Science in the British Colonies of America

"Plantations in their beginnings have work ynough,
& find difficulties sufficient to settle a comfortable way of subsistence, there beinge buildings,
fencings, clearinge and breakinge up of ground,
lands to be attended, orchards to be planted, highways & bridges & fortifications to be made, & all
thinges to doe, as in the beginninge of the world. . . ."

—*John Winthrop, Jr., November 12, 1668*

SCIENCE
in the
BRITISH COLONIES
of
AMERICA

Raymond Phineas Stearns

UNIVERSITY OF ILLINOIS PRESS

Urbana   Chicago   London

TO
JOSEPHINE

Contents

Contents

Illustrations

————

xi

Preface

THIS BOOK IS SET FORTH TO PROVIDE, within a single cover, a comprehensive overview of the scientific interests and activities of American colonials from early in the Age of Discovery to the end of the Old Colonial Era, or, roughly, from about 1520 to 1770. It has been done in the expectation that such a treatment would supply a basis for historical perspective, for comparison and contrast, and for the creation of a sense of growth and development of science in the colonial scene.

Some readers may find it unexpected to discover that the inquiring finger of science touched the New World in the sixteenth century and that the New World stimulated a scientific revolution already in progress in the Old. The "new science" that gradually emerged in Europe was of slow growth. The surprising feature, perhaps, is that it was present at all in the American colonies, especially when we consider the truth of John Winthrop's remark quoted above to the effect that in the New World there were "all thinges to doe, as in the beginninge of the world." It becomes more surprising when we realize that science in the New World was largely in the hands of amateurs, many of whom have hitherto escaped the scrutiny of history. They were persons with little or no formal scientific training. They operated in an atmosphere rarely favorable to scientific inquiry, working in isolation, without adequate libraries, apparatus, and the stimulation provided by able co-workers and by an informed, sympathetic public opinion. But their efforts led to more than a mere accumulation of crude scientific data. With the constant promotion, encouragement, and occasional correction by

Europeans, especially at the hands of the Royal Society of London, colonial men of science outgrew the role of field workers for European scholars and emerged as scientists in their own right.

But it was the eighteenth century, especially with the last generation of colonials prior to the American Revolution, before colonial scientific endeavor rose to uneven levels of achievement worthy of being classed alongside European scientific accomplishments. By this time colonial society offered a more favorable climate for scientific inquiry. It had become more populous, more urbanized, more wealthy, and many of its citizens had some leisure time. Communications had improved, and there was a widening popular interest in and appreciation of science. As these conditions developed, the cries of anguish were stilled which earlier colonial men of science had raised against their isolation, their paucity of books and apparatus, and their lack of sympathetic popular sentiment. Indeed, the emergence of colonial scientific *communities*, of scientific societies, both at the local and at an intercolonial level, with worldwide intercommunications among scientists and with colonial outlets for their scientific publications, was a major factor marking the relative maturation of colonial science. The role of the Royal Society of London, so vital in the earlier days, diminished in importance as American colonials were able to generate their own sources of inspiration, criticism, and leadership.

However, throughout the colonial era, European scientific achievement overshadowed that of the colonials, and this was still the case in 1770. By that time American colonials had begun to display an increasing independence of mind, an occasional capacity to challenge successfully their European mentors, and an ability to contribute a few scientific *ideas* which were original and arresting. But they stood, as yet, only on the threshold of independent scientific accomplishment. With the exhilaration generated by the success of the American Revolution, they were, like most adolescents, anxious to cast off the old scientific leadership and grope their way to scientific maturity. They were loath to admit that Europeans still led the way in science—and continued to do so for more than a century to come. To be sure, Americans made enviable progress during the nineteenth century in *applied* sciences—in engineering and technology—but it was the twentieth century before they equaled, and ultimately surpassed, the science of the Old World.

The relative newness of present-day scientific achievements is startling to contemplate. And so much of it has come to be taken for granted that many people—not the least being the scientists themselves—lose perspective and become impatient with our fore-fathers' "ignorance" of scientific knowledge nowadays looked upon as elementary, easily learned by schoolboys. It is jarring to realize that even by the end of the Old Colonial Era, Lavoisier and his associates had not yet organized chemistry as an independent scientific discipline; that Hutton and Lyell had not yet written their pioneer works in geology; that Cuvier's works in comparative anatomy and paleontology were a generation ahead; that Dalton, Faraday, Maxwell, Lobachevski, Darwin, and a host of other foun-ders of today's scientific outlook were unknown. In the art of medicine, the *materia medica* still wallowed in polypharmacy rem-iniscent of Galen, and surgeons had neither anaesthesia nor aseptic surgery. It is easy, from the standpoint of present-day scientific knowledge, to label our forefathers' beliefs as ignorance and super-stition; but, in their own time, their answers to scientific questions were no less "scientific" than those of today. The integrity of sci-ence at any moment of its history must be that of its own time; and the development of the sciences has been characterized by constant competition among different and often distinct views of nature, all of them "scientific" in their day. Those in competition in colonial times differed from those in competition today. During the latter years of the Old Colonial Era, there was a dispute in learned circles as to whether the culture of the Ancients or that of the Moderns was superior. It is interesting, and perhaps instructive, to note that the *Boston Evening Post* (No. 879, June 2, 1752) concluded that the Moderns were superior, especially because the sciences "flour-ish more at present, than anciently." But the writer was at pains further to point out that the Ancients had prepared the way for the Moderns, and quoted Clitus' comment to Alexander: " 'Tis true you have conquered, but it was with your Father's Soldiers."

In general, this book is postulated upon the belief that the history of science is properly the history of people, what they did and what they wrote, always with an effort to probe their real thoughts. Scientists study things; historians of science study scientists; and the history of science is part of intellectual and social history. Insofar as scientific principles, or laws, had been established, these are generally placed before the reader, usually as a kind of measuring rod whereby colonials may be compared to their European mentors. As science in Europe underwent considerable development during the era under survey, attention has been directed to the major European achievements behind which colonials generally lagged, although the distance closed during the latter years of the Old Colonial Era.

For the most part, this book is written from the original sources. I have, of course, consulted the many secondary treatments and have profited from many of them. Few of them, however, have gone beyond consideration of a single individual's scientific interests, and some of them have been founded upon an incomplete survey of original sources. I have been able, I believe, to avoid some of their errors and omissions; and I trust that I have committed no new errors. For those that have crept in, I apologize, resting secure only in the belief that subsequent studies will offer corrections.

I am deeply indebted to many persons and some institutions for aid in the preparation of this work. I owe particular thanks to the Graduate College of the University of Illinois, which provided a succession of research assistants to hasten my work and endowed me directly with six months' time uninterrupted by instructional duties as Associate Member of the Center for Advanced Study. The National Science Foundation, jointly with the University of Illinois, enabled me to spend a year on sabbatical leave in England, where research for this book was completed and a preliminary draft of the manuscript was drawn up. Among those research assistants who contributed to the preparation of this book, I wish especially to acknowledge the aid of George Pasti, Jr., now Professor of History at the State University of New York at Plattsburg; John Joseph Zacek, now Professor of History at the State University of New York at Albany; George Frederick Frick, now Professor of History at the University of Delaware, who also read a large portion of the typescript and offered helpful criticism and advice; Jack Richard Dukes, now Assistant Professor at Macalester College; John David Krugler, now Assistant Professor of History at Marquette University, who has been tirelessly helpful and creative in searching out obscure items and in proofreading the entire manuscript. David L. Coon and Mrs. Sally Kohlstedt, both graduate students in history at the University of Illinois, joined the procession near the end of the march, and both have been most helpful and obliging.

Archives and libraries have been, of course, indispensable. I wish to express my profound gratitude to the Royal Society of London, whose President and Council accorded me permission to quote freely from the materials in the Archives and Library of the Society, and whose librarians, the late Mr. Henry W. Robinson, and his son, Mr. Norman Robinson, at present the Assistant Librarian, together with the present Librarian, Mr. I. Kaye, extended to me every possible courtesy and all manner of assistance. Similarly, I wish to thank the Trustees and the staff of the British Museum, particularly those of the latter in the Division of Manuscripts; the staffs at the Bodleian Library and the Ashmolean Library, Oxford; of the Linnean Society, the London Society of Apothecaries, the Library of the Royal College of Physicians, the Wellcome Medical Library, and the Guildhall Library, all of London. Miss Phyllis I. Edwards and Dr. W. T. Stearn, of the Department of Botany in the British Museum (Natural History), also in London, were most cooperative in helping me ferret out items in the *Sloane Herbarium*. Similarly, the staff at the Library of the University of Leiden and of the Bibliothèque Muséum National d'Histoire Naturelle in Paris were helpful. In the United States, I am heavily indebted to the Library of the American Antiquarian Society (Worcester), to the Houghton Library of Harvard University, and to the Library of the Massachusetts Historical Society, Boston. The Newberry Library, Chicago, has furnished a few rare items, and the Library of the University of Illinois, particularly the Rare Book Room and the Natural History Library, has been unfailingly cooperative.

Three learned colleagues in historical studies, Professor I. Bernard Cohen, of the Department of the History of Science at Harvard University, Professor Winton U. Solberg, Chairman of the

Department of History, University of Illinois, and Arthur P. Whitaker, Professor Emeritus, University of Pennsylvania, have read the entire typescript of this book and offered valuable suggestions and corrections. My debt to them cannot be adequately acknowledged with words, but my sincere thanks to them are more than a mere formality. However, they share none of the responsibility for whatever errors may appear in this volume. These, alas, are my own.

To the staff of the University of Illinois Press, I extend my hearty thanks for their patient helpfulness and discriminating taste in seeing this book through the press.

Last, to my wife, Dr. Josephine Bunch Stearns, I am uncharacteristically speechless in attempting to express my gratitude for her constant assistance in research, literary criticism, typing from my crabbed longhand manuscript, and stimulating assistance at every turn.

R. P. S.
Champaign, Illinois
January, 1970

Abbreviations Used in Footnotes

Add. Mss.—the *Additional Manuscripts*, in the Division of Manuscripts of the British Museum, London.

Certificates—copies of the original signed certificates submitted to recommend persons to fellowship in the Royal Society of London, in the Royal Society Archives.

Classified Papers—bound volumes of original papers received by the Royal Society, in the Royal Society Archives.

Clokie—Hermia Newman Clokie, *An Account of the Herbaria of the Department of Botany in the University of Oxford*, Oxford, 1964.

C.M.—*Council Minutes*, manuscript records of the meetings of the Council of the Royal Society, in the Royal Society Archives.

Col. F.R.S.—Raymond P. Stearns, "Colonial Fellows of the Royal Society of London, 1661–1788," *Notes and Records of the Royal Society of London*, VIII, No. 2 (April, 1951), 178–246. Earlier versions of this article appeared in *William and Mary Quarterly*, 3rd Ser., III (1946), 208–268, and in a rather garbled form in *Osiris*, VIII (1948). Reasons for this unusual duplication are given in *Col. F.R.S.*, which is a revision of the earlier printings.

Coll. M.H.S.—*Collections of the Massachusetts Historical Society*, Boston. Series numbers are indicated by a digit preceding the abbreviation. Thus, *1 Coll. M.H.S.* refers to the first series.

D.A.B.—*Dictionary of American Biography*.

Dandy—J. E. Dandy, ed., *The Sloane Herbarium*, London, British Museum (Natural History), 1958.

D.N.B.—*Dictionary of National Biography*.

Guard-Book—bound manuscript materials from correspondents of the Royal Society, in the Royal Society Archives.

xix

Journal-Book—bound manuscript records of the meetings of the Royal Society, in the Royal Society Archives.

Letter-Book—bound volumes of manuscript letters from correspondents of the Royal Society, in the Royal Society Archives.

Notes and Records—*Notes and Records of the Royal Society of London.*

O.E.D.—*Oxford English Dictionary.*

Petiver—Raymond P. Stearns, "James Petiver, Promoter of Natural Science, *c.* 1663–1718," *Proceedings of the American Antiquarian Society*, LXII, Pt. 2 (Oct., 1952), 244–365. An offprint, with the original pagination, was published separately at Worcester, Mass., in 1953.

Phil. Trans.—*Philosophical Transactions.*

Proc. M.H.S.—*Proceedings of the Massachusetts Historical Society.*

Pulteney—Richard Pulteney, *Historical and Biographical Sketches of the Progress of Botany in England* . . . , 2 vols., London, 1790.

Register-Book—incomplete lists of members of the Royal Society and other data, in the Royal Society Archives.

Sloane—the *Sloane Manuscripts*, in the Division of Manuscripts of the British Museum.

Va. Mag. Hist. & Biog.—*Virginia Magazine of History and Biography.*

Wm. & Mary Qty.—*William and Mary Quarterly.*

Science in the British Colonies of America

Introduction

THE ACCEPTED PERIODIZATION OF POLITICAL and of intellectual history do not always coincide. However, the Old Colonial Era of American history, a political designation extending from the Age of Discovery from around 1500 to the recognition of American independence in 1783, closely paralleled the beginnings of early modern science, a phase of intellectual history reaching from the beginnings of "natural philosophy," as early science was often called, to its polarization into "the sciences" near the end of the eighteenth century, with the emergence of botany, chemistry, geology, and other more or less specialized scientific disciplines. It does no violence, then, either to political history or to the history of science to consider the rise and development of science in the American colonies as a topic for monographic treatment. And this is the focus of attention in this book.

However, natural philosophy was broadly European and international in origin whereas the American colonies which ultimately became the United States of America were mostly English in their institutional and intellectual development. Any consideration of the rise and development of early science in these American colonies must cast a glance both at the developing state of science in Europe as a whole and at that of England in particular, for it was from the latter that the colonists drew their principal sustenance. From the outset, as we shall see, European science was vitally and significantly affected by the discoveries in the New World. And from the outset, too, the English lagged behind their European contemporaries in some aspects of the new science. In the Americas

3

as a whole, and in the English colonies in particular, the colonists lagged behind their European fellows. Throughout the sixteenth century, Europeans who went to the Americas—and this was peculiarly true of the English—were so intent upon exploiting the gold, silver, and precious gems found there, upon Christianizing the natives, upon finding a passage through the American continents whereby to tap the commercial wealth of the Orient, and upon other dynastic designs, that they generally exhibited only a passing interest in the scientific problems posed by the New World. And the first generation or so of the English colonists in the seventeenth century either were similarly inclined (as in Captain John Smith's Virginia), or, as they settled down to the grim tasks of making homes in a virgin land and evolving a viable economy, they were too occupied with these matters to give serious attention to the new science.

Indeed, neither scientific knowledge nor the scientific method was *sui generis* in America. Both were tender plants introduced to the New World scene by European proponents, seeded, watered, and fertilized by European patrons, partly in the private and imperial interests of the Europeans themselves, and, as early modern science matured, increasingly promulgated in the more objective design of perfecting science for its own sake. Most of the American colonists who played a role in this development were relatively minor contributors of information, feeding American scientific data, with ever-increasing accuracy and growing sophistication, to English and occasionally European patrons, who in turn synthesized this data with similar information gathered around the world and gradually fashioned the whole into an organized body of knowledge—the corpus of early modern science. Not until the last colonial generation—that of the Founding Fathers of the American Republic—were there American colonials willing and able themselves to become active participants in this process of synthesis and organization and to become contributors of scientific *ideas* (as opposed to mere data) in their own right. In this manner science came to America.

From the vantage point of the third quarter of the twentieth century, it is difficult to accept the European mentality of the sixteenth, seventeenth, and eighteenth centuries when modern science was in its infancy. The greatest scholars of early colonial times subscribed to classical opinions regarding natural phenomena derived, for the most part, from ancient Greeks, medieval Arabs, and later medieval philosophers. Around the world today school children master scientific laws which the greatest minds of the colonial era were only beginning to comprehend. The great modern university, with its libraries, museums, experimental gardens, laboratories, and computers, is almost wholly a development of the post-colonial era, and the universities of that time often spurned the new science, so attached were they to the medieval curriculum. In consequence, early science developed outside the universities in public and private academies, usually financed (especially in the English world) by private assessments and donations. Only gradually, as the colonial era wore on, did colleges and universities, both in the Old World and the New, officially open their doors to "natural philosophy" and establish professorial chairs in science. Even then, the study of mathematics, especially in the colonial colleges, rarely went beyond simple arithmetic for everyday ciphering and bookkeeping.[1] Only a few scholars with patrons to support them in their work could devote the whole of their time to scientific endeavor. The vast majority of them made their daily bread as physicians, apothecaries, teachers, ministers of the gospel, or as merchants and tradespeople of one kind or another. At best they were amateur scientists, for there was as yet no means whereby a scientist, *qua* scientist, could earn his living. In this regard, the colonial was relatively no more disadvantageously situated than the bulk of his European fellows—except, perhaps, that the struggle for a living was more arduous in the New World than in the Old, and the colonial, living in a more rural environment, had far fewer library facilities to consult and fewer fellow scientists to fire his scientific efforts.

Moreover, none of the natural philosophers of this early day was a specialized scientist. Natural philosophy was general in its approach to natural phenomena as a whole. It covered the entire gamut of "the sciences" (as we know them today) from agronomy to zoology. Actually, few of the "sciences" had emerged as sepa-

---

[1] Samuel Pepys, the great diarist of seventeenth-century England, had to learn simple arithmetic to cope with his job in the navy in 1662. See Marjorie Hope Nicholson, *Pepys' Diary and the New Science* (Charlottesville, 1965), pp. 8–9.

rate, organized bodies of learning, such as chemistry, geology, and the like, and the two aforementioned did not spin off from the main body of natural philosophy until shortly after the American colonial period had ended. And although many of the natural philosophers of this early modern era demonstrated more interest and greater ability in some areas than in others, all of them were generalists. Thus, on both sides of the Atlantic we find physicians who are recognized as able botanists, zoologists, and astronomers; ministers of the gospel who serve as physicists, botanists, zoologists, and meteorologists; farmers, planters, merchants, ship captains, soldiers, and civil servants who have natural philosophy as an avocation, and so on. It is well-nigh impossible to cite a specialist in any one science in the present-day sense. The entire corpus of natural philosophy was still within the capacities of a single scholar.

This early age of science was devoted primarily to three broad objectives: the *collection* of data, in a valiant, if vain, effort to assemble together all of the varieties of nature's products from all parts of the world; the *classification* of data, in a long-sought attempt to discover a rational pattern, or system, into which the evidence of nature might be classified; and *nomenclature*, or the imposition of names which would win universal acceptance in the scientific world for all the various specimens of nature. As these involved the entire world (not to mention the stars, planets, and other objects of the firmament), it was an endless task. Indeed, the scientists of the colonial era left it unfinished and it still goes on, though shortly before the end of the colonial era the work of Carl Linnaeus (c. 1707–78) had furnished a fairly viable system for the world of natural history. As these tasks were undertaken on a worldwide scale, with attention to Europe, Asia, Africa, and the known island archipelagoes, as well as the Americas, the contributions of British colonials formed only a small proportion of the total. Moreover, the European scientists did not fix their attention upon the mainland colonies of North America to the exclusion of Central America, South America, and the West Indies, and English scientists, in particular, collected what they could from *all* of the English colonies in the New World, oblivious, of course, to the ultimate separation of the thirteen mainland North American colonies which became the United States. As their view was imperial, and as there was considerable traffic among all of the British colonies in the New World, we cannot exclude consideration of scientific activity in Canada, the West Indies, and parts of Central and South America. During the Old Colonial Era these efforts were all of one piece in the development of early European science.

The overwhelming evidence of scientific activity in the New World pertains to natural history. There were observations in astronomy, geography, geology, meteorology, electricity, and other areas, but the principal foci of interest lay in flora and fauna. To a wide extent, however, these discriminations among the sciences are valueless, partly because the early scientists themselves did not attend to them sharply and partly because their concept of natural history—for reasons which presently will become more apparent—envisaged a wide canopy which included under that denomination many of the more specialized studies into which the sciences fall today. And although some of the colonists were able to make scientifically meaningful observations in astronomy, most of those interested in natural philosophy could—and did—make useful collections of minerals, plants, shrubs, trees, birds, animals, fishes, and the like (including the seeds of flora), and supply helpful observations on these items together with occasionally useful data regarding geography, geology, mineralogy, medicine, drugs, meteorology, oceanography, the Indians, and a host of other things. These were subjects in which the gifted amateur could make his greatest contribution, serving, in effect, as the field collector and observer for the European synthesist. Of course, the better trained he was, the more useful his collections and observations would become. As the colonial era advanced, colonial collectors and observers became increasingly skillful and scientifically informed until, by the 1750's, some of them closely approximated their European correspondents.

Throughout the colonial era, scientists, both in the Old World and the New, subscribed in varying degrees to many fundamental scientific notions which had been widely accepted since ancient times. Some of these were modified or cast aside entirely as the new science developed during the sixteenth, seventeenth, and eighteenth centuries, but many of them persisted, especially in the more conservative or less well-informed minds, until the end of the colonial era and even later. Most of these were Greek in origin, for the

7

Greeks in the era extending from Thales of Miletus (early sixth century B.C.) through the early Roman Age (first century A.D.) were the earliest people to create a corpus of rational scientific learning relatively free from mystical conceptions. Some, too, were of medieval origin, largely from Arab scholars, who, in the years between the eighth and twelfth centuries A.D., revived much of the Greek learning and made additions of their own. Europeans slowly absorbed this stream of Arabic knowledge (and through the Arabs much of the ancient Greek science) and soon demanded the ancient knowledge directly from the Greeks themselves. This they accomplished in large part by means of the translations in the thirteenth century and beyond, at first by retranslation from the Arabic and then by direct translation of the Greek authors into Latin. The humanist scholars of the Renaissance era took a leading part in this salvage of Greek works, and by the sixteenth century a few of them were actually published in the vulgar tongues of western Europe. Inevitably, of course, the development of paper-making and the invention of printing by movable type played a vital part in this revival and spread of ancient learning. By the time of the discovery of the New World—itself a part of this classical revival and the creative application by Europeans of the "new learning" to problems of navigation, map-making, and overseas exploration—the scientific legacies of the Greeks and the Arabs largely had become a part of European learning. Europeans, thus reclothed in the best knowledge of the past, were ready and able to begin to build new scientific structures upon these ancient foundations. In the realm of natural history especially, the exploration of the New World supplied a wide and exciting opportunity for which the ancient syntheses, having been principally concerned with the Mediterranean world, soon proved inadequate.

Naturally, the first Europeans to explore the New World brought with them, as an inevitable part of their intellectual baggage, the fundamental notions in science which they had inherited from the Ancient and Medieval worlds. If some of these notions appear bizarre or ridiculous to the twentieth-century reader, he must remember that they were the results of logical deductions, based upon thoughtful observations (though few experiments), by the greatest minds of the past, by men who stood on the threshold of scientific

8

learning. Modern science still operates from scientific hypotheses. Many of these, in time, prove erroneous; but they serve a purpose until additional knowledge (nowadays usually discovered by experiments as well as thoughtful observation) makes possible the construction of new hypotheses to replace them. And this transposition in thinking is still sometimes accompanied by bitter disputes and mutual recriminations among the scientists themselves. The tempo of change has accelerated greatly during the past century, but the over-all developments, as extended in time, are similar to those of the earlier ages. Witness, for example, the present-day theory of matter as compared to that of the late nineteenth century. The ancient theory of matter was something else again, and it persisted for centuries into early modern times.

The Greeks conceived of matter in two different ways. The first, originating in the Pythagorean School (c. 530 B.C.), and ultimately passed on to subsequent generations by Aristotle (384–322 B.C.), held that different kinds of matter may be distinguished by their different properties or qualities. In their simplest forms, these properties can be presented as cold combined with dryness, as seen in earth; cold combined with dampness, as seen in water; heat combined with dampness, as seen in air; and heat combined with dryness, as seen in fire. These, then, were the four "elements," earth, water, air, and fire, arranged in order of heaviness and lightness. All other phenomena in the material world were combinations of these. The second Greek theory of matter derived principally from Leucippus (fifth century B.C.), who held that matter consists of ultimate, invisible, and indivisible particles scattered in a void. This theory, in origin very different from the atomic theory based upon experimental data set forth by Dalton and others in the nineteenth century, nonetheless resembled the atomic theory of modern science (before the atom itself was broken down at the turn of the present century).

A third theory of matter arose from the Arabs, who modified the Greek four-elements notion and held that the fundamental properties, or qualities, of matter are to be found in sulphur (the fire principle), in mercury (the liquid or water principle), and in salt (the earth principle). This theory passed into Europe during the fifteenth century and was endorsed, with modifications and addi-

9

tions, by Theophrast von Hohenheim (c. 1490–1541), a Swiss physician and mystic who called himself Paracelsus and was known by his envious colleagues as "Bombast." All of these theories of matter were severely castigated by Robert Boyle (1627–91), whose *Sceptical Chemist* (London, 1661) clearly set forth the modern conception of elements as concrete matter. But the name of Aristotle stood so high and the Aristotelian theory was so widely accepted in the learned world that the Greek four-elements theory of matter persisted long after Boyle had written. Several of the "natural histories" of the New World were projected along the lines of this classic theory, being organized around the four sections: "Of the Earth," "Of the Water," "Of the Air," and "Of Fire." José d'Acosta, in his *De Natura novi orbis* . . . (1589), was one of many to adopt these categories, although he omitted a treatment "Of Fire," arguing that it was the same in the New World as in the Old.[2]

The theory of "formed stones" was another legacy to early Americans from sixteenth-century Europeans. This theory, too, was Greek in origin, although it underwent progressive stages of modification and refinement during the American colonial era. Common to all these stages, however, was the belief that stones are stones, whether observed in the air, on the earth, in the earth, in the waters of the earth, or in animals, fishes, plants, or man. Little attention was paid to their physical and chemical structure; precious gems, common rocks, mineral ores, corals, and gall stones—all were stones. Stones also included all fossils and fossil-like substances; indeed, until the nineteenth century the term "fossil" in geological literature referred to anything dug out of the earth. Moreover, it was believed that stones, thus broadly conceived, grow after the manner of plants, in the earth, in the waters, in the air, in plants, in animals, and in man; some writers assigned sex to minerals and spoke of male and female stones. Between the time of Aristotle and the end of the American colonial era, three principal theories arose to explain the origins of rocks, stones, and minerals of all kinds. All three theories existed side by side, and virtually every writer on the subject associated himself with one or another of these. Aristotle

[2] E[dward] G[rimstone], tr., *The Naturall and Morall Historie of the East and West Indies* . . . (London, 1604), Bk. III, p. 117.

attributed the origin of stones and minerals to the celestial influences of the sun, the planets, and the stars upon the *qualities* of the four elements. If the *relative* proportions of these properties in one element (in association with the other elements) were altered, one body might be changed into another. Thus, as the sun warmed the land and the waters it gave off "exhalations," some dry and some moist. This in turn gave rise to clouds, causing thunder and lightning and other atmospheric phenomena which might alter the relative proportions of the properties of one or more of the four elements giving birth to new "fossils" or minerals. These might originate in the air (as a meteorite), in the waters (as coral or pebbles), on or in the earth (as boulders or stalactites), in vegetables (as amber or resins), and in man and animals (as gall stones, kidney stones, etc.). It was widely believed that as the rays of the celestial bodies fell more directly upon the earth's surface in tropical countries (especially those of the sun, that "king of the celestial orbs"), precious metals and gems were likely to be more abundant in those regions of the earth.

Agricola (1490–1555), the "father of mineralogy," observed the mines in Saxony, and he had little patience with the Aristotelian view of the origins of stones. He argued that stones and minerals are formed by a *succus lapidificus* (literally, lapidifying juice, or mineral-bearing solution). In the presence of heat, this *succus lapidificus* dries up and deposits the stony matter which it held in solution; similarly, in the presence of cold, it deposits its dissolved materials. Other sixteenth-century writers held that this *succus* would vaporize and infect the winds, to create stones in the air, in the waters, and in animals.

Last, there was the notion that arose largely in the seventeenth century to the effect that minerals and stones originate from a seed (the petrific or rocky seed), which, though no more visible than Aristotle's *pneuma*, or principle of life in general, produces stones after the manner that seeds grow into plants. Thus, "stones are generated from other stones, each after its own kind."[3] Like the other

---

[3] Quoted from M. J. C. Schweigger's *De Ortu Lapidum, Concerning the Origin of Roche* (1665) in Frank Dawson Adams, *The Birth and Development of the Geological Sciences* (Baltimore, 1938), p. 89. I am much indebted to Mr. Adams' volume for the discussion of the various theories of formed stones. Mr. Adams

theories, these "rocky seeds" may implant themselves in the earth, in the waters, in animals, and even in the air. The minerals which they produce may assume the shapes of fruits, plants, twigs, and the like, still sometimes confused with genuine fossil remains of once-living plants and animals. Indeed, before the close of the seventeenth century, scientists disputed whether "formed stones" found in the earth and resembling shellfish, animal skeletons, and plants were actually "stones bred in the earth," as the conventional ones maintained, or the actual organic remains of once-living animals and plants, as the newer school of scientists would have it. The discovery of fossil remains in the New World broadened this controversy, as we shall see.

All of the theories regarding the origins of stones and minerals held one thing in common, namely, that stones *grow*—in the air, in and on the earth, in the waters, in plants, in animals, and in man. By drawing no distinctions between the stones that grow in the earth, in the air, and in the waters, and those that grow in plants and animals, these theories continued the ancient practice of drawing no line of demarcation between the animate and inanimate objects of this world. Moreover, if stones are stones, whether grown in the inanimate(?) earth or in "living" plants and animals, then the line between mineralogy and geology, on the one hand, and botany and zoology, on the other, becomes badly blurred. It is for this reason, among others, that, as we noted above, sharp divisions among the sciences, physical and natural, and especially a narrow interpretation of natural history, simply did not exist in early modern science. As late as 1773, near the end of the American colonial era, Alexandre Savérein, writing his *Histoire des Philosophes Modernes*,

points out (p. 78): "The geologist who even now, after some four hundred years of additional study, finds that so much still remains to be learned, can indeed sympathize with these his predecessors, more especially when he remembers that they sought the solution of these difficult problems, not in the open field of nature, as he now does aided by all the appliances of modern science, but in the Aleian Fields of tradition and formal logic." M. J. S. Rudwick makes much the same point when he discusses the uncertainty that still remains regarding the organic versus the inorganic theories of the origin of fossils and the existence, still, of a small class of uncertain origin known as *Problematica*. See "Problems in the Recognition of Fossils as Organic Remains," *Ithaca 26 VIII 1962–2 IX 1962: Proceedings of the 10th International Congress of the History of Science* (Paris, 1964), pp. 985–987.

stated that natural history was composed of the mineral realm, the vegetable realm, and the animal realm, and he went on to add that the mineral realm includes waters, earths, sands, stones, salts, pyrites, demi-metals, metals, bitumens (a long list, including precious stones, petroleum products, granite, marble and other building stone, ambergris, belemnites, talcs, bezoars, and other stones found in animals), the products of volcanoes, and other petrifactions.[4] Actually, the line is still indistinct between natural history and the physical sciences.

A variety of somewhat related theories brought to the New World originated in the Christian philosophy of the Middle Ages, some of them so mixed with earlier, pre-Christian ideas that it is impossible to distinguish between those of Christian and of pagan beginnings. The Europeans who came to the New World, whether Roman Catholic or Protestant, were Christians, and they subscribed in common to a broad area of Christian percepts. Basic among these was the story of creation as set forth in the Book of Genesis. From this account Christian philosophers deduced not only that God had created the world in six days but also that He had created *all* the world. Nothing new or different had been created later, and nothing had been lost, or become extinct (Noah had rescued all living species when they marched, two by two, into the Ark). This belief in the absolute fixity of species led to disputes when the idea was broached in the seventeenth century that some fossils might be the remains of plants or animals which had become extinct (and it led to far more bitter disputes when Darwinism, with its possible evolution of new species, was set forth in the nineteenth century). Moreover, God's creation extended equally and without variation to the entire world: there were no plants, animals, or minerals anywhere in the world different from those known in Europe. Conservation was itself part of creation. This notion, of course, became patently absurd as the New World's creations became known, although it died very reluctantly in the minds of many sixteenth-century Europeans.

Yet another widely held belief was that of spontaneous genera-

---

[4] (8 vols., Paris, 1760–73), VIII, lxv. The *O.E.D.* states that natural history was *originally* "the systematic study of all natural objects, animal, vegetable, and mineral."

tion. Many of the Ancients had held that frogs were spontaneously generated in mud, and that flies, bees, and other insects were spontaneously generated in the putrefying flesh of dead animals. The latter belief even had Old Testament authority to support it.[5] Francesco Redi (1626–79), an Italian scientist, offered convincing experimental proof of the falsity of the belief in the spontaneous generation of insects during the 1660's. But in the eyes of many, Redi's conclusions were incompatible with the Scriptures, and the issue was inconclusive, arising repeatedly until late in the nineteenth century.

Sixteenth-century scholars generally accepted the belief, traceable as far back as Aristotle, that the torrid zone of the earth was a "burning zone" in which animal and plant life could not exist. This, too, became patent nonsense to early European explorers who observed the lush vegetation of the tropics and themselves had crossed the equator many times without suffering fatal effects. But the scholars at home often clung to the "burning zone" hypothesis. Closely allied to this concept of the tropics was the notion, also as old as the Greek sciences, that the climate of the world was essentially the same at any given latitude. Subscribing to this belief, many men on both sides of the Atlantic were puzzled to discover that the climate in North America, especially in Canada, New England, and Virginia, was much more severe both in winter and in summer than that of the corresponding latitudes of Europe. Isothermic lines, of course, were as yet unknown, and the climatic effects of the Gulf Stream on western Europe were not yet understood. Subscription to this belief caused bitter suffering to many early European colonists in North America who had anticipated a climate comparable to that of England or France.

The Doctrine of Signatures was another fancy of the sixteenth century carried to the New World, where it endures among superstitious folk to the present day. Briefly, the idea was that medicinal herbs are stamped by nature with a clear indication of their curative powers. This belief was prevalent among primitive peoples. The mystical relation between man and mandrake, with its forked, two-legged roots, was clearly depicted in early biblical times when

[5] Cf. Judges 14:5–6, 8–9. It was assumed that the bees had been spontaneously generated in the body of the lion which Samson had killed.

Rachel, desperate because of her barrenness, begged Leah for mandrake as a procreative drug.[6] The principle was supposed to function in either direction. Thus, pale, anemic women were given a ruddy complexion by consuming the blood-red beet, while the yellow tinge of jaundice was removed by imbibing the yellow celandine. Everything in nature, whether plant, animal, or mineral, either carried on its person, so to speak, or exhibited by its properties, its specific curative powers. Red medicaments were especially valuable for diseases of the heart; yellow, for the stomach; white, for the lungs; and black, for the kidneys. The upper parts of plants (leaves and fruits) were especially efficacious for treatment of the upper parts of the body, whereas the lower parts (roots and bulbs) were more proper for lower parts of the body. Perennial plants tended to impart long life, annuals operated to foreshorten it. To these self-obvious propositions was engrafted another very ancient philosophical principle, namely, that man and the universe bore toward one another the relationship of microcosm to macrocosm, so that in one was always to be found the counterpart of the other. This principle gave ancient peoples a feeling of oneness with nature, a sense that individual man was part of and in union with the universe; and Christians found that it fortified their belief in God, the Author of the universe in whose likeness every man himself had been created.

These ancient beliefs were given shape and an even greater degree of authority during the sixteenth century, especially at the hands of Paracelsus. Paracelsus applied primitive chemistry to medicine and proposed the addition of organic chemicals to the Galenic pharmacopoeia, consisting of inorganic concoctions derived from flora and fauna, thereby arousing bitter opposition. But Galenic medicine, derived from the Roman physician Galen (A.D. 129–c. 200), survived, and its bizarre prescriptions of various herbs and parts of animals, including their excreta, dominated the medical scene even beyond the colonial period of American history.[7] Still, Paracelsus had a few disciples, and his proposals were not entirely lost. His influence on the Doctrine of Signatures was carried to

---

[6] Genesis 30:14–16.

[7] Galenic anatomy and to some extent Galenic physiology were modified and corrected at points during early modern times.

further extremes by the Neapolitan, Giambattista Porta, whose *Phytognomonica* (1588) defended the proposition that long-lived plants lengthen and short-lived plants shorten men's lives, together with other notions extending further the Doctrine of Signatures. In England, William Cole and Nicholas Culpeper propounded the doctrine widely in the seventeenth century. Cole (c. 1625–62), a Fellow of New College, Oxford, published the *Art of Simpling* . . . (London, 1656), in which, among other things, he wrote:

> . . . the mercy of God which is over all his Workes Maketh Grasse to grow upon the Mountaines and Herbs for the use of Men and hath not only stamped upon them (as upon every man) a distinct forme, but also given them particular signatures, whereby a Man may read even in legible Characters the use of them. Heart Trefoyle is so called not onely because the Leafe is Triangular like the Heart of a Man, but also because each leafe contains the perfect Icon of an Heart and that in its proper colour viz a flesh colour. Hounds tongue hath a forme not much different from its name which will tye the Tongue of Hounds so that they shall not barke at you; if it be laid under the bottomes of ones feet. Wallnuts bear the whole Signature of the Head, the outermost green barke answerable to the thick skin [scalp] whereunto the head is covered, and a salt made of it is singularly good for wounds of that part, as the Kernel is good for the braines, which it resembles being environed with a shell which imitates the Scull, and then is wrapped up againe in a silken covering somewhat representing the *Pia Mater*.[8]

Culpeper (1616–54) was, perhaps, the major purveyor of the Doctrine of Signatures to the English colonies of North America. An apothecary who practiced medicine, Culpeper had much success and popular acclaim in eastern London, where he served his clients from Red Lion Street in Spitalfields. He also published several books in medicine, among which was his *English Physician Enlarged*. This volume, first published in 1653, passed through five editions by 1698 and was reprinted as late as 1809. It was widely used both in England and in the English colonies overseas as a home

---

[8] Cole also included a rare chapter on herbs useful in the treatment of animals. *The Art of Simpling* was enlarged and republished as *Adam in Eden* . . . in 1657. Cole bitterly disliked Culpeper and denied the latter's astrological botany described below.

guide to health problems. Its simple, clear instructions for the treatment of injuries and diseases with herbs and berries from the domestic garden made it a valuable aid to masses of people who had no physician at hand or could ill afford the fees of those available. And the book repeatedly set forth the Doctrine of Signatures in its manifold prescriptions. Culpeper may have performed a public service—and he certainly aroused the indignation of that medical guild, the Royal College of Physicians of London—when he published an English translation of the official college *Pharmacopoeia*, which had been first authorized by King James I in 1618.[9]

Culpeper was also a prominent advocate of astrological botany, the notion that there is a mystical relationship between herbs and celestial bodies.[10] Indeed, he practiced astrology along with medicine. The belief that herbs and vegetables are influenced by the moon and other heavenly bodies, and that their sowing and harvesting should be regulated by the signs of the zodiac, was a notion almost universally heeded in the sixteenth and seventeenth centuries.[11] This faith continued throughout the colonial era and well beyond, and the hundreds of almanacs which appeared on both sides of the Atlantic had, as a very practical part of their existence, the monthly appraisal of the moon phases and the positions of the other signs of the zodiac for the farmer and the gardener.

Such was the climate of opinion of the Europeans who first looked upon the New World. It was cast in the mold of ancient

[9] First published as *A Physical Directory, or a Translation of the London Dispensatory* in 1649. A second edition appeared in 1654. The official *Pharmacopoeia* was scarcely more modern in its scientific approach to medicine than Culpeper himself. See Sir George Clark, *A History of the Royal College of Physicians of London* (2 vols., Oxford, 1964–66), I, 227–230. ". . . not until 1788, in the sixth edition, was the London *Pharmacopoeia* cleared of the rubbish of the old polypharmacy." *Ibid.*, p. 230.

[10] He also extended this to a belief in a relationship between celestial bodies and the incidence of disease. See his *Semeiotica Uranica, or an Astronomicall Judgment of Diseases* (1651).

[11] Excellent accounts of the Doctrine of Signatures and of astrological botany appear in Agnes Arber, *Herbals: Their Origin and Evolution . . . 1470–1670* (2nd ed., rev., Cambridge, 1953), pp. 247–263; Eleanour Sinclair Rohde, *The Old English Herbals* (London, 1922), pp. 169–170; Redcliffe N. Salaman, *The History and Social Influence of the Potato* (Cambridge, 1949), pp. 110–112; and Pulteney, I, 180–183.

Greek philosophy, modified by the Christian scholars of the early church, with further reconstruction by the Arabs, the medieval scholastics, and the eager scholars of the Renaissance; but it was not wholly fixed and unyielding. Out of this old "science," with all its folklore, mysticism, and superstition, a new science was being born during the sixteenth century, and the explorers and early colonists of the New World, by their observations and specimens from the Americas, contributed to its delivery.

Old Science in the New World

THE EXCHANGE OF PLANTS AND ANIMALS between the Old World
and the Americas began at least as early as Columbus' second voyage (1493). By the end of the first two decades of the sixteenth
century this mutual traffic had reached wide proportions, with
profound effects upon the economies on both sides of the Atlantic
and equally profound effects upon the natural science of Europeans.[1] Puzzled by the unfamiliar specimens of plant and animal life
abounding in the Americas, Europeans naturally sought to incorporate them as much as possible into the varieties familiar to them
in Europe and already described in such accepted authorities as the

[1] Several useful books and articles on this subject are: U. P. Hedrick, *A
History of Horticulture in America to 1860* (New York, 1950); James A. Robert-
son, "Some Notes on the Transfer by Spain of Plants and Animals to Its Colonies
Overseas," *The James Sprunt Historical Studies*, XIX, No. 2 (Chapel Hill, 1927);
George Brown Goode, "The Beginnings of Natural Science in America," *Annual
Report . . . of the Smithsonian Institution, 1897. Report of the U.S. National
Museum*, Pt. II (Washington, 1901); Ira J. Condit, "Fig History in the New
World," *Agricultural History*, XXXI (1957), 19–24; Redcliffe N. Salaman, *The
History and Social Influence of the Potato* (Cambridge, 1949); Jerome E. Brooks,
*Tobacco: Its History Illustrated by the Books . . . in the Library of George
Arents, Jr.* (5 vols., New York, 1937–52); and Howard Mumford Jones, *O Strange
New World: American Culture in the Formative Years* (New York, 1964). Pro-
fessor Jones aptly points out (p. 61) that "In 1493 the New World dawned on
the European imagination as a few small, delectable islands, any one of which
was understandable in terms of a *hortus inclusus*, the walls of which sheltered the
Earthly Paradise, or a bower of bliss, or the garden of eternal youth and spring,
from the dark wilderness of the world. . . . But something more troubling than
the interminable extension of the land occurred. That something was the discovery of the terror of nature in the New World."

credulous, encyclopedic *Naturalis Historia* of Pliny the Elder or the *De Materica Medica libri quinque* of Dioscorides, dating from the first century A.D.[2] In general, when Europeans came upon a New World animal or plant, they tended to ascribe to it a name in one of four ways: first, to give it a generic name already attached to familiar European species thus extended to include the (supposedly) same genus in the New World, such as bears, wolves, goats, deer, oak, pine, and the like; second, to assign familiar European names to *somewhat similar* forms found in the New World in terms of general appearance, habits, use, cry, or note, so that European names were often given to American plants and animals later discovered to be different from European families from which the name had been chosen (thus the name of the lion was given to the American cougar, robin was given to the American thrush, partridge to the American grouse, crocodile to the American alligator, and others); third, they invented new names to describe striking features or peculiarities of American specimens whose forms were hitherto unfamiliar, such as pineapple, bluebird, cardinal, catfish, rattlesnake, and hummingbird—many of whose names have survived; fourth, they adopted the native Indian names of

[2] Of Pliny (A.D. 23–79) Gibbon later wrote that his *Natural History* was an immense register in which he "deposited the discoveries, the arts, and the errors of mankind." *The Greek Herbal of Dioscorides, Illustrated by a Byzantine A.D. 512, Englished by John Goodyear A.D. 1655, Edited and First Printed A.D. 1933,* Robert T. Gunther, ed. (Oxford, 1934), had circulated in manuscript form for years. Gunther aptly describes the seventeenth-century English edition as a Greek work of the first century as understood by a Hampshire scholar of the seventeenth century. The illustrations may have been based upon the beautiful drawings of Crateus, an herbalist who was Physician to Mithradites VI Eupator (120–63 B.C.), but some of the figures were confused, some had erroneous descriptions attached, and some were fictitious. *Ibid.,* p. vii. Dioscorides included about 500 plants and gave their healing virtues. Thomas Johnson, an English herbalist and physician, wrote in 1633 that Dioscorides' *De Materia Medica* "is as it were the foundation and grounde-worke of all that hath been since delivered in this nature." The elucidation of Dioscorides became the chief occupation of herbalists in the sixteenth century. To some it became a shackle beyond which their minds would not reach; to others it became merely a point of departure. Cf. Agnes Arber, *Herbals: Their Origin and Evolution . . . 1470–1670* (2nd ed., rev., Cambridge, 1953), pp. 10–11. For Latin translations of Dioscorides, Pliny, and Theophrastus during the Renaissance—the copies most available to New World commentators—see Pulteney, I, 34ff.

plants and animals (or what the Europeans *believed* the Indians had called them, for the linguistic differences were great and most of the native peoples had no written language), such as maize, papaya, iguana, potato, tobacco, caribou, moose, and others.[3] Many of these names, likewise, have survived in popular parlance.

From the time of Columbus onward, scores of Spaniards wrote about their adventures in and observations of the New World. Most of these were official reports, travelers' accounts, adventure stories, promotion literature, and descriptions written by seamen and soldiers of various ranks. Some of them, like Columbus, were learned in the new arts of ship-building and navigation, involving at least the application to seafaring of the newer knowledge of astronomy, geography, and cartography; but few of them were knowledgeable in natural science and, indeed, few could be recognized as natural philosophers. As a whole, their works are of immense value in political history and the history of geographical discovery; they are of small value in the study of the history of science. But three sixteenth-century Spanish authors wrote about the New World and its flora and fauna with an effort to approach these subjects in the light of the science of the day. They were widely read by their contemporaries and, as they were translated into English as well as other western European tongues before the end of the sixteenth century, they served the early English promoters of overseas colonies—and some of the colonists themselves—as pioneer studies of the natural history of the Americas by learned, on-the-spot observers. Their works also entered the stream of European learned literature and quickened the development of the natural sciences in Europe itself. Indeed, they still serve as foundation stones of the natural history of the New World. These writers were Oviedo, Monardes, and Acosta.

[3] This classification is adopted from W. F. Genong, "The Identity of the Animals and Plants Mentioned by the Early Voyagers to Eastern Canada and Newfoundland," *Proceedings and Transactions of the Royal Society of Canada*, 3rd Ser., III (1910), Sec. II, No. 5, pp. 197–242. Although this article treats the early French explorers primarily, the categories seem to apply equally to other Europeans who first came to the Americas. The Spanish sometimes employed a Spanish word, without regard to its previous Spanish meaning, if it approximated what they thought the Indians had said. See, for example, "avocado," *O.E.D.*

Oviedo

Gonzalo Fernandez de Oviedo y Valdés (1478–1557) was born in Madrid and early became a page in the service of Queen Isabella. He appears to have been largely self-educated at the Spanish court. He was a youthful friend of Columbus, whose voyages he favored. He entered the army and served in the Neapolitan War. In 1513 he was appointed Overseer of the mines in Hispaniola, to which he went in 1514. About 1523 he returned to Europe and saw the new Spanish king, Charles I, already elected Holy Roman Emperor as Charles V. Charles reappointed Oviedo to his post as Overseer of the mines and commissioned him to write a history of the New World. Probably when he returned to Spain on this occasion he brought his *Natural History of the West Indies*, for it was first published at Toledo in 1526.[4] Already he had traveled among the Spanish islands of the Caribbean and reached Central America, everywhere observing the natural products with a keen and speculative eye. Upon his return to New Spain he became Governor of Cartagena (1526) and later Governor of Santo Domingo (1535), posts he operated efficiently but tyrannically. About 1546 he returned to Spain to complete his *General and Natural History of the Indies*, Part I of which had been dedicated to Charles V and published at Seville in 1535.[5] Part II, published at Valladolid, followed in the year of Oviedo's death (1557). A large

[4] As *De la natural hystoria de las Indias,* illustrated with a few woodcuts too crude to have much scientific value. An excellent English edition was translated and edited by Sterling A. Stoudemire in the *University of North Carolina Studies in the Romance Languages and Literature,* No. 32 (Chapel Hill, 1959).

[5] As *La historia general de las Indias.* This work, with its second part (1557), is a general account of Spanish activities in the New World with less specific attention to natural history. It, too, was illustrated with woodcuts of an Indian dugout, a tomahawk, Indians panning for gold, a pineapple, and a few crude cuts of a lizard, nuts, pinions, cactus, and palm(?) leaves. In general, the natural history material appears largely to have been drawn from the *Natural History* of 1526. The *General and Natural History* was a revision, with additions, of both earlier volumes, growing to twenty volumes as *La Historia general y natural de las Indias occidentales.* A useful edition of the latter was published by the Real Academia de la Historia (2 vols., Madrid, 1851), with an account of Oviedo; a better and more recent edition is that of Juan Perez de Tudela Bueso, ed., in *Biblioteca de Autores Españoles,* Vols. CXVII–CXXI (Madrid, 1959)', with an account of Oviedo and his writings.

part of Oviedo's first book, the *Natural History of the West Indies*, was published in English by Richard Eden in his *The Decades of the newe world or West Indies, Containing the navigations of the Spaniards. . . .*[6] This was reprinted in 1577 and republished by Samuel Purchas in *Hakluytus Posthumus or Purchas His Pilgrims* (London, 1625), which in turn was reprinted by the Hakluyt Society in 1902. Thus, though Oviedo's work was originally published in Spanish, significant parts of it were accessible to English readers by 1555, thereby affecting English information and opinion before the time of such English promoters of overseas colonies as Sir Walter Ralegh, who is known to have read the English translation of the *Natural History of the West Indies*.[7]

For Oviedo, Pliny's *Natural History* was the measure of all things in the New World and he himself appears to have written in conscious imitation of Pliny. However, he recognized that New Spain produced animals and plants unknown to the Ancients. For example, when he described a jaguar, which he was inclined to call a tiger, he finally decided that it did not coincide with Pliny's account of tigers and therefore was not a tiger.[8] He observed that sharks were viviparous: "I mention this because Pliny in his *Natural History* does not include any . . . among animals producing living young."[9] He pointed out that whereas Pliny had described only five or six kinds of "trees that remain green and never lose their leaves," nearly all the trees in the islands of New Spain,

---

[6] (London, 1555), pp. 173–214.

[7] Eden's *Decades* displayed far more interest in Spanish exploits in the New World, and especially in the gold and silver found there, than in natural history. This is underscored by the minuscule selection he published from Venoccio Biringuccio's *De la Pirotechnica* (Venice, 1540). This work, the first printed book on metallurgy, was a full, illustrated account of the assaying, smelting, alloying, and working of metals, unfortunately lost sight of after it was overshadowed by Agricola's work in 1556. Eden saw fit to translate and publish only a small section on the origin and finding of gold. Cf. *The Decades*, pp. 326–343. Eden's work was republished by Edward Arber, ed., *The First Three English Books on America* . . . (Birmingham, 1885). Arber's edition includes Eden's translation of Oviedo, pp. 218ff. For confirmation that Oviedo's *Natural History* was read (probably in the Eden translation) by Sir Walter Ralegh, among others, see Stoudemire's edition of the *Natural History* cited above. Subsequent references to this work will be to this edition.

[8] *Natural History of the West Indies*, pp. 45–47.

[9] *Ibid.*, p. 112.

as well as those of Central America, were of this variety.[10] Thus, while Oviedo used Pliny's work as a constant guide, he readily accepted the new and unfamiliar plants and animals and sought to describe them in terms which would be meaningful to Europeans. He did not speculate about creation and how these strange beasts and plants came to exist at all, as some of his contemporaries were inclined to do, and he quickly disposed of the "burning zone" theory by pointing to himself as an eyewitness to the fact that the torrid zone is inhabitable by plants, animals, and men.[11]

Oviedo was the first field naturalist to prepare an organized account of the flora and fauna of the New World. But already, as he clearly recognized, difficulties arose in distinguishing between those plants and animals which were indigenous to the New World and those which Europeans, since the voyages of Columbus, had been introducing to the New World from Europe, Africa, the Levant, and the Far East.[12] By Oviedo's time, only thirty-odd years after Columbus' first voyage, New Spain was producing an astonishing list of things belonging to the second category. These included horses, cattle, and swine, all of which had gone wild and multiplied rapidly on several of the Caribbean islands; sheep, which ultimately were found to flourish only in the highlands of Chile and Peru; wheat, sugar cane,[13] cultured grapes, olives, oranges, lemons, limes, citrons, bananas, apples, pears, peaches, apricots, pomegranates, figs, and a number of vegetables and melons, including eggplant, squash, peas, cucumbers and watermelons.[14] Oviedo was thoroughly aware of these European transplantations and is himself

[10] *Ibid.*, p. 95.

[11] *Ibid.*, p. 30.

[12] *Ibid.*, pp. 48–57, 97ff. Europeans continued to introduce European plants and animals—and to experiment with others from Africa, the Levant, and the Far East —throughout the colonial era.

[13] Sugar cane was first introduced by one Pedro de Atienza at Hispaniola. Oviedo wrote that sugar was first produced by one Gonzalo de Vibora, also of Hispaniola, who erected a horse mill to extract the cane juice. "To him alone," said Oviedo, "are due thanks for the first manufacture of sugar in America."

[14] See Hedrick, *A History of Horticulture*, pp. 9–15; Robertson, "Some Notes on the Transfer . . . of Plants and Animals . . . Overseas," pp. 7–21; Condit, "Fig History in the New World," *loc. cit.* Bernabé Cobo, a Spanish Jesuit, remarked in 1652 that "the transfer of animals and plants has been more advantageous to the New World than the immense wealth of gold and silver sent thence to Spain." Robertson, p. 9.

the source of our knowledge of some of them. His concern, however, is primarily with the products of the New World hitherto
unknown in Europe—or at least not listed in Pliny's *Natural History*. His organization wanders somewhat, but in the main he
treats of them in the following order: animals, birds, insects, vipers,
trees, plants and herbs, precious stones, gold and other minerals, and
fishes. Beyond these natural categories, he makes no attempt to
classify or systematize the products of New Spain which caught
his eye, and his nomenclature is a mixture of those items he believed
had been described by Pliny, whose names he adopted; those that
were similar to items known in Spain (and with these he used the
common Spanish names); and those unfamiliar, for which he commonly employed the Indian names.

Obviously Oviedo displayed some confusion, as with the jaguar,
which, as we have noted, he first believed to be a tiger and later
identified with the leopard. Cougars he confused with lions, and
he saw no distinction between American deer and those of the Old
World. But he recognized differences between the European pig,
which had been transplanted to America, and the native American
peccary. Other American animals which he noticed included the
tapir, ocelot, anteater, rabbit, hare, armadillo, sloth, raccoon, various kinds of monkeys, "small dogs," and the opossum—although,
if he recognized the last named as a marsupial, he made nothing of
it, in marked contrast to later observers who did recognize it as a
marsupial and found it a most fascinating and unusual creature.
Birds he divided into two categories: those similar to birds of
Spain, including eagles and many others, and those that differ from
Spanish birds, among which he recognized many different varieties
of parrots, "man-of-war birds," "tropic birds," "boobies," petrels,
nighthawks, magpies, goldfinches, nightingales (mockingbirds?),
pelicans, cormorants, vultures, partridges, pheasants, toucans,
"crazy-birds," and hummingbirds—all too generally denominated
to be scientifically distinguishable with any accuracy.

The small section on insects was equally worthless, inasmuch as
Oviedo simply referred to flies, mosquitoes, wasps, bees, ants, and
the like in general terms, with almost no particulars. His treatment
of vipers was little better. He referred to the boa constrictor as a
very poisonous one and in such a way as to raise doubt that he ac-

tually had seen a specimen alive and in the wild. The iguana he described as "very frightful to look at but good to eat"; he referred to lizards of various kinds, including the alligator, which, in considerable advance of some succeeding European observers, he carefully distinguished from the crocodile.[15] And he referred to scorpions, crabs, toads, and huge spiders "as large as a sparrow" —probably tarantulas.

Of trees he mentioned first the "drumstick tree" (*Cassia fistula*),[16] an East Indian species introduced by the Spanish to the New World. A brisk and growing trade was developing in its seeds, from which a mild laxative was made. There followed a sizable list, many of them given only their native names, such as the mammee (*Mammea americana*), with a peach-like fruit as large as two fists; the guava (*Psidium*, probably *P. guajava*); the coconut (*Cocos nucifera*); various palm trees—though none of the native palms, said Oviedo, produced dates, although European date-producing palms had been successfully introduced; pines (*Pinus occidentalis*); evergreen oaks (species uncertain); various "grapes and vines," which "would do well here," as was soon discovered;[17] papaya (*Carica papaya*), later called "mammy tree" by the English, familiarly known as papaw; quince (*Cydonia*, variety uncertain, probably the mountain quince); avocado (*Persea gratissima*); calabash trees, from whose fruit the Indians made vessels; jobo tree (*Spondias lutea*, or hog plum); guayacan (two varieties later differentiated, *Guaiacum officinde* and *G. sanctum*), said to be a native specific for the cure of syphilis;[18] genipap (*Genipa americana*), from which the natives procured the black paint with which to

[15] *Natural History of the West Indies*, pp. 75–77. The distinctions between the crocodile and the alligator are technical, and early observers should not be criticized unduly for failing to make them. Cf. C. B. Lewis, "Alligators versus Crocodiles," *Glimpses of Jamaican Natural History* (2nd ed., Institute of Jamaica, 1949), pp. 13–15.

[16] The scientific names of the species were unknown to Oviedo, and have been added by later scholars.

[17] *Natural History of the West Indies*, p. 85.

[18] Here (*ibid.*, p. 89) Oviedo tells the story, often repeated (though erroneously), that syphilis had been introduced to America by Spanish sailors. The denotations of plants and animals according to their medicinal virtues, or "uses," was *a* system of classification; but as natural science developed it proved confusing and inadequate.

daub themselves; manchineel (*Hippomane mancinella*), from whose deadly fruit the Indians prepared poison for their arrows; and the huge ceiba (*Ceiba pentandra*, or *Copaifera homitomophylla*), from whose trunks the Indians made their dugout "canoes," and banana trees (*Musa sapientum*), which, although introduced to New Spain from Europe, were more fruitful in the New World than in the Old.

In the descriptions of plants and herbs Oviedo made much of maize, cassava, yams, and pineapples, all native to the New World. He supplied Europeans with their earliest illustration of maize (*Zea mays*), but he identified it with the millet of India described by Pliny, thus inaugurating a confusion in Europe which was not ironed out until the early seventeenth century.[19] Oviedo told of the preparation of cassava from the root of the yucca (*Manihot utilissima*), and he waxed enthusiastic over yams (of the genus *Dioscorea*) and "batatas" (sweet potatoes, *Batatis edulis*). The pineapple ("pinas," or *Ananas sativus*) was pictured in a recognizable woodcut in *La Historia General de las Indias* (1535 and later). He also found wild cress, larger than that of Spain but of similar taste; coriander, "like ours in taste" but with a larger leaf; a white clover similar to that of Spain; a strange herb which he called "Y" (*Ipomoea bona-nox*), which was a splendid purgative; hazelnuts (*Corylus americana*); large melons (either *Cucurbita pepo* or *Cucumia anguria*); prickly pear ("tunas," or *Opuntia tuna*); wild plantain ("bijaos," or *Heliconia bihai*), used by the natives, with other canes and grasses, to make baskets and thatches; and various undeterminable barks and leaves employed by the Indians to make dyes.

Oviedo traded with the natives to obtain precious stones, such as pearls, emeralds, cornelians, jasper, chalcedony, amber, and white

[19] See John J. Firmin, "Maize in the Great Herbals," *Annals of the Missouri Botanical Gardens*, XXXV (Galesburg, Ill., 1948), 149–191. Many European authorities followed Oviedo and spoke of its Oriental (East Indian) origin; others referred to it as *Frumentum turicum*, and held that it had been brought from Asia by the Turks; and a few stoutly insisted upon its American origin. A modern scientific dispute also exists about the origin of Indian corn. Similar confusions grew up in the sixteenth century over the origin of the American turkey (*Meleagris gallopavo*), whose English name still reflects the early belief in its Near Eastern derivation.

sapphires, and his descriptions of the gold mines emphasized his position in the New World as Royal Overseer of them. He argued that gold grew in the great hills and mountains from whence it was washed down by the rains into rivers and streams. Sometimes, however, it was found in the plains. He noted that coal and gold were often found together. Coal, he argued, had been formed from wood and that "in the passing of countless years a great deposit of earth accumulated on the coal."[20] The coal was above the gold and both were washed outward from the mountains by the rains. He also argued that the farther gold was found from its source, the purer it was. He described how the natives panned for gold in the streams and published a woodcut illustrative of the process.[21] He said that there were also fine silver and copper in New Spain, although evidently not much care had been taken to extract them in Oviedo's day.

Of the fishes in the New World he was brief, devoting most of his attention to turtles, sharks, manatees, and flying fish. He described huge turtles seen in Cuba, and bloodthirsty sharks in many parts of the Caribbean. The manatee (*Trichechus monatus*) was larger than a shark and very good to eat, with a flavor like beef. It had a bone, or stone, in its head, said Oviedo, which was found to be very useful, when burned and pulverized, "in curing pains in the side."[22] Flying fish, which Oviedo, in common with countless other voyagers to the New World, found very intriguing, would fly until they became dried by the air and then "fall again into the water."

In his final section Oviedo turned to a practical problem of geography and trade. Cosmographers and others, he said, believed that a navigable waterway existed from the Atlantic to the South Sea (the Pacific) across the barrier formed by the New World. On the basis of his studies and travels in the region, Oviedo denied the existence of such a passage. There is, he stated, a narrow land division, the Isthmus of Panama, as we know it today. However, Oviedo argued, a useful overland route could be constructed from

---

[20] *Natural History of the West Indies*, p. 109.
[21] The illustration first appeared in *La historia general de las Indias* (1535). For the first description see *Natural History of the West Indies*, p. 107.
[22] *Ibid.*, p. 113.

Nombre de Dios on the Atlantic to Panama on the Pacific, using the Chagre River part of the way on the Atlantic side. By this means, the trade of the East Indies, transshipped across the Isthmus along this combined overland and waterway route, could become a viable highway between the Spice Islands of the Orient to Europe, thereby saving "more than seven thousand leagues of sailing."[23] For a variety of reasons, the route never developed appreciably as a highway to the Far East, but, as the "Gold Road" between the eastern and western coasts, it became a vital link in the economy of New Spain a few years after Oviedo had returned to his homeland.

Thus ended Oviedo's *Natural History of the West Indies*. At the time of its first publication (1526), it was the most complete account of the natural history of the New World that had appeared. Although severely limited to the Caribbean area, it supplied the first significant survey of the natural products of the New World from the pen of an on-the-spot and (for his day) informed observer. No wonder that it was so widely translated and republished! European colonial promoters, persons interested in the exciting (and profitable!) colonial enterprises, and European scientists concerned with the perfection of natural philosophy found it a mine of exciting information, provocative of much speculation and thought. Subsequent visitors and settlers in America used it as a guide to what they might be expected to find there, and natural philosophers sought to incorporate its facts into a more systematic and scientific treatment of natural history. Its information—including its errors—became a part of the developing new science.

Monardes

Nicolás Bautista Monardes (c. 1493–1588) never set foot in the New World. The son of a Genoese bookseller and his Spanish wife, he was educated at the Alcalá de Henares, from whence he was graduated as a Bachelor of Medicine in 1533. There he became familiar with the *Lexico artis medicamentae*, an annotated work by Antonio de Nebrija (or Lebrija), combining the *materia medica* of Dioscorides and Pliny, and it became his favorite textbook for

---

[23] *Ibid.*, p. 120.

29

the rest of his life. He settled in Seville to practice medicine and soon became widely known both as a successful physician and as a medical scholar. He published his first medical treatise at Seville in 1536 and a number of later works followed, all of them subsequently translated into Latin, French, and Italian. His practice was profitable and he invested monies in colonial trade, including the slave trade, and became very wealthy. In 1547 he was granted the degree of Doctor of Medicine by the University of Seville. By virtue of his location at the port of entry for the trade with New Spain in America, he became interested in and familiar with the products of the New World, especially plants and herbs with a supposed medical value. Soon experimentation with these New World products became a major part of his medical practice. He developed a botanical garden of his own where he grew American plants for his experiments and corresponded with Spaniards in the New World in an effort to solicit their aid in collecting specimens of exotics and their seeds.[24] About 1565 he began a large book to describe the fruits of his experiments, *Dos libros, el uno que trata de todas que se traen de Nuestras Indias Occidentales, que sirven al uso de la medecina, y el otro que trata de la Piedra Bezoar, y de la yerva Escuerçonera*, first published at Seville in 1569. A second part appeared in 1571, and a further edition with a third part added was published in 1574 and republished in 1580. In the meantime, Monardes' work had excited the interest of one of the pioneers in scientific botany, Charles de L'Escluse (Carolus Clusius), who published a Latin edition of Monardes' work with illustrations, *De Simplicibus Medicamentis ex Occidentali India Delatis, quorum in Medicina usus est* (Antwerp, 1574, and later). In the same decade, an English merchant, who had wide connections in Spain, one John Frampton, prepared an English edition of Monardes' *Dos libros* (1574 edition) and published it as *Joyfull Newes out of the Newe Founde Worlde Written in Spanish by Nicholas Monardes Physician of Seville and Englished by John Frampton, Merchant*

---

[24] He mentioned one such correspondent, Peter de Osma, a soldier in Peru, who wrote to Monardes in December, 1568. Monardes compared him with Dioscorides, who was also a soldier-herbalist. See Stephen Gaselee, ed., *Joyfull Newes out of the Newe Founde Worlde* . . . (2 vols., London and New York, 1925), II, 145. The best life of Monardes is Fráncisco Guerra, *Nicholas Bautista Monardes, su vida y su obra, ca. 1493-1588* (Mexico City, 1961).

Anno 1577[25]—a title indicative of the hopefulness inspired in European medical circles by Monardes' work. In this manner the experiments of the Seville physician became the property of the European learned world, including the English-reading portion of it.

Monardes was immensely learned and highly esteemed by his contemporaries. He cited many of the authorities accepted in his day. If he leaned heavily upon Dioscorides and Pliny, he also had consulted many others. Of the Ancients, besides Dioscorides and Pliny, he cited Plato, Aristotle, Democritus, Hippocrates, and Galen; of the Arabic scholars, he referred to Rhazes, Averroes, and Gaber; and of the later medievalists, he mentioned Albertus Magnus and Agricola. But to a large extent he depended upon his own experiments,

> seeing that there is in the newe Spaine, so many Hearbes, and Plantes, and other things Medicinable, of so much importance, that there is not any that writeth of them, nor is it understoode, what vertues and formes they have, for to accorde them with ours, that if they had a desire for to searche out, and experiment so many kinds of Medicines, as the Indians doeth sell in their Market places and Faires, it would be a thing of great profit and utilitie: to see and to know their properties, and to experiment the variable and greate effects, the which the Indians doeth publishe, and manifest with great proofe, that amongst them selves, they have of them; and they of our part, without any consideration do refuse it, and such as doeth know their effects, they will not give us relation, nor knowledge what they are, nor write the efficacy and manner of them.[26]

These New World medicines might not only greatly enlarge the Europeans' knowledge of cures but also free them from dependence upon the medicines of the Levant and the Far East.[27] But Monardes' chief concern was with his experiments with the New World's "newe Medicines, and newe Remedies, wherewith they do cure and make whole many infirmities," as he himself had discovered by his experiments at Seville "this fourtie years."[28] Thus,

25 Subsequent references will be to the edition prepared by Stephen Gaselee noted above, cited simply as *Monardes*.
26 *Monardes*, I, 58.
27 *Ibid.*, p. 70.
28 *Ibid.*, pp. 10–11.

while he checked his findings against the old authorities, he did not hesitate to go beyond them when his own observations ran counter to or outside the collected wisdom of the past. In consequence, his work served, as Howard Mumford Jones has said, "to testify that a whole novel and miraculous pharmacopoeia was at last available to Europeans to heal them of their manifold diseases."[29]

Monardes made no attempt to systematize or classify the products of New Spain (other than in accord with their medicinal "vertues"), and his nomenclature was largely that of the natives, though now and then he identified, or believed he had identified, a New World product with one previously described by Dioscorides or some other ancient authority. He was pre-Copernican in his view of the universe, and his technology in preparing his medicines for human consumption was that of the sixteenth-century alchemist, although he did not accept the alchemists' belief in their ability to transmute baser metals into gold. He suggested a familiarity with the works of Paracelsus.[30] His penultimate section, "The Dialogue of Iron," argued that iron was, in reality, more precious to man than gold, both because of its usefulness in making tools and because of its virtues as a medicine. Citing an impressive array of ancient authorities, he held that all metals are engendered in the earth from substances placed there by God at the creation of the world, as St. Augustine had maintained. But iron, being compounded of contrary principles, both of heat and of cold, partakes of both. Hence its extraordinary medicinal virtues both for the outward disorders of the skin and inward disorders of the digestive tract.[31] His final treatise, "Of the Snow," was strictly Galenist. Snow is useful as a means of preventing putrefaction in foods and of cooling drinks. But snow water should never be used except as a medicine, for its continued use brings sickness and disease.[32]

Of the "newe medicines, and newe Remedies," Monardes began with gums and resins which, in many cases, the Indians used to

[29] *O Strange New World*, p. 37.
[30] *Monardes*, II, 153–187.
[31] *Ibid.*, pp. 113–130.
[32] *Ibid.*, pp. 105–116, 124–152.

burn as aromatics in sacrificial rites. Copal and animé he found to be different from those of the Levant but a great comfort for "griefes of the head." Tacamahac was useful for the "diseases of the Mother." Caranna was helpful as a plaster for swellings and "griefes of the joints." The "Oyle of the Fig Tree of Hell" was the same as Dioscorides had described, a good laxative. The bitumen, or tar, from Cuba was the same as the Ancients had described. Liquid-amber was good for colds in the head and brain afflictions. Balsam, derived from a tree, was different from that of Egypt, which was produced by a vine, but its medical virtues were equally good, useful inwardly for digestive troubles and menstrual irregularities and outwardly for colds, wounds, and surgical operations. But the greatest medicines from New Spain were guayacan ("lignum vitae," *Guaiacum officinde* or *G. sanctum*), china root (*Smilax china*), and sarsaparilla (several varieties, *Smilax officinalis, S. papyracea*, and *S. aristolochiaefolia*). Guayacan, found in Hispaniola, was said to be a sure cure for syphilis (here Monardes appeared to follow Oviedo), and useful for dropsy and bladder troubles. China root, which (he said) the Portuguese had previously introduced from China, was found in various parts of the New World and Monardes held it to be a veritable panacea, useful for treating fevers, syphilis, gout, sciatica, jaundice, and a variety of other human ailments. Sarsaparilla was almost equally efficacious in a wide variety of troubles. The bloodstone would stop bleeding of all kinds—here and elsewhere Monardes betrayed a faith in the Doctrine of Signatures; the bezoar stone, found in the bellies of wild sheep or goats in the mountains of Peru, was equal to those the Portuguese brought from the Levant, with the same qualities as an antidote to poisons.[33] Monardes stated that the Spanish had found them useful as antidotes for the poisonous arrows of the Indians.

Other medicines described by Monardes included kidney and reins stones, for kidney and bladder troubles and "for them that

[33] The original bezoar stones had been found in the bellies of ruminants in Persia. Monardes felt that Peruvian bezoars were somehow different, but they displayed the same medicinal virtues. He warned, however, that they must come from beasts which fed upon the "healthful Hearbes of the mountains," for those that pastured on the plains were useless. *Ibid.*, I, 134, 57–90.

doeth not pisse liberally"; various purgative nuts and pinions; mechoacan, a root from Mexico and Peru, sometimes known as rhubarb of the Indies, which excelled all other purgatives because of its gentle action and lack of unpleasant aftereffects—although Monardes goes on to claim much wider medicinal values. Tobacco was described and its plant illustrated, with an account of its medicinal virtues, especially as a poultice for wounds, headaches, and so on— although Monardes' claims for tobacco were modest compared with those made by others of his day.[34] A short section was devoted to "the Pep[p]ers of the Indies," already being grown in Spain and used both as a medicine and as a condiment. A magical herb, Escuerçonera, so called because its juice was used in New Spain as an antidote for the bite of the escorcu, a venomous toad, was hailed as a competitor with the bezoar stone as an antidote for poisons in general. Monardes said little about the key role played by the Spanish in the introduction of European plants and animals to the Americas, but he did state that Don Francisco de Mendosa had taken cloves, pepper, ginger, and other condiments from the Orient to New Spain, where all of them failed to grow except ginger, which, he said, was then being imported from New Spain.

Without question, Monardes was an unusually learned physician with a philosophic, experimental outlook. Without visiting the New World he told his contemporaries a great deal about its natural products. But his interest in natural history extended little beyond medical botany with its characteristic concern principally for the medical "vertues" of plants and its almost entire unconcern for such problems of early scientific botany as system, nomenclature, plant structure, ecology, and the like. Even the plant "vertues," as Monardes set them forth, generally approached those of the panacea rather than specific cures; they were good for anything and almost everything. He was often credulous in receiving accounts of the Indian remedies from New Spain, and he occasionally

[34] Tobacco had been brought to Europe in Monardes' day. Its seed was cultivated there as early as 1554. Two species were most widely used, *Nicotiana tabacum*, which originated in Brazil, and *N. rustica*, which originated in Mexico. When it first reached Lisbon, one Jean Nicot was French Ambassador to the Portuguese court. He wrote such extravagant accounts of its virtues to Catherine de' Medici, the French regent, that his name became identified with tobacco's active chemical agent, nicotine. Brooks, *Tobacco*, I, 13ff.; *Monardes*, I, xxvi.

assayed the role of the patriot, especially to further the Spanish Empire at the expense of the Portuguese. But no one in his century did so much to inform the learned world about the vast potential of the New World as a source of drugs and to excite their interest in the further enlargement of the pharmacopoeia by further exploration, experimentation, and study.

Acosta

José d'Acosta (1539?–1600) was born at Medina del Campo of an obscure family. At fourteen he entered the Society of Jesus and devoted his entire life to the Jesuit Order, as missionary, educator, administrator, and priest. In 1569 he was sent as a missionary to Peru, where he spent nearly twenty years teaching, founding schools and colleges, and converting the natives to the Christian faith. He also observed the flora and fauna, together with the political and social institutions of New Spain, and speculated philosophically on the differences between the New World and the Old.[35] He returned to Spain in 1588, bringing with him a Latin manuscript entitled *De Natura novi orbis libri duo*, which was published at Salamanca in 1589.[36] This book was subsequently enlarged to four volumes and published in Spanish as *Historia Natural y Moral de las Indias* at Seville in 1590.[37] This work was translated and published in French, Flemish, and German around 1600, and in 1604 one Edward Grimstone published an English translation as

[35] William Robertson, *The History of the Americas* (2 vols., London and Edinburgh, 1777), wrote in high praise of Acosta's work. Cf. I, 448n, 471n. According to a penciled note in the American Antiquarian Society's copy of Acosta (Grimstone's translation), Robertson said that "The Natural & Moral History of the New World, by the Jesuit Acosta, contains more *accurate observations*, perhaps, & *more sound science*, than are to be found *in any description* of remote countries published in the sixteenth century." I have been unable to identify the exact source of this quotation. The author of the penciling cites Robertson, *America*, VIII, 288, but this I have not found in the editions of Robertson's works that I have consulted.

[36] Republished in 1595. Acosta also published *De Promulgatione Evangelii apud barbaros* (8 vols., Salamanca, 1588; republished, Cologne, 1596); *De Christo revelato libri novem* (4 vols., Rome, 1590); and other theological works.

[37] Republished in 8 vols., 1591; revised and republished in 8 vols., Madrid, 1608, 1610.

The Naturall and Morall Historie of the East and West Indies . . .
(London, 1604.[38] For the remainder of his life Acosta remained in
Europe engaged in the affairs of his order. He became the Superior
at Valladolid for a time and was Rector of Salamanca when he
died.

Acosta was no scientist, but he was an acute observer, an informed thinker, and his *Naturall and Morall Historie* was a remarkably sane, wise, and informative book. Part of it had been
written while he was still in Peru. He was well grounded in the
works of the ancient and medieval scholars, was familiar with
Monardes' work, and approached the New World somewhat more
philosophically, though occasionally less accurately in matters of
detail, than his predecessors.

Acosta began with "the Opinions of some Authors, which supposed that the Heavens did not extend to the new-found world."
In this matter he combatted the claims of some of the early Christian philosophers such as Chrysostom, "a man better seene in the
study of holy Scriptures, than in the knowledge of Philosophie,"
who, even if they recognized "that the Heaven was round (as in
effect it is)," denied that it could "encompass another or New
World." St. Jerome approached this opinion and held that only
chaos and an "infinite gulph" were on the other side of the world.
Some of these men argued that the ancient, Aristotelian view of
the spherical form and circular motion of the heavens was contrary
to the Holy Scriptures. But whatever the older authorities said,
wrote Acosta, "it must not trouble us," for they knew little natural
philosophy. Even St. Augustine, "so well seene in all naturall
Sciences," doubted "whether the Heaven did compasse the earth
on all parts." He seemed to envisage Heaven as a kind of roof over
a house, covering only a portion of the earth while the remainder
hung, uncanopied, in the middle of the air. We must not censure the
early church scholars, said Acosta, for knowing no better, for they
had not seen the New World. Rather, he said, let us censure "those
vaine Philosophers of our age," who ought to know better regarding "the being and order of the Creatures, and of the course and
motion of the Heavens." For Acosta said "there is no doubt" that

[38] Grimstone used the Cologne edition (1596) of Acosta's *De Natura novi
orbis*. References hereafter will be to this English edition.

36

the Aristotelian views were correct in stating that "the figure of Heaven was round and did move circularly in his course . . . as we which doe now live in *Peru*, see it visibly." "Experience should be of more force, than all Philosophical demonstrations," continued Acosta, and this experience was sufficient to prove that "the Heaven is round, and comprehends and contaynes the earth within it of all parts." The heavens turn upon two poles, "as upon their Axeltrees." The circumnavigation of the world proved that there is no chaos or infinite gulf on the other side. Moreover, the earth is larger than it was formerly believed to be. It is round and it moves circularly, as an eclipse of the moon proves and as the darkness of nights also proves, "from the shadow which the earth makes." Acosta went on further to prove, from astronomical data which he had observed in Peru, from the church fathers, and from the Holy Scriptures, that the earth is round and stands "in the midst of the world."[39]

Acosta's cosmography is not wholly clear. He was concerned more with denying the contentions of early Christian philosophers than in setting forth a rounded cosmography of his own. As befitted a Jesuit scholar of his time, he denied the Copernican theory and evidently clung to a Ptolemaic view of the universe. But he clearly set forth a view of the sphericity of the earth, and he stoutly asserted that there were stars in the New World and "a Heaven in these parts of the Indies, which doth cover them as in *Europe, Asia,* and *Affricke.*"[40] Moreover, he asserted that there "is a firme Land on this Southern part as big as all *Europe, Asia,* and Affricke." The Eastern (Atlantic) and Southern (Pacific) seas join on the south at the Straits of Magellan, and he felt that there might be a Northwest Passage as well. He confuted St. Augustine's denial of the antipodes, "which is to say, as men marching with their feet opposite to ours." He argued that Aristotle and the others had denied the existence of the New World because their knowledge had been confined to a knowledge of the Mediterranean world alone.[41] He subscribed to the notion, with Aristotle, that "there is always one

39 Acosta, *Naturall and Morall Historie*, pp. 1–9.
40 *Ibid.*, pp. 17–18.
41 Plato, Seneca, St. Jerome, and others may have had a vision of the New World, however. Cf. *ibid.*, pp. 39–40.

37

temperature of the Heavens from East to West, being equally distant both from the Northerne cold and the Southerne heate"; but later, in his description of Peru, he pointed out that the winds on the seas differ from those on the land and that, in Peru, altitude caused the temperatures to vary.[42] He stoutly denied the ancient belief about the burning zone, and argued that the recent evidence regarding the habitability of Chile and Africa had laid this opinion to rest: "Contrarie to the antient and received Philosophy . . . the region which they call the burning Zone, is very moist, and in many places very temperate, and that it rains there . . ."; so that those who live under the equator "live a sweete and pleasant life."[43]

How the Indians got to the New World, and whence they came, was a problem to which Acosta addressed himself. All men, including the Indians, had descended from Adam, Acosta asserted, but how they migrated to the New World was unknown. They may have arrived by land or by sea, by accident or by design. In ancient times they might have come by sea, for there was no reason to suppose that ocean navigation was confined to modern times. But Acosta doubted that if they had arrived by sea, it was the result of design, as they had had neither compass nor knowledge of modern oceanic navigation. On balance, Acosta tended to hold that the Indians had arrived by land because he believed that the New World was joined to the Old at points of land not yet discovered. He did not subscribe to the opinion that the Indians were European, or Jewish, in origin and confessed that this was a question as yet unfathomable, especially as the Indians had neither written records nor certain remembrances from their founders. He believed that the Indians had descended from earlier migrants to the New World, and that their ancestors had lived "almost like savage beasts without coverings or houses, without tilled lands, without Cattell, without King, Law, God, or Reason." Only gradually had they improved their economy and political order to set up "a kinde of Commonweale, although it were very barbarous."[44]

After the first two books of his treatise, devoted to such philosophical speculations as are above recounted, Acosta began a "Nat-

[42] Cf. *ibid.*, pp. 29, 85–109, 117ff.
[43] *Ibid.*, "Advertisement to the Reader," and pp. 85–109.
[44] *Ibid.*, p. 503.

ural History of the Indies," setting forth an organization based upon "the three elements," air, water, and earth, and their compounds, "metals, plants, and beasts." He omitted fire, which, he asserted, was the same in the Indies as elsewhere. His treatment is neither as complete as Oviedo's nor as experimental and utilitarian as Monardes'. Like both of his predecessors, he simply described what he had seen, with no knowledge of scientific botany and zoology, and with little concern for scientific classification or system. If some of his observations are less accurate than those of Oviedo and Monardes, Acosta is even more emphatic than they in asserting that there are new flora and fauna in the Americas "whereof *Plinie* himselfe, *Dioscorides* and *Theophrastus*, (yea, the most curious) had no knowledge, not withstanding all their search and diligence."[45] At this point he challenged more pointedly than his predecessors the widespread sixteenth-century belief that nature was everywhere the same. He also issued a warning which should have placed European students of scientific nomenclature on guard: he pointed out that his predecessors in the New World had given European names to many New World products which, though they looked similar to the European counterpart, were in fact often very different.[46]

Acosta's treatment of the air was limited to the winds, which, he asserted, were very different in different places, some moist, some dry, and he was at pains to dispute Aristotle's contention that air temperatures were the same in all degrees of latitude. Winds were different, in the same latitudes, on sea and on the land; and mountain heights, as in Peru, gave rise to lower temperatures. His treatment of fishes is very sparse, merely noting that some are the same as found in Europe, others different. But, unlike Oviedo, he identified the American lizards and alligators with Pliny's crocodiles.[47] "Metals," he said, "are (as plants) hidden and buried in the bowels of the earth . . . and it seemes properly that these minerals grow like unto plants . . ."; in fact, "metals be plants hidden in the earth."[48] Plants and animals are superior creatures and have a more

[45] *Ibid.*, p. 280.
[46] *Ibid.*, p. 267.
[47] *Ibid.*, p. 165.
[48] *Ibid.*, pp. 203–204.

perfect being, "the inferior nature alwaies serving for the maintenance and sustentation of the superior, the lesse perfect yeelding
unto the more perfect. . . ." Metals are good for physic, defense,
ornament, and tools, but above all metals are useful for money, "the
measure of all things." Gold, other precious metals, and gems grow
in lands that are otherwise barren and unfruitful—yet, where the
mines are in America, because of the opportunities that arose in the
village settlements that grew up about them, the Indians have
the most, and the best, instruction in the Gospel of Christ. He described the mines, including those of Potosí, which he had visited,
and the methods of refining ores.

Of the flora of the New World, Acosta's account went little
beyond that of Oviedo and Monardes. He began with maize and
the uses which the Indians made of it; went on with the yucca,
from whose roots the natives made the bread-like cassava; and gave
the Indian names of about a dozen other American plants used for
food by the Indians, including "pattatres," perhaps the white potato (*Solanum tuberosum*, L.) which originated in the Andes and
later became an object of much confusion in historical botany.[49]
Acosta reported that the Indians had become very fond of garlic,
which had been introduced from Europe, together with beans
and melons.[50] He much admired the native American pineapple,
and pointed out, with Monardes, that the only spice native to
America was the chili pepper; ginger had been introduced by Europeans from the Far East. His reference to the tomato was probably to the *Physalis*, variously known as the husk tomato, cherry tomato, and strawberry tomato, which grew wild in the Americas and
was widely used both by the natives and the Europeans.[51] He also
spoke of coca (probably *Erythroxylon novogranatense*, of Peru;
perhaps *E. coca* of Mexican origin) and cacao (*Theobroma cacao*,
Mexico), the latter already much esteemed for making chocolate; of
coconuts; of West Indian balsam (for which he followed Monardes'

[49] See Salaman, *History and Social Influence of the Potato*. The white potato
was early confused with the Jerusalem artichoke and the West Indian sweet
potato and yam. Accounts of its Virginian origin were wholly erroneous. See
ibid., pp. 1–98.

[50] Acosta does not identify which species of beans and melons. Some were
indigenous to the Americas.

[51] Hedrick, *A History of Horticulture*, p. 22.

account, with the same distinction from Egyptian balsam); and of various flowers, of which, he said, the Indians were very fond of making nosegays. All in all, Acosta believed that the Indians received greater profit from those plants which the Spanish had brought from Europe than the Spanish received from American plants transferred to the Old World. The Indians, he said, produce in America as good—or better—wheat, barley, herbs, pulses, lettuces, oranges, lemons, citrons, peaches, apricots, pears, apples, figs, and the like as are cultivated in Spain, whereas American plants cultivated in Spain generally fail to thrive. He especially praised the fine wines produced in Peru and maintained that the most profitable plants to have been transplanted to America were grapes, olives, and sugar cane.

Except for the llama and the vicuña, from whose bellies were taken the best bezoar stones, Acosta did not enumerate the beasts of the New World extensively. But he asked the question so pertinent to sixteenth-century philosophy: "How it should be possible, that at the Indies there should be anie sortes of beasts, whereof the like are nowhere else?" In answer, he denied any new creations beyond those described in the Book of Genesis. But if they had been in Noah's Ark, how can one explain their absence in Europe? In America, he asserted, there are "a thousand different kindes of birdes and beasts of the forrest, which have never beene knowne, neither in shape nor name; and whereof there is no mention made, neither among the Latins, nor Greeks, nor any other nations of the world." All of them, he said, had been with Noah in the Ark, "yet by a naturall instinct, and the providence of heaven, divers kinds dispersed themselves into diverse regions, where they found themselves so well, as they would not parte; or if they departed they did not preserve themselves, but in processe of time perished wholly, as we do see it chaunce in many things."[52] This unorthodox explanation, so far removed from the current belief in the uniformity of nature the world over, must have jarred some of Acosta's European readers. But he also advanced another hypothesis, perhaps even more jarring: he proposed that perhaps these beasts did not differ essentially in kind, but by diverse accidents only appeared outwardly different, as men vary in size, build, color,

[52] Acosta, *Naturall and Morall Historie*, p. 307.

and other outward physical characteristics. But Acosta himself clearly doubted this latter explanation. Thus ended Acosta's attention to the natural history of the New World. The last 250 pages or so treated of the Indians, their history and religion, and their adjustments to Spanish overlordship. All this, Acosta assured his readers, was the work of God to insure the Christianization of the natives—a task still incomplete at the time of Acosta's departure from America.

The above survey of the works of three sixteenth-century commentators on the New World cannot be set forth as a definitive account of European literature of the Americas, nor does it pretend to be. Rather, it is a survey of the work of the three men who betrayed the greatest scientific interest in the New World during the century, who most excited the interests of European scientists, and whose works, partly because of this wide interest, were quickly translated into English—among other European tongues—and made the heaviest impact upon the English reading public at the time.

But their influence on scientific-minded Europeans was, perhaps, as much negative as positive. They went a good way to destroy several of the current learned opinions about cosmology and natural history and they posited a New World whose cosmology, geography, astronomy, native population, and natural history cried out for a new scientific nomenclature and a new scientific synthesis more adequate than those of Pliny, Dioscorides, Aristotle, Theophrastus, Ptolemy, and the other ancient and medieval authorities who had so long held sway in European scientific circles. Using the old science, they so demonstrated its inadequacies that the knowledge of the existence of the New World, with all its manifold new products, which they set forth so clearly, challenged the old science itself and created a demand for something more applicable to the new-found data. Indeed, before the end of the sixteenth century European minds were reaching out in this direction. A new science was slowly emerging out of the old, with a new scientific language and new scientific concepts that gradually rendered obsolete the language and the concepts of the old. Copernicus, Kepler, Galileo, Gilbert, and others, together with such practical achievements as those of Columbus and Magellan, created a new cosmology, a new astronomy, a new geography, and laid some of the foundations of a new physics; Vesalius improved upon the anatomy of Galen; and while these accomplishments were taking place, a host of northern European naturalists were struggling to create a new framework for natural science. Experiment and close scientific observation marked the new science more than they had the old, although the new scientists were still often tethered to the many remarkable logical deductions and the accumulated wisdom of ancient and medieval authorities. By 1600, however, a few of the observers of the New World were able to employ some of the tools of the new science, and slowly the old science of Pliny, Dioscorides, and the rest fell into obsolescence. The rise of this new science and its earliest applications to the New World scene, with primary attention to natural history, becomes the subject of the next chapter.

THREE

The New Science Comes to America

Herbals, Gardens, and Herbaria

SIMULTANEOUSLY, AS THE DISCOVERIES in the Americas opened up a new world of flora and fauna unknown to the Ancients, European scholars became increasingly aware of plants and animals native to western Europe which had escaped the attention of such authorities as Theophrastus, Dioscorides, and Pliny. The attentions of the latter had been principally confined to the eastern Mediterranean regions whereas those of the renaissance of the natural sciences were increasingly directed toward western and northern Europe, including England. The mere quantitative growth of the knowledge of flora during the sixteenth century was impressive. Dioscorides' great work, *De Materia Medica*, had described only about 500 plants; by 1623, when Gaspard Bauhin (1560–1624), a Swiss botanist and anatomist at Basel, published his πιναξ *theatri botanici . . . sive index in Theophrasti, Dioscoridis, Plinii et Botanicorum, qui a seculo scripserunt, opera, etc.* (Basel, 1623), better known as Bauhin's *Pinax*, he presented a methodical concordance of the names of plants then known totaling about 6,000.[1] In the meantime, Konrad von Gesner (1516–65), of Zurich, had published *Historia animalium* (1551–58), which performed a somewhat similar task for the knowledge of fauna. Each of these books has become, respectively, a groundwork in the early development of the modern sciences of botany and zoology.

Indeed, the modern natural sciences owe their origins principally to the efforts of a group of Teutonic scholars of Switzerland, Ger-

[1] Bauhin's plant collection, numbering about 4,000, is still extant at Basel.

many, the Low Countries, and England, who not only enlarged the knowledge of flora and fauna but also made scientific advancement in their descriptions, illustrations, nomenclature, the beginnings of new classification, and the establishment of system. Most of these men traveled extensively in Europe, their travels often becoming mandatory because of the religious upheavals occasioned by the Protestant Reformation.[2] They frequently knew one another personally, generally employed Latin as their common tongue, visited back and forth, exchanged specimens, information, and opinions, and, in short, formed one of the first *unorganized* "circles of botany" that dominated the scientific scene so widely in the Old Colonial Era. They also took advantage of the new technological processes available in their time, resorting not only to the printing press but also to the improved techniques of illustration by such Renaissance artists as Leonardo da Vinci, Albrecht Dürer, and others who employed the woodcut, and later the copper engraving, with wonderful effects.[3]

The principal medium of these men was the herbal, illustrative of the great debt which the natural sciences, especially botany, owed to medicine in their original impulse toward the systematic study not only of the "virtues" but also of the anatomy of plants and animals. At the outset, the herbal was often little more than an elucidation of Dioscorides, with an attempt to refine and enlarge upon the virtues of plants. But as time passed, as observations became keener, and as the number of new specimens so far exceeded

[2] Most of them were Protestants. It is important to note that neither Protestantism on the Continent nor Puritanism in England were inimical to the new sciences. Cf. R. Hooykaas, "Science and Reformation," *Cahiers d'Histoire Mondiale*, III, Pt. I (Paris, 1956–57), 109–141; T. K. Rabb, "Puritanism and the Rise of Experimental Science in England," *ibid.*, VII, Pt. I (Paris, 1962), 46–47. There is a large literature about English Puritanism and the rise of early modern science, most of which is cited by Rabb. As the issue does not seem particularly pertinent to my purpose, I shall ignore it hereafter. William T. Stearn shows clearly how the Continental Reformation redounded, in terms of scientific advancement, to the benefit of tolerant Leiden, in the Netherlands. See his "The Influence of Leyden on Botany in the Seventeenth and Eighteenth Centuries," *British Journal for the History of Science*, I, Pt. 2 (1962), 137–158.

[3] See Wilfrid Blunt and William T. Stearn, *The Art of Botanical Illustration* (3rd ed., London, 1955), pp. 1–90.

those known to the Ancients, the herbalist tended to become more "scientific" in the modern sense, giving more careful attention to the anatomy of his specimens, more careful observation of the functions of the various parts, more care to their nomenclature, and more experimentation with problems of classification. In consequence, after a century and a half of such studies, the herbalist ceased to exist. In place of the herbal arose the specifically medical pharmacopoeia, the specifically botanical flora, and the specifically zoological fauna. Pharmacy, botany, and zoology thus slowly emerged from the early efforts of the herbalists, although this process of evolution was by no means complete at the time of the founding of the first English colonies in America, and few, if any, of the early colonists were abreast of current knowledge in these areas.[4]

An important factor contributing to botanical progress in the sixteenth century and beyond was the botanical garden. Great formal gardens had adorned the mansions of emperors, kings, and wealthy noblemen since Egyptian times. The ancient Chinese emperors, the religious shrines of the East, the Aztec rulers of Central America, the Roman villas, the Italian villas of the Renaissance—all had had such formal gardens. Moreover, the apothecary and the physician often had his own "physic garden" in which to grow as much as possible his own supply of herbs for use in his business, and many simple folk had their own herb gardens to produce a supply of condiments and common home remedies for illnesses and injuries. Herb gatherings ("simplings") in the forests and common lands were common occupations of housewives and others every autumn. But the establishment of botanical gardens for the express purpose of testing new and exotic plants for both medicinal and botanical purposes was a development confined largely, if not entirely, to the sixteenth century and beyond. Some of these were privately maintained by wealthy men and women who often em-

---

[4] See Agnes Arber, *Herbals: Their Origin and Evolution . . . 1470–1670* (2nd ed., rev., Cambridge, 1953), especially pp. 264–268; Eleanour Sinclair Rohde, *The Old English Herbals* (London, 1922), pp. 1–120. Botany took the lead in emerging as a modern scientific discipline; and all three emerged more slowly in the English colonies of America than in Europe.

ployed gardeners who were knowledgeable in early botany. Others were set up in conjunction with universities whose medical faculties employed them for the instruction of students as well as for the empirical study of plants. The only botanical institutions of early modern times, the botanical gardens made available to students and botanists thousands of living plants from many lands. Thus they furnished, with the herbaria (collections of dried plant specimens, a means of preservation which became well known about 1550), the raw materials for the empirical study of plants, plant structures, plant similarities and differences, and the like, all contributing to the gradual establishment of an organized corpus of knowledge in botany. With the aid of the herbarium and, better, the garden, the botanist could become, with certain reservations as to ecology, his own field collector.

Although several "physic gardens" of note appeared earlier, the first botanical garden, as such, was the circular "Green Garden," established in 1545 by the Senate of Venice as an auxiliary to the medical school of Padua.[5] At about the same time, the Grand Duke of Tuscany, Cosimo de' Medici, founded a public botanical garden in conjunction with the medical school of Pisa and placed it under the direction of Andrea Cesalpino (1519–1603), Professor of Medicine whose later book, *De plantis libri XVI* (Florence, 1583), contained rudiments of the Linnaean system of plant classification. Soon afterward other botanical gardens were set up at Zurich (1560), where Konrad von Gesner and his friends were able to employ it for their studies; at Bologna (1547), where William Turner (1510–68), the "Father of English Botany," studied for a time before he moved to Zurich to become a close friend of Konrad von Gesner; at Leiden (1577), whose faculty, especially in the person of the peripatetic Charles de L'Escluse (1526–1609), had direct connections with early English botanists; at Paris (1591), whose early directors, Jean Robin and his nephew, Vespasien, grew some of the earliest plants from Canada and New England to find their way into European descriptive catalogues; at Montpellier (1598), whose erudite Guillaume Rondelet (1507–66) had already inspired

[5] Wolfgang Born, "Early Botanical Gardens," *Ciba Symposia*, XI (1949), 1099–1109.

a host of brilliant students in natural history; and others arose during the course of the Old Colonial Era.[6]

Early English Gardens and Herbalists

England lagged behind in the establishment of botanical gardens by public institution. However, there grew up a number of important gardens in private hands.[7] John Gerard (1545–1612), a barber-surgeon and lover of plants, obtained the patronage of William Cecil (1520–98), who became Queen Elizabeth I's chief Secretary of State, and later, as the first Baron Burghley, was Lord High Treasurer. Lord Burghley appointed Gerard as Superintendent of his gardens in the Strand, London, and at Theobald's, in Hertfordshire, and Gerard developed a private garden of his own in Fetter Lane, Holborn (1596). He superintended the noble lord's gardens for twenty years and after Burghley's death was, for a short time, Overseer of Queen Anne's (consort of James I) garden near Somerset House in London. Gerard was on close terms with Jean Robin, Director of the Paris garden, and they frequently exchanged plant specimens.[8] Matthias de Lobel (1538–1616), a Flemish botanist and physician who settled in England, became Superintendent of the garden maintained by Lord Zouche at Hackney in 1592 and later (1606) was appointed King's Botanist by James I.[9] John

[6] The most important were: Leipzig (1579), Heidelberg (1593), Giessen (1605), Strasbourg (1620), Uppsala (1657), Berlin (1679), Edinburgh (1680), Amsterdam (1682), St. Petersburg (1713), Vienna (1754), Dublin (1790). Cf. also Pulteney, I, 52–55.

[7] This situation was typical of the English encouragement to science in general. Whereas on the Continent there was wide support of science from a variety of public institutions (especially in France, where the royal purse lent liberal support to scientific researches), in England scientists generally were forced to pay their own way or seek a private patron, or both.

[8] Pulteney, I, 116ff.; Rohde, Old English Herbals, pp. 100–104.

[9] Pulteney, I, 96–98; Robert T. Gunther, Early British Botanists and Their Gardens . . . (Oxford, 1922), pp. 245–247. Lord Zouche (Edward Zouche, eleventh Baron Zouche of Harringworth, 1556?–1625) was said to have spent most of his fortune on his physic garden. It was fashionable for seventeenth-century English gentlemen to become amateur natural scientists. Lord Herbert of Cherbury, writing about 1643, said: "I conceive it is a fine study, and worthy of a gentleman to be a good botanic, that so he may know the nature of all herbs and plants, being our fellow-Creatures, and made for the use of man. . . ." Sidney Lee, ed., The

Tradescant (d. 1638), known as the Elder, together with John Tradescant the Younger (1608–62), made up a father-and-son team who did much to arouse interest in natural history in England during the first half of the seventeenth century. The elder Tradescant, possibly of Dutch origin, had traveled extensively in Europe and North Africa and had collected natural curiosities from many parts. With the patronage of Robert Cecil (1563?–1612), first Earl of Salisbury and son of Lord Burghley mentioned above, the elder Tradescant became Superintendent of the Earl's gardens at Hatfield about 1611. By 1629, with his son, he set up at Lambeth a garden and museum of natural history ("Tradescant's Ark") which, by the early 1630's, had become one of the sights of London, attracting many benefactors of high rank, including the King and Queen. The younger Tradescant made a voyage to Virginia (1637) to obtain new plants for the Lambeth garden and other rarities for the museum; and together the Tradescants introduced several new species of plants to the English public.[10] The *Musæum Tradescantianum: or, a Collection of Rarities preserved at South-Lambeth, neer London* (London, 1658), was a catalogue of the museum prepared by the younger Tradescant, and it listed flora, fauna (including fishes, insects, and shells), minerals, coins, medals, and other curiosities from various parts of the world, together with a list of the plants of the Lambeth garden. The younger Tradescant bequeathed the

Autobiography of Edward, Lord Herbert of Cherbury (2nd ed., rev., London, n.d.), p. 31. Lord Herbert also wrote (p. 28): "It will become a gentleman to have some knowledge in medicine, especially the diagnostic part. . . ."

[10] Pulteney, I, 176–179; Gunther, *Early British Botanists*, p. 333. The younger Tradescant was sent by Charles I "to gather all rarities of flowers, plants, and shells." He introduced to England the tulip tree (*Liriodendron tulipifera*), swamp maple (*Acer rubrum*), hackberry (*Celtis occidentalis*), sycamore (*Platanus occidentalis*), black walnut (*Juglans nigra*), and, perhaps, the butternut (*J. cinerea*) and bald cypress (*Taxodium distichum*). The garden also contained plants from Brazil, a "Cabbage tree from Barbados," and varieties of flowers from the West Indies, Peru, Virginia, New England, and Canada. The museum contained parrots from Brazil, hummingbirds from Virginia ("3 sorts"), a red-winged blackbird, a bluebird, a red bird (cardinal?), and a few others. The animals included a Virginia fox, muskrats, a wildcat, a flying squirrel, and a rattlesnake. Among other curiosities on display were American Indians' weapons (bows, arrows, darts, quivers, and "six sorts" of "Tomahacks"), thirty kinds of tobacco pipes, and Indian habits, one of bearskin, one of feathers, and one of shells said to have belonged to Powhatan, "the King of Virginia." Cf. Tradescant, *Musæum Tradescantianum, passim*.

museum to his friend, Elias Ashmole (1617–92), who later incorporated it with other collections which formed the nucleus of the Ashmolean Museum at Oxford. Other private English gardens included those set up at Long Acre by John Parkinson (1567–1650), a London apothecary; at Edgcombe, Surrey, by Sir John Tunstal, Gentleman Usher to Queen Anne; and at Droxford, Hampshire, by John Goodyer (b. 1592), a medical botanist of considerable merit who grew in his garden a variety of American plants, including the Jerusalem artichoke (*Helianthus tuberosus*), early confused with the potato (*Solanum tuberosum*), tobacco, and others.[11]

But the first botanical garden established in England under institutional (and hence more permanent) auspices was that set up for the advancement of the medical faculty at Oxford in 1621. In that year Magdalen College granted the university a five-acre plot of land and Henry Danvers, the first Earl of Danby (1573–1645), supplied a small endowment for the maintenance of the garden. However, the land was marshy and the funds were inadequate. With the disorders accompanying the English Civil Wars a few years later, the Oxford Physic Garden did not reach a position of importance until after the Stuart Restoration (1660). In the meantime, Jacob Bobart the Elder (1599–1680),[12] a Danziger, was appointed Horti Praefectus, or Keeper of the Garden, about 1640, and his catalogue of plants, first published as *Catalogus Plantarum Horti Medici Oxoniensis* (Oxford, 1648), listed about 1,600 species growing there.[13] Before the end of the seventeenth century, the Oxford garden had become one of the principal English botanical testing grounds for the observation of and experimentation in plants from the American colonies.[14] By this time, however, two more botanical gardens of import had been established in London. The first was that set up by the Worshipful Society of Apothecaries of

[11] The lesser gardens are cited by Pulteney and Gunther. See note 10 above, and Rohde, *Old English Herbals, passim.*

[12] Bobart's son, Jacob Bobart the Younger (1641–1719), succeeded his father as Horti Praefectus at Oxford in 1679.

[13] A second, much enlarged, edition appeared from Oxford in 1658. Prepared with the assistance of two Oxford Fellows, this catalogue enlarged the number of species to about 2,000. But while the book was improved in its nomenclature, much confusion is evident in the identification of species.

[14] Besides Gunther, *Early British Botanists*, p. 49, see Clokie, pp. 1–3.

London in 1673 along the Thames River in Chelsea.[15] The second
was that of Henry Compton (1632–1713), Bishop of London
(1675–1713), whose garden at Fulham became one of the show
places of London during the latter years of the seventeenth century. Under the guidance of a succession of learned, skillful overseers, those gardens, with that of Oxford, became the principal
English posts for the observation and study of American colonial
flora well into the eighteenth century.

Continental Forerunners of the New Science in England

A glance at the geographical locations of the botanical gardens,
English and Continental alike, together with their chronological
succession, suggests that developments in the natural sciences
broadly followed the course of change in the great trade routes
from the Mediterranean to the Atlantic seaboard during the fifteenth, sixteenth, and seventeenth centuries. This, in turn, emphasized anew the significance of overseas trade and colonization in
quickening the growth of natural science. Many of the scholar-
collectors were as interested in fauna as in flora, and just as the
traffic in commerce was international so was that in animals and
plants, until hundreds of scientists and amateur collectors made up
an informal international community of early natural scientists
bound together by mutual interests, travel, personal acquaintance,
correspondence, and the lively exchange of both live and preserved
specimens of flora and fauna.

It lies beyond the purpose of this study to trace in detail the rise
of the natural sciences during the sixteenth century. But as some of
the English scholars traveled and studied abroad, and as some of the
Continentals visited in England (or, as in the case of Matthias de

15 Henry Field, *Memoirs, historical and illustrative, of the Botanick Garden at
Chelsea* . . . (London, 1820), pp. 10–14; Cecil Wall and H. Charles Cameron, *A
History of the Worshipful Society of Apothecaries of London*, I (1617–1815),
rev. and ed. by Ashworth Underwood (*Publications of the Wellcome Historical
Medical Museum*, n.s., No. 8, London, 1963), 164–181; Sir George Clark, *A History
of the Royal College of Physicians of London* (2 vols., Oxford, 1964–66), I,
160; Pulteney, II, 99–100, 105–107; "Papers and Draughts of the Reverend Mr.
Banister in Virginia sent to Dr. Henry Compton, Bishop of London and Dr. Lister
from Mr. Petiver's Collection," *Sloane 4002*, 118 fols. Other references also occur
in the *Sloane Manuscripts*, especially *Sloane 2346, 3321, 3331*, and *3336*.

Lobel, settled there permanently), they contributed to the scientific outlook with which Englishmen first attacked the study of natural phenomena in their New World colonies. Moreover, their herbals had international effects, for they were almost all written in Latin, the common tongue of educated men everywhere in western Europe. At least seven of these sixteenth-century Continental naturalists played a role in fashioning English natural science. Leonhard Fuchs (1501–66), a German botanist, published, among other works, *De historia stirpium* . . . (Basel, 1542), which was frequently republished during the sixteenth century and translated into German, Flemish, French, and Spanish.[16] One of its finest qualities lay in the excellence of its woodcuts of plants, which were used again and again by various publishers and widely copied so that it became the model of plant illustration in the sixteenth century.[17] Konrad von Gesner (1516–65), a Swiss naturalist and "universal genius," had wide associations with Englishmen and wrote commentaries on medicine and natural history, including *Catalogus plantarum latine, Græce, Germanice, et Gallice . . . Adjectae sunt etiam herbarum nomenclature variarum gentium, Dioscoridi ascriptae* (Zurich, 1542), *De omni rerum fossilium genere, geminus, lapidibus, metallis, et hujusmodi, libri aliquot . . .* (Zurich, 1565), and others. But his most influential work was *Historia animalium*, in four volumes (Zurich, 1551–58), which included animals, birds, serpents, fishes, and insects of both the Old World and the New.[18] Andrea Cesalpino (1519–1603), an Italian teacher of medicine and natural philosophy and Director of the botanical garden at Pisa, published (among other works) *De plantis libri XVI* (Florence, 1583), one of the few sixteenth-century botanical studies to emphasize the importance of plant classification. And Gaspard Bauhin's

[16] The British Museum Catalogue lists Latin editions of 1542, 1543, 1546, 1547, 1549, 1551, and 1555; French translations of 1549, 1558; German, 1543; Flemish, 1550; and Spanish, 1557.

[17] Blunt and Stearn, *Art of Botanical Illustration*.

[18] Sir William Cecil [Lord Burghley] rewarded Gesner for his book on animals. Cf. Great Britain, *Historical Manuscripts Commission Calendar of the Manuscripts of the Hon. Marques of Salisbury . . . Preserved at Hatfield House, Hertfordshire* (London, 1833——), Pt. I, pp. 261, 835. See also L. C. Miall, *The Early Naturalists; Their Lives and Work, 1530–1789* (London, 1912), pp. 28–32; Diethelm Fretz, *Konrad Gesner als Gärtner* (Zurich, 1948).

DE PLANTIS
LIBRI XVI.
ANDREAE CAESALPINI
ARETINI,

Medici clarifsimi, doctifsimiq; atque
Philofophi celeberrimi, ac
fubtilifsimi.

AD SERENISSIMUM FRANCISCUM

Medicem, Magnum Aetruriæ Ducem.

FLORENTIAE,
Apud Georgium Marefcottum.
MDLXXXIII.

1. Andrea Cesalpino's early herbal (1583) contained rudiments
employed in later systems of plant classification.

Courtesy of the Missouri Botanical Garden.

ΠΙΝΑΞ
THEATRI BOTANICI
CASPARI BAVHINI
Basileenſ. Archiatri & Profeſſoris Ordin.

Colleg. Par. Soc. I.W.

ſive

INDEX
IN THEOPHRASTI DIOSCORIDIS
PLINII ET BOTANICORVM
qui à Seculo ſcripſerunt
OPERA:

PLANTARVM CIRCITER SEX MILLIVM
AB IPSIS EXHIBITARVM NOMINA CVM
earundem Synonymiis & differentiis

Methodicè ſecundùm earum & genera & ſpecies proponens.

OPVS XL. ANNORVM

Hactenus non editum ſummoperè expetitum & ad auctores
intelligendos plurimùm faciens.

GALENVS I. de ANTID. V.

Medicus omnium STIRPIVM, ſi fieri poteſt, peritiam habeat conſulo:
ſin minus, plurimarum ſaltem, quibus frequenter utimur.

M D C XXIII.

BASILEAE HELVET. *Sumptibus & typis Ludovici Regis.*

2. The Greek Word *Pinax* at the head of Gaspard Bauhin's title
meant "Catalogue" and implied all-inclusiveness.

Courtesy of the Rare Book Room, University of Illinois Library.

Pinax (1623) gave unusually clear descriptions of plants, early set forth the distinctions between genera and species, and employed a form of binary nomenclature widely retained by later students of botanical system down to the present day.[19]

Three Continental scholars had an even greater influence on the English scene: Rembert Dodoens (1517–85), a Flemish physician and botanist who, after serving as Physician to the Emperors Maximilian II and Rudolph II, became Professor of Medicine at Leiden; Charles de L'Escluse (1526–1609), a Walloon physician and botanist who traveled and corresponded widely over Europe, including at least three visits to England, before he settled down (1589) as Professor of Botany at Leiden; and Matthias de Lobel (1538–1616), a Flemish botanist and physician who, having studied with the famous Guillaume Rondelet at Montpellier, ultimately settled in England (1599) and became Royal Botanist and Physician to James I. These three men knew one another and to some extent worked together, their works being published, in part, by Christophe Plantin (1514–89), a French Huguenot printer who settled in Antwerp about 1555. Plantin was himself learned in botany and his publishing house collected a large number of the best woodcuts of plants, including those of Leonhard Fuchs, which were employed over and over in the botanical works issued under the Plantin imprint and lent or sold to other publishers for their use. Among the many published works of Dodoens was the *Cruijdeboeck. In den welcken de ghaheele historie dat es gheslacht, tfatsoen, naem, natuere, cracht ende werckinghe* . . . (Antwerp, 1554).[20] L'Escluse translated it into French (1557) and Henry Lyte used this to publish an English translation as *A Niewe Herball, or Historie of Plants . . . First set foorth in the Doutche or Almaigne tongue by D. Rembert Dodoens . . . and nowe first translated out of the French into English . . .* (London, 1578). Dodoens was a severe critic of the Doctrine of Signatures, although he arranged plants in accordance with their uses, dividing vegetation into six classes: plants remarkable for their flowers and seeds; plants of medicinal use; plants em-

---

[19] Arber, *Herbals*, p. 116.

[20] The first edition of the *Cruijdeboeck* was not published by Plantin. Later Dutch editions appeared at Antwerp, 1563, and Leiden, 1608, 1618. Dodoens' later works were published by Plantin.

ployed as cereals, legumes, and forage; herbs and fruits of culinary use; trees and shrubs; and a miscellany of plants that fall into none of the above categories. Within the groups each plant was described under six regularized heads: the name of the plant; the kind of the plant (or species); the description; the place of origin; the time of flower or fruit; the names—i.e., the various words used to denominate it in Flemish, German, Bohemian, French, English, Spanish, Italian, Arabic, Greek, and Latin; the nature of the plant, listing those similar in nature regardless of their names; and the virtues of the plant, with frequent protest against the notion that the appearance of a plant had anything to do with its uses.[21] Dodoens published his *Cruijdeboeck* in his native Flemish in order that it might be of greater use to those of his countrymen who knew no Latin. Its French and English translations proved of similar use to literate persons of those nations. It was illustrated with beautiful woodcuts from Fuchs, the best of the day, and the English edition by Henry Lyte included 870 of these, probably the best herbal in English in the latter part of the sixteenth century.

L'Escluse had a long and interesting life. Born (1526) in Arras (Artois) of a well-to-do noble family, his education was that of the typical medieval scholar—at Ghent, Louvain, Marburg, Wittenberg, Strasbourg, Geneva, Lyons, and Montpellier. He began in the study of law but later shifted to medicine and botany. After taking a degree in medicine at Montpellier, he returned for a time to Arras, then went on a series of wide travels, including a botanizing expedition in the Iberian Peninsula (1564–65), where he became acquainted with Monardes and his garden at Seville, a familiarity which impelled him to publish a Latin translation of Monardes' book (Antwerp, 1574).[22] He also learned of the works of Acosta, among others. In 1561 and again in 1571 he made his first visits to England. Returning to the Continent, he went to Vienna, where he was appointed Director of the gardens of the Emperor Maximilian

[21] P. J. Van Meerbeeck, *Recherches Historiques et Critiques sur la Vie et les Ouvrages de Rembert Dodoens* (Malines, 1841), pp. 89–124.

[22] His translation of Monardes was entitled *De Simplicibus Medicamentis ex Occidentali India Delatis, quorum in Medicina usus est.* Other editions appeared in 1579 and 1605. Another fruit of his Iberian visit was *Rariorum aliquot per Hispanias observantum Historia* (Antwerp, 1576).

II, a post he held for fourteen years. During this time he visited England again, becoming acquainted with Sir Francis Drake and other Elizabethan "seadogs," from whom he learned of their overseas voyages.[23] Also, while at Vienna, he was active in introducing new plants to western Europe, especially from the Turkish dominions in the Near East. Tiring of court life, he departed from Vienna in 1587 and settled for six years at Frankfurt am Rhein. In 1589 he was invited to become Professor of Botany at Leiden, where he spent his remaining years. He developed an immense correspondence with many of the leading botanical scholars in western Europe, had a prodigious memory, could read widely in both the ancient and modern languages, and possessed a genuine enthusiasm for the discovery of new and hitherto unknown plants. He was the principal factor in introducing (from Turkish sources) the bulb industry to the Low Countries, which became so profitable in Holland in the seventeenth century. He is credited with having introduced the potato from Spain to Germany, France, and the Low Countries. In all, he added about 600 new species to the European flora. Under his direction the Leiden botanical garden took new life and developed more than 1,000 varieties of plants, mostly of European or west Asian origin. L'Escluse was no systematist. His individual descriptions of plants were usually adequate and he helped to typify species for later scholars. But his principal contributions lay in his discovery of new plants and in his enthusiastic cultivation and accurate description of foreign species from Turkey, the Levant, and the Americas.[24]

Matthias de Lobel (or L'Obel) (1538–1616) was a Protestant Fleming who studied medicine and botany at Montpellier under Guillaume Rondelet. After receiving his medical degree, he botanized widely in southern France, Switzerland, Germany, and northern Italy, becoming familiar with the plant species of those regions. He then settled down to practice medicine, first at Antwerp

[23] He later published a Latin edition (Frankfurt am Main, 1595) of Thomas Hariot's *A briefe and true report of the new found land of Virginia* . . . (London, 1588), Hariot's account of Sir Walter Ralegh's "lost colony of Roanoke."

[24] T. W. T. Hunger, *Charles de L'Escluse (Carolus Clusius), Nederlandsch Kruidkundige, 1526–1609* (2 vols., The Hague, 1927, 1943); Blunt and Stearn, *Art of Botanical Illustration,* pp. 64–65; Arber, *Herbals,* pp. 82–89; Stearn, "The Influence of Leyden on Botany," *passim.*

and later at Delft. About 1569, disturbed by the religious troubles in the Low Countries and the confusion accompanying the approaching Dutch Wars for Independence, he went to England. There, with the cooperation of one Pierre Pena, he published his first work, dedicated to Queen Elizabeth I and entitled *Stirpium Adversaria Nova* . . . (London, 1570).[25] He returned to the Netherlands in the late 1570's, and, working jointly with Plantin at Antwerp, published *Plantarum seu Stirpium historia* . . . (Antwerp, 1576), with 1,486 woodcuts of plants.[26] Still later he served as Physician to William the Silent until the latter's assassination in 1584. Lobel then returned to England and became Superintendent of the garden maintained by Lord Zouche at Hackney. In 1592 he accompanied Lord Zouche on a diplomatic mission to Denmark, further enlarging his botanical knowledge by collecting plants there. After the accession of James I to the English throne (1603), he was appointed Royal Botanist and Physician to the King. He retired about 1606 and spent his last decade at Highgate in the home of his son-in-law, James Cole, devoting these years to work on *Stirpium Illustrationes* . . . , a work left incomplete at his death in 1616. The work fell into the hands of John Parkinson (1567–1650), who incorporated much of it in his *Theatricum Botanicum* . . . (London, 1640), and portions of it were later published by William How under Lobel's title (London, 1655). Lobel was criticized for a coarse style in his writings, but he contributed greatly to the botanical system. His considerable travels, both on the Continent and in England, together with his keen observations and knowledge of the works of his predecessors from the Ancients to his own day, enabled him to approach a natural system of classification in botany. This, though filled with many inaccuracies, was an improvement upon arrangement by mere alphabetical order and upon a classification based upon plants' supposed medicinal virtues. Lobel organized plants into forty-four orders and, although his tribes were ill-defined and the arrangement sometimes incongruous, he was clearly groping toward a better system. He was the first to distinguish between plants with primary single

---

[25] Later editions appeared in the Netherlands, 1571, 1572, 1576; again, in London, 1605; and Frankfurt, 1651.

[26] A later edition enlarged this number to 2,191.

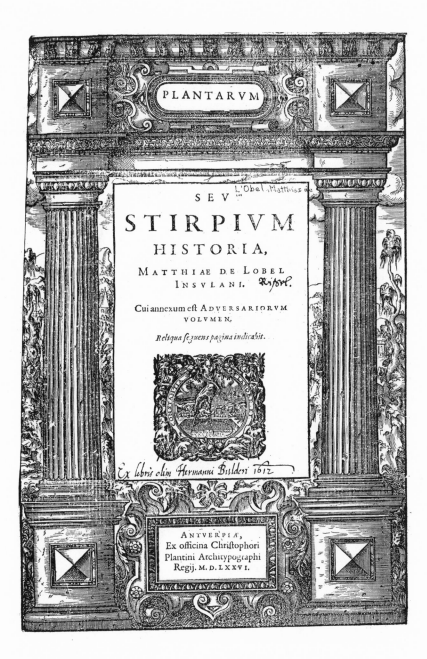

PLANTARVM

SEV
STIRPIVM
HISTORIA,
MATTHIAE DE LOBEL
INSVLANI.

Cui annexum est ADVERSARIORVM
VOLVMEN,

Reliqua sequens pagina indicabit.

ANTVERPIÆ,
Ex officina Christophori
Plantini Architypographi
Regij. M.D.LXXVI.

3. Matthias de Lobel's work introduced some of the first permanent
contributions to plant classification, and the author's settlement in
England coincided with an upturn in early English studies of flora.

Courtesy of the Rare Book Room, University of Illinois Library.

IAC. CORNVTI
DOCTORIS MEDICI
PARISIENSIS

CANADENSIVM PLANTARVM,
aliarúmque nondum editarum

HISTORIA.

Cui adiectum est ad calcem

ENCHIRIDION
BOTANICVM PARISIENSE,

Continens Indicem Plantarum, quæ in Pagis, Siluis, Pratis, &
Montosis iuxta Parisios locis nascuntur.

PARISIIS,
Venundantur apud SIMONEM LE MOYNE, viâ Iacobeâ.

M. DC. XXXV.
CVM PRIVILEGIO REGIS.

---

4. Jacques Cornuti, a Paris physician, adapted Lobel's nomenclature
to describe Canadian plants (see pp. 81-82 below).

Courtesy of the Rare Book Department, McGill University Library.

seed leaves (monocotyledons) and those with primary double seed leaves (dicotyledons). In general, his efforts at classification excelled those of Dodoens. In his English travels he discovered about eighty new plants which hitherto had escaped the notice of English scholars. He was familiar with the English botanical gardens and personally knew nearly all the English naturalists of his generation. How much he contributed to their botanical expertise, beyond his published works, is impossible to assess.[27]

The First English Scholars in the New Science

In England, the natural sciences at the opening of the colonial period were more derivative than original. In general, the works of Continental scholars overshadowed those of their English contemporaries. Although English scholars were abreast of the Continentals, and occasionally in the vanguard, in mathematics, geography, astronomy, cosmography, cosmology, cosmogony, and the arts and technology related to ship-building and navigation,[28] they made

[27] Pulteney, I, 96–107; Gunther, *Early British Botanists*, pp. 245–252.

[28] See James Orchard Halliwell [Halliwell-Phillips], ed., *A Collection of Letters Illustrative of the Progress of Science in England from the Reign of Queen Elizabeth to That of Charles the Second* (London, 1841); Foster Watson, *The Beginning of the Teaching of Modern Subjects in England* (London, 1909); Lambert Lincoln Jackson, *The Educational Significance of Sixteenth Century Arithmetic* . . . (New York, 1906); *Shakespeare's England: An Account of the Life and Manners of His Age* (2 vols., Oxford, 1917), a cooperative work of great value; Louis B. Wright, *Middle-Class Culture in Elizabethan England* (Chapel Hill, 1935); E. G. R. Taylor, *Tudor Geography, 1485–1583* (London, 1930), *The Mathematical Practitioners of Tudor and Stuart England* (Cambridge, 1954), and *Ideas on the Shape, Size and Movements of the Earth* (London, 1943); Bois Penrose, *Travel and Discovery in the Renaissance, 1420–1620* (Cambridge, Mass., 1952); Walter I. Trattner, "God and Expansion in England: John Dee, 1527–1583," *Journal of the History of Ideas*, XXV (1964), 17–34; Ernest A. Strathmann, "The History of the World and Ralegh's Skepticism," *Huntington Library Quarterly*, III (1939–40), 265–287; Francis R. Johnson, "The Influence of Thomas Digges on the Progress of Modern Astronomy in Sixteenth-Century England," *Osiris*, I (1936), 390–410; Johnson and Sanford V. Larkey, "Robert Recorde's Mathematical Teaching and the Anti-Aristotelian Movement," *Huntington Library Bulletin*, No. 7 (April, 1935), pp. 59–87; R. A. Skelton, "Ralegh as a Geographer," *Va. Mag. Hist. & Biog.*, LXXI (1963), 131–149; W. D. Walters, "The Sea Chart and the English Colonization of America," *The American Neptune*, XVII (1957), 28–37; and, for a more general overview, Raymond P. Stearns, "The Scientific Spirit in England in Early Modern Times (*c.* 1600)," *Isis*, XXXIV, Pt. IV (1943), 293–300; reprinted in R. S. Schuyler and H. Ausubel,

comparatively few original contributions to the natural sciences. The early Anglo-Saxons excelled in their knowledge of medicinal herbs and their uses, but this decayed after the Norman invasion, and the principal herbal extant in the English language in the early sixteenth century was a printer's flyer compiled from Continental sources and published as *The Grate Herball* . . . (London, 1526, and later sixteenth-century editions), a strictly medieval compilation innocent of the early scientific strivings of the day.[29]

The English pioneer in the natural sciences was William Turner (c. 1510–68), called the "Father of English Botany." Turner was a physician and divine whose Protestant learnings sometimes caused him to run afoul of the law. He was educated at Pembroke College, Cambridge, where, he said, "I could learn never one Greke, neither Latin, nor English name, even amongst the physicians, of any herbe or tree; such was the ignorance at that time; and as there was no English Herbal, but one [*The Grate Herball?*] all full of unlearned cacographies and falsely naming of herbes."[30] After he had graduated, he preached and botanized over England, but his reforming zeal soon led to imprisonment. Upon his release, he went into voluntary exile on the Continent, where he resided until the accession of Edward VI. Returning to England, he was incorporated Doctor of Physic at Oxford, appointed Physician to Edward, Duke of Somerset (who maintained a fine botanical garden), and multiple offices in the English church, including the Deanery of Wells. During the reign of Mary Tudor, he again retreated to the Continent, returning to England after the accession of Queen Elizabeth I in 1558. Elizabeth reinstated him in his church preferments and he remained in England for the rest of his life.

During his enforced exiles Turner had traveled widely on the Continent, served as Physician to the Dutch Earl of Embden, Lord of East Friesland, acquired a degree of Doctor of Medicine at

---

The Making of English History (New York, 1952), pp. 225–231. Two recent publications of the Folger Shakespeare Library are useful. See F. D. and J. F. M. Hoeniger, *The Development of Natural History in Tudor England* (Charlottesville, 1969), and *The Growth of Natural History in Stuart England from Gerard to the Royal Society* (Charlottesville, 1969).

29 Rohde, *Old English Herbals*, pp. 3–70; Pulteney, I, 46, 50.
30 *Ibid.*, I, 60.

Ferrara, Italy, studied natural science at Bologna under Lucas Ghinus, the organizer of a botanical garden there, and resided at Basel, Strasbourg, Bonn, Cologne, and elsewhere. In the course of these peregrinations, he became personally acquainted with many of the Continental scholars in the natural sciences, including Leonhard Fuchs and Konrad von Gesner, to the latter of whom he referred as a "lifelong friend." He exchanged scientific specimens with some of these men, became familiar with their work, their botanical gardens, and their natural history collections. Thus he gradually drew abreast of Continental scholarship, especially in botany, in which he had displayed an interest since his youth. Back in England after Elizabeth's accession to the throne, he developed botanical gardens of his own, at Wells, where he was Dean of the Cathedral, at Crutched Friars, in London, where he had a house, and at Kew.

Turner published several works in medicine, religion, and natural history. Gesner considered him an authority in ornithology and English fishes. But his greatest efforts were in botany, in which field he published the earliest original scientific work in English. One of his books, perhaps an attempt to supply the want he had found at Cambridge, was *The Names of Herbes in Greke, Latin, English, Duche, and French with their Commune names that Herbaries and Apothecaries use* (London, 1548). His greatest contribution, however, was his *Herball*. This book appeared in piecemeal fashion in three parts. The first part, though written while Turner was in exile, was published in London as *A New Herball* . . . (1551); the second part was published in Cologne (1562); and these were combined with the third part as *The first and seconde partes of the Herbal of William Turner . . . enlarged with the Thirde parte . . .* (London, 1568). It was illustrated with 502 woodcuts, upwards of 400 of which were Fuchs's figures and about 90 were new, cut especially for this volume. Turner listed plants alphabetically, using their Latin names, with no other attention to botanical system; and he used Dioscorides as his guide wherever possible. However, he included plants "whereof is no mention made nether of ye old Grecians nor Latins," including a few plants from the Americas and 238 plants native to England. To some of the English plants he attached scientific names for the first time.

He proposed several corrections to the works of Fuchs and others of his contemporaries, especially in their application of the names employed by Dioscorides. On the whole, and in spite of errors in the application of figures to specific plants, the work was carefully and critically done, with several corrections to the Ancients' views regarding the virtues of plants where Turner's own experience proved different. But it was soon outdated. Henry Lyte's English translation of Dodoens' *Cruijdeboeck, A Newe Herball, or Historie of Plants* . . . (London, 1578), supplied the English public with a better arrangement (also using Fuchs's woodcuts), albeit with less attention to plants of specifically English provenance.[31]

In the years between the publication of Turner's *Herball* (1568) and the end of the sixteenth century a variety of events combined to transform the English scene. It was in this generation, the "glorious days of Good Queen Bess," that the scientific spirit took firm root in England, that England turned to overseas adventure, and that New World products first excited English consciousness. Sir John Hawkins engaged in the slave trade to the West Indies and the Spanish Main and made subsequent voyages to the New World; Sir Martin Frobisher sought the Northwest Passage to Cathay in Canada and later sailed to the West Indies; Sir Walter Ralegh voyaged to North America in 1584, established the short-lived English colony at Roanoke (1585–86), and explored parts of South America in the mid-1590's; Sir Francis Drake circumnavigated the globe (1577–80) and returned to North America in 1586; Thomas Cavendish again circumnavigated the earth (1586–88); and the Spanish Armada was turned back in 1588 amid wild excitement in England. Many of these men enlisted the technological aid of both English and Continental scientists in problems of ocean navigation,[32] and some of them, Sir Walter Ralegh in particular, adopted something closely approximating a scientific attitude of mind and patronized several of the scientists of the day.[33]

---

[31] *Ibid.*, I, 56–76; Rohde, *Old English Herbals*, pp. 75–119.

[32] See, for example, E. G. R. Taylor, "Instructions to a Colonial Surveyor in 1582," *The Mariner's Mirror*, XXXVII (1951), 48–62, *Tudor Geography*, pp. 75–96, 256, and *Mathematical Practitioners*, pp. 35–45.

[33] John William Shirley, "The Scientific Experiments of Sir Walter Ralegh, The Wizard Earl, and the Three Magi in the Tower, 1603–1617," *Ambix*, IV (1949), 52–66.

Their overseas exploits were made widely known to the English public by the promotional literature published by Richard Hakluyt, the first edition of whose *Principall Navigations, Voiages, and Discoveries of the English Nation* appeared in 1589. In the meantime, hundreds of specimens of the natural products of the New World had been brought into England. A number of wealthy patrons of the sciences emerged.[34] The works of Monardes and Acosta had become known, and the medicinal and commercial opportunities to be exploited in the New World aroused the interest of adventurous Englishmen to a considerable heat.[35] Henry Lyte had published his English translation of Dodoens' *Cruijdeboeck* (1578); one of Charles L'Escluse's visits to England was to help celebrate Ralegh's return from a voyage and to view the natural products which the latter had brought back from the New World; and Matthias de Lobel settled permanently in England about 1590, thus enhancing English studies in the natural sciences to a considerable degree. Dynastic politics, commercial expansion, and the new science conjoined to place England once again among the Great Powers.

It was in this atmosphere that there arose in England a new generation of scientists, some of whom equaled and even excelled the achievements of their Continental contemporaries.[36] Such men as John Dee (1527–1608), a mathematician, geographer, astrologer, and alchemist; Thomas Digges (d. 1595), mathematician and military engineer; Thomas Hariot (1560–1621), algebraist and physicist—all went far to bring England abreast of the Continent in mathematics, astronomy, physics, and the art of navigation. William Gilbert (1540–1603) published *De Magnete* (1600), the first great English scientific work of lasting value; and William Harvey's *Exercitatio de Motu Cordis et Sanguinis* (1628) revolutionized anatomy and physiology with the earliest account of the role of the heart in the circulation of the blood.

But these achievements were not equaled in the realm of natural

34 D. Nichol Smith, "Authors and Patrons," *Shakespeare's England*, Vol. II, Chap. 22. The Earl of Leicester, Lord Burghley, the Earl of Essex, and Sir Philip Sidney were generous patrons.

35 Garcia de Orta (or del Huerte), a Portuguese physician in Goa, had written about the Far Eastern products in much the same manner as Monardes had done for Spanish-American products. This, too, excited European interest.

36 See Stearns, "The Scientific Spirit in England," for details and references.

science, although marked improvements were made. After William Turner, the most famous of the English herbalists was John Gerard (1545–1612). Gerard was a barber-surgeon who, as has been noted, became Superintendent of Lord Burghley's gardens and also developed a private botanical garden of his own in London. He had something of a classical education, although his knowledge of the ancient languages was faulty. But he had a "feel" for plants, was widely acquainted with the botany circle that was arising in England in the 1590's and beyond, and had friendly intercourse with Jean Robin, Director of the Paris garden. In 1596 Gerard published *Catalogus Arborum, Fruticum, ac Plantorum, tam indigenarum quam exoticarum, in horto Johannes Gerardi . . .* , listing about 1,100 sorts of plants in his garden. During the following year he published his principal work, *The Herball, or Generall Historie of Plantes . . .* (London, 1597), a massive volume with some 1,800 woodcuts, most of them reused from the Dutch edition of Johann Teodor Tabernaemontanus' *Eicones Plantarum . . .* (1588). Gerard's *Herball* was not wholly an original work. For the most part it was plagiarized from an English translation of Dodoens' *Pemptades*,[37] with generous borrowings from Lobel's *Stirpium Adversaria* (London, 1570),[38] especially Lobel's system of organization. Scientifically, it was inferior to Continental works of the time. Lobel was indignant about it and was said to have offered more than a thousand corrections to the early portion of the book which he had seen in manuscript, but Gerard refused to accept them. The book consisted of more than 800 chapters divided into three parts: the first containing grasses, grains, rushes, seeds, flags, and bulbous-rooted plants; the second, all herbs used in medicine, food, and ornamentation; and the third, trees, shrubs, fruit-bearing plants, resins, gums, roses, mosses, mushrooms, and sea plants. With a few exceptions, the plant descriptions were indifferent to poor from a scientific point of view. He had described a potato in his *Catalogus* of 1596, and his *Herball* devoted a whole chapter to it, with the first woodcut of it ever to be published.[39] His description, however,

---

[37] . . . *Stirpium historiæ pemptades sex, sive libri XX* (Antwerp, 1583), commonly referred to as *Pemptades*.

[38] For other editions, see note 25 above.

[39] The cut, one of the few new ones in Gerard's *Herball*, is one of his best.

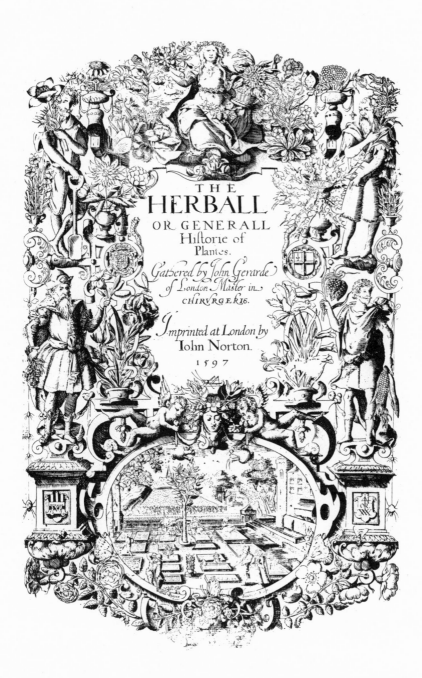

THE
HERBALL
OR GENERALL
Historie of
Plantes.

Gathered by John Gerarde
of London Master in
CHIRVRGERIE.

Imprinted at London by
Iohn Norton.
1597

5. John Gerard's famous English *Herball*, in spite of grave shortcomings,
aroused English interest and enthusiasm in plant studies.

Courtesy of the Rare Book Room, University of Illinois Library.

6. The title page of "Gerardus Immaculatus" wherein Thomas Johnson corrected and enlarged Gerard's *Herball*.

Courtesy of the Rare Book Room, University of Illinois Library.

did not make clear what variety of "potato" he meant. Most likely he referred, in fact, to the Jerusalem artichoke (*Helianthus tuberosus*), native to North America, a common food of the Indians of the eastern seaboard, and long confused—thanks to Bauhin, Gerard, and others—with the white potato (*Solanum tuberosum*), which originated in the Andes of South America.[40] Still, with all its faults, Gerard's *Herball* became a widely used and even beloved book. In large part this popularity arose from Gerard's fine literary style, with which he imparted to his readers his own love of plant life. Moreover, Gerard's compilation was, in spite of errors, a useful book for its day. Pulteney's judgment (1790) holds good still:

> In this manner, Gerard, with Dodoens for his foundation, by taking in also many plants from Clusius [L'Escluse], and from Lobel, by the addition of some from his own stock, published a volume, which, from its being well timed, from its comprehending almost the whole of the subjects then known, by being written in English, and ornamented with a more numerous set of figures than had ever accompanied any work of the kind in this kingdom, obtained great repute. . . . And notwithstanding his manifest inferiority to Lobel in point of learning, it must yet be owned that Gerard contributed greatly to bring forward the knowledge of plants in England.[41]

In 1633 Thomas Johnson (c. 1600–1644) published a revised and much-improved edition of Gerard's *Herball*.[42] This volume, which came to be known as "Gerardus Immaculatus," was the work of a highly respected apothecary-physician-botanist who later served as Lieutenant Colonel with the Royalist forces in the first English Civil War, losing his life at Basing-House in 1644. Prior to his death Johnson had earned a splendid reputation as a botanist, having botanized widely in England and Wales, and published two catalogues of local English plants before his revision

[40] See Redcliffe N. Salaman, *The History and Social Influence of the Potato* (Cambridge, 1949), pp. 36–85. If Gerard got his "Virginia potato" from Ralegh, as seems likely, it was certainly a sweet potato or, possibly, a Jerusalem artichoke.

[41] Pulteney, I, 122. Benjamin Daydon Jackson, ed., *A Catalogue of Plants Cultivated in the Garden of John Gerard, 1596–1599* (London, 1876), is a useful commentary.

[42] *The Herball, or Generall Historie of Plants . . . Enlarged and Amended by Thomas Johnson . . .* (London, 1633). Another edition was published in 1636.

of Gerard's work. He was a more learned and careful scholar than Gerard, and he had the advantage of the considerable new knowledge of botany that had accumulated since the time Gerard's *Herball* had appeared. He added about 800 plants not in Gerard, used cuts from Plantin's collection, adding about 700 figures not included by Gerard (a total of 2,766 cuts in all), and made innumerable corrections to Gerard's text. He was materially aided by John Goodyer (see below), who, with Johnson, was one of the most able botanical scholars England produced before the Stuart Restoration. In short, Johnson's "Gerardus Immaculatus" was a vast improvement over Gerard's *Herball* and it became the guidebook of American colonial botanical collectors for many years.

John Goodyer (1592–1664) appears to have left no published works although his library and manuscript remains (in Magdalen College, Oxford) are considerable and revealing.[43] About 1616 he became Steward to Sir Thomas Bilson of Mapledurham and studied botany on the side. Within five years he had become a good descriptive botanist, familiar with Johnson, James Cole (Lobel's son-in-law), and others interested in natural history, as well as with the principal gardens and gardeners, including the Tradescants and John Parkinson (see below). At Droxford, Hampshire, he developed a garden of his own where he grew, among others, several Virginian plants, including the Virginian "potato," "pompions" (or "macocks," *Cucurbita pepo*, L.), "no bigger than a great apple," and tobacco. He drew up yearly plant lists which, lying among his manuscripts, form some of the earliest histories of English gardens with scientific names attached. He appears to have been familiar with plant mutations and did not subscribe to the ancient idea of the fixity of species. But his influence was limited to his own day through the assistance and advice he gave to his friends, particularly to Thomas Johnson in his revision of Gerard's *Herball*.

John Parkinson (1567–1650) was the last remaining member of the small English "circle" of naturalists in the years before the English Civil Wars.[44] An apothecary and herbalist, he became

[43] These have been carefully studied by Gunther, *Early British Botanists*, pp. 16–83, 100ff., 197–232.

[44] Pulteney, I, *passim*, cites a number of lesser figures.

Royal Apothecary and Herbalist to James I and Botannicus Regius Primarius to Charles I. His interest lay in gardens and gardening, although he combined a shrewd knowledge of botany mixed with medieval superstition and Galenic medicine. His first book was about gardening, dedicated to Queen Henrietta Maria and entitled *Paradisi in Sole Paradisus Terrestris. A Garden of all sorts of pleasant flowers* . . . (London, 1629). His chief volume, *Theatrum Botanicum, the theatre of plants, or, an herball* . . . (London, 1640), was heavily indebted to the previous works of L'Escluse, Lobel, Bauhin, and Dodoens, but gave, in addition, the earliest descriptions of some English plants. With descriptions of some 3,800 plants arranged mostly after the pattern of Dodoens, with much attention to their supposed "virtues," and with hundreds of new (but inferior) woodcuts, the *Theatrum* was the most copious herbal in England at the time. Parkinson had a private garden in London (Long Acre) in which he grew a considerable number of New World plants from Peru, Brazil, Bermuda, and Virginia, New England, and Canada. It was said that the first yucca (*Yucca gloriosa*, L.) to flower in England (1604) grew in Parkinson's garden. He leaned much upon Monardes, Acosta, and Jacques Philippe Cornuti (see below) for descriptions of American products. But. in spite of his dependence upon the works of others, Parkinson was more original than Gerard or Johnson, and excelled Bauhin in his lists of synonyms for plants. His descriptions were fresh, sometimes vivid, and he was careful to point out the places of growth. But in treating the virtues of plants, he depended upon ancient and medieval authors with little attention to the more recent discoveries. The net result was a curious mixture of the old science and the new. On balance, Parkinson appears to have been the last important herbalist in England.[45] Civil strife interrupted English scientific endeavors during the 1640's, and after the Stuart Restoration (1660) English science took on a new and more vigorous development.

Until the Restoration, however, the best achievements of English natural science were little more than digests of current works, with

[45] *Ibid.*, I, 138ff.; Gunther, *Early British Botanists*, pp. 265ff. Gunther also includes (pp. 369–371) lists of seeds, nuts, and fruits from New England and from Virginia (1636) evidently known to Parkinson.

very little originality. No English naturalist had gone significantly beyond the Continental scholars. But, indeed, the Continental scholars had advanced very slowly, after a century of endeavor. With the emphasis upon virtues in the herbals, botany still remained a handmaiden of medicine—although it must be admitted that botany (and natural history as a whole) was much indebted to medicine, from which it had obtained its original impulse.[46] Perhaps the sudden expansion of natural history with the discovery of the New World and the commerce with the East had been a handicap to the development of classification and system in the natural world. Scholars had become so excited in incorporating into the old science the new products that captured their notice that they gave too little attention to problems of natural science as a whole. Descriptions were still poor, and scholars could do no better until the essential parts—the anatomy and the physiology of the products of nature— were more fully comprehended. They still depended largely upon general observations, with only vague distinctions. Guiding principles, universally acknowledged, were still lacking. No accurate lines of definition had been drawn between genera and species. Lobel had recognized a sound principle for distinguishing between classes of plants when he called attention to the differences between monocotyledons and dicotyledons; Bauhin had formulated a system of binary nomenclature, but he was inconsistent in its application and was not followed by others until years later; Gesner had set forth the idea of genera, had stated that they should be founded upon the structure of the flower, especially the seed, rather than the leaf of plants, and he had proposed some substantive names, but his proposals were not yet widely heeded; Lobel, Bauhin, and a few others had proposed groupings of plants which progressed from simple to more complex forms, from grasses to trees, whereas the older systematists had started with trees, but this, too, failed to win wide acceptance. Many herbalists were still content to list their specimens in alphabetical order, using their Latin names, or to group them according to their supposed medicinal virtues. In fact, until clear lines of distinction between genera and species were

[46] Thomas Johnson wrote in his "Gerardus Immaculatus" (1633) that the *materia medica* "is as it were the foundation and ground-worke of all that hath been since delivered in this nature."

drawn, until careful analysis of anatomy and functions of the essential parts of plants had been made, and until the roles of soil, situation, and culture were appreciated, nothing better could be accomplished in botany. And much the same could be said for zoology—and for natural history as a whole. Scholars also suffered in their capacity to illustrate their specimens. Woodcuts were far better than no illustrations at all, but even the finest of them could not equal the precision in detail which copper plates made possible later on.

English Scholars in the New World

Such, then, was the state of scientific knowledge in the homeland when the first English colonists set foot in the New World. Few, if any, of them, however, were abreast of their most advanced fellow Englishmen in any branch of scientific learning. Nor did they go to the New World on a scientific expedition per se. What scientific observations they made were generally incidental to public and private commercial, religious, or dynastic interests. Hawkins, Ralegh, Drake, Cavendish, and other Elizabethan "seadogs" who sailed forth in the later sixteenth century brought back many natural specimens which soon filled the gardens of English herbalists and museums of curiosities like that of the Tradescants. But such studies of these items as were made were done by the English naturalists; to the seadogs they were, for the most part, merely New World curiosities.[47] When Sir Walter Ralegh established his settlement at Roanoke in 1585, as a base for privateering, he took with the expedition John White, a painter, and Thomas Hariot (or Harriot),

---

[47] Accounts of these voyages, mostly published in the volumes of the Hakluyt Society, are not unlike those of the Spanish written many years before. They have great value in the history of discovery and international politics, but their contributions to the history of science are relatively slight. They point out, as the Spanish had done, the proofs of the sphericity of the earth, the fallacies of the burning zone theory, some vague comments on weather, ocean currents, and the like, the union of the Atlantic and Pacific oceans at the southern tip of South America, and excited references to flying fishes, strange birds, animals, natives, and some of the flora. But their comments are not couched in scientific terms and show little familiarity with the scientific work of their own day. One of the best examples is that of Master Francis Fletcher, a chaplain with Sir Francis Drake, in his *The World Encompassed . . .* (London, 1628) (Hakluyt Society, *Works*, Vol. XIV, London, 1854).

a servant, friend, and mentor of Ralegh's, as well as a prominent English mathematician, surveyor, and astronomer. These men worked together to some extent to describe the flora and fauna in and about the Roanoke settlement. Hariot published *A briefe and true report of the new found land of Virginia* . . . (London, 1588).[48] Richard Hakluyt subsequently persuaded Theodore de Bry, a Flemish goldsmith and engraver, to reprint Hariot's *Report* with several of John White's paintings (as black-and-white engravings) and a sketch map of coastal topography. This was published at Frankfurt in 1590, with a London edition following in 1593. White's paintings were very valuable. He produced many identifiable pictures of American birds and fishes, but some of his views tended to be stylized in the European fashion of the time, with Indians hardly identifiable as such, looking more like idealized portraits of Greek gods and goddesses playing Indian.[49]

Hariot traveled into the interior of Carolina (and probably in Virginia) and made observations with his "mathematical instruments," but records of these were lost. His *Report*, however, bore more the marks of a promotion pamphlet than those of a scientific treatise. Organized under three heads—"Merchantable Commodities," "Commodities for Victuals," and "Commodities for Building Materials"—together with an essay on the nature and manners of the natives, it gave glowing accounts of the healthful climate, the commercial potentials, and the need for greater financial support of the colony. Hariot had armed himself with Monardes' book, and

[48] This has been repeatedly reprinted. A facsimile reproduction, with an Introduction by Randolph G. Adams, was published at Ann Arbor, 1931. See also Increase N. Tarbox, *Sir Walter Ralegh and His Colony in America* (*Publications of the Prince Society*, Boston, 1884), and others. The best edition is by David Beers Quinn, ed., *The Roanoke Voyages 1584–1590* (Hakluyt Society, *Works*, 2nd Ser., Vols. CIV–CV, London, 1955). Mr. Quinn cites the modern scientific names of flora and fauna described and for that reason I have omitted repeating them.

[49] White's paintings have since been published in full, in color. In 1946 Stefan Lorant published some of them in *The New World* . . . (New York, 1946). But Lorant's work was unsatisfactory. Later, Paul Hulton and David Beers Quinn edited, with the help of others, *The American Drawings of John White, 1577–1590* (2 vols., London and Chapel Hill, 1964), with all of White's works in color. White's birds have also been reproduced in Thomas P. Harrison, *John White and Edward Topsell, the First Watercolors of North American Birds* . . . (Austin, 1964).

one suspects that he viewed much of the natural scene about Roanoke Island through the pages of John Frampton's English translation of Monardes.[50] Still, he made an accurate report of the geology of the region, for he recognized the differences between the Coastal Plain and the Piedmont, and learned that mountains existed to the westward. He described the mineral resources of the Coastal Plain, and they were assayed by Thomas Vaughan, the "mineral man" who accompanied the expedition. They found alum, white "co-presse" (copperas), "nitrum" (saltpeter), "allumen plumen," and other commercial and medicinal earths; they also reported iron, copper, and silver—although Hariot was somewhat misled by the copper and silver ornaments worn by the Indians of the interior.[51] He reported medicinal plants such as sassafras (*Sassafras offinale*), for the virtues of which he referred to Monardes, "Sweet Gummes," and "many other Apothecary drugges. . . ." The Indians used various herbs, seeds, roots, and barks for coloring their bodies, but "their goodnesse for our English clothes remayne to be proved." But other dyestuffs, such as madder and woad, could doubtless be produced in America. Also sugar cane and citrus fruits were likely products. Oils were obtainable from walnuts, berries, and bear's fat; and pearls were found in mussels, though the quality was poor. Fur-bearing animals abounded—deer, bears, martens, otters, "civet catts," and others. And the prospects for wine-making were excellent, for Hariot found two kinds of grapes which he assumed would make potable wines. Under the second head, "Commodities for Victuals," Hariot gave a useful account of Indian agricultural techniques, praised maize as a "graine of marvelous great increase," described Indian beans, "potatoes," gourds, melons, sunflowers, "pompions," and tobacco, whose virtues "woulde require a volume by its selfe." He found that oats, barley, and peas would do well in Carolina, but did not try wheat or rye. He described various roots, but as he used only their Indian names they are of uncertain identity and, he said, their virtues were unknown. Leeks, however, were similar to those in England. Of nuts and fruits, he found acorns (five sorts), chestnuts, two kinds of walnuts, two kinds of grapes,

[50] *Joyfull Newes out of the Newe Founde Worlde* . . . (London, 1577).
[51] George W. White, "Thomas Hariot's Observations on American Geology in 1588," *Transactions of the Illinois Academy of Science*, XLV (1952), 116–121.

strawberries, mulberries, crab apples, hurtleberries, "medlars" (which from the description were probably persimmons), and two or three others for which only Indian names were given. He listed several small beasts, but only squirrels and conies were mentioned by their English names. In all, he mentioned twenty-eight species of mammals. He had the names of eighty-six birds, giving pictures of eight "strange sortes of water fowle" and seventeen of land fowls, to which he gave English names on the basis of their resemblance to English birds. Similarly, he gave English names to fishes and pictured twelve, but the "Seekanauk" defied his nomenclature—probably the land crab which he pictured from White's drawing.[52] Of building materials, Hariot listed trees for ship-building and homes, (oyster) lime, clay for brick-making, and stone—although the latter was upcountry and the Indians had no quarries. Trees included oak, elm, ash, walnut, fir, cedar, maple, witch hazel,[53] holly (holly plant?), willow, beech, sassafras, and others for which only Indian names were given. His account of the nature and manners of the Indians was sometimes perceptive. He noted their lack of metallic weapons, but underrated their arts and crafts. He felt that they were easily defeated in war, and would easily become dependent upon the English, "brought to civilite, and embracing of true religion." He dazzled them with his mathematical instruments, sea compass, loadstones, "perspective glass whereby was shewed manie strange sights," burning glasses, guns, spring clocks, writing, books, and other European gadgets, but he found them skillful and ingenious. All in all, Hariot's *Report* was a straight job of reporting with promotional overtones and few indications of current scientific knowledge in the natural sciences. He betrayed a belief in the ancient idea that the same plants grew on the same parallels of latitude all over the world, and while he accurately distinguished between American and European deer, he also told of having killed a lion (mountain lion?) in North Carolina. As the first English attempt to describe in detail the nat-

[52] All the illustrations were from White's drawings—one of the most original features of Hariot's book. For Hariot's birds, see Elsa G. Allen, "The History of American Ornithology before Audubon," *Transactions of the American Philosophical Society*, n.s., XLI, Pt. 3 (1951), 444–448.

[53] *Hamamelis virginiana* was not a European tree or shrub, and it is gratifying to see that Hariot recognized it.

ural products of the New World by an observer in the field, Hariot's *Report* was a valuable historical account; but as a scientific treatise it is disappointing. Its effect was clearly more discernible in the realm of English colonial promotion than in scientific circles.[54]

Captain John Smith (1580–1631) is far better known as a seaman, soldier, adventurer, and colonial administrator than as a scientific observer. Like Hariot, with whose work he was thoroughly familiar, he was a promoter of English colonization in America. But by his discovery of Chesapeake Bay, his investigations of the rivers flowing into it, his similar observations along the New England coast, his publication of *A Map of Virginia. With a Description of the Countrey . . .* (Oxford, 1612), *A Description of New England . . .* (London, 1616), *New England Trials . . .* (London, 1620), *The Generall Historie of Virginia, New England, and the Summer Isles* (London, 1624), and his other voluminous works, he emerged as a geographer, cartographer, and geologist of no mean stature. His works were frequently reprinted during his lifetime and became well known to the English public.[55] Smith was widely read in the works of discovery preceding his own, including Spanish authors (his acquaintance with Richard Hakluyt probably contributed to this); but his references were often too imprecise to enable a reader to pin down his exact source. That he held to a theory of colonization at variance with many of the leaders in the Virginia Company is well known, and it colored his works throughout. The capacity of a colony to support itself in terms of food and most other necessities of life, together with its ability to enter into profitable trade with the homeland, was of far greater importance, in Smith's eyes, than the discovery and exploitation of gold, silver, and precious gems: ". . . to thinke that gold and silver

---

[54] Fulmer Mood's "The English Geographers and the Anglo-American Frontier in the Seventeenth Century," *University of California Publications in Geography,* VI, No. 9 (Berkeley, 1944), 365–395, surveys English geographical literature from George Abbot's *A Briefe Description of the Whole World* (1599) to John Ogilby's *America . . . the New World* (1671) and finds them generally more promotional in nature than scientific.

[55] Smith's works have been republished over and over since his death. They were collected together in Edward Arber, ed., *Travels and Works of Captain John Smith* (2 vols., Edinburgh, 1910), with a bibliography of the various editions and reprints up until that time. I have used Arber's edition throughout.

Mines are in a Countrey otherwise most rich and fruitfull, or the greatest wealth in a Plantation, is but a popular error; as is that opinion likewise that gold and silver is now the greatest wealth of the West Indies at this present."[56] Accordingly, while he struggled against Captain Christopher Newport and others who wildly sought gold in Virginia and elsewhere,[57] he directed his attention to soil types, to agricultural resources, and to minerals other than gold and silver, especially to iron ore.[58] He gave considerable attention to flora and fauna but not with the eye of a naturalist. He noted the vast forests as resources for colonial home-building, ship-building, fuel for smelting iron ore; sassafras and other flora as a source of drugs and dyestuffs;[59] mulberry trees for feeding silk-worms;[60] fruit and nuts for food and oil; and vines for a future wine trade. He observed the waters choked with fishes as a basis for the establishment of profitable fisheries; and he noted the animals with an eye to the fur trade. Even the Indians were no barrier to English colonization; like Hariot, he felt that they were docile,

[56] *The Generall Historie of Virginia*, Arber, I, 581.

[57] "There was no talke, no hope, no worke, but dig gold, wash gold, refine gold, load gold. Such a brute of gold, as one made fellow desired to bee buried in the sandes, least they should by their art make gold of his bones." *A Second Part of the Map of Virginia* (1612), Arber, I, 104. Smith's personal interest in gold appears to have been merely to get capital to support later voyages. Newport and his eager followers were inspired by the Spanish example and, perhaps, by Richard Eden's *Decades* (1555). See Chapter 2 above.

[58] Actually, iron manufacturing began in Smith's time and later grew into a considerable industry in Virginia. Cf. Charles E. Hatch, Jr., and Thurlow Gates Gregory, "The First American Blast Furnace, 1619–1622: The Birth of a Mighty Industry on Falling Creek in Virginia," *Va. Mag. Hist. & Biog.*, LXX (1962), 259–296.

[59] Sassafras was madly sought by early travelers and colonists for its sale in England, where its medicinal values were highly regarded. Cf. John Brereton, "A Briefe and true Relation of the Discoverie of the North part of Virginia . . . 1602, by Captain Bartholomew Gosnold . . . ," in George Parker Winship, ed., *Sailors' Narratives along the New England Coast, 1524–1624* (Boston, 1905), pp. 32–50. Wyndham B. Blandon, *Medicine in Virginia in the Seventeenth Century* (Richmond, 1930), pp. 108ff., describes the interest in sassafras and other Virginia flora with supposed medicinal values.

[60] The red mulberry (*Morus rubra*) was indigenous to Virginia and the southern colonies; the Spanish introduced the white mulberry (*Morus alba*). The early efforts to produce silk in Virginia are well described by Charles E. Hatch, Jr., "Mulberry Trees and Silkworms: Sericulture in Early Virginia," *Va. Mag. Hist. & Biog.*, LXV (1957), 3–61.

"a more kinde and loving people cannot be."[61] In general, though, his attentions to flora and fauna were incidental to his topographical, geographical, and geological observations. They added little to Hariot's *Report* from North Carolina, which, at many points, Smith appeared to follow closely.

Smith's *Map of Virginia*,[62] whether made by Smith or by others working at his direction, was a remarkable production. It was reasonably accurate, showing the Indian territories and towns. It distinguished between the Coastal Plain and the Piedmont, as Hariot had recognized in Carolina, and it showed landfalls whereby mariners might recognize the region as they approached from the sea. It described the climate, the topography, the rivers—many of which Smith had navigated up to the Fall Line—the soils, and the mineral resources. He recognized that Virginia was no island, as many in England had been led to believe, and he gathered reports from the Indians about "Countreys beyonde the Falles," with accounts of the mountains and resources of the interior parts.

> The vesture of the earth in most places doeth manifestly prove the nature of the soile to be lusty and very rich. The Colour of the earth we found in diverse places, resembleth *bole Armoniac, terra sigillata ad lemnia*, Fuller's earth, marle, and divers other such appearances. But generally for the most part the earth is a black sandy mould, in some places a fat shiny clay, in other places a very barren gravell. But the best ground is knowne by the vesture it beareth, as by the greatnesse of trees or abundance of weeds, &c.
>
> The country is not mountanous nor yet low but such pleasant plaine hills and fertle valleyes, one prettily crossing another, and watered so conveniently with their sweete brookes and christall springs, as if art it selfe had devised them.[63]

---

61 *The General Historie of Virginia*, Arber, I, 309.

62 Smith sent a "Mappe of the Bay and Rivers, with an annexed Relation of the Countries and Nations that inhabit them" to the Council of London in the fall of 1608, as a report of his explorations of Chesapeake Bay in the previous summer. His *Map of Virginia. With a Description of the Countrey . . .* (1612) is an expansion of this earlier report.

63 *A Map of Virginia. With a Description of the Countrey*, Arber, I, 49. "Bole Armoniac" and "terra sigillata" are varieties of clay used for pigmentation and medicinal purposes. Smith clearly recognized the variation of vegetation with

Captain John Smith left Virginia in 1609 never to return. After several years in England, during which he published his valuable book on Virginia, he returned to North America, this time to explore the New England coast (1614), which, as he said, was "still but euen as a Coast vnknowne and vndiscovered." His return to England (1615) was prevented by storms at sea and he fell a prisoner of the French. During his captivity, which lasted only a few months, he began *A Description of New England* . . . , published in 1616 with a map almost as remarkable as his earlier *Map of Virginia*. In fact, it is the first real map of this part of North America, and Smith first named the region "New England." As Smith spent less time in New England than in Virginia, his map does not show the interior as well as the former, and there is less geological information in it. Obviously he made his observations largely from a coastal view only. He contrasted the rocky coast of Maine with the sandy shores of Massachusetts. Of the latter, he observed that they are "so indifferently mixed with high clayie or sandy cliffes in one place, and then tracts of large long ledges of diuers sortes, and quarries of stones in other places so strangely diuided with tinctured veines of diuers colours: as, Free stone for building, Slate for tiling, smooth stone to make Fornaces and Forges for glasse or iron, and iron ore sufficient, conveniently to melt in them. But the most parte so resembleth the Coast of *Deuonshire*, I think most of the cliffes would make such lime stone."[64] He had no "mineral man" with him on the voyage, and he confessed that "I am no Alchymist." He tended, therefore, to generalize on mineral resources from insufficient direct evidence. But he was confident that there were minerals of value in the country, and he stated that "Who[ever] will vndertake the rectifying of an Iron forge, if those that buy meate, drinke, coals, ore, and all necessaries at a dear rate gaine; where all these things ar to be had for the taking vp, in my opinion cannot lose."[65]

William Strachey (1572–1621), a onetime student, but not grad-

---

soil types. See George W. White, "Geological Observations of Captain John Smith in 1607–1614," *Transactions of the Illinois Academy of Science*, XLVI (1953), 124–132.

[64] *A Description of New England*, Arber, I, 193.
[65] *Ibid.*, p. 201.

uate, of Emmanuel College, Cambridge, arrived in Virginia the year after Captain John Smith left (1609). Strachey was a literary aspirant who went to Virginia to recoup his fortune. His account of a hurricane in the Bermudas, in which he, with Captain Christopher Newport (Smith's nemesis), were caught became the basis for Shakespeare's play *The Tempest*. He became Secretary and Recorder of the colony in the time of Lord de la Warr (1610–11), and prepared *The Historie of Travaile into Virginia Britannia* . . . in 1612. This book, though circulated in manuscript, was not published until 1849.[66] Strachey's work depended heavily upon others, and although he gave a fulsome description of the flora and fauna of Virginia, it added little to those of his predecessors with no more of a scientific flavor than they. In one particular, however, he added to Captain John Smith's account of the geology of Virginia. "All of the Low Land, of South and North Virginia," he wrote,

> is conjectured to have been naturally gayned out of the Sea; for the Sea through his impetuous, and vast revolution (who knowe not), swayinge vpon every Coast, in some places wynes, and in other places looseth . . . bouches of Oysters and Scallops, which ly unopened, and thick together, as if there had been their naturall Bed, before the Sea left them. . . . Moreover, the Mould and sward of the earth is not 2. foot deep all along neare the Sea, and that which is, comes only by the grasse and leaves of Trees, and such rubbish, rotting upon yt in continuance of tyme; for in digging but a fathome or 2, we commonly find quick sand, againe under the crust of the Surface, we fyne not any stones, quarryes nor rocks (except near the high-Land). . . .[67]

Strachey's calm acceptance of fossils in the modern sense was notable, and his theory of the rising coastline of Virginia was unique in its day.

Strachey referred to a contemporary of his in Virginia, Dr. Lawrence Bohun (or Bohune, or Boone), who arrived in Virginia in 1610. Dr. Bohun was a physician trained in the Low Countries, and although he was not the first physician in Virginia, he was the first

---

[66] By the Hakluyt Society. The best edition is that edited by Louis B. Wright and Virginia Freund in the Hakluyt Society, *Works*, 2nd Ser., Vol. CIII (London, 1953).

[67] *Ibid.*, pp. 39–40.

to be appointed by the London Company for service in the colony. He appears to have had an experimental bent of mind and some scientific knowledge in medical botany. He set up the first botanical garden in the English colonies of the New World, imported Europe-pean seeds, fruit trees, and shrubs, and experimented considerably with native Virginia plants, gums, berries, and earths in an effort to establish their medicinal value.[68] Very likely his work encouraged the London Company to urge its colonists to search the country for drugs, gums, and medicinal earths and to make trial, as one writer proposed,

> by boaring holes in divers trees, whose vertues wee are yet ig-norant, and collecting the juice thereof, a scrutiny be made which are fit for Medicinall liquor and Balsomes; which for gummes, Perfumes, and Dyes, and heere I may justly take occasion to com-plain of our owne sloth and indulgence, if compared to the laborious Spanyard, who by this very practice have found out many excellent Druggs, Paints and Colours meerely by bruizing and grinding Woods. . . . The French relations of their Voyages to Canada, tell us that the Indians and themselves falling into a contagious disease, of which Phisitians could give no Reason or Remedy, they were all in a short space restored to their health meerly by drinking water, in which Saxifrage was infused and boyled, which was then discovered to them by the Natives, and wee justly entertaine beliefe that many excellent Medicines either for conservation of Nature in her vigor or restauration in her decadence may be communicated unto us. . . .[69]

Strachey also reported that Dr. Bohun made excellent wine of Virginia wild grapes: "I dare say yt we have eaten there as full and lusheous a Grape as in the village between Paris and Amiens, and I

---

[68] Dr. Bohun's experimentation was, in part at least, the result of a critical shortage of drugs in the colony. Blandon, *Medicine in Virginia*, pp. 11–14, 101–104; Alexander Brown, *The Genesis of the United States* . . . (2 vols., Boston, 1891), I, 412; *D.A.B.*

[69] Edward Williams [ghost writer for John Ferrar by self-appointment], *Virgo Triumphans: Or, Virginia richly and truly Valued* . . . (London, 1650), p. 45, quoted in Blandon, *Medicine in Virginia*, pp. 103–104; reprinted in Peter Force, ed., *Tracts and Other Papers, Relating Principally to the Colonies in North America* (4 vols., Washington, 1836–48; reprinted, 1947), III, 15. This is a promotion pamphlet with purple passages. The company's interest was clearly commercial, only incidentally scientific.

have drunck often of the . . . wyne, which Doctor Bohoune and other of our people have made full as good as your French-Brittish wyne, xx. gallons at a tyme. . . ."[70] Unfortunately, Dr. Bohun's garden expired with him and there are few records of comparable experimentation in Virginia until many years afterward.

William Wood's *New Englands Prospect. A true, lively, and experimentall description of that part of America, Commonly called New-England* . . . (London, 1634)[71] was one of the best topographical accounts of New England by a temporary settler in the region. Little is known of Wood other than that he was obviously a man of wide liberal education. He emigrated to New England in 1629, probably settled at Saugus (Lynn), Massachusetts, was well received among the Puritans, and returned to England in 1633. Details of his later life are uncertain. His book is ample testimony to his fine descriptive style and his close, systematic observations. But there was little "experimentall" in the treatise other than the fact that the author made on-the-spot observations and perhaps experimented somewhat in planting English seeds in New England gardens to find that "whatsoever growes well in England, growes as well there, many things being better and larger. . . ." There was more than a hint of promotionalism in the book, and the author gave directions regarding the needs of those Englishmen who might be considering settlement in New England. The book contained a map of the coastal area of New England showing English settlements from Long Island Sound to southern Maine. Wood was familiar with Captain John Smith's *Description of New England*. He praised Boston Bay for its many snug harbor facilities and, although he recognized that the winter cold of New England sometimes exceeded that of Old England, he lauded the New England climate as more healthful than that of the mother country. His praises of the soil of New England were somewhat misleading in that the reader might easily conclude that the fine soils were more extensive than they actually were. He felt that the country

[70] Wright and Freund, eds., *Historie of Travaile*, pp. 121–122. The colonists became very thirsty for wine!

[71] Later editions in 1635 and 1639. The book was republished in whole or in part in 1764, 1865, and 1898. The best reprint edition is that edited by Charles Deane in the *Publications of the Prince Society* (Boston, 1865).

was splendidly adapted by nature for the production of forage for cattle, for all manner of garden stuffs, herbs, small fruits, hemp, and flax. The forests were filled with good timber for fuel, home-building, mill construction, and ship-building. He recognized three kinds of oak trees and that American ash, walnut, and cherry differed from those of England. He believed that the wild grapes would produce wines equal to those of Bordeaux. His account of the fauna of New England was both perceptive and remarkably complete. He gave a good description of moose, recognized that the deer were larger than in England, gave a splendid account of the beaver and their dams, and raised a question as to whether there were lions in New England. He had seen none, "but others affirm they had seen one at Cape Ann," and Wood concluded that lions certainly existed on the North American continent, especially in Virginia. (Actually, the mountain lion ranged over most of North America.) On the whole, Wood's account of the fauna of New England contained fewer credulities than was common at the time. He recounted the popular belief that porcupines darted their quills at their enemies; and while he believed, contrary to many of his day, that the poison of the rattlesnake "lyeth in her teeth, for she hath no sting,"[72] he also affirmed that "snakeweed" was a competent antidote for rattlesnake venom. He said that he had little knowledge of minerals but he knew of the existence of iron and, on the basis of Indian reports, he believed that the mountains of New England hid rich mineral deposits of an undisclosed nature. The second part of Wood's *Prospect* was devoted to the Indians. And in spite of the fact that he said the Indians' "carriage and behaviour hath afforeded more matter of mirth, and laughter, than gravity and wisdome," his lighthearted treatment of the natives was instructive, with attention to the different tribes, the differences among them, their various arts and crafts, and their relations with the English. He also appended a brief vocabulary of Indian words and their English equivalents, a list helpful to later students of the Indian tongues. Over-all, Wood's *Prospect* is that of an intelligent, lay observer, with no expressed attention to the scientific problems of the day, no attempt at scientific collection, analysis, or classification, and no reference to scientific works or scientific contempo-

[72] No one, as yet, had dissected a rattlesnake.

raries in the homeland. The book is a traveler's account, not a
scientific treatise. It is a most readable description with acute lay
descriptions of the topography and the flora and fauna of New
England designed to still the voices of those who would criticize
the region as a place for English colonization by emphasis upon
the potentialities of New England for future trade, commerce, and
economic growth. Capable and industrious Englishmen, with a
small amount of capital, Wood concluded, could live far better in
New England than in Old.

Thomas Morton (d. c. 1646), the gay and licentious "Morton of
Merry Mount," whose questionable efforts to establish a "plantation"
at Quincy, Massachusetts, ran him into trouble with the authorities
both of New Plymouth and the Massachusetts Bay Colony, wrote
the *New English Canaan or New Canaan Containing an Abstract of
New England* . . . (Amsterdam, 1637). This book is similar,
though inferior, to William Wood's *New Englands Prospect*. Morton arrived in New England in 1622, liked what he saw, and set
out to promote English colonization there before the Dutch or the
French had taken over the region. His book, in consequence, belongs clearly and specifically to promotional literature with no
more of a scientific approach to the New England scene than
William Wood demonstrated. Indeed, Morton's account is less
complete than Wood's, although he approached his subject differently, and in general presented an even more favorable picture
of New England's climate and resources. Much of the detail of the
two books, especially with reference to flora and fauna, closely
coincides. Morton agreed with Aristotle that the frigid zones of the
earth are uninhabitable, and he held a low opinion of the torrid
zone. New England, being in the temperate zone, was ideally
situated with a climate, Morton affirmed, "warmer in winter, than
some parts of France." He praised the Indians, "with their tractable
nature and love towards the English," believed that their language
was a mixture of ancient Greek and Latin, and, after speculation
as to their origin, concluded that they were Trojans dispersed by
the ancient wars. He noted the vast forest resources of New
England and praised the other plant resources, with herbs "of a more
maskuline vertue than any of the same species in England." Of
fauna, he took note of the stores of fish, fowl, and animals. He

especially remarked about the hummingbird, "no bigger than a great beetle," and concluded that they fed upon insects ("bees") caught among flowers. He felt that the moose held possibilities as a domesticated draught animal, gave an excellent description, similar to Wood's, of beaver and beaver dams, and concluded that there were no lions in New England.[73] He was more specific than Wood in describing the mineral resources. These included, said Morton, marble and limestone for building, chalk, slate, whetstones, load-stones, iron, lead, black lead, red lead, "Bol armoniack," vermilion, brimstone, tin, copper, and, perhaps, silver and gold. The soil was fertile, with no boggy ground(!); the air was aromatic, perfumed with sassafras, balsam, marsh roses, violets, and the like—a very healthful climate; the water was sweet, with many medicinal springs of wondrous curative properties; and, altogether, New England was truly a New Canaan, a land of milk and honey. This promotional note characterized much of Morton's treatise, which, while it holds much value as political history and travel literature, was marked neither by a scientific approach couched in scientific language, nor concerned with scientific problems.

French Scholars in the New World

It is somewhat ironic that the earliest work on the natural products of North America to be prepared in the spirit and language of the new natural science was of French origin. French interests in the geography and natural products of the New World dated from the voyages of Jacques Cartier in the 1530's. René de Landonnière and Jacques Le Moyne, officers associated with the Huguenot attempts to set up a French colony in Florida and South Carolina (1562), described the flora and fauna of that region. But the ablest promoter of French colonization in North America and the first to give wide and prolonged attention to natural products was Samuel de Cham-

[73] Morton's comments, however, do not justify the remark that he had "banished" the lion from eastern North America. Cf. George Brown Goode, "The Beginnings of Natural Science in America," *Annual Report . . . of the Smithsonian Institution, 1897. Report of the U.S. National Museum*, Pt. II (Washington, 1901), p. 373. What Morton said was that "Lyons there are none in New England," because it was contrary to their nature "to frequent places accustomed to snow." *New English Canaan*, p. 83.

plain (1567?–1635). In his several voyages to the West Indies, Central America, Mexico, and North America, Champlain made constant observations of the flora and fauna of these regions. After he entered French service (1603), he became vitally concerned with French colonization in the St. Lawrence Valley. He explored and mapped much of this area as far south as the lake which bears his name, including much of Maine and western New England.[74] Champlain "aimait à herboriser." He experimented with European garden plants, cereals, and fruit trees in Quebec, setting up a garden near Mount Royal (Montreal) for this purpose in 1610.[75] He also raised Indian corn (maize) and tried to domesticate the native grapevines. More important than these, however, Champlain shipped back to France untold numbers of the seeds of plants, shrubs, and trees growing in the St. Lawrence Valley and upper New England. These were placed in the hands of Jean and Vespasien Robin, who planted them in the Paris garden. By the 1620's they had a considerable number of North American flora growing there.[76] They attracted the attention of various herbalists, and Bauhin listed several of them in his *Pinax* (1623). They also attracted Dr. Jacques Philippe Cornuti (d. 1651), a physician and Professor of Medicine at Paris.

Cornuti made a study of the plants in the Paris garden with special attention to those of North American origin. This was published as *Canadensium Plantarum, aliarúmque nondum editarum Historia. Cui adiectum est ad calcem Enchiridion Botanicum Parisiense, Continens Indicem Plantarum, quæ in Pagis, Siluis, Pratis, &*

[74] For Champlain's interest in natural history, see H. P. Biggar, ed., *The Works of Samuel De Champlain (Publications of the Champlain Society*, 6 vols., Toronto, 1922), I, 1–80, 83–189, 192–469; II, 1–112, 157–308; III, 114–168, 174, 177, 203–205, 235ff.; IV, 35ff.; V, 121ff., 141ff., 257ff.; VI, 253–346.

[75] Louis Hébert (d. 1627), an apothecary, moved from Paris to Quebec in 1617. He, too, had a garden, and helped spark Champlain's interest in botanical affairs.

[76] Jean Robin's catalogues of the Paris garden list Canadian plants as early as 1601. At least fifty were being cultivated by 1635. Jacques Rousseau, "Michel Sarrazin, Jean-François Gaultier, et l'Etude Prelinnéenne de la Flora Canadienne," *Les Botanistes Français en Amérique du Nord avant 1850 (Colloques Internationaux du Centre National de la Recherche Scientifique, Paris, 11–14 Sept., 1956,* Paris, 1957), pp. 150–151. See also Ph. Guinier, "Ce que les Jardins et les Forêts de France Doivent aux Forêts Américaines," *ibid.*, pp. 329–337; George S. Bonn, *Science-Technology Literature Resources in Canada: Report of a Survey of the Associate Committee on Scientific Information* (Ottawa, 1966).

Montosis iuxta Parisios locis nascuntur (Paris, 1635).[77] Written in excellent Latin, the volume contained sixty-eight illustrations made from copper plates, including thirty-eight pictures of Canadian plants. The plates were made from living plants, and they frequently illustrated the root structures, with occasional insets to show seeds and flowers—although these were not always adequate for the modern determination of species.[78] Although the volume included plants from Spain, Africa, the West Indies, and the environs of Paris, Cornuti focused upon Canadian specimens, of which he described forty-three in all. His approach was that of the herbalist, with much attention to the medicinal virtues of the plants described. After an attempt to fit the Canadian specimens into the old frameworks of Dioscorides and Pliny, he became impressed by "un tres nombre de plantes inconnues," and, being familiar with the works of the new scientists, ended by adopting names patterned after the nomenclature of Lobel. Thus, though Cornuti himself never set foot in North America, his work was the first treatise on North American plants couched in the terms of the new science as it existed in his day.[79]

In scientific quality, Cornuti's book excelled all others on North American flora before the Stuart Restoration in England in 1660. Neither the Dutch in New Netherland nor the English in Virginia, Maryland, and New England produced anything comparable to it. Indeed, the first generation of New England Puritans, with a high percentage of university-trained men, including some physicians, showed comparatively little concern for natural science.[80] Their

---

[77] Later editions in 1651 and 1672.

[78] Mgr. C. La Flamme has listed all of Cornuti's American plants, with modern scientific nomenclature where possible, in "Jacques-Philippe Cornuti—Note pour Servir á l'Histoire des Sciences au Canada," *Proceedings and Transactions of the Royal Society of Canada*, 2nd Ser., VII (1901), Sec. IV, pp. 57–72.

[79] Cornuti included no aquatic or marine plants and no large trees. Both the Recollets and Jesuit missionaries to New France wrote on the natural products of the region, but their early treatises have little scientific interest. See William J. Robbins, "French Botanists and the Flora of the Northeastern United States . . . ," *Proceedings of the American Philosophical Society*, CI (1957), 362–368; Walter Hanns, *Die Verdienste der Jesuitenmissionare um die Erforschung Kandas: Ein Beitrag zur Entdeckungsgeschichte von 1611–1759* (Jena, 1910).

[80] Several New England physicians had had medical training in European universities. But the poverty of European drugs, together with the plethora of

accounts of the natural products of New England were promotional literature, couched in lay terminology with little or no attention to the new science developing in the homeland. Obviously, this fact reflected no anti-scientific attitude per se, for in Old England the Puritans were in the vanguard of the new scientific developments. But in New England a preoccupation with political, economic, and religious affairs shoved aside, for the moment, concern with scientific matters. Post-Restoration colonials, both Puritan and non-Puritan, displayed a more active—and occasionally a more scientific—attitude. To some extent this was traceable to the influences of the Royal Society of London, founded on the heels of the Stuart Restoration and dedicated to the promotion of the new science. The Royal Society becomes, then, the subject of the next chapter.

fascinating "Indian remedies," led them to experiment with native New England herbs and other remedies, often falling into the trap of the Doctrine of Signatures. The early Puritan theory of disease was anti-scientific (disease was a visitation of the wrath of God for the victim's sins) and this choked out more scientific practice in some cases. Even so, they were not basically unscientific for their day. Some of them were familiar with Gerard's *Herball* (1597), a few did some notable clinical work, and Giles Firmin lectured on anatomy at Boston in the 1640's and performed an autopsy in 1647. Clearly, the first generation of New England Puritans was cognizant of aspects of the new science although not actively engaged in its furtherance. See Malcolm Sydney Beinfield, "The Early New England Doctor: An Adaptation to a Provincial Environment," *Yale Journal of Medicine*, XV (1942–43), 99–132; Samuel Eliot Morison, *The Intellectual Life of Colonial New England* (Ithaca, 1960), pp. 241ff.

FOUR

The Advent of the Royal Society of London

Forerunners

PRIOR TO THE STUART RESTORATION, English scientists had no formal organization—nor, in fact, had their Continental fellows. However, as has been noted, informal "circles" had arisen by virtue of the natural philosophers' mutual correspondence, travels, visitations, and exchanges of seeds, specimens, and information. Toward the end of the sixteenth century these informal groups tended to congeal into more formal organizations, and several societies of them rose and fell in Italy, France, and England before permanent association could be achieved.[1] Obviously, the impulse for organization was in the air by the first half of the seventeenth century.

In England, the existence (since 1518) of the Royal College of Physicians of London may have served as a spur toward further scientific association, but the college operated principally as a licensing agency for physicians in the London area and more meaningful cooperation was in demand.[2] Many proposals for scientific foundations and associations were set forth from the latter years of the reign of Elizabeth I, some of them advocating royal sanction and

[1] See Martha Ornstein, *The Role of the Scientific Societies in the Seventeenth Century* (New York, 1913); Harcourt Brown, *Scientific Organizations in Seventeenth Century France, 1620–1680* (Baltimore, 1934); Raymond P. Stearns, "The Scientific Spirit in England in Early Modern Times (*c.* 1600)," *Isis,* XXXIV, Pt. IV (1943), 293–300.

[2] Sir George Clark, *A History of the Royal College of Physicians of London* (2 vols., Oxford, 1964–66); Charles C. Gillespie, "Physick and Philosophy: A Study of the Influence of the College of Physicians of London upon the Foundation of the Royal Society," *Journal of Modern History,* XIX (1947), 210–225.

state support.[3] Among these, that of Francis Bacon (1561–1626), first Baron Verulam, was the most imaginative and the most fruitful.

Bacon was a philosopher, not a scientist. In a series of interrelated works, including his *Advancement of Learning* (1605), the *Novum Organum* (1620), and *De Augmentis Scientiarum* (1623), he proposed to replace Aristotle's "Organon" with a "New Organon," or new instrument, for the advancement of human knowledge. Bacon deplored the state of learning in his own day, arguing that Aristotelianism, through no fault of its founder, had become constrictive at the hands of the Schoolmen whose academic sterility had made all philosophy a mere slave to Aristotelian logic. New discoveries were merely engrafted upon old systems of thinking—as has been evident with reference to the effects of New World discoveries on the old science of the sixteenth century. As Bacon said, "We must begin anew from the very foundations, unless we would revolve for ever in a circle of mean and contemptible progress."[4]

In order to break out of this repressive atmosphere, Bacon urged greater reliance upon the impartial, irrefutable evidence of experimental research and a union in one person of the empirical and rational faculties of man:

> The practical skill and diligence which characterized the approach of the artisan, freed from an attachment to immediate ends, must be carried into the realm of the intellect, itself freed from sub-

[3] The precise sources of inspiration and leadership in the founding of the Royal Society of London recently have been in considerable dispute. The standard, older accounts are in Thomas Birch, *The History of the Royal Society of London* . . . (4 vols., London, 1756–57, cited hereafter as Birch), and C. R. Weld, *A History of the Royal Society* . . . (2 vols., London, 1848, cited hereafter as Weld). These have been challenged by such works as those of Dorothy Stimson, "Dr. Wilkins and the Royal Society," *Journal of Modern History*, III (1931), 539–563; R. F. Young, *Comenius in England* . . . (London, 1932); Francis R. Johnson, "Gresham College: Precursor of the Royal Society," *Journal of the History of Ideas*, I (1940), 413–438; R. H. Syfert, "The Origins of the Royal Society," *Notes and Records*, V (1948), 75–137; G. H. Turnbull, "Samuel Hartlib's Influence on the Early History of the Royal Society," *ibid.*, X (1953), 101–130; and Douglas McKie, "The Origins and Foundation of the Royal Society of London," *ibid.*, XV (1960), 1–38. These views have been canvassed and yet another added by Margery Purver in *The Royal Society: Concept and Creation* (London, 1967, cited hereafter as Purver).

[4] Quoted from Bacon's *Novum Organum* in Purver, pp. 33–34.

servience to preconceived opinion. In this way the range of man's research might become the whole universe, and the object of his investigation the *causes* of natural phenomena. To have discovered a cause in nature was to have perceived a law, which in its turn would give rise to other discoveries and further laws.[5]

Bacon's plan for the "new Philosophy" envisaged a slow, continuous process extending over generations of men. The first step was to get the facts straight, because particular facts must be known before deductions could follow from them. Minor laws would proceed from a knowledge of particular facts, and they, in turn, would proceed to major laws: "I propose to establish progressive stages of certainty," he said, "starting from the simple sensuous perception."[6] Thus began a continuous and expanding process of discovery, in which an axiom derives from "the senses and particulars, rising by gradual and unbroken ascent, so that it arrives at the most general of axioms last of all."[7] In this gradated process, hypotheses became laws only when they had been proved in practice by producing new information itself verifiable by experiment. The experimenter must always be tentative in his thinking, and even when he arrived at a conclusion, he should record the steps taken in reaching it, so that any possible error or omission might be seen and corrected by others. Bacon condemned the impatient experimenter, anxious for quick "results" and willing to arrive at premature conclusions. A vast quantity of "spade work" was required. Bacon set forth a "Catalogue of Particular Histories" as a means of defining the fields of endeavor. A hundred and fifty "Histories" were listed, involving practically every modern scientific discipline. Within these categories, the "Phenomena of the Universe; that is to say, experience of every kind" must be constituted to serve as "a foundation to build philosophy upon." By means of such a gradual, long-term building-up process, Bacon envisaged the development of "a true and active philosophy," a new and generative science, in place of the dead hand of Greek science as rendered up by the logic-chopping Schoolmen. With this new engine of the mind, men would speed up scientific discovery (as compared with the tempo

[5] *Ibid.*, p. 35.
[6] *Ibid.*, p. 38.
[7] *Ibid.*

of the past thousands of years), and create, from its foundations, a new organized body of related inductive knowledge capable of continuous, unlimited development in the future. Thus, too, man could develop a new imperium over nature, "for," as Bacon said, "the empire of man over things depends wholly on the arts and the sciences."[8]

Bacon saw that the realization of his proposals must rest on a corporate undertaking whose efforts would extend over successive generations in the future. In his *De Augmentis Scientiarum* he wrote: "I take it that all these things are to be held possible and performable, which may be done by some persons, though not by every one; and which may be done by many together, though not by one alone; and which may be done in the succession of ages, though not in one man's life; and lastly, which may be done by public designation and expense, though not by private means and endeavour."[9] In short, Bacon proposed a planned scientific revolution to be effected at the hands of an approved public institution supported by state funds.

None of the early Stuart monarchs could afford, or could be persuaded to afford, such an undertaking as Bacon proposed. Criticisms of his plan arose, too, especially from scholars wedded to Aristotelianism. The long-term efforts involved in Bacon's proposals caused many persons to throw up their hands in despair. His latitudinarianism offended religious persons who were more bent upon the salvation of men's souls than the creation of an exciting new materialistic philosophy. But among the several groups of scholars that arose in England during the Civil Wars and the Interregnum, one association consciously sought to approximate Bacon's ideals. This was the Oxford experimental science club which arose about 1648 under the leadership of John Wilkins (1614–72), at that time Warden of Wadham College.

This club consisted of thirty or more men of extraordinary ability. It included Seth Ward, Oxford Professor of Astronomy and

---

[8] *Ibid.*, p. 52. I am deeply indebted to Miss Purver's book for this brief reconstruction of Bacon's ideal.

[9] James Spedding, R. L. Ellis, and Douglas Heath, eds., *The Works of Francis Bacon* (14 vols., London, 1857–74), IV, 291.

later Anglican bishop; Robert Boyle, physicist and chemist; William Petty, political economist and statistician; Dr. John Wallis, Oxford Professor of Geometry; Dr. Thomas Willis, anatomist, physician, and later Physician to Charles II; Christopher Wren, astronomer and architect; and others. Dr. Wilkins, the leader of the group, was a mild Puritan (he married Oliver Cromwell's sister in 1656), but in the political flux enveloping the Stuart Restoration he managed to ride out the storm, and later was appointed Bishop of Chester. During the critical years, 1658–59, when the Cromwellian Protectorate tottered and fell from its foundations, the Oxford Club dispersed. Several of its members moved to London, where they soon foregathered with Londoners of a similar bent of mind. For a dozen years or more, a group of Londoners had been meeting periodically at Gresham College to witness scientific experiments and to discuss problems of natural philosophy. In the course of 1659–60, the erstwhile Oxonians—notably Boyle, Petty, Willis, Wren, and Wilkins—joined with such Londoners as John Evelyn, the famous diarist, architect, antiquary, and landscape gardener; Robert Hooke, an extraordinarily versatile and ingenious experimental philosopher, who previously had been associated with members of the Oxford Club; Sir Robert Moray, a Scottish statesman-philosopher who, having joined the King in exile in France, had kept his political reputation unsullied in the Royalist camp; and several of the professors at Gresham College.

Whether the impulse for the formation of the Royal Society of London originated in the Oxford Club or in one of the several London groups that had sprung up since the mid-1640's is, for the purposes at hand, a matter of indifference. The immediate facts in the origin of the Royal Society are clear and well substantiated. On November 28, 1660, only a few months after the restoration of the Stuart monarchy, a group of men interested in the new science met at Gresham College, London, to hear a lecture by Christopher Wren. Dr. Wilkins was in the chair. After the meeting Dr. Wilkins joined with eleven other persons present at the occasion to propose the foundation of a "College for the promoting of Physico-Mathematicall Experimental Learning." The members felt that they could improve their regular meetings "according to the manner of other Countries, where there were voluntary associations of men into academies for the advancement of various parts of learning, so they might do something answerable here for the promoting of Experimental Philosophy."[10] So they agreed to continue weekly meetings and for the future to charge each member a ten-shilling admission fee as well as a weekly assessment of one shilling to defray the costs of experiments and other expenses. They also drew up a tentative list of forty-one names of individuals to be invited to join the association. At this point, in fact if not in name, the Royal Society of London had its beginnings.

A week later (December 5), 112 persons signed the "Obligation" (as it came to be called) to "consent and agree that we will meet together weekly (if not hindered by necessary occasions) to consult and debate concerning the promoting of Experimental Learning. . . ."[11] They also agreed to pay the admission fee and the weekly shilling assessed at the previous meeting. Sir Robert Moray reported to the assembly that the King had been informed of their plans to form a voluntary association, approved of them, and indicated his readiness to encourage them. The assembly then pushed forward to perfect a temporary constitution for the association, to appoint a committee to arrange for experiments to be conducted at future meetings, and to enter into a scientific correspondence with interested, like-minded, and able persons both at home and abroad. The following week (December 12), temporary officers were set up (President, Treasurer, and "Register," or Secretary), and membership rules were discussed. No one, save those with the rank of baron or above, was to be admitted without scrutiny. Elec-

---

[10] *Journal-Book*, I, 1. This, and all other materials in the Archives of the Royal Society of London, is cited with the gracious permission of the President and Council of the Society. Cf. *Add. Mss. 4447*, fol. 13 (Birch Mss.). The twelve were William, Viscount Brouncher, Robert Boyle, Robert, Lord Bruce, Sir Robert Moray, Sir Paul Neile, Dr. John Wilkins, Dr. Jonathan Goddard, Dr. William Petty, William Ball, Laurence Rooke, Christopher Wren, and Christopher Hill. The "manner of other Countries" probably referred to the short-lived Florentine Accademia del Cimento founded in 1657 by Prince Leopoldo de' Medici. Thomas Sprat's official *History of the Royal-Society of London For the Improving of Natural Knowledge* (London, 1667, cited hereafter as Sprat) sets forth the Wilkins-Oxford origin of the Society. Miss Purver presents most of the pertinent evidence. Sir Harold Hartley, ed., *The Royal Society: Its Origin and Founders* (London, 1960), brings together useful essays about the founding and the founders of the Society.

[11] *Add. Mss. 4447*, fols. 12, 14.

tion was to be by ballot, two-thirds of the members present must approve or the election was invalid, and twenty-one members were to constitute a quorum. No new member was to be elected on the same day that his name was proposed. Future meetings, at least for the time being, would be held at Gresham College or, during the time of the college's vacation, in William Ball's room in the Middle Temple. Two servants were also employed, an amanuensis to assist the Secretary and an "operator" to assist in the conduct of experiments.[12]

Charters, Statutes, and Objectives

In the course of the following year, a variety of experiments were performed at the association's meetings, membership was somewhat expanded, scientific correspondence was initiated, plans were drafted to test, by empirical means, the scientific validity of a number of classical works hitherto given wide credence (such as Virgil's *Georgics*), and committees were appointed "to consider of proper Questions to be enquired of in the remotest parts of the world." During this time, too, John Evelyn suggested the name (the Royal Society) for the association, and Sir Robert Moray, carefully selected as one in good standing with the King, successfully negotiated a royal charter, issued as of July 15, 1662. By its terms the King expressed a desire "to extend not only the boundaries of the Empire, but also the very arts and sciences," especially those philosophical studies "which by actual experiments attempt either to shape out a new philosophy or to perfect the old." To this end, the King granted corporate existence to the Royal Society, "whose studies are to be applied to further promoting by the authority of experiments the sciences of natural things and of useful arts."[13] In addition to corporate status, the charter also extended to the Society various privileges including the ordering of its own internal affairs, the right to meet in London or within a radius of ten miles of the city, the right to correspond abroad on scientific matters, and the imprimatur, or privilege to

[12] *Journal-Book*, I, 1–2; *Add. Mss. 4447*, fol. 15v.

[13] *The Record of the Royal Society of London for the Promotion of Natural Knowledge* (4th ed., London, 1940, cited hereafter as *The Record*), p. 226. The charters of 1663 and of 1669 repeated the wording. *Ibid.*, pp. 250, 274ff.

print items relating to the work of the Society. A second charter, issued in 1663, tidied up several administrative matters, altered the Society's legal title to "the Royal Society of London for Promoting Natural Knowledge," and declared the King to be founder and patron. A third charter (1669) corrected some legal omissions of the second charter and reasserted the purposes of the Society in the same words. By these documents the Royal Society acquired legal status and the prestige which was associated with royal approval.

In the meantime, the Society drew up, in 1663, its original statutes. They proclaimed: "The business of the Society in their ordinary Meetings shall be, to order, take account, consider, and discourse of philosophical experiments and observations; to read, hear, and discourse upon letters, reports, and other papers, containing philosophical matters; as also to view, and discourse upon, rarities of nature and art; and thereupon to consider what may be deduced from them, or any of them; and how far they, or any of them, may be improved for use or discovery."[14] There were limitations set, however. The Society was specifically designed to promote "Physico-Mathematicall Experimental Learning," and it refused to meddle with "Divinity, Metaphysics, Moralls, Politicks, Grammar, Rhetorick or Logick." The Society set out to "examine all systems, Theories, principles, Hypotheses, Elements, Histories & Experiments of things Natural, Mathematicall & Mechanicall invented, recorded, or practised by any Considerable Author, Ancient or Modern, in order to the compiling of a Compleat system of Solid Philosophy, for explicating all phenomena produced by Nature or Art, & rendering a rational account of the causes of things."[15] It would own no "hypothesis, system, or doctrine of the principles of Natural Philosophy . . . till by mature debate & clear arguments, chiefly such as are deduced from legitimate experiments, the truth of such positions be demonstrated invincibly." The time of the assembly was "to be employed in proposing and making experiments, discoursing of the truth, manner, grounds & use thereof; reading & discoursing upon Letters, reports, and other papers concerning philosophicall & mechanicall matters; Viewing

14 *Ibid.*, p. 289.
15 *Add. Mss. 4441*, fols. 1–1v; Weld, I, 146.

and discoursing of curiosities of Nature and Art; and doing such other things as the Council or the President alone shall appoint."[16] The limitations doubtless helped the Society to avoid sterile controversies and to steer clear of many of the rocks upon which similar organizations foundered. By this token, also, they enabled the Society more rapidly to advance experimental learning in the years ahead. But at the same time, they may well have contributed to that unhappy separation of science and morals which many scholars find deplorable in the present day.

The Royal Society, as chartered, was a self-governing and self-perpetuating body. Every year, on St. Andrew's Day (November 30), the Fellows of the Society elected their own officers by ballot. The officers consisted of a President, Vice-President, Treasurer, two Secretaries, and a Council of twenty-one members. The President and Council constituted the governing body of the Society, attending to all questions of policy, business, and finances. The latter, especially in the early years, was often troublesome. Although the King was generous in the privileges granted by the charters, he did not endow the Society with viable lands or other resources from the royal purse, and the Society, unlike the French Academy of Sciences founded by Louis XIV only a few months later (1662), had no access to state funds and was left to its own private resources. At the outset, these consisted principally of the admission fees of Fellows and the weekly assessments of one shilling each. As time passed, these sums were increased by action of the Council, which also encouraged wealthy Fellows and other well-wishers to make gifts to the Society. But endowments accrued slowly, and until well into the eighteenth century the Society's funds were often inadequate to meet its needs, especially for the publication of scientific treatises. However, the Society's lack of access to the royal purse largely left it free from royal dictation, and in this respect the Royal Society was better off than its French counterpart. With the exercise of judicious policies as set by the President and Council, the Society was free to set its own course without royal interference.

It is evident from the records of the Royal Society that its objec-

16 *Ibid.*

tives were broadly Baconian in nature. But for the realization of the Baconian ideal—to promote the development of a new experimental philosophy looking toward the enunciation of scientific laws underlying natural phenomena—some of the greatest barriers lay within the Society itself. There was an ever-present strain of utilitarianism in the objectives of the Royal Society. This was evident in many of the English forerunners of the Society, even in the works of Francis Bacon himself. It was referred to in the charters, which coupled the "useful arts" with the object of "further promoting by the authority of experiments the science of natural things." It was given concrete expression by Robert Boyle's book, *Some Considerations Touching the Usefulness of Experimental Natural Philosophy* (London, 1663, followed by a second part in 1671).[17] And it permeated the minds of many of the Fellows from the Society's beginning. The danger lay, as Bacon had foreseen, in "the little cells of human wit" content with one of the foundation stones of the larger edifice of science which Bacon envisaged. Not all of the men elected into the fellowship of the Royal Society comprehended the Society's larger objectives or were able to sustain that objective constantly before their eyes. In consequence, while the larger "wits" of the Society pursued the long-term philosophical ends of Bacon's works, some Fellows, together with forces outside the Society, sought the more proximate end of private advantage.

As the Society officially sought to have "all the proposals that shall be made concerning Mechanical Inventions be referred to the Council of this Society, to be by them Examined, whether they be new, true, and usefull,"[18] it was forced from time to time to fend off applications for its approval sought for private advantage. In 1663 Sir William Petty brought his double-bottomed ship to the Society for formal approval, and the Council, after entertaining "the sense of the Society thereupon," ordered "that the Matter of

---

[17] Boyle himself comprehended Francis Bacon's larger objectives, but his book was subject to a purely utilitarian interpretation. Actually, the book was a promotion pamphlet for the Baconian philosophy, pointing up the strictly utilitarian benefits as a by-product of the pursuit of Baconianism.

[18] *C.M.*, I, 58 (March 9, 1664); cf. *ibid.*, I, 60. Queen Anne later ordered that all patent applications be referred to the Royal Society "before the Patent be passed." *Journal-Book*, XI, 326 (Jan. 22, 1712/13).

Navigation, being a State-Concern, was not proper to be managed by the Society; And that Sir William Petty, for his private satisfaction, may, when he pleases, have the sense (if he hath not already) of particular Members of the Society, concerning his new Invention."[19] Later the same year, in a debate on whaling occasioned by a report on the whaling industry by a representative of the Greenland Company, the Society was led by a motion from Sir Robert Moray to consider "whether there might not be devised an Engin fit to strike the Whales with more ease and surnesse, and at a greater distance," and it was ordered that "Dr. Wilkins and Mr. Hook should think upon such an Engin that might be cheap and easy to be managed."[20] In the 1670's the importunities of John Beale of Yeovil, Somerset, a valued correspondent of the Society and an agricultural reformer of substance, to meddle in policies to improve English cider-making and free England from its dependence upon French wines, went beyond the real interests and powers of the Society.[21] In 1693 the Society mistakenly gave a testimonial to John Marshall, a glass-grinder, for a new method of grinding glasses and soon found itself embroiled with all the glass-grinders of London.[22] This experience led to greater caution in 1695, when John Davys sought approval of "ye Invention of an Engin to weave stockings &c." This time the Society stated that a detailed account of the machine "will be entered in our Registers wch will be a means to assert the Invention to yorself. . . ."[23]

As many Fellows were investors in and officers of the great

[19] C.M., I, 8 (May 27, 1663).

[20] Journal-Book, I, 243 (Nov. 4, 1663). Such a machine was presented in 1752. Ibid., XXII, 115–118.

[21] Oldenburg-Beale Letters, Letter-Book "O," No. 137 (Royal Society Archives); ibid., Supplement, I, 315ff.

[22] Journal-Book, VIII, 206, 215, 216, 260, 261–262, 264, 267; Stowe 747, fol. 39 (British Museum).

[23] Letter-Book, XI, Pt. II, No. 40. In 1712 one Mr. Evrard, a brewer, sought the Society's approval on the wholesomeness of his malt extract. The Society demurred, instructing its Secretary to "tell him by word of Mouth, That the Society, as such, cannot give any such Certificate as he desired. . . . That tho' some of the Members might attest they did not find it Unwholesome, yet that would not answer his Design; but more they could not do." Journal-Book, XI, 286 (April 24, 1712).

English trading companies of the time, and as the Society itself invested funds in these companies and sought their aid in scientific commerce with distant parts of the world, it is not surprising that the Society was occasionally plagued by private interests.[24] Sometimes, too, the Society could not escape experiments which it was asked to undertake. In 1740, for example, His Majesty's Board of Ordnance requested the Society to set up a series of experiments to test three new sorts of gunpowder recently received from Danzig. These experiments occupied a committee of the Society for nearly two years before it was concluded that English gunpowder was superior to them all.[25] With such pressures arising upon its Fellows from both within and outside the Society, it is not surprising that from time to time they lost sight of Bacon's majestic *Organon* and fell into consideration of matters purely utilitarian.[26] And if such lapses from the Baconian ideal were possible among the Fellows of the Royal Society of London, it is hardly surprising if English

[24] The Society bought stock in the East India Company, Royal African Company, and South Sea Company. See *C.M.*, II, 25 (1683), 100 (1699), 238 (1717), 253 (1720). The Society was also closely tied in with the Hudson's Bay Company.

[25] *Journal-Book*, XVIII, 166–167, 448–449 (Dec. 11, 1740–Nov. 4, 1742).

[26] The tug of the old science was also apparent from time to time. Indeed, the experimental approach did not immediately destroy the old scientific beliefs, and the Society itself did not propose to cast them aside until careful experimentation had demonstrated their errors. Thus, on May 31, 1699, the Society spent much of its meeting time solemnly discussing the medicinal values of "cows piss" (*Journal-Book*, IX, 153–154—no experimentation was reported), and it was 1727 before the Fellows, having been repeatedly presented with plans for perpetual motion machines, finally stated: "That such subjects as that of an artificial perpetual motion not falling under the Society's Enquiry, They cannot hold any Correspondence concerning it." *Ibid.*, XIV, 23 (Oct. 26, 1727). In 1759 Dr. William Stukeley, "as a Senior member," protested against the Society's refusal to consider problems involving longitude, squaring the circle, perpetual motion, the philosopher's stone, and the like, arguing that "tho these points are not proved or found out yet in the pursuit of them many valuable and excellent discoveries have been hit upon by these mad projectors, as is fully known in all the Litterary world." There was added an "N.B. however just and worthy the Drs. observations are, I find the Society will go its old way. . . ." *Egerton 2381*, fol. 87 (March 22, 1759) (British Museum). As late as 1785 the Council of the Society was constrained to order "that no member of the Society who may be engaged in any Trade or Mechanical Profession, shall upon any occasion affix the addition of Fellow of the Royal Society to his name upon any article of Commerce or Bill published for the purpose of promoting his said Trade or Profession." *C.M.*, VII, 238 (Nov. 24, 1785).

colonials across the wide Atlantic sometimes confused the goals of experimental philosophy.

On the average, the Fellows of the Royal Society assembled for their weekly meetings about thirty times a year. A recess of two or three weeks was usually taken at Christmas and again at Easter, and it was customary for the Society to hold a "long recess" from early July until late October. Attendance at the meetings varied greatly. During the first ten years, the meetings were heavily attended, with upwards of eighty Fellows present. The number fell off rapidly in the 1670's,[27] grew "very full" in the early 1680's, fell again as the political confusions of the "Glorious Revolution" approached, and remained at a low level until the last years of the seventeenth century.[28] Membership increased after 1700 and weekly attendance grew heavier. The number of visitors grew exceptionally large in the 1740's (sometimes being as many as forty or fifty), so that an attendance of 200 or more persons was not unusual during the third quarter of the eighteenth century.[29]

What took place at these meetings? A chart of the activities at the weekly meetings during the first century of the Royal Society's existence (1660–1760) as recorded in the *Journal-Books* of the Society is revealing. During this interval, the Society held about 3,000 regular meetings. Some 1,677 experiments were conducted, 9,876 papers were read, 1,265 *consignments* of specimens were received into the "Repository of Rarities," or the Society's museum (these varied from a single rattlesnake from William Byrd II of Virginia to three consecutive gifts by the Hudson's Bay Company in the early 1770's, constituting the "most valuable Collection of Natural Productions from Hudson's Bay by the favour of the Directors of that Company"[30]), and scores of book reviews

[27] Meetings were dismissed because of sparse attendance. *Journal-Book*, III, 263; V, 58, 62, 133, 149, 186, 214, 223, 233.

[28] No meetings were held in June, 1687; April 18, June 6, July 25, 1688; and too few were present to hold a valid election on Nov. 30, 1688. *Ibid.*, VII, 41, 44, 105, 121, 138, 152. The Society made conscious efforts to improve its status in the late 1690's. See *Journal-Book*, IX, 118, 124, 126, 148; *C.M.*, II, 103, 104, 106, 107.

[29] The information in this paragraph derives from systematic perusal of the *Journal-Book*, I–XL, *passim*. Cf. also Margaret 'Espinasse, "The Decline and Fall of Restoration Science," *Past and Present*, No. 14 (Nov., 1958), pp. 70–89.

[30] *Journal-Book*, XXVIII, 197.

were presented. The number of experiments performed at meetings declined after the first five years. For example, between 1661 and 1665, 362 experiments were recorded; from 1666 to 1670, 276; after 1685 the number never reached 100 for an entire five-year period; and from 1756 to 1760 there were none at all. Indeed, after 1750, very few experiments were performed at meetings of the Society.

The subjects with which the experiments were concerned varied widely, in part corresponding to the interests of the curators or of prominent members at any given time, and in part corresponding to the incidence of unusual events in the natural world. Robert Boyle, Robert Hooke, Denis Papin, Francis Hauksbee, and Stephen Gray, for instance, generally produced experiments in physics; Nehemiah Grew, Patrick Blair, and Stephen Hales would lead the Society into experiments in the anatomy or physiology of plants; and James Douglas, James Stuart, and William Cheselden presented anatomical dissections of birds, beasts, and, in a few instances, human bodies. An earthquake, a comet, or an outbreak of smallpox generally provoked much discussion and experimentation.[31] The decline of experiments after 1750 is difficult to explain. Individual Fellows and groups of Fellows still performed experiments, especially in electricity and in chemistry, and many of these were reported in detail to the Society. But very few were actually performed in the presence of the Society as a whole.

As the number of experiments made at the Society's meetings declined, the number of papers increased. During the first five years, only 245 papers were read; during the second, 304. The number fell throughout the parlous decade of the 1670's, but rose to 441 in the period between 1681 and 1685. After the "Glorious Revolution," it rose to more than 500 each five years, reached a peak of 701 for the years 1736–40, and slowly decreased after 1740. By the last decade of the eighteenth century, the number of papers

---

[31] An eclipse of the sun in 1715, a comet in 1716, and other extraordinary celestial events up through 1719, resulted in 163 papers on astronomy between Nov. 30, 1714, and Nov. 30, 1719; an earthquake in England in 1750 brought forth 63 papers on the subject before the end of the year. For the Royal Society and smallpox, see my paper (with the assistance of George Pasti, Jr.), "Remarks upon the Inoculation for Smallpox in England," *Bulletin of the History of Medicine*, XXIV (1950), 103–122.

read was less than fifty per year. The reduced number of papers, together with the relative absence of experiments, was an index of the decline in the Society, and it was accompanied by a marked decrease in scientific correspondence both with the British colonies and with the European Continent. The American and the French revolutions appear to have affected the Royal Society more adversely than the earlier wars of Louis XIV, the War of Austrian Succession, and the Seven Years' War. The number of members continued to increase, to be sure, but a larger and larger proportion of them were retired military and naval men, with a large amount of "back-scratching" evident among the servants of the East India Company. The Council of the Society was aware of the eclipse, and commented at the end of 1791 that the Society had become "unusually deficient in the quality of its produce."[32]

However, the decline was not as great as the lessened number of papers and experiments seemed to indicate. For as their numbers decreased, the length of individual papers increased, and they showed a marked difference in nature. Whereas the earlier papers had generally dealt with the analysis or description of particular things, such as a specimen in natural history, an ancient inscription, a surgical operation, or, at most, a small group of related experiments in physics, and so on, there was a rapidly developing tendency after about 1760 for the papers to become large syntheses of data, interpretative rather than analytical or descriptive in nature. These were the more mature accomplishments of such men as Joseph Priestley, Henry Cavendish, Benjamin Thompson, and James Hutton. They marked the beginning of the end of that period in the history of science devoted principally to collection of data and simple experimentation. They indicated the rise of greater specialization and the emergence of new scientific disciplines, such as chemistry and geology. Experimental philosophy, in short, had reached a new stage in its development, and the fewer—but longer—papers presented to the Royal Society were, in this regard, a mark of new achievement rather than a signal of decline.

The proceedings of the Royal Society covered a wide range of scientific activities, and the survey of its papers and experiments

[32] *Journal-Book*, XXXIV, 437 (Nov. 30, 1791).

shows where the emphasis lay and furnishes, in a large measure, an index of the scientific life of the British Empire for the period under survey. Speaking broadly, natural history and the medical sciences led the field. Of the 9,876 papers read to the Society during its first century, 2,174 dealt with natural history and 2,171 related to medicine. Physics ran third, with 1,560 papers; astronomy was fourth, with 1,134 papers; and antiquities and meteorology almost tied for fifth place, with 471 and 470 papers respectively.

Turning to experiments conducted at the meetings of the Society, however, the sequence is altered. Physics held a clear first, claiming nearly half (780) of the 1,677 experiments performed during the first century of the Society's existence. The medical sciences were second, with 353; natural history ran third, with 234; and chemistry was fourth, with 144.

Taking papers and experiments together, it is clear that problems concerned with medicine and natural history consumed a major portion of the time and efforts of the Royal Society during its first century and that problems of physics ran third. When it is considered that the vast majority of the Fellows of the Society who had any formal training in scientific subjects were physicians and apothecaries, and that the amateurs could make their greatest contributions in such subjects as botany, entomology, and zoology, it is not surprising to find such heavy emphasis upon medicine and natural history.

The "Repository of Rarities" of the Royal Society, founded upon the motion of Daniel Colwall, rapidly grew into a scientific collection which far outshone that of the Tradescants.[33] Its dried plants, seeds, shells, serpents, birds, insects, minerals, fossils, and the like were collected from all over the world, being donated by far-flung correspondents of the Society (or of its members), or by the East India Company, the Royal African Company, or others of the English trading companies of the time. It early included a variety of gifts from the New World—the first of any size being a collection of plants, acorns, ores, minerals, and shells from New England given in 1669 by John Winthrop, Jr., Governor of Connecticut and, as we shall presently see, the first of several

[33] See pp. 48–50 above.

colonial Fellows of the Royal Society.[34] Throughout the remainder
of the Old Colonial Era the Royal Society's Repository was augmented by gifts from the New World—although many of the
insects, animals, fishes, and birds became so putrified, owing to
inadequate means of preservation, that in time they had to be
thrown out.[35] Many of the Fellows of the Society used the Repository for study, and the collections became the source both of
many papers and of experiments presented to the Society and of
scientific treatises published by the Society's Fellows. In 1681
Nehemiah Grew, acting upon the authority of the President and
Council of the Society, published *Musaeum Regalis Societatis, Or
a Catalogue & Description of the Natural and Artificial Rarities
Belonging to the Royal Society and Preserved at Gresham Colledge. . . .*[36] Grew's *Catalogue* described nearly a hundred specimens which had come from the New World, and these were
multiplied many times over before the end of the Old Colonial
Era.[37] On June 3, 1779, the Council decided to close out the
Society's Repository and to present its contents to the recently
founded British Museum "for the public use." Delivery was effected in 1781, and the specimens of the "Repository of Rarities"
were incorporated into the collections of the British Museum,
where many of them are still extant, especially in the Department
of Natural History.[38]

The Royal Society did not hide its light under a bushel. It ex-

[34] Listed in detail in *Journal-Book*, IV, 112–117; Birch, I, 67ff.

[35] In March, 1690/91, the Council ordered a house-cleaning of "the Animals
that are decayed in the Repository. . . ." *C.M.*, II, 83. A second cleaning-up in
1729–30 cost the Society £120. *Ibid.*, III, 30–31, 74. A third took place in 1763.
Add. Mss. 5166, fol. 240. By this time the Society had professional taxidermists in
charge of their soft specimens.

[36] Grew was the official Caretaker of the Repository at the time. See *Journal-
Book*, VI, 192; *C.M.*, II, 24–25. A new catalogue was prepared in 1730, but so far
as I can discover, never printed. *C.M.*, III, 30–31, 47–48.

[37] See, for example, the gift of John Winthrop III (1681–1747) in 1734. This,
constituting the largest single gift to the Repository up to that time, consisted of
"above Six hundred curious specimens, chiefly of the Mineral Kingdom." Cf. my
article, "John Winthrop (1681–1747) and His Gifts to the Royal Society," Publications of the Colonial Society of Massachusetts, *Transactions, 1952–1956*, XLII
(1964), 206–232.

[38] *C.M.*, VII, 27, 30, 70; *Journal-Book*, XXX, 78, 606. Dandy lists many of the
botanical specimens still extant.

erted great efforts to inform the world about its work and to enlist the active interest, understanding, and cooperation of persons of all ranks everywhere. These efforts can be treated under three principal heads: correspondence, publications, and patronage.

Interest in the Colonies: "Enquiries for Foreign Parts"

Even before it had become a chartered corporation, the Society had set up a Secretary and appointed a Committee for Foreign Correspondence.[39] In the course of 1661, correspondence was established with scientists in France and Italy, and the Committee began to draw up "Enquiries for Foreign Parts" to be dispatched to the Far East, the Levant, Muscovy, and the Americas (see, for example, Appendixes I, III, IV, and V herein). The charters provided for two Secretaries of the Society, and in 1686 an assistant was added to help with foreign correspondence. In 1719 Robert Keck died, leaving the Society a bequest of £500 "to support their foreign Correspondence."[40] With these funds an endowed office of Foreign Secretary was set up, and for the remainder of the eighteenth century the Society had four Secretaries almost constantly in its employ.

The volume of the Society's overseas correspondence varied over the years in direct proportion to the energies displayed both by individual Fellows, whose correspondence was often brought before the Society, and to those of its Secretaries. The first two Secretaries named after the Society had been chartered were Dr. John Wilkins and Henry Oldenburg. Oldenburg was possibly the most energetic and able Secretary the Society had during the Old Colonial Era. Oldenburg (c. 1615–77), a German who migrated to England during Cromwell's Protectorate, was associated with the founders of the Royal Society from its beginnings. By his extraordinary linguistic achievements (he had command of German, English, Latin, French, and Italian), he was admirably equipped to direct the Society's foreign correspondence. For an unsalaried position—he earned his living by teaching, translating,

[39] Dr. William Croone, Professor of Rhetoric at Gresham College, was the first Secretary. *Journal-Book*, I, 10, 16, 35.

[40] *C.M.*, II, 250 (Oct. 26, 1719); *The Record*, pp. 110, 142.

and writing—he devoted himself unstintingly to the service of the Society: "He attends constantly the meetings both of the Society and the Councill; noteth the observables said and done there; takes care to have them entered in the Journal- and Register-Books; reads over and corrects all entrys . . . writes all letters abroad and answers returns made to them, entertaining a correspondence with at least 50 persons; employs a great deal of time and takes much pains in satisfying forran demands about philosophical matters. . . ."[41]

After Oldenburg's death in 1677, the Society's correspondence fell into serious disarray and remained in a parlous state until Richard Waller became Secretary in 1710—although the relative weakness of the Secretaries was, in part, overcome by the private correspondence of such Fellows as Dr. Martin Lister, James Petiver, and Dr. Hans Sloane. Unfortunately, Waller died after a few years. Dr. James Jurin (Secretary, 1721–27) and Dr. Cromwell Mortimer (Secretary, 1730–51) were the only other Secretaries whose energy in the position approximated that of Henry Oldenburg during the Old Colonial Era. But Oldenburg had laid a firm foundation for the Society's correspondence with the British colonies overseas, with the principal scientists and academies of Continental Europe, and with the Dutch, Portuguese, and Spanish far-flung empires in the Far East and the New World.[42] Through the good offices of John Evelyn and Sir Joseph Williamson, His Majesty's Principal Secretaries of State, permission was

[41] R. K. Bluhm, "Henry Oldenburg, F.R.S.," *Notes and Records*, XV (1960), 187. See also the Introduction to Vol. I of A. Rupert Hall and Marie Boas Hall, eds., *The Correspondence of Henry Oldenburg* (in progress, Madison and Milwaukee, 1965——).

[42] The Royal Society had little success in penetrating the Portuguese and Spanish empires. Cf. *Letter-Book*, I, 377–379; II, 218–219; IV, 345–346; *Journal-Book*, III, 194; IV, 3, 6. Other efforts to extend the Society's correspondence are evident in *Letter-Book*, I, II, III, *passim*; XV, *passim*; XVI, 32; *Guard-Book*, W-3, Nos. 72, 74, 75–76, 78, 80, 83, 85 (Waller); *C.M.*, VII, 97–98, 100–101, 105–106, 111; VII, 150, 152, 158, 161, 165; *Journal-Book*, VI, 232, 234, 235, 246, 320; VIII, 45, 50; IX, 80, 100, 142, 145, 204; *Classified Papers, 1660–1740*, XVII ("Miscellaneous"), Nos. 46, 48; *Add. Mss. 4811*, *passim*; *Sloane 4036*, fols. 266–267, 370; *4037*, fol. 189. American colonial correspondence dwindled badly on the eve of the American Revolution. Cf. *Journal-Book*, XXIX, XXX, XXXI, *passim*.

given to send and to receive the Society's letters abroad in the packets of the Secretaries of State.[43] And the ships of the East India Company, the Hudson's Bay Company, and subsequently those of other English trading companies also carried without charge the Society's correspondence—and sometimes its agents, with their instruments and specimens—to and from England.[44]

To supplement and insure the continuance of its foreign correspondence, the Society often enlisted the assistance of Fellows stationed abroad on diplomatic business and other affairs. Such men as Dr. Edward Brown at Vienna in the late 1660's, Sir Robert Southwell at Lisbon in the 1670's, Sir Thomas Dereham at Rome in the 1720's, Dr. James Burnet at Madrid in the 1730's, Dr. Murdock Mackenzie at Constantinople in the early eighteenth century, Jakob Theodor Klein at Danzig in the mid-century, and Dr. John Amman and some seven or eight others at St. Petersburg in the course of the eighteenth century—all these and others too numerous to relate served to promote the philosophical interests of the Royal Society abroad.[45] They distributed—and sometimes collected—scientific books and information for the Society, and they enlisted the cooperation and correspondence of foreigners. Several of them were rewarded by remission of their payments as Fellows and by paid-up life membership in the Society.

The private correspondence of Fellows was often laid before the Society, and occasionally it was fully as significant as that of

[43] *Journal-Book*, VI, 217 (Dec. 4, 1679); *C.M.*, I, 303 (Dec. 10, 1679); Weld, I, 420–421.

[44] *Letter-Book*, II, 1–2.

[45] *Sloane 1911–13, passim; 1908*, fols. 1–85; *1922*, fols. 48ff.; *Letter-Book*, III, 38–39, 95; *Supplement*, V, 233–278. Much of the correspondence of these men survives in the Royal Society Archives and in the *Sloane Manuscripts* of the British Museum. Oldenburg's letter to Dr. Edward Browne, Dec. 3, 1668, will serve to illustrate the Society's expectations. Hearing that Browne was on the point of going to Rome, Oldenburg wrote "that your being a Member of the R. Society would easily engage you to doe them what service you could in those parts for the advancement of their dessein in inquiring after philosophical Experiments and Discoveries, and Communicating them to that Illustrious Body, as also in establishing such a Correspondence betwixt the Italian Philosophers and us whereby we may be made participant from time to time of what passes and is discovered by them . . . as we shall be ready to return unto them. . . ." *Sloane 4062*, fols. 170–171.

the Secretaries. Martin Lister, Tancred Robinson, John Wood-ward, Hans Sloane, James Petiver, Peter Collinson, and John Ellis maintained a wide correspondence in scientific matters, much of which was presented to the Society.[46] All of these men held correspondence with American colonials, some of which had special significance. Notable examples were John Woodward, with whom Cotton Mather corresponded so voluminously; John Ellis, through whom Alexander Garden of Charles Town was intro-duced to Linnaeus; and Peter Collinson, who not only put John Bartram in touch with Linnaeus but also aroused Benjamin Frank-lin's interest in electrical experiments and helped to bring Franklin to the attention of the Royal Society.

Manuscript remains of the Royal Society's scientific corre-spondence repose in many collections of manuscripts around the world. They supply overwhelming documentation for the con-clusion that the Society fulfilled with energy, enthusiasm, and de-termination its chartered purpose to promote experimental philosophy. And they render Thomas Sprat's statement of 1667 no idle boast when he wrote that the Fellows pursued a policy designed to make the Royal Society "the general Banck" of experimental learning and that its correspondence was so ordered "that in a short time, there will scarce a Ship come up the *Thames*, that does not make some return of *Experiments*, as well as of Merchandize."[47]

The publications of the Society further advanced their cause. Chief among them was, of course, the *Philosophical Transactions*, the first number of which appeared in March, 1665. At the outset, and for many years afterward, the *Philosophical Transactions* was a private undertaking, at first by Henry Oldenburg and subse-quently by later Secretaries—all, however, operating with the knowledge and formal approval of the President and Council of the Society.[48] The contents consisted almost entirely of papers, letters, and other proceedings drawn directly from the Society's

[46] The correspondence of these men is scattered, but the core of what remains exists in the Royal Society Archives, the Bodleian Library, Oxford, the British Museum (especially the *Sloane Manuscripts*), and the Library of the Linnean Society, London.

[47] Sprat, pp. 64, 86.

[48] *The Record*, p. 36, *passim*; Weld, II, Appendix, 563–565.

weekly meetings and from the Society's correspondence.[49] In consequence, the *Philosophical Transactions* was generally—and properly—received as an official organ of the Society. Financially, however, it was an undertaking by Oldenburg and his successors. As time passed, transitions from one Secretary to another occasioned interruptions in the publication, and as disputes sometimes arose over the policies of the journal, the Royal Society assumed full and direct charge of the *Philosophical Transactions* in 1752.[50] Already it had become the world's foremost publication in experimental philosophy. Practically every Fellow of the Society received it regularly; the President, Council, Secretaries, and many Fellows circulated it widely among their friends and correspondents in many parts of the world—to Pekin, Manila, Fort St. George (India), Boston, Philadelphia, Charles Town, Kingston (Jamaica), Danzig, St. Petersburg, Stockholm, and elsewhere, until the virtuosi on the frontiers of European civilization sometimes complained if they failed to receive it regularly. Articles from it were lifted and republished in the *Gentleman's Magazine*, in Continental journals, and occasionally in colonial newspapers. The *Philosophical Transactions*, together with the correspondence of the Society and its members, became the principal mastic, philosophically speaking, which bound the American colonies to the homeland.

Besides the *Philosophical Transactions*, the Royal Society occasionally assumed the financial burden of publishing scientific books, or assisted in the publication of them, either by directly assuming the entire cost, by holding subscription of funds among the Fellows, or by the guaranteed purchase of fifty or more copies for distribution among the Society's benefactors. The works published at the Society's expense—either in whole or in part—included several of the classics of scientific literature, such as Robert Hooke's *Micrographia* (1665); John Evelyn's *Sylva: Or a Discourse of Forest-Trees* (1664, 1670); Marcello Malpighi's *Anatome Plan-*

---

[49] Not all the papers etc. read at the Society's meetings found their way into the *Philosophical Transactions*. Many of these appear in the *Journal-Books* and in the *Guard-Books* in the Royal Society Archives. A few were published separately.

[50] *C.M.*, IV, 49–70; *Journal-Book*, XXII, 29–30, 62–69, 100–102; *Add. Mss. 4441*, fols. 21–23; *4445*, fols. 46, 150; *6229*, fols. 32v–35.

tarum (1675); Denis Papin's *A New Digester, or Engine for Softning Bones* (1681); Nehemiah Grew's *Anatomy of Plants* (1682); John Ray's *Historia Plantarum* (1686–88); and Isaac Newton's *Principia* (1687).[51]

The patronage of the Society extended far beyond the publication and distribution of scientific journals and books. Indeed, a full account of its further undertakings lies beyond the purpose of this book. We can only suggest something of its extent and attempt to portray its significance. The core of the matter lies in the fact that the Society sponsored, directly or indirectly, in whole or in part, more than forty scientific expeditions into various parts of the world during the Old Colonial Era. Some of these are well known, such as Edmond Halley's voyage to the South Seas in 1700, the various expeditions to observe the transits of Venus in 1762 and 1769, and the voyages of Captain James Cook. But many of them are little known, except to burrowers in the Archives of the Royal Society.[52] In addition to expeditions of this kind, the Society furnished scientific instruments to correspondents in remote places in order that these men might carry on experiments and observations, the results of which were then sent to the Society. This was particularly common in regard to thermometers and barometers, as the Society collected meteorological data systematically for many years in order to improve methods of weather forecasting. But it extended as well to sextants, clocks, and telescopes for astronomical observations and cartography, all of which were supplied at one time or another to correspondents in the East Indies, in Jamaica, and in Hudson's Bay.[53]

Statutes Governing Fellowship for Colonials

At the time of its first charter there were 119 original Fellows of the Royal Society. The membership rose to 187 by 1671, not in-

[51] *The Record*, pp. 36–38. Actually, although the Society undertook to defray the expenses of Newton's *Principia*, its funds proved inadequate, and the cost was borne by Edmond Halley. The Society continued its practice of publication throughout the Old Colonial Era.

[52] Pulteney, I, 347–349.

[53] *Journal-Book*, III, 219; *C.M.*, I, 96, 303; II, 284; III, 120; IV, 11–15; *Sloane 4052*, fol. 20.

cluding twelve foreign members. It diminished during the later years of the seventeenth century, falling to 121 in 1699. But the membership list grew throughout the eighteenth century, numbering 195 in 1721 and 479 in 1781. The membership "on the Foreign List" also expanded greatly, having reached 64 in 1721 and 154 in 1761. In 1765, however, because of protests against the caliber of several foreign members, the list was limited to 80 in that year. This limit was advanced to 100 in 1776 and remained at that figure for the rest of the eighteenth century.[54]

Although the London area was more heavily represented on the "home list" than the provincial areas of the United Kingdom, there was a considerable number of Fellows from the outlying districts. The university towns (Oxford, Cambridge, Edinburgh, Dublin) almost always could boast of residents who were Fellows of the Royal Society, and Yorkshire, East Anglia, the Midlands, Devonshire, Hampshire, and Gloucestershire usually contained several Fellows in each. The "home list" also embraced a liberal sprinkling of colonial Fellows as well as others in the trading companies' "factories" in the Far East, the Levant, Hudson's Bay, the South Sea Company, and the Muscovy Company. As noted above, John Winthrop, Jr., Governor of Connecticut, was an original Fellow of the Society, having been elected on December 18, 1661, before the first charter was issued.[55] Between 1663 and 1783, there were 53 Fellows of the Royal Society elected from the British colonies of North America. Twenty of these were chosen primarily because of their position as governors in the colonies (although some of these contributed modestly to experimental philosophy) and the remaining ones were elected solely on the basis of their interest in and contributions to the scientific program of the Society.[56] Although

[54] *The Record*, pp. 15, 45, 49–50, 568.

[55] After the Society received its second charter (1663), the Council (May 20, 1663) "Considered again who should be received as Fellows of into the Society." It then resolved to admit 115 of the "Original Fellows," including "John Winthrop Esqr." *C.M.*, I, 3–6; *The Record*, pp. 16–18.

[56] See the complete list in my *Col. F.R.S.* An earlier version of this article appeared in *Wm. & Mary Qty.*, 3rd Ser., III (1946), and in *Osiris*, VIII (1948). Of the thirty-three colonial Fellows (exclusive of the colonial governors), ten were from New England, three were from the middle colonies, four were from the southern colonies, two were from Canada (including the Hudson's Bay region), and fourteen were from the British West Indies.

the Royal Society tended to be an elite group, neither social status, religion, nor nationality was a stay to membership, for, as Thomas Sprat had written in 1667, the Society sought "not to lay the Foundation of an *English, Scotch, Popish,* or *Protestant* Philosophy, but a Philosophy of *Mankind.*"[57] Ship captains, merchants, ministers, physicians, apothecaries, civil servants, and others mingled with peers of the realm without regard to creed or nationality in the fellowship of the Royal Society. And they were distributed widely over the globe.

Even before the Society was chartered its members determined that no one should be admitted to the association (other than persons with the rank of baron or above) without scrutiny. The original statutes of the Royal Society (1663) provided that candidates for membership were to be proposed by one or more Fellows at a regular meeting of the Society; balloted for at a subsequent meeting (provided a quorum was present); notified of election and of the moral and financial obligations incurred by membership; required to subscribe to the Obligation, which provided that the candidate promised to *"endeavour to promote the good of the* Royal Society *. . . to pursue the ends for which the same was founded . . .* [to] *be present at Meetings of the Society, as often as conveniently we can . . .* [and to] *observe the Statutes and Orders of the said Society."*[58] This done, the candidate agreed to pay weekly contributions of one shilling "toward the charges of experiments, and other expenses of the Society"; and, within four weeks after election, "or within such further time as shall be granted by the Society or Council, upon cause shewed to either of them," the candidate was required to appear in person before the Society, to pay an admission fee of forty shillings, and to be admitted formally into the Society by the President, who, taking the candidate by the hand, said, *"I do by the authority and in the name of the* Royal Society of London for improving natural knowledge, *admit you a Fellow thereof."*[59]

---

[57] Sprat, p. 63.

[58] Weld, I, 145–146. Any Fellow who refused to subscribe to the Obligation was to be ejected from the Society; any newly elected candidate who refused would thereby void his election. *C.M.*, I, 12 (June 17, 1663).

[59] *The Record*, p. 291.

Candidates residing outside of London, especially colonials, found it difficult, if not impossible, to comply with these regulations. To ameliorate their situation, the Council passed (April, 1664) a new statute: "When any person residing in Forreign parts shall be elected into the Society, in due and accustomed forme and manner; the said person shall be registered among the Fellows of the Society, and to be reputed a Fellow thereof, *without Subscription and admission in the usual Forme, any thing contained in the Statutes requiring Subscription and Admission to the Contrary notwithstanding*. And the said person may have an Instrument under the Seal of the Society testifying him to be elected and reputed a Fellow of the Society accordingly."[60] This statute, though designed primarily for foreign members, was extended to colonial Fellows as well. In the same year (1664) another statute provided that colonial Fellows and others who derived no direct benefit from the experiments performed before the Society should pay no weekly contributions.[61] Almost ten years later (October 30, 1673), the Council provided that "such Members of the R. Society as are or shall go abroad & continue absent from England aboue three Months shall not be obliged to pay their weekly Contribution after those 3 months are expired, but every such Fellow shall be left at his own Liberty to pay or not after the said time, untill he shall be return'd againe unto England."[62] Thus, after April, 1664, a colonial subject of the English Crown was elected a Fellow of the Royal Society merely by being propounded candidate by a Fellow at a regular meeting of the Society and by being chosen by secret ballot at a subsequent meeting, without subscription to the Obligation, payment of fees, or formal admission.

On August 5, 1682, however, "The Statute for Election of Fellows having by long Experience been found insufficient for bringing in persons qualified for the ends of the Institution of the Royal Society," the Council passed a new statute:

Every person that would propose a Candidate shall first give in

[60] *C.M.*, I, 62–64 (italics added).
[61] *The Record*, p. 95. Difficulties in the collection of weekly fees led the Society, by 1674, to require of all candidates "a Bond for payment of the contribution." Later a bond was required of colonial candidates too.
[62] *C.M.*, I, 219–220.

his name to some of the Councell, that so in the next Councell it may be discoursed *vivâ voce* whether the person is known to be so qualified as in probability to be usefull to the Society. And if the Councell return no other Answer but that they desire further time to be acquainted with the gentleman proposed, the Proposer is to take that for an Answer. And if they are well assured that the Candidate may be usefull to the Society then the Candidate shall be proposed at the next meeting of the Society and ballotted according to the Statute in that behalf, and shall immediately sign the usual Bond[63] and pay his admission money upon his Admission.[64]

Thereafter all candidates proposed were referred first to the Council, which, if it approved, recommended them back to the Society, which then balloted for and elected them according to the former statutes. Other regulations applying to English candidates who resided abroad or in the colonies remained unchanged, and such candidates were still exempted from subscription, payment of fees, and formal admission.

Although alterations occurred in the statutes relating to admission of Fellows, the exemptions in favor of colonial Fellows existed until 1753. Indeed, as the number of colonial Fellows increased, a more precise statement of their exemptions and of their exact status in the Society appeared in the statutes. Jarred, perhaps, by the misunderstandings which arose about the election of Cotton Mather to the Society,[65] the Council enacted (January 4, 1727/28) a new statute which stated:

---

[63] See note 62 above.

[64] *The Record*, p. 90.

[65] Details of this unpleasant matter, too intricate for consideration here, are given by G. L. Kittredge in "Cotton Mather's Election into the Royal Society," *Publications of the Colonial Society of Massachusetts*, XIV (1912), 81–114. Whether Mather's case led the Council to effect the clarifying statute of 1727–28 is impossible to say. The only unique feature of the Mather incident was the fuss stirred up by Mather's jealous enemies. The names of several other bona fide Fellows residing in the colonies were more than once omitted from the printed list of Fellows. For example, a search among the printed lists in the Royal Society Archives reveals that the list for 1715 omitted William Brattle (elected in 1714), those for 1722, 1724, 1725, and 1727 omitted Paul Dudley (elected in 1721), and that for 1727 omitted Thomas Robie (elected in 1725). The fact is, these lists were not intended to be complete and official lists of all the Fellows of the Society. Rather, they were printed, as Secretary Waller told Mather in 1713, "for the Con-

Every Person who is a Foreigner and every one of his Majesties Subjects whose habitation or usual place of residence is more than forty miles distance from London, shall be and be deemed as a Fellow immediately after he shall be Elected, and shall be registered in the Journal Book of the Society as such: Provided always, that no such person shall have liberty to Vote at any Election or meeting of the Society before he shall be qualified pursuant to the Statutes. And if he shall neglect so to qualify himself the first time he comes to London when he may be present at a meeting of the Society and can be admitted; his election shall be declared Void, and his Name shall be cancelled in the Register.[66]

At the same time, the Council passed another statute which, though it did not deprive colonial Fellows of their exemptions, altered the manner of electing them and all other candidates for admission to the Society. This statute, which replaced that of August 5, 1682, read as follows:

Every Person to be Elected Fellow of the Society shall first at a meeting of the Society be propounded as a candidate to be approved by the Council, and shall be recommended by three members, one of which at least shall be a member of the Council, and one of them shall at the same Time mention and Specify the qualifications of the said Candidate. And afterwards such Person shall be at another meeting of the Society (where at there shall be a competent Number for making Elections) be referred back from the Council, if approved, and shall then be propounded and put to Vote for Election. Saving and Exempting that it shall be free for every one of his Majesties Subjects who is a Peer or the Son of a Peer of Great Britain or Ireland, and for every one of his Majesties Privy Council of either of the said Kingdoms to be propounded

venience of the Fellows present in choosing the Council, President, and officers for the ensuing year" at the yearly elections on St. Andrew's Day. Mather and other colonial Fellows were omitted because they were not present at the elections and, indeed, until they were formally admitted into the Society, had no vote and were not eligible for office.

[66] *C.M.*, III, 5–6; *The Record*, p. 91. In the original draft of this statute the wording was more explicit: "That any Person residing out of Great Britain, who shall be Elected into the Society, shall be a Fellow of the Society, without Subscribing the Obligation, Giving Bond for Contributions, paying his Admission Money, or being Admitted. . . . That however no Person shall have the right to be present at the Meetings of the Society or to give his vote in any Election till he shall have been Admitted." *C.M.*, II, 304, 308.

by any single Person and to be put to the Vote for Election on the same Day, there being present a competent Number for making Elections.[67]

The above statute survived only about three years. Certain Fellows, suspecting that it contained elements contrary to the Society's charters, brought about a review of the statute by the Attorney General, who, after deliberation, declared the statute to be opposed to the charter and therefore void.[68] Accordingly, the Council prepared a new statute. This one, which was set forth on December 10, 1730, introduced two new practices in the election of candidates for fellowship, namely, the use of certificates and the lapse of a stated interval between propounding a candidate and balloting for his election. Each of these innovations is fully described in the statute, which ran as follows:

> Every person to be elected a Fellow of the Royal Society, shall be propounded and recommended at a meeting of the Society by three or more members; who shall then deliver to one of the Secretaries a paper, signed by themselves with their own names, specifying the name, addition, profession, occupation, and chief qualifications; the inventions, discoveries, works, writings, or other productions of the candidate for Election; as also notifying the usual place of his habitation.
>
> A fair copy of which paper, with the date of the day when delivered, shall be fixed up in the common meeting room of the Society at ten several ordinary meetings, before the said candidate shall be put to the ballot. . . .[69]

Beginning in 1731, then, for each candidate propounded for fellowship there was a certificate describing the candidate's qualifications and bearing the signatures of at least three Fellows of the Society.[70] Further, in order to prevent abuses, the Council set forth (June 19,

---

[67] C.M., III, 5–6; The Record, p. 90.

[68] The entire case is treated in Weld, I, 458–461.

[69] C.M., III, 77, 79; The Record, p. 91. The remainder of the statute makes the same reservations regarding Peers of the Realm, Privy Councillors, and "every foreign Prince or Ambassador" that were included in the statute of Jan. 4, 1727/ 28. See above.

[70] The certificates for all the colonial Fellows are printed in my Col. F.R.S.

1738), "as a useful Precept," that "no Person sign a Certificate in recommendation of a Candidate unless he be personally acquainted with him."[71] Later (February 19, 1756) the "precept" was somewhat modified: "That no Fellow of the Society, Do Sign any Certificate of recommendation of any Candidate for Election into this Society, unless Such Candidate be known to the person Signing Such Certificate, *either in person or by his Works*, and that such Knowledge be particularly expressed in every Such Certificate of Recommendation."[72] Until after the close of the Old Colonial Era this form of election into the fellowship of the Royal Society remained unaltered.

However, one further change in the statutes of the Society affected colonial Fellows. This was the withdrawal, in 1753, of exemptions from the payment of fees, which since 1664 had been extended to colonial Fellows and others residing at places remote from London. Financial needs of the Society, especially after the formal taking over of the *Philosophical Transactions* in 1752, appear to have prompted the reversal of policy. In any case, on July 5, 1753, the statute of January 4, 1717/18 (officially cited as the eighth statute of the sixth chapter of the printed statutes, published in 1752, which specifically granted the exemptions), was repealed. In its place the Council provided a new statute by unanimous vote. The latter read:

> Whereas by the Eighth Statute of the Sixth Chapter of the Statutes of this Society, It is Enacted 'and Declared, That every Person who is a Foreigner, and every one of his Majesty's Subjects whose habitation or usual place of residence is at more than forty miles distance from London, shall be and be deemed a Fellow of the Society immediately after he shall be Elected, and shall be Registered in the Journal Book of the Society as Such. . . . And Whereas the said Statute, so far as the same concerns His Majesty's Subjects, has been found to be detrimental to the Society, The Council therefore have thought proper to Order and Enact, And

---

[71] *C.M.*, III, 194.

[72] *Ibid.*, IV, 171 (italics added). That the "precept," even as modified, was not strictly observed by the Fellows is witnessed by the complaint of the Vice-President, Sir James Burrow, in 1774. Cf. *Journal-Book*, XXVIII, 334–335 (March 17, 1774).

do hereby Order and Enact, That the said Eighth Statute of the said Sixth Chapter, so far as the Same concerns any of his Majesty's Subjects, or any person residing in His Majesties Dominions be, and the same hereby is repealed and made void. And it is further declared and Enacted, That no one of his Majesties Subjects, or any person residing in his Majesties Dominions, who Shall be Elected a Fellow of the Society Shall be deemed an actual Fellow thereof, nor shall the Name of any such person be registered in the Journal Book, or printed in the List of Fellows of the Society, until Such Person shall have paid his admission Fee, and given the usual Bond, or paid the Sum of Twenty-one pounds for the use of the Society in lieu of Contributions. But that upon Such payment or giving Bond as aforesaid, It shall be lawful for the Society to give leave for the name of any Such person so elected as aforesaid to be entred in the Journal Book, and printed in the List of the Fellows of the Society, Provided Always that no such person Shall have liberty to Vote at any Election or Meeting of the Society, before he shall be duly admitted as a Fellow pursuant to the former Statute.[73]

Thereafter, having been elected a Fellow of the Royal Society in accordance with the statute of 1730 governing elections to the Society, no colonial candidate was properly deemed a Fellow until he had paid the admission fee[74] and had either given bond to insure payment of contributions or had paid twenty-one pounds "for the use of the Society in lieu of Contributions."[75] Moreover, as former statutes stated, no colonial Fellow, having been duly elected and having paid his fees, could attend meetings or vote in elections of the Society until he had been formally admitted in person according to the ancient ceremony of admission. This, of course, required his presence in person at a meeting of the Society in London. The "free ride," which colonial Fellows had enjoyed in the Royal Society for ninety years, had ended.

---

73 *C.M.*, IV, 120–122, 124–125.

74 Originally forty shillings, the admission fee was increased to two guineas sometime before 1752, at which time it was raised to five guineas and remained at this level to the end of the Old Colonial Era. After 1752, however, all Fellows received gratis copies of the *Philosophical Transactions*. *The Record*, p. 93.

75 This "composition fee" was increased to twenty-six guineas on Dec. 11, 1766. *Ibid.*, p. 95.

The Role of the Royal Society in the Colonies

The Royal Society effected a revolution in English science. Gone were most of the scattered fumblings of pre-Restoration days. In their place was an organized, cooperative, voluntary effort to explore the entire empire of nature, to seek the underlying laws of nature by experimental design and the exchange of ideas through rational discourse. Gradually the Society succeeded in replacing the dead hand of ancient science, stultified by the Schoolmen, with a dynamic new instrument of planned research in which the Society or its members acted as the synthesizing agent, yet free from the regimentation and central dictation which characterized the work of the French Academy of Sciences. The goals were long-range, involving the works of many hands over unknown generations of men. They required unremitting effort, patience, persistence, and constancy of faith for their ultimate attainability. Small wonder if individual Fellows—and occasionally the Society as a whole—lost sight of the ends in view, their visions temporarily beclouded by more immediate, utilitarian objectives. After all, the Society consisted of men, most of them amateur scientists, some of them mere dilettantes, curiosity-seekers, and other species of dead wood incapable of sustaining the vision of the Baconian ideal. Still, in spite of these "little cells of human wit," as Bacon had called them, the Society succeeded in the long run in retaining its grand objectives and, in time, their partial realization. Before 1700 it was difficult to find anywhere in the British Empire a person interested in any branch of scientific knowledge who was not either a Fellow of the Royal Society or a correspondent of the Society directly or indirectly. The Royal Society had become the most illustrious body of scientific men in the world, and to place "F.R.S." after one's name was a matter both of personal pride and of public prestige.

Of equal importance, especially to the colonial scene, the Society succeeded in imparting its objectives, its experimental approach, and its spirit to countless persons in the English colonies of North America. By its unrelenting promotion in the New World in the way of personal appeals, publications, patronage in a variety of forms, and constant encouragement, it went far to reproduce in the colonies men and women imbued with the spirit of experimental philosophy. Colonials who exhibited an interest in any phase of science or technology were welcomed and encouraged. Their communications were eagerly sought, their letters and papers were read before the Society, and upwards of 260 of them were published in the *Philosophical Transactions*. An impressive number of colonials were elected into the fellowship of the Society and, for most of the Old Colonial Era, were given the privileges and prestige of being Fellows without incurring any pecuniary expenses. Those colonials who visited London, whether Fellows, correspondents, or merely interested observers, were welcomed as vistors to the meetings of the Society.[76] If they needed books, instruments, money, or other supplies in the conduct of experiments and observations, the Society stood ready to assist them. No available opportunity was overlooked to seek to impart to the colonist the objectives of the work of the Society and to assist him in identifying himself with its vast program. Many colonists took part in the effort. Before the end of the Old Colonial Era, and in the midst of sharp political controversies, colonials themselves were able to muster enough persons sufficiently imbued with the spirit of experimental philosophy to launch their own organization more or less in imitation of the Royal Society of London. This was the American Philosophical Society at Philadelphia, created in 1768 and early in 1769.[77] To be sure, these accomplishments—or something similar to them—might have been effected in other ways; the fact is, however, they were achieved to a major extent under the aegis of the Royal Society of London.

[76] They attended as guests of Fellows present, who introduced them to the Society as a whole at the outset of the meeting. The meeting place of the Society in London varied during the Old Colonial Era. The Royal Society met at first in Gresham College. After the Great Fire of London (1666), it was forced to find new quarters. These were found (1667) at Arundel House in the Strand until 1673. Temporarily, then, the Society again met in Gresham College until new facilities were found at Crane Court (Fleet Street) in 1710, where the Society continued until after the end of the Old Colonial Era. It is now (1969) housed at 6 Carlton House Terrace, London, S.W. 1, just off Pall Mall.

[77] For the background of the American Philosophical Society and its final emergence, see Brooke Hindle, *The Pursuit of Science in Revolutionary America, 1735–1789* (Chapel Hill, 1956), Chaps. 6, 7.

FIVE

The Royal Society's Early Promotions: New England

The First Colonial Fellow: John Winthrop, Jr.

THE PRESENCE OF JOHN WINTHROP, JR., among the founders of the Royal Society of London excited high hopes among the Fellows that he would serve as a valuable promoter of the new science in New England. Obviously, they held Winthrop in high regard as an exponent of experimental philosophy and believed that New England, with its considerable reservoir of educated men, would both accept the new philosophy with open arms and supply the Society with all manner of scientific observations and other data to enrich its store of natural knowledge. Dr. Cromwell Mortimer, Secretary of the Society, writing in the late 1730's, placed Winthrop among "the highest rank of learned men" in his day and went on to say:

> His distant abode from *London* and his not putting his name to his writings made him not so universally known as the *Boyles's*, the *Wilkins's*, or the *Oldenburg's* of his days, nor his name handed down to us with such general applause. In concert with these and other learned friends (as he often revisited *England*) he was one of those who first formed the plan of the *Royal Society;* and had not the civil wars happily ended as they did, Mr. *Boyle* and Dr. *Wilkins* (as may appear in letters from Mr. *Boyle*, Dr. *Wilkins*, Sir *Kenelm Digby*, &c, to Mr. Winthrop), with several other learned men, would have left England, and out of esteem for the most excellent and valiant Governor, John Winthrop the Younger, would have retir'd to his new-born Colony, and there have established that Society for *Promoting Natural Knowledge*, which these gentlemen had formed as it were in *embryo* among them-

selves, but which afterwards, obtaining the protection of King Charles II, obtained the style of Royal, and hath since done so much honour to the British nations. . . .[1]

How seriously the founders considered the establishment of the Society in Connecticut, in the event their hopes could not be realized in England, is impossible to say. But that Winthrop "was in a particular manner invited to take upon him the charge of being the chief Correspondent of the Royal Society in the West" is unquestionably true.

John Winthrop, Jr. (1606-76) attended Trinity College, Dublin, and entered the Inner Temple, London. But he left Trinity before he had finished for a degree and relinquished the study of law before he had completed his apprenticeship. He was early attracted to the study of alchemy and medicine, and in the course of time became recognized as a self-made chemist and physician of no mean ability. After he gave up the study of law, he served as secretary to the captain of a ship engaged in the ineffective expedition led by the Duke of Buckingham for the relief of French Protestants at La Rochelle (1627). Then he shipped with a vessel of the Levant Company plying in the Mediterranean trade. He spent fourteen months in the Mediterranean, visiting Italy and Turkey. In Italy he seized upon opportunities offered to enlarge his scientific interests and associations by making the acquaintance of some of the Italian savants at Pisa, Florence, Venice, and elsewhere, and to visit the famous Italian botanical gardens. In Turkey he viewed the wonders of the Turkish capital and made friends among the English and other Europeans stationed in Constantinople. He returned to England (August, 1629) just in time to become caught up in his family's association with the Massachusetts Bay Company enterprise. He followed his father to New England in 1631 and became the principal planner of Agawam (Ipswich) in 1633. He returned to England for a brief visit in 1634-35, becoming associated, on this

---

[1] *Phil. Trans.*, XL, for the years 1737, 1738 (London, 1741), unpaged. This volume was dedicated to Winthrop's grandson, John Winthrop (1681-1747). For the circumstances see my article, "John Winthrop (1681-1747) and His Gifts to the Royal Society," Publications of the Colonial Society of Massachusetts, *Transactions, 1952-1956*, XLII (1964), 206-232. Cf. R. F. Young, *Comenius in England* . . . (London, 1932), pp. 8-9.

occasion, with English plans for a settlement at Saybrook, Connecticut. When this enterprise languished for want of English support, Winthrop returned to Ipswich, where he lived until 1641. Although his family position inevitably placed him in the ruling circles of the Bay Colony, he shunned extreme measures both in politics and in religion. To many of his friends he became known as the "gentle Mr. Winthrop," in contrast to his father, the dedicated Governor of the Puritan colony. During the late 1630's and early 1640's, when the Bay Colony suffered economic depression and struggled to evolve a viable economic system, Winthrop sponsored salt works and iron works, and spent much time and effort prospecting for useful mineral deposits in New England. He returned to Europe in 1641 primarily to seek capital and expert workmen for his manufacturing enterprises. But he interrupted these efforts with a visit to the Continent (Hamburg, the Netherlands, and Brussels), where he cultivated his scientific interests in association with such various promoters and scientists as Samuel Hartlib, Sir Kenelm Digby, Jan Amos Komensky (Comenius), and Johann Rudolf Glauber (of Glauber's salt fame). After his return to Massachusetts (September, 1643), he continued to promote iron and salt works. But he became increasingly interested in Connecticut affairs, especially after his foundation of New London in 1646. Land hunger and a kind of Connecticut expansionist policy seized his imagination. If, as appeared to be true, he entered upon a political career with reluctance, nevertheless he became the architect of Connecticut Colony. The charter which he received in 1662 from Charles II was his master stroke, indicative of great political skill, especially by one of such indelible Puritan origins treating with the recently restored Stuart king. He was Governor of the newly chartered colony continuously until his death in 1676.[2]

By virtue of his knowledge of alchemy, early chemistry, and medicine—as well as an enthusiasm for experimental philosophy as a

[2] Robert C. Black III, *The Younger John Winthrop* (New York and London, 1966), *passim;* Richard Slater Dunn, *Puritans and Yankees: The Winthrop Dynasty of New England, 1630–1717* (Princeton, 1962); "John Winthrop, Jr., and the Narragansett Country," *Wm. & Mary Qty.*, 3rd Ser., XIII (1956), 68–86; and "John Winthrop, Jr., Connecticut Expansionist: The Failure of His Designs on Long Island, 1663–1675," *New England Quarterly*, XXIX (1956), 3–26. I know of no full monographic treatment of Winthrop as a scientist.

whole—together with his wide travels and personal acquaintances, his warm personality, his moderate and sensitive spirit, and his flexible and wide-ranging intellect, John Winthrop, Jr., was well qualified to become the Royal Society's "chief correspondent" in New England. Like many of his Fellows in the Society, he was prominently placed and many-sided. He was a colonial governor and promoter, an entrepreneur in salt-making, iron-making, and mining,[3] a land speculator and farmer, an administrator and diplomat, a physician, an alchemist, and a mineralogist. His medical practice, evidently balanced between the herbalism of the Galenists and the chemical specifics of the Paracelsus school, became famous in New England. Patients came—and wrote—to him from far and wide in New England to get his secret remedy, "Rubila powder" (formula unknown), a powerful purgative considered to be a panacea for ague, worms, smallpox, colds, and other disorders. He possessed a considerable amount of chemical apparatus and a sizable library, and while the contents of the latter are unknown in full detail, the collection contained books on medicine, chemistry, physics, astronomy, mathematics, natural history, navigation, geography, metallurgy, military science, agriculture, and politics.[4] He had the first telescope in New England, a ten-footer which he had acquired sometime before 1660, possibly during his visit to the Continent in the early 1640's; and a three-and-a-half-foot 'scope which he evidently brought from England in 1663.[5] As a scientist Winthrop's chief fault lay in his tendency to spread himself too widely. Per-

[3] Winthrop's mining went beyond bog iron. In 1644 he acquired control of a black lead (graphite) deposit at Tantiusques, near Sturbridge, Massachusetts, about a mile from the present Connecticut boundary. Winthrop worked this mine (with the help of other "mineral men") and it became a Winthrop family enterprise until about 1750. George H. Haynes, "The Tale of Tantiusques: An Early Mining Venture in Massachusetts," *Proceedings of the American Antiquarian Society*, n.s., XIV (1901), 471–497. For years it was erroneously believed that the lead ore contained silver.

[4] C. A. Browne, "Scientific Notes from the Books and Letters of John Winthrop, Jr.," *Isis*, II (1928), 324–342; Herbert Greenberg, "The Authenticity of the Library of John Winthrop the Younger," *American Literature*, VIII (1937), 448–452; Ronald S. Wilkinson, "The Alchemical Library of John Winthrop, Jr. (1606-1676) and His Descendants in Colonial America," *Ambix*, XI (1963), 33–51.

[5] Ronald S. Wilkinson, "John Winthrop, Jr., and America's First Telescope," *New England Quarterly*, XXXV (1962), 520–523.

haps, as Professor Black has argued, this was a counterpart of his lifelong tendency to undertake tasks left abandoned and incomplete. But saying this, especially in regard to many of Winthrop's New England enterprises, is to discount the insuperable obstacles which he faced. His various commercial enterprises, all plagued with the general colonial malaise of inadequate capital and labor supply, were seldom successful and often abandoned. These, together with his political activities, so engulfed his time and energies that his scientific observations and experiments were correspondingly curtailed. And the raw colonial scene, where his neighbors and friends were necessarily preoccupied with the details of day-to-day living and the problems of colony-building, robbed him both of much of the leisure which his English fellows enjoyed and of the exchange with other scientifically-minded men which often proves intellectually stimulating.

Winthrop had been in correspondence with some of the founders of the Royal Society before the Society was established.[6] His voyage to England (1661–63) in behalf of the Connecticut charter placed him once more in direct association with these men. He was proposed as a candidate for the new Society on December 18, 1661, by William Brereton and admitted on January 1, 1661/62.[7] Subsequently he was appointed to two of the committees of the Society, the first called "Mechanical. To consider and improve all mechanical inventions"; the second, "For Histories of Trade."[8] And he threw himself actively into the business of the Society. On February 12, 1661/62, he "promised to bring in an account of strange tides against the next day," but there was no record that he did so.[9] On April 23, 1662, the minutes recorded that "Mr. Winthrop brought a time Lamp, call'd a bladders Lamp, burning high like a

[6] "Correspondence of the Founders of the Royal Society with Governor Winthrop of Connecticut," *Proc. M.H.S.*, XVI (1878), 212–251.

[7] These, and subsequent details regarding Winthrop's activities in the Royal Society during his stay in London, are recorded in Thomas Birch, *The History of the Royal Society of London . . .* (4 vols., London, 1756–57, cited hereafter as Birch), I, 67–68, 76, 80, 82, 84–88, 99, 102, 111–112, 162, 166–167, 171, 178, 198, 205, 206–207, 212. As Secretary of the Society, Birch lifted most of his information directly from the *Journal-Book* of the Society.

[8] Birch, I, 406–407.

[9] *Journal-Book*, I, 50.

Candle continually feeding itself: & [the Society] Ordered that a Diagram thereof bee made & Registered. He also produced Malleable Mineral Lead, & a piece of a Rock of Granats [garnets?]."[10] The latter specimens were from New England, the lead probably from his mine. The following week (April 30) he produced further specimens from New England, "a little stone of which one part was as it came from the rock of an amethist colour & the other was after calcination of a flesh colour," and "a piece of wood called Lignum Columbrinum, or in Dutch Schlangen holtz, whose taste is bitter."[11] On June 25 Winthrop was "intreated to bring in writing the manner of making Pitch and Tarr." The manner of the request suggested that Winthrop had discussed this process, probably in one of the committees on which he served. The following July 9, "Mr. Winthrop read the History of Making Pitch and Tarr; and thanks [were] returned to him for his pains, and [he was] intreated to prosecute it."[12] A week later (July 16) he presented to the Society the finished draft of a paper entitled "A Description of ye Artifice and making of Tarr and Pitch in New England, and ye Materialle of wch it is made."[13]

This paper, Winthrop's first formal presentation to the Society—and the first given by any colonial—was a detailed description of the process of making tar from the knots of pitch pines which had blown down over the years, the trees rotted and burned by the annual fires set by the Indians, and the knots lying about in great profusion. Pitch pines, he said, grew in various parts of New England, generally "vpon ye most barren plaines" and upon "rocky hills" among those plains. "Ye most Tarr is made about Coneticutt above 50 miles vp ye river where there be great plaines of those pines on

[10] *Ibid.*, I, 55.

[11] *Ibid.*, I, 57. Robert Hooke reported in 1686 that he had "true Amethists and transparent" from Winthrop. *Ibid.*, VIII, 112 (Dec. 12, 1686).

[12] *Ibid.*, I, 68.

[13] A copy of this paper is in *Classified Papers, 1660–1740*, III (1) ("Mechanicks"), No. 23. It is endorsed "Read July 9:1662. Ent'd. R.B. I, 179." The "R.B." was the *Register-Book*, I (1661–62), 179–184, where a copy of the paper is given. Its inclusion under "Mechanicks" suggests that it arose out of the Mechanical Committee of which Winthrop was a member. It was printed in substance in John Evelyn, *Sylva: Or a Discourse of Forest-Trees . . .* (2nd ed., London, 1670), pp. 110–113; and in Birch, I, 99–102.

both sides of ye river," but tar was also similarly made in Massachusetts and New Plymouth. Winthrop observed that the pine knots
were impregnated with "yt Terebinthin or resinous matter" whose
"balsamick" nature "can preserve ye wood from rotting it may be
many hundred years, tho exposed to all weathers." After the tar
had been extracted in a covered hearth made of stones and clay,
Winthrop noted that the pine knots had been "made into excellent
charcoal, wch ye smiths find to be ye best sort of coale for their
use." Split into small slivers, the knots were also used by the colonists "instead of candles, giving a very good light; and thence they
call it candle-wood," although it gave off an offensive smoke and
was usually burned "in ye Chimney-corner upon a flat stone or
iron." There were "millions" of living pitch pines with similar tar-
laden knots, but "ye labour of falling ye trees, and cutting out those
knots would far exceed ye value of ye Tarr . . . especially in yt
Country, where labour is very deare." A few pines were found
"whose whole body of ye lower part towards ye roote is as full of
turpentine, as ye knots." But those were few in number and were
"commonly preserved to split out into Candlewood." Some persons
held that the lower part "of any living pine-tree" may be "as fully
impregnated with Turpentine as ye knots," but if this were true no
effective way to harvest the tar had been found in New England,
although there were reports that "such artifice is used in Norway."
Whether it was that the "right season of ye year" had not been
found, "or what else omitted, is not knowne to those yt have missed
in such trials, it were worth inquiring whether any such art be used
in Norway or Sweden, or elsewhere, and what it is." Winthrop
concluded with accounts of the boiling—and burning—of tar to
make pitch and of the boiling of the pine knots, split into small
pieces, to make resin. The paper promoted a considerable discussion
about the method of making tar in Sweden, and Mr. Brereton
"intreated Mr. Winthrop's account of making Pot-ashes."[14] There
is no evidence that Winthrop responded to Brereton's entreaty.

On September 17 "A paper about the improvement and planting
of Timber-Trees brought in by Sir Robert Moray, Sent by His
Maties Commissioners for the Navy, was read and referred to the

14 *Journal-Book*, I, 69 (July 16, 1662).

Consideration of Mr. Evelyn, Dr. Goddard, Dr. Merret, & Mr. Winthrop."[15] This paper set off a series of discussions in the Royal Society which continued long after Winthrop's return to New England. The shortages of timber for building, especially ship-building, including mast timbers, had been troubling the English since the late sixteenth century, and they feared the consequences of their dependence upon the Scandinavian countries for ship-building materials. The Society's committee to which His Majesty's Commissioners of the Navy's paper was referred decided that reforestation of England, particularly of the royal lands, together with conservation of the use of existing resources, were promising solutions of the problem. In the course of the committee's discussions, "Mr. Winthrop's Consideration about ye planting of Trees for Timber" was offered. The written report of this ran as follows:

> Yt it may be probably a good way to plant 4. or 5. within a little distance and to leave some of them to grow together so long, till they do apparently hinder ye growth one of ye other; and then to leave only that wch is of ye best growth.
>
> Yt though it is probable it may be but to raise plants by setting of Acorns, yet in respect of gayning time in ye growth of ye Timber it may be best to remouve young Trees of as great bigness as may be fit to be remouved; whereof there may be enough found that grow in ye woods too thick together.
>
> Yt it is considerable whether to plant them in ye natural Earth . . . than in Earth that is made too rich whiles they are young.
>
> Yt it were good to make full enquiries into ye way used in Biscay, where they have very strict orders and constant practise about planting their trees.
>
> Yt care should be had about ye planting ye best sort of timber &c.[16]

---

[15] *Ibid.*, I, 85. The Commissioners' paper consisted of "Quaeries Touching ye Preserving of Timber now growing. And planting more in His Maties Dominions of England and Wales." *Classified Papers, 1660–1740*, X (3) ("Agriculture"), No. 21.

[16] *Ibid.* John Evelyn's *Sylva* (see note 13 above) grew out of this discussion when, on Oct. 15, he presented a paper "by way of answer to the Queries of the Commissioners of the Navy." The Society desired him to print his paper and the *Sylva* was the result. *Journal-Book*, I, 93.

As a colonial promoter, Winthrop could not withhold pointing out the resources of New England for ship-building, and the following week (September 24) he read to the Society propositions "concerning ye building of Shipps in New England."[17] It began: "The honorable Society may be pleased to Consider whether it maye not be fit to propose to his Matie or his honorable Commissioners for the Navy the Conveniency of building of ships in some of the Northern parts of America there being severall reasons yt may be propounded as motives incouraging thereunto." There followed nine numbered paragraphs enumerating the advantages: the great store of oak timber for ship-building; the supply of spruce and fir for masts; pitch pine for tar and pitch; plenty of sawmills for sawing "planck and board of all sorts"; timber cheap and easily available "upon, or neare good harbours and Navigable Rivers"; much experience in ship-building in New England: "there were this summer divers here at London that were built there . . . there is one now of an hundred tonnes in this River"; trial could easily be made, for New England has "all ready very good Artists for Master workmen and other ordinary workmen . . . also kalkers and smiths and all other trades necessary towards yt worke"; a ship built in New England can easily be freighted with ship timbers, masts, etc., "all of which will be of good use for supply of building ships here"; if more workmen are needed, housing can be provided "and plenty of all sorts of provisions"; and the only things "necessary to be transported thence" are "Riggin[g], Sayles, Anchors, Cables and Spanish Iron for spikes, & bolts &c. which is best to be heere allso ready made into bolts and other Iron worke necessarye: for though there be Iron workes and good Iron made there, yet Spanish Iron is most approved for that worke."

---

[17] *Ibid.*, I, 86. The paper is among *Classified Papers, 1660–1740,* VII (1) ("Travels, etc."), No. 11; copy in *ibid.,* X (3) ("Agriculture"), No. 21. Printed in Birch, I, 112–113. The Royal Society expressed interest in this anew in 1668, this time aroused by Dr. John Beale, a zealous Fellow from Yeovil, Somerset, who was concerned about England's rapid deforestation because of the consumption of timber for charcoal for the rapidly multiplying iron mills. If New England made "a good gain by ye sale of boards, wainscoat, planks, joyners work, coopers work, carpenters work, ready fram'd [buildings]," and the like, perhaps, with "an encouragement" from the homeland, it might be "a kind of redemption to England."

It was about two months later when John Winthrop, Jr., again appeared in the Royal Society's minutes. On December 17, 1662, it was recorded that "Mr. Winthrop, shewing the Society some Indian Corne, whereof some grains were blewish, promised to give them in writing the History and Ordering [of] it in the West Indies. He shewed also the taile of a rattle-snake, wch he said did increase every year by one ring, whence ye people concluded ye age of this animal. Dr. Merret took it home with him to make some tryall of the powder of it."[18]

It seems likely that Winthrop's display of maize led to questions and discussion and that it was recalled that the English herbalists, Gerard and Parkinson, had expressed differing opinions with regard to the food value of it. Gerard had written that maize "doth nourish far lesse than either wheat, rie, barly, or otes. The bread which is made thereof is meanely white, without bran: it is hard and dry as Bisket is, and hath in it no clamminess at all; for which cause it is hard of digestion, and yeeldeth to the body little nourishment. . . ."[19] Parkinson, on the other hand, had reported:

> Many doe condemne this Maiz to be as dry and of as little nourishment as Millet or Panicke, but they doe not as I thinke rightly consider the thing, for although the graine be dry, yet the meale thereof is nothing so dry as [that] of the *Turkie Millet*, but hath in it some clamminesse, which bindeth the bread close and giveth good nourishment to the body, for wee finde both the Indians and the Christians of all Nations that feede thereon, are nourished thereby in as good manner no doubt, as if they fed on Wheate in the same manner. . . .[20]

Whatever the background may have been, Winthrop presented to the Society a paper, "Of Maiz," on December 31, 1662.[21] He stated that maize was "Generally used in many parts of America

[18] *Journal-Book*, I, 122; Birch, I, 162.

[19] John Gerard, *The Herball, or Generall Historie of Plants . . . Enlarged and Amended by Thomas Johnson . . .* (London, 1636), p. 82.

[20] John Parkinson, *Theatricum Botanicum . . .* (London, 1640), pp. 1138–39.

[21] Several copies of this paper are in the Royal Society Archives: *Classified Papers, 1660–1740*, X (1) ("Botany, etc."), No. 3; *Journal-Book*, II, 81–92; *Boyle Papers*, Bl. V, fol. 197. The copies are essentially alike but vary in detail. It was subsequently published, with editorial dressing, in *Phil. Trans.*, XII, No. 142 (Dec., Jan., Feb., 1678), 1065–69. And Professor Fulmer Mood published the copy from the *Boyle Papers* in the *New England Quarterly*, X (1937), 121–133.

for food," by the English and the Dutch as well as by the Indians, even where there "is plenty of wheat & other grain"; "Mr. Gerrard wrote of it in his Herball as hard to digest, & of little nourishment; now [it] is found both wholesome & pleasant." Having thus confuted Gerard and aligned himself with Parkinson, Winthrop went on to describe the ear and the stalk. The ear, he said, is "very beautiful," and "Nature hath delighted it selfe to beautify this Corne with great Variety of Colours." The stalk grows to a height of "6 or 8 foot," varying according to the condition of the soil and the kind of seed. Virginian seed grows taller than that of New England, and there is a shorter variety grown by "the Northerne Indians farr up in the Countrey."[22] The stalk is jointed "like to a Cane and is full of sweete Juice like the Sugar Cane, and a Syrrop as sweete as Sugar Syrrop may be made of it which hath beene often tryed." Winthrop suggested that perhaps cornstalks might become a source of sugar if "some trialle may make it knowne whether it may be brought into a dry Substance like Sugar." He went on to describe the planting and cultivation of maize, its harvest, and its storage. With the help of the plow, the English colonists had much improved upon the earlier Indian methods of planting and cultivation, especially for quantity production, and Winthrop also described how this was done, though, he said, they still found it helpful to fertilize with fish, as the Indians had done, or with cow dung, as the English sometimes did, especially "where the Ground is not very good, or hath beene long planted and worne out." Beans, squashes, and pumpkins were also planted amongst the corn, "loading the Ground with as much as it will beare." Many English also, "after the last Weeding their Ground sprinkle Turnep-seed between the hills, and so have after Harvest a good crop of turneps in the same Field."

The stalks of corn made good winter fodder for cattle, and the Indian women often used the husks for basket-weaving. Winthrop described in detail both the Indian and the English manners of preparing the grain for human consumption. He especially

---

[22] Later Winthrop referred to this as "Mohawkes Corne, which though planted in June wil be Ripe in Season, the Stalkes are shorter than the other Sorts, and the Eares grow neerer the bottom of the Stalke and are generally of divers Colours."

favored maize made into "Sampe," a kind of pudding which, he asserted, "is of a nature Divertical and Clensing and hath no Quality of binding the Body, as the Herball supposeth, but rather to keepe it in a fit temperature. . . ." Sampe, with a good "Quantity of Milke" mixed in, was "the most common diet of the planters." Indeed, a London physician, the late Dr. [Edmund?] Wilson, "had every yeare some Quantity brought over ready beaten, and fitt to be boyled, and did order it to such Patients as he saw cause for it." Winthrop concluded with an account of the New Englanders' discovery of "a way to make very good Beere of this Graine which they doe either out of Bread made of it, or by Maulting of it. . . ." But because of the difficulties encountered in malting maize without its moulding in the process, Winthrop stated that the beer was most commonly made from cornbread, broken or cut into "greate Lumps, as bigg as a mans Fist or bigger," which are mashed and "proceed every way about brewing of it, as is used in Brewing Beere of Mault, adding hopps to it as to make Beere."

Winthrop's paper "Of Maiz" was undoubtedly the most complete description of Indian corn, its cultivation, and its uses that the English public had seen. His account of beer brewed from cornbread excited the interest of his Royal Society Fellows and they demanded a demonstration with Winthrop as curator of the experiment "as soon as he could be provided with material, Dr. Goddard to provide the accommodations."[23] But it was March before this could be done. In the meantime, Mr. Winthrop "shewed a swimming Earth, wch he had brought out of New England, wch did swim for about ½ houre, and yn sunk to ye bottom"; mentioned "yt there was no right black lead any where, but in England and New England"; reported that "a ship built in N. Eng. had arrived here; and was desired to bring in account of its size & shape"; and on March 4, 1662/63, was again reminded of the beer-brewing experiment.[24] Finally, on March 11,

> Mr. Winthrop presented the Company with some bottles of beer brewed of maiz-bread, which was a pale well-tasted middle Beer.

[23] *Original Minutes of the Royal Society*, I (Aug. 13, 1662–March 23, 1663/64), unpaged (Jan. 7, 1662/63); *Journal-Book*, I, 132, 142, 145.

[24] *Ibid.*, I, 124, 126, 132, 145.

He had the thanks of the Company, and was directed to keep some of his liquor for a while, to see how it would bear age. Being asked what proportion of Maiz there had been taken to the 8 Gallons of Beer which was brewed, he said about halfe a Bushell. Dr. Merrett suggested that some Spirit might be drawne out of this Maize-beer, a fortnight hence.[25]

A week later (March 18, 1662/63),

Mr. Winthrop acquainting the Company with his intention to returne shortly into New England, was desired to take with him a Copy of the Directions for Seamen, and to make as many of the Observations and Experiments therin as he could. The Amanuensis was Commanded to have the said Copy ready for him against next meeting: And the Operator to have a dozen Sounding Leads and 3 or 4 Balls made against the same time, as also the cylindricall Vessel with Valves to fetch up water from the bottom of the Sea.[26]

Winthrop received the "Directions" the following week and the operator was ordered to have ready the equipment for the experiments.[27] But the experiments failed. On July 22, 1663, Secretary Oldenburg read to the Society a letter from Winthrop

giving an Account of the Tryalls made by him at Sea with the Instrument for Sounding of Depths without a line; and with the Vessel for drawing of Water from the bottom of the Sea: both of which proved successless; the former, by reason of too much wind at the time of making Sounding; the latter, because of the leaking of the Vessel. Captain [Silas] Taylor, being to goe shortly to Virginia, and offering himself to make the same Experiments: the Company recommended to him the trying of the one in Calme weather, and of the other with a staunch Vessel.[28]

---

[25] *Ibid.,* I, 146.

[26] *Ibid.,* I, 147. The "Directions for Sea-men, bound for far Voyages" are printed in full in Appendix I below. They include a description of the instruments provided and how to use them. Subsequently (Oct., 1663), the Society provided Winthrop with a thermometer. Birch, I, 322.

[27] *Journal-Book,* I, 50.

[28] *Ibid.,* I, 203–204; Birch, I, 280. The leaking vessel was not the ship, but rather the cylindrical vessel used in the experiment. See Appendix I below, "To take up water near ye bottom of ye Sea."

Winthrop's services to the Royal Society after his departure from England got off to a poor start.

Indeed, Winthrop's activities in behalf of the Royal Society suffered one misfortune after another upon his return to New England, and his friends in the Society protested against his supposed inaction. But several of his letters were lost in transit, others arrived long after they normally should have reached their destination, and shipments of specimens of New England life were lost at sea. It was more than five years after his departure from England before his Royal Society friends received any significant correspondence from Winthrop—and longer before his collections of New England curiosities began to arrive. From those letters which have survived, it is clear that he was not inactive during this time, nor unmindful of his obligations to the Society. However, the loss of his letters and the absence of other sources of information make it impossible to reconstruct his scientific efforts during this period. He wrote to Robert Boyle in November, 1663, to send two Latin papers "composed by two Indians now scollars in the Colledge in this Country," and while he suggested that the papers might be of interest to the Royal Society, the burden of the letter related more to the progress of the Christianizing of the Indians at the hands of the agents of the New England Company, an English corporation for that purpose of which Boyle was the President.[29] Henry Oldenburg wrote to Winthrop on March 26, 1664, asking him to make astronomical observations of the conjunction of Mercury with the sun, due to have occurred "here in England about London on the 25th of October of this present year." As a postscript, Oldenburg added the reminder, "Remember ye History of New England, & particularly the matter of the mines, of Tydes, of making salt in your Compendious & cheap way; these things well accounted of to ye Society, will set you very high in their esteem, to my knowledge."[30] If Winthrop made the requested astronomical observations his report of findings appears to have been lost, although, as he made other observations in 1664 which are a matter

[29] Many of Winthrop's letters to Fellows of the Royal Society were published in *Proc. M.H.S.*, XVI, 212ff. In April, 1664, Boyle wrote Winthrop regarding the presentation of John Eliot's Indian Bible to Charles II. *Ibid.*, V (1862), 376–377.

[30] *Letter-Book*, I, 137–138.

of record, it seems likely that he did observe the eclipse to which Oldenburg referred. He made observations of a comet seen in New England in 1664 (although his written account did not reach the Royal Society until early in 1670),[31] and it was in October of this year that he reported his supposed discovery of a fifth satellite of Jupiter.[32] On September 29, 1664, Oldenburg wrote to Boyle saying that he had recently had a letter from Winthrop "but almost 11. months old, wch I much wonder'd at." The letter referred to another, earlier, letter which Oldenburg had never received. Winthrop reported little of consequence. He was busy looking for useful ores in New England and thought he had an Indian relation which promised copper, but it was not yet confirmed. His work up-country, far from the sea, precluded further trials of salt-making, observation of tides, and the like, for the time being.[33] The following December Sir Robert Moray gently chided Winthrop for his supposed neglect of his English friends:

> But I would gladly be allowed to complain to you that in so long a time as you haue been in those parts, and haue, I doubt not, acquired new knowledge as well as practised the old, you do not acquaint none of your friends hereaway with any thing you have don, found out, or do designe nor giue them any account of such matters as you are very well able to do; that is the peeces of Naturall History, philosophicall matters, Inventions, & Mechanick practises; nor any thing of Minerall businesses. . . .

According to Winthrop's endorsement of this letter, he did not receive it until December, 1667.

Already, in 1665, Winthrop had sent a shipment of New England specimens "for the view of the Gentlemen of the Royall Society," but it was lost at sea. In October, 1667, both Henry Oldenburg and Theodore Haak wrote to Winthrop to remind him of his obligations to the Royal Society. Oldenburg was rather sharp in his admonition:

---

31 See Appendix II below.

32 Wilkinson, "John Winthrop, Jr., and America's First Telescope." Actually the fifth satellite of Jupiter was not observed until 1892.

33 Oldenburg-Beale Letters, *Letter-Book "O,"* No. 14 (Royal Society Archives).

So good an opportunity as this I could not let passe w[t]hout putting you in mind of y[r] being a member of ye Royall Society, though you are in New-England; and that even at so great a distance you may doe that Illustrious Company great service. You cannot but remember both yr general Obligation to them, when you were received, of contributing what you could to promote the dessein and end of their Institution; and also yr particular Engagements, of communicating to them all the Observables both of Nature and Art, yt occurr in the place, you are, and especially such as concern the Mines of that contry, and yr ingenious way of making salt out of sea-water by a cheap and speedy method; wch, if I mistake not very much, you seemed here resolved to try in New England. I am persuaded the R. Society, who retains still a particular respect and kindnesse for you, will receaue what shall come from you . . . wth no ordinary affection and thankfulnesse. Sr, you will please to remember that we have taken to taske the whole Vniverse, and that we were obliged to doe so by the nature of our Dessein. . . . We know yr ingenuity, experience, and veracity, ye best qualities of a man and a Philosopher. . . . And since you have now been from us severall years, give us at last a visit by a Philosophicall letter.

Oldenburg sent with his letter copies of the *Philosophical Transactions* and Thomas Sprat's *History of the Royal-Society*—the latter just off the press. Winthrop received the packet by the hand of Peter Stuyvesant of New Amsterdam, who had received it from Amsterdam, where it had gone from London. He was delighted with the books, whose receipt he acknowledged with thanks on November 12, 1668. Embarrassed by Oldenburg's reference to his obligations to the Royal Society, Winthrop acknowledged them, reaffirmed his esteem for the Society, regretted that because of accidents at sea he had been unable to perform his duty, and assured Oldenburg that

It is my constant sorrow that (*penitus toto orbe divisus*) my great remotenesse makes [me] so little capable of doing them that service to wch my desires & indeavours have beene, and are greatly fixed & devoted. Had former letters & collections of such mean things as could be had in such a wilde place as this arrived, it might have appeared, yt I had beene gathering frō many parts of this wildernesse: and there had beene the relation of some observables fallen out in these parts, & of other matters wch were then thought of, as yt might have beene considerable. . . .[34]

He would try to make amends. He had found few minerals of any significance, and lately Indian wars had rendered them difficult to collect in remote places.[35] Both his plans for mines and his salt works had been retarded for want of capital. He continued:

Plantations in their beginnings have work ynough, & find difficulties sufficient to settle a comfortable way of subsistence, there beinge buildings, fencings, cleeringe and breakinge up of ground, lands to be attended, orchards to be planted, highways & bridges & fortifications to be made, & all thinges to doe, as in the beginninge of the world. Its not to be wondered if there have not yet beene *itinera subterranea.* . . .

He summarized the contents of previous letters which had failed to reach the various Fellows (Boyle, Lord Brereton, Moray, Charles Howard, Dr. Goddard, Dr. Merret, and others) to whom they had been sent. He had written "some additions" to his essay "Of Maiz," had written about a new way to make charcoal "of any and all sorts of wood," a new way of "malting potatoes," and about "a speciall kind of that Indian graine Mais, wch may be planted very late in the summer . . . & yet be ripe as soon as that [other], wch may probably ripen well in England, if planted there." He had procured "an ear or 2 of it from a remote Northern part of this Continent, and had it planted every year since to preserve the seed of it." He related several "curious occurrences" in Connecticut, of the use of rotten creek thatch and seaweed as fertilizer for the land, and of steeping wheat seed in salt water mixed with oyster-shell lime to rid it of smut, which had harassed their wheat production in recent years. He had some "new notions about findinge the longitude at sea," but they were not mature and had not been tested in "a long sea voyage." In short, Winthrop went far to vindicate himself of the charges of slothfulness and his Royal Society Fellows did not renew the criticism.

[34] Winthrop had written to Sir Robert Moray in somewhat the same vein, Aug. 18, 1668, *Proc. M.H.S.*, XVI, 232–234. The other letters quoted above, and the one immediately at hand, are printed in *ibid.*, pp. 224–225, 229–230, 234–239.

[35] Winthrop evidently depended heavily upon friendly Indians to bring him specimens of minerals from the interior parts of New England.

The following year (1669), Winthrop went further to vindicate himself in the eyes of his Fellows in the Royal Society. On October 4 he addressed a letter to Oldenburg to cover a shipment of specimens of the products of New England nature and arts shipped from Boston to London in Captain Christopher Clark's ship under the watchful eye of Adam Winthrop, "a kinsman of mine" (nephew). The shipment consisted of four boxes, including upwards of fifty specimens, groups of specimens, and two copies of Eliot's Indian Bible, an Indian grammar, works of piety translated into the Indian language, and "Two Astronomical Descriptions of the Comet of 1664."[36] The shipment was, in part, an effort to replace items formerly lost at sea. It contained "Three dwarf oaks, with cupps of Acorns on them," together with several other natural products, both of flora and of fauna; various items of Indian craftsmanship, including several strings of different kinds of wampum; mineral products of New England and other colonies; pieces of candlewood which Winthrop had described in his paper to the Royal Society on the making of tar and pitch; five ears of the special kind of Indian corn which ripened "a month at least before other kinds"; and a fish, resembling a starfish "yet differing from it in divers particulars." Of the dwarf oaks, Winthrop wrote that

> the roots wilbe seene vpon them being plucked up so that it might appeare certainly that they are not slipt off a greater tree: It would seeme a strange hyperbole to report of a country, where the swine are so tall, that they eat acornes upon the tops of standing growing oakes, but it will appeare, not to be a riddle to those that shall see acornes, or the cups of them, vpon these little oakes . . . and of this sort of oakes, there are whole forrests in many parts of the inland country. [They make the land very difficult to break up for farming] so that ten or 12 oxen with a strong plough can scarce stir them, without the helpe sometymes of an ax, or such like toole.[37]

Winthrop's letter was read to the Royal Society on February 10, 1669/70, and his boxes of specimens were viewed with obvious

[36] The entire list, as given in *Journal-Book*, IV, 112–117 (Feb. 10, 1669/70), is printed in Appendix II below.

[37] Extracts of this letter were published in *Phil. Trans.*, V, No. 37 (March 25, 1670), 1151.

enthusiasm by the Fellows. Indeed, as the Curator was absent, the Society dispensed with experiments, and devoted the entire meeting to a perusal and discussion of "divers Curiosities of Nature, sent partly out of New England by Mr. Winthrop, partly out of Hungary and Transylvania by Dr. Edward Brown, [and] partly out of Warwickshire by Mr. [Francis] Willoughby." But Winthrop's was the largest of the three—the largest collection, in fact, that the Royal Society had received to date in a single shipment.

The Society ordered the Secretary to give thanks to Winthrop, and Oldenburg wrote on March 26, 1670. He reported that Adam Winthrop had been invited to the Society, "being an eye-and-ear witnesse of the kind reception" accorded to Winthrop's gifts. Moreover,

> His Matie himselfe hearing of some of the rarer things, would see them, and accordingly the extraordinary fish, the dwarf-oaks, the gummy fragrant bark with knobs, the silken podds,[38] the bags with little shells in them &c. were carried to Whitehall, where the King saw them with no common satisfaction, expressing his desire in particular to have the stellar fish engraven and printed; Wee wish very much, that you could procure a particular description of the said fish viz: whether it be common there; what is observable in it when alive; what colour it hath then; what kind of motion in water; what use it maketh of all that curious workmanship wch nature hath adorned it with? &c.

With the thanks of the Society, Oldenburg sent—"that the returne may not bee altogether verbal"—a few books, "lately printed by several Fellows of the R. Soc., Viz. M. Boyles Continuation of Experiments concerning the spring and weight of the Aire. 2. Dr. Holders Philosophy of Speech. 3. Dr. Thruston de Respirationis usu primario. 4. The Transactions of the last yeare." Oldenburg went on to add:

> I hope that the New-English in America will not bee displeased with what they find the Old-English do in Europe, as to the matter of improving and promoting usefull knowledge by Observations and Experiments; and my mind presages to mee, that within a little time wee shall heare that the ferment of advancing

---

38 The King asked for more of the "silken podds" (probably milkweed pods), enough to make a pillow.

reall Philosophy, wch is very active here, and in all our neighbouring Country, will also take in your parts, and there seize on all
that have ingenuity and Industry, for the further spreading of the
Honour of the English Nation, and the larger diffusing of the
manifold advantages, and benefits, that must proceed from thence.
I am persuaded, Sir, that you will lay out your talent for that purpose, and instill the noblenesse of this institution and worke with
your best Logick and Oratory, into the minds of all your friends
and acquaintances there, especially of those pregnant youths, that
have begun to give proofe of their good capacityes for things of
that nature. . . .

Even the Indians will doubtless subject themselves more cheerfully
to the English, who become as addicted to "all sorts of Ingenuityes,
pleasing Experiments, usefull inventions, and practices" as they are
to piety and virtue![39] If any doubt exists about the Royal Society's
design to promote experimental philosophy in the American colonies, let Oldenburg's words above be reread.

Winthrop's "stellar fish" (*Astrophyton agassizii*) excited much
attention. A figure of it was printed in the *Philosophical Transactions*, together with a long description by the Royal Society savants.[40] Winthrop was pressed to find out more about it, but it was
some time before he could comply. He was also pressed to prepare
a complete natural history of New England. Winthrop approved
of this project but argued that the time was not yet ripe to undertake it: too little was known as yet, too much unexplored and
untested by experiment. But in the meantime Winthrop sent a second collection of New England curiosities to the Royal Society.
Shipped on August 26, 1670, it was received by the Society the
following October 27, placed in the Repository, and catalogued.[41]
It consisted of three boxes containing two stuffed rattlesnake skins,
four roots of Virginia snakeweed, more quick-ripening Indian corn,
a bag of mixed hazelnuts and walnuts, more dwarf oaks gathered

[39] *Letter-Book*, III, 313–315.
[40] *Phil. Trans.*, V, No. 57 (March 25, 1670), 1152–53.
[41] *Guard-Book*, W-3, No. 23; *Letter-Book*, IV, 53–56; *Journal-Book*, IV, 153.
Nehemiah Grew's *Musaeum Regalis Societatis, Or a Catalogue & Description of
the Natural and Artificial Rarities Belonging to the Royal Society and Preserved
at Gresham College* . . . (London, 1681) listed many of Winthrop's gifts. See
passim.

early so that the acorns were intact, a new kind of fish, and a piece of crystal-like rock, which the Indians said "fall from thunder" (belemnite?). In the meantime, on October 11, Winthrop sent two letters, one to Henry Oldenburg with gifts "for the Repository of the Royal Society," the other to Lord Brereton, who presented it to the Society. The former consisted of a new kind of shellfish, known locally as a "Horse-foot" (*Similus polyphemus*), a hummingbird's nest with two eggs in it, "a fly with feathers," some "small shells, and bullets, and hardened clay" from a hill in Maine (near Wells) that had "blown up"; and some "silke-down of those silke-pods, for Sir Robert Moray," who had requested them.[42] The letter to Lord Brereton gave further details about the "hill near Kennebeck, Me, that turned over in summer last (June or July) into the River. Told by Major Wm Philips & Mr. Herlskenden Symons." The specimens were minerals from the "inside" of the hill after the event had taken place. The letters and the curiosities were displayed to the Society on March 23, 1670/71. The hummingbird's nest, whose eggs were still intact, was ordered "to be first shewn to his Maty," before it was placed in the Repository. Oldenburg wrote his letter of thanks on the Society's behalf (April 11, 1671) and indicated that Winthrop's gift had excited considerable comment in the Society and that the King was mightily pleased to see the hummingbird's nest. But the Society desired more information about the overturned hill in Maine. He also sent Winthrop copies of the *Philosophical Transactions* for 1670, Boyle's "New Tracts about the wonderfull Rarefaction, and condensation of the Air," and "Monsieur Charras New Experimts upon Vipers."[43] Even before Oldenburg had written, the Society had

---

[42] *Guard-Book*, W-3, Nos. 24, 25; Birch, II, 473–474. Evidently no one was eyewitness to the "blowing-up" of the hill in Maine. Josselyn later described it as follows: "a piece of Clay Ground was thrown up by a mineral vapour (as we supposed) over the tops of high Oaks that grew between it and the River, into the River, stopping the course thereof, and leaving a hole two yards square, wherein were thousands of Clay Bullets as big as Musquet Bullets, and pieces of Clay in shape like the Barrel of a Musquet." John Josselyn, *New-England's Rarities Discovered* . . . (London, 1672), pp. 113–114. The "Clay Bullets" etc. were probably concretions in the late Pleistocene marine clay. See George W. White, "John Josselyn's Geological Observations in 1672–1674," *Transactions of the Illinois Academy of Science*, XLVIII (1956), 180.

[43] *Journal-Book*, IV, 182; *Letter-Book*, IV, 274–275.

received (March 23, 1670/71) Winthrop's letter (dated October 26, 1670) giving a few more details about the "stellar fish."[44] He had seen the fisherman and said that "I asked all the questions I could thinke needfull concerning it." But he learned little of value. The fisherman had caught the starfish while fishing for cod near Nantucket. He had seen six or seven in all, all taken near Nantucket. As there was no English name for it, Winthrop proposed that it might be known as a "basket fish," because when it seized bait or other food it placed its "arms" together forming a kind of basket.

John Winthrop's last shipment of New England rarities to reach the Royal Society was sent from Boston, October 17, 1671, under the supervision of Waitstill Winthrop. Wait wrote to Oldenburg that his father was remote from Boston and had asked him to oversee the shipment, which consisted of "An Indian bow and quiver of arrowes, the Sword of a fish, and a smale fish with a horne on the back, alsoe among the arrowes there is a Sea thorn. The quiver is made of an Indian doggs skin, the arrowes are headed according to the Indian manner, some with horsefoot tayles, some with stones and others with deere horn and shark's teeth. . . ."[45] These were produced at the Royal Society on December 7 following and placed in the Repository. Nearly two years later (September 25, 1673), Winthrop wrote to Oldenburg indicating that some of his letters had again gone astray and that a packet of things sent from New York had not been acknowledged.[46] He thanked Oldenburg for the books, for information sent about a comet seen at Danzig by Johannes Hevelius, the remarkable astronomer there, and promised to make inquiries whether the comet had been seen in New England. He had "several things" at Boston to send to the Society and relied upon a friend there to find shipping. But there is no record that those items arrived at the Royal Society. Nearly three years before his death (1676), Winthrop's active intercommunication with the Royal Society ceased. Although marked by disappointments on both sides, it had been a mutually beneficial

[44] *Guard-Book*, W-3, No. 26; printed in *Phil. Trans.*, VI, No. 74 (Aug., 1671), 2221–24.

[45] *Guard-Book*, W-3, No. 28; *Journal-Book*, IV, 219.

[46] *Guard-Book*, W-3, No. 27.

relationship. And there is evidence, which will be reviewed later in this chapter, that the Royal Society's design to employ its first colonial Fellow to erect a "cell" in New England for the future promotion of experimental philosophy there was at least temporarily successful.

John Josselyn, a Source outside the Society

Henry Oldenburg's reiterated appeal for a "good History of New England . . . containing ye Geography, Natural Productions, and Civill Administration"[47] was answered in the early 1670's by John Josselyn—although the tone of Josselyn's works could hardly have been music in the ears of the Puritan colonists nor their contents wholly satisfying to the rising generation of experimental philosophers in the Royal Society. John Josselyn (c. 1608–75) was the second son of Sir Thomas Josselyn, Kt., of Essex, by his second wife. Sir Thomas held appointment as a Councillor of the Province of Maine, and he and John visited at the house of John's older brother, Henry Josselyn, at Black Point (Scarborough), Maine, in 1638–39. John returned to England in October, 1639, but he made a later and much longer visit which extended from July, 1663, until August, 1671. It was during this second sojourn that he gathered most of the materials for his books, the first entitled *New-England's Rarities Discovered: In Birds, Beasts, Fishes, Serpents, and Plants of that Country. Together with the Physical and Chyrurgical Remedies wherewith the Natives constantly use to Cure their Distempers, Wounds and Sores . . .* (London, 1672); and *An Account of two Voyages to New England* (London, 1674; 2nd ed., London, 1675), dedicated to the President and Fellows of the Royal Society of London. The first book included (pp. 103–114) "A Chronological Table of the most remarkable passages in that part of America, known to us by the name of New England" from 1492 to 1672.[48]

47 To Winthrop, April 11, 1671, *Proc. M.H.S.*, XVI, 250–251.

48 Accounts of Josselyn appear in the *D.N.B.* and the *D.A.B.* See also Fulmer Mood, "Notes on John Josselyn, Gent.," *Publications of the Colonial Society of Massachusetts*, XXVIII (1932), 24–36; Moses Coit Tyler, *History of American Literature during the Colonial Period* (2 vols., Boston, 1879), I, 180–185. The best edition of his *New-England's Rarities* is that by Edward Tuckerman in

Most of what is known about John Josselyn is derived from his own writings. He appears to have been a genial, educated gentleman with training as a physician. He was familiar with the works of Aristotle, Pliny, and other Ancients as well as with Johnson's "Gerardus Immaculatus," Parkinson's *Theatricum Botanicum*, Captain John Smith's works, William Wood's *New Englands Prospect*, and others of his day. Although outwardly on friendly terms with the leaders of the Massachusetts Bay Colony, he was, politically, a Royalist; and this, together with his family's close support of the interests of the heirs of Mason and Gorges in Maine, led him to cast a jaundiced eye upon the people and policies of the Bay Colony. He was a born *raconteur*. And while he sought to soothe the doubts of an incredulous reader,[49] some of the accounts of remarkable cures and wondrous beasts which he occasionally interpolated into his catalogue of rarities must have lifted the eyebrows of his most credulous follower. Obviously he loved a good story.[50]

Archaeologia Americana: Transactions and Collections of the American Antiquarian Society, IV (Boston, 1860), 105–238. It was given a descriptive, noncommittal review in the *Philosophical Transactions* on July 15, 1672, and Robert Hooke brought it to the attention of the Royal Society again on March 23, 1686/87, by reference to Josselyn's report of "a Scarlet-Muscle [mussel]" in Maine waters used by the colonists as a purple dyestuff. *Journal-Book*, VIII, 135. In Oct., 1692, John Ray referred to Josselyn's account of New England barberries. Robert T. Gunther, ed., *The Further Correspondence of John Ray* (Oxford, 1928), p. 179. Obviously, Josselyn's book, with all its scientific shortcomings, was used by the natural philosophers of the late seventeenth century. The papers of James Petiver include a note, undated (c. 1696–1700), "Observations on Mr. Josselyn his New England Rarities, discovered. sent these with ye book to Mr. Bullyvant & Mr. H. Usher." *Sloane 3332*, fols. 231–232. Benjamin Bullivant, a Boston apothecary, and Hezekiah Usher were Boston correspondents of Petiver. Cf. *Petiver*, pp. 321–326. Josselyn's *Account of two Voyages* has been republished in *3 Coll. M.H.S.*, III (1833), 211–396.

49 "It was a good *proviso* of a learned man, never to report wonders, for in so doing, of the greatest he will be sure not to be believed, but laughed at, which certainly bewraies their ignorance and want of discretion. Of Fools and Mad-men then I shall not care. I will not invite these in the least to honour me with a glance of their supercilious eyes; but rather advise them to keep their inspection for their fine tongu'd Romances and playes. This homely piece [*An Account of two Voyages*] I protest ingeniously, is prepared for such only who well know how to make use of their charitable constructions towards works of this nature. . . ." *An Account of two Voyages*, *3 Coll. M.H.S.*, III, 235. Subsequent references to this work will be to this edition.

50 *Ibid.*, p. 251.

His descriptive prose hardly rivaled that of William Wood, although he covered a much wider range of subjects. Indeed, Josselyn's works, with all their extravagances and other scientific shortcomings, constituted the most complete natural history of New England produced during the Old Colonial Era. Manasseh Cutler's "An Account of some of the Vegetable Productions naturally growing in this part of America, botanically arranged," published in the first volume of the *Memoirs of the American Academy* in 1785, was the first major successor to Josselyn, but it lay in the national period of American history, after the end of the colonial era.

Josselyn's prime concern was with flora: I "made it my business," he said, "to discover, all along, the natural, physical, and chyrurgical rarities of this new-found world";[51] "Johnson hath added to Gerard's *Herbal* 300. [plants] and Parkinson mentioned many more; had they been in *New-England* they might have found 1000 at least never heard or seen by any Englishman before. . . ."[52] His *New-England's Rarities* was

> cast . . . into this form: 1. Birds. 2. Beasts. 3. Fishes. 4. Serpents and Insects. 5. Plants, of these, 1. such Plants as are common with us, 2. of such Plants as are proper to the country, 3. of such Plants as are proper to the Country and have no names known to us, 4. of such Plants as have sprung up since the *English* Planted and kept cattle in New England, 5. of such Garden Herbs (amongst us) as do thrive and of such as do not. 6. Of Stones, Minerals, Metals, and Earths.[53]

Josselyn's account of New England birds was rather brief in his *Rarities* and was enlarged upon considerably in his *Voyages*. In the former he mentioned only ten varieties, interspersed with accounts of several medical "cures." These included the hummingbird (*Trocilus*), the chimney swift, the "Pinkannow" (crested eagle?), the "Turkie," and three sorts of geese (with an interpolation regarding the value of goose grease as a cure for the bloody flux!). Among birds "not to be found in New England" he listed, quite

51 *New-England's Rarities* (Tuckerman ed.), p. 3.
52 *An Account of two Voyages*, p. 251.
53 P. 6.

erroneously as Mrs. Allen has pointed out,[54] about a dozen birds, including, strangely enough, the sparrow, blackbird, woodcock, quail, and robin. In his *Voyages* he speculated that there must be about 120 kinds of birds in New England. He drew distinctions between land birds and waterfowl and, in particular, referred to the partridge, pigeon (carrier pigeon), thrush, and a few others. But his treatment of birds is more suggestive than complete, and was wholly lacking in scientific description and terminology. He said nothing of nesting habits, feeding, migration, form, color, differences between the sexes, or anatomy.

His treatment of beasts was mixed with concern for their value in the fur trade and their supposed medicinal virtues. In his *Rarities* he mentioned only ten: the black bear, the wolf, the "Ounce" or lynx, the raccoon, the porcupine, the beaver, the moose, the "maccarib" (elk?), the "Jaccal" (grey fox?), and the hare (with obvious confusion between the grey rabbit and the northern hare). In his *Voyages* he added the skunk, the otter, the musquash, the martin, and three sorts of squirrels. Altogether, the treatment is incomplete and, again, lacking in scientific description. His concern for the medicinal value of bear's grease for the treatment of aches and swellings was significant, for bear's grease was long valued by American colonials and frontiersmen as an ointment with great curative properties. He said that the horns of a moose had a better use in physic than hartshorns. He believed, with others, that porcupines "shoot" their quills, and his. account of the beavers' "dams" was inferior to that of William Wood.

"Pliny and Isadore," said Josselyn, "write there are not above a hundred and forty-four kinds of fishes; but, to my knowledge, there are nearer three hundred. I suppose America was not known to Pliny and Isadore." His catalogue of fish "seen between the English Coast and America, and those proper to the Country" listed but twenty varieties, including a few mollusks. There was no evidence that he had consulted with experienced fishermen about the varieties of fishes in the waters in and about New England, and again he concerned himself with the supposed medicinal virtues

[54] Elsa G. Allen, "The History of American Ornithology before Audubon," *Transactions of the American Philosophical Society*, n.s., XLI, Pt. 3 (1951), 457–458.

of the stones found in codfish and the cure of the toothache by picking the gum with a dogfish fin (an osprey beak was equally effective). His accounts of the spermaceti whales found in New England waters, of spermaceti oil and its uses, were informative, but he believed that whales feed upon ambergris, which, he said, "is no other than a kind of mushroom, growing at the bottom of some seas." In "An Addition of some Rarities Overslipt" at the end of his book, Josselyn listed a few more fishes, including the starfish (*Asterias rubens*, L.), probably not the same variety as that sent by Winthrop to the Royal Society, and the sea horse (*Hippocampus*), whose medicinal virtues were highly extolled as an antidote to poison, a diuretic, and a cure for tumors, hemorrhoids, cramps, bruises, wounds, and other disorders.

"Of Serpents and Insects" Josselyn had little to say and what he wrote had more relevance to their supposed "virtues" than to natural history. He spoke of large frogs in the ponds, and said that the Indians had reported an up-country frog "as big as a child of a year old." Frog fat was a palliative for burns, scalds, and any inflammation. Rattlesnake fat was also helpful in the treatment of frostbite, aches, and bruises, and the "heart of a Rattle-snake dried and pulverized & drunk with wine or beer is an approved remedy against the biting and venome of a Rattle-snake. Somebody will give me thanks for discovering these secrets and the rest: *Non omnibus omnia conveniant.*" This was but one of many of Josselyn's bows to the theory of a special *vis naturae curatrix* in the New World. Of insects he mentioned several intermediate worms, an unidentifiable beetle-like bug, the "flying gloworm" (firefly), and wasps which built strange nests of a paper-like substance, "like a great pineapple."[55] Honeybees throve in New England, having been introduced by the English: "in time they may be produced from Bullocks when the wild Beasts are destroyed"—thus did he reveal his belief in their spontaneous generation.

Josselyn's greatest contribution lay in his account of the flora of New England. But he was no botanist in the sense that John

---

[55] These wasp nests were a great curiosity in France, having been sent over from Canada. See the illustration of the *Vesparum quarundam Canadensium* in Martin Lister, *A Journey to Paris in the Year 1698*, ed. Raymond P. Stearns (Urbana, Ill., 1967), Fig. 1 (opp. p. 248) and pp. 265–266.

Ray and others in England and on the Continent were developing into at the time. Indeed, his technique fell short of that of Cornuti, the Paris physician who had described Canadian plants growing in the Paris garden. There is no evidence that he collected seeds or prepared an herbarium to share with fellow students in England or elsewhere. Obviously he collected plants and seeds for his own immediate use and experimentation. He practiced medicine among the people of Maine, and he delighted in discovering the secrets of Indian medicine, often taking part with Indians in the brewing of herbs, barks, and other native concoctions. As a reporter of folk medicine, both English and Indian in origin, he was without parallel in his day.[56] But his instincts were those of the herbalist, his primary concern being the virtues of plants, his principal guide being Thomas Johnson's revision of John Gerard's *Herball*. He had little, if any, conception of scientific botany, no ecological sense, little concern for plant structure, botanical system, or nomenclature. The organization of his catalogue was utilitarian rather than scientific; but his recognition "of such Plants as have sprung up since the English planted and kept Cattle in New England" was a shrewd observation making useful distinction among plants indigenous to New England, plants consciously introduced by the English, and plants accidentally introduced by the chance of English livestock droppings, fodder from abroad, and the like. He published ten rather crude woodcuts of plants, mostly "of Such plants as are proper to the Country, and have no Name." These included the first illustration and description of skunk cabbage (*Symplocarpus foetidus*, L.), and, while a few included helpful illustrations of the plant leaf, none of them showed root structure in any useful detail. Still, with all his faults, John Josselyn published the most complete account of the flora of New England—with obvious emphasis upon upper New England—of the Old Colonial Era.

His first category of plants, "of such as are common with us in England," included ninety-three items with extensive attention to their virtues and uses as employed in New England. He noted some

[56] See Harvey W. Falter, "The Genesis of American Materia Medica, Including a Biographical Sketch of John Josselyn, Gent.," *Bulletin of the Lloyd Library of Botany, Pharmacy, and Materia Medica*, No. 26 (1927), Reproduction Ser. 8.

differences in the plants in New England, especially "Dragon's [lily; *Arum*, L.]. Their leaves differ from all the kinds with us. They come up in June"; and he had high praise for the native strawberry (*Fragaria vesca*, L.). His second category, "of Such Plants as are proper to the Country," described forty plants and trees, including maize, the food value of which, like Winthrop, he praised, although he made no distinctions as to varieties as Winthrop had done; wild sarsaparilla (but whether *Aralia nudicaulis*, L., or *A. hispida*, Michx., is impossible to say); white pine (*Pinus strobus*, L.); fir, referred to as balsam fir and the "pitch tree" (*Abies balsamea*, L.); black spruce (*Picea mariana*); white spruce (*Picea glauca*); hemlock (*Tsuga canadensis*, L.); and cranberries (*Vaccinium macrocarpon*, Ait.). In his *Account of two Voyages*, Josselyn enlarged upon his descriptions of trees, with particular attention to their value in the building trades, medicinal uses, and ship-building, with a description of the making of tar and pitch similar to Winthrop's account. Considering his interest in medicinal virtues, he gave strangely little attention to sassafras, but emphasized the value of turpentine as a cure for wounds, resin "good as Frankincense," pine cones boiled in beer as treatment for scurvy, and sumac, valuable both as a dye and as a medicine.[57] His *Account of two Voyages* also referred to American beans of various colors, pumpkins, watermelons, tobacco (of which little was grown in New England), and various small fruits (elderberries, juniper berries, currants, gooseberries, etc.).

Josselyn's third category of plants in the *New-England's Rarities*, "of Such plants as are proper to the Country, and have no

---

[57] It is strange that Josselyn did not mention the maple tree or maple sugar. Indeed, English scientists apparently first learned of maple sugar from samples sent to England from France early in 1685 and referred to as "Canada sugar." The Royal Society received a "large piece of brown Sugar, delivered . . . by Monsʳ [Henri] Justel" on March 4, 1684/85 (*Journal-Book*, VII, 300), with a brief account of its production "by the Savages of Canada." It was reported "to be as secret as the Sugar from the Sugar Canes." About the same time, Tancred Robinson sent a sample to John Ray. Edwin Lankester, ed., *Correspondence of John Ray* (London, 1848), pp. 162–163. It excited wide interest. Attempts were made to extract sugar from other trees and plants, resulting ultimately in the distinction between the sycamore maple (*Acer pseudoplatanus*, L.) and the sugar maple (*A. saccharum*, Marsh.). See "An Account of a sort of Sugar made of the Juice of the Maple, in Canada," *Phil. Trans.*, XV, No. 71 (May 20, 1685), 988.

Name," listed eight items, with cuts of them all. He gave them the popular New England names—wintergreen (*Pyrola*); skunk cabbage (see above); touch-me-not or "Humming-bird tree" (*Impatiens fulva*, Nutt.), which he said was a favorite for hummingbirds to feed on; snakeweed (species uncertain); the "American marigold" or sunflower (*Helianthus strumosus*, L.); and two plants of uncertain identification. The fourth part, "of such Plants as have sprung up since the English planted and kept Cattle in New England," listed twenty-two varieties, including several grasses and such "weeds" as the ubiquitous dandelion and plantain. In his final list "of Such Garden-Herbs amongst us as do thrive here, and of such as do not," Josselyn gave a useful guide for New England planters, including sixty items in all. They embraced forty-nine common garden vegetables, herbs, flowers, roots, and cereal crops[58] which, he said, throve in New England, such as parsnips "of a prodigous size" and "Pease of all sorts, and the best in the world." Eleven items, including rosemary, bay, lavender, pennyroyal, and some other herbs did not, he said, do well in New England.

The sixth section of *New-England's Rarities* treats of "Stones, Minerals, Metals, and Earths." Like Captain John Smith, Josselyn believed that New England abounded in precious stones and ores; but unlike Smith, he had traveled widely over parts of the country and had what he believed to be acceptable evidence for his opinions. He gave the earliest printed description of the White Mountains, and his detailed account of Mt. Washington could only have been by one who had climbed it.[59] His physiographic observations that the "gullies" on Mt. Washington (the gulfs or ravines) had been eroded by melted snow was an early recognition in America of the fluvial origin of valleys—a view not yet universally accepted by English students of geology.[60] But Josselyn did not refer to rainfall

[58] Wheat, rye, barley, and oats. Josselyn subscribed to the old belief that barley would degenerate into oats. In the *Account of two Voyages* Josselyn enlarged this list considerably. He included hemp, flax, various common fruit trees, and domestic livestock and poultry.

[59] Josselyn was not the first New Englander to climb Mt. Washington; that honor belonged to Darby Field, who had ascended it in 1642.

[60] With regard to Josselyn's geological observations, I am deeply indebted to my colleague, Professor George W. White, whose article "John Josselyn's Geological Observations in 1672–1674" is the source of what follows on this

as a source of the erosion, and he believed, with most of his contemporaries, that the "pond of clear water" (Lake of the Clouds) near the top of Mt. Washington, as well as springs and other "sources" of streams in the hills, were the consequence of an upward subterranean movement of sea water which, desalted in some mysterious way, broke to the surface. Like his contemporaries, further, he also believed that the mountains were hollow, wherein the waters rose, and wherein the "trees" of ores grew in the earth. He found what he believed to be emerald, "which grows in flat Rocks, and is very good," rubies, diamonds, crystal, Muscovy glass, black lead, "Bole Armoniack," "Red and yellow *Oker*," "Terra Sigilla," vitriol, antimony, "*Arsnick*, too much," lead, tin, "Tin Glass," silver, iron "in abundance, and as good bog Iron as any in the World," and copper. But his emeralds were probably beryl; his rubies, garnets; his diamonds, quartz; his Muscovy glass, muscovite or biotite; his "Arsnick," arsenopyrite; and the "Bole Armoniack" and "Terra Sigilla," varieties of clays. In his *Account of two Voyages*, Josselyn enlarged upon his list of New England minerals and chided the colonists for not exploiting the deposits of lead, silver, copper, tin, and the like in their midst. The younger Winthrop and other New Englanders already were realizing that these hopes of valuable mineral deposits and precious stones were probably vain. Bog iron and black lead were the only ones found in such quantities as to be commercially exploitable, and the lack of capital and labor proved to be handicaps to these enterprises.

Josselyn had much to say about the waters in and about New England. The springs, the lakes, the rivers, and the sea all called forth comment. The springs had all the medical properties "ascribed to the best in the world," and a sulphur spring at Black Point was "*purgative and cures scabs and Itch* &c." The lakes were many, some of them large, all of them filled with fish, beaver, and other fur-bearing animals. Of the rivers he wrote:

> All the Rivers of note in the Countrey have two or three desperate falls distant from one another for some miles, for it being

aspect of Josselyn's works. Professor White concluded that "John Josselyn deserves a place in the history of geology in America." Cf. also Gordon L. Davies, "The Eighteenth-Century Denudation Dilemma and the Huttonian Theory of the Earth," *Annals of Science*, XII (1966), 129–138.

rising ground from the Sea and mountainous within land, the Rivers having their Originals from great lakes, and hastning to the Sea, in their passage meeting with Rocks that are not so easily worn away, as the loose earthlie mould beneath the Rock, makes a fall of the water in some Rivers as high as a house. . . .[61]

In his description of the location of New England and its boundaries he mentioned the many fine harbor facilities of New England and discussed the sea. His statements about the depths of the sea are very early ones, although English sea captains since Sir Francis Drake's time had sounded the seas about New England. Josselyn concluded that "this is certainly true, that there is more Sea [i.e., deeper sea] in the Western than the Eastern *Hemisphere*," and at the spring tides, "in more places than one," the sea "riseth 18 foot perpendicular." This was specific information which the Royal Society had been seeking for some time. Probably Josselyn had heard, as Winthrop had reported he had been told, of even higher tides in the Bay of Fundy. But the "reason of this great flow of waters," said Josselyn, "I refer to the learned." He had seen icebergs and a "Triton or Merman" on his first voyage to New England in 1638. The icebergs, he said, "congealed in the North" and were brought down in the spring by ocean currents "to the banks on this side of New-found-land, and there . . . they dissolve at last to water." He made no further comment on the merman.

As he had approached Boston in late June, 1638, Josselyn had been told of an earthquake which had shaken New England on June 1. This, the first recorded in the list of earthquakes in the British colonies of North America, was a moderately violent shock, and was extensively reported in New England.[62] Josselyn speculated in a remarkably "modern" manner as to the possible relationship between earthquakes and mountain building in New England and, in a less advanced way, as to their influence in producing the supposed "hollowness" of the mountains. He also reported an eclipse of the moon between eight and eleven at night on December

61 *An Account of two Voyages*, p. 306.
62 W. T. Brigham, "Volcanic Manifestations in New England . . . ," *Memorials of the Boston Society of Natural History*, II (Boston, 1871), 1–28; N. H. Heck, *Earthquake History of the United States* (*U.S. Coast and Geodetic Survey*, Ser. 609, Washington, 1938).

10, 1638, adding that seamen could correct their observations as to latitude and longitude if they would mark accurately the times of beginning and ending of such eclipses, "in all places they happen to be, and confer their observations to some Artist."[63] Later he referred to the comet seen in New England in 1664, regretting that he had had no telescope with which to observe it. But he believed that comets proceed from natural causes and concluded, with Tycho Brahe, that their preternatural consequences were "fancie." His meteorological comments were confined largely to the observation that whereas in England the coldest weather came from the sea (i.e., the North Sea), in New England the reverse was true, the coldest weather approaching from the north and west while the sea winds exercised a warming effect.

Josselyn held the Indians in high regard and evidently was on close terms with many of them. He ended his book on *New-England's Rarities* with a description of an Indian squaw "trick'd up in all her bravery." He described the men as "somewhat Horse Fac'd, and generally . . . without Beards." Many of the women had "very good Features," and though they were "seldom without a *Come to me*, or *Cos Amoris*, in their Countenance," they were, "saving here and there one, of a modest deportment." It is unfortunate, in view of Josselyn's associations with Indians, that his account of them was so brief.

Taken together, John Josselyn's *New-England's Rarities* and *An Account of two Voyages* complemented one another. They constituted, as Professor Mood has remarked, "a rudimentary sociological description of New England," approximating the account which Henry Oldenburg implored John Winthrop to prepare.[64] Josselyn described the flora and fauna, the physiography and scenery, the Indians, the English settlers together with their distribution and frames of government, and the history of New England by way of

[63] Josselyn also referred to an eclipse of the moon during the evening of July 15, 1668.

[64] April 11, 1671: "I cannot yet desist from recommending to you the composure of a good History of New England, from the beginning of ye English arrival there, to this very time; containing ye Geography, Natural Productions, and Civill Administration thereof, together with the notable progresse of yt Plantation, and the remarkable occurences in the same." *Proc. M.H.S.*, XVI, 250–251.

a chronological table of important names, dates, and events. If he was critical of church and state in the Massachusetts Bay, he was also critical of the economy and want of enterprise of the colonists of Maine. His works constituted a sort of handbook of New England, larded with "cures," credulities, and an occasional racy quotation worthy of English Restoration literature. His works lacked the scientific advances in natural history demonstrated by John Ray, Francis Willughby, Robert Hooke, and Nehemiah Grew, but he belonged to a somewhat older generation and shared none of the direct benefits derived from fellowship in the Royal Society.[65] Indeed, other than the dedication of his second book and the review of his works in the *Philosophical Transactions,* there is no record that Josselyn had any connection with the Royal Society or its Fellows. He belonged with the pre-Restoration generation of English herbalists. As such, he was a cultivated, forward-looking gentleman, scholar, physician, and descriptive writer of his travels.

Increase Mather's "Philosophical Society"

The life spans of John Josselyn (c. 1608–75) and of John Winthrop, Jr. (1606–76) were almost coterminous, although Winthrop spent a longer portion of his life in New England and his influence was felt more widely. Born into the elite society of Massachusetts Bay, he moved freely and upon terms of wide friendship among the mighty personalities of both church and state in the Bay Colony, New Plymouth, Connecticut, and early New York. Infected with the spirit of the new experimental philosophy and prodded by Henry Oldenburg and other Fellows of the Royal Society to promote this spirit,[66] Winthrop strove from his frontier home in Con-

[65] The English publications most comparable to Josselyn's works were Dr. Robert Plot's *Natural History of Oxfordshire* (Oxford and London, 1676; 2nd ed., Oxford, 1705) and *The Natural History of Staffordshire* (Oxford, 1679; 2nd ed., 1686). Plot (d. 1696) was a F.R.S., served as Secretary of the Royal Society (1682), and later became Keeper of the Ashmolean Museum at Oxford and Professor of Chemistry there.

[66] In addition to the letters quoted above, Oldenburg wrote to Winthrop in 1667: "Sr, I persuade myself yt you, who know so well the vselesnes of ye notional and disputacious School philosophy, will make it a good part of yr business to recommend this reall Experimental way of acquiring knowledge, by conversing with, and searching into the works of God themselves; and that you

necticut, where there were "all thinges to doe, as in the beginninge of the world," to diffuse the knowledge and the methods of experimental science in New England. He told Oldenburg that by the *Philosophical Transactions* "you have engaged thousands by the publication of them."[67] He lent his copies to friends (who sometimes lent them to *their* friends),[68] and he built up a small circle of men interested in experimental science with whom he exchanged books and information.

The limits of this circle are impossible to establish. The records show that the group included Dr. Robert Child (M.D., Padua), the Remonstrant, who was expelled from the Bay Colony in 1647 for his Presbyterian sympathies; George Starkey, a Bermuda-born Harvard graduate (1646) who subsequently returned to England, where he became a widely known physician and experimental alchemist; Jonathan Brewster, son of Elder Brewster of the Pilgrim church at Plymouth, who had a trading post at Moheken (Norwich), Connecticut, and maintained a private laboratory for chemical experiments; the Reverend Mr. Gershom Bulkeley, who had a

will endeavour, and yr ingenious and sober friends, to season and possess the youth of New England with ye same." *Proc. M.H.S.*, XVI, 230. Two years later Oldenburg urged Winthrop to devise ways to insure that Harvard College might be "furnisht wth the modern books of ye most Ingenious and famous Philosophers and Mathematicians, as Descartes, Gassendus, Ricciolo, Helvetius, Cassini, Fabri, Ward, Wallis, Boyle, Pell, Hugenius, Willis, Hook, Merret, Wilkins, Evelyn, Vossius, etc." *Ibid.*, pp. 239–240. John Beale urged a somewhat similar policy upon Robert Boyle and pleaded with Boyle to use his position in the Company for the Propagation of the Gospel in New England to encourage the ministers of New England to print monthly bills of mortality, collect information on the soils and natural history, and promote the new science generally: "If New England would now begin it, probably ye English would follow yr Example in all parts of yt large Continent & ye Islands about it." Beale to Boyle, Yeovil, Somerset, Feb. 16, 1680/81, March 2, 1680/81, and June 26, 1682, *Letter-Book "O,"* Nos. 141, 142, 143.

[67] Winthrop to Oldenburg, Boston, Oct. 4, 1669, *Guard-Book*, W-3, No. 22.

[68] See the letter of Thomas Shepard to Winthrop, Charlestown, March 8, 1668/69, thanking Winthrop for copies of the *Philosophical Transactions*, and saying that he had showed them to Samuel Danforth of Roxbury and others. 3 *Coll. M.H.S.*, X (1849), 70–72. Shepard himself was interested in scientific learning and supplied Winthrop with astronomical observations, and Danforth, in publishing *The New England Almanack For . . . 1686* (Cambridge, 1686), betrayed his modern view in the doggerel:

In Joshuah's *Solstice* at the Voyce of man
The *Rapid sun* became *Copernican*.

similar establishment at his parsonage on the Connecticut River; Richard Nicolls, the first English Governor of New York after New Amsterdam was taken over from the Dutch in 1664, and was highly valued by Winthrop as a congenial neighbor and natural philosopher; Winthrop's son Waitstill, who developed an interest in science and ultimately became a practicing physician; and the Reverend Mr. Thomas Shepard of Charlestown, Massachusetts, Samuel Danforth of Roxbury, and no doubt others of the Boston area. It was a small and undistinguished group, but it widened into a far more significant number of scientific observers soon after Winthrop's death.

During the winter of 1671/72, as increasing age and the cares of office diminished Winthrop's capacities to carry on scientific observations, he presented his three-and-a-half-foot telescope, with its various attachments, to Harvard College. He accompanied the gift with "a large and learned letter" of instructions about its use, and obviously intended, as Henry Oldenburg had urged upon him, "to season and possess the youth of New England" with experimental philosophy. The gift was enthusiastically received by Harvard, one of whose tutors, Alexander Nowell, had already written on comets.[69] Indeed, the Boston area embraced several persons interested in astronomy. Samuel Cheever's *An Almanack For the year of our Lord 1661* . . . (Cambridge, 1661) contained a remarkable account entitled "A brief Discourse of the Rise and Progress of Astronomy," beginning with the Egyptians, discussing Ptolemy's system, and ending: "Whereupon in this last age, *Galilaeus, Bulliadus, Keplerus, Gassendus* and sundry other Mathematicians, have learned[ly] confuted the Ptolemaick & Tychonick Systeme, and demonstrated the Copernican Hypothesis to be the most consentaneous to truth and ocular observations, as the *Ephemerides* composed from thence clearly evince. . . . Thus, after many sore and tedious throwes . . . at length *Experientia scientiam, scientia veritatem, sed veritas* odium *perperit.*" All of which placed Mr. Cheever firmly in the vanguard of natural philosophers who accepted Copernicanism as refined by Kepler, Galileo, and Gassendi. Indeed, this attitude of mind appears to have prevailed among the New Eng-

[69] Samuel Eliot Morison, *The Intellectual Life of Colonial New England* (Ithaca, 1960), p. 252.

land intelligentsia—an attitude, as Professor Morison has pointed out,[70] far removed from that of the Roman ecclesiastics who had condemned Galileo in 1633 for unholding a hypothesis said to be contrary to Holy Writ.

One of the earliest recorded uses of Winthrop's telescope at Harvard was the work of a young Harvard graduate, Thomas Brattle (1658–1713), in observing the comet of 1680 ("Newton's Comet"). Brattle's observations were printed in the local almanac for 1681 and sent to John Flamsteed, the Astronomer Royal at Greenwich, via the Royal Society.[71] Flamsteed sent them to Isaac Newton, and Newton, in his *Philosophiae Naturalis Principia Mathematica* (London, 1687), after citing several "rude" observations of this comet, said that "those made by Montenari, Hooke, Ango, and the observer in New England, taking the position of the comet with reference to the fixed stars, are better."[72] Professor Morison has summed up Brattle's contribution as follows:

> Considering his limited equipment, Brattle's observations are remarkable. Independently he recognized the fact that a comet which disappeared at perihelion into the glare of the sun's rays and reappeared on the other side of the sun was one and the same comet. His observations were valuable to Newton because, being based on fixed stars rather than on azimuths and altitudes, they helped him to determine the orbital elements of a comet moving in an ellipse. Hence our little college telescope, used by a careful and intelligent observer, contributed its mite toward helping Newton to test Kepler's three laws, to work out the law of gravitation, and to write the great *Principia*.[73]

Brattle continued to make astronomical observations until his death. He contributed to the *Philosophical Transactions* observations he and Henry Newman made by a brass quadrant with telescopic sights of a solar eclipse at Cambridge in 1694. Near the turn of the century, Harvard College acquired a new telescope, a foot longer

[70] *Ibid.*, p. 254.

[71] *Journal-Book*, V, 234; read to the Royal Society April 13, 1681. The Society had official oversight of the Royal Observatory.

[72] Newton also referred to observations made in Jamaica and by Arthur Storer on the Patuxent River, near Hunting Creek, in Maryland. Cf. Newton, *Philosophia Naturalis Principia Mathematica* (London, 1687), pp. 495–497, 507.

[73] Morison, *Intellectual Life*, p. 253.

than Winthrop's; with this Brattle made observations of lunar eclipses in 1700, 1703, and 1707, the last of which was read to the Royal Society (October 22, 1707).[74]

In the meantime, in 1682, Halley's Comet made its periodic appearance [and received its name after Edmond Halley (1656–1742), the English astronomer who paid for the publication of Newton's *Principia* when the Royal Society's treasury ran low]. How many New Englanders viewed this phenomenon through Harvard's "optic tube" is unknown, but several of them published observations and comments about it. Increase Mather (1639–1723), soon to become President of Harvard (1685–1701), and his two sons, Cotton (later to be elected Fellow of the Royal Society) and Nathaniel, all observed the comet and wrote about it. Cotton made his comments in *The Boston Ephemeris, An Almanack for MDCLXXXIII* (Boston, 1683), and Nathaniel, still a Harvard undergraduate, published in the same *Almanack* for 1685 a list of astronomical discoveries which demonstrated that he was keeping abreast of the latest astronomical works as reported in the *Philosophical Transactions*. Increase, who had preached a sermon on the occasion of the comet of 1680, now published two sermons about *Heaven's Alarm to the World* (Boston, 1682) and a longer treatise entitled *Kometographia, or A Discourse Concerning Comets* (Boston, 1683). This was an erudite study which revealed not only that Increase Mather had kept up with the *Philosophical Transactions* but also that he had read widely in the recent European literature on astronomy.[75] He had viewed the comet with the telescope on September 12, 1682, and he recognized that comets proceed from natural causes, and that they move like planets with orbits greater than the planets; but he insisted that the appearance of comets could not be predicted ("Comets then are Temporary, whereas the true *Planets* are per-

---

[74] *Journal-Book*, XI, 120. In 1711 Brattle sent a report to the Society about the incidence and virulity of smallpox in New England. He argued that epidemics of it occurred at twelve-year intervals and that the next would come in 1714. *Ibid.*, XI, 275 (March 6, 1711/12).

[75] Besides Stanislai de Lubienietz's *Theatrum Cometiarum*, he cited Erasmus Bartholinus, Johannes Kepler, Johannes Hevelius, whose *Cometographia* (Danzig, 1688) may have borrowed its title from Mather, Giovanni Battista Riccioli (an anti-Copernican Jesuit), Johannes Heinrich Alsted, and Christian Severin (Longomontanus), Tycho Brahe's assistant.

petual"). He gave a history of all comets since the creation of the world and the remarkable or calamitous events their appearance had portended. For, like John Evelyn and other Fellows of the Royal Society of London, Mather was able to accept what the scientists had observed; but as historical events appeared to prove a sequential correlation between the appearance of comets and extraordinary events in human affairs, either for good or for evil, Mather clung to the age-old belief that there was an interrelationship. Wise men of the past had viewed comets as evil omens, and although events had sometimes proved their predictions "false and foolish," Mather chose the safe side and urged people to prepare for calamity.

The interest stimulated by the comets of 1680 and 1682 brought together in and about Boston a number of New Englanders who in 1683 formed a scientific club or, as they called it, the Philosophical Society. This was the first child of the Royal Society of London, and it demonstrated a considerable degree of success for the latter's policy of promoting experimental philosophy in the New World. The Boston group met fortnightly "for a Conference upon Improvements in Philosophy and Additions to the Stores of Natural History." Perhaps, as Professor Morison has suggested, they planned to compile a natural history of New England such as Henry Oldenburg had urged Winthrop to do, and to plan it according to the "rules and method described by that learned and excellent person Robert Boyle, Esq."[76] Obviously, they continued to make astronomical observations. Increase Mather submitted some of their work to Wolferdus Sanguerdis (1646–1724), Professor of Philosophy at Leiden, including descriptions and drawings of a parhelion

[76] Morison, *Intellectual Life*, p. 255. Boyle's "rules and method" had been published as "General Heads for a Natural History of a Countrey, Great or small," in *Phil. Trans.*, I, No. 11 (April 2, 1666), 186–189. Cotton Mather told of the Boston Philosophical Society in *Parentator. Memoirs of Remarkables in the Life and Death of . . . Dr. Increase Mather . . .* (Boston, 1724), p. 86. Robert Boyle and some of the men associated in the Boston Philosophical Society, including the Mathers, H. Usher, Morton, Danforth, Sewall, and possibly others, were also bound together by their participation in the work of the Company for the Propagation of the Gospel in New England, of which Boyle was the energetic Governor from 1662 until 1689. The *Boyle Papers* in the Royal Society Archives contain evidences of correspondence relating both to science and to missionary activities. See also William Kellaway, *The New England Company, 1649–1776* (New York, 1962), *passim.*

and a double rainbow which Sanguerdis published in the second edition of his *Philosophia Naturalis* . . . (Leiden, 1685; 1st ed., 1680). Thus, the Philosophical Society had at least one foreign correspondent. It also gathered "Collections" of observations and descriptions of natural phenomena, some of which Cotton Mather later employed in the preparation of his *Curiosa Americana* (1712–24) for the Royal Society.[77] But if any formal records of the Society were kept they have not come to light, and the scanty evidence regarding the Boston Philosophical Society suggests only that it was a group of men who voluntarily met periodically to confer upon topics of mutual interest in the broad spectrum of natural philosophy. Its leadership evidently sprang from Increase Mather; its membership can only be surmised; and it ceased to meet about 1688, after about five years' existence. But some of its members continued to exhibit a lively interest in natural—and experimental—philosophy; and these men, with other New Englanders of the early eighteenth century, traced their initial interest in science to the work of the ephemeral Boston Philosophical Society.

Who might have participated in this work? To seek the answer to this query is to survey the men of known scientific interests in the Boston area—men who, in spite of religious, political, and private differences, were sometimes able to meet together for friendly intercourse in the neutral atmosphere of experimental philosophy. Increase Mather and his two sons, Cotton and Nathaniel, were doubtless regular attendants. Their good friend Samuel Willard, minister of the Old South Church and Vice-President of Harvard College, was another. Thomas Brattle and his brother William, both of Cambridge, were likely to have been present occasionally: William Brattle, minister at Cambridge and Harvard tutor, was author of a textbook on logic long used at Harvard, and of *An Ephemeris of Coelestral Motions, Aspects, Eclipses &c. For the Year of the Christian Aera 1682* (Cambridge, 1682), a curious mixture of modern astronomy, with excellent popular descriptions of how eclipses

[77] Otho T. Beall, Jr., "Cotton Mather's Early 'Curiosa Americana' and the Boston Philosophical Society of 1683," *Wm. & Mary Qty.*, 3rd Ser., XVIII (1961), 360–372. See also Michael G. Hall, "The Introduction of Modern Science into 17th Century New England: Increase Mather," *Ithaca 26 VIII 1962–2 IX 1962: Proceedings of the 10th International Congress of the History of Science* (Paris, 1964), pp. 261–264.

take place, and astrological lore on the signs of the zodiac. Charles Morton, who had attended Oxford at the time of the meetings of the Oxford Club and had been a scholar at Wadham College when Dr. John Wilkins was Warden,[78] was minister at Charlestown after 1686. His academy at Stoke Newington, near London, had developed a splendid reputation for its teaching of natural philosophy before it was closed for religious reasons in 1685; and his *Compendium Physicae*, adopted at Harvard shortly after his arrival in New England, though hardly up-to-date in its contents, had a wide influence upon Harvard students for more than thirty years, stimulating student theses on heat, light, motion, and other problems of physics.[79] Dr. William Avery of Boston and his son, Dr. Jonathan Avery of Dedham, were both interested in chemistry and mining and had chemical books, apparatus, and other scientific instruments. The elder Dr. Avery was described as "a man of pretty ingenuity: who from the Ars veterinaria [i.e., farriery] fell into some notable skill in physick and midwifery & invented some usefull instruments for that case & besides was a great inquirer & had skill in Helmont & chemical physick. . . ."[80] Hezekiah Usher, Boston speculator in mines and mineral lands and brother of John Usher, later Lieutenant Governor of New Hampshire, was interested both in mineralogy and natural history. He corresponded with Robert Boyle and

[78] The Reverend Samuel Lee, minister at Bristol, Rhode Island, was another Oxonian from Wilkins' time, and though he was too distant to attend the Boston Society with any regularity, he may have been present at times, for he knew and was known well by the Bay Colony leaders. His interest in and knowledge of scientific matters were manifest in his letters to Nehemiah Grew. Cf. G. L. Kittredge, ed., "Letters of Samuel Lee and Samuel Sewall Relating to New England and the Indians," *Publications of the Colonial Society of Massachusetts*, XIV (1912), 142–186.

[79] Charles Morton, *Compendium Physicae* (*Collections of the Colonial Society of Massachusetts*, XXXIII, Boston, 1940). This volume has a biographical sketch of Morton by S. E. Morison and the *Compendium* is edited by Theodore Hornberger. I agree with I. Bernard Cohen, "The Compendium Physicae of Charles Morton, 1627–1698," *Isis*, XXXIII (1942), 657–671, that Morton's work was overpraised as an up-to-date work of experimental philosophy and that the book's importance lay in its influence rather than in its contents.

[80] *Pubs. Col. Soc. Mass.*, XIV, 147. The original manuscript by Lee is in *Sloane 4062*, fols. 235–236. Avery included Thomas Brattle and Joseph Dudley among his friends, and he corresponded with Robert Boyle about chemicals as medicines in his "earnest desire to be a comfort to the sick." *The Works of the Honourable Robert Boyle in Six Volumes* (new ed., London, 1772), VI, 610–614.

later with James Petiver.[81] Both he and Waitstill Winthrop may have attended meetings of the Boston Philosophical Society. Further, Joseph Dudley and Edward Randolph, by virtue of their positions as well as an interest in natural philosophy, were likely attendants, and Randolph communicated accounts of New England natural phenomena to Boyle, Flamsteed, and Sir Robert Southwell, some of which were reported to the Royal Society.[82] And surely Samuel Sewall, the jurist and diarist, must have attended meetings of the Boston Philosophical Society although his *Diary* is curiously silent on the subject.

Sewall was on terms of close familiarity with the Mathers, the Brattles, the Winthrops, Charles Morton, Hezekiah Usher, Dr. William Avery, and, indeed, nearly all the principal ministers and magistrates of New England and New York. *The Diary of Samuel Sewall*[83] revealed much about the Boston area and its people during the last quarter of the seventeenth century. It presented a society of surprising mobility with considerable traveling about, private gatherings, meetings at taverns and coffeehouses, and much intercommunication by verbal exchange. It demonstrated that Samuel Sewall had a lively interest in science. He took part (with a physician and five others) in the dissection of an Indian who had been executed; he went on herb-gathering expeditions; he made note of extraordinary high tides, unusual weather, and a variety of astronomical phenomena, including Halley's Comet, several eclipses, and an aurora borealis which bespangled New England skies on January 18, 1686/87. Moreover, in 1688 Sewall, in the company of Increase Mather, Thomas Brattle, and others, visited England to defend the interests

[81] *Boyle Papers*, Bl. VII, No. 36; Usher's revealing will was published in *The Historical Magazine*, 2nd Ser., IV (1868), 120–126. See also *Petiver*, pp. 321–326.

[82] *Boyle Papers*, Bl. VII, No. 38; *Letter-Book*, IX, 139–140; *Journal-Book*, VI, 118, 136; VII, 218: "Sir Robert Southwell gave in a Paper communicated by Mr. Randolph, concerning the strang effect of a Thunder Storm upon severall Companies, which were in a ship Called the Albemarle. Mr. Edward Lad Capt. being then in 48ᵈ North Latitude, upon the Coasts of America, the 24ᵗʰ July 1681" (March 5, 1683). One compass "went dead," two reversed their polarity. "One of those Compasses which had quite changed the Polarity from North to South is still extant in the Country in the hands of Mr. Increase Mather." Letter by Sir Robert Southwell, London, March 8, 1683/84, *Letter-Book*, IX, 129–130.

[83] Published in 5 *Coll. M.H.S.*, Vols. V–VII (1878–82). The first volume is of special value for the period under discussion.

of the Bay Colony during the "Glorious Revolution" and to negotiate for a new charter. He was in England from January 13, 1688/89, when he landed at Dover, until the following October 10, when he sailed from Plymouth for New England. From Dover he visited Canterbury, with "time enough to view the Cathedral, and Kentish Husbandry as [we] went along" to London. With Thomas Brattle he visited Gresham College and was shown the Royal Society rooms by Charles Dubois, F.R.S. He made two visits to Greenwich, one with Increase Mather and Thomas Brattle, when John Flamsteed, the Astronomer Royal, "shewed us his Instruments for Observation, and Observed before us, and let us look and view the Stars through his Glasses." They visited the "Physick Garden" at Chelsea and Sewall commented upon several exotic plants they saw there. They made the acquaintance of—and dined with—Dr. Nehemiah Grew, plant physiologist, onetime Secretary of the Royal Society, and author of *The Anatomy of Plants* (London, 1682). They visited Dr. William Avery's cousin near Deptford, who took them "to a Gentleman who showed us many Rarities." They visited Cambridge, especially Emmanuel and Trinity colleges, and in the midst of their political business they managed a great amount of sightseeing, much of it to visit persons and places prominently associated with the promotion of experimental philosophy.

New England's Receptivity to the New Science

Obviously, then, during the last quarter of the seventeenth century New England developed a small circle of men devoted to experimental philosophy. They were able and willing to contribute to its development, and their contributions were incorporated in some of the works of the great English natural philosophers of the "century of genius," notably those of John Evelyn and Isaac Newton, not to mention the *Philosophical Transactions*. They enthusiastically accepted "the new hypothesis" of the stellar universe as described by Copernicus and Galileo, and set forth their own astronomical observations within this new framework. Indeed, it appears evident that they had confidence in the experimental method and were ready to accept both the evidence acquired from observation and experiment and the rational conclusions derived therefrom.

Cotton Mather spoke (1693) of the wonders revealed by the microscope: "There is not a Fly, but what would confute an Atheist; and the *Little* things which our Naked Eye cannot penetrate into, have in them a *Greatness* not to be seen without Astonishment. By the Assistance of *Microscopes*, have I seen *Animals*, of which many Hundreds would not Aequal a Grain of Sand. How exquisite . . . must the structure of them be!"[84]

Still, although the New Englanders readily accepted experimental science as a means of discovering the laws of nature, they were not ready to accept a wholly mechanistic view of the universe such as Robert Hooke and Nehemiah Grew envisaged in the seventeenth century and many of the philosophers of the Enlightenment adopted in the eighteenth. They retained a conventional belief in the Christian religion and merely extended their theology to admit the experimental method. Natural laws were part of the law of God. It was the duty of Christian man to seek to know God and His laws as fully as possible, as Charles Morton had stated in his *Compendium Physicae:*

> Though Man can't fully know what God hath done
> Yet 'tis his duty still to think thereon.

Between the knowledge of God revealed in the Bible and natural knowledge obtained by rational, experimental means there was an essential unity, as both derived ultimately from God. Natural knowledge was, then, a partial manifestation of the mind of God. There was no conflict between science and religion, nor were there any controversies of this nature either in colonial America or in the homeland. To the adherent of this natural theology, as it soon became known, experimental science was an instrument of theology whereby man might enlarge his knowledge of God's laws and the wonders of God's universe. It was an additional weapon of the Christian against the unbeliever. "There is not a Fly, but what would confute an Atheist," as Cotton Mather said. Still, while God might choose to operate the universe by natural laws much of the time, there was no natural law that God could not and did not occasionally set aside as He, in His inscrutable ways, saw fit. In-

---

[84] Quoted by Theodore Hornberger in "Cotton Mather's Interest in Science," *American Literature*, VI (1935), 413–420.

crease Mather's *Essay for the Recording of Illustrious Providences* (Boston, 1684), Cotton Mather's more famous *Magnalia Christi Americana* (London, 1702), and a host of other New England publications were testimonies to God's intervention in the affairs of New England and of New Englanders. These "Things Preternatural" included tales of demons and witches, monstrous births, terrible storms, and other horrors occurring outside the natural course of events. The *Philosophical Transactions* published similar accounts during these same years, and this dualistic, natural theology was as common in England as in New England.[85] Indeed, as Professor Morison has remarked,[86] this dualism, if not the whole fabric of natural theology, still exists. Renowned scientists, convinced from Monday through Saturday that the universe operates solely according to the inexorable and unchanging laws of nature, attend church on Sunday and pray for such things as prosperity, rain, and deliverance from "all the deceits of the world, the flesh, and the devil" as well as from "plague, pestilence, and famine; from earthquake, fire, and flood; from battle and murder, and from sudden death." Similarly, New Englanders accepted experimental philosophy without sacrificing their faith in "the special providences" of God. Indeed, they treasured it as a new instrument with which to combat the forces of evil and learn the ways of God.[87]

[85] Robert Boyle's *Usefulness of Experimental Natural Philosophy* (London, 1663) sounded the keynote of natural theology; and John Ray's *The Wisdom of God manifested in the Works of Creation* (London, 1691, with revised editions in 1692, 1701, and 1704, and many later editions) was a most learned and eloquent defense of the view. John Locke and some of the Cambridge Platonists also supported it. Cf. Charles E. Raven, *John Ray, Naturalist, His Life and Works* (Cambridge, 1950), Chap. 17.

[86] *Intellectual Life*, p. 270.

[87] I cannot agree with Frederick G. Kilgour, "The Rise of Scientific Activity in Colonial New England," *Yale Journal of Biology and Medicine*, XXII (1949), 123–138, when he states that the Puritans never gave science any real place in their culture. On the contrary, they adopted it enthusiastically—and gave it a prominent place in their theology! I also believe that its incidence in New England was almost a generation earlier than Mr. Kilgour presented it.

SIX

The Royal Society's Early Promotions: Virginia

Refinements in English Natural Science

Between the publication of John Josselyn's *New-England's Rarities* (1672) and the end of the seventeenth century remarkable advances were achieved in the study of natural sciences. A major impetus to new accomplishments derived from the posthumous works of Joachim Jungius (or Jung, 1587–1657), a German philosopher at Hamburg, whose *De Plantis Doxoseopiae Physicae Minores* (1662) and *Isagoge Phytoscopia* (1678) gave new directions to the study of botany. The latter work, especially, created a new scientific terminology for the more adequate description and processes of plants, such as expressions for the different kinds of flowering (i.e., *spica*, a cluster of flowers growing directly from the stem; *panicula*, a loose cluster of flowers; *umbella*, a cluster of flowers arising from the same level of the stem; *carymbus*, a cluster from different levels of the stem; and others). Many of Jungius' terms are still in use. He also clarified simple leaves from compound leaves (often mistaken for branches), and his full descriptions of the forms of flowers enabled him to distinguish among classes, such as the Compositae (daisy family), the Labiatae (deadnettle family), and the Leguminosae (bean family). Thus, while Jung himself set forth no system, he clearly advanced fundamental concepts of plant morphology upon the bases of which better plant classification could be made. At the same time, he dismissed such secondary features as scent, taste, color, and medicinal uses (upon which the herbalists had often depended), and scoffed at the ancient classification of plants into such categories as trees, shrubs, and herbs.[1]

A dozen or so English scholars contributed in varying degrees to the development of the natural sciences in the later seventeenth century, the most influential of them taking Jungius' works as a point of departure. Robert Morison (1620–83), however, was a law unto himself. A Scot, educated at Aberdeen, he espoused the Royalist cause during the English Civil Wars. Wounded at the Battle of Brig o' Dee, he went into exile in France. Here he was employed as a tutor and applied himself assiduously to the study of zoology, medicine, and botany, the latter under Vespasien Robin, the King's Botanist. Upon the recommendation of Robin, he was taken under the patronage of the Duke of Orleans, uncle of Louis XIV, and placed in charge of the royal gardens at Blois, a position he held from 1650 to 1660. The Duke sent him into various parts of France to search for new plants, and Morison not only enriched the gardens at Blois but also became widely familiar with the French gardens and French botanical learning as a whole. In 1660 Charles II invited him to England, appointed him King's Physician, and gave him charge of the royal gardens. He remained in this position until 1669, when he was chosen Professor of Botany at Oxford and incorporated Doctor of Physic. In the same year was published his *Hortus Regius Blesensis* (London, 1669). This work, while it improved Morison's reputation as a scholar, aroused criticism of him as a man. The book listed plants he had studied at Blois, some of them new and rare, advanced the rudiments of a "new" system of classifying plants, and demonstrated that Morison was an accurate, diligent worker. But it angered readers by unbecomingly severe indictments of errors alleged against two illustrious writers, the Swiss brothers Jean and Gaspard Bauhin, and Morison's boasted "wonderful system," "known only to myself," based upon characteristics of fruit and flower, was little more than a revival and extension, without credit, of that set forth by Andrea Cesalpino in 1583. Still, Morison helped to focus attention upon

[1] F. W. Oliver, ed., *Makers of British Botany* (Cambridge, 1913), pp. 16–17; A[braham] Wolf, *A History of Science, Technology, and Philosophy in the 16th and 17th Centuries*, with the cooperation of Dr. F. Dannemann and Mr. A. Armitage (new ed. by Douglas McKie, London, 1950), pp. 399–400.

system, his subsequent work at Oxford enlisted new scholars in the study of natural history, and, with the Bobarts (Jacob and his son of the same name), he brought the Oxford Physic Garden to new standards of development. He published the first monograph devoted to a single family of plants (the Umbelliferae, or parsley family), a detailed account, with subdivisions arranged according to the character of the fruit,[2] and he projected a vast history and classification of all the world's known plants under the title *Plantarum Historiæ Universalis Oxoniensis*. But of this he lived only to publish the second part (Oxford, 1680), his life having been brought to an end by a London traffic accident late in 1683. The first part, on trees and shrubs, was never printed. The second part treated herbaceous plants, dealing with only nine of the fifteen classes of his own system, which, however, did not strictly adhere to his proposed principles, namely, to found the system on the characteristics of the seed, or fruit, and he borrowed heavily from Jean Bauhin and others. The third part, or more properly the second volume, was completed in 1699 by Jacob Bobart the Younger (1640–1719), who succeeded his father as Keeper of the Oxford Physic Garden after his father's death in 1679, and later succeeded Morison as Acting Professor of Botany at Oxford. But Bobart had the assistance of several others, notably John Ray, England's greatest naturalist of the time, whom Morison had treated shabbily, Samuel Dale, an apothecary, physician, and disciple of Ray, and William Sherard, who will appear again later as a skilled botanist of the early eighteenth century. This second, published volume of the work undertaken by Morison was so much the product of other hands that it can scarcely be credited to him. Yet Morison had the advantage of powerful patronage. The university had liberally supported his work. Draughtsmen and engravers had been employed to complete illustrations for the work to an extent which Bobart could not have commanded. And had Morison been less vain, less impelled to quarrel with his contemporaries, more inclined to give credit to others from whom he borrowed freely, he might well have won a high reputation. But in spite of his faults he exerted a considerable influence, and suc-

[2] *Plantarum Umbelliferarum Distributio Nova* (Oxford, 1672). The subdivisions, in some cases, were also determined by the form of the leaf.

ceeding scholars, including both Joseph Pitton de Tournefort, the great French botanist, and Linnaeus, the Swedish professor of system in natural history of the eighteenth century, professed their indebtedness to him.[3]

Robert Hooke's *Micrographia* (London, 1665) opened the field of microscopy and was first to point out the cellular structure of plants. Dr. Nehemiah Grew (1641–1712), the English physician trained at Leiden, carried the work further and became the founder of the science of plant anatomy.[4] Beginning his work in 1664, Grew presented to the Royal Society his first study, *The Anatomy of Vegetables Begun* (1670), which was printed by the Society in 1672. This was followed by a succession of papers: "An Idea of a Philosophical History of Plants," "The Anatomy of Roots . . . ," "An Account of the Vegetation of Roots . . . ," "The Anatomy of Trunks," "The Anatomy of Leaves, Flowers, Fruits and Seeds," and others. These were revised, drawn together, and published as *The Anatomy of Plants With an Idea of a Philosophical History of Plants* . . . (London, 1682), with copious and splendid illustrations of the parts of plants as viewed under the microscope. Like the herbalists, Grew began his studies of plants from a medical viewpoint, but his training in animal anatomy led him to apply similar principles to plants. He went beyond Hooke's discoveries to describe the vascular anatomy of plants, sexuality in plants,[5] and aspects of ecology. His work established sound foundations, often confirmed and extended by subsequent scholars. No alert botanist, seeing Grew's book, could be contented with such descriptions

[3] Pulteney, I, 298–312; Oliver, ed., *Makers of British Botany*, pp. 17–28; Clokie, pp. 10–14. Morison left a considerable herbarium at Oxford. See S. H. Vines and G. C. Druce, *An Account of the Morisonian Herbarium . . . of Oxford* (Oxford, 1914). The second volume of his *Plantarum Historiæ Universalis Oxoniensis* (1699) was one of the earliest sources to question the Virginian origin of the potato, but whether this originated with Morison, or with Bobart, or with some of Bobart's assistants is impossible to say. It seems likely that John Ray was the source.

[4] Working contemporaneously, Dr. Marcello Malpighi (1628–94), an Italian physician elected to the Royal Society in 1668, applied the microscope to animals and became a co-founder of microscopic anatomy. His *Anatome Plantarum* was published by the Society in 1675.

[5] Grew regarded the flowers of plants as their sex organs, with the stamen as the male organ, pollen the seed, and pistils the female organs. He thought all plants were hermaphrodites.

of plants as the herbalists had made, "Prosecuted with the bare Eye."

Henry Compton (1632–1713), Bishop of London from 1675 to 1713, became an enthusiastic patron of botany in the latter part of the seventeenth century. He studied at Oxford (1649–52) but left without a degree to travel abroad. Returning to England at the Restoration, he entered Cambridge, was graduated M.A. in 1661, and entered holy orders. He rose in the English church to become Bishop of Oxford (1674) and was translated to London in 1675. His strong anti-Catholic views got him into trouble with James II and he was suspended from all ecclesiastical functions in 1686, although his revenues were untouched. At this point he retired to Fulham Palace and devoted himself to the development of the Fulham Palace gardens, which for a time became one of the most celebrated in England. The Bishop not only took great personal delight in the gardens but also made them freely accessible to others for observation and study. He became personally acquainted with many of the English botanists of his day, and entered into a considerable scientific correspondence with collectors in the American colonies and elsewhere. He employed his own funds to support overseas collectors and, after he was fully reinstated as Bishop by William and Mary, and the Society for the Propagation of the Gospel in Foreign Parts was being organized (late 1690's, formally chartered in 1701), there is some evidence that he directed some of the society's funds to the purpose of botanical collecting in the colonies. He was particularly interested in trees and shrubs, and his gardens became a showplace of exotic varieties. In this he was ably assisted by his gardener, George London (d. 1713), probably the most renowned English gardener of his day. London had been trained by John Rose, Chief Gardener to Charles II, and he had visited several of the gardens on the Continent. He was said to be "perfectly well skill'd in Fruit, which seemed to be his Masterpiece. . . ." When William and Mary appointed him Superintendent of all the royal gardens, Bishop Compton acquired William Milward as his successor. Milward, though less famed than London, continued to maintain the Fulham gardens at a high level. From the late 1680's until Bishop Compton's death in 1713, the Bishop was a generous patron of botanical studies, and his garden at Fulham

Palace became a nursery of exotic trees and shrubs gathered from many parts of the world, a worthy competitor to the Oxford Physic Garden and the Chelsea Physic Garden.[6]

Dr. Martin Lister (1638–1712) was another Englishman who contributed to the development of natural history during the last quarter of the seventeenth century. Educated at Cambridge (B.A., 1658; M.A., 1662), he studied medicine, including anatomy and botany, at Montpellier (1663–66). After his return to England, he practiced medicine at York, moving to London in 1683, where he stayed the remainder of his life.[7] Between the time of his return to England (1666) and the end of the seventeenth century, he took an active part in the study of natural history. He worked closely with John Ray (see below) and the latter's wide circle of friends, and he assisted Ray in the preparation and publication of the latter's work on fishes, birds, plants, and insects.[8] He was elected Fellow of the Royal Society in 1671, and after his removal to London took an energetic part in the Society's affairs. Invited by Henry Oldenburg (and later by Nehemiah Grew) to contribute to the works of the Society, he sent more than eighty articles and letters published in the *Philosophical Transactions*.[9] His scientific correspondence included collectors and scholars in Virginia, Maryland, South Carolina, Barbados, and the Far East, from all of whom he received specimens and observations that appeared in his works. Although

[6] Dandy, pp. 114–115, 157–159; *Petiver*, pp. 293–295; *D.N.B.* After Compton's death, his garden fell into "the management of ignorant persons," many of the exotic trees were removed, and when William Watson reported on it in 1751, it had fallen into neglect. See *Phil. Trans.*, XLVII (1751), 241–247; *Journal-Book*, XXI, 682–685 (June 27, 1757).

[7] He was awarded the Doctor of Medicine degree by diploma at Oxford in 1684.

[8] For Lister's relations with Ray, see Edwin Lankester, ed., *Memorials of John Ray* (London, 1846), and *Correspondence of John Ray* (London, 1848); Robert T. Gunther, ed., *The Further Correspondence of John Ray* (Oxford, 1928); Charles E. Raven, *John Ray, Naturalist, His Life and Works* (Cambridge, 1950); and Martin Lister, *A Journey to Paris in the Year 1698*, ed. Raymond P. Stearns (Urbana, Ill., 1967), Introduction, pp. xxii–xxiii, xxix–xl, xliv.

[9] A full bibliography of Lister's works appears in *ibid.*, pp. 292–308. James Petiver later described a Carolina crab which he had received from Lister, who in all likelihood received it from Madame Hannah Williams, of Goose Creek, South Carolina, one of Lister's correspondents. See *Phil. Trans.*, XXIV, No. 299 (May, 1705), 1952–60.

his interests spanned the whole of natural history, including geology, his early attention centered upon insects (especially spiders and flies), snails, and other mollusks. His first major work, published by the Royal Society in 1678–81, was *Historiæ Animalium Angliæ Tres Tractatus: Unus de Araneis, alter de Cochleis tum terrestribus tum fluviatilibus; tertius de Cochleis Marina*. . . . This was a pioneer study, with figures and systematic descriptions, of spiders, snails, and fossils, heavily drawn upon later by John Ray in his *Methodus Insectorum* (1704) and *Historia Insectorum*, published posthumously in 1710. In 1684 Lister published in the *Philosophical Transactions*[10] "An Ingenious proposal for a new Sort of Maps together with Tables of Sands and Clays." The proposal was for a geological map, a "*Soil* or *Mineral* map," and he emphasized the correlation between different kinds of rocks and their fossil contents, pointing out that "quarries of different stone yield us quite different sorts or species of shells." If one were with "great care . . . very exactly to note upon the *Map* where such and such *Soiles* are bounded . . . I am of the opinion that such upper *Soiles* if natural, infallibly produce such *under Minerals*, and for the most part in such order." He demonstrated an excellent knowledge of living conchylia, and although he was reluctant to identify fossils with living animals, he illustrated living and fossil conchylia side by side in order to show their resemblances. He observed that certain rocks are present over a definite extent of the earth's surface, so that maps might be constructed with regard to the distribution of different kinds of rock. And as the fossil bivalves and snails differed in various kinds of rocks, he set forth for the first time the basic geological principle that different rocks might be distinguished from one another according to their particular fossil contents. Lister's proposal was not acted upon immediately. It was sixty-two years before the Frenchman, Jean Etienne Guettard, presented his famous *Mémoire et Carte Minéraloque* (1746), and 131 years before William ("Strata") Smith constructed the first geological map of England. But Lister had anticipated a basic principle in modern geology.

[10] XIII, No. 164, 739–746.

He had also forecast his greatest works. These appeared piece-meal between 1685 and 1696. A major portion of them were drawn together as *Historia sive Synopsis Methodica Conchyliorum* (2 vols., London, 1685–92) and followed by three lesser volumes on the anatomy of shellfish. These works, containing more than a thousand illustrations from copper plates, including some of specimens from his correspondents in the American colonies, established Lister as the foremost authority of his day on mollusks. He became known as the "Father of British Conchology." Lister's scientific work tapered off about 1700. His health was failing, and after his appointment in 1702 as Second Physician in Ordinary to Queen Anne, he did nothing of scientific significance. His acceptance of experimental philosophy was always tempered by caution and an inborn regard for the learning of antiquity. He would never fully accept John Ray's opinion that "petrified shells" and "Stones figur'd like Plants" were the fossilized remains of once-living specimens. Instead, he tended to cling to the old formed stones tradition, always leaving the matter unresolved, however, pending the discovery of additional evidence.[11] Still, although he was sometimes cautious in the adoption of new scientific hypotheses, he generally accepted the fruits of experimental philosophy to which his own works contributed much in the realm of natural history as a whole. His works in chonchology, in particular, remained the standard of reference for many years.

A London neighbor of Dr. Lister was Leonard Plukenet (1642–1706), though there is no evidence that they were particularly close friends.[12] Plukenet was the son of a Westminster gentleman. He studied at Oxford[13] and may have acquired the degree of Doctor of Medicine at a Continental university, although he does not appear to have practiced the art. He returned to Westminster, where he had a small garden, and devoted himself to botany. He also had a farm in nearby Buckinghamshire, the income from which, together with that of other rental properties, he expended in publishing engravings of plants, a costly undertaking and one for

[11] See Lister, *A Journey to Paris*, Introduction, pp. xxxviii–xxxix, lvi.
[12] They both lived in Old Palace Yard, Westminster.
[13] Joseph Foster, *Alumni Oxonienses* . . . (4 vols., Oxford, 1891), III, 1172.

which he was unable to obtain adequate patronage. In 1689, on the recommendation of Bishop Compton, Queen Mary placed him in charge of the Hampton Court gardens with the title of Regius Professor of Botany. He entered into a wide scientific correspondence and exchange of plant specimens from many parts of the world, including the American colonies, and developed a herbarium of about 8,000 plants. Plukenet's catalogues of his collections were published at different times between 1691 and 1705. The first four parts, constituting the first volume, were published as *Phytographia* . . . (1691–96); the second volume, *Almagestum Botanicum* . . . (1696); the third, *Almagesti Botanici Mantissa* . . . (1700); and the last, *The Amaltheum Botanicum* . . . (1705). The volumes included about 2,750 illustrations of very unequal quality, most of them too small to be of great value. Plant descriptions were often inadequate to establish true generical characters, and varieties were occasionally set forth as real species. The catalogues were alphabetical in arrangement, with no attempt at system. Nonetheless, Plukenet made shrewd observations, with useful criticisms of his predecessors (and sometimes of his contemporaries!), and Pulteney stated that "no work before published by one man, ever exhibited so great a number of new plants."[14] Plukenet recorded the names of many benefactors, including several from the American colonies, by whom he had been enabled to enlarge his collection. John Ray was indebted to him for assistance in the preparation of his *Historia Plantarum Generalis* . . . (3 vols., 1686, 1688, 1704), but, while the two men remained friends, Ray ultimately found Plukenet to be "reserved," "jealous of his reputation," and "ill-natured and liable to mistakes, however confident and self-conceited he may be."[15] In the third volume of his work Plukenet severely censured the early works of two rising young scholars, Hans Sloane (1660–1753) and James Petiver (c. 1663–1718), both of whom will appear at greater length in a later chapter. The works of the three men abound in reciprocal criticisms and abuse—with little credit to any of them. However, as the younger men, especially Sloane, were rising rapidly in the estimation of their fellows, particularly in the

14 Pulteney, II, 27.
15 Lankester, ed., *Correspondence of John Ray*, pp. 307–371.

Royal Society, which Sloane lifted out of its late seventeenth-century decay, the conflict redounded to Plukenet's temporary loss of reputation.[16] Still, Plukenet's works survived these attacks, and in spite of the fact that he worked largely from dried specimens of plants in his vast herbarium, he excelled both of his younger critics as a botanical scholar. He added greatly to the knowledge of new plants, especially from the New World and the Far East, and later botanists held his works in high esteem. Indeed, Linnaeus later called Plukenet's *Phytographia* an "Opus incomparabile."[17]

All of the above scholars were overshadowed by the manifold works of John Ray (1627–1705), England's greatest naturalist before Charles Darwin. Born of humble origin, Ray was educated at Cambridge (B.A., 1647/48; M.A., 1651), where he subsequently held the offices of lecturer, tutor, and other appointments until 1660. He was ordained in 1660 and evidently had intended to make his career in the church. But he was unwilling to subscribe to oaths required by the Bartholomew Act, and in consequence lost his fellowship at Cambridge and all hope of preferment in the church (1662). He was cast, as he said, "upon Providence and good friends." Fortunately, Providence was kind and he had many loyal friends. As early as 1650 Ray had become interested in botany, and his enthusiasm fired many of his friends and students at Cambridge. Together they collected and studied plants, and in 1660 Ray published *Catalogus Plantarum circa Cantabrigiam nascentium*, a small volume which at once demonstrated Ray's familiarity with the works of both English and Continental scholars (especially the Bauhins, Dodoens, Lobel, Jungius, Johnson's revision of Gerard, and Parkinson), his freedom from fanciful, legendary, and superstitious accounts,[18] his capacities for accurate observation, and his fine

---

[16] It may have been a factor in the fact that Plukenet was never elected F.R.S.

[17] Many of Plukenet's papers and much of his herbarium later fell into the hands of Sir Hans Sloane, and through him passed to the British Museum. See Dandy, pp. 183–187. See also Pulteney, II, 18–29; Benjamin Daydon Jackson, "Leonard Plukenet," *Journal of Botany*, XX (1882), 338–342.

[18] Even at this early date Ray rejected the Doctrine of Signatures. As time passed, he also rejected the still widely accepted belief in witchcraft, the doctrine of formed stones, the novity of the earth as set forth by Ussher, the notion of spontaneous generation, and questioned the fixity of species.

Latinity. But this book was only a harbinger of Ray's works for the next forty-odd years.

In 1662, cast out of employment at Cambridge and in the church, Ray and one of his friends and former students, Francis Willughby (1635–72), "finding the History of Nature very imperfect, agreed between themselves . . . to reduce the several tribes of things to a method and to give accurate descriptions of the several species from a strict view of them. And forasmuch as Mr. Willughby's genius lay chiefly to animals, therefore he undertook the birds, beasts, fishes, and insects, as Mr. Ray did the vegetables."[19] Willughby, from Middleton Hall, Warwickshire, was a talented amateur with means. He had already accompanied Ray on tours for observation and collection, and he financed Ray's work for several years, leaving Ray an annuity upon his early death in 1672. Theirs was a partnership of mutual esteem and affection, and when Willughby died leaving his work incomplete, Ray felt a keen obligation to bring his friend's efforts to a conclusion. But Willughby's studies of animals, birds, fishes, and insects had not been far advanced. Moreover, Ray's was the guiding intellect of the partnership, and his completion of Willughby's studies in the animal world bore the marks of Ray's overshadowing genius.

John Ray devoted upwards of forty years to the observation, collection, dissection, and study of natural products. Between 1658 and 1662, and again later, with various friends, he traveled widely over England, Scotland, and Wales. Between 1663 and 1666 he traveled and collected in the Low Countries, Germany, Austria, Italy, Malta, Switzerland, and France—and Willughby visited Spain. Thus, Ray was a field naturalist who early collected first-hand information about the habitat and natural environment of his specimens. Further, he was a thorough student of his predecessors, and because of his travels and generous, friendly disposition, he became widely acquainted and esteemed by his contemporaries both in England and on the Continent. He maintained an enormous scientific correspondence and exchange of information, opinion, and specimens with them, and indeed, when one studies Ray's life and works he cannot but be impressed by the assistance which the

[19] Lankester, ed., *Memorials of John Ray*, p. 33; quoted in Raven, *John Ray*, p. 123.

man had from his friends and contemporaries.[20] It was gladly and freely given, and generously acknowledged; and John Ray, by his widely recognized genius and gently compelling personality, mobilized the forces of almost his entire generation to supply him with specimens, information, opinion, and criticism, out of all of which, together with his own wide experience, observation, experiment, and collection, he fashioned some of the most remarkable works of his age. Personally, Ray had few correspondents in the New World, but by means of the Royal Society, to which he was elected Fellow in 1667, and the correspondence of his friends, including Lister, Bishop Compton, Plukenet, Sloane, Petiver, and others, he kept abreast of New World developments.

Ray's scientific works[21] were marked by their unusual accuracy and the author's extraordinary capacity to master his data and set forth significantly advanced suggestions for methodization. His botanical works went far to prune out the duplicated species erro-

[20] Raven's life of Ray is learned, thorough, and judicious; and it demonstrates well Ray's indebtedness to others in the conduct of his work. For Willughby, see the eulogistic "Memoir of Francis Willughby, Esq., F.R.S.," *The Naturalist's Library*, V (Edinburgh, n.d.), by Sir William Jardine.

[21] His principal scientific works were, in addition to the catalogue of Cambridge plants already mentioned: *Catalogus Plantarum Angliae* (1670; 2nd ed., 1677), followed by *Synopsis Methodica Stirpium Britannicarum* (1690; 2nd ed., 1696), a remarkably "modern" handbook of British flora, though, of course, without Linnaean names; *Ornithologiae Libri Tres . . .* (1676, with an English version in 1678), Ray's revision and extension of Willughby's work on birds; *Methodus Plantarum Nova* (1682, extended and improved in *Methodus Emandata*, 1703); *Historia Piscium Libri quator . . .* (1686), Ray's extension of Willughby's work on fishes; *Historia Plantarum Generalis . . .*, I (1686), II (1688), III (1704), his major work on flora, unfortunately not illustrated because of lack of funds; *Synopsis Methodica Animalium, Quadrupedum, et Serpentini Generis* (1693); *Methodus Insectorum* (1704), followed by *Historia Insectorum*, published posthumously in 1710. The latter two, Ray's loyal attempt to complete Willughby's beginnings on insects, were, in spite of significant additions by Ray, especially in the butterflies and moths, and the efforts of his Braintree friend, Samuel Dale, never wholly completed. In addition to these scientific works, Ray published books in theology, etymology, geology, and philosophy. His *Collection of English Proverbs* (1670; 2nd ed., 1678) and *Collection of English Words* (1673) were pioneer studies of great value; his *Wisdom of God manifested in the Works of Creation* (1691, with enlarged editions in 1692, 1701, and 1704) and *Three Physico-Theological Discourses* (1692; 2nd ed., 1693; 3rd ed., 1713) were widely read, learned interpretations of the new science embracing the new religious approach to science which became known as natural theology and, as has been noted, was common in New England.

neously listed by his predecessors, and his *Methodus Plantarum Nova* introduced a feasible limitation of the term "species."[22] He held that plant classification must consider the way of life, the function as well as the structure, and the growth and behavior of plants. He paid tribute to the works of Cesalpino, the Bauhins, Jungius, and other predecessors, and he showed a close acquaintance with those of his contemporaries, especially those of Grew, Malpighi, and Morison. But his "General Method" was more than a compilation of other sources. It was drawn from additional original observations and dissections of seeds and their embryos, together with analogies between the growth of vegetable seeds and of animal ova. In consequence, though Ray could not include all genera of plants (for many were not yet known or described), his "General Method" provided the foundations upon which a natural system of the classification of plants was ultimately erected, using the fruit, and later the flower, in classifying.[23] Ray's *Historia Plantarum Generalis* attempted, as Bauhin's *Pinax* had attempted, to set forth all the botanical knowledge of his time, arranging plants not in alphabetical order, or according to their supposed "virtues," but in accord with their natural characteristics as outlined in his "General Method." Assisted by his many friends, and embracing new materials supplied from the New World and the Far East, he dealt with about 19,000 plants, arranged into 125 classes including about 6,100 species. It was a masterly work, pregnant with suggestions for his successors. For as his biographer has stated, John Ray was "the most accurate in observation, the most philosophical in contemplation, and the most faithful in description of all botanists of his own or any other time."[24] He gave new meaning to the concept of "species," and by his emphasis upon generic principle he devel-

[22] Ray wrote that "forms which are different in Species always retain their specific natures, and one species does not grow from the seed of another species." However, Ray did not hold as rigidly to the notion of the fixity of species as most of his contemporaries. He added to the above statement, "Although the mark of unity of species is fairly constant, yet it is not invariable and infallible."

[23] Following Cesalpino, Ray divided seed plants into dicotyledons and monocotyledons, which all subsequent botanists have adopted, and his tables of classification set forth several natural orders with accuracy. A full abstract of Ray's system is in Oliver, ed., *Makers of British Botany*, pp. 33–34. See also Raven, *John Ray*, pp. 196–200.

[24] *Ibid.*, p. 335.

oped the basis of every clearly arranged system for the future clas-sification of plants—and, to a lesser extent, of animals as well.

For Ray was a close student of fauna as well as of flora. In bringing to completion the works undertaken by Willughby on birds, fishes, animals, and insects, he contributed significantly to the principles of their classification. His observations of birds were as keen as were those of plants, and it is evident that he and Wil-lughby, both singly and together, had dissected many different kinds of fowl to study their anatomical structure. Previous treatments of birds were few, and as a rule they had arranged birds according to their habitat. Ray introduced new scientific principles which com-bined habitat with anatomical structures and feeding habits. He divided birds into land fowl and waterfowl, each subdivided by anatomical structures and dietary characteristics, i.e., flesh-eating, insect-eating, and seed-eating. This system of classification was subsequently followed by Linnaeus and Buffon and remained the basis of scientific ornithology for the remainder of the Old Colonial Era.[25] Ray's *Historia Piscium* was of less value. Although he had studied the earlier literature on the subject, collected and dissected fishes both in England and on the Continent, and had generous as-sistance from his friends, his principal contribution lay in the defi-nition of a fish, based largely upon anatomical features and, unlike the widely held popular view, rejecting squids, crustaceans, and mollusks from the fish family. His *Synopsis Methodica Animalium, Quadrupedum, et Serpentini Generis* was, however, a pioneering study in the scientific classification of animals. Beginning with the Aristotelian division of animals into those which have red blood (vertebrates) and those which do not (invertebrates), Ray further divided the vertebrates into those which breathe through lungs and those which breathe through gills. The former were later sub-divided into viviparous and oviparous animals (reptiles, birds), and the viviparous animals were further classified according to the na-

[25] *Ibid.*, Chap. 12. See also Elsa G. Allen, "The History of American Ornithol-ogy before Audubon," *Transactions of the American Philosophical Society*, n.s., XLI, Pt. 3 (1951), 421–426. Page 424 summarizes Ray's arrangement of birds. Ray's *Ornithology* included a few American birds, and his *Synopsis Avium et Piscium*, completed in 1694 but not published until 1713 by William Derham, included other American and West Indian fowl, including mention of John Josselyn's American turkey.

ture of their teeth or their toes, or hoofs (Ungulata, with hoofs, subdivided into single, double, and quadruple hoofs; and Unguiculata, with nails, two-toed, five-toed, etc.). Much of Ray's system of zoological classification was later adopted by Linnaeus. Willughby had done much work on insects, and Ray supplemented this from the works of Martin Lister and other friends, adding much of his own in the collection of Lepidoptera. He studied the generation of insects, especially metamorphosis, and used these observations as a basis for classification. But the *Historia Insectorum* was left incomplete, cut short by Ray's death. In some respects it fell short of Jan Swammerdam's *General History of Insects*, originally published in Dutch in 1669, including microscopic studies of insect anatomy which Ray had consulted but, lacking a proper microscope, had not emulated.[26]

Virginia: The Earliest Correspondents

The scientific advances made in England during the last quarter of the seventeenth century were only partially reflected from the New World colonies. In New England, as the previous chapter has shown, this reflection was revealed to some extent in the realm of astronomy but almost none in natural history. New York and Maryland scarcely entered upon the scientific scene in the seventeenth century, but Virginia had responded to a comparatively generous degree by 1700. The Royal Society of London played a prominent role in these developments. It promoted experimental philosophy directly in Virginia as it did elsewhere in the colonies; and its Fellows, individually and collectively, furthered these efforts by their scientific correspondence, supply of books, and exchange of scientific data—all with the enthusiastic approval of the Society, which constantly served as a clearing house for their correspondence and specimens and frequently published their correspondence in the *Philosophical Transactions*.

At its outset in 1660, the Royal Society had no such entrée to Virginia as it possessed in New England through John Winthrop, Jr. One Captain Silas Taylor, probably a shipmaster, contributed an

[26] Besides the works of Raven and Oliver cited above, Pulteney, I, 189–281, gives a valuable account of Ray.

account "Of the way of killing Rattle-snakes" in Virginia,[27] but the first correspondent from Virginia whose letters were reported to the Society was Edward Digges. Digges was probably in London when, on September 29, 1663, he addressed a letter "for ye Society" to Dudley Palmer, F.R.S., who read it before the Royal Society the following day.[28] Digges accompanied his letter with "two sorts of Silk, one Course, the other fine, sent from Virginia and made there;[29] together with some written Observations of the said Silke, Contrary to the received opinion thereof." Digges reported: "The great skaine is a sort that my man easily windeth six pounds a day; The little skaine in the paper, beeing of some seed that I procured from languedock yeilded bottoms as big againe as wee usually had before, soe that my man saved twenty pound of those bottoms to make seed for next year, and onely wound off eight bottoms to that extraordinarily finess you see." He held that mulberry leaves gathered the previous day were as good to feed silkworms as freshly gathered ones; that neither thunder, nor tobacco smells, nor the odor of "women having theire courses" disturbed silkworms; that he was hoping to be able to produce two silk crops a year. These opinions caught the eye of the Royal Society, for this enterprise was highly treasured in Restoration England and the techniques were imperfectly known. Early in 1666 Digges again visited London and presented the Society with further samples of his Virginia silk.[30] Three years afterward, upon occasion of yet another visit to London, Digges "offered his service for philo-

27 *Phil. Trans.*, I, No. 3 (May 8, 1665), 43. The method, said to have been observed by Captain Taylor in 1657, was to use bruised leaves of dittany (or "Wild Penny royal") tied to a cleft stick and held "to the nose of the Rattlesnake," which died "in less than half an hours time." This myth persisted in Virginia for many years. In 1666 the Society received four specimens of plants from Virginia, including dittany, said to kill snakes. The donor was one Mr. Green, who had lived "many years" in Virginia and had visited Carolina. *Journal-Book*, II, 243–244 (March 28, 1666).

28 *Ibid.*, I, 228. The original of the letter is in *Guard-Book*, D-1, No. 1. It was published with some omissions and editorial changes in *Phil. Trans.*, I, No. 2 (April, 1665), 26–27. It was also reprinted in Henry Barham, *An Essay upon the Silk-worm: Containing many improvements upon this Curious Subject* (London, 1719), pp. 98–99.

29 After Fellows of the Society had inspected the silk it was turned over to Robert Hooke for microscopic study.

30 *Journal-Book*, II, 235 (March 14, 1665/66).

sophicall purposes" and was presented by the Royal Society with its "Directions and Inquiries Concerning Virginia" (July 22, 1669).[31] But to these detailed instructions and requests the Society apparently received no response. To be sure, they imposed upon Edward Digges more than a busy Virginia planter ordinarily had the time, the energy, and the skill to accomplish. But they demonstrated what the Royal Society hoped to learn about Virginia: "a good History of the Virginian Plantation, concerning its Beginning, Increase, misfortunes and the present state thereof"; to give "a perfect account of ye Planting and ordering of Tobacco" and its uses in Virginia; to supply "a full account of the progress of the Silkwork"; to experiment in raising rice, coffee, olives, wine grapes, "and the like"; to report on "the Varieties of Earth," minerals, drugs, and medicinal baths; to describe the rivers of Virginia, including those west of the mountains, their sources and whither they flow; to check on the timber resources for ship-building and ship supplies (resin, tar, pitch); to inquire whether Virginia could produce hemp; to acquire "all sorts of Berries, Grasses, Grains and Herbs, growing in Virginia," with instructions about their collection, preservation, and shipping to England; to discover the fauna of Virginia; to foster the collection of data about the tides and the weather; "To make a good Map of Virginia"; and a number of other items, some of them designed to check upon doubtful points reported by previous travelers. Obviously, few colonists, even given the abilities, could spare the time to respond fully to such requests: who, for example, could keep an accurate "register of all changes of wind and Weather at all hours by night and day, shewing ye Point, ye wind blows from . . ."?[32]

[31] These are printed in full in Appendix III below.

[32] Fellows of the Royal Society were much interested during the 1660's in collecting data about tides and the weather in various parts of the world. Dr. John Wallis, of Oxford, was seeking to formulate a theory of tides based upon observations and measurements; and others, urged on by John Beale, sought meteorological information in the hope of improving weather forecasting. Robert Boyle perfected a "portable Barascope" (a kind of primitive barometer) which was supplied to observers in the American colonies. Cf. *Letter-Book*, I, 308, 320, 332, 410, and elsewhere; *Journal-Book*, III, 219 (June 4, 1668, when the Society agreed to distribute Boyle's baroscope); and *Phil. Trans.*, I, No. 16 (Sept., 1665), 263–289, No. 17 (Oct., 1665), 297–300, No. 18 (Nov., 1665), 311–314, No. 21 (Jan., 1665/66), 378–379; III, No. 34 (March, 1668), 652–653, No. 35 (April, 1668),

A second correspondent from Virginia in the 1660's was the Reverend Mr. Alexander Moray from "Ware-river in Mockjack bay." A Scottish minister who evidently ran into difficulties in establishing a church in Virginia, Mr. Moray corresponded with Sir Robert Moray, probably a kinsman, who presented extracts of his letters to the Royal Society in 1666 and 1668.[33] It is not clear how long Mr. Moray had resided in Virginia, although in the summer of 1668, because of heavy losses on his plantation, he shipped his family to England and prepared to return himself in the following spring.[34] He, too, may have been provided with the Society's "Directions and Inquiries Concerning Virginia." Edward Digges was urged to get in touch with him,[35] and his letters contain information coinciding, in part, with that requested by the Society. But he had been disappointed in collecting "rarityes of stones, minerals and mettals," having been promised them "by a gentleman of good esteeme here" who lived "above 100. miles distant . . . upon the freshes at the falls of the mountains, and there is but seldome oc-

656–659, No. 37 (July, 1668), 726–727, No. 41 (Dec., 1668), 813–817. The interest in meteorological data for weather forecasting continued throughout the Old Colonial Era. In 1683 Dr. Martin Lister proposed a "compendious way" of observing and recording weather observations by means of a chart whereby wind directions, barometric pressures, and temperatures would appear side by side, a method deemed "very ingenious and usefull." *Letter-Book*, X, 212–221; Lister, *A Journey to Paris*, Introduction, pp. xlii, xlvi.

33 *Journal-Book*, II, 244 (March 28, 1666). The extract of the first letter, dated Feb. 1, 1665/66, is in *Letter-Book*, I, 241–242; and *Guard-Book*, M-1, No. 36a. It was published in *Phil. Trans.*, I, No. 12 (May 7, 1666), 201–202. A second letter, dated June 12, 1668, is in *Guard-Book*, M-1, No. 37.

34 "My wife will acquaint you of my troubles; and what hopefulness of providing a settlement of a church." Moray was evidently a loyal Scot and a Presbyterian who spoke of "differences betwixt us and the English." The Scots in Virginia suffered many "disappointments in justice," although their trade with the English prospered and many "from so mean a beginning as being sold slaves here, after Hamilton's engagement and Worcester fight are now become great masters of many servants them selfs. . . ." It is not clear whether he had been a victim of Cromwell's severe measures after "Worcester fight." If so, he had been in Virginia for fifteen years or so. He was "very happy in living in this country; being so pleasant, so fertile & so plentifull," and he proposed to return to England "with a designe of recommending to our country men a settlement and plantation to the southward of this: which may be the hopefullest business, that has been aimed at." Letter of June 12, 1668.

35 See Appendix III below.

casion of meeting with him."[36] Moray had experimented with rice-growing and French barley, would be pleased to try coffee if seed were sent by "some of those marchants, yt are of your Society, who keepe a correspondency there." He requested a receipt for making "common white salt, and how they make bay-salt at Rochel in France; for salt is very dear here; and what else you can recommend to me for anything worth the improving here." He would "willingly be at the charges to improve art and virtue." He raised tobacco, and was experimenting with the new sweet-scented variety. But his major interest lay in silk production. He reported that he had planted 10,000 mulberry trees, and was experimenting with their cultivation and with "a new way, for few hands to serve many worms, and yt more cleanly than before." It was about these experiments in particular that Edward Digges was asked by the Royal Society to inquire. There is no evidence that he did, and after 1668 Mr. Moray was heard from no more. Perhaps he returned to England in 1669 as he had planned.

In 1673 the *Philosophical Transactions* published "An Account of the Advantage of Virginia for building ships: communicated by an observing Gentleman."[37] The author remains unidentified. The article was a promotional piece, with enthusiastic affirmatives to the Society's inquiries about ship-building materials (oak for hulls, pine for masts) and ship supplies (resin, pitch, tar, hemp, iron). However, the most valued communication about Virginia in the 1670's was that read before the Society on March 9, 1675/76, having been communicated to the Secretary by John Ray's former pupil and friend, Sir Philip Skippon. The author was Thomas Glover, "an ingenious Chirurgeon of these Parts, who lately came from our Western Plantations," having lived some time in Virginia and nine months in Barbados.[38] How long Glover had resided in Virginia is not recorded. Since he described a windstorm of August, 1667, as if he had witnessed it, he may have been there

---

[36] Possibly William Byrd I.

[37] VIII, No. 93 (April 21, 1673), 6015–16.

[38] William Derham, ed., *Philosophical Letters Between the late Learned Mr. Ray and several of his ingenious Correspondents* . . . (London, 1718), pp. 135–136; *Journal-Book*, V, 148.

for several years.[39] Glover's account referred to several of the same points raised by the Society's "Directions and Inquiries Concerning Virginia," but the information tended to be general and incomplete, and there is no evidence that he had seen the Society's document. He described the geographical position of Virginia, with special praise for its "Commodious" navigable rivers flowing into Chesapeake Bay. The tides flow "as they do in England, only they appear not so large," being diffused in the spacious rivers. These rivers, he said, rise from the Appalachian Mountains and these mountains had been explored to some extent. Some five years before, a German surgeon had obtained a commission from Governor William Berkeley to explore southwestern Virginia. He had gone along the mountains "as far as ye lake of Usherre," had discovered two passes over the mountains, and had brought back an emerald and some Spanish money.[40] The rivers of Virginia abounded in fish, waterfowl, beavers, and other creatures. Glover mentioned fish common to England "and divers others whose names I know not." He told of seeing a sea monster like a merman, and praised the drumfish, in the head of which, he said, "There is a Jelly, which being taken out & dried in ye sun, then beaten to powder and given in broath, procureth speedy delivery to women in labor." In general, however, his account of the fauna of Virginia was inferior to those of his predecessors. He hinted broadly at the existence of precious stones and praised the possibilities for exploiting iron, but Virginians were shy of erecting iron works because of the lack of capital. However, Glover added, "I believe the true reason is, their being so intent on their Tobacco Plantations that they neglect all other more noble & advantageous improvements";

[39] See "An Account of Virginia, its Scituation, Temperature, Productions, Inhabitants and their Manner of Planting and Ordering Tobacco etc. . . . by Mr. Thomas Glover . . . ," *Classified Papers, 1660-1740*, VII (1), No. 18; published in *Phil. Trans.*, XI, No. 126 (June 20, 1676), 623-626; reprinted, Oxford, 1904.

[40] Glover referred to the explorations of John Lederer, who was commissioned by Governor William Berkeley and made three trips into western Virginia (including part of North Carolina) between May and Sept., 1670. The accounts of his travels were published in English by Sir William Talbot of Maryland as *The Discoveries of John Lederer* . . . (London, 1672; reprinted, 1891, 1902). Lederer gave some attention to geography, geology, flora, fauna, and the Indians.

and they found the "greatest encouragement from England by reason of ye vast revenue it brings into ye Exchequer."

Glover drew an excellent, but not uncritical, picture of Virginian soils and agriculture. He described the soils as high, low, and marshy. All of them contained sand, and in general they were as fertile as those of England. The lowlands consisted of a blackish loam over a foot deep that "will hold its strength for seaven or eight Crops successively without manuring. . . . When ye strength of their ground is worn out they never manure it to bring it in heart, but let it lie for pasture for all mens Cattell to grase upon and clear more ground out of their woods to plant in." Many kinds of fruits flourished in Virginia. There were "few plantations but that have fair and large Orchards, some whereof have twelve hundred trees and upward. . . ." Besides "all sorts of English Appells," from which much cider was made, there were peaches, quinces, apricots, plums, pears, cherries "as plentifull as they are in Kent," gooseberries (but no currants), walnuts, chestnuts, and hazelnuts. There were also figs as good "as those do in Spain but there are few planted as yet." But no oranges or lemons grew in Virginia. Mulberry trees had been planted about the houses to feed silk-worms, but "that design fayling, they are now of little use amongst them." Wild vines supplied grapes from which some planters made wine. Kitchen gardens flourished, with all sorts of English "pot herbs and sallads"; various English flowers, including roses, did well. Glover was critical of Virginian animal husbandry. There were great stocks of cattle, "which are greater than ours." Indeed, they might be larger still but for their lack of care, especially in winter, when the colonists gave them only "ye huskes of their Indian Corne, unless it be som of them that have a little wheat straw; neither doe they given them any more of these then will serve to keep alive, by reason whereof they venture into ye Marshy ground and Swamps for food, where very many are lost." Sheep were few because of the hazards from wolves. Horses were as good and as many as in England. Glover failed to mention swine. The highlight of his account of Virginian agriculture, however, was his detailed description of the planting, cultivation, harvest, cure, and packing of tobacco. Few, if any, writers had supplied such a succinct, yet complete, description.

Of the wild herbs that grew in the woods, Glover mentioned a few and confessed that there were many "whose names, nature, virtues, and operations are altogether unknown to us in *Europe;* neither have there been any Physitians in these parts that have made it their business to understand much of them." His approach to natural history, with regard to both flora and fauna, was destitute of all the newer advances being made in the natural sciences in Europe. But it was too early to expect much, and Glover, being a "chirurgeon," had the medicinal approach of the older generation. He drew attention especially to the black snakeroot (*Radix serpentaria nigra*), which, he said, had been so widely used in London during the Great Plague "that the price of it advanced from ten shillings to three pounds sterling a pound." He also mentioned dittany, or pepperwort, which "biteth upon the Tongue" like pepper, and distilled in water "is one of the best things I know to drive worms out of the Body." Two very useful purges were the roots of turpeth and mechoacan, and the fever and ague root were also described.[41] As a whole, however, Glover's treatment of the flora and fauna of Virginia was disappointing. He told nothing that was new. He saw a merman—but reported no lions. He said little about the Indians except to praise their medicine and to point out their diminishing numbers in Virginia.[42] His account contributed more largely toward the study of Virginia's economy than to its natural history.

The First John Clayton

John Clayton (1657–1725) arrived in Virginia in the spring of 1684 and served as minister at Jamestown for two years. Born of a good Lancashire family, he graduated from Oxford (B.A., 1678; M.A., 1682) and took orders in the church. He early demonstrated an interest in science and became well versed in medicine, chemistry, anatomy, zoology, meteorology, geology, and geography. He had an inquiring, experimental bent of mind, remarkably free from credulities.

[41] Wyndham B. Blandon, *Medicine in Virginia in the Seventeenth Century* (Richmond, 1930), pp. 109–111.

[42] He estimated about 3,000 Indians left in Virginia, and said they were more numerous in Maryland.

Before he went to Virginia (1684) he had been in communication with Robert Boyle and had taken part in experiments with Dr. Denis Papin's "New Digester," or primitive pressure cooker, upon which Clayton was said to have made improvements in design. He also had begun experiments relating to the specific gravities of liquids, particularly urine; and on a visit to relatives in Lancashire, he distilled coal to produce a gas similar to that of the natural gas "spring" at Wigan, demonstrated that the artificial gas was flammable, that it was insoluble in water, and that it could be stored, thereby becoming the discoverer of gas lighting more than a century before the process was put to practical use.

Clayton, then, rated as something of a scientist, and though he went to Virginia in a ministerial capacity, he was known among scientific men at Oxford and to some extent at the Royal Society. Certainly he went well prepared, for his equipment included, besides scientific books, various glasses and chemical instruments, tools for grinding glasses, microscopes, a barometer, and a thermometer. Probably no one previously had gone to the New World with so much scientific gear. Unfortunately, however, he shipped his equipment in a separate vessel which was wrecked and he lost it all. This accident disheartened him, for he could not replace his instruments in Virginia, and he was reduced to making what observations he could without them. Moreover, the Royal Society was in decline, its records were in temporary disarray, the events of the "Glorious Revolution" interrupted some of its meetings, and its correspondence and publications (including the *Philosophical Transactions*) were temporarily curtailed.

Clayton returned to England (May, 1686) on the eve of these events. Soon he was in London, where he met Boyle in person and became acquainted with Nehemiah Grew, Edmond Halley, and various London apothecaries, some of whom became his warm friends. During the winter of 1687/88, he joined forces with Dr. Allen Moulin to make a series of anatomical studies of birds to seek a correlation between birds' feeding habits and the nerve supply to their beaks. Dr. Moulin presented the results of their study to the Royal Society (February 1, 1687/88), and Clayton attended meetings of the Society as a visitor. He was elected Fellow on November 30, 1688. In the meantime, his friends in the Royal

Society urged him to report his observations in Virginia. Clayton demurred at first because, as he wrote, having lost his books and instruments,

> I was discouraged from making so diligent a scrutiny as otherwise I might have done: so that I took very few Minutes down in Writing. And therefore since I have only my Memory to rely on, which, too has the disadvantage of its own Weakness, and the distance of two years since now I left the Country; if future Relations shall in some small Points make out my Mistake, I thought this requisite to justify my Candor: for I ever judged it villanous to impose in matters of Fact. But descriptions of things that depend on memory may be liable to mistakes, and yet the Sincerity of the person that delivers them entire. But thereof I shall be as cautious as possible, and shall rather wave [waive] some things whereof I have some doubts, & am incapable now of satisfying my Self, than in any sort presume too far.[43]

Clayton's hesitation did him credit. But he succumbed to the importunities of the Society, and in 1688 began a series of communications expounding his observations in Virginia. These were interrupted by the "Glorious Revolution" and Clayton stopped sending them until, upon the application by Richard Waller, Secretary of the Royal Society, in 1693, they were resumed and brought to a conclusion.[44]

Clayton had been well received in Virginia.[45] He was critical of

[43] Clayton to the Royal Society, Wakefield, May 12, 1688, *Letter-Book, Supplement*, II, 463–475; *Phil. Trans.*, XVII, No. 201 (June, 1693), 781–789; Peter Force, ed., *Tracts and Other Papers, Relating Principally to the Colonies in North America* (4 vols., Washington, 1836–48; reprinted, 1947), III, 4. Force's *Tracts* was the best printed source for Clayton's letters until Edmund and Dorothy Smith Berkeley edited them with a few additions, excellent annotations, and a biographical sketch of Clayton under the title *The Reverend John Clayton: A Parson with a Scientific Mind* . . . (Charlottesville, 1965, cited hereafter as Berkeley and Berkeley).

[44] Waller's letter, dated Oct. 13, 1693, is in *Letter-Book*, XI, Pt. I, 163; Clayton's reply is in *Letter-Book, Supplement*, II, 498–511, and in *Guard-Book*, C-2, No. 22. He resumed his series with the account of birds in Virginia.

[45] His first letter after his arrival stated that "tis now our great Assembly & on Sunday by a peculiar order from ye Governor & Councell I am to preach so yt something peculiar is expected & I must mind my hits to preserve yt blooming repute I have got. I have had ye happinesse to be cried up farr beyond my deserts. Ye people are peculiarly obliging. . . ." Clayton to Nehemiah Grew, James City,

many of the people and customs, but he evidently made friends widely and spoke especially well of William Byrd I, "who's one of the most intelligent Gentlemen in all *Virginia*, and knows more of *Indian* Affairs than any man in the Country,"[46] and of John Banister, a naturalist in Virginia who will appear again below. Evidently he did not travel widely over Virginia, but he knew the most informed persons who had, and he brought to Virginia the most up-to-date and scientifically informed mind, with the exception of John Banister, that had looked upon the Virginian scene. He made some notes of his impressions and collected some specimens, although it appears likely that very few of the latter were taken to England. With both Clayton and Banister in Virginia in the 1680's, the Fellows of the Royal Society had two of the most able observers that they had had in the colonial field.

Some years after his return to England, Clayton was appointed (1708) Dean of the Protestant Cathedral of Kildare, Ireland, where he stayed until his death in 1725. Some of his unpublished papers were subsequently transmitted to the Royal Society. One of these

---

Va., April 24, 1684, *Sloane 1008*, fol. 335; published in *Wm. & Mary Qty.*, 2nd Ser., I (1921), 114–115. Other materials about Clayton are in *ibid.*, 1st Ser., XXIV (1916), 218–219; XXV (1917), 209; 2nd Ser., XVII (1937), 499; XIX (1939), 1–7; Conway Zirkle, "John Clayton and Our Colonial Botany," *Va. Mag. Hist. & Biog.*, LXVII (1959), 284–294; Joseph Foster, *Alumni Oxonienses . . . 1500–1714* (Oxford, 1895), p. 287; and a bibliography of Clayton's works in Earl G. Swem, *A Bibliography of Virginia*, Pt. I (Richmond, 1916), p. 132.

46 Force, ed., *Tracts*, III, 20. The elder Byrd was a patron of Banister and himself had connections with the Royal Society, to which he later sent Virginia specimens. See "Capt. Byrd's Letters Continued," *The Virginia Historical Register and Literary Companion* (6 vols., Richmond, 1848–53), I, 114–119; "Letters of William Byrd First," *Va. Mag. Hist. & Biog.*, XXV (1917), 128ff.; XXVI (1918), 247ff. In Clayton's early impressions of Virginia, he stated that it was "a place where plenty makes poverty. Ignorance, ingenuity, & covetousnesse causes hospitality, yt is thus every one covets so much & there is such vast extent of land yt they spread so far they cannot manage well a hundred pt. of wt they have. Every one can live at ease, & therefore they scorne & hate to work to Advantage ymselves so are poor wth abundance. They have few Schollars so yt every one studys to be halfe Physitian halfe Lawyer & with a naturall acutenesse [that] would amuse yee, for want of bookes, They read men ye more. Then for ye third thing Ordinarys, i.e. oʳ Inns, are extreeme expensive wherefore wth common impudence they'll go to a man's house for diet & lodgings tho they have no acquaintance at all rather yn be at ye expence to lie at an Inn & being grown into rank custom it makes ym seem liberal. . . ." Letter to Grew quoted above.

fell into the "Papers Relating to the Royal Society" collected by the Secretary, Thomas Birch, who published *The History of the Royal Society of London*. Entitled "A Letter from the Rev^d Mr. John Clayton afterwards Dean of Kildare in Ireland to Dr. [Nehemiah] Grew in answer to several Quaerys sent to him by that learned Gentleman. A.D. 1687," it failed to reach the Royal Society in time to be included with Clayton's other letters read before the Society and published in the *Philosophical Transactions* in the early 1690's.[47] Chronologically speaking, however, it is the first of Clayton's series of letters about Virginia that has survived. It consisted of replies to twenty-five queries[48] submitted by Grew relating to the organization and practice of medicine among the Indians. Clayton did not reveal the sources of his information, but from his other works it may be concluded that he obtained most of it from the nearby Pamunkeys. Obviously, he had difficulty in prying information from them for, as he said, "they are a sullen close people & will answer very few questions." The Wiochist, who was both priest and physician, was held in high honor next to the War Chief. His services were free to members of the village, although they accepted fees when treating an Englishman, usually in articles of clothing, rum, and the like. He was his own apothecary, depending for his remedies largely upon herbs, roots, and the leaves, buds, and bark of various trees and shrubs, together with turpentine and the oils of various fish, reptiles, and animals. His treatments were empirical, often based on trial and error, with no scientific knowledge of anatomy or physiology. He gave no attention to the pulse or urine of his patients, and he never bled them. He employed purges, emetics, blistering plasters, and sweating stones. Clayton indicated a high regard for the medicine man's knowledge of *materia medica* as gathered in the Virginia wilderness, especially for his capacities to cure wounds and snake-bite. He described their methods and their medicines in some detail. He believed that the Indian's life span was about the same as that of the English; if they

[47] This letter, a copy of the original, is in *Add. Mss. 4437*, fols. 85–97. It has been published by Bernard G. Hoffman in "John Clayton's Account of the Medicinal Practices of the Virginia Indians," *Ethnohistory*, XI, No. 1 (1964). Cf. also Berkeley and Berkeley, pp. 21 ff.

[48] They are numbered to twenty-four, with two numbered 13.

survived to the age of thirty-three, they usually reached an old age. Their common complaints were intermittent fevers, sore throats, dry "griping of the guts," and pains in the legs, often dropsical in nature. The Wiochist passed on his medical arts by a system of apprenticeship, usually, if not always, only to his own children. Clayton's account of Indian medical practices was a rare item in its day, and it still has value for the student of Indian culture.

Between May 12, 1688, and May 26, 1694, Clayton sent six communications to the Royal Society, all of which were read before the Society and five of which were published in the *Philosophical Transactions*.[49] In his first letter he proposed an outline of the entire series, which was "to give an Account of the Air, and all such Observations as refer thereto; then of the Water, the Earth and Soil; the Birds, the Beasts, the Fishes, the Plants, the Insects; and lastly, the present State of the Inhabitants."[50] Actually, he said very little about the fishes, plants, and insects, and "the present State of the Inhabitants" was scattered through various letters rather than concentrated in a single section. In discussing the air of Virginia he lamented the loss of his barometers and thermometers, for the temperatures were much affected by the winds, "both as to heat and cold, driness and moisture, whose Variations

[49] Of those published in the *Philosophical Transactions*, four of the originals are in the Royal Society Archives, and four are noted as having been read to the Society: 1. Dated May 12, 1688, *Letter-Book, Supplement*, II, 463–475; read before the Society May 16, 1688, *Journal-Book*, VIII, 207; published in *Phil. Trans.*, XVII, No. 201 (June, 1693), 781–789. 2. Original not found; read before the Society June 13, 1688, *Journal-Book*, VIII, 212–213; published in *Phil. Trans.*, XVIII, No. 201 (June, 1693), 790–795, 941–948. 3. N.d., *Letter-Book, Supplement*, II, 475–483; no record found of its being read before the Society, but the *Journal-Book* is poor during the "Glorious Revolution"; published in *Phil. Trans.*, XVII, No. 205 (Nov., 1693), 941–948. 4. Dated Nov. 24, 1693, *Letter-Book, Supplement*, II, 498–511; read before the Society Jan. 3, 1693/94, *Journal-Book*, IX, 148; published in *Phil. Trans.*, XVII, No. 206 (Dec., 1693), 978–999. 5. Dated May 22, 1694, *Guard-Book*, C-2, No. 23, and *Letter-Book, Supplement*, II, 511–538; read before the Society June 6, 1694, *Journal-Book*, IX, 162; published in *Phil. Trans.*, XVIII, No. 210 (May, 1694), 121–135. All were published by Edmond Halley in *Miscellanea curiosa* (3 vols., London, 1705–07), III; and reprinted in Force, ed., *Tracts*, III, No. 12, 1–45, and in Berkeley and Berkeley, pp. 21–127. Subsequent references will be to Force's reprint unless otherwise indicated. Garbled portions of Clayton's letters are among the *Boyle Papers* (Misc.), XLIV (Royal Society Archives).

[50] Force, p. 3; Berkeley and Berkeley, p. 41.

being very notable . . . there being often great and suddain Changes." North and northwest winds were "very nitrous and piercing, cold and clear, or else stormy." South and southeast winds were "hazy and sultry hot." The winter was fine, clear, dry, and pleasant, with short, sharp frosts which sometimes froze over the rivers. Snowfall rarely lasted more than a day or two. Spring was earlier than in England, with frequent rains in April. May and June were pleasant, but in July and August the air became stagnant with oppressive heat. In September the weather usually broke suddenly, with heavy rains. This was the season of sickness, and Clayton was of the opinion that the physicians of Virginia were a very poor lot, ill-prepared to practice their art.[51] He believed that changes in weather had great influence upon health, and that thunder was often attended by fatal circumstances (speaking "confusedly of Thunder and Lightning"). He described severe damage caused by lightning in Virginia and compared it with sulphurous spirits such as he had drawn from coals, stored in bladders, and found flammable even after having been passed through water— a very early experiment with coal gas even if it had no valid connection with lightning. He ended his treatment of air in Virginia with a description of "frequent little sorts of Whirl-winds" that he had observed.

Clayton's second letter treated "Of the Water" in Virginia.[52] He began with a description of the location of the colony between Cape Henry and Cape Charles, Chesapeake Bay, and the four great rivers which empty into the bay and "plentifully water all the other Parts of *Virginia*"—the James, the York, the Rappahannock, and the Potomac. These interlocking water systems were, in Clayton's opinion, "the greatest Impediment to the Advance of the Country, as it is the greatest Obstacle to Trade and Commerce. For the great Number of Rivers, and the Thinness of the Inhabitants, distract and disperse a Trade." Ships loaded up and down the rivers, and "the best of Trade that can be driven is only a Sort

[51] On another occasion he wrote that the physicians did not understand medicine, for there were "none except such as have been Surgeons, or Apothecaries boys in England (& indeed untill there be mony here as in other places, it can hardly be expected, yt any able man will come hither." *Boyle Papers* (Misc.), XXXIX, No. 3. See also Blandon, *Medicine in Virginia*, pp. 78ff.

[52] Force, p. 11.

of *Scotch* Peddling." The number of rivers was "one of the chief Reasons why they have no Towns," which Clayton deplored, feeling that it disposed the colonists to look to their private interests with too little concern for the public welfare. He observed that the tide regularly ebbed and flowed "about two Foot perpendicular at *James Town*," and stated that "as far as the salt Waters reach the Country is deemed less healthy"—a result, Clayton speculated, of the air becoming impregnated with marine salt. Virginia fresh water required more malt to make strong beer than that of England, and "will not bear Soap." Clayton had tested Virginia water from various wells and springs and found them all very much alike, but he was unable to test their specific gravity for lack of equipment. Most of them had petrifying qualities, and "nothing is more common than petrify'd Shells, unless you would determine that they are Parts of natural Rock shot in those Figures, which indeed I rather think." Both at this point and later on Clayton identified himself with the conservative view regarding the origins of fossils.

Following his second letter, Clayton sent to the Royal Society a communication which was not published in the *Philosophical Transactions.* Entitled "The Journal and Relation of a New Discovery made behind the Appalachian Mountains to the West of Virginia," it was read before the Society on August 1, 1688.[53] The journal was a copy of that of Thomas Batt, Thomas Woods, and Robert Fallam, who had been commissioned by Major General Abraham Wood "for the finding of the ebbing & flowing of the Waters on the other side of the Mountains, in order to the discovery of the South Sea." Accompanied by Perecute, "a great Man of the Apomatock Indians," and Jack Neason, a former servant of Major General Wood, they had set out from "Apomatock Town" on September 1, 1671, and returned a month later to Fort Henry. They reached streams west of the mountains flowing west-north-west "to empty themselves into the great River," and took formal possession of the region for Charles II. The journal excited much

[53] *Journal-Book*, VIII, 223. Copies of the journal are in *Classified Papers, 1660–1740*, VII (1), No. 43; *Boyle Papers* (Misc.), XLIV; and *Add. Mss. 4432*, fols. 27–32. It has been published in Clarence W. Alvord and Lee Bidgood, eds., *The First Explorations of the Trans-Allegheny Region by the Virginians* (Cleveland, 1912), pp. 184–193, and in Berkeley and Berkeley, pp. 68ff.

speculation about the geography of the interior of North America, and it was believed that it might be "of some use to rectifie the Mapps of America." Some held that the streams must empty into the South Sea (the Pacific Ocean), "not thinking it rational ye Continent should be so broad opposite to Virginia being so narrow at Panama."[54] The explorers reported that the streams west of the mountains ebbed and flowed with the tide, but John Clayton, after discussing the matter with William Byrd, refused to accept this and concluded that the streams emptied into "Laque petit where the greatest number of Beavours is caught" and where the French had recently established themselves.[55] Obviously, the Fellows of the Royal Society had much to learn about the geography of the interior of North America.

Clayton's third letter began a discussion of the earth and soil which was continued in the fourth letter. He commented upon the vast quantities of oyster shells in the earth, sometimes "many Yards perpendicular" and "being petrified seem to make a Vein of a Rock." Of those that "are not so much petrified" the Virginians burned their lime. Many persons thought that these had been oyster beds and "that all that Tract of Land, now high Ground, was once overflowed by the Sea." As to whether they had once been oysters or "truly Stones, *sui Generis*," Clayton left "to the honourable Society to determin."[56] But his own opinion was that they were formed stones grown in the earth. However, he admitted that among them were sometimes found "perfect Teeth petrified," probably from fishes; that a backbone of a whale, and some of its

[54] *Boyle Papers* (Misc.), XLIV, unpaged.

[55] *Journal-Book*, VIII, 225 (Oct. 24, 1688); *Guard-Book*, C-2, No. 21; Force, p. 20. Byrd assured Clayton that the streams west of the mountains were not tidal rivers, and that they must run into a fresh-water lake. It is ironic that the English referred to the Great Lakes as "Laque Petit." Dr. Daniel Cox, using sources from the Hudson's Bay Company voyages, had reported to the Society (Jan. 12, 1686/87) that the English had surveyed the Great Lakes and found them to be "a great Mediterranean sea of above 5000 miles round." He felt that it was "a great probability" that the Susquehanna and perhaps the Delaware rivers had their source in this lake. Cox spoke of the "great Lakes" as one lake (*Journal-Book*, VIII, 121), but Clayton recognized that there were "several large Lakes" and stated Byrd believed these lakes had no communication with one another and that no river had its source in Lake Petit.

[56] Force, p. 14.

ribs, had been found; and that John Banister had showed him "the Joynt of a Whale's Back-bone, and several Teeth . . . found in Hills beyond the Falls of *James River*, at least, a hundred and fifty Miles up into the Country." The soil was generally sandy, and Clayton speculated about the difference in soils as they "seem appropriated to the several Sorts of Tobacco." Again he deplored the loss of his microscopes, for he felt that with microscopic examination of tobaccos, together with chemical experiments of soils, he could help to improve tobacco culture.[57] He made experiments of his own in tobacco culture and offered suggestions for its improvement, but he made little impression upon Virginian tobacco growers, who were wedded to their own methods. In general, he was highly critical of Virginia agriculture, for, although he believed the soil of Virginia was more fertile than that of England, the planters made poor use of their soil resources.[58] They only cultivated the highlands, whereas the richest soil was in the boggy marshes and swamps with "a very deep Soil, that would endure planting twenty or thirty Years, and some would scarce ever be worn out." Clayton argued that if the planters would learn to drain the marshes, they could raise better tobacco and other crops, obviate the practice of clearing new ground, reduce the need for such large plantations, tighten their society, and overcome living such "solitary and unsociable lives" with "Trading confused and dispersed; besides other Inconveniences." Farms of 200 or 300 acres would be adequate and "would make the Country much more healthy," especially if planters would learn to fertilize their lands with manure and marl and produce more fodder such as "*Saintfoin*" (sainfoin) for their livestock.[59] Clayton even oversaw

---

[57] *Ibid.*, pp. 14–18. Clayton believed tobacco "to be a Plant abounding with nitro-sulphureous Particles," that the "Goodness of Tobacco . . . consists in the Volatility of its Nitre: And hence the sandy Grounds that are most impregnated therewith, and whose nitrous Salt is most volatile" would produce the best tobacco. "Nitre" refers to potassium nitrate (saltpeter) or to sodium nitrate (Chile saltpeter).

[58] This portion of Clayton's comments was in his continuation of his account of the soil in the fourth letter. Force, pp. 20ff. Cf. Zirkle, "John Clayton and Our Colonial Botany," who takes issue with Clayton's view.

[59] Clayton tried to persuade planters to provide winter shelters for their livestock and keep milk cows in winter, using their empty "Tobacco Houses" for the purpose.

the drainage of "a good large Swamp," but he was unable to convince the generality of Virginians of the merits of his plans. They said scornfully that he "understood better how to make a Sermon, than managing Tobacco." He also described and mapped the location of Jamestown,[60] with severe criticisms of its "silly sort of Fort" and suggestions for its better security against attack. He found clays in Virginia suitable for pottery-making; great quantities of iron ore (although capital was not available for its exploitation, "and for Persons in *England* to meddle therewith, is certainly to be cheated at such a Distance"); black lead which had been brought to William Byrd by the Indians; chalk which made "a delicate Whitewash"; and other earths alluded to vaguely.

At this point in his fourth letter, Clayton launched upon a treatment "Of the Birds." By virtue of his anatomical work with Dr. Moulin it was a subject for which he was well qualified, and he was familiar with the studies of Nehemiah Grew and with those of Willughby and Ray as published in the latter's *Ornithology*. Indeed, his account of the birds of Virginia was by far the best that had been produced. He had made a collection of birds while in Virginia "but falling sick of the Griping of the Guts" had given it up in despair.[61] In all, he mentioned forty-five birds, which Mrs. Allen has called "an outstanding list" and "though informal and interrupted in many places, is good, showing that he observed critically with close attention to detail."[62] Some of his anatomical studies were difficult, and although they have not borne out some of the facts as they are known today, they were indicative of a scientific point of view. As of his day, Clayton was the best bird observer who had reached the American colonies.

Clayton's last letter referred to the "Beasts of Virginia."[63] He averred that "There were neither Horses, Bulls, Cows, Sheep, or

---

60 The map is in *Guard-Book*, C-2, No. 21 (Aug. 17, 1688). Reproduced in Berkeley and Berkeley, opp. p. 84.

61 Force, pp. 27ff. On Clayton's birds see W. L. McAtee, "North American Bird Records in the 'Philosophical Transactions' 1665–1800," *Journal of the Society for the Bibliography of Natural History*, III, Pt. I (1955), 46–60.

62 Allen, "History of American Ornithology," p. 460. It may have been Clayton who supplied the data for "The Description of the American . . . Humming Bird . . ." that appeared in *Phil. Trans.*, XVII, No. 200 (May, 1693), 760–761.

63 Force, pp. 35ff.

Swine, in all the Country, before the coming of the *English*," and began his discussion by reference to domestic livestock. There was a "good Store of Horses," though the colonists were "very negligent and careless about the Breed," seldom stabled them nor kept them shod. Many cattle had been allowed to go wild and were difficult to shoot, "having a great Acuteness of Smelling." Sheep were of "a middling Size," well fleeced, and planters were beginning to keep flocks now that the danger from wolves had abated. Swine were in great abundance, and Virginia hams were highly praised, "as good as any *Westphalia*, certainly far exceeding our *English*." Of wild animals, Clayton mentioned about eighteen different sorts of beasts, vipers, and fish.[64] He said he saw no lions, although there was a cat-like creature like a "Pard [leopard] or Tyger" which he had not himself seen. He described the opossum, "the Skin of its Belly is very large, and folded so as to meet like a Purse, wherein they secure their Young whilst little and tender," but he speculated no further upon its nature as a marsupial, as Englishmen did for several years. His account of the flying squirrel was accurate: its "flight" was "rather skipping than flying." He dissected a "Musk-rat" to investigate the source of its scent and tore apart one of their lodgings to examine its interior arrangement. His description of a bullfrog was less exaggerated than that of John Josselyn, although he found it to be "of a prodigious largeness." He refused to believe John Banister when the latter asserted that the noise heard in the forests, "much like that which our shrew-Mouse makes, but much sharper," was produced by "a sort of *Scarabœus* Beetle [cicada] . . . as big as the humming-Bird," but he offered no other explanation for it. He mentioned seven sorts of snakes, with special attention to rattlesnakes. He believed that poisonous vipers "kill so very speedily by injecting the Poison" through fistulous teeth in contrast to mad dogs, whose poison "sticks only to the outsides of the Teeth." Probably he was aware that Dr. Edward Tyson had dissected a rattlesnake at the Royal Society early in 1683 and had found that its fangs were hollow and

---

[64] The fish were Portuguese men-of-war, mentioned in his first letter describing his voyage to Virginia. Force, pp. 4–5. He considered them "to be a sort of Sea-Plant," which seamen called "Carvels." He dissected a turtle and found that it fed upon the "Carvels." See Berkeley and Berkeley, pp. 42–43.

connected with the poison sacs.[65] He expressed justifiable doubts about the value of dittany in the cure of snake-bite and evidently believed, as he had said in his letter to Grew about Indian medicine, that the Indian cure—applying a tourniquet, sucking out the poison, and cauterizing the wound—was the best treatment.[66]

John Clayton had little communication with the Royal Society after his last letter had been sent on May 22, 1694.[67] Evidently he devoted himself to his professional career in the church. His letters about Virginia provoked considerable discussion in the Society at the time they were read, but their publication in the *Philosophical Transactions* did not reach a wide audience outside the Society, and his failure to send or to bring back specimens of the natural products of Virginia left his scientific friends with nothing on which to experiment and make further observations. In consequence, his contributions made a less lasting impression than they deserved, and they were soon lost from the sight of experimental philosophers.

John Banister

Moreover, John Clayton's contributions were overshadowed at the time by the collections, observations, and correspondence of John Banister (1650–92). Born of humble parentage at Twigworth, Gloucestershire, Banister was educated at Magdalen College, Oxford (B.A., 1671; M.A., 1674), where he remained as clerk (1674–76) and chaplain (1676–78).[68] Here he became acquainted with Robert Morison, Robert Plot, and Henry Compton, who became Bishop of Oxford in 1674 before being translated to London in the

---

65 *Phil. Trans.*, XII, No. 144 (Feb. 10, 1682/83), 25–47. Clayton stated that the specimen had been sent by one Colonel Claybourn (William Claiborne) from Virginia to Henry Loades, a London merchant, who presented it, still alive, to the Society. See also *Journal-Book*, VII, 109, 116, 119, where it was related that the snake had come from the West Indies.

66 Robert Hooke had reported to the Royal Society (July 25, 1678) that he had received from a merchant who had lived in Virginia a substance called "punk" found in "the wilde Walnut tree." It caught and held fire easily and was often carried by planters to cauterize snake-bites. Hooke concluded that "punk" was a fungus or mushroom. *Journal-Book*, VI, 121–123.

67 He wrote to Secretary Waller, June 20, 1694, to send a fossil from Yorkshire and to inquire how to refine sulphur. *Letter-Book, Supplement*, II, 539–540.

68 *D.A.B.*; Foster, *Alumni Oxoniensis*, I, 66.

following year. It was probably under the influence of these men that Banister became devoted to natural history. In any case, he became familiar with their works and widely *au courant* with the works of other students in the field. He prepared the *Herbarium siccum Jo. Banister*, a collection of 374 folios of pressed plant specimens gathered in the Oxford Physic Garden and from the fields about Oxford.[69] Robert Plot appropriated it in his *Natural History of Oxfordshire* (1676). In 1678 Banister sailed to Virginia via Barbados and other West Indian isles and prepared an initial "Catalogus Stirpium Rariorum in Virginia sponte nascentium. Inchoatus. Do^{mi} MDCLXXIIX."[70] Obviously, Banister was supported in Virginia by English patrons who subscribed funds for that purpose, but the details, especially of Banister's early years in Virginia, are blurred. The presumption that he went as a missionary for the Church of England has no direct evidence in its support, although it cannot be ruled out completely in view of Bishop Compton's enthusiasm for botany, Banister's efforts in his behalf in Virginia, and the Bishop's later actions in support of Hugh Jones, one of Banister's successors in Virginia.[71] By 1684 the evidence is clear that Banister was being supported in his Virginian researches by the contributions of a number of English patrons. William Byrd I served as treasurer and bookkeeper for the group, and Banister long made his headquarters with Byrd, first at James River Falls and later at Westover.[72] Byrd himself was in communication with English naturalists and Fellows of the Royal Society. He received bulbs, seeds, shrubs, and trees from Jacob Bobart the Younger, and

[69] Dandy, pp. 50 (H.S. 168), 86.

[70] *Sloane 4002*, fols. 7–29. Banister struggled to acquire a botanical vocabulary in Latin, sometimes gave English and Indian names to trees and plants, and revealed close knowledge of earlier descriptions of American flora by both English and Continental scholars, whom he corrected in some instances. A similar catalogue dated 1679 is in *Sloane 2346*, fols. 101v–107v.

[71] *D.A.B.;* see Bishop Compton's part in the appointment of Hugh Jones in 1695, in *Petiver*, pp. 293–295. Jones's association with a particular parish appears to have been more nominal than real, and Banister, though he apparently held no church at the outset, ultimately held some connection, possibly as pastor, with Appomattox (later Bristol) Parish. *Wm. & Mary Qty.*, 1st Ser., VII (1898), 162.

[72] Byrd to Jacob Bobart, May 20, 1684, in "Capt. Byrd's Letters Continued," p. 114. Subsequent letters by and to Byrd further demonstrate his role. See "Letters of William Byrd First," *loc. cit.*

books from various sources; and he sent specimens of Virginia minerals, flora, and fauna to his English correspondents.[73] In 1687 Byrd was in London, and shortly before his return to Virginia Leonard Plukenet asked him to meet with himself, Dr. Martin Lister, Tancred Robinson, George London (Bishop Compton's gardener), William Charleton (or Courten), a collector and generous friend of natural history, and other "Worthy Gentlemen" at the Rainbow Coffee House near Temple Bar.[74] The purpose of the meeting was to consider the "Encouragemt. of yt worthy and Reverend Gentleman Mr. Banister," and to ask Byrd to urge Banister not to be discouraged by the apparent lack of financial support he had received. Byrd was asked also to "Addresse in my Lord of London's [i.e., Bishop Compton's] Name to several persons of Honour of whom my Lord has had a promise of Contribution. . . ." In 1690 Byrd appealed to Mr. Arthur North, a London merchant who served as Byrd's agent: "I . . . pray lett me know wt money you have recd & given mee Credit for, on Mr. Banisters Acco^t that hee & I may bee able to Reckon."[75] From Plukenet's letter mentioned above and from other manuscript sources bearing upon Banister's career, it appears that Banister's English patrons included Bishop Compton, Dr. Martin Lister, Leonard Plukenet, Robert Morison, Robert Plot, Jacob Bobart the Younger, Tancred Robinson, George London, Dr. Hans Sloane, James Petiver, and the latter's close friend, Samuel Doody, Keeper of the Chelsea Physic Garden (1693–1706) and a specialist on the study of English mosses. Edward Lhwyd, who succeeded Robert Plot

[73] Byrd sent many of his specimens, especially the minerals, for assay purposes. See his letters cited above and Richard M. Jellison, "Scientific Enquiry in Eighteenth-Century Virginia," *The Historian*, XXV (1963), 292–293; Maude H. Woodfin, "William Byrd and the Royal Society," *Va. Mag. Hist. & Biog.*, XL (1932), 26ff.

[74] Plukenet to Byrd, n.d., *Sloane 4067*, fol. 105. Byrd's acceptance, dated Dec. 14, 1687, is in *Sloane 4062*, fol. 226. The former letter has been published in *Wm. & Mary Qty.*, 2nd Ser., IV (1924), 224–226. Another letter by Plukenet addressed to an unnamed "Worthy Sir" (n.p., n.d., *Sloane 4067*, fol. 99v) illustrates the solicitations on Banister's behalf: "I have Inclosed one of the Proposals that relates to a promotion of a Natural History of Virginia if it lies in your way to Add any thing to Mr. Banister's Encouragemt. You will oblige all partys concern'd especially your humble servant L.P."

[75] Byrd wrote on Aug. 8, 1690, and again on Oct. 25, 1690, to the same effect. *Va. Mag. Hist. & Biog.*, XXVI (1918), 389; XXVII (1919), 274.

(1690) as Keeper of the Ashmolean Museum at Oxford and was a close friend of Dr. Lister and John Ray, may have contributed, together with other "Worthy Gentlemen" such as Mr. North and Daniel Horsmanden, named in Byrd's letters.[76] Whether John Ray knew Banister before he went to Virginia is not clear, but in the early 1680's he entered into active correspondence with the Virginia naturalist. In view of Ray's slender resources, however, it is unlikely that he gave material support to Banister's work. The majority of Banister's patrons were Fellows of the Royal Society and many of his communications were read before the Society, some of them being published subsequently in the *Philosophical Transactions*.

Until near the end of his career, it is evident that Banister, too, had slender resources. There was no mention of such scientific instruments and equipment as John Clayton had had at the outset of his voyage to Virginia, although Banister had a considerable library of scientific books. He was forced to appeal to his patrons for brown paper with which to press the plant specimens in his herbarium, and on one occasion he begged for "a good microscope."[77] Except for the assistance afforded by William Byrd, he could "expect no helps" in Virginia.[78] However, his situation im-

---

[76] For Horsmanden, see *Va. Mag. Hist. & Biog.*, XXIV (1916), 237, 356. Horsmanden was a relative of Byrd, whose wife was the daughter of Warham Horsmanden.

[77] See his letters in *Sloane 3321*, fols. 4–7v, n.p., n.d., addresses uncertain.

[78] William Byrd I remained Banister's loyal friend and patron to the end. He also continued to appear in the records of the Royal Society during the 1690's. On Dec. 5, 1694, the Society received a live rattlesnake with fifteen rattles from "Coll. Bird of Va." (*Journal-Book*, IX, 175); Sir Robert Southwell presented "The Method the Indians in Virginia and Carolina use to Dress Buck and Doe-Skins," probably from Byrd, which was published in *Phil. Trans.*, XIV, No. 194 (July–Sept., 1691), 532–533; in 1694 and in 1695 Southwell had lists of seeds from Virginia (*Sloane 3343*, fols. 244, 251–252); on Nov. 21, 1694, beeswax which "smelt of tobacco" was exhibited to the Society (*Journal-Book*, IX, 173); and on July 13, 1697, Southwell "related that Mr. Bird told him that the Indians were very patient in hunting & that they stalke on all foure with horns & some of the Skin on their head to deceive deer and that they used ashes in want of common Salt" (*ibid.*, X, 41). On April 30, 1695, Dr. John Woodward read a letter from Virginia (source not identified) with an account of the yaws "having the appearance of the Pox but it is gotten without Copulation" (*Journal-Book*, VIII, 296). Other Virginia data received in the 1690's are recorded in *ibid.*, VIII, 242, 248, 252; X, 18; and *Sloane 3328*, fol. 88.

proved after 1688 when he married "a young Widow."[79] Two years later, he imported persons to Virginia, including two Africans whom he evidently appropriated as his own slaves, and, on the basis of the headright system, patented a plantation in Bristol Parish, Charles City County. He had become a landowner and the founder of a family which later became highly esteemed in Virginia.[80] And he was named among the founders of William and Mary College.[81] Unfortunately, however, his life in this world soon ended. In May, 1692, while botanizing along the Roanoke River, he was accidentally shot and killed by one Jacob Colson. Colson was later granted bail and acquitted "for the death *per misadventure* of Mr. John Banister."[82] But the act was irrevocable and Banister's friends and patrons in England mourned his loss deeply. Every effort was made to salvage Banister's papers, and they continued to be the source of English scientific publications for years afterward. Martin Lister, after deploring the death of "my friend," whom he described as "a verie learned & sagacious naturalist," wrote that

I understand after his death ye Government of yt place tooke special care of his papers to seal ym up, & transmit ym to my Lord Bishopp of London here: in whose hands they are & therefore I doubt not, but they will be carefullie & speedlie publisht unlesse possiblie, it may not be thought as proper, being posthumous & incoharent pieces, to have them printed in yor Transactions. . . . You are best able to conduct & manage such papers. . . . I thinke it will be verie obliging in you, to offer to my Lord of London yor service in yt kind before my Lord shall have otherwise disposed of them. . . .[83]

[79] Byrd to Bobart, April 16, 1688, *Va. Mag. Hist. & Biog.*, XXV (1917), 255–256.

[80] Banister had a son, John, and his grandson of the same name was prominent in Virginia during the American Revolution. Lyon G. Tyler, "Virginia's Contribution to Science," *Wm. & Mary Qty.*, 1st Ser., XXIV (1916), 218.

[81] "Papers Relating to the Founding of the College," *ibid.*, 1st Ser., VII (1898), 162.

[82] *Va. Mag. Hist. & Biog.*, XI (1904), 164–165. The story originally was that Banister had fallen from rocks and was killed. It was perpetuated by Pulteney, II, 56, and repeated in 1950 in Raven, *John Ray*, p. 301.

[83] "The Extracts of four Letters from Mr. John Bannister to Dr. Lister Communicated by him to ye Publisher" (of the *Philosophical Transactions*), *Classified Papers, 1660–1740*, XV (1) ("Zoology"), No. 43, dated April 18, 1693; read before

To be sure, some of Banister's papers, together with his herbarium, were collected together and sent to England, where they ultimately fell into the hands of Sir Hans Sloane and lie among his collections in the British Museum.[84] But many had passed through the hands of James Petiver, whose careless and disorderly habits did much damage to Banister's scientific collections and literary remains.

Many of Banister's papers were prepared with an eye toward their eventual publication, and some of them appeared in print at the hands of his English friends. But Banister himself published nothing. Still, the widely admired ability of the man, and the collection of Virginia flora and fauna, often accompanied with competent black-and-white figures, drawn to scale, and with scientific descriptions and observations remarkably abreast of the latest scientific opinion in England, captured the imagination of English scholars as no other colonial field naturalist had done. John Ray, with Banister's patrons, and the Royal Society as a whole, awaited

---

the Royal Society April 26, 1693. Extracts from the letters were published in *Phil. Trans.*, XVII, No. 198 (March, 1693/94), 667–672.

[84] "A List of diverse Physiological Collections, or writings, made by the Reverend Mr. John Banister of Virginia, lately deceased," *Sloane 3321*, fol. 8. The list is dated "from Charles City, June 27, 1692," and signed by the Virginia officials, Hugh Davis, Robt. Bolling, and Ri. Blandon. On June 29 Davis added to the box some late items received from "the Honorable Edward Hill, Esq." *Ibid.*, fol. 8v. These papers are mere lists of catalogues and observations. *Sloane 4002* is richer, containing 118 folios of "Papers and Draughts of the Reverend Mr. Banister in Virginia sent to Dr. Henry Compton Bishop of London. and Dr. Lister from Mr. Petiver's Collection." Other Banister items appear elsewhere in the *Sloane Manuscripts*. In both collections, however, most of the items are copies of the originals, usually in Petiver's crabbed hand. A long letter by Banister to Morison, dated The Falls, April 6, 1679, is in the *Fulham Palace Papers 14, S.R. 652* (Lambeth Palace Library). See William Wilson Manross, comp., *The Fulham Papers in the Lambeth Palace Library, American Colonial Section, Calendar and Indexes* (Oxford, 1965), p. 159. A microfilm copy is in the Alderman Library at the University of Virginia. The remnants of Banister's herbarium are in the Department of Botany at the British Museum (Natural History) and listed in Dandy, H.S. 90–92, 98, 114, 168, with a biographical sketch on pp. 84–87. Other remains of Banister are at Oxford. See *Sherard Letters*, M.S. B26 (Bodleian Library); George Pasti, Jr., "Consul Sherard: Amateur Botanist and Patron of Learning, 1659–1728" (Ph.D. thesis, University of Illinois, 1950); Clokie, p. 182. Martin Lister's gift of shells and drawings to the Ashmolean Museum in 1683 included items which he had received from Banister. Lister, *A Journey to Paris*, Introduction, p. xliii.

with eager anticipation "the many discoveries of North America from Mr. Bannister by the next Shipping."[85]

Among the literary remains of Banister, one of the earliest letters from Virginia still extant was dated from "The Falls, Apr. 6, 1679," and was addressed "To my much esteemed Friend Dr. Robert Morison."[86] At the outset, it demonstrates Banister's esteem for and close association with Morison:

> The kindness you have shewn me when at Oxford, & the benefit & satisfaction I ever received from yo[r] free & communicative Nature deserves more than an Annual tribute of thanks, & I should be very ungratefull if I neglected any opportunity of paying it. But I should adde stupidity to ingratitude, & show my self a fool as well as a Clown, if I gave you not some Acct. of the Country I am now in; & so (tho I fail in the attempt) endeavour at least in some measure to make good the Character you were pleas'd to give of me to my Ld of London; who (as I am inform'd by Capt. Bird, a gentleman to whom I am in many ways besides this exceedingly oblig'd, had thereupon effectually Recommended me to my Ld Culpepper whose arrival not only I but all men else that love Peace & a Settle[mt] of Affairs impatiently desire.

Banister went on to relate the northern Indian attacks on Virginia and the damages and confusion in the colony. He described Virginia as "a Country excellently well water'd, & so fertile yt it does or might be made yield any thing yt may rendure to the pleasure or necessity of life. But want of Peace too much land & ye great Croppe of Tobacco men strive to make hinders Virginia from improving." He told of plants he had seen: "a sort of *Marshmallow*," growing on "Our clear'd grounds" and probably useful for cordage; "a sort of milky Plant call'd Silk-grass. It is I take it Apocynum Americanum Latifolium," and the Indians dyed it and

---

85 T. Robinson to Sloane, London, April 8, 1688, *Sloane 4036*, fol. 16. Ray had written to Robinson in 1684 that he awaited word from Banister, who "might make a great addition to it [Ray's *History of Plants*] by communicating the nondescript species of the country." See Gunther, ed., *Further Correspondence of John Ray*, p. 141.

86 The original of this letter is among the *Fulham Palace Papers 14*, S.R. 652. I quote from it with the permission of Mr. E. G. W. Bill, Librarian of the Lambeth Palace Library.

wove it into baskets; "most sorts of English Grain besides those proper to the Country, as Maize, of wch there are divers kinds," red, blue, yellow, and mixed, but the two most commonly planted were known as "Flint-corn," very hard, and "She-corn," "more soft & feminine, on whose Superfices Nature has impressed the Signature ♀." The Indians had yet other varieties. There were several kinds of legumes,

> besides Pumpkins and Macocks [gourds] of several sorts, Mush-
> & Water-Melons wch are a large very pleasant & innocent fruit.
> I have eaten near half a score of them in an afternoon. Most of
> these I suppose grow naturally somewhere or other in this
> Continent, for the Natives had them before this was a Colony &
> We from them. We have also Potatos—not of that kind you call
> Virginienses, wch is Erect: Ours run on a Vine & are the same
> they plant in Barbados: We bury them near our hearths to keep
> them from ye frost; The great Roots we eat, reserving the little
> Ones, wch we call Plantings for ye next Years increase.[87]

He distinguished between the two kinds of tobacco raised in Virginia, the sweet-scented, "wch is that usually piped in England," and "Arinnocoe," which was commonly sent "in Holland & other Places beyond the Seas"—although "that wch is Dark & not fit for the Market we sell to Irish & West Country men, or to the Barbadians for Rum & Sugar." The Indian trade consisted largely of furs—otter, fox, cat, raccoon, deer, and beaver—and he described again the beaver dams and their curious organization. The Virginia deerskins were very good and "Their flesh sear-rosted sweet as Mutton." He described the Indian method of hunting deer in the fall and winter by firing the woods ("fire-hunting"). Besides domestic animals, Banister referred to bears, panthers, elk, "three sorts of Squirrels," "Hedge-Conies," and opossums, "a sort of creature with a false belly, into wch it receives its young when in danger."

[87] Here Banister made a clear distinction between the sweet potato (*Batatis edulis*) and the Andean white potato (*Solanum tuberosum*), and pointed out that the latter was not native to Virginia—a confusion already noted in botanical literature. Morison had known the white potato at Blois, and the second volume of his *Plantarum Historiæ Universalis Oxoniensis* was the first to clear up the distinction, in part, no doubt, because of Banister's observations. See Redcliffe N. Salaman, *The History and Social Influence of the Potato* (Cambridge, 1949), p. 98.

He mentioned the huge flocks of carrier pigeons that sometimes "darken the Sky," wild turkeys—"I have heard some old men affirm they have known Cocks weigh 60 lb. but we commonly meet with those of 30-odd"—and waterfowl, such as "Swans, geese, ducks, &c." The rivers were filled "wth very good Fish: sturgeon, Shade, Herrings, Rocks, an excellent Fish, Catts, garrs, Perches, &c." Of the salt-water varieties "I am yet ignorant." There were "multitudes of Vegetables, many of them unknown to me," peaches, "more & better kinds than yo^{rs}," and Banister speculated that they may have been indigenous to the country. Vines of several sorts produced "very good Wine." There were both black and white mulberries, and the "Pascimmon a sort of fruit between the Plum & Medlar," pleasant enough "when rotten . . . but green it is of an exceeding harsh Tast wch draws the Mouth into a Purse." He referred to the "Chinquapin Bush," the black walnut, the "Pick hickkory or Hickkory," nine or ten kinds of oak, two or three sorts of laurustine, three kinds of lady's-slipper, a dwarf iris, and "several sorts of rare Capillaries, Funguses, &c." Of the latter, he told of the dog prick mushroom, which he described as "Phalloidea sive Penis Caninus," whose glans "was covered over with a shiny kind of Substance wch stunk egregiously. I have never smelt any thing Vegetal or animal like it." He had "met with a great Number of Trees, Shrubs & herbs yt I know not wt to make of for want of Books & other helps & Assistances pertinent to him yt undertakes & intends to go through with such a Work." Little could be said of the minerals of Virginia "because We seldom break the Earth any further than wth ye Plow or Hoe." He longed to see "wt Naturals of this & other kinds the great Ridge of Mountains . . . does produce." When the Indian forays stopped, he planned to visit the natives "to take a View of their Towns, Forts, Manner of Living, Customs &c. And also to informe my self of ye Names, but especially the Virtues of Plants, wch Nature has taught them to a Miracle." But he had evidently visited the nearby peaceful villages and described their "Sweating Houses" in detail, much after the manner of John Clayton a few years afterward. Banister's letter revealed that he had been alert during the few months that he had spent in Virginia, and he struggled to invent Latin descriptions of the nondescript plants that he had observed.

During the following years, Banister labored hard to enlarge his observations, prepare a herbarium of Virginia plants, collect seeds, roots, shells, fossils, insects, animals, and birds, and increase his knowledge of the Virginia air, waters, climate, soil, and Indians. He taught himself to draw so that he could make figures of the flora and fauna. He did some dissections, and described the anatomy of a land snail for Lister in 1690. And he sent an unknown number of collections of the natural life of Virginia to his patrons in England. Indeed, such is the state of Banister's specimens and literary remains that it is impossible to reconstruct his activities with assurance as to accuracy because many of the lists and letters bear no date and no clear indication as to whom they were sent. There is no evidence that he reached the mountains of Virginia, although he may have accompanied William Byrd on trading expeditions to the interior; but his collections appear to have been limited to the Coastal Plain, with excursions into the Piedmont from Byrd's house at the James River Falls. Repeatedly he demonstrated his knowledge of the current literature in natural history, both English and Continental, and showed himself to be a capable representative of the new post-Restoration generation of English naturalists.

Banister was familiar with some Virginia plants before he went to Virginia. He had seen those growing in the Oxford Physic Garden, had read of them and some others in works of the Bauhins, Johnson (the "Gerardus Immaculatus"), Parkinson, Morison, Cornuti, and others. His letter of April, 1679, indicated that he had collected and sent specimens to Morison, and his "Catalogus Stirpium Rariorum" was prepared in the same year. Unfortunately, Morison's accidental death (1683) precluded his use of Banister's data other than that which appeared in the posthumous second volume of his *Plantarum Historiæ Universalis Oxoniensis*, completed by Bobart and others in 1699. Leonard Plukenet saw Banister's early catalogue of plants sent to Bishop Compton and wrote asking for drawings to be included in his book *Phytographia* . . . (1691–96). By 1680 Banister had prepared a catalogue of insects, probably sent to Lister,[88] and Ray had received seeds from him via Bishop Comp-

---

[88] *Sloane 4002*, fol. 94. Followed by drawings of insects and shellfish, *ibid.*, fols. 110–118.

.66

*Phyllitis parva saxatilis
per summitates foliorum.
prolifera.*

7. John Banister's drawing of the walking fern, *Camptosorus
rhizophyllus*, as it is now known. Banister's caption indicates that he
invented a different name. Leonard Plukenet published
a revision of this drawing in his *Phytographia* (1691).

Sloane 4002, fol. 37. *Courtesy of the British Museum.*

NB. Let ye graver make the
Stalk 2; or 3 inches Shorter:
for, by inadvertence, it is
made so much too long

·5

Auricula Ursi floribus albis
Cyclamineorum more reflexis.

8. That Banister intended to publish his materials is illustrated
by his instructions to the engraver
in the above drawing of *Dodecatheon meadia* (shooting star).

Sloane 4002, fol. 42. *Courtesy of the British Museum.*

ton. But Banister was somewhat overwhelmed by "a new world of plants where in as yet very small discovery has been made." In one of his early letters, he said that some of the plants were

> so strange and monstrous that I am affraide that they may be thought chameras to be found no where but in his braine that drew them. last yeare I rec'd a letter from the worthy Mr. J. Ray who desires me to give him a more full account . . . [of a plant] and some other rare capillarryes . . . in a Catalogue I some years since sent to my Ld London. He also requested me to send some . . . [nondescript seeds] which I have done. Wch considering that dryed plants tho illustrated with never so plaine descriptions will but lead the Limner or designer into many errors . . . I betook my self therefore to drawing. . . .[89]

Besides the figures of things, Banister proposed to send "annually . . . the things themselves whether plants roots seed or figures to your selfe Mr. Ray or other gentlemen of your Society" interested in "that part of Natural Hystory." He further offered the proposal: ". . . supposing that some member of your Society (which I fancy may be no great difficulty to do) can prevaile so far with the royall affrican company as bestow me 4 or 5 young Negros it will be but a small matter to them yet with what flock I can procure my selfe may in time make me a pritty livelyhood." There is no evidence that the Royal African Company responded, although, as mentioned above, Banister did ultimately manage to obtain slaves of his own.

Meanwhile, he kept very busy. In 1680 he had made a collection of insects with figures and observations, probably for Lister; the observations were subsequently (1701) published in the *Philosophical Transactions* "with Remarks by Mr. James Petiver."[90] Mixed in with these manuscripts were Latin descriptions of Virginia plants,

[89] N.p., n.d., *Sloane 3321*, fols. 3–3v. Copy in Doody's hand. I would guess that the date was 1685, the addressee, Martin Lister. Two pages of "Plants Delineated by Mr. Banister" follow. *Ibid.*, fols. 4–5.

[90] "Collectio Insectorum atque Aliarum Rerum Naturalium in Virginia: Inchoata ult. Maij A.D.mi MDCLXX," *Sloane 4002*, fols. 94–109. There are notes on insects in *ibid.*, fols. 92–93, and illustrations, *ibid.*, fols. 111–118. Another copy of the "Collectio Insectorum" is in *Sherard Letters*, M.S. B26. See also *Phil. Trans.*, XXII, No. 270 (March–April, 1701), 807–814.

an account of the seashore, and lists of shells, soils, stones, sands, and fossils. "In divers places of the Freshes of our River," he wrote, "The Banks I am informed are composed of multitudes of Cockles, Scallops, & other sea-shells mixed with the Earth," and he concluded—as his predecessors had done—that the sea "in former Ages came farther up in the Country." In his descriptions of the shells, he referred to the works of Lister, Christopher Merret,[91] and other recent authors. Of the stones in Virginia he said "some of these may prove Indices of a Mine below," suggesting familiarity with Lister's hypothesis inherent in his proposal for a geological map (1684).[92] He continued to supply Lister and Lhwyd with fishes, shells, land and fresh-water snails, mussels, insects, and fossils up to the end of his life.[93] His papers as collected by the Virginia officials and sent to London included treatises entitled "Some Observations on Beasts, Byrds, Fishes, Insects &c Natural to Virginia," and "A Full rough Treatise in 4° and diverse loose leaves treating of diverse sorts of Stones, Shells, Vegetables & Insects in Virginia."[94] Banister was the first field naturalist to cast a scientifically trained eye upon the insects, the mollusks, and the fossils of Virginia, and his English patrons profited mightily from his collections, figures, and observations. Unlike Clayton, with whom he viewed (1686) "the Joynt of a Whale's Back-bone, and several Teeth . . . found in Hills be-

---

[91] Author of *Pinax Rerum Naturalium* (1666), "Dr. Merret's bungling Pinax," as Ray called it. Lister's work was *Historiæ Animalium Angliæ Tres Tractatus . . .*, published by the Royal Society, London, 1678–81. John Woodward included a Virginia óyster shell contributed by Banister in his "Catalog of the Foreign Fossils in the Collections of J. Woodward, M.D. . . . ," in *An Attempt towards a Natural History of the Fossils of England . . .* (London, 1729), Pt. II, p. 8.

[92] *Sloane 4002*, fols. 2–4. With these manuscripts is an invoice of furs consigned to London, suggesting that Banister added to his income by fur-trading.

[93] His last letter to Lister was dated May 12, 1692, only a few days before his death. See note 83 above. Many of the snails were figured. The consignment of Aug. 2, 1690, with the anatomy of a land snail, was probably that referred to in Lister's letter to Lhwyd, Feb. 8, 1691/92, *Ashmole 1816*, fol. 88 (Bodleian Library): "I have received verie latelie 8 new Species of Land & Fresh water Snails from *Virginia* of Mr. Banister, who hath industriously searcht yt Country for you, besides a Confirmation of verie many others formerlie sent me."

[94] This last manuscript may have been the basis for "Some papers . . . read of Mr. Bannister, concerning the Insects, Shells & Stones of Virginia" before the Royal Society, March 19, 1700/01, *Journal-Book*, IX, 250.

yond the Falls of *James River*, at least, a hundred and fifty Miles up into the Country," Banister hedged on the question of formed stones, evidently inclining to the opinion that skeletal parts of animals found in the soils and sedimentary rocks of Virginia were in some cases fossilized remains of creatures once living, while at the same time nature continued to produce formed stones in the earth. Besides the article in the *Philosophical Transactions* noted above, Petiver published bits and pieces from Banister's works on insects and plants in his *Musei Petiveriani . . .* (1695–1703), in *Gazophylacii Naturae et Artis* (1702–06), both with "grateful" acknowledgments to his various donors, and in *The Monthly Miscellany: or, Memoirs for the Curious*, a short-lived excursion into popular science journalism (1707). With somewhat less attention to his sources, Lister published engravings of seventeen of Banister's mollusks in his *Historia sive Synopsis Methodica Conchyliorum* (1685–92).[95]

By 1688 or 1689 Banister had advanced enough in his collections and observations to plan a natural history of Virginia. As funds were wanting for its publication, he planned to prepare it piecemeal for serial publication. This created problems of organization, for the method was "inconvenient for a history . . . nor can I think of any other that this annual printing will not interrupt, break to pieces, & confound."[96] He rejected "so nice & particular a Method as that of ye Learned Mr. Ray (tho it be necessary in a General History) would fill a Particular one too full of Divisions." Accordingly, he wrote,

> I shall therefore dispose my matter according to 3 Elements. In that of Fire [I] shall give a Relation of ye Violent and suddain Thunder & lightnings we have here, wth accidents relating thereto: as also ye Indians way of fire-hunting, sweating &c. In that of ye Aire, I shall endeavour to give an acct. of ye healthfulness of ye Climate, wth ye diseases incident thereto. In that of ye Water, I shall relate wt great Rivers we have, how far Navigable; wt quantity of Ships trade hither yearly; how we are stored wth

95 Petiver later referred to plants, shrubs, and trees growing in English gardens about 1712. See *Phil. Trans.*, XXVII, No. 333 (Jan.–March, 1712), 416–426; XXVIII, No. 337 (1713), 33–64, 177–221.

96 Banister to [?], n.p., n.d., *Sloane 3321*, fols. 7–7v.

fowles, Fishes, &c; & of Those, wch are Undescribed, wch not. I shall also give some reasons I have to think that ye now habitable part of ye Country was heretofore Sea. In that of ye Earth, of ye Nature & fertilitie of ye soile; also somewt of ye Nature, Customes, & qualities of ye Natives & of ye Trade we have wth theme; of the Birds, beasts, reptiles, Insects that are here; together with a more particular Acct. of ye Plants of this country, viz. Such as are Cultivated & manured, or Wild & Spontaneous: of both which, what are already described, I shall barely Name; Unless I meet with any thing New worth Observation touching in medicinal, culinary, or other use. Such as are not well described, I shall endeavour to rectify; & to resolve ye Doubtfull; Those also that are Nondescript I shall delineate & describe & rank ym: all under these Heads:

1. Such as are More or Less imperfect: viz. ye Fuci, Algi, Fungi, Musci, Filici, &c; all those of ye 2^d & 3^d Book of ye fore mentioned incomparable Hist. [i.e., Ray's]. 2. Such as are Less perfect, as Grasses, hops, Hemp, Nettles, Spinage; all yt are in ye 22nd & 4th Books. 3. Such as are More-perfect; whose Seeds are, 1st Rais'd; 2^d Capsulate. 4. Of Trees, & Shrubbs.

In his description of plants "their particular place" would be noted. He planned to number the new plants consecutively to those of the previous year, "very willing to cast it into any Other [form] as you shall think fit." He begged the favor "of you to send me some Instructions towards ye finding of Mineral Waters, several sorts of Earth, &c. & to buy me Willughby or some other that has writ at large of Birds & Beasts yt I may know wt are & are not described." He also asked for a good microscope "and anything else" thought necessary for his design. Here was the plan for a natural history of Virginia such as the Royal Society had long promoted.

But it was not to be fulfilled. The weaknesses of the Royal Society at the moment and the shortages of funds occasioned by the turmoils and uncertainties of the "Glorious Revolution" rendered it impossible to publish any part of Banister's natural history of Virginia before his regrettable death in 1692. Some of his manuscripts were about ready for the press. Besides those mentioned above, so widely appropriated in the works of Petiver and Lister, his manuscripts included "A Treatise of the ffertility of ye Waters, & Soyl in Virginia," "A full Catalogue of some of ye more rare Shrubbs &

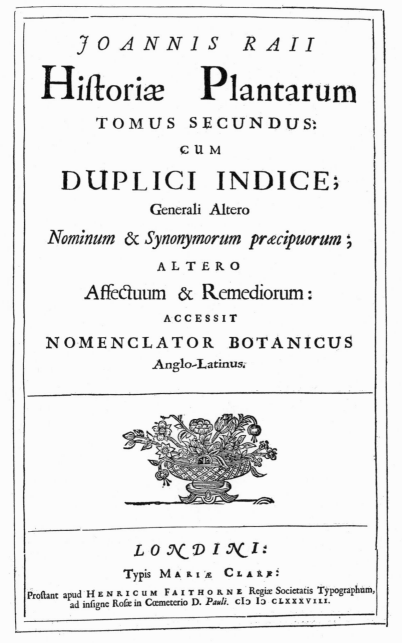

JOANNIS RAII

Hiſtoriæ Plantarum

TOMUS SECUNDUS:

CUM

DUPLICI INDICE;

Generali Altero

Nominum & *Synonymorum præcipuorum* ;

ALTERO

Affectuum & Remediorum :

ACCESSIT

NOMENCLATOR BOTANICUS
Anglo-Latinus;

LONDINI:

Typis Mariæ Clark;

Proſtant apud HENRICUM FAITHORNE Regiæ Societatis Typographum,
ad inſigne Roſæ in Cœmeterio D. *Pauli.* cIɔ Iɔ cLxxxvIII.

9. The title page of the second volume of John Ray's *Historia Plantarum*
(1688) which published much of Banister's Virginia material.

Courtesy of the Rare Book Room, University of Illinois Library.

E Catalogo huc tranſmiſſo *Anno* 1680. quem compoſuit eruditiſſimus Vir & conſummatiſſimus Botanicus D. *Johannes Baniſter* Plantarum à ſeipſo in Virginia obſervatarum.

A.

ALſine Spergula latifolia reptans.
Becabungæ folio.
Althæa lutea Pimpinellæ majoris folio, floribus parvis, ſeminibus roſtratis. Folia hujus plantæ pediculis inſident.
Althæa magna Aceris folio, cortice Cannabino, floribus parvis ſemina rotatim in ſummitate caulium, ſingula ſingulis cuticulis roſtratis co-operta ferent.
Althæa magna quinquecapſularis, cortice Cannabino, foliis integris ſubtus albicantibus, floribus magnis ex fundo ſaturatè rubro albis.
Alth. magna quinquecapſularis, cortice Cannabino, foliis Malvarum modo diviſis, ſubtus viridibus.
Ambroſia inodora foliis non diviſis.
gigantea inodora foliis aſperis trifidis. ,
Anchuſa lutea minor, quam Indi Paccoon vocant ſeipſos ea pingentes.
Anemone latifolia ſylveſtris alba.
Apocynum erectum non ramoſum folio ſubrotundo, umbellis florum rubris.
Apoc. erect. non ramoſ. latiore folio, umbellis florum albicantibus.
Apoc. erect. minus, umbellâ florum candida.
Apoc. erect. non ram. Aſclepiadis folio, umbellis florum rubentibus.
Apoc. minus non lactescens, caule & foliis hirſutis, floribus ſaturatè luteis.
Apoc. erect. non ram. Roris marini foliis umbellis florum candidis.
Apoc. petræum ramoſum Salicis folio.
Apoc. ſcandens, capſulis brevibus ſpinis aſperis.
Apoc. ſcand. capſulis alatis.
Apoc. ſcand. capſulis planis.
Hæc omnia ſiliquas ferunt tumentes.
Apoc. erect. ramoſum, caule rubente, foliis oblongis parvis, ſiliquis [ex floſculis albis] tenuiſſimis, binatim ad extremitates conjunctis.
Ariſarum triphyllum, pene viridi.
Ariſ. triph. minus, pene atro-rubente.
Ariſ. Dracontii foliis pene longo acuminato.
Arum aquaticum, foliis in acumen definentibus, fructu viridi.
Arum fluitans, pene nudo.

C.

Carduus Jaceoides purp. foliis ſubtus incanis, capite viſcoſo.
Caryophyllata flore ſemper albo.
Caſtanea pumila racemoſo fructu parvo, in ſingulis capſulis echinatis unico, **The Chinquapin.** Autor deſcriptionis Carolinæ ex hac nuce Chocolatam fieri refert non multò inferiorem ei quæ ex Cacao fit.
Centaurium minus caule quadrato alato, flore carneo amplo, umbilico luteo.
Centaurium luteum Aſcyroides.
Clematis purpurea repens petalis florum coriaceis.
Clem. erecta, humilis non ramoſa, foliis ſubrotundis, flore unico ochroleuco.
Cochlearia flore majori In locis udis à ſalſis procul remotis.

Conyza cærulea acris Americana.
Cucumis fructu minimo viridi, ad maturitatem perducto nigricante. Fructus Bryoniæ albæ baccâ non multo major eſt, cujus primo aſpectu ſpeciem eſſe putaveram.

D.

Dens caninus flore luteo.
Digitalis flore pallido tranſparenti, foliis & caule molli hirſutie imbutis.
Digit. rubra minor, labiis florum patulis, foliis parvis anguſtis.
Digit. lutea elatior Jaceæ nigræ foliis.
lutea altera, foliis tenuiùs diſſectis thecis florum foliaceis.
parva comis coccineis.

E

Eryngium campeſtre Yuccæ foliis, ſpinis tenellis hinc inde marginibus appoſitis.
Euonymus capſulis eleganter bullatis.

F.

Filix mas foliis integris auriculatis.
mas rachi ſeu nervo medio alato.
fœmina foliis per margines pulverulentis, ſeminibus fimbriatis.
Fumaria ſiliquoſa lutea.
Siliquoſa altera grumoſa radice, floribus gemellis ad labia conjunctis.
Fungus (ex ſtercore equino) capillaceus capitulo rorido, nigro punctulo in ſummitate notato. Ex recenti fimo noctu exoritur cauliculis erectis, vix digitum longis, capillorum inſtar tenuibus nec minùs denſis ſeu confertis. Singuli Cauliculi parvulo globulo aqueo coronantur, qui in ſumma ſui parte macula parva nigra Limacis oculo ſimili inſignitur.

G.

Gentianæ affinis foliis glabris ſerratis, floribus Ranæ referentibus.
Gladiolus cæruleus hexapetalos, caule etiam gladiato.
Gratiola foliis latioribus ſerratis.

H.

Hedera trifolia Canadenſis foliis ſinuatis.
Helleborine flore rotundo luteo, purpureis venis ſtriato. **The Mockaſine flower.**
Helxine latè ſcandens ſeminibus majoribus.
Helxine fruteſcens Bryoniæ nigræ foliis, capſulis triquetris amplis Pergamenis.
Hieracium fruticoſum latifolium foliis punctulis & venis ſanguineis notatis.
Hyacinthus Occidentalis flore pallidè cæruleo.
Hypericum parvum caule quadrato ſeu Aſcyron minimum.
Hyper. pumilum ſemper virens caule compreſſo ligneo ; ad bina latera alato, flore luteo tetrapetalo, ſeu Crux S. Andreæ.
Hyper. fruteſcens luteum Phillyrrheæ foliis.

Jacea

10. The first page of Banister's catalogue of Virginia plants (1680) as it appeared in Ray's *Historia Plantarum.*

Plants of ye growth of Virginia," "A Discourse of the Natives, their Habitt, Customes, & manner of living," and numerous herbaria, catalogues, figures, and descriptions of Virginia flora.[97] Like his works on insects and mollusks, these were largely appropriated by others. John Ray recognized Banister's contributions to the flora of Virginia in the first volume of his *Historia Plantarum* (1686), published the new species mentioned in Banister's "Catalogus Stirpium Rariorum" (1679) together with Banister's seed list of 1687 in the second volume (1688), and gave notice to yet other Banister materials in the third volume (1704).[98] Leonard Plukenet made even more extensive use of Banister's materials and included about eighty of his figures of plants in his *Phytographia: sive, Stirpium Illustriorum et Minus Cognitorum Icones* (1691–96).[99] Jacob Bobart and the other scholars who completed Robert Morison's *Plantarum Historiæ Universalis Oxoniensis* (1699) published accounts of Virginia plants from Banister.[100] Professor Joseph and Mrs. Nesta Ewan have demonstrated that Robert Beverley's *History and Present State of Virginia* (1705) borrowed parts verbatim, without credit, from Banister's treatises "of the ffertility of ye Waters, & Soyl in Virginia" and the "Discourse of the Natives, their Habitt, Customes, & manner of living," and that John Oldmixon's *British*

[97] *Ibid.*, fols. 8ff. Banister's other catalogues etc. of Virginia flora (often undated) appear in *Sloane 2346*, fols. 1, 25v–27; *3331*, fols. 58–59; *3336*, fols. 34v–38, 51–65. For his herbaria see note 84 above.

[98] Ray, *Historia Plantarum*, I, 145, 148, 270, 278–279, 674; II, 1857, 1914, 1919, and Appendix, pp. 1928ff., mispaged, 136 items listed; III, 22ff. Ray's third volume (1704) included more than fifty references to Banister's collections and descriptions of Virginia flora, some of which referred to items previously cited in Plukenet's works. See also Joseph and Nesta Ewan, "John Banister and his Natural History of Virginia, 1679–1692," *Ithaca 26 VIII 1962–2 IX 1962: Proceedings of the 10th International Congress of the History of Science* (Paris, 1964), pp. 927–929. The Ewans have in press an excellent and very full study of Banister under substantially the same title. I have seen this manuscript, but as its mass of detail is more than I could incorporate into this book, I fell back upon my own rather full notes, previously collected, and covering much of the same ground as the Ewans' manuscript, though including less bibliographical and scientific data.

[99] Published in four parts, consisting entirely of figures, some of which were probably from Banister although the author was not consistent in citing his sources.

[100] The volume cited at least forty-eight Banister items (pp. 19ff.). Some of the figures at the back of the book may have been copies of Banister's drawings, but the sources were not given with uniformity.

Empire in America (2 vols., London, 1708) appropriated materials from both Banister and Beverley.[101] Other eighteenth-century writers, including a few Continentals, borrowed from Banister's materials, and Linnaeus, in his *Species Plantarum* (1753), referred to well over a hundred of Banister's Virginia plants and drew heavily upon his figures as published in Plukenet's *Phytographia*. Dr. William Houstoun, who will appear later as a collector in the service of the South Sea Company and the Georgia trustees, collected seeds in the West Indies to which he gave the generic name *Banisteria*. Philip Miller, the gardener at the Chelsea Physic Garden, grew them at Chelsea and published an account of the plants in his *Gardener's Dictionary* (7th ed., 1759). Linnaeus adopted the name in his *Genera Plantarum* (1737) and *Species Plantarum*, and today *Banisteria* is still recognized as a tropical and subtropical genus of many species. Though Banister himself had published nothing, his works were widely employed in the publications of others, and his name has been perpetuated in botanical nomenclature.

In the meantime, his works enriched the studies of countless others, both within the Royal Society and outside, many of whom, like Banister, published nothing. The gardens of England, both public and private, enlarged their American species from the many seeds and roots which Banister sent to his patrons. Bishop Compton, who doted especially on exotic trees, received from Banister many Virginia specimens for the Fulham Palace gardens, including the flowering dogwood, black haw, willow oak, chinquapin chestnut, northern hemlock, red oak, and sweet bay.[102] In all, Banister sent specimens of about 340 species of plants, 100 insects, 20 mollusks, and some fossils and stones to his English patrons. About two-thirds of the plants were new species, accompanied in many cases by careful descriptions and dried specimens from his herbarium. He drew sketches of about eighty plants, besides his figures of insects and mollusks. He contributed little toward a wider knowledge of other fauna, and his scattered references to birds were inferior to those of John Clayton. But no other field naturalist had contributed so

101 See note 98 above.

102 U. P. Hedrick, *A History of Horticulture in America to 1860* (New York, 1950), pp. 111–112; John Claudius London, *Arboretum et fruticetum Britannicum* (2nd ed., 8 vols., London, 1854), I, 49–51.

much, and so well, to the natural history of Virginia, or to any other British colony in North America. A generation passed before others of his ability carried forward his work.[103]

[103] For this summary of Banister's achievements, I am indebted to the works of Joseph and Nesta Ewan as cited in note 98 above.

SEVEN

The Royal Society's Early Promotions: The West Indies and Hudson's Bay

Barbados

AT THE OUTSET OF THE SECOND CIVIL WAR in England, Richard Ligon, a sturdy Royalist, lost all his English possessions "by a Barbarous Riot." In consequence, he left England in June, 1647, and made his way to Barbados. There he acquired an interest in a plantation in conjunction with one Colonel Modiford and lived in Barbados for three years, helping to manage the property. Illness forced his return to England, where he was soon cast into prison, probably as a political victim of the Commonwealth. While in Barbados, Ligon had taken an interest in natural history and had made sketches of some of the flora of the island. His close friend, Dr. Brian Duppa, Anglican Bishop of Salisbury, encouraged him to write a history of Barbados. Evidently written in prison, Ligon's *A True & Exact History of the Island of Barbadoes* was published in London in 1657. It was a small book (122 pages), devoted as much to the economy of the island as to its natural history. The author was strictly an amateur, with no evident familiarity with the literature of the pre-Restoration herbalists. But he had an interest in natural history and possessed some ability to describe and figure plants and trees. He described the geography and climate of Barbados, gave considerable attention to its products of commercial value (indigo, cotton, tobacco, sugar, ginger, and dye-woods), and he concentrated with some detail upon some of the vegetables and trees, especially the "Physick-nut," the "Poyson Tree," "Cassavie" (cassava), "Coco," and the "Palmetto Royal" (*Sabal palmetto?*, also called the "cabbage tree"), for the last of which he published a picture. Of great interest, especially to post-Restoration English-men, was his account of "Sugar Canes, with the manner of planting; of their growth, time of ripenesse, with the whole process of Sugar Making,"[1] for the English in the West Indies were poorly informed about the culture of sugar cane and the technological processes of sugar refining. Ligon's thin volume became a starting point for sub-sequent English studies of the West Indies, and a basis upon which the Royal Society of London formulated various sets of "Inquiries" directed to its correspondents there.[2]

A second volume of wide interest to the Royal Society was Charles de Rochefort's *Histoire Naturelle et Morale des Isles An-tilles de l'Amerique* . . . (2nd ed., Rotterdam, 1665), published in English as *The History of the Caribby-Islands* . . . (London, 1666) by "John Davies of Kidwelly," with plates illustrating fauna but not flora. The second book included a section devoted to "la manière de faire le Sucre." After study of both Ligon and Rochefort, the Society's Committee drew up "Inquiries for ye Antiles: out of ye French Naturall History of those iles and Ly-gon's hist. of the Barbadoes," published in the *Philosophical Trans-actions* early in 1668.[3] Similar to the earlier "Inquiries" noted, the Society sought information about and specimens of the flora and fauna, including *"Earths and Minerals"* of the West Indies, with attention to medicinal and other commercial values. Early in 1667, Edward Littleton set out for Barbados armed with these "Inqui-ries," but there is no evidence that he responded to them.[4] During

---

[1] Ligon, *A True & Exact History*, pp. 84–96. The description included an illus-tration of a sugar mill.

[2] John Oldmixon, *The British Empire in America* (2 vols., London, 1708), II, 5ff., relied heavily upon Ligon's work. See also Appendix IV below.

[3] Original in *Classified Papers, 1660–1740*, XIX, No. 64; *Phil. Trans.*, III, No. 33 (March 16, 1667/68), 634–639.

[4] *Journal-Book*, III, 58 (Jan. 23, 1666/67). Possibly Edward Littleton was a kinsman of Sir Charles Littleton, Deputy Governor and later Governor of Jamaica (1662–63); see Arthur Collins, *Early History of the Littleton Family* (London, n.d. [1931?]). Edward Littleton later served on the Council of Barbados, as a judge in the courts, and as Agent for Barbados. *Calendar of State Papers, Col. Ser., Am. & W. Indies, 1669–1774*, pp. 433, 496, 546, 626; *ibid., 1675–1676*, pp. 206, 363; *ibid., 1677–1680*, pp. 89, 151; *ibid., 1681–1685*, pp. 22, 24, 334; *ibid., 1689–1692*, p. 519; *ibid., 1693–1696*, pp. 54, 397. He also published *The Groans of the Plantations: Or A True Account of Their Grievous . . . Suffer-ings By the Heavy Impositions Upon Sugar . . .* (London, 1689; reprinted, 1698).

the same year, when Charles Howard requested (October 3) an "Account for ordering Sugar-Canes to make sugar," Henry Oldenburg responded with a lengthy manuscript for the Society about "The History of ye Culture of ye Sugar-Cane and ye making of Sugar."[5] In 1667, too, Dr. Henry Stubbes, "a Curious and Learned Person," sailed to the Caribbean Islands, and although he wrote more about Jamaica than Barbados, one of his letters to Sir Robert Moray related to the latter. Dr. Stubbes had read Oviedo, Monardes, and Ligon, but he appeared to be little better informed than they, and fully as credulous. His letter to Moray, dated March 30, 1668, was read before the Royal Society on June 4, 1668, and was published in the *Philosophical Transactions*.[6] He told of plants "with stony roots" (coral) in the harbors at Barbados; of dissecting a tortoise to disprove Ligon's story that tortoises have three hearts;[7] and of the stones found in the belly of a "Crocodile" (alligator?) which, he affirmed, are "for digestion" only and have no medicinal value, "whatever *Monardes* relateth." He noted that the manatee (*Trichechus monatus*) differed from the shark; argued that the cabbage tree was a species of palm, very edible; reported that tobacco raised in nitrous soil flashed when it was smoked and spoiled more quickly than tobacco grown in other soils; held that sugar cane cured much faster in Jamaica than in Barbados; told of a "Bastard-Cedar" whose wood was so porous that cups made of it leaked out brandy and wine in a short time; described tanning barks in Barbados far superior to those in use in England; and stated that fireflies in Barbados continued to emit light after their death. Much of what Dr. Stubbes reported raised questions in the minds of Fellows of the Society, and the Society's "Inquiries" for Jamaica, prepared a short time later, reflected these doubts.[8]

[5] *Journal-Book*, III, 125. Oldenburg's "History," based upon "an acct., or reply, furnished him by one Mr. Drarce, long resident in Barbados," is in *Classified Papers, 1660–1740*, XXIV, No. 49. The account from Mr. Drarce (spelling uncertain) is lost.

[6] *Guard-Book*, S-1, No. 91; *Phil. Trans.*, III, No. 36 (June 15, 1668), 699–709. John Ray referred to a plant he had had from Dr. Stubbes in *Historia Plantarum* (1688), II, Appendix, p. 1856.

[7] Paul Bussière later anatomized "the Heart of Land Tortoises from America" (Jamaica) and found but one. *Phil. Trans.*, XXVII, No. 328 (Oct.–Nov., 1710), 170–185.

[8] See Appendix IV below.

In the early summer of 1674, Dr. Thomas Townes, "a learned and ingenious Physician" who had "practis'd long in Cleveland," wrote to Dr. Martin Lister, "I goe shortly to ye Barbados, where if I can serve yᵣ curiosity in inquiring after anything yt is rare to these Northern parts, you may command me."[9] Lister was at the time engaged with John Ray in the completion of Francis Willughby's *Ornithology*, and he evidently urged Dr. Townes to be attentive to birds. Townes arrived in Barbados on August 7, and soon dispatched a letter to Lister to tell of the "Booby" (*Sula fusea*, a species of gannet), the "Tropic-Bird" (a sea tern, family of Phaethontidae), of a hurricane that "whirled us like a top about all points of the compass," with great damage in Barbados, and of fishes and seeds which he had collected to send.[10] Lister passed on the information to Ray (December 13, 1674), and Ray requested additional details about the birds, although he stated that he had seen a tropic bird in the Repository of the Royal Society.[11] On March 26, 1675, Townes wrote again to Lister. He expressed gratitude for Lister's letters: ". . . now I am so remote from the learned world & I hope & heartily beg you would continue your kindness this way, if for nothing else but out of pity to an American . . . most men here being wholly intent upon riches. . . ."[12] He described the ample game and fish of the island. Beef and veal were scarce, but "Truly here's sufficient belly provision." He praised the climate and was building a house on a nearby plantation. He described plants similar to those of England, and had collected dried specimens and made drawings which he sent to Lister.

[9] Townes to Lister, n.p., n.d., *Lister 3*, fol. 228 (Bodleian Library). See also *Lister 2*, fols. 22–23. Townes thanked Lister for "ye Short directions" and asked for a copy of Ray's *Ornithology* when published. Dr. Townes was born in Barbados, educated at Cambridge (B.A., 1667/68; M.D., 1674). Both Ray and Lister may have known him there. John and J. A. Venn, *Alumni Cantabrigiensis*, Pt. I (4 vols., Cambridge, 1927), Vol. IV. When he corresponded with Lister, his address in Barbados was "at Mr. Henry Harding's Plantation, St. George's parish."

[10] *Lister 2*, fol. 28, dated Aug., 1674.

[11] William Derham, ed., *Philosophical Letters Between the late Learned Mr. Ray and several of his ingenious Correspondents* . . . (London, 1718), pp. 128–130. See also Edwin Lankester, ed., *Correspondence of John Ray* (London, 1848), pp. 111–112, 124, 126.

[12] *Lister 2*, fols. 32–33.

The soil is fertil, though not above a foot or two thick upon a white & spongie limestone rock, wch affords good quarries here & there, that serve very well for building. Every dwelling house, with the sugar-worke, & other our housing looks like a handsom town, most being now built with stone & covered with pan-tile or slate (brought hither in the ballast of shipps; as are likewise sea coal for forges, & soe are brought cheap enough). Indeed, the whole Island appears in a manner like a scattered Towne, wch with perpetual green fields & woods makes this place very pleasant.

He promised to send an account of the indigo plantations, and he was confident that "many good gums may be found here," specimens of which he would send to Lister. But he ended with an account of observations which aroused considerable debate for some years afterward:

It will not be unwelcome to you perhaps, If I tell you that the blood of Negroes is almost as black as their skin. I have seen the blood of at least 20 both sick & in health, drawn forth, and the superficies of it all, is as darke as the bottome of any European blood, after standing a while in a dish; soe that the blackness of Negroes is likely to be inherent in them, & not caused (as some imagine) by the scorching of the Sun, especially when other Creatures here that lie in the same Clime & heat with them, have as florid blood as those that are in a cold latitude viz. England.[13]

Lister (who still resided at York) sent a copy of Dr. Townes's letter to Henry Oldenburg. It was read to the Royal Society on July 1, 1675, and subsequently published in the *Philosophical Transactions*.[14] Only one other letter from Dr. Townes appears to be extant. It was undated, addressed to one Mr. Dent, "and by him Communicated to Mr. Ray."[15] With it he sent the stuffed skin of a young shark which, he said, "you may receive . . . from Mr.

[13] Sir Philip Skippon wrote to John Ray, Feb. 11, 1675/76, to say that he had conferred with Thomas Glover, who had spent nine months in Barbados as well as a longer time in Virginia, and Glover denied Dr. Townes's allegations regarding the color of Negroes' blood. Skippon urged that the question "ought therefore to be farther examined by Correspondents in that and other Places (where Blacks inhabit)." Derham, ed., *Philosophical Letters*, pp. 135–136.

[14] *Journal-Book*, V, 118–119; *Phil. Trans.*, X, No. 117 (Sept. 26, 1675), 399–400.

[15] Derham, ed., *Philosophical Letters*, pp. 132–133.

[Samuel] Penn, if it come safe to him." He enclosed his observations on the shark, but had been cut short by "the Seamen urging me to rid him out of hand, thinking it a very childish Thing to sit poring on a Carcass." The shark, he said, "has usually the Attendance of about 2 or 3 *Pilot-Fishes* about a foot long, which are commonly seen to taste of the Bait immediately before their Master." He referred to strange plants and seeds in Barbados which "I would have long ago sent you, could I have got a Friend that would take care of them; but as soon as may be you shall have them." There is no evidence that Dr. Townes supplied Ray with any plants or seeds; most of his collections went to Dr. Lister, who, when he moved from York to London in 1683, gave them to the Ashmolean Museum at Oxford. However, both Ray and Lister profited by Dr. Townes's collections and observations. His contributions were too late to be acknowledged in Ray's publication of Willughby's *Ornithology* (1676), but Ray cited his assistance in Willughby's *Historia Piscium* (1686),[16] and Lister gave recognition to his aid in the *Historia sive Synopsis Methodica Conchyliorum* (1685-92).[17]

Only scattered references to Barbados appear in the scientific correspondence of the seventeenth century after Dr. Thomas Townes withdrew from the scene. On December 6, 1677, it was reported to the Royal Society that Sir Peter Colleton had sent several baroscopes to Barbados "to examine whether they would be of any use for the foretelling the seasons & mutations of the weather as they are found to doe here, especially concerning harricanes."[18] At least one of the baroscopes was sent to Colonel William Sharp, to whom a thermometer was also sent in 1680.[19] On August 16 of the same year, Colonel Sharp sent a long "Barometric Calendar for Barbados" with daily observations of the barometric pressure, wind direction, and weather from April 8 to August 14,

16 Charles E. Raven, *John Ray, Naturalist, His Life and Works* (Cambridge, 1950), p. 364.

17 Martin Lister, *A Journey to Paris in the Year 1698*, ed. Raymond P. Stearns (Urbana, Ill., 1967), Introduction, p. li.

18 *Journal-Book*, VI, 20.

19 *Ibid.*, VI, 259 (Nov. 11, 1680).

1680.[20] Two communications from Barbados in the 1680's excited high hopes among the Fellows of the Royal Society, but neither of them brought forth the expected returns. On December 28, 1680, William London wrote to John Evelyn to say that although he did not know Evelyn personally he admired him greatly. London said that he had been a farmer in England, had migrated to Barbados, where he had found more profitable undertakings, had read all the modern authors on husbandry, agriculture, and descriptions of the West Indies (including the Royal Society's "Inquiries" for those parts), had made a beginning of a natural history of Barbados, enclosed a catalogue of Barbados plants said to include many that were new, hoped to experiment with East Indian products in Barbados (tea, nutmeg, pepper, cloves, coffee, and cinnamon), and offered his services as a scientific correspondent from Barbados.[21] The letter bore the marks of an informed and ambitious man, and Evelyn quickly dispatched it to Robert Hooke at the Royal Society, saying that it came "from an unknown person . . . seeming by his exact Enumeration of particulars, to be the project of a most industrious Man. . . ."[22] Evelyn suggested that it be communicated to the Royal Society "not only for their Approbation, and to encourage his undertaking, but to consider what Rarities and Exotics we might desire him to send us for the Repository, and to propagate in this Climate." Evelyn would attempt to procure East Indian seeds for London's experiments, and, although the latter's "odd style" did "not seem to be so well fitted with English, for the writing an History of this Comprehension," yet the design appeared worthwhile, and "I would therefore faine encourage this ingenious Drudge (whoever he be) and receive your Commands. . . ." At the Society's behest, Hooke drew up detailed instructions with regard to the Society's requests, to involve the entire gamut of natural history, including pictures, specimens, and observations,

[20] *Letter-Book*, VII, 127–134; *Journal-Book*, VI, 258. These observations were compared with similar records from England, St. Helena, and Tangiers.

[21] London's letter is in *Guard-Book*, L-5, No. 114. I have found no further identification of William London. Was he a kinsman of George London, Bishop Compton's gardener?

[22] Evelyn to Hooke, Sept. 23, 1681, *Letter-Book, Supplement*, III, 132–133; read Oct. 19, 1681, *Journal-Book*, V, 251, 253; VII, 28.

and sent it to Evelyn to forward to London in Barbados.[23] At this point the correspondence ceased, and nothing further was recorded of William London's plans. The second letter was written by John Frank, from Bridgetown, to one Mr. Hunt, on May 5, 1687. Frank wrote:

> Tho' I am altogether a stranger to your Person, yet have Adventured to give you the trouble of Looking upon the following Observation which was taken with a 9 foot Telescope, and a Quadrant 30 Inches Radius, which had a Telescope of about 28. Inches affixt and fitted to the graduation of the quadrant by previous Observations. I am doubtfull whether or no it may be worth your while to peruse it, however I shall rest satisfyed in my selfe, that I endeavour to contribute my Mite to the Advancement of so noble a Study as Astronomy. . . . But that I may not increase my Crime by being overtedious, I shall only request that you would be pleased to intimate wherein I may be serviceable in Communicating any thing home to you, that we have here, you shall always find a willing minde to oblige. . . . In the Interim earnestly requesting your Correspondence. . . .[24]

The observations related to an eclipse of the sun taken at Bridgetown on May 1, 1687. Hunt forwarded the letter to Edmond Halley, who reported his own and Frank's observations on the eclipse to the Royal Society on June 29, 1687.[25] Nothing further was recorded from John Frank.

Other occasional notices of Barbados appeared in the minutes of the Royal Society. On June 26, 1679, Hooke presented the Society with a coconut "newly brought from the Barbados." The Curator cut it in half. Fellows tasted the "milk" and the "kernell, wch was much of the Same Tast with the liquor but pretty hard & Tuff to bite."[26] John Houghton, F.R.S., presented (June 29, 1681) a va-

---

[23] The "Heads" of the Society's instructions are in *Letter-Book, Supplement*, III, 134. The Society reported that "we cannot send any thing of the East India." A committee of the Society, headed by Hooke and including Robert Plot, drew up the instructions.

[24] *Letter-Book*, XI, Pt. I, 66.

[25] *Journal-Book*, VII, 49.

[26] *Ibid.*, VI, 203.

riety of specimens of the flora for the Society's Repository.[27] Halley described the use of the prickly pear for fences in Barbados, and Sloane added that the dividing line between the French and English sectors of St. Christopher's Island was marked by "a triple hedge of this pricklepear."[28] The most significant additional event, however, was the voyage to Barbados by James Reed,[29] the Quaker gardener, sent by King William to gather seeds, plants, shrubs, and trees for the royal gardens, as well as insects, fishes, shells, serpents, and birds for English scientists. Reed was a highly respected field naturalist who had made collections in England for various English patrons. Financed by the King, he was sent out with glasses, bottles, paper, spirits of wine, and other items required for collecting, and with specific directions drawn up by William Charleton (or Courten).[30] His voyage took him to several islands, but his principal collections were made in Barbados in 1690. Moreover, his collections appear to have been confined to plants, trees, and shrubs, which he catalogued with a short description of each with particular attention to the nature of the soil in which they grew in Barbados.[31] Reed shared his collections, especially seeds and nuts, with Charleton, who, in turn, shared with William Sherard, who planted them in the Oxford Physic Garden. In consequence, Reed's collection supplied English scholars with Barbados flora for observation and study for several years thereafter.[32] But in the eighteenth century new sources arose in Barbados and to these attention will be directed in a later chapter.

[27] *Ibid.*, VII, 23.

[28] *Ibid.*, VIII, 290 (Feb. 5, 1689/90).

[29] Variously spelled Reed, Reid, Rhede, Rheed, and Rede.

[30] *Sloane 3962*, fol. 188.

[31] Several manuscript lists of Reed's Barbados collections (which do not conform to one another) exist among the *Sloane Manuscripts*. See *Sloane 2346*, fols. 14–14v ("Directions Concerning ye Describing of Plants," probably by Charleton), 121–125 (93 items), 197v–199v (86 items); *3962*, fols. 83–83v (114 items); *4070*, fols. 17v–18 (129 items), 19–21 (93 items). Pulteney records that the Earl of Portland repeatedly sent Reed to the West Indies to collect exotic plants. Pulteney, II, 21.

[32] William Sherard wrote, "I have raised above 60 Barbados plants sent by Mr. Charleton brought over by James Reed the Quaker." Clokie, pp. 79, 230; Petiver, *Musei Petiveriani Centuria Octava* (1700), p. 72; *Phil. Trans.*, XIX, No. 220 (Jan., 1696), 393–400; XXVIII, No. 337 (1713), 216.

The Bermuda Islands

Richard Norwood (1590–1675) went to sea as a youngster and made the earliest surveys of Bermuda in 1616 and 1622. Subsequently he settled in London as a teacher of "the Mathematicks" and published a number of textbooks about navigation. In 1635 he measured an arc of the meridian with results very close to those of the Frenchman, Jean Picard, some years later. After the English Civil Wars began, Norwood returned to Bermuda, "hoping," as he said, "here to find much quietness and freedome for my studyes and practises (though it hath fallen out much otherwise)." He kept a small school for a time, worked on the completion of Briggsian logarithmic tables to fifteen decimal places, and practiced as a surveyor. He revisited England a few times and was well known and respected by the Fellows of the Royal Society.[33] On March 6, 1664/65, Secretary Oldenburg wrote in behalf of the Society to invite Norwood "to a Philosophical Correspondence," with special attention to "the Motion of Mercury [and] . . . the said planets conjunction with the sun, which, according to the calculations of best Authors will happen here in England on the 25 October of this present year. . . ." The Society assumed that Norwood was "furnished with Instruments necessary for it," and hoped that his "generousness" would extend to "what in and about your Island occurs considerable for the inriching of the History of Nature. . . ."[34] Unfortunately, Oldenburg's letter miscarried. Whereupon, on October 24, 1666, Oldenburg renewed his plea, this time to urge Norwood to make "Observations and Experiments, as conveniently as you may your selfe in those parts where you are, concern-

---

[33] E. G. R. Taylor, *The Mathematical Practitioners of Tudor and Stuart England* (Cambridge, 1954), pp. 347ff. Norwood gave autobiographical data in a letter to Oldenburg, July 26, 1668. See *Guard-Book*, N-1, No. 27. John Parkinson, in his *Theatricum Botanicum* (1640), p. 1671, published a list of Bermudan plants, the source of which is unknown. See also Robert T. Gunther, *Early British Botanists and Their Gardens* . . . (Oxford, 1922), pp. 368–369. It seems unlikely that Norwood was the source.

[34] *Letter-Book*, I, 136–137. This and other correspondence with Norwood are published in A. Rupert Hall and Marie Boas Hall, eds., *The Correspondence of Henry Oldenburg* (in progress, Madison and Milwaukee, 1965———), II, 123, 146–147; III, 276–278, 442–444; IV, 166–169, 547–550.

ing natural and artificial things" and to excite persons in other English plantations "to joyne in the same with you."

> Which being so, I am particularly to recommend to you and your Ingenious friends in America, that you would impart to the Royall Society, what ever in the Bermudas, and other Colonys occurs considerable for the enriching of ye History of Nature (a faithfull composure whereof is one of the main things they have in their eye) and more especially of the History of the Tydes, the particulars whereof, if well observed about such Islands . . . would probably give much light for the finding out of a good Theory to solve those puzling Phaenomena. . . . Besides, we being informed of a new whale-fishing, undertaken about the Bermudas, we should be very glad to receive from you the truth, method and success of that enterprise, with a description of the kind and qualities of those whales, and whether any of that substance called Sperma Ceti, be found in them, and if so, in what part, and quantity, and how 'tis ordered;[35] to which if you would please to adde, what observations you make of Eclipses, of the Motions of the Satellites of Jupiter, and such like, you would thereby exceedingly oblige the publicke, and gratifie the R. Society. . . .[36]

In a long postscript, Oldenburg revealed that he enclosed two copies of the *Philosophical Transactions* with the printed "Directions for Sea-men" and other particulars "in the matter of Tydes." The Society had been informed, further, that Norwood possessed "many accurate maps, as well of other parts of the world, as of America, and of the particular Plantations of the English, Duch, &c"; and it requested that "they may all be carefully preserved" and would be highly pleased if Norwood would "lodge them in the repository, or with the books and writings of the R. Society," which, if Norwood would consent, would "cause them to be printed, as yours, as indeed they are, with a character due to your Person and merits."

To this letter Norwood replied on June 18, 1667. His reply was

---

[35] Two articles about "the New American Whale-fishing about the Bermudas" appeared in *Phil. Trans.*, I, No. 1 (March 6, 1664/65), 11–13, and No. 7 (Dec. 4, 1665), 132–133. They were written by the same author but are unsigned, and the author has not been identified.

[36] *Letter-Book*, I, 352–354.

read before the Society on December 5 following and extracts from it were published in the *Philosophical Transactions* on December 9.[37] He had not previously heard of the Royal Society but he promised what assistance he could muster. However, he had grown old and had largely laid aside philosophical affairs. He had planned to observe the conjunction of Mercury with the sun in 1665 without Oldenburg's prodding (which he had not received), but the weather was cloudy and he could observe nothing. He said that the tides at Bermuda rose to four feet and in calm weather settled from southwest to northeast. Water in wells near the sea ebbed and flowed with the tides—a report which the Society was loath to accept without confirmation.[38] Whale-fishing in Bermuda only recently had been revived after former attempts had proved fruitless. Within recent years, however, the whalers sometimes caught two and three a day during spring months. The whales were smaller and more lively than those of Greenland. No spermaceti whales had been found in Bermuda waters, although they were found in the Bahamas.

Norwood's communication enlarged the hopes of Oldenburg, who, on February 10, 1667/68, addressed a long letter to the aging mathematician enclosing "some printed Inquirys, both generall for all Countreys, and particular for the West Indies, and the island you reside in. . . ."[39] The Royal Society requested more precise information regarding the tides about the Bermudas, "the Sperma Ceti, said to be plenteously found in the heads of whales, caught about the Bahama Islands," reports of remarkable longevity of Bermudans and "what you and other men doe conceive to be the cheife cause of such a great age," observations on "the Eclipses of the Moon," and the usual plea for specimens, seeds, roots, and the like of the natural produce of the islands.

[37] *Guard-Book*, N-1, No. 26; *Letter-Book*, II, 37–40; *Journal-Book*, III, 154; *Phil. Trans.*, II, No. 30 (Dec. 9, 1667), 565–567.

[38] See Appendix IV below.

[39] *Letter-Book*, II, 147–149. Oldenburg enclosed a copy of Thomas Sprat's *History of the Royal-Society* (1667). The inquiries included those in Appendix I below, together with others designed for both Virginia and the Bermudas (*Classified Papers, 1660–1740*, XIX, No. 65), somewhat similar to—and equally demanding as—those sent to Edward Digges of Virginia (see Appendix III below).

I hope [concluded Oldenburg], the English Plantations in America, and every where else, when they shall know the end and work of this Institution, how it aimes at the Improvement of all usefull sciences and Arts, not by meer speculations, *but* by exact and faithfull observations and Experiments, will not onely congratulate this felicity to England, but emulate the same method everywhere for attaining the same end. By it, I am persuaded, England may becom the glory of the Western world, making it selfe the seat of the best knowledge, as well as it may be the seat of the greatest Trade: And that Almighty God would Crowne it with his blessings for that end. . . .

The Royal Society's demands were too much for the old man. Unable to respond to the inquiries, Norwood wrote (July 16, 1668) that he had passed them on to "my loving friend Mr. Richard Stafford now sheriff of this Country, an Ingenious gentleman and his Father was very industrious in prying into the mysteries of Nature, who is most likely to satisfy you of any man in the Country that I know."[40] This letter evidently was delivered personally to Oldenburg by Norwood's son.[41]

A letter from Richard Stafford arrived bearing the same date and accompanied by curiosities from Bermuda, including cedar berries, palmetto berries, palma Christi seeds, a "Poysen weed" (poison ivy, *Toxicodendron radicans*), and some "Silken Spiders webbs." Stafford said that the season was wrong for collecting Bermuda flora, and he postponed further collections of this kind until a more propitious time. The spider webs had been made by large spiders which spun them between trees and they were strong enough to catch a bird "as big as a thrush." He reported that the tides of Bermuda rose to five feet between Michaelmas and Christmas, were about three feet high in calm weather the remainder of the year, and flowed northwest to southeast. Spermaceti whales occasionally were caught there, but he said nothing about where the "Sperma Ceti" was found in the whale. Bermudans were said to be migrating rapidly to New Providence Island in the Bahamas, whose climate and resources Stafford praised highly. People in Bermuda often lived to become centenarians. Their only distempers were colds in

40 *Guard-Book*, N-1, No. 27.
41 Oldenburg to Norwood, Nov. 16, 1668, *Letter-Book*, II, 319–320.

hot weather, and they generally succumbed to old age. Mr. Norwood would endeavor to supply observations of any eclipses that took place in Bermuda, but if he failed, "it will bee his age and weakness," for he "hath a great desire to serve the R. Society in every respect."[42] Oldenburg thanked Stafford (November 16, 1688), reimbursed him for expenses incurred in the transport of the Bermudan curiosities, and sent further inquiries with regard to the flora, the spiders, the tides, whales, the variation of the compass in the Bermudas, and hurricanes.[43] But if further communications to the Royal Society were sent by either Norwood or Stafford, they were not recorded in the minutes of the Society.

In March, 1670, Oldenburg prepared another set of "Inquiries Concerning Bermudas recommended to Mr. Hotham going thither," urging him "To Salute very kindly from me Mr. Norwood, and Mr. Stafford, and to let ym know, yt I wrote largely to them by ye ship yt sailed thither last summer, but receaved hitherto no answer from them." Nor was any reply from Mr. Hotham recorded.[44] Indeed, excepting a few observations about and curiosities from the Bermudas—the latter by the hand of London merchants trading there—the Society had no direct communications with the islands for the remainder of the seventeenth century.[45]

[42] Stafford to Oldenburg, July 16, 1668, *Guard-Book*, S-1, No. 106; *Letter-Book*, II, 241–245; *Journal-Book*, III, 248 (Oct. 28, 1668); IV (Nov. 19, 1668). Extracts published in *Phil. Trans.*, III, No. 40 (Oct. 19, 1668), 792–795. Both Norwood and Stafford were cited in John Matthew Jones, *The Naturalist in Bermuda . . .* (London, 1859), pp. 17–20, and in Oldmixon, *The British Empire in America*, II, 361–379.

[43] *Letter-Book*, II, 316–318; Hall and Hall, eds., *Correspondence of Henry Oldenburg*, IV, 550–553.

[44] *Classified Papers, 1660–1740*, XIX, No. 54. The inquiries were similar to those previously sent, requesting information about the variations of the compass, more exact and complete data about the tides, whales (especially spermaceti whales caught "at Eleutheria, and other Bahama-ilands"), the natural products of New Providence Isle (evidently inspired by Stafford's letter), the flora of the Bermudas, and "ye cause of the longevity of ye inhabitants of ye Bermudas, and the Bahama ilands."

[45] *Boyle Letters*, Bl. IV, fol. 146 (Royal Society Archives). Sir Robert Southwell presented to the Royal Society comments about the difference between Bermudan whales and those of Greenland. His information appears to have been derived from Sir Robert Clayton, who had formerly had an interest in Bermudan whale-fishing. *Journal-Book*, VIII, 97 (March 26, 1690), 298 (April 2, 1690), 299 (April 9, 1690); IX, 20 (Jan. 14, 1696).

Jamaica

If both quantity and quality of scientific communicants to the Royal Society be considered, Jamaica excelled all of the English colonies in the western hemisphere during the seventeenth century. Scientific exchanges began early. On May 8, 1661, Thomas Povey, a friend and correspondent of both Robert Boyle and Sir Robert Moray, sent Boyle "a small quantitie of Pepper of Jamaica" to "thoroughly examine it, by making such Experiments, that the qualities of it may bee well discovered and understood; for I am readie to believe that it is not of a lowe or ordinarie degree."[46] He had also sent a sample to Moray "that Hee alsoe will attempt something upon it." Povey had a brother in Jamaica, and he may have visited the island himself. In the course of the next five years, the two men presented the Royal Society with several curiosities from Jamaica, including shells, minerals, and cloth dyed with "yellow wood," probably *Madura tinctoria*. In 1667 Dr. Henry Stubbes, noted above in relation to Barbados, sent the first of three letters to Henry Oldenburg and Sir Robert Moray, to relate his observations in Jamaica.[47] These letters became the foundations of three articles which appeared in the *Philosophical Transactions*.[48] Stubbes had sailed from Deal to Jamaica, evidently in 1667. He contested Ligon's report that iron rusted more in the West Indies, holding the cause of rust to be the salinity of the air rather than rain or ocean water; that there was less "Burning of the Sea" (St. Elmo's fire?) than Ligon had related; and that tortoises had but one heart instead of the three that Ligon had accorded them. However, he added to the confusion about tortoises, saying that the blood of Jamaican tortoises was "colder, than any water, I ever felt there," and that the urine of men who had eaten tortoise was a curious shade of

---

[46] *Ibid.*, II, 44 (Feb. 17, 1663/64), 129 (Aug. 10, 1664); III, 43 (Nov. 28, 1666).

[47] Two to Oldenburg, dated May 27, 1667, and March 24, 1667/68; one to Moray, March 30, 1668, all written from Norwich, after Stubbes had returned from his voyage to the West Indies. See *Guard-Book*, S-1, Nos. 89, 90, 91. All were read to the Society, the last on June 4, 1668. *Journal-Book*, III, 219.

[48] II, No. 27 (Sept. 23, 1667), 493–500; III, No. 36 (June 15, 1668), 699–709; IV, No. 37 (July 13, 1668), 717–722.

yellow. He said that every mountainous island of the West Indies developed night winds that blew outward from the islands in all directions whereas those without mountains had no such winds. He told of waters in the wells at Jamaica that ebbed and flowed with the tides, of corals in the harbor, of animals, both domestic and wild, that drank water but rarely, and of rain that fell on a plain called "Magotti Savanna" and turned the seams of garments "in half an hour to Magots." He referred vaguely to the sandy soil of Jamaica, good for raising melons, and "In some ground that is full of Salt peter, your *Tobacco*, that grows wild flasheth as it is smoked." He spoke of crocodiles' and alligators' eggs, of swallows, ducks, teal, and other Jamaican fowl, of insects, fishes, trees, and roots—all, however, after a manner likely to satisfy the curious rather than the scientific reader. And he described the various colors of the sea, said that sugar cured as thoroughly in ten days in Jamaica as in six months at Barbados, and reported that Jamaica had no hurricanes so far as he had heard. On the whole, Dr. Stubbes's reports from the West Indies gave rise to more questions in the Royal Society than they supplied acceptable answers.[49]

Richard Norwood's son, referred to simply as "Mr. Norwood the Younger," supplied additional comments about Jamaica published in the *Philosophical Transactions* in 1668.[50] Described as "an Eye-witness," Norwood reported that alligators were shaped "like Lizzards," that if the blood of tortoises became heated they died (for it must not become hotter than the elements they live in), and that fireflies continued to shine after death ("he hath applyed them dead to a Printed and Written paper in the dark, and read it"). He gave a vivid account of the insect called "chegos" (chigoes, jiggers, or chiggers), how they bred and the discomforts they caused. And he described the beautiful but poisonous "Manichenel-Apples" (*Hippomane mancinella*), whose wood, if green or damp, raised blisters on the skin.

The reports of both Stubbes and the younger Norwood gave rise to several of the queries included by the Royal Society when it pre-

[49] Oldmixon later cited Stubbes extensively. See *The British Empire in America*, II, 323–338.
[50] III, No. 41 (Nov. 16, 1668), 824–825.

pared "Inquiries Recommended to Colonel [Sir Thomas] Linch going to Jamaica, London, Decem. 16, 1670."[51] Sir Thomas went forth as Governor of the island, directed to deliver up for trial the previous Governor, Sir Thomas Modiford (or Muddiford), a task which he executed with guile after having stayed at Modiford's house as a guest for a fortnight.[52] Sir Thomas Lynch, however, accomplished more than his political errand. He also sent to the Royal Society (via Sir Robert Moray) "an Account Concerning Cacao-Trees, their Planting & culture, the way of cureing them, the observables in their fruit &c." The tree described was an old one growing on Sir Thomas Modiford's estate near Port Royal, and Lynch sent a picture of it "painted to the life." His report was read before the Royal Society on May 22, 1672, and printed in the *Philosophical Transactions* the following spring.[53] The trees were planted twelve feet apart under shade, in a "fat moist soil." They began to bear fruit in the third year, after which the shade was cut down. They reached their prime in the tenth to twelfth year. Recently they had suffered from an unknown disease called "blast," and the crop was down "very severely." They bore fruit both in December-January and in May, the larger crop being in December-January. Each tree bore two to eight pounds of cods, each cod containing twenty to thirty nuts. The cods were cured in the sun three or four days, then the kernels were removed, allowed to "sweat" sixteen to twenty days, then dried in the sun for three or four weeks, after which they were ready for the market. Sir Thomas' account of the cacao tree was probably the most complete eyewitness description that had been received in English circles of the source of chocolate, which was becoming so popular in London.

George Ellwood sailed from the Downs to Jamaica on October 12, 1671, armed with a request to send an account of his voyage. Little is revealed about him other than that he set out to become a

[51] Printed in Appendix IV below. The extravagant stories about the light emitted by fireflies may well have come from John Davies' translation of Rochefort's *Histoire naturelle*, which relates similar tales (pp. 80–81).

[52] This according to George Ellwood, Jamaica, June 15, 1672, *Guard-Book*, M, No. 14.

[53] *Journal-Book*, IV, 252; *Phil. Trans.*, VIII, No. 93 (April 21, 1673), 6007–09.

planter in Jamaica. He visited many of the West Indies before he finally arrived at his destination, and his single long letter to his father, a copy of which was sent to the Royal Society, gave a useful overview of the West Indian scene.[54] He described Nevis as a "pittyfull & inconsiderable place, ye growth of it is chiefly Sugars but exceeding brown, ye buildings are poore much like to a Hogstie in England." Small Dutch islands to the leeward of St. Christopher's were used as a rendezvous for Dutch slave ships "& from thence furnish these windward Islands." St. Christopher's itself was divided between English and French settlements. Hispaniola was "a fine levell Country, & esteemed ye best & richest Island, & most desired by ye English of any part of ye West-Indies, it is abundantly stocked with Cattle, insomuch yt they kill them for their fat & hides." The northern part of it contained many French settlers, who were likely to envelop the entire island unless the English prevented it. Jamaica was a healthful island except during July and August, when it was very hot, and the people suffered from the flux. Port Royal, the principal harbor, was hot and unhealthful, but across the harbor a short distance away lay Spanish Town, the seat of the governor and a splendid community of about 200 families and 7,000 soldiers, both horse and foot, all well equipped. Jamaica was well stored with wild hogs, wild cattle, fresh fish, and turtles. Salt cod and "Bisket" were supplied from New England, beef from the Bermudas, "but within a few years ye Island will be sufficiently stocked with Cattle; there being above 10,000 head of cattle kept for breeding." Ellwood continued with a considerable description of the diet and economy of Jamaica. Cacao was the "richest" commodity, although during the past five years the crop had been seriously curtailed because of the "blast." Sugar was the second largest commodity, and indigo, "which doth farre exceed that of Barbados," was the third. Logwood "fetch'd from Campeach by our Privateers" was also a source of great profit. Ellwood estimated that "it will require at least £1000 to raise a Sugar-work, £250 for an Indico-worke." He planned ultimately to purchase "ye former." Ellwood drew an informative panorama of the West Indies, with particular attention to the topography and economy of Jamaica,

[54] See note 52 above.

but his account is of greater value to the social and economic historian than to the historian of science.[55]

William Hughes published the first English book devoted to the flora of the West Indies, entitled *The American Physitian; Or, A Treatise of The Roots, Plants, Trees, Shrubs, Fruit, Herbs, &c. Growing in the English Plantations in America. Describing the Place, Time, Names, Kinds, Temperature, Vertues and Uses of them, either for Diet, Physick, &c. Whereunto is added A Discourse of the Cacao-Nut-Tree. And the use of its Fruit; with all the ways of making chocolate* (London, 1672). Hughes was a physician stationed with the English fleet in the West Indies, principally at Jamaica, with which his book is primarily concerned. The volume was not designed for the botanist, being written "only in plain and easie Terms." The author stated that it was not *"written in a Closet or Study, in the corner of a house, amongst many Books; but the most of it, some time since, was taken, with many other Observations, rather in travelling the Woods, and other parts, (when I had the leisure at odd times to go on shore, being then belonging to one of his Majesties Ships of War) especially in that praise worthy Island of Jamaica. . . ."*[56] He endeavored to make his account *"as nigh to truth as I could possibly draw it, if my eye-sight failed me not,"* and though he hoped to make it yet more complete upon another voyage there to bring back seeds, roots, and the like to grow in English gardens, there is no indication that he did. Someone who signed himself "H. E." contributed the following eulogistic verse "On Mr. Hughes's Treatise of American Plants":

> Our *Lovel*,[57] *Gerrard, Johnson,* and learn'd *Ray,*
> Did travel far in the *Botanick* way:
> But this our Author hath out-went them clear,
> As by the following lines it doth appear:

[55] Ellwood promised to send a larger, fuller relation in his next letter. If he did, it evidently did not reach the Royal Society. A description of Jamaica similar to Ellwood's, written by R. Blome and dated 1672, appears in *Sloane 1394*, fols. 1–28. If it reached the Royal Society, the fact is not recorded in the Society's minutes.

[56] *The American Physitian,* "To the Reader."

[57] Robert Lovel, author of *Panbotanologia . . . or, A compleat Herbal . . .* (Oxford, 1659; 2nd ed., 1665), described as a "misapplication of talents" by Pulteney, I, 181.

> In which the Plants of *India* may be found,
> And their *Vertues*, to keep our Bodies sound.
>
> . . . .
>
> What here's disclos'd, *Columbus* did not see
> In his American Discoverie.

Hughes's method was to describe the specimen, the place where it grew, its uses or "Vertues" in different seasons, and its popular names. As a whole, the book resembled the older herbals far more than the more scientific works of the "learn'd Ray." It included accounts of thirty-odd plants, trees, and shrubs, with somewhat greater attention to items of diet than to items of "Physick." The section "of the Wheat of America, or Maiz" described it well, though less learnedly than John Winthrop, Jr., and reported that it was grown on nearly all the plantations in Jamaica. The description "of the Sugar-Cane, or Reed" resembled that of Henry Oldenburg in 1667, although it was less detailed; but the methods employed in its cultivation in Jamaica were reported, and Hughes believed that Jamaica "will produce the best Sugar in time, and the most." However, he said, the Jamaicans who produce sugar used it to excess, resulting in widespread scurvy there. Salt was made by evaporating sea water let into ponds, as the French did at La Rochelle.[58] A common drink used widely to entertain visitors consisted of the spirit of wine, or rum, mixed into a punch with water, sugar, and lime juice. The author devoted a large portion of the latter part of his book to a description "of the Cacao-Tree and Fruit," but he added little to the account previously given by Sir Thomas Lynch except to give specific recipes for the American way of making chocolate, which Hughes praised highly. He said little about the fauna of Jamaica, with only short references to the "sea star-fish" and the alligator. He described the white coral rocks and the dangers they held for ships in the harbors of Jamaica. On balance, however, Hughes's book appears to have promised more in its title than its contents

[58] An unpublished letter in the *Classified Papers, 1660–1740*, VII (1), No. 19, described the way of salt-making in Jamaica and the general want of salt in all American colonies together with suggestions for improving English fruit, flowers, and vegetables "by sowing or setting ye Seeds of our best" from the colonies. It cited Richard Ligon, John Josselyn, and William Hughes, and described Jamaica as a "Gentleman's Paradise." It had merit mostly as a promotion piece, and I suspect was compiled by Governor Lynch.

delivered. The *Philosophical Transactions* printed a descriptive review of it in a noncommittal vein;[59] and John Beale, the avid West Country horticulturist, gave the only other notice of it to appear among the Royal Society's papers when he wrote to Robert Hooke to comment that "Honest Mr. Hughs the American Physitian reports strang things of the Spanish pear in Jamaica," suspecting that it "is no native of Jamaica, but brought thither by the Spaniard" because of "strang alterations" in it, possibly resulting from the change in soil and climate.[60] Generally, however, Hughes's *American Physitian*, possibly because of its incompleteness and lack of attention to the newer approaches to botanical studies, made little impression on the scholarly world. Possibly, too, because William Hughes does not appear to have "belonged" to the Royal Society set, and he sent no collections from the West Indies with which Society Fellows could make observations and experiments.

The avid hopes of the Society were to be raised repeatedly in subsequent years, only to be imperfectly fulfilled. Lord John Vaughan, Governor of Jamaica, wrote two letters to Oldenburg in 1675–76 promising a scientific correspondence and collections from Jamaica, but nothing further was heard from him.[61] The Governor had taken with him Thomas Willisel. Willisel had been a foot soldier in General John Lambert's regiment during the Civil Wars. But he developed an interest in natural history, and after the Restoration became one of the most highly valued field collectors in England. Christopher Merret, Robert Morison, John Ray, and others had employed his services, and he accompanied Ray on

[59] VII, No. 83 (May 20, 1672), 4078.

[60] Nov. 29, 1679, *Letter-Book*, VIII, 88. Near the end of the seventeenth century there was published in England a competent popular natural history which cited Ligon, Hughes, the *Philosophical Transactions*, Grew, Gerard, Ray, Josselyn, Stubbes, and a variety of other recent works. Written by Sir Thomas Pope Blount, it was entitled *A Natural History: Containing many not Common Observations: Extracted out of the best Modern Writers* (London, 1693).

[61] Lord Vaughan to Oldenburg, Jamaica, Aug. 12, 1675, and Aug. 1, 1676, *Guard-Book*, V, Nos. 30, 31. It was probably Lord Vaughan who wrote to Martin Lister from Port Royal, Jan. 18, 1675/76, promising to act upon Lister's instructions. But aside from this letter, which briefly described beasts and reptiles, with a note about a "very malignant distemper" called the "Dry-belly-Ache" which affected Jamaicans disastrously, he delivered nothing more. The signature to the letter is "John [?]." It is *Lister 2*, fol. 31.

some of his botanical tours of England.[62] In 1669 the Royal Society itself employed Willisel at thirty pounds a year to collect "such Plants, Fowle, Fishes, and Minerals . . . in such parts of his Maties Kingdomes, as they shall think best for the use of the R. Society."[63] Everyone held Willisel in high esteem, and was confident that when he went to Jamaica he would reap a rich harvest for the students of natural history. Unfortunately, however, Willisel had been in Jamaica only a short time when he died (1675). His death was a great loss, both personally and professionally, to English scholars. As late as 1692, Ray lamented "Dr. Moulin's & poor Tom Willisel's losses wch I cannot remember without some trouble,"[64] and he went on to add: "Had God granted them life & health, they would have made great discoveries, & highly improved natural History. Very few species would have escaped their Notice; especially T. Willisel, who was indefatigable & could endure any hardship & live as well as Oatcake & whig [buttermilk] as another man upon flesh & wine, & ramble over hills & mountains & woods & plains."[65]

Charles Bouchar, whom Edmond Halley described as a "Coadjutor in my studies," did somewhat better. He left for Jamaica sometime before March, 1675, and Halley was confident that "he will not neglect to make what [astronomical] observations he can but especially those of Mercury for which that horizon will be most convenient."[66] Bouchar reported to Halley from Jamaica (March 10, 1675/76) in a letter read before the Royal Society on June 8 following. He described the climate and topography of Jamaica, and gave a brief attention to flora, insects, birds, and fishes, adding a promise of specimens for the future. "As for Astronomy," he said, "no Country better. For you shall have scarce

---

[62] *D.N.B.*; Lankester, ed., *Correspondence of John Ray*, pp. 5, 48, 61, 85, 87; Raven, *John Ray*, pp. 66, 77, 151, 153, 389, 442.

[63] The Society supplied Willisel with a "Certificate" recommending him to "All whom it may concerne. . . ." *C.M.*, I, 187–188, 193–194 (April–Oct., 1669).

[64] Ray to Edward Lhwyd, Dec. 28, 1692, in Robert T. Gunther, ed., *The Further Correspondence of John Ray* (Oxford, 1928), p. 233.

[65] Ray to Lhwyd, March 22, 1692/93, in *ibid.*, pp. 234–235.

[66] Halley to John Flamsteed, Oxford, March 10, 1674/75, in Eugene Fairfield MacPike, ed., *Correspondence and Papers of Edmond Halley (Publications of the History of Science Society*, n.s., Vol. II, Oxford, 1932), pp. 37–39.

two cloudy nights in a year. . . ." However, he lacked good instruments, and was forced to use a three-foot telescope belonging to Lieutenant Colonel William Beeston, with whose aid he sent observations of two eclipses of the moon which had occurred on June 26, 1675, and December 21 following. He also reported rebellions of the Negro slaves in Jamaica, "and ye biggest part is not yet taken: those that were taken are executed. They have done much mischief, and killed many people: But I hope they will all be taken in a short time, their gang having been broken already 5 or 6 times."[67]

About fifteen years passed before the Royal Society had further communications from Jamaica. In 1691 the *Philosophical Transactions* published "Observations on the making of Cochineal, according to a Relation had from an Old Spaniard at Jamaica, who had lived many years in that part of the West-Indies where great quantities of that rich Commodity are yearly made."[68] The author of the article was not revealed. He stated that cochineal was made from an insect (*Coccus cacti*) which "at first appears like a small blister or little knob upon the leaves of the Shrub on which they breed [the cochineal cactus, or *Opuntia* (*Nopalea*) *coccinellifera*] which afterward . . . becomes a live Insect . . . or small Grub." These were allowed to mature, then smothered in smoke, gathered on cloths, and spread in the sun until they were dry. Their wings were then removed and only their bodies were used, when crushed, to prepare the dye.

Three further communications from Jamaica appeared in the *Philosophical Transactions* during the 1690's. The former Lieutenant Colonel William Beeston, whose telescope had been used by Charles Bouchar, was now Sir William Beeston, Governor of Jamaica. In 1696 was published a letter from him "containing some

[67] Bouchar wrote from "Liguanea," twelve miles from Port Royal. *Letter-Book, Supplement*, II, 174–178; *Guard-Book*, B-2, No. 33; *Journal-Book*, V, 161; Thomas Birch, *The History of the Royal Society of London* . . . (4 vols., London, 1756–57), III, 318. About five years later (June 29, 1681) Sir Christopher Wren observed in the Royal Society that it had been found that West Indian Negroes who were fed only upon potatoes were likely to die of dropsy, and that the planters had been forced to allow them milk and bread to curb their losses. *Journal-Book*, VII, 23.

[68] XVI, No. 193 (March–June, 1691), 502–504.

observations about the Barometer, and of a Hot Bath in that Island."[69] His barometric observations puzzled him because weeks passed without observable changes. The hot mineral spring in Jamaica was very inaccessible, but he had persuaded two persons to try it and they found it very helpful in the treatment of belly-ache, pox, ulcers, and other disorders. One Mr. Robert Tredwey wrote from Jamaica, February 12, 1696/97, to Dr. Leonard Plukenet with "an Account of a great piece of Ambergriese thrown on that Island; with the opinion of some there about the way of its Production,"[70] but there appeared nothing remarkable about the communication other than the obvious ignorance of the Jamaicans as to the origin of ambergris. In 1698 one Captain Langford, a ship captain, related his way to predict hurricanes in the Caribbean area. He had been in five between 1657 and 1667, but his prognostications derived from an Indian in the island of Dominica. All hurricanes, he said, arrived on either the day of a full moon, or on the day of a change in the moon's phases. If, on these days, the sky was turbulent, the sun redder than usual, there was a great calm, and there were "burrs" about the stars, the sun, and the moon, then a hurricane was likely.[71]

In the midst of these relatively insignificant communications from Jamaica, however, occurred one of the most exciting events in the natural history of Jamaica during the Old Colonial Era. This was occasioned by the visit of Dr. Hans Sloane for about fifteen months between December 19, 1687, and March 16, 1688/89. Dr. Sloane (1660–1753) was born in Ireland of Scottish parentage. From his youth he found delight in the study of natural history, and at nineteen he went to London where he studied medicine, which, of course, included botany, pharmacology, chemistry, and anatomy. Here, too, he won the friendship of Robert Boyle, Dr. Martin Lister, and other luminaries of the Royal Society. In 1683 he went to Paris to further his studies of medical botany with

69 XIX, No. 220 (March–May, 1696), 225–228; Beeston to Charles Barnard, Jamaica, April 18, 1696, *Letter-Book, Supplement*, II, 190–194; *Journal-Book*, VIII, 364 (July 8, 1696). On March 15, 1698/99, Sir William presented the Royal Society with "a piece of Rock Gold in Spar" found in a new mine in Panama. *Ibid.*, X, 110.

70 *Phil. Trans.*, XIX, No. 232 (Sept., 1697), 711–712.

71 *Ibid.*, XX, No. 246 (Nov., 1698), 407–416.

Joseph Pitton de Tournefort and of anatomy with Joseph Guichard Duvernay. In Paris he met Dr. Tancred Robinson, who later introduced him to John Ray, and William Charleton, later to become a great English patron of science and collector, whose collections, comprising about 30,600 items, Sloane himself acquired in 1702 as the principal nucleus of his own expanding "cabinet." He received his medical degree at Orange (July 27, 1683), and moved on to Montpellier for work with Pierre Magnol. However, his stay at Montpellier was cut short in 1684 owing to disturbances growing out of Louis XIV's persecutions of the French Protestants, and he returned to London, bringing collections of plants and other items to his English friends.[72] He was assisted in establishing a medical practice by the renowned English physician Thomas Sydenham, and he rapidly made a place for himself in the London scene. He entered upon a scientific correspondence with John Ray and other English and Continental scientists, was elected Fellow of the Royal Society (November, 1684) upon the nomination of Dr. Lister, and to the Royal College of Physicians (April, 1687). Few young men won such speedy recognition in London at the time, a tribute both to his abilities and to his pleasant personality—although it was insinuated that Dr. Sloane was something of a "climber," with a disarming capacity to cultivate the "right people." To be sure, his ultimate reputation rested more upon his philanthropic promotion of science and aid to the poor (both of which were considerable), his collections (which he generously placed at the disposal of his friends), and his wide and fashionable medical practice rather than upon his achievements in science and the art of medicine—although in these he was more than ordinarily competent. His enthusiasm for science sometimes outran his judgment.[73]

Dr. Sloane went to Jamaica as physician to the new Governor, the Duke of Albemarle. His decision to accept the appointment and

[72] He presented fifty mineral specimens to the Royal Society in Feb., 1686. *Sloane 1968*, fol. 174.

[73] The principal secondary accounts of Sloane are: *D.N.B.*; Pulteney, II, 65–85; G. R. de Beer, *Sir Hans Sloane and the British Museum* (London, 1953); E. St. John Brooks, *Sir Hans Sloane: The Great Collector and His Circle* (London, 1954). Raven, *John Ray*, p. 209, states that Sloane "was an enthusiastic but not always judicious naturalist, a born collector . . . a liberal and energetic friend of science."

make the voyage was made only after long hesitation and the encouragement of Ray, Bobart, Robinson, and others, and he wrote to Ray that "I have talked a long while of going to Jamaica with the Duke of Albemarle as his physician, which, if I do, next to serving his grace and family in my profession, my business is to see what I can meet withal that is extraordinary in nature in those places. I hope to be able to send you observations from thence. . . ."[74] Ray replied: "If you go to *Jamaica*, I pray you a safe and prosperous Voyage. We expect great things from you, no less than the resolving all our doubts about the names we meet with of Plants in that part of *America*. . . ."[75] Sloane embarked with the Duke on September 12, 1687, and they arrived in Jamaica on December 19, sailing by way of the Madeiras, Barbados, Nevis, and St. Christopher's, at all of which Sloane went ashore and made hasty observations and collections in natural history.[76] From his collections, the plaudits of his friends, and his subsequent writings, it is evident that Dr. Sloane worked intensely while he was stationed in Jamaica. He traveled widely about the island, made notes on the soil, climate, topography, flora and fauna, and the people, together with their ways of living, and he employed the Reverend Mr. Garrett Moore to draw pictures of plants, birds, fishes, animals, and insects, some of them colored in crayon. And he sent seeds and specimens to some of his friends in England.[77] But his stay in Jamaica was foreshortened by the untimely death of the Duke of Albemarle and the Duchess' subsequent haste to return to England.

[74] Derham, ed., *Philosophical Letters*, p. 206 (Jan. 29, 1686/87).

[75] *Ibid.*, pp. 209–210 (n.d.). For Sloane's duties and privileges in Jamaica, see *Sloane 4069*, fols. 200–201.

[76] He collected at Barbados a rare species of the "Masticke Tree" (*Mastichodendron sloaneanum*), now on the verge of extinction. De Beer, *Sir Hans Sloane*, p. 39.

[77] See the letters of thanks from Tancred Robinson, April 8, 1688, *Sloane 4036*, fols. 32–33, William Charleton, July 29, 1688, *ibid.*, fols. 39–40, and Jacob Bobart, Oct. 1, 1688, *ibid.*, fols. 43, 55. Robinson wrote that "I will never let you rest" until Sloane had prepared an accurate natural history of the West Indies, for "Oviedo, Margrave, Piso, Rochfort, Hughes, etc., are so far from satisfying mee that they starve me. . . ." And Bobart implored Sloane to "remember Oxford." Martin Lister received shells from Sloane. See illustrations in *Lister 9*, fols. 14, 23, 45. Letters from Jamaica to Sloane after his return to England indicated that the colonists deplored the loss of Sloane's medical skills. *Sloane 4036*, fols. 47–50ff.; *4078*, fol. 116.

Sloane embarked with the Duchess on March 16, 1688/89, arriving in England on May 30, uncertain as to the person of the reigning monarch, so recent had been the "Glorious Revolution" by which William and Mary had risen to the English throne.

During the next four years Sloane served as household physician to the Duchess of Albemarle in London and worked on his Jamaican collections. He had brought with him a collection of about 800 plants, most of them new to English scholars, together with his drawings of the flora and fauna and a record of the Jamaican weather from May, 1688, to March, 1688/89. Three live specimens of the fauna—a large yellow snake, an iguana or "great Lizard," and an alligator—had been lost at sea. As early as July 31, 1689, he began to show his Jamaican collections and to make observations about them to the Royal Society.[78] Soon some of these began to appear in the *Philosophical Transactions*. The first of these contained detailed descriptions of the "*Jamaica*-Pepper, or All-Spice-Tree" (probably *Eugenia aromatica* or *Pimenta officinalis*) and of the "Wild-Cinamon-tree, *commonly but falsely called* Cortex Winteranus" (*Canella alba* or *Myrcia acris*), with handsome illustrations, accounts of their medicinal uses, and a collection of the names attributed to them since the time of Monardes.[79] The second was "*An Account of the true* Cortex Winteranus, *and the Tree that bears it*" (*Wintera aromatica*, Soland), with an illustration and description of its history and virtues.[80] The false "Cortex Winteranus" came from the West Indies, said Sloane. It was often sold by apothecaries as the true "Cortex Winteranus," which came from the Straits of Magellan, and Sloane emphasized that they derived from wholly different trees growing in widely separated places. Sloane's next article was entitled "An Account of a prodigiously large Feather of the Bird Cundur, brought from Chili, and supposed to be a kind of Vulture; and of the Coffee-Shrub, with a

---

[78] *Journal-Book*, VIII, 273 (July 31), 274 (Aug. 7, Oct. 16), 277 (Oct. 30); IX, 20 (Dec. 17, 1690), 25 (Jan. 14, 1690/91). On Oct. 30, 1689, Sloane showed the figure of the plant and fruit "called in Jamaica Soursop" (*Annona muricata*). On Jan. 14, 1690/91, he gave a description of corals found on the coast of Jamaica, dividing them into seven groups.

[79] *Phil. Trans.*, XVI, No. 192 (Jan.–Feb., 1690/91), 462–468.

[80] *Ibid.*, XVI, No. 204 (Oct., 1693), 922ff.

Figure thereof drawn by the Plant itself."[81] John Ray had previously considered the condor to be a fabulous creature but now, upon Sloane's testimony, he inserted it among the vultures.[82] Widespread ignorance of the nature of the coffee tree (which the Arabs had guarded closely) was only partially dispelled by Sloane's figure of the "Coffee-Shrub," for its flower was still unknown and its berry had only been seen in its roasted state.[83]

Sloane had been badly frightened when, on February 19, 1687/88, an earthquake shook the houses of Jamaica: "I wondered what it was at first, till looking out, I saw the Turtle-Doves in an Aveary to be in as great a concern, as myselfe, and then I concluded what it was, and betooke me to my heales, and designed to make an honorable retreat; but by the time I was got to the Staircase, we were all safe, and the house stood still. . . ."[84] The episode prompted him not only to write an account of the quake for the Royal Society but also to gather relations of earthquakes by others. In all, he collected eight accounts, including his own, which were read to the Royal Society and published in the *Philosophical Transactions* in 1694.[85] His last formal report to the Society relating to Jamaica was "Some Observations . . . Concerning some Wonderful Contrivances of Nature in a Family of Plants in Jamaica, to perfect the Individuum, and propagate the Species, with several Instances analagous to them in European Vegetables."[86] The observations related to parasitical plants, principally the "wild pine" (club moss?), which he grouped with the mistletoe family (Loranthaceae) and remarked about the sticky nature of its seeds, which fastened themselves to the bark of trees where they grew.

[81] *Ibid.*, XVIII, No. 208 (Feb., 1693/94), 61–64.

[82] Raven, *John Ray*, p. 332.

[83] Sloane's figure was made from a dried branch brought "from Moka in Arabia" by an English merchant.

[84] *Letter-Book*, XI, Pt. I, 65, dated from Jamaica, April 17, 1688; *Sloane 4068*, fol. 7.

[85] The original letters are in *Letter-Book*, XII, 290–291. They describe earthquakes in Lima, Peru, Oct. 20, 1687, and two in Jamaica, Feb. 19, 1687/88, and June 7, 1692, the latter far more severe and destructive than Sloane had experienced. Extracted in *Phil. Trans.*, XVIII, No. 209 (March–April, 1694), 78–100.

[86] *Ibid.*, XXI, No. 251 (April, 1699), 113–120.

In the meantime, Sloane had made several lesser reports to the Royal Society about his West Indian observations, and continued to study his Jamaican collections in the preparation of published accounts. He was exceedingly generous in sharing his collections with his friends, and Ray, Bobart, Sherard, and others lent material aid to the work.[87] The task was both aided and enlarged by the addition of James Harlow's large Jamaican collections of 1692. Harlow was the gardener of Sir Arthur Rawdon, who, at the instigation of William Sherard (who in turn had been inspired by Sloane's Jamaican discoveries), dispatched Harlow to gather Jamaican flora for his garden at Moyra in Ireland. He returned in the spring of 1692, and Sherard wrote to Sloane from Oxford (May 3, 1692) that

> James Harlow is at last return'd from Jamaica, he came last week
> . . . with 20 cases of shrubbs & trees each containing above 50,
> well condition'd . . . he has brought little else, not above 6 shells
> & but one new . . . his collection of ferns very large, but as for
> herbacious things and grasses very few, seeds he has not above
> 100 (of wch I will send next week for London) having sent his
> whole collection last fleet, wch we have not yet heard of, & must
> desire ye favʳ of you to enquire after. . . .[88]

Harlow's collections were widely dispersed to Sloane, Bobart, George London, Sir Arthur Rawdon, and others in both England and the Netherlands, although Sherard exercised care lest any fall into "Dr. P. [Plukenet's?] hands (especially ye ferns) lest he prejudice you [Sloane] in yr noble design."[89] Parts of the collection

[87] *Journal-Book*, IX, 90 (June 29, 1692); X, 40 (July 16, 1697); *Sloane 3341*, fol. 14 (Oct. 28, 1696); *4068*, fols. 9, 11 (June 12, July 18, 1692). He reported that rice grows well in the West Indies but that the difficulty of threshing it rendered it "not worth their while to sow this grain." He presented a root of "Hipeco-acanha" (ipecacuanha) from Brazil and told of its medicinal uses by the Indians. He argued that Jamaican cochineal was identical to that of the Middle East. And he reported that the china root of Jamaica (ginseng) had properties similar to that of the East.

[88] *Sloane 4036*, fols. 119–120.

[89] See Ray's letter to Sloane, July 22, 1696, in which he expresses confusion about the enormous number of Jamaican ferns referred to in Plukenet's works and seeks Sloane's aid in denominating them. Lankester, ed., *Correspondence of John Ray*, p. 298.

are still extant at Oxford.[90] Further West Indian data were derived from a remarkably versatile French priest, Father Charles Plumier (1646–1704), who made three voyages to the French West Indies, two of them at royal expense, as Botanist to Louis XIV, and published *Description des Plantes de l'Amerique* (Paris, 1695). Father Plumier made hundreds of drawings and paintings of West Indian plants and shared many of his discoveries with English scholars (and the Royal Society) with whom he was in friendly correspondence.

Dr. Sloane's first book, appropriately dedicated to the late Duke of Albemarle, was published in London in 1696. Entitled *Catalogus Plantarum Quae in Insula Jamaica Sponte proveniunt . . . Seu Prodromi Historiæ Naturalis Jamaicæ Pars Prima*, it was often referred to by his friends as his *Prodromus*, or introductory volume.[91] John Ray wrote a laudatory Preface to the volume, pointing out its chief merits, namely, that Sloane had described *living* specimens from Jamaica; that he had severely reduced the numbers of names and species; that he had given exact definitions; that he had included remarkably inclusive lists of synonymous names of plants, reaching back to the early Spanish nomenclature; and that he had discovered some West Indian plants common to Europe (including England) and the Americas. Indeed, Sloane did demonstrate unusual attention to the various names attached to West Indian plants by Oviedo, Monardes, Acosta, Bauhin, L'Escluse, Eden, Hakluyt, Purchas, Ligon, Hughes, Ray, Plukenet, and others reaching into his own generation. But subsequent scholars have found that in his enthusiasm for the reduction of the number of species Sloane sometimes overdid it and fell into error. His attention concentrated upon the flora of Jamaica, although he gave some space to the products of other islands which he had visited. He

90 Clokie, pp. 79–80, 177–178.

91 On June 5, 1696, the Royal College of Physicians objected that "ye License of ye Royal Society being placed before yt of ye President & Censors of ye College of Physicians, ye Censors Board was very angry & ordered him either to have it reprinted & place ye License of ye President & Censors before yt of ye Royall Society or leave both out, wch he promised to do." *Horsman Mss.*, I, n.p. (Library of the Royal College of Physicians, London). Evidently Sloane arranged to have both omitted in later printings.

employed Ray's system of classification, dividing his book roughly into two equal parts, the first devoted to plants, the second to trees and shrubs. The volume was written in Latin with only an occasional inclusion of English popular names. It included no figures, although Sloane's later works made up for this defect. As a pioneer study of the times, it was a first-rate work, and it won wide acclaim. No other English scholar with Sloane's ability had visited the West Indies, and no one (save possibly Father Plumier) had produced so valuable a study in the spirit and manner of the new science. For Jamaica, Sloane's study was unique at the time.[92]

Sloane planned further works to extend his *Prodromus*, but he was slow in completing them. When they appeared, they were cast in a wholly different fashion, written in English, dedicated to Queen Anne, and sumptuously printed and illustrated. The first volume was entitled *A Voyage to the Islands Madera, Barbados, Nieves, S. Christophers and Jamaica with the Natural History* . . . (London, 1707). The front pages included a fold-out map of Jamaica with a less detailed map of the West Indies as a whole. A long Introduction was devoted to the history of discovery in the West Indies, with accounts of their situation, air, waters, soils, economy, diet, and the diseases of the islands. Sloane's meteorological record of Jamaica from May, 1688, to March, 1688/89, was included here. Again attention was focused upon Jamaica, although subsequent pages included a journal of Sloane's voyage with accounts of the other islands visited, the fishes seen (some of which he had dissected to describe), and the birds and plants collected at Madeira, Barbados, Nevis, St. Christopher's, and Jamaica. Sloane described his travels in Jamaica, including the "Magotty Plain" which Dr. Henry Stubbes had described. It was here that he discovered the rare specimen of *Mimosa viva*, but he did not mention the alleged effects of the rainfall there to which Dr. Stubbes had referred. He told of his employment of the Reverend Garrett Moore to figure the natural products of the island, and the majority of the first volume (pp. 49–264) was devoted to the flora of Jamaica, excluding trees and shrubs, arranged according to Ray's

[92] Bobart borrowed from it somewhat in the completion of Morison's *Plantarum Historiæ Universalis Oxoniensis* (1699), and Ray borrowed extensively from it in the third volume of his *Historia Plantarum* (1704).

system. In the back of the book were 156 large plates to illustrate the plants described. Sloane's second volume, with the same title and format, was not published until 1725. It was devoted solely to Jamaica and contained a brief introduction and some figures of trees and shells, together with an account of the production of cochineal. There followed eight books devoted respectively to trees and shrubs, insects, shells, crustaceans, fishes, birds, quadrupeds, and serpents, and stones, earths, sands, minerals, and the like. A Table of Contents and Index to both volumes, with plates numbered consecutively from the first volume (Nos. 157–274) completed the book. The illustrations showed Jamaican trees, insects, shellfishes, fishes, birds, and serpents, but no quadrupeds. On the whole, Sloane's treatments of the fauna were inferior to and less complete than his accounts of the flora—although the section devoted to birds (Book VI), describing twenty-eight in all, appears to have been somewhat better than the rest. In the absence, however, of any other illustrated natural history of Jamaica, it was a masterly accomplishment.

Indeed, Sloane's three volumes about the natural history of Jamaica constituted a pioneer effort in the writing of the natural history of the New World, fulfilling all the major criteria previously laid down by the Royal Society for such undertakings. Where John Banister and a host of others had failed to bring their efforts to a conclusion, Sloane had lived to complete and to publish a work such as the Royal Society had been promoting for many years. Students of the natural history of the West Indies still turn to Sloane's volumes as one of the earliest works of value, especially with reference to the flora of Jamaica.[93] Similarly, whether in political, economic, or social history, historians still find Sloane's works, especially his two-volume *Voyage to the Islands,* to be a useful and reliable source of information.

The Lesser Islands

John Locke (1632–1704), the well-known English philosopher and empiricist, was a Fellow of the Royal Society with various

---

[93] See A. H. R. Grisebach, *Flora of the British West Indies Islands* (London, 1864), p. vii.

colonial associations. As a Royal Society Fellow, he appears to have taken relatively little part in the Society's activities. On May 27, 1675, however, Henry Oldenburg presented to the Society several relations from Locke, giving an account of poisonous fishes found in the waters about New Providence Island in the Bahamas.[94] Of these, the most significant was extracted and published in the *Philosophical Transactions*.[95] Locke did not reveal the name of his correspondent. His letter to Oldenburg, as copied in the Society's *Letter-Book*, ran as follows:

> I herewith send you an account I lately received from New Providence, one of the Bahama Islands, concerning fish there, wch is as followeth:
>
> I have not met with any rarities here worth yr acceptance, though I have been diligent in enquiring after them. Of those wch I have heard of, this seems most remarkable to me.
>
> The fish, wch are here, are many of them poisonous, bringing a great pain on their joints who eat them & continue for some short time & at last with 2 or 3 days itching the pain is rub'd off. Those of the same species, size, shape, Colour, taste, are one of them poyson, the other not in the least hurtfull; and those that are, only to some of the company. The distemper to men never proves mortall: Dogs & Cats sometimes eat their last. Men who have once had that disease, upon the first eating of fish, though it be those wch are wholesome, the poisonous forment in their body is received thereby & their pain increased.
>
> Thus far the Ingenious person from whom I had this relation, who having been put a very little while upon the Island when he writt this, could not send so perfect an account of this odde observation as one could wish, or as I expect to receive from him in answer to some Quaeries I lately sent him by a ship bound thither. When his answer comes to hand, if there be any thing in it wch may gratifie yr curiosity, I shall be glad of that or any other occasion to assure you that I am, Sir, your most humble Servant.[96]

If Locke received any further information from his Bahamas

[94] *Journal-Book*, V, 10.
[95] X, No. 114 (May 24, 1675), 312.
[96] *Letter-Book*, VII, 234-235. The date is derived from the *Journal-Book*.

correspondent, he does not appear to have communicated it to the Society.[97]

One Mr. Crisp, a correspondent of Sir Robert Southwell at St. Christopher's Island, sent barometric observations in 1680. But he had had troubles. He had set up the barometer supplied by Sir Robert "according to the Rules writt and demonstrated to me at Gresham Colledge but I conceive it will not do in this climate. It never alters where it first fixt at that degree opposite to Stormy for 3 months together although ye weather Changed 100 times in that time."[98] Subsequently, on June 13, 1683, the Society's Curator experimented with the magnet on the sands from St. Christopher's, though there was no record by whose hand the sands had been received.[99] Two years later (December 2, 1685), minutes were read from the Oxford Society of the previous November 17 "mentioning a Catalogue of severall Leaves, and Seeds, brought from St. Christopher's. . . ."[100] But again, the collector of the leaves and seeds in St. Christopher's was not identified. In 1685 a variety of curiosities from the West Indies was exhibited to the Society. These were recorded as "A Sample of Silk from the Antilles finer than the East-Indian. A Curious Plant at St. Christophers called Echinomelocardus, and another called Frangipane. A sort of Miratelaus at Martinique, A piece of West Indian wood smelling like human Exrement. Virgin sugar that condensed in the Canes."[101] The identities of neither the donor nor those of several of the specimens appear to be known.

In January, 1690, Dr. Clopton Havers presented "a West-Indian Animal called a Possome or Pirson" which, it was reported, "hangs itself bending by the tayle into a hook on boughs of Trees, and has a pouch under its belly to harbour its young. . . ." Dr.

---

[97] Locke made few, if any, other communications to the Royal Society, although he wrote to Sloane March 15, 1703/04, to send a register of the weather he had kept at Oates from Dec., 1691, to Dec., 1692. *Sloane 4039*, fol. 259.

[98] Crisp to Southwell, St. Christopher's I., Nov. 10, 1680, *Guard-Book*, C-2, No. 7; *Letter-Book, Supplement*, II, 449–450; read Feb. 23, 1680/81, *Journal-Book*, VI, 272.

[99] *Ibid.*, VII, 159.

[100] *Ibid.*, VIII, 19.

[101] *Ibid.*, VIII, 1–2 (June 24, 1685).

Havers proposed to dissect the creature and to give a report at the next meeting. The source of Dr. Haver's opossum was not revealed.[102]

Three communications from Nevis reached the Royal Society in the 1690's. On July 23, 1690, Robert Hooke read a discourse about a recent earthquake at Nevis and the Leeward Islands. He explained that it was the consequence of an explosion of underground "vapours," "like subterranean lightning."[103] Sir Robert Southwell read two letters (October 20, 1697) from "Mr. Tovey of Nieves" describing hummingbirds there.[104] And the last seventeenth-century communication from the West Indies—more or less a fruit of the Royal Society's promotions there—was "An Account of a Foetus, voided by the Ulcered Navil of a Negro in Nevis, by Mr. James Brodie; Communicated by Dr. Preston." This appeared in the *Philosophical Transactions* in 1697.[105] It is not for the reader who is easily nauseated by unpleasant medical details, for the fetus was, of course, dead and so badly decayed that it was remarkable that the woman survived at all.

New sources both for observations and collections developed in the West Indies in the eighteenth century, although the geographical emphasis was somewhat altered. Besides the continued promotional efforts of the Royal Society, there arose also additional patronage from private groups, some of them stemming directly from Fellows of the Society. As a whole, these activities had both the assent and approbation of the Society, and their fruits were generally laid before it. Subsequent chapters will treat of these developments.

[102] *Ibid.*, VIII, 289 (Jan. 29, 1689/90). The opossum may have come from Virginia or Maryland, for the term "West Indies" was often used loosely to refer to North American mainland colonies as well as the island colonies. The Society had had descriptions of the opossum before, but this appears to have been the first specimen received. Dr. Haver's report on his dissection was given to the Society on Feb. 19, 1689/90 (*ibid.*, pp. 293–294), but it was not published. It was overshadowed in 1698 by Dr. Edward Tyson's illustrated "Anatomy of an Opossum" published in *Phil. Trans.*, XX, No. 239 (April, 1698), 106–164. Tyson's specimen came from Virginia, supplied by William Byrd II.

[103] *Journal-Book*, IX, 2.

[104] *Ibid.*, X, 47.

[105] XIX, No. 229 (June, 1697), 580–582.

The Hudson's Bay Area

Although the royal charter of the Hudson's Bay Company (May 2, 1670) stated that territories granted to "the Governor and Company of Adventurers of England trading into Hudson's Bay" should "bee from henceforth reckoned and reputed as one of our Plantations or Colonys in America called Ruperts Land," such were the vicissitudes of the company that, throughout the Old Colonial Era, the area contained little more than a number of relatively small—and occasionally ephemeral—fortified trading posts. More than six years before the company was chartered, the Royal Society had exhibited interest in the Hudson's Bay area, primarily, it would seem, because of its concern about the geography of the region and the likelihood of there being a Northwest Passage through the continent to the Orient. On February 17, 1663/64, John Beale read a letter to the Society "concerning Capt. [Thomas] James's Voyages and wintering in charleton island in Hudson's Bay." Again, in December, 1665, when the artful reports of Pierre Esprit Radisson and Médard Chouart, Sieur des Groseilliers, were heard in England, the interest of the Society was heightened. After Captain Zachariah Gillam returned from his voyage to Hudson's Bay in 1668–69, Henry Oldenburg addressed twenty-two queries to him, and these, with Gillam's replies, were read before the Society on May 19, 1670.[106] They furnished considerable information about the voyage, the location of places visited, the weather at different seasons, the variations of the compass during the voyage, the soil, the flora and fauna, the behavior of the tides in Hudson's Bay, and the language, government, religion, and manner of life of the natives. This report to the Royal Society occurred only about a fortnight after the charter of the Hudson's Bay Company had been issued.

Sir John Clapham once remarked upon the extent to which the personnel of the Royal Society and that of the early Hudson's Bay Company interlocked.[107] Four of the seven adventurers in the en-

---

106 *Classified Papers, 1660–1740*, XIX, No. 19.

107 E. E. Rich, ed., *Minutes of the Hudson's Bay Company, 1671–1674* (Toronto, 1942), Introduction, p. xxvii.

terprise before the charter was granted were Fellows of the Society; six of the eighteen adventurers named in the charter were Fellows; and Sir Christopher Wren, who was a heavy investor in the company, was President of the Royal Society from 1680 to 1682.[108] To a lesser extent this interlocking membership existed throughout the Old Colonial Era. It clearly underlay the amicable, and sometimes close, relationships between the company and the Society and enabled the latter to penetrate deeply into the natural history of the Hudson's Bay region.[109] As of April 18, 1672, the minutes of the Society recorded: "There were read Some Observations Concerning the Voyage lately made to East Hudsons Bay, and the Constitution of that Country and its Inhabitants, Communicated to the Secretary, upon his Enquirys by two of the chiefest persons Employed in that Voyage who had wintered there."[110] The report derived from a second interview by Henry Oldenburg with Captain Zachariah Gillam and Charles Bailly, first Governor of the Hudson's Bay Company forts in Hudson's Bay, and the queries, together with "the chiefest persons" answers, are preserved in the Royal Society Archives.[111] It stimulated discussion in the Society about the possibility of the discovery of a Northwest Passage. Someone pointed out that José d'Acosta had suggested the likelihood of it, and the speculations gave rise to an article in the *Philosophical Transactions*.[112] A few years later (1678), Dr. Daniel Cox related "a description of certaine Animals that were found in a mighty large plaine or champion Country lying between Hudsons Bay & California in vast great numbers or heards." One Fellow (Mr. Henshaw) hazarded the opinion that these were a "Sort of Indian Goats call'd by the Spaniards Vac-

[108] Much of the material in this section appeared in my article, "The Royal Society and the Company," in the 275th anniversary number of *The Beaver* (Outfit 276, June, 1945), pp. 8–13. *The Beaver* is a quarterly journal published by the Hudson's Bay Company at Winnipeg.

[109] In 1680 the company instructed its Governor to collect seeds in Hudson's Bay for experimental purposes. *Hudson's Bay Co. Archives*, A-6, I, 5–7, 10 (May 21, 29, 1680) (Hudson's Bay House, London).

[110] *Journal-Book*, IV, 245.

[111] Printed in full in Appendix V below.

[112] IX, No. 109 (Dec. 14, 1674), 207–208.

cuneos or Quereanadoes [vicuña]."[113] This was the Royal Society's first introduction to the American buffalo. Henshaw also affirmed that there was no Northwest Passage, and that "he could demonstrate that Hudson bay and the South Sea were 1000. miles distant." In 1681 the President, Sir Christopher Wren, gave an account of the way of life of the natives of the Hudson's Bay region, with particular emphasis upon their alleged longevity and their manner of tanning hides. He said that the people live to the age of 130 or 140 years "without the use of spectacles," and that they go on hunting trips to the very end of their long lives. The discussion about the Indians' methods of tanning furs "soe as to be like a peice of Cotton [when] they lye upon it in the night" continued into the following week, and Robert Hooke reminded the membership of earlier accounts of the Indians' methods of tanning leather in Virginia.[114] Early in 1687, Dr. Cox produced maps and descriptions of the Great Lakes region "which he affirmed some Englishman had surveyed." Whether his information stemmed from the Hudson's Bay region or from the reports of discoveries to the west of the Virginia settlements (which were reaching England about this time)[115] is impossible to say. He reported that the Great Lakes were "a great mediterranean sea of above 5000 mi. around," that the sources of the Susquehanna River—and possibly the Delaware River—were in the lake, or lakes (he appeared to be uncertain as to how many lakes there were), and he felt that the English could make an "advantageous settlement of the beaver trade . . . on these Lakes."[116]

From the outbreak of the Wars of the Palatinate until after the War of Spanish Succession (1688–1713) the Royal Society received little information from Hudson's Bay. In 1690 Edmond Halley presented the Society with "a white Hudson-bay Partridge whose feet were all overgrown with a thick Down to preserve them from the Cold." Halley had received it from a sea captain,

[113] *Journal-Book*, VI, 117 (July 11, 1678). See also the following week's discussion, *ibid.*, p. 119.

[114] *Ibid.*, VII, 22–25 (June 29, July 6, 1681).

[115] See Chapter 6 above.

[116] Forty years before Oswego was established! *Journal-Book*, VII, 5–6 (Jan. 12, 1686/87).

who reported that the partridge was decked in its winter feathers, whereas "In the Summer he turns Gray like the ends of two feathers on his breast, & the down or hair weares off his feet."[117] The captain also told Halley of the "Island of Ice" between Iceland and Newfoundland, which Halley related to the Society in the following June. During January, 1694, the Society received two minerals from the Hudson's Bay area, neither of which excited much comment.[118] In 1699 "a girdle from Hudson Bay & also a lower jaw of a Beaver" were presented; and in 1701 Sir Robert Southwell, on the authority of James Young, a captain in the company's employ, reported on the habits of Indians west of Hudson's Bay.[119] Nothing more arrived until the end of the War of Spanish Succession, when a pair of snowshoes used by the natives of Hudson's Bay was presented and placed in the Society's Repository (July 16, 1713).[120] The following November, one Mr. Callet, writing from York Factory on September 19, desired a correspondence on philosophical matters, but no records of further commerce with him exists in the Royal Society Archives.

Captain Christopher Middleton (d. 1770), one of the most able employees of the Hudson's Bay Company in the early eighteenth century, won much attention from the Royal Society. On February 17, 1725/26, Edmond Halley presented to the Society "A New and Exact Table, Collected from several Observations, taken in from Voyages to Hudson's Bay . . . from London: Shewing the Variation of the Magnetic Needle, or Sea Compass, in the Pathway to the said Bay, according to the several Latitudes and Longitudes, from the Year 1721, to 1725. By Mr. Christopher Middleton." The tables were published in the *Philosophical Transactions* in 1726.[121] In the course of the next fifteen years, Captain Middleton supplied the Royal Society with similar tables which were the fruits of his observations during a dozen voyages to Hudson's Bay. Most of them included meteorological observations and records as well, and the report of the voyage of 1736–37 included astronomical

---

117 *Ibid.*, VIII, 296 (March 19, 1689/90).

118 *Ibid.*, IX, 149, 152.

119 *Ibid.*, X, 212.

120 *Ibid.*, XI, 375.

121 *Ibid.*, XIII, 551–553; *Phil. Trans.*, XXXIV, No. 393 (March–April, 1726), 73–76. A biographical sketch of Middleton is in the *D.N.B.*

data gathered during an eclipse of the moon at Hudson's Bay (September 8, 1736). All of these were published in the *Philosophical Transactions*.[122] In the meantime, Captain Middleton was invited to attend a meeting of the Royal Society as a guest of the Secretary, Cromwell Mortimer (February 24, 1731/32), and in recognition of his services to the Society was elected Fellow on April 7, 1737.[123] In the same year he aroused further interest by becoming one of the first navigators to employ Hadley's quadrant for the determination of longitude at sea, and, at the request of Secretary Mortimer, he performed an experiment at Hudson's Bay to discover whether "sea water discharged its Salt by the act of freezing." He melted forty quarts of frozen sea water, and after distillation found about six ounces of salt, which he estimated to be one one-hundredth part of the whole.[124]

In 1740 Captain Middleton left the service of the Hudson's Bay Company and accepted a post as Commander in His Majesty's Navy. In this capacity he sailed again to Hudson's Bay in 1741, this time in a fruitless search for the Northwest Passage. After his return in 1742, he communicated to the Royal Society "some very curious Observations made in 1741–42 on his voyage to discover a NW Passage to the East Indies while he resided at Prince of Wales Fort in Churchill River. . . ." These were published in the *Philosophical Transactions* in 1742.[125] During this year, also, Com-

---

122 Records of Middleton's report of the Royal Society are in *Journal-Book*, XIV, 632 (June 24, 1731); XV, 76, 82 (Feb. 24, 1731/32). See also *Phil. Trans.*, XXVII, No. 418 (March–May, 1731), 71–78; XXXVIII, No. 429 (July–Oct., 1733), 127–133; XXXIX, No. 442 (July–Sept., 1736), 270–280.

123 *Certificates, 1731–1750*, p. 124. Signed by Sir Hans Sloane, Edmond Halley, R. Middleton Massey, and Thomas Birch. Another (unpublished) record of his voyage of 1736–37 was reported in *Journal-Book*, XVII, 237–239 (April 20, 1738).

124 *Journal-Book*, XVII, 355–356 (Jan. 25, 1738/39); *Phil. Trans.*, XLI, No. 461 (Aug.–Dec., 1741), 806–807.

125 *Journal-Book*, XVIII, 445–446 (Oct. 28, 1742); *Phil. Trans.*, XLII, No. 465 (Oct.–Nov., 1742), 157–171. Middleton's later associations with the Hudson's Bay Company were clouded by the suspicion that he cooperated with Arthur Dobbs in the latter's attacks upon the company, alleging that company lethargy and avarice had delayed efforts to discover the Northwest Passage. See *Journal-Book*, XX, 251–253 (April 9, 1747); Arthur Dobbs, *An Account of the Countries Adjoining to Hudson's Bay in the North-west Part of America* . . . (London, 1744; reprint, Johnson Reprint Corp., New York, 1967); Christopher Middleton, *A Vindication of the Conduct of Christopher Middleton* . . . (London, 1743; reprint, Johnson Reprint Corp., New York, 1967); and the *D.N.B.*, "Arthur Dobbs."

mander Middleton was awarded the Copley Medal by the Royal Society, thus becoming enrolled with such scientists as Stephen Gray, Jean Théophile Desaguliers, Sir William Watson, and, a few years later, Benjamin Franklin. The remainder of his naval career was largely confined to European waters. For nearly twenty years he had served the Royal Society and had supplied it with much valuable data in navigation, geography, oceanography, astronomy, physics, and the natural history of the Hudson's Bay region.

Captain Middleton's contributions to the Royal Society heralded a period of scientifically rich relationships with the Hudson's Bay region. The occasional small gifts continued to come in to the Society—as, for instance, George Edwards' presentation of "a large white owl from Hudson's Bay, with a description of its way of life."[126] The English Parliament's offer (1746) of a prize of £20,000 to the English ship commander who would discover and sail through the Northwest Passage to the Pacific led to intensified efforts in this direction, although Henry Ellis (1721–1806), the hydrographer, mineralogist, and later Governor of Georgia and Nova Scotia, was skeptical about the existence of such a passage after a voyage to the Hudson's Bay region in 1746–47.[127] However, he brought back a report to the Royal Society and was elected Fellow (1750) for his pains.[128] The scientific excitement generated by the transit of Venus in 1761 led the Society to make careful preparations for the short-cycle repeat performance in 1769. Years before, Edmond Halley had urged more exact observations in 1761 and 1769 for accurate determination of the solar parallax in order to calculate more closely the distance between the earth and the sun.

Conscious of the value of observing the parallax from several widely separated places on the earth, the Royal Society, with the cooperation of the Hudson's Bay Company, planned to locate an observation post—among others—in Hudson's Bay. William Wales (c. 1734–98) was selected as the Society's observer, to be assisted by Joseph Dymond. They were supplied with appropriate instruments and, with all their gear, transported to Prince of Wales'

---

126 *Journal-Book*, XIX, 251–252 (May 24, 1744).

127 Ellis, *A Voyage to Hudson Bay* . . . (London, 1748).

128 *Journal-Book*, XXI, 163, 237, 246.

Fort at the mouth of the Churchill River in 1768. The company refused remuneration for their transport but accepted £250 "for the accommodation of the observers."[129] By early November, 1769, Wales and Dymond had returned to England, and on November 9 they handed in their astronomical observations to the Society.[130] The Society's Council, however, requested them to draw up a more extended report with the full particulars of their voyage. Two weeks later (November 23), they presented a fifty-page manuscript entitled "Observations on the State of the Air, Winds, Weather &c at Prince of Wales' Fort on Churchill River Hudson's Bay."[131] This was referred, with the previous astronomical tables, to the Astronomer Royal at Greenwich. William Wales, however, went further still. He drew up a manuscript entitled "Journal of a voyage made by order of the Royal Society to Churchill River on the Northwest coast of Hudson's Bay; of thirteen months residence in that Country; and of the voyage back to England, in the years 1768 & 1769." This account was read to the Society at two successive meetings (March 8 and 15, 1770) and, according to the minutes, contained the following:

> A variety of new and useful matter; such as latitudes and longitudes of Harbours and Capes, and similar occasional investigations at Sea; Observations for determining the variations of the needle in different latitudes and longitudes; Remarks upon the nature, formation, course, and Navigation of the numerous Islands of floating Ice; concerning the temperature; fossil, vegetable, and animal productions of the Country about Hudson's Bay; of the genius of the inhabitants, and of the Eskimaux, Indians, &c.

This report was published in the *Philosophical Transactions*.[132] Wales subsequently accompanied Captain James Cook on his second and third voyages and was elected Fellow of the Royal Society upon his return from Cook's second voyage in 1776.[133]

The presence of the Royal Society's observers in Hudson's Bay

---

129 *C.M.*, VI, 60 (Dec. 7, 1769).

130 *Ibid.*, VI, 54–55.

131 *Ibid.*, VI, 59; *Journal-Book*, XXVII, 247; *Phil. Trans.*, LX (1770), 137–178.

132 *Journal-Book*, XXVII, 323, 325–326; *Add. Mss.* 5169C, fols. 68–70; *Phil. Trans.*, LX (1770), 100–136.

133 Wales and Dymond were each paid £200 by the Royal Society for their services in Hudson's Bay. *C.M.*, VI, 55, 57.

appears to have inspired the Hudson's Bay Company to unprecedented action. On November 29, 1771, the Secretary of the Royal Society announced receipt of a letter dated the previous day by the Secretary of the company and written to advise that the company was presenting the Society with eight boxes of dried, stuffed skins of quadrupeds and birds, together with a collection of stones and fossils, and a paper of "Descriptive & Historical remarks on several arts sent from Severn River in Hudson's Bay by Mr. Andrew Grahame," Chief Factor at Severn Fort.[134] This was the first of four large collections presented by the Hudson's Bay Company to the Royal Society in as many successive years.

The Council of the Society appointed a special committee (March 26, 1772) to arrange and preserve the specimens.[135] The committee appointed Johann Reinhold Forster (1729–98), the ornithologist and naturalist, to name, describe, classify the specimens, and prepare a complete catalogue of them. On May 21, 1772, Forster read a paper, "Quadrupeds from Hudson's Bay," describing for the Society twenty of the animals; and on June 25 he read a similar paper about the birds of the collection.[136] The subsequent gifts of the company (1772, 1773, 1774) contained insects, fishes, and plants, reported to be "in excellent preservation." The Royal Society reciprocated by presenting the company with a variety of instruments for the use of the company's agents in registering the weather in Hudson's Bay—obviously in the expectation that the observations would be sent to the Society from time to time.[137] The Society's committee, which by 1773 had become a standing Committee on Natural History, was hard pressed to dispose of the items from Hudson's Bay—both in terms of room in the Repository and scientific allocation. It deplored the disorder of the

---

[134] *Ibid.*, VI, 119.

[135] Chairman of the committee was Samuel Wegg, F.R.S., later Governor of the Hudson's Bay Company.

[136] Both papers appeared in *Phil. Trans.*, LXII (1772), 370–381, 382–433. Forster also prepared a paper on roots used by Hudson's Bay Indians to dye porcupine quills (*ibid.*, pp. 54–59) and "An Account of some Curious Fishes, sent from Hudson's Bay" (*ibid.*, LXIII, Pt. I (1773), 149–157). Daines Barrington also published "Observations on the . . . Ptarmigan" from the Hudson's Bay collection in *ibid.*, Pt. II (1773), pp. 224–230. For fuller treatment, see C. S. Beals and D. A. Shenstone, *Science, History and Hudson Bay* (2 vols., Ottawa, 1968).

[137] *C.M.*, VI, 203 (Dec. 23, 1773).

Repository, gave duplicate (second-best!) specimens to the new British Museum, and assigned names after the naturalists who prepared the best scientific descriptions of the specimens—a practice which provoked a plethora of papers on the Hudson's Bay collections.

Despite these difficulties, the Royal Society's scientific curiosity was insatiable. The generosity of the Hudson's Bay Company inspired the Committee on Natural History to recommend that the Society should appeal to the East India Company, the Muscovy Company, the Levant Company, and others for similar collections. On December 23, 1772, Secretary Mortimer, acting upon instructions, addressed a letter to the Earl of Dartmouth, F.R.S., Secretary of State for the Colonies,

> to inform your Lordship that we are desirous of procuring for our Musaeum the more uncommon natural productions of his Majesty's most extensive dominions in America.
>
> Whether your Lordship will think it proper to lay this matter . . . before his Majesty, or in what other manner these natural productions may be procured, is with great deference submitted to your Lordship.
>
> We take the liberty of enclosing a Copy of some short instructions which may be useful in forming (and preserving) such a collection. . . .

The Earl replied three days later, promising to lay the matter before the King and "to take any other steps that may be necessary to forward their wishes." Had the responses been commensurate with the Hudson's Bay Company's contributions, it is difficult to imagine what the Committee on Natural History could have done. Perhaps it was fortunate that the disorders attendant upon the approaching American Revolution precluded any such eventuality!

In the meantime, the Royal Society, acting through the Committee on Natural History, sought further to repay the Hudson's Bay Company in a practical manner. On May 5, 1773, the Committee reported to "the Governor and Committee of the Hudson's Bay Company" its attempts "to find out whether some of the natural productions which you have been so obliging as to present to the Royal Society may not furnish materials for our manufactures. . . ."[138] It found that buffalo hides "seem to be as good a material as the Skin of Russian Buffalo for Bookbinding"; that buffalo hair mixed with rabbit made acceptable stockings; and that the down of the "wild Swan" (Canada goose) made excellent powder puffs. Whether or not these suggestions were viable in the marketplace, they illustrate the cordial relations that existed between the Royal Society and the Hudson's Bay Company on the eve of the American Revolution. This cordiality continued throughout the war, for Hudson's Bay was not as seriously cut off from commerce with the homeland as the American colonies were elsewhere during the American Revolution. Efforts were mounted to discover the North Pole, further attempts were made to discover the Northwest Passage, and Thomas Hutchins, Chief Factor and Governor at Albany Fort, in cooperation with the Royal Society, made "experiments on the Dipping Needle," meteorological observations, and "A Paper entitled, Experiments for determining the freezing point of Quicksilver." These were sent to the Royal Society and published in the *Philosophical Transactions*.[139] And in recognition of his work, Hutchins was presented with the Copley Medal (December 1, 1783) by the Society, the second of the Hudson's Bay Company employees to be honored in this manner.

Thomas Hutchins' award was a handsome termination to a century and a quarter of reciprocal association between the Royal Society and the Hudson's Bay Company, a period marked by more meaningful association than any like number of years after. Tangible results are difficult to assess. But surely it is evident that the cooperation of the Hudson's Bay Company materially assisted Fellows of the Royal Society and their scientific colleagues throughout the world to improve their knowledge of geography, to correct maps of the Hudson's Bay area and the Arctic regions, to enlarge their knowledge of oceanography, astronomy, meteorology, natural history, and their information regarding the native peoples of the New World—all of which added priceless increments to learning as a whole, to the problems of classification and

138 This letter is published in full in my article "The Royal Society and the Company," p. 13.

139 LXV, Pt. I (1775), 129–138; LXVI (1776), 174–181; LXVIII, Pt. II (1783), 303–370.

nomenclature of the natural productions of the Arctic regions of North America, and to the formulation of natural laws which regulate them. And while the Hudson's Bay Company contributed to the advancement of knowledge, it also reaped profits in the form of improved maps, safer navigation, wider and more economical uses of the raw materials of the Hudson's Bay region, and infinite good will.

The Southern Mainland Colonies

A New Generation of Fellows in the Royal Society

A NEW GENERATION OF FELLOWS WAS ARISING in the latter years of the seventeenth century. Although by no means the greatest scientist, one of the most eminent personalities of this new generation was Dr. Hans Sloane. From the time of his return from Jamaica (1689) until his retirement to Chelsea (1741), Dr. Sloane was a central figure in the scientific promotions and collections of the time. From his popular, widespread medical practice, which reached into the highest circles of English society, he made a fortune and won a deserved reputation as a philanthropist.[1] He made many contributions to the poor, both in money and in services, and his collections of scientific specimens, medals, coins, and other antiquities, together with scientific books and rare manuscripts, were matchless in their time. At his death, in 1753, his will provided that his collections and papers should become an instrument for further education and research. To this end he proposed that the nation should purchase his collections from his daughters for £20,000 and make them into a national museum. The British Parliament accepted the terms, and Sloane's collections thus became the nucleus of the British Museum. In 1712 Sloane had purchased the Manor of Chelsea as a country home and place of retirement. The manor included the site of the Chelsea garden and, finding that the Society of Apothecaries was in financial difficulties in supporting it, Sloane

[1] A scrutiny of some of Sloane's "Medical cases" in *Sloane 4078* and *4075* demonstrates that his practice included all classes of society, both at home and abroad. Much of it was by mail, often including consultation with other physicians in France, Portugal, Ireland, the colonies, and elsewhere. Evidences of his private philanthropy are widely scattered among the *Sloane Manuscripts*.

conveyed the garden to the society (1722) for five pounds per year together with a yearly presentation to the Royal Society of London of fifty plants, all different, grown in the garden each year until 2,000 had been supplied.[2] With Sloane's aid, the Society of Apothecaries was able to employ as gardener the gifted botanist Philip Miller (1691–1771), whose popular work, the *Gardener's Dictionary* (1731), ran through eight editions in Miller's lifetime, and the Chelsea garden emerged as the most famous English botanical garden during the remainder of the Old Colonial Era, performing valuable services to the sciences of botany and pharmacy.

Dr. Sloane terminated his services to the Duchess of Albemarle in 1693, married (1695) the wealthy widow of a Jamaican sugar planter, and purchased a house in Bloomsbury, near the present site of the British Museum. Here he practiced medicine for forty-seven years. Here, too, he entertained his friends from the Royal Society and the Royal College of Physicians, setting a splendid table with "the best chocolate in London."[3] Doubtless his social charm contributed to his public acceptance. He was probably a Fellow of the Royal Society longer than any other person, his membership having extended from his election in 1684 until his death in 1753. He was chosen Secretary of the Society in 1693 and served until 1712. In the meantime, he was repeatedly elected a member of the Council, Vice-President in 1713, and, upon the death of Isaac Newton in 1727, Sloane was elected President, a position he held continuously until his retirement in 1741. He was also President of the Royal College of Physicians from 1719 to 1735. As Secretary and Council member of the Royal Society he was one of the chief instruments for the resuscitation of the Society after its disastrous decline in the 1680's. He personally maintained a worldwide scientific correspondence throughout his active life, and few scientific promotions took place during these years without his participation. In 1716 he was raised to a baronetcy by George I, thereafter, of course, being known as Sir Hans Sloane.

It was probably Dr. Sloane who, soon after his return from

[2] Terms of the bequest are given in Henry Field, *Memoirs, historical and illustrative, of the Botanick Garden at Chelsea* . . . (London, 1820), pp. 36–38.

[3] Sloane was author of a recipe for milk chocolate, used by the Cadbury firm in the nineteenth century. G. R. de Beer, *Sir Hans Sloane and the British Museum* (London, 1953), p. 60.

Jamaica in 1689, was the moving force in the assembly of the Temple Coffee House Botany Club, which came into being in the autumn of that year. Several such private, unofficial, and voluntary associations sprang up in the early eighteenth century both within the Royal Society and outside it, and between those who were Fellows and those who were not.[4] Some of the groups existed for only a few months, others continued for years. The members assisted one another in their private collections, engaged in correspondence overseas, performed experiments, demonstrated and shared their discoveries with one another at their meetings, and frequently carried their findings to the Royal Society, which welcomed their efforts and gave them tacit approval. Such private groups as the Temple Coffee House Botany Club supplemented the promotional activities of the Royal Society, which, renewed from time to time, remained essentially unchanged in the eighteenth century.[5] However, as the new century moved on, the promotions of individual Fellows, working privately or in conjunction with small groups, rivaled and even surpassed the efforts of the Secretaries of the Royal Society in establishing fruitful scientific correspondents in the American colonies.

The Temple Coffee House Botany Club was a voluntary and unofficial group, social as well as scientific in its *raison d'être*, and, as it apparently had little formal organization, there are no records of it in any collected sense. Only from the private correspondence of its members are we informed of its, existence. It met on Friday evenings, and it mixed social intercourse with botanical discussions, the exchange of botanical specimens, and communications

---

[4] One was the Botanical Society, 1721–26. See John Nichols, ed., *Illustrations of the Literary History of the Eighteenth Century* (8 vols., London, 1817–58), I, 217n.

[5] Nearly every newly elected Secretary of the Society promptly sent out letters of inquiry and invitations to begin, or to continue, scientific correspondence with the Society. Richard Waller in 1713–14, James Jurin in the early 1720's, and Cromwell Mortimer in 1730 were especially active in this matter. See some of Waller's letters, including invitations to Cotton Mather and the President of Yale College, in *Letter-Book*, XV, 4ff. (to Mather, pp. 44–47; to Yale, pp. 14–15); for Jurin, *Journal-Book*, XIII, 210–211, 538–539; *C.M.*, II, 284, 290, 305 (Jurin distributed eighteen barometers and thermometers at the Society's expense to correspondents to enable them to make meteorological observations, 1725–26); for Mortimer, *Letter-Book*, XIX, 494–495 (1731).

from correspondents—all, as James Petiver (one of its most active members) put it, for the benefit of "the World in the Recreative Science of Botany."[6] On Sundays and holidays during the summer months members of the club occasionally embarked upon botanizing expeditions to Greenwich, Hampton Court, Primrose Hill, the Fulham Palace gardens, and elsewhere. By the spring of 1691 the club was said to consist of forty members.[7] These included persons already familiar—Dr. Sloane, Dr. Martin Lister, Dr. Tancred Robinson, Nehemiah Grew, William Sherard, William Charleton, George London, James Petiver, Samuel Dale, and Dr. Leonard Plukenet, who soon quarreled with Sloane and Petiver. In addition there were Charles Hatton, an aristocratic promoter who had encouraged John Ray in his various works; James Ayrey, a Quaker merchant; Charles Dubois, an officer of the East India Company who soon was elected Fellow of the Royal Society; Adam Buddle of Henley, the accepted authority on English grasses and mosses; and occasional visitors from Oxford, Cambridge, the American colonies, and the European Continent. As time passed, new and younger men were added to the group, and it appears likely that the occasional visitors included William Byrd II before his return to Virginia in 1704 after the death of his father. The club continued to hold its weekly meetings until about 1720, perhaps longer, although, as has been noted, its personnel was obviously fluid.[8] To some extent the correspondence of its members enlarged the vision of colonial correspondents. Sloane, Petiver, and Sherard, in particular, by their correspondence and exchange of specimens with British and Continental scholars—including many items of American colonial origin—excited further the interest of Europeans in the natural history of the Americas and sometimes placed their American correspondents in direct exchange with Scottish and European scholars.[9] In turn, before many years had passed—by 1740 or shortly

[6] Petiver to J. P. Breynius, c. Dec., 1692, *Sloane 4067*, fol. 81.

[7] *Sloane 3961*, fol. 41.

[8] *Sloane 4039*, fol. 80; *4020*, fol. 107; *3336*, fol. 11v; *4066*, fols. 277–291; *4067*, fols. 144, 146. Professor George Pasti, Jr., now at the State University of New York at Plattsburg, first described the Temple Coffee House Botany Club in his unpublished Ph.D. thesis, "Consul Sherard: Amateur Botanist and Patron of Learning, 1659–1728" (University of Illinois, 1950).

[9] M. Jean Jacquot, "Hans Sloane and the French Men of Science," *Notes and*

after—colonial scientists were themselves in correspondence with Scottish, Dutch, French, Swedish, German, and Italian scholars. In their exchange with non-English scholars, colonials soon became aware that they were no longer solely dependent upon English sources for their scientific education, direction, and inspiration.[10] Through the efforts of the members of the Temple Coffee House Botany Club, in conjunction with the Royal Society, Dr. Sloane and his associates became a vital center of scientific promotion for half a century—and led American colonials to extend their reach outside the British Empire.

Maryland

A problem important in the minds of English naturalists in the 1690's was to find an able successor to John Banister in the southern mainland colonies. To this problem members of the Temple Coffee House Botany Club—some of whom had been patrons of Banister—addressed themselves with vigor, although many years passed before a suitable man was found. They were assisted in their efforts by William Byrd II, who was elected Fellow of the Royal Society in 1696 and continued his interest in the Society until his death in 1744. Francis Nicholson also contributed. Nicholson, though not always popular in the colonies, was a "professional colonial governor," being appointed Lieutenant Governor of the Dominion of New England (1688), Lieutenant Governor of Virginia (1690–92), Governor of Maryland (1694–98), Governor of Virginia (1698–1705), commander of the expedition against Port

---

Records, X (1953), 265ff.; see also: the correspondence of Sloane, Petiver, and Sherard with John Philip Breynius of Danzig, Sloane, passim, and the Sherard Letters in the Royal Society Archives; with Dr. Patrick Blair of Dundee, Charles Preston of Edinburgh, Fredrik Ruysch, Anton van Leeuwenhoek, Hermann Boerhaave, and various members of the Gronovius family at Leiden, Sloane, passim; and others.

[10] One index of this development is the number of colonial medical men who studied abroad at non-English medical schools, especially at Edinburgh and Leiden. Between 1744 and 1776, fifty-two colonials took medical degrees at Edinburgh, thirty-four from the mainland colonies. List of Graduates in Medicine in the University of Edinburgh from MDCCV to MDCCCLXVI (Edinburgh, 1867), passim. A smaller but still significant number studied at Leiden.

Royal (1710), Governor of Nova Scotia (1713), and Governor of South Carolina (1720–25). By his extensive travels and successive offices in the colonies, Nicholson made a wide circle of acquaintances. He was elected Fellow of the Royal Society in 1706.[11] His enthusiastic support of the English Society for the Propagation of the Gospel may have lent support to Bishop Compton's doubtful use of that society's funds to support a scientific correspondent in Maryland with monies earmarked for Church of England missionary activities.[12]

This correspondent was the Reverend Mr. Hugh Jones (d. 1701/02), a Welshman who previously had been Edward Lhwyd's "servitor and Deputy at the [Ashmolean] Museum" in Oxford.[13] He was only twenty-four years old and had matriculated at Gloucester Hall in 1694—hardly, as yet, a proper candidate for ordination in the English church. But in their eagerness to find "a fit man to succeed Mr. Banister,"[14] Lhwyd, Bobart, Lister, London, Bishop Compton, Nicholson, and the members of the Temple Coffee House Botany Club pushed through Jones's ordination in 1695, arranged for him to be paid twenty pounds in advance from public funds,[15] and sent him off to Maryland.[16] The original intent had been for the appointee to serve as chaplain to Governor Nicholson and as Commissary for the Bishop of London, a position to which Thomas Bray was subsequently appointed.[17] But Jones was too immature for the latter post, and although he stayed with Governor Nicholson for five weeks upon his arrival in Maryland, he ultimately settled as minister of Christ Church Parish in Calvert

11 *Col. F.R.S.*, pp. 197–198.

12 *Petiver*, pp. 293–295.

13 Lhwyd to Lister, Aug. 26, 1695, *Lister 36*, fol. 128 (Bodleian Library). Care must be exercised not to confuse this Hugh Jones with several others of the same name at the time, especially with the Hugh Jones (d. 1760) who wrote *The Present State of Virginia* (London, 1724). See *Petiver*, p. 292n.

14 Lhwyd's words to Lister, Jan. 1, 1693/94, *Lister 36*, fol. 83.

15 *Calendar of Treasury Books, 1693–1696*, p. 1266 (Dec. 27, 1695).

16 Jones had toyed with the idea of going to the East Indies. See Lhwyd's letters to Lister in *Lister 3*, fols. 140, 142, 144, 145. Lhwyd assured Lister that Jones would "be no lesse industrious than Mr. Bannister."

17 Lister to Lhwyd, Jan. 10, 1693/94, *Ashmole 1816*, fols. 107–108 (Bodleian Library).

County. He remained in this position until his death early in 1702.[18]

Before he sailed for Maryland in May, 1696, Jones was warmly courted by members of the Temple Coffee House Botany Club, especially by the ambitious apothecary, James Petiver. They supplied him with equipment for collecting flora and fauna, a thermometer for weather observations (unfortunately broken on the voyage), and showered him with pleas for specimens of the natural productions of the New World. Unfavorable winds delayed the departure of Jones's ship, and he was marooned in the Downs for more than a month. However, he fished up some shells and seaweeds to the delight of his London friends, and one plant specimen was sent to John Ray, who included it in his *Synopsis Methodica Stirpium Britannicarum*.[19] He arrived in Maryland in early August, 1696, and promptly started making collections for his English friends. On March 26, 1697, he addressed letters to Messrs. Petiver, Doody, and Ayrey, and sent two boxes, one with trees, plants, seeds, and berries for Doody and Ayrey, and the other with fossils, shells, and insects for Petiver.[20] Governor Nicholson had provided him with an assistant to "help him in the part of his greatest drudgery, as carrying the box, basket, &c," and his English friends spoke of endowing him with another, more useful helper, "being a Batchelor of Arts allready," although nothing came of the proposal.[21] To brighten Jones's life in Maryland, James Ayrey sent him bottles of beer "for yr morning draughts," and Petiver sent him a Cheshire cheese, medicines, newssheets, copies of some of his own "Centuries" (i.e., *Musei Petiveriani*), Sloane's *Catalogue of Jamaica Plants* (*Prodromus*), copies of the current *Philosophical Transactions*, long lists of American plants extracted from Banister's catalogue and from Plukenet's *Phytographia* (so that Jones could compare

[18] His will is listed in Jane Baldwin, ed., *The Maryland Calendar of Wills*, II (Baltimore, 1906), 228. See also *Wm. & Mary Qty.*, 1st Ser., X (1902), 203.

[19] (3rd ed., London, 1724), p. 39.

[20] *Sloane 3333*, fols. 65–67, 71v. The "large letter," relating particulars of Jones's voyage and his early impressions of Maryland, is published in *Petiver*, pp. 297–300.

[21] J. Bobart to Lhwyd, March 5, 1697/98, *Ms. English History C. 11*, fol. 13 (Bodleian Library).

these with what he found in Maryland), and additional supplies of brown paper, wide-mouth bottles, and spirits for preserving specimens of flora and fauna.[22] In 1697 Jones was unhappily caught in the crossfire of a running dispute between Dr. Leonard Plukenet and Dr. John Woodward, on the one hand, and Petiver, Sloane, and other members of the Temple Coffee House Botany Club on the other. Dr. Woodward (1665–1728), F.R.S., was an able but contentious physician, geologist, and botanist, and Plukenet, who was said to be "very intimate" with Woodward, has already been noted as a severe critic of Petiver and Sloane. These men—Woodward in particular—protested because Jones supplied Maryland curiosities and specimens to Petiver, Sloane, Sherard, London, Lhwyd, Bobart, and others without also supplying them. Indeed, Woodward appeared to imply that Jones should collect for him and Plukenet to the exclusion of all the others, to whom he applied a variety of uncomplimentary names. Tancred Robinson wrote to Edward Lhwyd that "It seems Dr. W[oodward] had the Impudence to write to the Governour [Nicholson] complaining of Mr. Jones his sending naturall things out of the Country to Rogues and Rascalls here, but the Governour only ridicul'd him for his pains. . . ."[23] Plukenet died in 1706, but Woodward lived on for many years to insult many of his contemporaries and to win the reputation of being an insolent, jealous, and thoroughly unpleasant fellow. But he suc-

---

[22] See various letters from Petiver to Jones, 1697–98, in *Sloane 3333*, fols. 73ff. Petiver's medicines were for Jones's landlord, the landlord's wife, and Jones himself. See *Petiver*, pp. 300–302.

[23] Sept. 20, 1698, *Ms. English History C. 11*, fol. 90v. *Sloane 3333*, fol. 91v (Dec. 31, 1697): "Our good ffriend Dr. Woodward is mightily disgruntled yt you should send petrifications to me & Mr. Lloyd [Lhwyd] with some other reflexions, wch I wonder not at since he has not spared ye best of men, nay he has villified all. . . . This I thought fit to acquaint you withall yt forewarned you may know who are yr ffriends & yt Dr. Woodward's anger will not I hope frighten you from sending by all opportunitys to me & ye rest of yr friends. . . ." Petiver referred to the quarrel again in a letter of March 10, 1697/98 (*ibid.*, fols. 119v–120), imploring Jones to hasten his shipments lest Woodward (who was reported to have sent his own collector to Maryland) steal a march on Jones's correspondents. He said that Woodward somehow had won possession of "All Mr. Banister's Collection of Plants &c . . . so yt I despair seeing of them." Doody mistakenly had given Woodward "all ye greatest part of those ffossils you sent him," and both Robinson and Lhwyd were agitated about it.

ceeded in getting some fossils from the Reverend Hugh Jones's collections.[24]

Hugh Jones served his English friends until illness overtook him in 1700, and death intervened only little more than a year afterward. On January 23, 1698/99, he wrote a long letter to Benjamin Woodroffe, F.R.S., his former Principal at Gloucester Hall, Oxford. Woodroffe sent the letter to the Royal Society, where it was read January 17, 1699, and extracts were printed in the *Philosophical Transactions*.[25] Evidently Woodroffe had solicited an account of Maryland, and Jones's letter gave a general description of the colony, its geography, climate, soil, varieties of timber, animals, birds, Indians, Indian medicine, tobacco culture, and the success of the Church of England, with Governor Nicholson's assistance, in overcoming Roman Catholics and Quakers.[26] Jones's other communications and shipments of specimens excited the admiration of a wide circle in England, although Jacob Bobart complained that "the trees he sent to London were soe ill & ignorantly packt, that they were all spoiled."[27] As a whole, however, Jones's specimens arrived in good condition, and they were widely distributed among his English friends. Bishop Compton and George London received plants and trees for the Fulham Palace gardens. William Sherard cultivated Maryland plants and seeds in the Duchess of Beaufort's gardens at Badminton, where he was in tutorial charge of the grandson of the Duchess, and set out to improve the Duchess's garden, which was rapidly developing into one of the finest in England.[28] Jacob Bobart experimented in the Oxford Physic Garden with seeds supplied by Jones.[29] Dr. Sloane, James Ayrey, Tancred Robinson, and Samuel Doody shared seeds, berries, plants, and

[24] John Woodward, "A Catalog of the Foreign Fossils in the Collections of J. Woodward, M.D. . . . ," in *An Attempt towards a Natural History of the Fossils of England* . . . (London, 1729), Pt. II, pp. 5, 7, 11, 14, 15, 24, 29.

[25] XXI, No. 259 (Dec., 1699), 436–442.

[26] For the original of Jones's letter, see *Guard-Book*, J, No. 183; *Letter-Book*, XI, Pt. II, 237–245. See also *Journal-Book*, IX, 188 (Jan. 17, 1699/1700).

[27] *Ms. English History C. 11*, fol. 14.

[28] See Sherard's letters in *Sloane 4038*, fols. 58, 332; *4063*, fol. 97; *4070*, fol. 175; *4071*, fol. 175. In June, 1699, the Duke of Beaufort received a box weighing 400 pounds containing eleven Virginia trees with earth to preserve them. See *Sloane 4062*, fol. 305.

[29] *Ms. English History C. 11*, fol. 14.

trees from the Maryland collector. Dr. Lister, Edward Lhwyd, and James Petiver received Maryland fossils, mollusks, minerals, birds, and insects at the hands of Mr. Jones. John Ray and his friend Samuel Dale discussed Maryland specimens received from Petiver and others among Jones's English correspondents.[30] But of all Jones's correspondents, James Petiver gave him the greatest encouragement, received the lion's share of his Maryland collections, and published the most regarding them. In 1698 Petiver published in the *Philosophical Transactions* an article describing crustaceans, shells, insects, and plants "sent to him from Maryland, by the Reverend Mr. Hugh Jones."[31] Further, both in his *Musei Petiveriani* and in his *Gazophylacii*, Petiver acknowledged the contributions of Mr. Jones, referred to as a "very curious Person in all parts of Natural History; particularly in Fossils; some of which he hath sent me from *Maryland*, with several *Volumes* of *Plants* very finely preserved; with divers *Insects* and *Shells*. From this obliging Gentleman, I am promised frequent remittances of whatever those Parts afford, as well *Animals* & *Fossils* as *Vegetables*."[32] Petiver's manuscript remains testify further to the collections of Mr. Jones by way of catalogues and descriptions of hundreds of Maryland plants and insects.[33] The *Sloane Herbarium* in the British Museum still has several of the Maryland specimens which Jones sent to his English patrons.[34] However, although he proved to be an industrious field collector, Jones did little to describe what he gathered or the places where he gathered his specimens, leaving all matters of description, classification, and nomenclature to the English recipients of his collections. He published nothing, and his manuscript

---

[30] Petiver to Dale (several letters), *Sloane 3321*, fols. 57ff.; *3333*, fols. 251ff.

[31] XX, No. 246 (Nov., 1698), 393–406. Some of the items were from Virginia by the hands of others. Cf. *Sloane 3324*, fol. 4.

[32] *Musei Petiveriani Centuria Quarta & Quinta* (1699), Nos. 418–419. See also *Musei Petiveriani* (1695–1703), pp. 39, 50, 56, 59, 69; *Gazophylacii Naturae et Artis Decas Prima* (London, 1702), pp. 5, 7, 8, 10, 41. Tancred Robinson gave the specimens "adept Latin names" at Petiver's request. See Petiver to Robinson, n.d. [Jan., 1699?], *Sloane 3333*, fol. 142v.

[33] Petiver's catalogues of Maryland plants received from Jones are in *Sloane 3330*, fols. 771, 772, 825–826; *3324*, fols. 4–5; *3331*, fol. 50. A list of Maryland butterflies is in *Sloane 3324*, fols. 75–78.

[34] Dandy, pp. 142–143 and *passim*. See also George Edwards, *Gleanings of Natural History* (3 pts., London, 1758–64), Pt. I, pp. 54–58.

remains appear to include nothing by way of a scientific catalogue, paper, or treatise beyond a few rather general descriptions of the Maryland scene such as might have been produced by almost any literate traveler. No wonder, then, if his English friends found him to be disappointing as a successor to John Banister.[35]

The Maryland collections of the Reverend Hugh Jones were supplemented and reinforced by those of Dr. David Krieg and William Vernon. Both of these men journeyed to Maryland in 1698, with the enthusiastic support of members of the Temple Coffee House Botany Club, although while Dr. Krieg appears to have been primarily the protégé of James Petiver, Vernon was patronized by the Royal Society and Governor Nicholson. They sailed separately, Vernon in January and Krieg in March, although Krieg arrived in Maryland more than a month before Vernon. But they worked together to some extent, and both men appear to have made the acquaintance of Hugh Jones.

William Vernon (d. 1706) was a Fellow of Peterhouse, Cambridge. He was well known among the English naturalists and contributed largely to John Ray's studies, both in botany and in insects, especially butterflies.[36] His voyage to America arose from a motion made by William Byrd II at a meeting of the Royal Society on November 10, 1697. Mr. Byrd "moved the Society that if they would think of a Fitt person to send over to Virginia in order to make observations and Descriptions of all the Naturall productions of those parts and to write the History thereof, and that for the Encouragement of such a fitt person the charge of his passage and £25 per Ann. would be allowed him by the Governor of Maryland."[37] The motion met with the Society's favor, and on December 31, 1697, William Vernon was approved as the "fitt person" after a testimonial signed by Sir John Hoskins, Vice-President (and

[35] In 1721 Sir Hans Sloane wrote that he and his friends were "very much disappointed and losers" by the "undertakings" of Hugh Jones and William Vernon, "yet I intend to encourage all these undertakings, notwithstanding they cannot promise better than they did." Nichols, ed., *Illustrations*, I, 278–279. For Vernon, see below.

[36] Vernon to Lister, Feb. 7, 1694/95, *Lister 36*, fol. 111; J. Archer to E. Lhwyd, Feb. 8, March 21, 1695/96, *Ashmole 1829*, fols. 12, 14; Pulteney, I, 264–265; II, 57–58; Charles E. Raven, *John Ray, Naturalist, His Life and Works* (Cambridge, 1950), pp. 257, 301–302, 393, 409–411.

[37] *Journal-Book*, IX, 70.

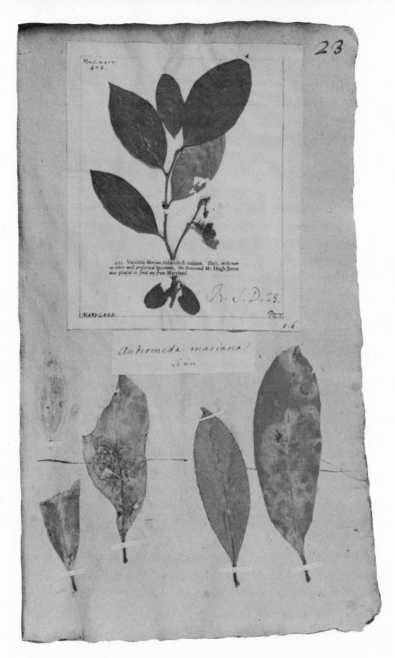

11. A plant specimen sent from Maryland by the Rev. Hugh Jones to James Petiver (c. 1698) and published by the latter. As Petiver's collections were later purchased by Sir Hans Sloane, the specimen remains in the *Sloane Herbarium*, a vast collection of plant specimens still preserved in the British Museum (Natural History). Botanical scholars have studied these specimens from time to time since the eighteenth century without firm agreement on the precise identification of many of them.

Sloane Herbarium, H.S. 249, fol. 23. *Courtesy of the British Museum (Natural History).*

12. Plant specimens sent from Maryland by William Vernon to James Petiver (1698).

Sloane Herbarium, H.S. 37, fol. 102. *Courtesy of the British Museum (Natural History).*

former President) of the Royal Society, and by Dr. Hans Sloane, one of the Secretaries.[38] The following month (January, 1698), Vernon set out for Maryland to claim Governor Nicholson's allowance, arriving there in early April.

Vernon went forth with high expectations and evidently with the intent to stay in America for three or four years. Petiver, pressing faster action upon the Reverend Hugh Jones, wrote that Vernon had said "he will make more discoveries of this nature than any man yt ever was in these parts & will bring over as many Plants, Shells, Insects, Fossills, Serpents, &c as will take up our Botanic Club & Royall Society a twelve month ye looking over."[39] This enthusiasm, however, was short-lived. By mid-July it was reported in England that Mr. Vernon "thinks he will return for old England this winter, he not liking those parts so well as he expected."[40] Indeed, on July 24, Vernon wrote to Dr. Sloane:

> I met severall curious parts of Naturall knowledge, which I'd rather refer to you in the Temple Coffee-House, than in Scriptis. I've a collection of plants for you & any other part of my collection is at your service. When I return which I expect will be the later end of October, I shall bring Every Fryday night a collection of plants to be discussed by you, & that Honourable Club; to whom my service. Mr. Krieg will be back about that time. . . . Out of the 100 Correspondents Woodward told me he had in America, I'm sure since I've not met One, or heard of.[41]

It may have been that Dr. Woodward's machinations affected Vernon's decision to leave Maryland after a single summer there. Even before he left England, Vernon had reported to Sloane that Woodward had impeached his character to Bishop Compton and that he

[38] *Sloane 4068*, fol. 16. Already (Dec. 23, 1697) Vernon had been granted leave by the Visitors of Peterhouse "to be absent for three or four years to improve his Botanick Studies in the West Indies." Christopher Wordsworth, *Scholae Academica* (Cambridge, 1877), p. 207. Vernon evidently had a reputation for perseverance, it having been jocularly reported that he once "followed a butterfly nine miles before he could catch him." John Aubrey, *Letters Written by Eminent Persons . . . and Lives of Eminent Men* (2 vols., London, 1813), II, Pt. I, 100–101.

[39] Oct. 6, 1698, *Sloane 3333*, fols. 170–171.

[40] Petiver to John Ray, July 16, 1698, *ibid.*, fol. 149.

[41] From "Anapolis," *Sloane 4037*, fol. 102.

"told me he was satisfi'd I Kept Company with you Dr. Robinson, Dr Lister etc &c but that wou'd not doe, I must Expect dealings accordingly. In Short he's an abominable Villanous & Silly fellow. . . ."[42] Perhaps Woodward's charges against Vernon led Governor Nicholson to entertain doubts about the recipient of his grant.[43] Whatever may have been the reason for it, Vernon returned to England in the autumn of 1698.

He had been industrious, however, and he made large collections of plants, seeds, insects, and fossils for himself, the Royal Society, and his English friends. The Royal Society was so impressed by Vernon's work that on January 11, 1698/99, it subscribed "in the name of the Society 20 1. for Mr. Vernon . . . towards an encouragement for his voyage to the Canaries."[44] But as Vernon spent all the rest of the winter and the spring following in a vain attempt to find acceptable passage to the Canaries, the venture was abandoned.[45] Subsequently Vernon returned to Peterhouse, where he actively solicited subscriptions for Petiver's works and gave valued assistance to many other English naturalists of the day. His collections were shared widely with the Royal Society, Sloane, Petiver, Lhwyd, and Dr. Richard Richardson of North Bierley, and some were kept by Vernon and subsequently presented to the University of Cambridge. Many were handed about to Ray, Buddle, Lister, and others for study, and Vernon's correspondence for several years after his return from Maryland demonstrated the efforts being made to digest his materials.[46] Petiver published accounts of

[42] However, Governor Nicholson received two quires of dried plants from Vernon which he passed on to William Blathwayt in 1698. See Nicholson to Blathwayt, Annapolis, Aug. 13, 1698, *Blathwayt Papers*, XV, No. 3 (Colonial Williamsburg, Inc.). I am indebted to Professor Bruce T. McCully of the College of William and Mary for this reference.

[43] N.p., Dec. 28, 1697, *Sloane 4061*, fol. 234.

[44] *C.M.*, II, 101–102.

[45] Letters of Vernon to Sloane, Margate, Feb. 5, 1698/99, and Deal, May 23, 1699, and a third, n.p., n.d., in *Sloane 4037*, fols. 209, 274, and *Letter-Book*, XII, 248–249. The failure, alleged to have been the result of Vernon's mismanagement, led to a motion in the Royal Society Council, Nov. 8, 1699, to recover the money from Vernon. The motion failed, and the record of it was omitted in the final draft of the Council minutes. *Sloane 4026*, fol. 364v.

[46] See, for example, the letters to and by Vernon in *Ms. Radcliffe Trust C. 1*, fols. 38ff. (Bodleian Library); *Sloane, passim;* Dawson Turner, ed., *Extracts from*

some of Vernon's Maryland butterflies;[47] Woodward referred to about a dozen of his fossils in "A Catalog of the Foreign Fossils . . .";[48] and John Ray included about forty Maryland plants from Vernon's collections in the third volume of his *Historia Plantarum* (1704). But Ray was unhappy with Vernon's dried plants. He wrote to Dr. Sloane (June 2, 1694):

> Since I wrote last to you, I have, with what care and considera-tion I could, viewed and compared your Maryland dried plants, and find that I can make but poor work with them. They all, save one or two, want of roots, and I can determine the stature of but few of them certainly. The figure and number of petals in the flower can clearly be discerned scarce in any, nor the colour. The Seed-vessel and seed are wanting or imperfect in most; so that it is a very hard matter for me, who am a stranger to them, to determine the genus.[49]

Ray's words reflected much of the disappointment felt regarding Vernon's Maryland venture. Some of his plant specimens are still extant among the herbaria at Oxford,[50] but the larger portion of them fell to Dr. Sloane and rest in the British Museum.[51]

Dr. David Krieg (d. c. 1712) was a Saxon physician who, in the service of various noblemen and military officers, drifted from Leipzig to Riga in the early 1690's and then to London in 1697.[52] He was very skillful in drawing and coloring specimens in natural history, and as he was also learned in these matters, he was wel-comed among the members of the Temple Coffee House Botany

the Literary and Scientific Correspondence of Richard Richardson (Yarmouth, 1835), *passim.*

[47] *Musei Petiveriani Centuria Quarta & Quinta*, Nos. 304, 382. See also Petiver's printed list of butterflies in *Sloane 3324*, fols. 75–78.

[48] In *An Attempt towards a Natural History of Fossils*, Pt. II, *passim.*

[49] Edwin Lankester, ed., *Correspondence of John Ray* (London, 1848), pp. 364–365.

[50] Clokie, pp. 79, 86, 259; S. H. Vines and G. C. Druce, *An Account of the Morisonian Herbarium . . . of Oxford* (Oxford, 1914), pp. 207, 213. John Jacob Dillenius used some of these in the preparation of *Historia Muscorum . . .* (Oxford, 1741), p. viii, *passim.*

[51] Dandy, pp. 226–228, *passim.*

[52] Data reconstructed from dates and places mentioned in his "Album Amicorum, 1691–1697," *Sloane 2360, passim.* See also Pulteney, II, 58.

Club. He lodged with James Petiver while he was in London, made sketches of some of Petiver's collections (especially insects and birds), and the two men became fast friends. When Dr. Krieg set out for Maryland (where he arrived in late March, 1698), Petiver wrote a letter to introduce him to Hugh Jones, saying that Dr. Krieg was

> a most ingenious . . . German Gentleman, he hath lodged at my house about a year in wch time he hath painted most of our English Insects & several other things admirably well. . . . I heartily recommend him to your friendshipp & favour & whatever service you can do him, I shall esteem it as done to my selfe. This Gentleman hath made large Collections & wonderful observations of all parts of Nature that hath come in his way this year whilst in England & I know will do no lesse in your parts, it being the great and only motive of his going over. I have also sent my Butterflie Catcher to attend him, a poor boy I tooke on purpose to run on Errands &c.[53]

"Isaack the Butterfly Boy"—for whom Petiver left no surname—accompanied Dr. Krieg, and Petiver had given the following "Directions for Isaack":

> Whenever you goe ashore take with you a Quire of Brown Paper or Collection Book. An Insect Box, Pins & a small Viall halfe fil'd with *Spirit* in which drown all your supernummery [i.e., unusual] Flies, Beetles, Catterpillars, & other Insects especially such you shall find in water. Also a Booke for Butterflies & Moths of each of wch get al you can find, with a paper bag or two to put all ripe seed, ffruit & berries as also all ye shells you meet with both land & water & as many of each sort as you can find; such as are thin & brittle you must put into a Pocket by themselves with moss or any soft leaves to keep them from breaking.[54]

Laden with this gear, "Isaack" would have cut a rather ludicrous figure to the colonists and Indians,[55] but, unfortunately, Dr. Krieg

[53] N.d. [c. Jan. 1, 1697/98], *Sloane 3333*, fols. 91v–93.

[54] *Ibid.*, fols. 94–94v. Petiver drew up more complete printed directions for his amateur collectors. One is printed in *Petiver*, pp. 363–365. See also *ibid.*, pp. 279–281.

[55] Jezreel Jones, brother of the Reverend Hugh Jones, in collecting for Petiver in Spain in 1701, wrote that "I have been suspected for one that Studys

reported on May 7, 1698, that "as for the boy Isaac he has little opportunity to get something because he is most forct to stay on board."[56] No explanation was offered as to why or upon whose authority Isaac was confined to the ship, but it is evident both that he collected nothing of consequence and that he gave little or no assistance to Dr. Krieg.

Dr. Krieg made the voyage to Maryland as physician on the ship *John and Thomas,* and he also practiced medicine while in Maryland.[57] Like Vernon, he stayed in Maryland for the summer of 1698 and arrived back in England in November, laden with specimens, and drawings of specimens, of Maryland plants, birds, and insects for his English admirers. He remained in England and was elected Fellow of the Royal Society in January, 1699, upon the recommendations of Petiver, Sloane, and others. But in May he accepted a position as physician to a nobleman with whom he returned to Riga. He subsequently traveled in Scandinavia, Holland, and France, returning to Riga permanently in 1703. He continued a lively scientific exchange with his English friends, with frequent nostalgic references to the Royal Society and the Temple Coffee House Botany Club. Indeed, he hoped to return to England and to undertake further scientific journeys for his English patrons, but he was caught up in personal matters and problems posed by the Northern Wars and spent the remainder of his life in Riga. His "Collectanea Curiosa" was subsequently purchased by the Czar of Russia for "an inconsiderable sum."[58]

As with Vernon's collections, Dr. Krieg's Maryland specimens were widely distributed and studied by English scholars. Jacob Bobart received a share at Oxord.[59] John Ray published descriptions of about twenty of them in the third volume of his *Historia Plan-*

---

witchcraft, Necromancy, and a Madman by some who observed me following butterflies, picking of herbs and other lawful exercises and I have had much to do to escape the censure of the higher powers." To Petiver, April 2, 1701, *Sloane 4063*, fol. 76.

56 Krieg to James Ayrey, *Sloane 3333*, fol. 144.

57 For which he was promised payment in tobacco, although it seems unlikely that he ever received the payment. Krieg to Ayrey, May 7, 1698, *Sloane 3333*, fols. 144–144v. See also Krieg's later inquiries about the tobacco in *Sloane 4063*, fol. 112; *4067*, fol. 7.

58 Dr. Nicholas Martini to Sloane, Riga, Dec. 20, 1717, *Sloane 4045*, fols. 83–84.

59 Petiver to Bobart, *Sloane 3333*, fols. 250–250v.

tarum (1704) and some of Krieg's Maryland insects in his posthumous *Historia Insectorum* (1710). James Petiver exhibited several of Krieg's drawings and paintings to the Royal Society,[60] made extensive notes on Krieg's Maryland insects, especially butterflies,[61] and published a few notices of them both in his *Musei Petiveriani Centuria Quarta & Quinta* (1699) and in his *Gazophylacii* (1702–06). But Ray had the same criticism of Krieg's Maryland plants that he had expressed regarding Vernon's collections,[62] although the former's birds, butterflies, and other insects appear to have drawn no adverse comments. Dried specimens of Maryland plants collected by Dr. Krieg are still extant in the herbaria of both Oxford University and the British Museum.[63]

After the activities of the Reverend Hugh Jones, William Vernon, and Dr. David Krieg at the turn of the century, scientific communications between Maryland and England waned. Evidently no articulate scientists of consequence developed in Maryland before the end of the Old Colonial Era. On August 4, 1697, Dr. Sloane exhibited to the Royal Society a fossil from Maryland, communicated to him by Dr. Robinson and thought to be "the tongue or chops [mandibles] of a Pastinaca marina which he [Sloane] had seen in Jamaica."[64] The specimen was supposed to be akin to a thornback, or skate, and Dr. Sloane published an account of it in the *Philosophical Transactions*.[65] One Stephen Bordley, writing from Chester River, Kent County, to James Petiver (March 26, 1706), offered his services in return for "wt News Europe affords."[66] He had seen Petiver's directions for collections, but he had no equipment for the preservation of specimens and inquired whether there were some method to preserve "fleshy bodies" other than in spirits, which were very dear in Maryland. But Petiver's reply miscarried, and when he sought to renew the exchange in 1709, with gifts of brown paper, copies of Banister's "Catalogus," and other information about American plants, shells, and insects, Bordley's enthusi-

---

[60] *Journal-Book*, IX, 291 (Jan. 14, 1701/02) (birds and butterflies).
[61] *Sloane 3324*, fols. 4–12, 75–78.
[62] See note 48 above, and Lankester, ed., *Correspondence of John Ray*, p. 363.
[63] Clokie, pp. 86, 195; Dandy, pp. 151–152 and *passim*.
[64] *Journal-Book*, X, 45.
[65] XIX, No. 232 (Sept., 1697), 674–676.
[66] *Sloane 3321*, fol. 194.

asm was lost, and nothing of consequence resulted from the exchange.[67]

Later scientific relations with Maryland were equally desultory and generally unproductive. Philemon Lloyd, listed in the minutes of the Royal Society as "Secretary to the Province of Maryland," was elected Fellow of the Royal Society on November 9, 1727, but there are no records of any communications he sent to the Society.[68] Similarly Charles Calvert, Lord Baltimore, was elected Fellow (December 9, 1731) and formally admitted to the Society (January 27, 1731/32);[69] and the Honorable Benedict Leonard Calvert, Governor of Maryland, was elected Fellow (March 25, 1731/32), but he died the following June and was never formally admitted to the Society. Neither of the Calverts left any correspondence with the Royal Society indicative of active scientific interest. On May 14, 1733, Mr. Gilbert Falconer, a Quaker, of Kent County, Maryland, having been solicited by the Society's Secretary, Cromwell Mortimer, wrote a long letter to the Royal Society, "giving an account of the extraordinary virtues of some Vegetables in Maryland."[70] He especially extolled the virtues of the Jerusalem oak (*Chenopodium botrys?*), whose stock and leaf (specimens of which he sent), steeped in rum, were useful in the treatment of "hysterick fitts" and fevers.[71] He also praised the "Quinzie Root" (throatwort, probably button snakeroot) which the Indians used both as a gargle and as a poultice for sore throats. He greatly admired Indian medicine and he wrote that "I hope thou wilt give me leave to tell thee as Secretary of the Royal Society, that we in America wonder that they have several times sent Gentlemen over at their expence to take Account of our Insects, but never any to

[67] See the correspondence in *ibid.*, fols. 247–247v; *Sloane 3337*, fols. 62–62v. James Miller and Charles Combs, surgeon, were other Petiver correspondents from Maryland, but their letters do not appear to have survived. Perhaps they were ship surgeons or other seagoing men who touched Maryland only periodically.

[68] *Journal-Book*, XIV, 17, 140; *C.M.*, II, 312. Lloyd was never formally admitted F.R.S. *Ibid.*, III, 238.

[69] *Journal-Book*, XIV, 586.

[70] *Letter-Book*, XX, 332–337; read Jan. 17, 1733/34, *Journal-Book*, XV, 369.

[71] William Byrd II had previously (1708) recommended the virtues of Jerusalem oak, but he had found the seeds, mixed in honey, to be a sure cure for worms. *Letter-Book*, XIV, 239–240.

learn from the Indians the Use of our plants and herbs. . . . The Indians have many valuable Secrets among them, that are not yet communicated to the English, and perform several notable cures, especially in Surgery. The use of the Jerusalem Oak was learned from them. . . ." He urged the Society to send over men who would live with the Indians, learn their languages, and discover their medical secrets: "This we think would be a more usefull enquiry, but perhaps not so pleasant an amusement as the discovery of things curious and new." Having thus betrayed his lack of comprehension of what the Royal Society was seeking to accomplish, perhaps it was no great loss that Mr. Falconer did not continue the correspondence.

Peter Collinson (1694-1768), a Quaker merchant and influential promoter of science in the American colonies (of whom more will appear later), communicated to the Royal Society three letters and several specimens from Richard Lewis, of Annapolis, in the years between 1731 and 1735. Lewis' first letter, reported to the Society on March 4, 1730/31, was an account of the aurora borealis and sunspots as seen with the naked eye on October 22, 1730.[72] His second, read to the Society on January 4, 1732/33, described a "Male Mulberry" or "fly-tree" (so called because it supposedly bred flies in its leaves), an earthquake felt in Maryland and Pennsylvania on September 5, 1732, and a fiery meteor observed at Annapolis on October 22, 1725.[73] Lewis appears to have believed that the "fly-tree" generated flies spontaneously, which sent the Fellows of the Society to Francesco Redi's works, where they could find no reference to it. Lewis' last communication, accompanied by two specimens, was a description of Maryland insects which he called "man-gazers." Dr. Sloane said this was a corruption of "mantis," perhaps the well-known praying mantis.[74]

Secretary Mortimer, in an effort to check further on Gilbert Fal-

---

[72] *Journal-Book*, XIV, 575-576; *Phil. Trans.*, XXXVII, No. 418 (March-May, 1731), 69-70. Isaac Greenwood, of Harvard College, sent a more sophisticated account, based upon telescopic views, that was read at the same time. See *ibid.*; *Guard-Book*, G, Pt. II, No. 10, and *Phil. Trans.*, XXXVII, No. 418 (March-May, 1731), 55-69.

[73] *Journal-Book*, XV, 213-214.

[74] *Ibid.*, XVI, 108; *Letter-Book*, XXI, 308-309; *Phil. Trans.*, XXXVIII, No. 429 (July-Oct., 1733), 119-121.

coner's fulsome account of the virtues of Jerusalem oak and the "Quinzie Root," or throatwort, wrote to Dr. Richard Hill of Londontown, Maryland. Dr. Hill replied on March 8, 1735/36, and his letter was presented to the Royal Society on June 3, 1736.[75] He reported:

> I'm afraid that Jerusalem Oak and the Throatwort, as I call it, have had the credit of Cures they've not altogether deserved: altho' I'm satisfied that they are both usefull Remedies . . . those who are fond of it [Jerusalem oak] say that a glass of it suddenly relieves in a depression of Spirits. . . . But this effect I believe is partly owing to the Rum. Gilbert Falconer, a Gentleman of great veracity, and some others less to be depended on, are very fond of it, and say it is effectual in the above cases. But I've no instance of it within my own observations.

Dr. Hill found Jerusalem oak far more useful in scorbutic cases and in the treatment of worms in children (by mixing the powdered seeds in honey or treacle). He said that throatwort was an Indian remedy widely used by them for sore throats, quinsy, and even tumors. But he made no comment as to its efficacy.

Sir Hans Sloane brought to the Royal Society letters and specimens he received from two Maryland correspondents in 1739–40. Andrew Scott sent a letter (dated August 6, 1739), two live rattlesnakes, and a specimen and short account of a turkey buzzard, which he described as a carrion fowl.[76] He had supplied Maryland seeds to Robert James, eighth Lord Petre (1713–42), a gentleman with splendid gardens in Essex, on which he lavished great pains and treasure to procure seeds and plants from all over the world. And he promised to supply Sir Hans with a collection of seeds and fruits.[77] Edward Lloyd was Sir Hans's second Maryland correspondent. He had seen the Sloane collections in London and had been told by Sir Hans that "an Oyster with a Pearle as it grew in it was a curiosity that would readily find a place among your vast Collection."[78] He had found an oyster with two pearls in it and

---

[75] *Letter-Book*, XXII, 352–354; *Journal-Book*, XVI, 344.
[76] *Letter-Book*, XXVI, 38–40; read Nov. 8, 1739, *Journal-Book*, XVII, 482.
[77] Dandy, p. 202.
[78] Lloyd to Sloane, Wye River, Md., May 14, 1740, *Sloane 4056*, fol. 235.

presented it to Sir Hans. He also enclosed a copy of the Lord's Prayer and one of the Psalms of David written in Arabic by a Maryland Negro slave "who seems very perfectly to understand that Language." Lloyd promised to add further to Sloane's collection at every opportunity, but there appears to be no further communication from him.

After Sir Hans Sloane's retirement (1741), only three persons from Maryland communicated with the Royal Society before the end of the Old Colonial Era. On March 12, 1752, Mr. Richard Brooke, "a gentleman present by leave" at the meeting of the Society, told of having witnessed a strange phenomenon in Maryland on May 3, 1749, when there were three concentric rings around the sun. He also presented an undescribed insect, a specimen of asbestos, and the proboscis of an amphibious creature said to have been killed by an Indian on the banks of the Mississippi River.[79] The following month (April 23), Dr. James Parsons, F.R.S., presented to the Society a paper about "Medico-Physical Observations . . . in Maryland," by Mr. Brooke, now described as a physician and surgeon there. These were mostly notes about Indian remedies, with a specimen of a medicinal root which, as it was neither named nor described, cannot be identified.[80] In May Dr. Parsons communicated another paper from Mr. Brooke, this time relating his experiences with smallpox inoculation in Maryland, which he did without incision on very young persons "by rubbing." He reported wide success, and Fellows of the Society confirmed the method.[81] The following year, Dr. John Pringle, F.R.S., read a letter from Mr. Brooke to praise the efficacy of Maryland pokeweed (*Phytolacca decandra*) as a cure for cancer. Cadwallader Colden of New York was quoted to confirm the value of the treatment.[82] The remainder of Mr. Brooke's communications to the Royal Society consisted of "Thermometrical" accounts of the weather in Maryland from Sep-

[79] *Journal-Book*, XXII, 90. A description of the amphibious creature is in *Wellcome Library Ms.* 67457 (London) among Sir Joseph Banks's papers, but I cannot identify it.
[80] *Journal-Book*, XXII, 119–120.
[81] *Ibid.*, XXII, 134–135 (May 14, 1752).
[82] *Ibid.*, XXII, 285 (March 8, 1753).

tember, 1753, to the end of 1757, with an account of the diseases prevalent there during the last three years, 1754–57.[83]

James Bate, another Maryland surgeon, reported via the Reverend Mr. Alexander Williamson, of St. Leonard's Town, to Thomas Birch, Secretary of the Royal Society, that a Negro maid, a cook owned by one Colonel Barnes of Virginia, had, over a period of fifteen years, changed color from black to "⅘ white." She was in good health, and she was said to blush and flush like a white person. Mr. Bate offered "to try any Experiments that shall be suggested to him to the better examining of this matter."[84] And the last communication recorded from Maryland to the Royal Society before the outbreak of the American Revolution was an "Observation of the transit of Venus taken in Maryland by Mr. John Leeds," Surveyor General of the colony.[85] It was addressed to Nevil Maskelyne, the Astronomer Royal, and passed on to him at Greenwich, where it was studied in relation to other similar observations of the event promoted by the Royal Society in many parts of the world. Leeds's account was published in the *Philosophical Transactions*,[86] although it was relatively worthless, inasmuch as Leeds had no accurate knowledge either of the exact time of the event or of the longitude of the place near Annapolis, where his observations were made.[87] Thus Maryland's colonial contributions to experimental science, never impressive when compared to those of many of her sister colonies during the Old Colonial Era, ended on a characteristic note.

Virginia

Governor Francis Nicholson was transferred from Maryland to Virginia in 1698 and remained in Virginia until 1705. During this

---

[83] *Ibid.*, XIII, 80 (March 13, 1755); *ibid.*, pp. 284–286 (March 8, 1759); *Egerton 2381*, fol. 84v (British Museum); *Add. Mss. 4445*, fol. 304; *Phil. Trans.*, LI, Pt. I (1759), 58–69, 70–82.

[84] *Journal-Book*, XXIV, 311–313 (May 10, 1759); *Phil. Trans.*, LI, Pt. I (1759), 175–178.

[85] *Journal-Book*, XXVII, 269 (Dec. 21, 1769); *Add. Mss. 5169B*, fol. 260.

[86] LIX (1796), 444–445.

[87] Brooke Hindle, *The Pursuit of Science in Revolutionary America, 1735–1789* (Chapel Hill, 1956), Chap. 8.

time he sent several consignments of Virginia trees, nuts, and seeds to Dr. Sloane, George London, and others.[88] After his return to England, he was elected Fellow of the Royal Society (December 4, 1706), and he retained an active interest in the Society until his death in 1728.[89] Other communications from Virginia reached the Society at the turn of the century, and William Sherard's brother Sampson, in Virginia on business (1700–1704), collected seeds for his brother and his English friends.[90] On July 20, 1697, James Petiver displayed a large female snake, which he opened before the Society to divulge seventeen "young ones all perfectly formed & covered with tunicles by which he said it appeared to be a true viper bringing forth live ones."[91] It was not related what variety of snake it was. Again, on December 3, 1701, one Mr. Haistwell presented "A Box of Curiosities from Virginia" containing "Maiz or Indian Corn of several Colors, the same with 5 Spikes from the same Stalk, Sorgum, an Eagles foot, the Exuviae of a Serpent, Virginia Snake-weed, a Rhinoceros Beetle, a Silk-worm in his Bag, some Shells and seeds."[92] The seeds were distributed, in part, to the Duchess of Beaufort, Bishop Compton, Samuel Doody, and others.

However, the most constant link between the Royal Society and Virginia during the first half of the eighteenth century was William Byrd II (1674–1744), although the records of his association with the Society indicate a curiously uneven performance on his part. It is possible that Byrd had been introduced to Royal Society Fellows by his father, William Byrd I, although the younger Byrd had been sent to England for his education at a tender age, and had himself spent nearly fifteen years abroad before he was called to the English bar on April 15, 1695. About a year later (April 29, 1696) he was elected Fellow of the Royal Society, and visited Virginia soon afterward.[93] On July 20, 1697, the Secretaries of the Society recorded that

---

[88] *Sloane 3343*, fols. 105ff., 204–205; *4072*, fol. 49.

[89] *Journal-Book*, XI, 103, 104, 106; XIV, 52. He served on the Council of the Society in 1726–28. *C.M.*, II, 288, 292–299; III, 1, 6.

[90] *Sloane 4063*, fols. 44, 83; *Va. Mag. Hist. & Biog.*, XXV (1917), 70. See also *Journal-Book*, X, 84, 134, 162–163, 227, 253.

[91] *Sloane 3341*, fol. 55.

[92] *Journal-Book*, X, 232.

[93] He was formally admitted on the same day. *Ibid.*, VIII, 348. He was elected

Mr. Bird present to the Society a live Rattlesnake brought from Virginia by him in a Box, wherein he had lain 7 months without food, the members present saw him in a Box verry lively and ordered Mr. Hunt should take care of him.

The Same presented another Strange creature called an opossum from the Same place; itt had a pounch under its belly wherein ye young ones entered when ever they were put to itt; it was Committed to the same persons care.[94]

There is no record of Mr. Hunt's experiences with the rattlesnake, but the opossum died early in 1698 and was turned over to Dr. Edward Tyson, who dissected the creature, made anatomical drawings, and published the results in the *Philosophical Transactions*.[95] The 'possum's classification was much in doubt: John Ray thought it was a canine, Dr. Tyson held that it was a variety of rodent. A search of old authorities led to a discussion of "monsters," but Tyson believed that this was only "Ignorance . . . for Nature in his regular Actings, produces no such Species of Animals." Former authorities consulted had declared that the pouch on the animal served as its uterus in reproduction, but Dr. Tyson found that it had two uteri (ovaries?) in addition to its pouch, and wished that he had a male specimen to dissect for comparison. He got one in 1704 and found that it had no pouch.[96] Later in 1697 Mr. Byrd read to the Society an account of a Virginia Negro boy who "is dappled in several Places of his Body with White Spots." The boy had been born of black parents and was himself all black until he was three years old, although he had had no illness to account for his changes in pigmentation. Byrd's relation was published in the *Philosophical Transactions*.[97] During the next four years Mr. Byrd's name appeared several times in the Society's minutes. He showed

to the Council of the Society in 1698 and re-elected in 1699 and in 1703. *C.M.*, II, 98–112, 123–124. His friendship with Sir Robert Southwell, President of the Royal Society (1690–95), doubtless contributed to his cordial reception in the Society. See Alden Hatch, *The Byrds of Virginia* (New York, 1969), *passim*.

[94] *Journal-Book*, X, 44. See also *ibid*., pp. 59 (Feb. 23, 1697/98), 60 (March 2, 1697/98).

[95] XX, No. 239 (April, 1698), 105–164.

[96] *Phil. Trans.*, XXIV, No. 290 (March–April, 1704), 1565–90. The donor of the male opossum was not disclosed.

[97] *Journal-Book*, X, 52 (Nov. 17, 1697); *Phil. Trans.*, XIX, No. 235 (Dec., 1697), 781–782.

"an Artificiell Cupp made to resemble agat" (February 15, 1698/99), told the Society that Indian arrowheads were made of flint (March 1, 1698/99), presented the exuviae of a rattlesnake (February 7, 1699/1700), explained how the Indians made glue (February 5, 1700/01), and exhibited "the Skeleton of an Oak-leaf, which was all eaten away except the Fibres."[98] Then, for reasons unknown, Mr. Byrd's name does not appear in the Royal Society's records until 1706.

The death of his father (December 4, 1704) led William Byrd II to hasten back to Virginia. He wrote to Dr. Sloane on April 20, 1706:

> The news of my Father's Death hurry'd me so suddenly from England that I had not time to receive the commands of the Society or of yourself, so laborious a Member of it. However I think my self obliged to offer my Service by this first oppertunity, and should be very ambitious to do any thing for you that might make me worthy of ye honour I have of being of that illustrious Body that are ever at work for ye good of ungratefull mankind.
>
> The Country where fortune hath cast my lot is a large feild for Natural inquiry, and tis to be lamented that we have not some people of Skil and curiosity amongst us. I know no body here capable of making very great discoveries so that Nature has thrown away a vast deal of her bounty upon us to no purpose. Here be some men indeed that are called Doctor but they are generally nothing above very common Remedys. They are not acquainted enough with Plants or the other parts of Natural History, to do any service to ye World, which makes me wish that we had some missionary Philosopher that might instruct us in ye many usefull things which we now possess to no purpose. . . .[99]

[98] *Journal-Book*, X, 104 (the "Cupp" was not of Virginia origin); *ibid.*, pp. 108, 164, 210, 234. It was probably about this time that Byrd gave sassafras berries and other seeds to Dr. Sloane. N.p., n.d., *Sloane 4058*, fol. 99.

[99] Byrd to Sloane, April 20, 1706, *Sloane 3335*, fols. 77–99. In another copy of this letter (*Sloane 4040*, fol. 151), Byrd is reported to have written that Virginia physicians were "generally discarded Surgeons of ships." Read Dec. 11, 1706, *Journal-Book*, XI, 105, 108. See also "Letters of William Byrd II and Sir Hans Sloane Relative to Plants and Minerals of Virginia," *Wm. & Mary Qty.*, 2nd Ser., I (1921), 186–200; and Sarah P. Stetson, "The Traffic in Seeds and Plants from England's Colonies in North America," *Agricultural History*, XXIII (1949), 45ff.

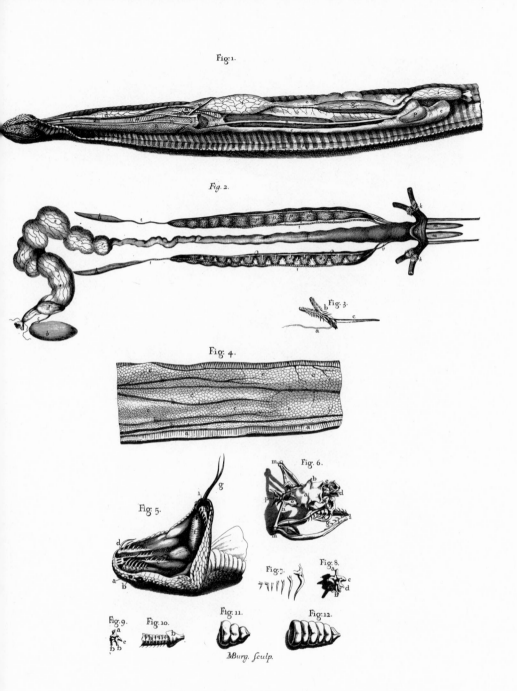

13. Dr. Edward Tyson's drawings to illustrate the anatomy of a rattlesnake (1693).

Phil. Trans., XX, No. 144 (1693), 25-54.

14. Sir Hans Sloane (1660-1753), simultaneously President
of the Royal College of Physicians
and of the Royal Society of London, collector,
active promoter of science in the colonies,
and founder of the British Museum.

Courtesy of the National Portrait Gallery, London.

He sent a box of snakeroot[100] which, he said, Indians and Indian traders found to be a cure for rattlesnake-bite, and he sent other herbs in hopes that the Society would advise him regarding their uses. The Society referred them to James Petiver, who reported to Byrd (May 1, 1707), as ordered by the Society, that the snakeroot was new to him, and its specific quality, as related by Byrd, was very wonderful; but if it were to be properly classified, Byrd must supply the entire plant, with leaves, flowers, and seed. The same was true of the other herbs, although Petiver believed that one other root sent by Byrd was the ipecacuanha root identical to that commonly received from Brazil and used as an emetic, purgative, and diaphoretic. If so, Petiver believed that Byrd might profit by collecting it for sale in England.[101]

Byrd's next letter to Sloane was dated September 10, 1708. He was doubtful whether ipecacuanha would be profitable to gather for sale, for "if it should be never so dear in Europe, I'm confident the Quantity that can be sent from hence will hardly make it cheaper, for it grows in very few places. . . . I planted some in my Garden, but it does not thrive. I should be glad to hear how much it will sell for a pound however, that I may Judge whether it be encouragement sufficient to employ anybody about it. . . ."[102] He sent more roots and seeds, "that the Society may try if there be any Vertue in them." These included: "a Root which I think very like Jalop, we call the Plant here Poke; it bears a purple Berry which would dye an admirable Colour if we understood the right way of fixing it," and he begged for instructions on this point; seeds of the Jamestown weed (Jimson weed): both the seed and the plant were a "rank poison if taken inwardly," but as a poultice both the seeds and the leaves were helpful in the cure of burns; the seed of the Jerusalem oak which, mixed with honey, "kills worms better than

---

[100] The roots of any of several American plants reputed to be antidotes for snake venom: *Polygala senega*, *Aristolochia serpentaria*, or *Ascyrum hypericoides*, L. (St. Andrew's cross).

[101] *Sloane 3336*, fols. 10v–11. Petiver also sent Byrd three tables of Virginia plants from Banister's drawings. When he heard nothing from Byrd, Petiver wrote again (April 1, 1708) to repeat his report, send the tables, this time with a printed copy of Banister's catalogue, and a plea for Byrd to collect for him in Virginia. *Sloane 3336*, fols. 48v–49.

[102] *Letter-Book*, XIV, 239–240; presented to the Royal Society Dec. 7, 1709, *Journal-Book*, XI, 179. See also *Sloane 3337*, fols. 64–65.

any Worm-Seed I ever heard of"; stickweed root, "the great Leaves of which never fail to stop bleeding either at the Nose or else where"; and another root "for which I have no Name, but by the Taste I judge it to have a great deal of Virtue." He deplored his ignorance of plants and returned to the virtues of snakeroot. Powdered in Canary wine, it was effective in the cure of "a Violent Diarrhoea." He also spoke of Virginia mines and minerals and asked Dr. Sloane to send "Samples of several Ores, that I might by Comparing them with those which I find be able to make some Judgement of them."

As Byrd's letter was read before the Royal Society, both Dr. Sloane and James Petiver replied to it, Sloane on December 7, 1709, Petiver on the following January 3. Dr. Sloane's letter was official in the sense that it was read and approved by the Royal Society before he sent it. Ipecacuanha sold at thirty shillings a pound in London. Sloane suspected that it grew more widely in Virginia than Byrd had indicated: "I dare say You will find plenty of it, and you will save so much Money as goes from hence to Portugall and Spain on this occasion." Byrd's "Poke" was not jalap. Sloane had seen it in Jamaica, and John Ray had named it "Solanum racemosum Americanum" (*Phytolacca decandra*). Sloane gave brief instructions how to fix the dye with alum and potash. Likewise, Sloane had met with the Jamestown weed (*Datura stramonium*) in Jamaica; Gerard long since had described it, its poisonous qualities, and its use in the treatment of burns. Similarly, Gaspard Bauhin had described the Jerusalem oak: "It may very likely kill Worms given with Honey." Sloane observed drily, "Honey is certainly one of the greatest Killers of Worms of any medicine we have . . . !" The stickweed root had been lost en route, and the unknown root would remain unknown until its leaves, fruit, and flower could be examined: "When you send any other Herbs, pray send their Leaves and Flowers dryed between Papers, and their Seeds that they may be known and raised here." As to minerals, "it is much easier to you to send over what you want to be informed of, in which case you shall receive the best Satisfaction I can give you." Shocked by Byrd's proclivity to take medicines made from unknown or untried plants, and sensing, perhaps, that Byrd was something of a hypochondriac (as he was), Dr. Sloane advised him to do "what I practice my self, never to take Physick when I am well, and not to make use of any Medicines but such as are very well tryed when I am ill, Observation and Experience being the best way to find out the Virtues of Plants."[103] Mr. Petiver, having heard Sloane's letter above, confirmed much of what Sloane had written. He had hoped for a supply of ipecacuanha from Virginia (which, he said, was "much different from ye true Brasile one of Piso"), but he had since found a source of supply in Maryland, where it grew plentifully. He had seen specimens of the stickweed root from Maryland also and had told Dr. Sloane about it. He sent two more tables of Virginia plants, some from "Mr. Banister's own designs," to aid Byrd in plant identification. And he renewed his plea to Byrd to employ his servants and Negroes to collect specimens of Virginia natural products:

> I have as yet observed very few Plants, Insects, Shells, or ffosills from Virginia but from Maryland & Carolina several hundreds & am by every shipping in Expectation of many more . . . particularly from one who hath lately written a Naturall History of Carolina [John Lawson]. I am very sensible many of ye same plants grow with you yet nevertheless wtever I can obtain from Virginia & more Northward convince me they may more probably be ye very same Mr. Banister had himself observed who lived upon that spott.[104]

Mr. Byrd replied to Sloane on June 10, 1710. Encouraged by Sloane to collect ipecacuanha, he had gathered thirty pounds the previous year, only to run into customs troubles that destroyed his incentive. He sent some more "Raritys" to the Royal Society, but the shipment did not arrive, probably lost or seized at sea in a wartorn Atlantic. He asked for books, including Sloane's account of Jamaica, travels, and others on "curious" subjects.[105] However, after this letter, nothing further was heard from Mr. Byrd at the Royal Society until he arrived in person in 1715, bearing, as gifts to the Society, "the bones of the penis of a Bear and of a Raccoon, which were ordered to be put in the Repository."[106] Indeed, al-

[103] *Letter-Book*, XIV, 241–243.
[104] *Sloane 3337*, fols. 66–67.
[105] *Sloane 4042*, fols. 143–143v.
[106] *Journal-Book*, XII, 74 (July 21, 1715).

though he was in England much of the time between 1715 and 1726, the minutes of the meetings of the Royal Society record little participation by Mr. Byrd.[107] He did not resume scientific correspondence with the Society until 1737.

In the meantime, an important new naturalist had appeared on the Virginia scene, and although his ultimate worth had not yet been appreciated when Mr. Byrd returned to England in 1715, Byrd had made his acquaintance, had entertained him royally at Westover, and to some measure had patronized and encouraged him in his work. This man was Mark Catesby (1683–1749). Born and educated in Essex, Catesby had early developed a passion for the study of nature, and his interest had been sharpened further by such mentors as John Ray, who lived nearby, and Ray's helpful friend Samuel Dale (c. 1658–1739), the Braintree apothecary and author of the widely admired *Pharmacologia* (London, 1693, and later).[108] With slender means and as yet no wealthy patrons, Catesby went to Virginia in April, 1712. He was drawn to Williamsburg by his sister Elizabeth, whom he accompanied to Virginia. Elizabeth had married Dr. William Cocke, who had emigrated to Virginia in 1710 and had become a friend of Governor Alexander Spotswood, had been chosen Secretary of the colony, and soon (1713) became one of Her Majesty's Councillors. Dr. Cocke was well received among the best families of Virginia, and a week after his arrival at Williamsburg, Catesby met William Byrd II. During the summer and early autumn of that year (1712), Catesby was entertained for two rather lengthy visits at Westover, and the two men became fast friends. By means of Byrd, Dr. Cocke, and

---

[107] Byrd nominated Sir Wilfred Lawson as a Fellow of the Royal Society on Nov. 20, 1718, and Lawson was elected. *Ibid.*, XII, 251. His political opposition to Governor Spotswood led Byrd to London in 1715. He returned to Virginia in 1720, was sent to England in 1721 as Agent for the Virginia Assembly, remarried there (his first wife having died of smallpox in London in 1716), and went back to Virginia in 1726. Cf. Richmond C. Beatty, *William Byrd of Westover* (Boston and New York, 1932), pp. 83–126; Louis B. Wright, *The First Gentlemen of Virginia* (San Marino, Calif., 1940).

[108] Professor George Frederick Frick, now at the University of Delaware, has uncovered further, more persuasive, evidence of Catesby's early association with John Ray since the publication of our book *Mark Catesby: The Colonial Audubon* (Urbana, Ill., 1961). See pp. 6–10 therein. For further materials on Catesby's first visit to Virginia, see this book.

Governor Spotswood himself, Catesby won rapid acceptance and ready hospitality among several of the great families of Virginia.

Almost immediately he began botanizing. About a year after Catesby's arrival, Governor Spotswood sent to Henry Compton, the Bishop of London, seeds collected by Catesby, and additional collections were sent to Samuel Dale, to Thomas Fairchild, a prominent nurseryman at Hoxton, and to other English friends and relatives. By 1715 Catesby had attracted the attention of James Petiver, who published accounts of specimens from "that curious Botanist Mr. Mark Catesby of Virginia." He also lent assistance to various Virginia planters in the cultivation of their gardens, especially Byrd, John Custis, and Thomas Jones, who later married Catesby's niece, Elizabeth Cocke. He traveled extensively in Tidewater Virginia, and in 1714 went westward to the mountains. In the same year he made a voyage to Jamaica and the Bermudas, supplementing Dr. Sloane's collections in these islands, and sending specimens to Samuel Dale. Dale shared these with William Sherard, who after John Ray's death was, as Catesby said, "one of the most celebrated Botanists of this Age." Sherard, as has been noted, was one of the rising young scientists in the latter years of the seventeenth century. In 1703 he accepted a lucrative position as Consul at Smyrna, hoping to travel in the Levant and collect the flora there. War, the prevalence of bandits in the Smyrna area, and other difficulties foiled many of his botanical hopes, and he returned to England in 1717, a disappointed but wealthy man. Before he had gone to the Near East, Sherard, upon the suggestion of Joseph Pitton de Tournefort, the widely heralded professor at Paris, undertook to revise Gaspard Bauhin's *Pinax*. He had the task well under way before he went to Smyrna. Now, in 1719, he planned to complete it. Dale's gifts of American plants from Catesby excited Sherard's interest. On October 15, 1719, Dale wrote to Sherard: "Mr. Catesby is come from Virginia, he hath brought me about 70 Specimens of which about half are new, these I shall likewise send up in the box for your view, and where there are duplicates, they are at yours, and Mr. [Charles] Dubois service; He intends againe to return, and will take an opportunity to waite upon you with some paintings of Birds &c which he hath drawn. Its pitty some encouragement can't be found for him, he may be very usefull for perfecting of Natural History."[109] Dale's thinly veiled promotion of Catesby bore fruit. Catesby was introduced to Sherard, Sir Hans Sloane, and others in the Royal Society. On October 19, 1720, Colonel Francis Nicholson, about to depart to assume his last gubernatorial position as first Royal Governor of South Carolina, notified the Council of the Royal Society that "he would allow Mr. Catesby, recommended to him as a very proper person to Observe the Rarities of that Country for the uses and purposes of the Society, the Pension of Twenty Pounds per Annum during his Government there [1720-25], and at the Same time to give him Ten pounds by way of advance for the first half years payment and so for the future a Years pay beforehand."[110] Governor Nicholson's act, together with the approval of the Royal Society, loosened the purse strings of others, and after some delay Mark Catesby embarked for South Carolina to begin a task which, in its ultimate outcome, was one of the greatest achievements in natural history during the Old Colonial Era. (Further attention to Catesby's work there will be postponed to the section devoted to South Carolina.)

After William Byrd's return to Virginia in 1726, he was occupied with the survey of the boundary between Virginia and North Carolina, several literary undertakings (few of which were published in his lifetime), and various projects, as he put it, to "be very rich; but if I can benefit my Country, and make it usefull to Great Britain, it will be a greater satisfaction."[111] Between May, 1737, and April, 1741, he wrote six "philosophical" letters which were presented to the Royal Society.[112] Taken as a whole, these letters asserted, and reasserted, against the opinions of Sir Hans Sloane, Mark Catesby, and others of the Royal Society, that Virginia gin-

[109] *Sherard Letters*, II, fol. 211.

[110] *C.M.*, II, 324.

[111] Byrd to Sloane, May 31, 1737, *Letter-Book*, XXIV, 17-20; *Sloane 4055*, fols. 112-113; read Nov. 3, 1737, *Journal-Book*, XVII, 30.

[112] These included the above (note 111); to Catesby, June 27, 1737, *Letter-Book*, XXIV, 115-118; read Jan. 12, 1737/38, *Journal-Book*, XVII, 172; to Peter Collinson, July 5, 1737, *Letter-Book*, XXIV, 217-219; read March 2, 1737/38, *Journal-Book*, XVII, 205; to Sloane, Aug. 20, 1738, *Letter-Book*, XXV, 101-103; *Sloane 4055*, fols. 367-368; read Dec. 7, 1738, *Journal-Book*, XVII, 326-327; to Collinson, July 10, 1739, *Letter-Book*, XXVI, 21-25; read Nov. 1, 1739, *Journal-Book*, XVII, 476; and to Sloane, April 10, 1741, *Sloane 4057*, fols. 20-20v.

seng was identical to that obtained from China,[113] and that the snakeroot of Virginia possessed wide medicinal properties which the Fellows of the Royal Society were unable to confirm in spite of large quantities shipped to them by Mr. Byrd for their experimentation.[114] Of the American ginseng, Byrd wrote that "I believe ever since the Tree of Life has been so strongly guarded, the Earth have never produced any Vegetable so friendly to Man as Ginseng." It revived men from great fatigue, warmed the blood, frisked the spirits, strengthened the stomach, and comforted the bowels—all without "those naughty effects that might make Men too troublesome and impertinent to their poor wives." In his high opinion of the virtues of snakeroot, Byrd was supported by Dr. John Tennent of Virginia. Dr. Tennent (c. 1700–1748), author of *Every Man his own Doctor* (2nd ed., Williamsburg, 1734; reprinted, Philadelphia, 1736) and an *Essay on Pleurisy* (Williamsburg, 1736),

---

[113] More recent botanists agree with Sloane: American ginseng is classified as *Panax quinquefolium;* Chinese ginseng as *Panax schinseng.*

[114] Thomas Cooper, a graduate of Wadham College, Oxford (B.A., 1720), went to Charles Town, South Carolina, where he set up a very successful medical practice and became known as "Doctor of Phisick," though whether he held a doctor's degree is uncertain. In Charles Town, during the early 1720's, he took part in a series of experiments on rattlesnake-bite to demonstrate the "certainty & speediness of this poison on many kinds of animals," and to test the value of various antidotes, all of which were found to have little or no effect. A ship captain, Fayer Hall, later communicated an account of these experiments to Sir Hans Sloane, who, in January, 1728, confirmed that Charles Town experiments by others of his own, and dissected a rattlesnake to confirm the anatomical structures previously reported by Dr. Edward Tyson. All this was reported to the Royal Society in 1727–28 (*Journal-Book*, XIV, 123, 155–157, Oct. 26, 1727; Jan. 11, 1727/28), and Captain Hall published "An Account of Some Experiments on the Effects of the Poison of the Rattle-Snake . . ." in *Phil. Trans.*, XXXIV, No. 398 (April–June, 1727), 314. Mark Catesby, who became a close friend of Cooper in Charles Town, must have learned of these experiments before Sloane's confirmation of them in 1727–28. Catesby also had ample opportunity to observe rattlesnakes during his own travels. He gave no credence to the value of snakeroot in the treatment of rattlesnake-bite, for "where a Rattle-Snake with full Force penetrates with his deadly Fangs, and pricks a Vein or Artery, inevitably Death ensues, and that, as I have often seen, in less than two Minutes. The *Indians* know their Destiny the Minute they are bit, and when they perceive it mortal, apply no Remedy, concluding all efforts vain. If the Bite happenneth in a fleshy Part, they immediately cut it out to stop the Current of the Poison." *The Natural History of Carolina, Florida, and the Bahama Islands* (2 vols., London, 1729–47), II, 41. Catesby had written in the same vein in 1724. See Frick and Stearns, *Mark Catesby*, pp. 27–30.

was an enthusiastic advocate of the curative powers of Senega snakeroot (*Polygala senega*). With a letter of introduction from Byrd to Sir Hans Sloane (1737), Tennent turned up at the Royal Society with a sample of his Senega root "decoction," which he claimed was a specific for the cure of pleurisy and "peripneumonick fevers." He hoped for public acclaim and large financial rewards if the Society would place its stamp of approval upon his "discovery." When the Society failed to respond, Tennent returned two years later, this time offering as proof of his claims testimonials from "the upper and lower Houses of the Legislature of Virginia," which, he said, had been "transmitted to the Board of Trade here." Again the Royal Society withheld approval, and, as the truth turned out, the Virginia houses, divided in their counsels, had grudgingly voted Tennent a reward of £100 Virginia currency, a sum which had disgusted Tennent as a parsimonious recompense. Somehow, too, Dr. Tennent, in place of Royal Society approval, found himself in a London debtors' prison. Sir Hans Sloane gave financial assistance to bail him out, and Tennent went free, protesting eternal gratitude to Sloane and promising to make a fresh start in Jamaica, where he had a "recommendation" and "no enemies."[115] Byrd endorsed Dr. Tennent's claims for snakeroot, and prophesied that "our Assembly will reward him very handsomely." He continued to sing the praises of snakeroot, claiming to the end that it would "operate either by Purge, Vomit or Sweat, according to the present disposition of the Person who takes it," and recommending it for fevers, worms, hydrophobia, rattlesnake-bite, smallpox, pleurisy, pneumonia, gout, and dropsy. Indeed, he said, "I am not without hopes, that it will disgrace the Peruvian Bark, and put the Jesuit quite out of Countenance." When the Royal Society refused to uphold his claims for ginseng and snakeroot, Byrd con-

[115] Dr. Tennent did go to Jamaica, and died there Oct. 27, 1748. *Gentleman's Magazine*, XVIII (1748), 524. I mistakenly identified him with Dr. John Tennant, F.R.S., of New York in *Col. F.R.S.*, pp. 223–225. I have seen no first edition of Tennent's *Every Man his own Doctor*. Benjamin Franklin published the Philadelphia edition, as well as extracts from his *Essay on Pleurisy* and testimonies for the Senega snakeroot. Cf. Leonard W. Labaree, ed., *The Papers of Benjamin Franklin* (in progress, New Haven and London, 1959———), II, 155–158, 239–240. Tennent had better luck with the French Academy of Sciences, where a committee reported favorably on the Senega snakeroot. *Ibid.*

cluded that they lost all their virtues "by passing the Sea," and clung to his opinions.

Byrd also confirmed the popular stories in circulation about the mysterious powers of snakes, especially of rattlesnakes, to "charm" their victims so that they fell motionless awaiting consumption. To test this ancient superstition of the "evil eye," Sir Hans Sloane, with others, conducted experiments reported to the Royal Society on October 22, 1730. After observing a rattlesnake with its prey in the garden of the Royal College of Physicians, Sloane told the Society that

> In my Opinion the whole mistery of their enchanting or charming any Creature is only this, that when such as are their proper prey, namely small Quadrupeds or Birds &c., are surprized by them they bite them, the poison allowing them time to run a small way as our Dog abovementioned did, or perhaps a bird to fly up into the next tree, where the Snakes watch them with great earnestness till they fall down, or are perfectly dead, when having licked them with their Spawl or Spittle they swallow them down.[116]

But Mr. Byrd would accept none of this. He wrote to Catesby: "There is something in the steady Look of that Reptile, that fascinates any of the little Animals designed for its food. After some Struggle with their Fate, they fall down before their Charmer senseless and without motion. . . . I assure you, I have ogled a Rattlesnake so long, till I have perceived a Sickness at my Stomach." And, in spite of Dr. Tyson's published report on the anatomy of a female opossum (for which Byrd had supplied the specimen), Mr. Byrd clung to the opinion that the pouch of the animal "is the actual matrix wherein the young are bred." He quoted his own

[116] *Sloane 1966*, fol. 208v; *1968*, fols. 208–213; *Journal-Book*, XIV, 500; *Phil. Trans.*, XXXV, No. 401 (Jan.–March, 1728), 377–381; XXXVIII, No. 433 (July–Aug., 1734), 321–331. After Byrd insisted upon the charming powers of rattlesnakes, Peter Collinson inquired of Isham Randolph for another opinion. Randolph, on the basis of personal observations and reports of others, opined "that their [the snakes'] Victories are not gained by their Charms. . . ." Randolph to Collinson, Aug. 22, 1737, *Letter-Book*, XXV, 123–125; reported Dec. 21, 1738, *Journal-Book*, XVII, 338–339. Randolph, though not a correspondent of the Royal Society, was on close terms with several men interested in natural history. "Letters of John Clayton, John Bartram, Peter Collinson, William Byrd, Isham Randolph," *Wm. & Mary Qty.*, 2nd Ser., VI (1926), 303–325.

observations on the creature "from my Journal of the Line" (*History of the Dividing Line*) as proof, and offered to send a pair of opossums so that "If therefore you can . . . perswade them to generate, you may be convinced of the Truth of what I write by ocular demonstration."

On balance, William Byrd II appears to have been a backward student of experimental science. Still, his diaries, letters, and other works betray a constant interest in the "curious" phenomena of nature. He generously supported the scientific endeavors of others and he had a utilitarian concern about experimentation, although he was not always able to cast off old opinions when experiments proved them to be baseless.[117] He once dissected a muskrat. He experimented with hemp, vineyards, and other products in the hope of finding profitable new crops for Virginia. He appealed to Peter Collinson for mangoes and nectarines to try them in Virginia and Carolina. And he repeatedly pleaded with his English correspondents to send competent observers to America to explore the rich opportunities for finding new drugs for the cure of human ailments: "What signifies the Tour of France and Italy [he wrote to Collinson] wherein such young Fellows have a greater chance to grow Fops, than they have to grow Physicians. Here everything wou'd be perfectly new, and a large Field of knowledge wou'd discover itself to the ingenious Enquirer. In doing this he wou'd perform a more pious work, than by endowing twenty Hospitals, or building twice that number of Churches. . . ." But the nature of Byrd's concern was more utilitarian than scientific, and he was

---

[117] Most of Byrd's works have been published posthumously. See John Spencer Basset, ed., *The Writings of Colonel William Byrd of Westover in Virginia Esqr.* (New York, 1901); Louis B. Wright and Marion Tinling, eds., *The Secret Diary of William Byrd of Westover, 1709–1712* (Richmond, 1941); Maude H. Woodfin and Marion Tinling, eds., *Another Secret Diary of William Byrd of Westover, 1739–1741* (Richmond, 1942); William K. Boyd, ed., *William Byrd's Histories of the Dividing Line betwixt Virginia and North Carolina* (Raleigh, 1929). Percy G. Adams has shown that in all likelihood *William Byrd's Natural History of Virginia, or the Newly Discovered Eden*, originally published in German and translated and edited by Richmond C. Beatty and William J. Mulloy (Richmond, 1940), was actually written by John Lawson. See Percy G. Adams, "The Real Author of William Byrd's Natural History of Virginia," *American Literature*, XXVIII (1956), 211–220. By his subscription to Eleazar Albin's *A Natural History of English Insects* . . . (London, 1720), Byrd was probably the first colonial to lend support to the scientific works of others. See also p. 520 above.

more given to accept tales of "wondrous cures" than the less spec-
tacular results of experimental medicine. With his literary ambitions, William Byrd was more of a late seventeenth-century
virtuoso than an eighteenth-century scientist.

The Carolinas

The Royal Society of London received relatively few messages
from North Carolina throughout the Old Colonial Era, and it recorded little from South Carolina until the turn of the eighteenth
century. On July 11, 1678, Dr. Daniel Cox told the Society of a
"certain Powder" recently brought from Carolina. The natives
made of it a drink called "Cassenj," which "did strangely exhilerate
those that drank it and freed them from troublesome fears for 24
houres or 30 houres after but that then they were more flat."[118] He
promised a more extensive account of it for the future, but nothing
further was recorded. John Aubrey, at the same meeting, related
that Sir Peter Colleton had spoken of a Carolina herb called "Ter-
rara" highly valued as an antidote against "all manner of Poysons."
It was said to be kept a closely guarded secret by the Indians, but
its seed had been procured "by one that married an Indian King's
daughter," and was grown in "Mr. Johnson's Garden." On December 14, 1692, Dr. Robert Plot presented "An Account of a Colony
of Welsh in the West Indies by Mr. Morgan Jones." Dated from
New York, March 10, 1685/86, this was the familiar story of the
Welsh minister who reported that he had voyaged from Virginia to
Carolina in 1669, and, having been captured by the "Tuscorara
Indians," was approached by a "War Captain belonging to the
Sachim of the Doegs (whose original as I afterwards understood
was from the Welsh)." To his amazement, Jones discovered that
the "Doegs" spoke the Welsh tongue, and upon this mutual discovery, he was treated civilly by the natives, conversed with them "in
the British Tongue," preached to them in Welsh, conferred with
them "Easily," and was ultimately freed to go on his own way.[119]

---

[118] *Journal-Book*, VI, 116.

[119] *Classified Papers, 1660–1740*, VII (1), No. 54; *Register-Book*, VII (1687–94),
259–261; *Journal-Book*, IX, 104–105; Robert T. Gunther, ed., "Dr. Plot and the
Correspondence of the Philosophical Society of Oxford," *Early Science in Oxford*, XII (Oxford, 1939), 327–328. The rather doubtful account of the alleged

Dr. Plot, of course, argued that this was evidence that the medieval Welsh prince, Madoc, had reached America and founded a settlement there. Robert Hooke presented to the Society two curiosities, a strange sort of cloth said to have been made by Indians "about 500 Miles to the Northwest of Carolina"[120] and three arrows "such as are used in Carolina made of Reeds and bounded with a piece of sharp Cane,"[121] together with an extract of an account of a journey "made by one of that place to 900 miles distance towards the S.W. from the Savanna Town in Carolina."[122] And on October 17, 1694, a "Relation" was read "of ye Maner of Burials of ye Indians in South Carolina by Mr. Richard Warwick whoe hath lived theire Several years[123] [and] Part of a letter was read from Mr. Marshall at Carolina, giving an account of Rice and silk thriving in Carolina, but yt they could not know a way to hull it easily. Nor had they Industry enough to carry on the Silk trade."[124]

More substantial scientific exchanges began in the late 1690's between Carolinians and members of the Temple Coffee House Botany Club. Martin Lister maintained correspondence with Robert Steevens of "Goose Creek in Carolina," near Charles Town, in 1696, and the evidence extant indicates that it had been in progress for at least a year or more earlier.[125] James Petiver joined Lister in the exchange with Steevens in 1698 and in the course of the next two years received both Carolina plants and animals for his collections and for publication in his *Musei Petiveriani*.[126] Between 1700

---

discovery of America by Prince Madoc in 1170 has recently been set forth exhaustively in Richard Deacon, *Madoc and the Discovery of America* . . . (New York, 1967).

120 *Journal-Book*, VIII, 283 (Dec. 4, 1689).

121 *Ibid.*, IX, 87 (June 1, 1692).

122 *Ibid.*, IX, 131 (June 28, 1693). Hooke promised to try to get the entire journal of the trip, but no evidence exists to show that he did. Distances mentioned in these reports must be received with skepticism.

123 *Classified Papers, 1660–1740*, VII (1), No. 49; *Letter-Book*, XI, Pt. II, 29–30.

124 *Journal-Book*, IX, 140 (March 29, 1699).

125 Steevens to Lister, Oct. 24, 1696, *Lister 36*, fol. 161 (Bodleian Library). Steevens regretted "to find you were disappointed in the Shells I sent you." He sent more, adding that few land shells were available "by reason of ye yearley burning of ye Woods [by the Indians?] which destroyeth them." See also Lister to Steevens, Sept., 1697, *Sloane 3333*, fol. 108.

126 *Ibid.*, fols. 266v–267, 846. The latter reference lists thirty-eight plants which Petiver had received from Goose Creek. In 1701 Petiver displayed to the Royal

and 1713 Petiver enlarged the number of his Carolina correspondents. At the turn of the century, five of these worked together in Petiver's behalf, and at least two of them were personally known to Petiver, Sloane, Buddle, and others of the Temple Coffee House Botany Club before they went to South Carolina. The most energetic of them was Edmund Bohun, son of the Chief Justice of South Carolina of the same name who, falling victim to political disputes, was discharged in 1698 and died the following year.[127] About 1696 the younger Bohun settled in Charles Town as a merchant, although he occasionally returned to England in the course of his mercantile affairs. On at least one of these occasions, early in 1699, he was invited to join with Petiver and others of the Temple Coffee House Botany Club on a "moss-cropping" walk to Hampstead.[128] In Charles Town he lived with Robert Ellis, who cooperated with him in collecting for Petiver.[129] These two were also associated with George Franklin, an apothecary-physician who had known Petiver in England, with Robert Rutherford, a ship surgeon, and with William Halsteed, a ship captain, sometimes referred to as "Major Halsteed." Arriving in Charles Town on one of his voyages, in April, 1700, Halsteed wrote to Petiver (May 1) that "This morning I dranck with Mr. Bohun and Mr. Ellis; they tell me they have sent you home a very fine collection. I am preparing to doe the same. . . ."[130] None of these five, however, served Mr. Petiver in Carolina for an extended period. Halsteed and Rutherford, by the nature of their occupations, were itinerants; Franklin was captured by the French on a return voyage to England in 1703;[131] Bohun returned to England in 1701 to take posses-

Society a bullfrog, crab, sturgeon, hummingbird, and other specimens from Carolina. *Journal-Book*, X, 222 (June 18, 1701). In 1704 Robert Ellis reported that Steevens balked at collecting for Petiver, partly because Petiver paid too little (five shillings per volume for dried plants) and that he now collected seeds and plants "only to please himself, & obleidge one or two particular friends." Ellis to Petiver, April 25, 1704, *Sloane 4064*, fol. 2.

127 S. Wilton Rix, ed., *The Diary and Autobiography of Edmund Bohun Esq.* (Beccles, Eng., 1853), *passim*.

128 Petiver to Bohun, n.p., n.d. [c. Feb. 15, 1698/99], *Sloane 3333*, fols. 274v, 275–275v.

129 Ellis to Petiver, Charles Town, April 20, 1700, *Sloane 3321*, fol. 41.

130 *Sloane 4063*, fol. 18.

131 Ellis to Petiver, April 25, 1704, *Sloane 4064*, fol. 2. Franklin (or Francklyn) later settled at Downton, Wiltshire. In the *Sloane Herbarium* at the British

sion of estates left to him by his father;[132] and Ellis lost interest in collecting after his friends had departed.

For a few years at the turn of the century, however, these men supplied Petiver, Sloane, and others of the Temple Coffee House Botany Club with rather large collections, including seeds, plants, shells, fish, minerals, butterflies, and other insects. In April, 1700, Bohun sent shells, seeds, and two volumes of dried plant specimens, and Ellis sent "above 100 butterflies."[133] In July of the same year, Bohun sent "a small parcel of Tulip tree seed," and "a very pretty collection of Insects & about 6 or 7 Vollums of plants with 6 or 7 hundred butterflies & moths some very fine & scarce."[134] Indeed, between 1699 and 1705 these five men added materially to Petiver's hoard of natural products from Carolina.[135] Petiver shared some of the specimens among his friends, displayed them to the Royal Society, and sometimes read to the Society the communications from his Carolina correspondents.[136] Further, the correspondents were advertised as donors in Petiver's published works.[137] By their travels in Carolina, by persuading colonists to make collections— usually in exchange for garden seeds which Petiver supplied from London—and more especially by enlisting the cooperation of Indian traders, Petiver's correspondents extended their collections beyond

---

Museum (Natural History) is a list of twenty-eight different plant specimens (many in multiples of up to eight each) and five insects sent to Petiver by Franklin from Charles Town, June 28, 1700. See H.S. 159 (in back, unpaged).

132 Bohun returned to Charles Town in 1709–10. Petiver hoped that he would continue "your wonted Favours," but Bohun did not respond. Petiver to Bohun, March 20, 1709/10, *Sloane 3337*, fols. 98v–99.

133 Bohun to Petiver, April 18, 1700, *Sloane 3321*, fol. 39.

134 Bohun to Petiver, July 16, 1700, *ibid.*, fol. 51.

135 See the correspondence in the *Sloane Manuscripts:* Ellis (1700–1705), *Sloane 3321*, fols. 41, 45; *4063*, fols. 81, 82, 191; *4064*, fols. 2, 55; Franklin (1700–1701), *Sloane 3321*, fols. 44, 80; *4063*, fols. 34, 53; Halsteed (1700), *Sloane 4063*, fol. 18; Bohun (1699–1710), *Sloane 3321*, fols. 33, 39–40, 42, 51, 61, 74. Petiver's replies are mostly in *Sloane 3334*, fols. 74–75v; *3337*, fols. 97v–99. Rutherford appears only in Petiver's published works.

136 *Journal-Book*, IX, 265; X, 222.

137 See *Gazophylacii Naturae et Artis* (1702–06), pp. 4, 9, 11, 12, 40; *Phil. Trans.*, XXIV, No. 299 (May, 1705), 1957–60; *Musei Petiveriani* (1695–1703), especially *Centuria Octava* (1700), pp. 67–69, 71, 76, 79–80; and *A Classical and Topical Catalogue of all the things Figured in the Five Decades, or First Volume of the Gazophylacium Naturae & Artis* . . . (1706), pp. 92–96.

the immediate environs of Charles Town. In a letter of April 25, 1704, Robert Ellis referred to some of these efforts:

> I have delivered some of your Bookes [for collecting dried plants and butterflies] out to Severall Gentlemen of my acquaintance, to fill for me. One I have delivered to Capt. Nerne, who lives up to the South Ward among the Indians, who promised me to fill it, and to set down of what use each Plant is amongst the Indians, with their Names, another I have delivered to one Capt. Jonathan Fitch . . . who trades with another Nation of Indians, and has promised to do the like; and would desire you to send him Ten Potts of your Lucatillio Balsam . . . he will make you returnes in Skins, or Rice, or any thing else that this country affords, that you shall send for, I would desire you to send me one Pott of your Lucatillio Balsam, and . . . something that will disperse Wind. . . .[138]

Ellis and Bohun visited Silk Hope, the plantation of Sir Nathaniel Johnson, later Governor of South Carolina, and enlisted him as a Petiver correspondent, in return for which Petiver sent seeds for Sir Nathaniel's garden.[139]

Captain William Halsteed appears to have been the intermediary who enlisted for Mr. Petiver the cooperation of two other correspondents in South Carolina. One of these was Madame Hannah Williams; the other was the Reverend Mr. Joseph Lord. To these were added, in 1706, Mr. Daniel Henchman. As these three persons were mutually acquainted, and to some extent cooperated in their

---

[138] *Sloane 4064*, fol. 2. Samuel Eveleigh wrote to Petiver from Charles Town, June 13, 1706, saying that he craved his advice about the collection and curing of herbs as medicines. He offered to correspond with Petiver and promised to send among the Indians for collections. He had also written to "my Coz Eveleigh, druggist in Exon," for directions and was especially interested in collecting medicinal plants that would "bring profit." No further correspondence appears to have followed. *Sloane 3321*, fols. 200–200v. Another potential Carolina correspondent with Petiver was one Mr. Cooper, but he, too, did not respond. Petiver to Cooper, n.p., n.d., *Sloane 3332*, fol. 49v.

[139] Johnson to Petiver, Feb. 14, 1700/01, *Sloane 4063*, fol. 82; Bohun to Petiver, April 18, 1700, *Sloane 3321*, fol. 39. When Bohun was about to sail for England in 1701, he warned Petiver that if he expected Ellis to continue to supply collections from Carolina, "you must take care to send him some Garden seeds every year to supply his ffriends with or else you cannot expect yt ye Country people heere will be verry ready to give him any assistance for without some help from them a man can doe but little." April 28, 1701, *ibid.*, fol. 74.

endeavors, they formed a second group of Carolinians that, between 1701 and 1713, contributed richly to Mr. Petiver's collections.

Madame Williams sent a collection of butterflies to Petiver by Captain Halsteed in 1701. As the collection contained several that were new to Petiver, he was delighted with it and promptly returned his printed directions for collecting and preserving specimens with a request for more.[140] With this exchange began a correspondence that continued for several years. Madame Williams expanded her field to include snakes, lizards, scorpions, insects, plants, shells, and Indian items, such as an Indian "King's tobacco pipe & a Queens Petticoat made of moss," and Indian herbs and medicines.[141] In return, Petiver sent collector's supplies, newssheets, copies of his "Centuries" and "Decades," catalogues of plants and shells as guides for Madame Williams' efforts, and drugs from his apothecary shop. In 1705 Madame Williams complained of trouble with her spleen, asked for medicine, and proposed that she might profitably serve as an agent to sell Petiver's prepared medicines among her friends and neighbors in Carolina. Petiver responded with "some Hysterick pills & others for your head and stomach," sent medicinal items for possible sale, added three tables of West Indian shells and one of his published accounts of Carolina objects received from Madame Williams, in which "you will see ye Picture of your . . . Kings Tobacco Pipe."[142] The exchange languished thereafter for a time, possibly because of interference in the shipping lanes caused by the War of Spanish Succession ("Queen Anne's War"), and also because Madame Williams heard a premature report of Petiver's death. But in 1713 Madame Williams' son visited London, met James Petiver in person, and prompted Petiver to seek a resumption of the exchange.[143] There is no evi-

140 *Sloane 3334*, fols. 67v–68. Petiver's printed directions are published in *Petiver*, pp. 363–365.

141 Madame Williams to Petiver, Charles Town, Feb. 6, 1704/05, *Sloane 4064*, fol. 63.

142 Nov. 15, 1706, *Sloane 3335*, fols. 39v–41. The published account was probably "An Account of Animals and Shells from Carolina," *Phil. Trans.*, XXIV, No. 299 (May, 1705), 1952–60.

143 Petiver to Madame Williams, Dec. 12, 1713, *Sloane 3339*, fols. 87–88.

dence, however, that Madame Williams sent further collections.

Joseph Lord (1672–1748) was the son of a cordwainer of Charles-town, Massachusetts, and graduate of Harvard College (1691). He was a schoolmaster at Dedham and later at Dorchester before he was ordained (October 22, 1695), having studied for the ministry under Samuel Danforth, already noted as a member of the younger Winthrop's Boston circle of advocates of the new science.[144] The new Reverend then accompanied five other New Englanders as a missionary to Charles Town, South Carolina, in December, 1695. Soon he settled in a town later named Dorchester, about twenty miles from Charles Town on the east bank of the Ashley River.[145] In 1701, probably at the instigation of Madame Williams, he sent five quires of dried Carolina plants to England via Captain William Halsteed. Halsteed delivered them to James Petiver, who acknowl-edged their receipt in a letter to Lord of May 20, 1701, expressing delight with the collection, "wch I understand by him [Halsteed] I am obliged to you for as also Collecting most of them yourself." Lord had included some notes on the specimens, and Petiver pro-posed to print them, as "There hath as yet been very little published concerning ye Animals, Vegetables & Fossills of America." He an-nounced his pleasure at the prospect of a correspondence with a person of such obvious "character & Merrit," especially because to-gether they might be "better able to perform greater things for Carolina." He added: "You have already, Sir, so well begun there needs noe further directions but that you will be pleased to make your Collections monthly by which we shall know the true Seasons when every Tree & Herb produces its Flower & Fruit." To aid and encourage Lord in further collections, Petiver sent him collecting materials, a copy of the *Philosophical Transactions*, newssheets, and the "8th Century of my Museum." And to reassure his ministerial friend of his religious motivations, he told Lord that he often "con-templates on ye wonderful works of ye Creation."[146]

---

[144] See Chapter 5 above, p. 152.

[145] C. K. Shipton, sketch of Joseph Lord, in the continuation of J. L. Sibley, *Biographical Sketches of Graduates of Harvard University*, IV (Cambridge, Mass., 1933), 101–106; *South Carolina Historical and Genealogical Magazine*, XXI (1920), 4.

[146] *Sloane 3334*, fols. 63v–64.

To the above letter Mr. Lord replied (January 6, 1701/02):

In September I rec'd your Letter Dated May 20^{th,} 1701: In which I find you expressing your apprehension of Obligation to me for Collecting of Some Plants & my Remarks upon them; which, tho I did indeed collect, & make remarks upon them, yet I think not worthy the Appellation of unmerited Favor; especially considering that as you were unknown to me, & I reckoned not on being this way make known to your self, (haveing never, that I know of, seen Major Halstead) my design in collecting them was, only that I might make known what Plants were here that so such whose Vertues [as] are not yet known might be enquired into; & therefore your Printing the Rarities that you receive answer my Design in collecting them; only I desire it may be done without the mention of my name. . . .[147] [Lord had not yet received all of the supplies that Petiver had sent him, although he had been told that Robert Ellis had them.] Nor have I any Books for Collections, nor could I get any Quires of Brown Paper at Charlestown; which is the reason I send so few herbs or Plants. . . . If your Design (as you hint) be to write a Natural History of Carolina, I suppose whatsoever grown Naturally in the Country may be acceptable to you. . . . [He had two pine cones with seeds and asked] whether those seeds be the Pine-kernals usefull in Physick, & what Pine-nuts be. . . . In your letter I found enclosed an Extract of a Letter from Dr. [Benjamin] Bullivant to yourself, containing some remarks he made in New England, upon which (being myself a New-Englander by Birth) I shall make some remark. . . .[148] I should be glad to obtain several things (as Gum Fragaranth, Galbanum, Mastick, Rhubarb, Elaterium, Agarick &c). . . . [I send] some of the Moss of our Countrey, which I suppose to arise of seed sown on the Branches of trees, where the Bark is Rough, by means of the wind blowing it off from other trees. . . .[149]

Mr. Lord wrote yearly to Petiver until 1711, when, having heard the same rumor of Petiver's death that Madame Williams had cred-

[147] Lord withdrew this qualification in a letter of May 3, 1704, saying that "you may use your pleasure as to publishing my name." *Sloane 4064*, fol. 4.

[148] Lord's remarks made clear that the copy of the *Philosophical Transactions* which Petiver had sent to him included "Part of a Letter from Mr. Benjamin Bullivant, at Boston . . . to . . . Petiver." XX, No. 240 (May, 1698), 167–168.

[149] *Sloane 4063*, fol. 132.

15. Joseph Lord labeled this specimen (1704) in his beautiful, neat, and miniscule hand:

"A sort of Fern yt is green wth us all ye year. If I mistake not, it was the root of yt wth wch an Indian trader told me yt he cured himself of ye Lame distemper (as it is here called). But what [it] is, it is hard to know. For some call *Lues Venerea* so; others call another Distemper wch is a more yn ordinary Malignant Scurvey."

Sloane Herbarium, H.S. 158, fol. 17. *Courtesy of the British Museum (Natural History)*.

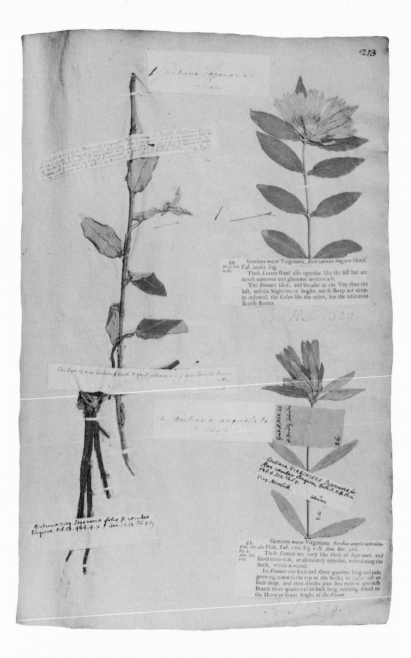

16. Two species of gentian sent by Joseph Lord to James Petiver. The one on the left is labeled: "Gathered Dec. 12, 1707."

Sloane Herbarium, H.S. 158, fol. 213. *Courtesy of the British Museum (Natural History).*

ited, he ceased the correspondence.[150] The activities of the French and Spanish fleets during the War of Spanish Succession constantly imperiled the exchange, and some of the collections were lost, especially items sent by Petiver to Carolina. The exchange was remarkable, however, because of the faithfulness of the correspondents, because of the variety and quantity of Carolina flora and fauna sent by Mr. Lord, and because of the unusual quality of the latter's observations and remarks on what he sent. Mr. Lord had little scientific knowledge of botany, but he hoped to extend his knowledge, and in 1702, with "some hundreds of species" of Carolina plants sent to Petiver, he expressed the desire "that if you know the names & vertues of any which I give no name unto, that you would please to let me know them also. I have often thought it a very difficult [task] to find the Tribe to which plants belong. . . ."[151] Again, in 1705, he wrote that "I have but little skill in Natural Productions, & have very few advantages to increase my skill (Books of that nature not being here to be bought, & Gerrard's Herbal which I had borrowed, & was the only considerable help I had to get such skill by being called for by the owner) & by that means I am not capable of ranking them under their Proper Heads."[152] Two years later he wrote that "the only book that I have that describes Herbs is [Nicholas] Culpepper's English Physician."[153] Mr. Lord's interest stemmed, in part, from the fact that he served as physician to the bodies of his flock as well as to their souls. But it went further than that, for he was a good Latinist who read with a critical eye the accounts of natural history which Petiver sent to him; and he sought to amplify and correct statements he read in Petiver's "Centuries," in the *Gazophylacii*, and in the other works that he re-

---

[150] Petiver had also heard that Lord had died. But learning his error, he wrote to Lord, Dec. 14, 1713, and begged to renew the exchange, "which Peace now I hope will give us both a more frequent Opportunity to Cultivate." *Sloane 3339*, fol. 88. But Lord did not respond. As a dissenter he was pressed by the South Carolinian government, and after the Yamasee War broke out he returned to New England, where he served the churches of Eastham and Chatham before his death in 1748.

[151] May 11, 1702, *Sloane 4063*, fol. 155.

[152] April 10, 1705, *Sloane 4064*, fol. 69. Lord did not specify whether he had borrowed the original edition of Gerard (1597) or Thomas Johnson's revision of 1633.

[153] Aug. 7, 1707, *Sloane 4064*, fol. 146.

ceived by Petiver's hand. His acute powers of observation and his eagerness to learn went far to outweigh his antiquated reference books. Few of Petiver's correspondents "talked back" as did Mr. Lord, and from the observations, queries, and comments found in his communications to Petiver, it is likely that Petiver found him to be one of his most stimulating correspondents among the permanent residents of the colonies.[154]

Mr. Lord sent Petiver specimens of Carolina plants, shells, fossils, butterflies, and other insects. As was his custom, Petiver added these to his collections, shared duplicates with his friends of the Temple Coffee House Botany Club and elsewhere, displayed them before the Royal Society, and used them in his publications, with flowery acknowledgments. His "Account of Animals and Shells sent from Carolina," published in the *Philosophical Transactions* in 1705,[155] included accounts of items sent by Mr. Lord; his *Classical and Topical Catalogue of all the things Figured in the Five Decades, or First Volume of the Gazophylacium Naturae & Artis* (1706) contained a section devoted to "Carolina Plants from Mr. Job Lord &c"; and his *Pterigraphia Americana* . . . (1712) listed other items, with one table dedicated to "that Curious Observer of Nature *Mr. Joseph Lord* at Carolina." Some of Lord's plants are still extant in the British Museum, and the editor of *The Sloane Herbarium* comments that the specimens, "which are in good condition . . . have a descriptive label in Lord's beautiful neat hand, giving information as to habitat, characters and date of collection, and showing that Lord was possessed of some botanical knowledge."[156]

When, in December, 1705, Petiver called upon Mr. Lord to send him specimens of the fauna of Carolina, especially animals, birds, fish, and reptiles, Lord begged off, claiming "no acquaintance" with

---

[154] The correspondence extant between Lord and Petiver is found in *Sloane 3334*, fols. 63v–64; *3335*, fols. 41–42v; *3337*, fols. 97v–98v; *3339*, fol. 88; *4063*, fols. 132, 155; *4064*, fols. 4, 11, 69 (printed in *S.C. Hist. & Geneal. Mag.*, XXI, 50–51), 146, 148, 150, 155, 192 (printed in *ibid.*, pp. 6–8), 233 (printed in *ibid.*, pp. 8–9), 258. Until Pinkney's death about 1704 or 1705, Lord instructed Petiver to address him at "Mr. Thomas Pinkney at Charleston." Pinkney himself forwarded one of Lord's shipments to Petiver with a covering letter. *Sloane 4064*, fol. 11.

[155] XXIV, No. 299 (May, 1705), 1952–60.

[156] Dandy, p. 159.

such natural history productions "that go far into the Countrey." However, he added that his "next neighbor," Mr. Daniel Henchman, "has shewed me several things which he has collected having been himself more than 300 miles in the Countrey, which he tells me he intends to send to the Royal Society," and had "acquaintance with those that go to trade with the Indians." Henchman (1677–1709) was the son of a Boston schoolteacher and Indian fighter. He attended Harvard College, where, in addition to some scholastic ability, he displayed much incorrigibility, and after he had stolen a silver spoon in 1696, he was dismissed before graduation. Subsequently he shipped off to South Carolina, where he served Governor Nathaniel Johnson as an Indian agent.[157] When he wrote to Petiver, who approached him soon after he had received Mr. Lord's letter, Henchman instructed him to direct his letters to Sir Nathaniel's home at Silk Hope, "where I reside." Actually Petiver received only one letter from Henchman, dated June 1, 1706.[158] It was labeled "Letter the 4th" but, as Petiver pointed out, he never received the first three—if, indeed, Henchman had written them.[159] It was a long, rambling, boastful letter, praising the regime of Governor Johnson, especially the enlargement of the Empire by alliances with the Indians (the Creeks) and extension of the Indian trade "for near Eight Hundred miles from the most frontier Plantation in this Province directly West." Henchman said that he himself had traveled 500 to 600 miles "beyond any English plantation into the Continent." He belittled the Indian factors, who were generally ignorant and passed over remarkable things, but Henchman "with Indefatigable pains both day & Night labour'd to make such Discoveryes and Observations as may render me in some respect serviceable to my Country." He had kept "jornals" of his travels, and he had seen such admirable things that he "resolved to lay aside

---

[157] Shipton, *Biographical Sketches*, IV, 297–298.

[158] *Sloane 3321*, fols. 196–199; read Dec. 18, 1706, *Journal-Book*, XI, 106.

[159] Joseph Lord suggested that the earlier letters had been addressed to the Royal Society. If so, it is unlikely that they were delivered, as the Royal Society Archives do not contain them, nor were they mentioned in any of the minutes of the Society's meetings. It is likely that they were lost en route, the dangers of ocean passage being as they were at the time. And it is possible that Henchman, being the none-too-truthful braggadocio that he appears to have been, invented them to impress Petiver with his promise as a correspondent—although the evidence is far too scanty to justify any such unkind conclusion.

all by-Ends of Trade," even if it cost him £150 to £200 per year, so that he could reveal "for the Publick Benefit, what by a more than ordinary Providence I had been the spectator of." He had learned much about the Indians and was writing a general history "of these Nations." He had collected plants and minerals, had made experiments regarding their medicinal virtues, and had found them "very wonderfull and almost past belief." In consequence, he was preparing a "Medicus Occidentalis." He also planned a book about his travels, for he had studied natural and moral history everywhere he had gone and had found almost everything written about the Indians and the country to be "very imperfect and mostly conjectured or Romantick." He had perfected a map of the country as well. He chose James Petiver "as my Patron," craved his directions for "better proceeding," urged him to communicate to the Royal Society whatever "you think will favour my Designs," begged Petiver, if he printed any of Henchman's letters, to edit them so as to free them from "tautologies" and "barbarisms," and stated that "were I but, sufficiently, encouraged, I would lose my Head without regrett" to natural history. He sent Petiver a copy of an address made by an Indian "King" at a council, and promised to deliver herbs, roots, stones, soils, and other minerals, "some of which I have brought 6 or 700 miles by Land out of the Continent." He would like to go to England "but that I fear being presst and carried into the Service," and he believed that he could be of far greater service to Queen and country in other ways. However, if Petiver would get him a pass from the Lord High Admiral of the Fleet, with a guarantee that he would not be pressed into service, he would embark on the first ship for England.

There can be little doubt that Henchman had traveled more widely on the Carolina frontiers and among the southern Indians than any other Petiver correspondent. But his account of his travels is difficult to follow, partly because of uncertainty as to the location of some of the Indian towns said to have been visited, and partly because Henchman's statements as to the distances traversed are both contradictory and incredible. It is difficult to believe that any Englishman, in 1704 or 1705, had gone 500 to 600 miles "beyond any English plantation into the Continent" and had opened up a "safe trade" with a variety of Indian tribes "for near Eight

Hundred miles from the most frontier Plantations in this Province directly West"! Mr. Lord's statement that Henchman had been "more than 300 miles in the Countrey" is more plausible. Did Henchman, perhaps, refer to his round trip when he described it to Petiver?

Petiver read Henchman's letter to the Royal Society on December 18, 1706.[160] Two days earlier he had sent thanks to Henchman for his letter, expressed pleasure at the prospect of further correspondence, and advised Henchman that none of his promised shipments of specimens had arrived.[161] As he had received nothing further from Henchman nearly two years later, Petiver wrote again in November, 1708, pressing for "more frequent Correspondence with you both of Letters & things." He sought to sweeten the appeal with a gift of dried English plants, "as a Pattern," together with John Ray's *Methodus Plantarum*, some of his own tables of American specimens, and remarks on shellfish—with an added appeal for fossil shells and "formed stones . . . supposed to have been there ever since the Flood."[162] But he received nothing further from Daniel Henchman. During the winter of 1709–10 Petiver heard that Henchman had died. He inquired of Mr. Lord, and requested him to seek information regarding "some Collections of Nature which he [Henchman] formerly promised to send me & other Papers concerning his Travells up into the Countrey, which if to be procured, I should be glad to have Coppies of which I pray inquire after." Mr. Lord replied (August 9, 1710), confirmed the report of Henchman's death (November, 1709), promised to confer with Henchman's wife, and, if permitted, to copy any papers about natural history and forward them to Petiver. That Petiver received nothing only enhances the suspicion that Daniel Henchman had grossly exaggerated his actual accomplishments.

Evidently, it was by means of the promotions of Edmund Bohun, Robert Ellis, and Captain William Halsteed, in Charles Town, that James Petiver developed an exchange with John Lawson (d. 1711), the only important scientific observer in North Carolina in correspondence with English scholars during the Old Colonial Era. If, as

160 *Journal-Book*, XI, 106.
161 *Sloane 3335*, fols. 46v–47.
162 *Sloane 3337*, fol. 26.

has been suggested, Lawson had been an apprentice in the London Society of Apothecaries before he went to the American colonies,[163] it is probable that he knew Petiver, at least by reputation, before their correspondence began. Little is known about Lawson's antecedents before he went to the colonies in 1700, arriving in New York and proceeding to Charles Town. At Charles Town he met Bohun, Ellis, and Halsteed, by whom he was persuaded to open a correspondence and exchange with Petiver, as his following letter to Petiver, dated from "Bath County on Pamphlicouph [Pamlico] River, North Carolina, April 12, 1701," would appear to indicate:

> I have sent you a Letter dated from Albemarle County in North Carolina, by wch I desyre y[r] advertisem[ts] in order to ye collections of Animalls Vegitables etc.[164] I shall be very Industrious in that Imploy I hope to y[r] satisfaction & my own, thinking it a more than sufficient Reward to have ye Conversation of so great a Vertuosi. I shall shortly goe to ye Sea Board which voyage I hope will furnish me with shells according to yr Request to Mr Bohun & Mr Ellis whose letters of yors to them I have. As for Cortex Elutheria [eleuthera bark?] I have laid oute for, & will use wt means I possibly can to procure it butterflies, & other Insects you may depend of wt our new Settlement affords fish likewise. I came aboute Xmas last to that place by land, from Ashley River in 11th lat. 32:45 to Okakock Inlet wch this River runs to. Sr my Service to Major Halsteed. . . . [Postscript] pray send us yr Opinion concerning ye distillation of Spirits from Malt Molasses figs apples Cherries pearrs &c wch fruit we have plenty. My journal of my Voyage through Carolina I shall send wth ye rest. Adeu. direct for me att pamphlicough River in North Carolina to be left for me att Collon[l] Quarmes in Philadelphia Pensilvania. I shall (God willing) send you some collections by Octob[r] next, by way or Pensilvania or Virginia.[165]

[163] *Petiver*, p. 335. Lawson's familiarity with medicines and botany is evidence in support of this conjecture, and his appreciation of John Banister's earlier work in Virginia suggests familiarity with the London circle of natural history scholars. See his *History of Carolina* . . . (London, 1714), p. 78.

[164] If this refers to an earlier letter to Petiver, it does not appear to be extant.

[165] *Sloane 4063*, fol. 79. Petiver endorsed it "Recd. Aug. 4, 1701." In the previous publication of this letter (*Petiver*, p. 336), I misread "Ellis" as "Elliot." Was it his "Journal of A Thousand Miles Travel among the Indians, from South to North Carolina" that Lawson proposed to send to Petiver? See Lawson, *History of Carolina*, pp. 6–60.

17. Although John Lawson gathered more trees than plants
in North Carolina, he sent to James Petiver the above plant specimens
gathered in June, 1710.

Sloane Herbarium, H.S. 145, fol. 62. *Courtesy of the British Museum
(Natural History).*

18. Lawson labeled the specimen at the top left "The celebrated snake root," but which of several plants called snakeroot is not clear. The item on the right is one of several varieties of silk grass, the fibers of which the Indians wove into a rough cloth, baskets, hammocks, matting, rope, etc.

Sloane Herbarium, H.S. 242, fol. 117. *Courtesy of the British Museum (Natural History).*

No further communications have been preserved until those of 1709. In the meantime, Lawson, who had been appointed by the Lords Proprietors to make a reconnaissance survey of Carolina, traveled extensively in North Carolina and, on the basis of his observations, prepared *A New Voyage to Carolina*, first published in 1709 and republished posthumously under the title *The History of Carolina: Containing the Exact Description and Natural History of that Country: Together with the Present State thereof. And A Journal of a Thousand Miles, Travel'd thro' several Nations of Indians, Giving a particular Account of their Customs, Manners &c.* (London, 1714). Lawson also acquired lands and political interests in North Carolina, and was co-founder of Bath (1705), the oldest town in the colony. In 1708 he returned to England, arranged the publication of *A New Voyage to Carolina*, met with the Lords Proprietors, who appointed him (with Edward Moseley) as one of the commissioners to settle the long-disputed boundary between North Carolina and Virginia (March 14, 1708/09), and was commissioned as Surveyor General of North Carolina (April 28, 1709). He also touched the Lords Proprietors for twenty pounds to prepare a map of North Carolina (probably the one already prepared and published in *A New Voyage*), and became associated with a new project to settle Swiss emigrants and German Palatine refugees in North Carolina. And, in the summer of 1709, he consulted in person with James Petiver.[166]

Petiver referred to Lawson's visit in a letter to George London, September 7, 1709:

I have lately obtained an Acquaintance with one Mr. Lawson Surveyor General of Carolina. He is a very curious person & hath lately printed a Natural History of Carolina wherein he hath treated ye Quadrupeds, Birds, Fishes, Reptiles, & Vegitables particularly the Trees with a great deal of Judgment & Accuracy. He suddenly designs to return for Carolina & as he may be serviceable to you is very ambitious of being known to you. If therefore you will as soon as possibly you can appoint a time I will wait on you

---

[166] *Calendar of State Papers, Col. Ser., Am. & W. Indies, 1708–1709,* Nos. 402, 479, 480, 675, 813, 828; William L. Saunders, ed., *The Colonial Records of North Carolina* (30 vols., Raleigh, 1886–90), I, 703, 717; *Collections of the South Carolina Historical Society,* I (1857), 178–180; A. T. Dill, Jr., "Eighteenth Century New Bern," *North Carolina Historical Review,* XXII (1945), 11, 15, 152, 171.

with him. He is very desirous of procuring what variety of Grapes & Plumbstones he can before his departure for which I doubt not but he will make you a suitable return from thence of whatever you desire from those parts.[167]

That London met Lawson before the latter's return to Carolina is indicated in Lawson's later correspondence. He did not sail with his Palatine refugees until January, 1710, and in the interim he consulted frequently with Petiver, who gave him encouragement, instructions, and collecting supplies for further work in natural history. He also furnished medicine and medical directions for the care of Lawson's sickly emigrants, who suffered much in their overcrowded vessel. At least three of Petiver's letters to Lawson survive from the autumn of 1709 before Lawson's departure with his human cargo for Carolina.[168] He sent Lawson John Ray's printed tract on Banister's Virginia plants, "a few pinns yt your Insects may not fly away after you have once caught them," some of "ye cholick Root" with directions for its use, and other drugs with a price list of "ye lowest ready money Prices, for the incouragement of those who practise with you or can otherwise dispose of them." He assured Lawson that many ship surgeons "get great reputation by ye Practise of these medicines, they being all very choise & such as will not change or decay in your climates." And he renewed his plea for collecting natural products, both for himself and George London. Writing from Portsmouth, where his ship was preparing for its voyage, Lawson acknowledged Petiver's

kind present of Mr. Rays book & the Physick, which I hope to make suitable return for soon after my arrival in America. . . . As for the citron laudanum I have given some few drops to a poor Palatine which did him a great kindness in relieving him of a delirium. I think them to be choice remedies and fit for transportation to our parts. . . . Therefore if you please to send some parcels for sale & ye price, I doubt not but they will do a great deal of good amongst us & answer the expectation in ye returns. . . . If ye last months memoirs be out pray send me one & a half

---

[167] *Sloane 3337*, fol. 56v. In publishing this letter in my article on *Petiver* (p. 337), I erroneously printed the name of the addressee as "William London." Professor Hugh Talmage Lefler, of the University of North Carolina, kindly brought this error to my attention.

[168] Oct. 12, 14, Nov. 14, *Sloane 3337*, fols. 63, 74-75.

dozen corks for the vials, seeing quarts are too small to stop them. . . . Our Palatine children dye apace, we have lost above 40 already, they are chiefly very younger ones. The men & women are generally in health. . . . This is wt I can think on att present until my arrivall in America where you shall [have] my monthly observations with ye Collections. . . .[169]

Upon his arrival in America, Lawson was busily engaged with the settlement of the Swiss and Palatines at New Bern, with the disputed boundary between North Carolina and Virginia, and with his duties as Surveyor General. In the management of all these affairs he made enemies: the Germanic colonists charged him with sharp practices in land transactions; the Indians and the Virginians accused him of self-interest, the use of inadequate surveying instruments, and obstruction during the abortive efforts to settle the boundary problem; and an explosive political situation in North Carolina at the time enhanced the difficulties of Lawson's doubtful dealings.[170] Mr. Petiver wrote to him (March 14, 1709/10), sending "more physick," several books on natural history, and an appeal to sell colonials subscriptions to his languishing publishing venture, *The Monthly Miscellany: or, Memoirs for the Curious*, at five shillings per month.[171] Lawson found no opportunity to reply until December 30, 1710. He had sent a collection of insects, plants, bird skins, a snakeskin, and some fossils the previous July, and he needed more brown paper, white paper, larger bottles for snakes, and "preserving liquor."[172] The season had been good, with excellent crops; he held high hopes for the Swiss at New Bern, was less enthusiastic about the Palatines, "seeing they are the most sloathfull people I ever saw." More than half of them had died: "one reason & the chiefest was seeing they were twice as many as ought to have been on a healthful ship." Lawson had been unable to keep "an

[169] Nov. 11, 1709, *Sloane 4064*, fol. 214.
[170] Vincent H. Todd, ed. and tr., *Christoph von Graffenried's Account of the Founding of New Bern* (Raleigh, 1920), pp. 225, 226, 293, 360–362, 373, 392; Saunders, ed., *Col. Records of N.C.*, I, 723, 725, 727–728, 735–738, 747, 750, 752, 757, 786.
[171] *Sloane 3337*, fols. 97–97v.
[172] It appears from this letter that Lawson's collections were shared with (besides Petiver and London) Fettiplace Bellers (1687–1750?), F.R.S., philosophical writer and minor dramatist. *D.N.B.*

exact dyary as you & I proposed until July last, when I began & shall very strictly commence it the year 1711 from the first of January with the weather & whatever happens worthy of notice in this Collony." Moreover, he had made further resolves, and "if God prolongs my dayes my Intention is this:

1st. To make a strict Collection of all ye plants I can meet withall in Carolina, always keeping one of a sort by me giving an account of ye time & day they were gotten, when they first appear in what sort of ground, with the flower, seed & disapear [dispermous?] & what medicinall uses the Indians or English make thereof to have ye Accompt of the same & let me know how near they agree with the European plants of the same species & wherein they differ. Besides I would send seeds of all ye physicall plants & flowers to be planted in England. As for the trees the time they bud, flower, bring their ripe fruit & soil. I hope to comply with most of them this year 1711.

2ndly. Beasts the most easy to discover fully their kinds &c. seeing they are not many as the other tribes are.

3dly. Birds to procure all of this place both land & water fouls from ye Eagle to the wren to know if possible the age they arrive to, how & where they build their nests, of what material & form. The colour of their Eggs and time of their Incubation, & flight, their food, beauty, & colour, of what medicinall uses if any, if rarely designed & to ye Life this would illustrate such a history very much, their musicall notes & cryes must not be omitted, wch of them abide with us all ye year & those that go away, and wt strange birds tempestuous weather, winds, unusual seasons & other accidents affords us.

4thly. Fishes their species, names here, how far they agree with those of Europe & wherein not, their food, usuall haunts, taste, virtues, time of brood & running to scawne as they call it, wt are all ye year when in season, which of them delight in salt water & of others in fresh streams, Lakes, ponds &c. Wt species are found in ponds that are often destitute of water in dry seasons, whether spawn may be transported for propagating strange species of fish & how far.

5th. Insects. the months they appear to us in the places of their resort, how they breed & what changes they undergo, their food, make, & parts: this may be very well done by having a many small Phyols or boxes wth descriptions of every insect contayned in each bottle & when you receave them you may rank them on wyer pins in little drawers as you think fitt, having ye notes constantly by you.

Fossills, as earth, shells, stones, mettalls, mineralls, stratas, paints, Physicall Earths, where & when found & what subterranean matters are yet discovered & wt methods has been theretofore taken to discover & work mines of all sorts.

Of ye Nature of soyles & Land wt agriculture they are most fitt for as to orchards, Gardens, grass, pasturage, Exotick plants, how they thrive & what species may be expected to agree best with this Climate, as fruits, lumber, trees for pleasure, grain of all sorts, garden seeds & seeds of all usefull sorts of Grasses, of the nature and disposition of our spontaneous Vines: if they are to be meleorated & made profitable and how. Wch is ye best way to bring in forraign vines that they may thrive, whether by the seed or by slips or roots. How fruit came first from ye seed to be meleorated to the perfection they are at now in Europe. I think grafting not so absolutely necessary in hott climates as in cold since ye best peaches & apples I ever saw are eat in Carolina come from ye seed wch will not do (I presume) in great Brittain. As to ye Vine a great help might be had from ye Event the planters of Madera, Cape of Good Hope, & other new Collonies had in planting them were we acquainted therewith. Of the making wine, its ripeness & fermentation, the making of fruit as raisins, figgs, prunes &c an account of this kind would be extreamely pleasant & profitable.

The present state of such a new country as this wth weather each day & month affoards in season for fruits, fish, fowl, etc, the way of living, cookery, dayry, seasons of planting, reaping, advancement of trade, increase of stock of cattle, sheep, &c., wt designs are in hand to Inrich ye Country & wt experiments have been made and how far they have answered ye design; then buildings, fencings, fortifications, Improvements of barren lands, dreyning of marshes and other improvements, what rivers, mountains, valleys, nations of Indians, mineral waters, springs, cataracts and other natural rarieties are discovered, what other accidents are of moments as to the sick, who died and who recovered and what means were used for the same either by the Christian or Indian practitioners. These transactions being faithfully communicated to you and such ingenious gentlemen of the Royal Society with their remarks on these same will lay a foundation towards a compleat history of these parts which I heartily wish I may live to see publisht. I shall be ready and glad at all times when I can with my mite be any ways serviceable towards compleating so good an undertaking. I shall be glad at all times when opportunity serves to hear from you and Mr. [Fettiplace] Bellers and to have your instructions. I have some more plants collected but the books being not full I omit sending them until compleated. Sir, pardon this freedom I take with you. I only tell you my intentions and beg your advice and am and shall ever remain to the utmost of my power sir your most humble servant to command.[173]

Lawson's "intentions" bespoke both a broad grasp of problems involved in being an effective collector in natural history and a lively, enthusiastic imagination. Their accomplishment would have involved a full-time occupation hardly possible in view of his other interests and duties. Had he been able to undertake the tasks he set himself over an extended period of time, his value as a collector —and possibly even as a natural historian in his own right—might have equaled that of John Banister in Virginia or Dr. Hans Sloane in Jamaica. But God did not prolong his days to carry out his plans. On an expedition trip up the Neuse River in the following September (1711), Lawson was seized by the Tuscarora Indians, who, enraged by his insolent attitude and ill treatment in land transactions, put him to death on September 22. Scarcely two months earlier he had sent to Petiver "by our Govr's Lady one book of plants very Lovingly packt up," and he had expressed the hope that he would soon hear again from his London friend.[174] Petiver wrote to Jacob Bobart (March 8, 1711/12) that "I very lately rec'd a Collection of Virginia [Carolina] Plants from Mr. John Lawson, a very curious Person who wrote a Naturall History of Carolina, but at ye same time rec'd advice of his unfortunate death, being Kil'd by ye Indians of ye Continent so yt ye Death of these good ffriends & ye dilatory performances of many living ones quite disheartens me, so that nothing but ye glimmerings of Peace & a South Sea Trade gives me hope of a faint recovery. . . ."[175]

[173] *Sloane 4064*, fols. 249–250. A partial copy in *ibid.*, fol. 267, adds an appeal for grape seeds from Petiver's correspondents in Spain and elsewhere, together with other fruits and acorns of cork oaks. It also includes regards to George London, "whom I am collecting for."

[174] *Sloane 4064*, fol. 271.

[175] *Sloane 3338*, fol. 37.

John Lawson's *A New Voyage to Carolina* (1709) was one of the best travel accounts of the southern colonial scene.[176] Its value, however, is greater for the history of the political, social, and economic aspects of the region than it is for the history of science. Lawson undoubtedly had a sharp eye for natural history, and he possessed rudiments of botanical and pharmacological knowledge; but he was not *au courant* with the scientific advances in the study of natural history of his generation. It would be scarcely true to label his book as promotional literature, although there was a clear note of promotionalism running through it, with exaggerated praise of Carolina's climate and soil and its potential for the production of citrus fruits, silk, and wine. More accurately, it may best be described as a travel account, and a very sprightly one at that, with sections devoted specifically to topography (including a useful map of the Carolinas), vegetation, beasts, insects (including with them, curiously enough, alligators and snakes), birds, fishes, and the Indians. Lawson's approach to the natural productions of North Carolina was little, if indeed any, in advance of the herbalist. His descriptions of flora and fauna, though occasionally good, were uneven in quality, sometimes verbose, generally lacking in exact details, and couched in popular, sometimes quaint, language seldom associated with the scientific terminology that was developing at the time. Though less credulous than John Josselyn had been in describing New England, Lawson betrayed much credulity, espe-

---

[176] Lawson's work has often been republished. The latest and best edition, with an introductory sketch, annotations, and index, is that of Hugh Talmage Lefler (Chapel Hill, 1967). Professor Lefler lists all the previous editions of Lawson's work (pp. xliv–lii) and provides a map of Lawson's "Thousand Miles Travel" (p. x), estimated to have been nearer to 550 miles. His annotations are particularly helpful in identifying the names and locations of Indian towns and villages mentioned by Lawson. Lawson's work has been plagiarized several times. John Brickell, *The Natural History of North Carolina* . . . (Dublin, 1737, 1743; Raleigh, 1911), has been called a "shameless" plagiarization of Lawson, but Professor Lefler softens the charge by showing that Brickell included valuable materials of his own (pp. lii–liii). More clearly, the book commonly listed as the *Natural History of Virginia, or the Newly Discovered Eden* (1737) and attributed to William Byrd II (ed. and tr. from the German version by R. C. Beatty and W. J. Mulloy as *William Byrd's Natural History of Virginia* . . . , Richmond, 1940) was taken verbatim, in considerable part, from Lawson's book. See Percy G. Adams, "The Real Author of William Byrd's Natural History of Virginia," *loc. cit.;* Lefler, ed., *A New Voyage to Carolina,* pp. liii–liv.

cially in his estimate of snakeroot as a cure for snake-bite, the capacity of rattlesnakes to charm their victims (in both cases he echoed opinions akin to those of William Byrd II), and his attitudes toward the remarkable cures attributed to Indian medicine. He appears to have accepted the ancient Doctrine of Signatures. Still, his lists and descriptions of beasts, insects, birds, and fish of Carolina (though lacking illustration) generally excelled those of his predecessors, even granting that the descriptive terms were innocent of scientific accuracy, terminology, and evidences of dissection. He listed twenty-seven beasts (exclusive of snakes, vipers, and alligators, the latter of which he declared to be identical with crocodiles); he spoke of "tygers," but said they differed from "the Tyger of *Asia* and *Africa*"; and he specifically laid to rest the long-prevailing belief in the existence of lions in the region. His bird lists (fifty-six land birds and fifty-three waterfowl) exceeded in quantity if not in quality those of John Clayton—though he set forth a curious account of the bat, which he thought was a biological cross between a bird and a mouse.[177] Unlike many of his predecessors, he did not assume that the fauna of North Carolina were all of the same species as those of the Old World, recognizing distinctive features in the wildcat, wolf, skunk and opossum— though he asserted that American deer were a source of the occidental bezoars, which he distinguished from the oriental bezoars, but not in distinct fashion. His accounts of the fish of Carolina were largely confined to the salt-water varieties. They were unusually complete for the time, but too frequently Lawson identified them with European varieties. Lawson held a good opinion of the Indians. Indeed, his informative and sometimes lusty descriptions of Indian ways of living, social and political organization, traditions, and languages (including a brief Indian vocabulary) are one of the best portions of his book, with considerable value for the anthropologist of today. In comparison with his predecessors, however, Lawson suffered a relative lack of attention by his contemporaries. Both his book and his meager collections (mostly trees) were too

177 Elsa G. Allen, "The History of American Ornithology before Audubon," *Transactions of the American Philosophical Society*, n.s., XLI, Pt. 3 (1951), 461–463; Waldo Lee McAtee, *The Birds in Lawson's New Voyage to Carolina (1709)* (Raleigh, 1955–56).

late to attract such great scholars of natural history as Plukenet, Bobart the Younger, and Ray; and no great English publicists rose up to succeed them in the field. Only James Petiver gave attention to Lawson's collections in his publications.[178] No doubt they were studied by others—Sloane, Sherard, London, for example—and some of them still lie in the British Museum.[179] But probably the greatest blow to Lawson's reputation was the attention given to Mark Catesby, who, entering upon the scene within a short time after Lawson's death, soon overshadowed him and all his predecessors in the mainland colonies.

Catesby, already known for his work in Virginia, arrived in Charles Town, South Carolina, on May 3, 1722. He was comparatively well sponsored by a dozen patrons, led by William Sherard, Sir Hans Sloane, Charles Dubois, four noble peers of the realm, and Governor Francis Nicholson of South Carolina, whose grant of twenty pounds, it is well to recall, had been given "for the uses and purposes" of the Royal Society.[180] At least three of these men had been members of the Temple Coffee House Botany Club, whose tradition of patronage for colonial collectors they were carrying on. Catesby was cordially received by Governor Nicholson, and as he had previously done in Virginia, the Governor quickly introduced his protégé to the best families in and about Charles Town, with whom Catesby soon found cordial hospitality and cooperation.[181]

Catesby spent about three years in Carolina. He collected specimens and made notes, drawings, and watercolors of the natural products of the country, ranging from the borders of North Caro-

[178] See Petiver's acknowledgment of gifts from "My Late curious Friend Mr. John Lawson" in *Phil. Trans.*, XXIX, No. 346 (Nov.–Dec., 1715), 355.

[179] Dandy, p. 154.

[180] The patrons were listed in the Preface of the first volume of Catesby's *Natural History of Carolina* (1731). The peers were: the Duke of Chandos, the Earl of Oxford, the Earl of Macclesfield, and Lord Perceval. Another patron was the well-known collector and philanthropist, Dr. Richard Mead, F.R.S. At about the same time, four of these patrons, including Sherard and Sloane, with three new contributors, financed a similar collecting expedition to New England by Thomas More, the elusive "Pilgrim Botanist." But More's voyage proved to be of comparatively little value. See the account of More in the Appendix of Frick and Stearns, *Mark Catesby*, pp. 114–127.

[181] For more complete details and references, see *ibid., passim.*

lina into Florida (including areas soon to be Georgia), and west-
ward into the Piedmont. Generally, he was adequately supplied
with collecting materials by his sponsors. He had, for a time at least,
the assistance of a Negro boy, and he was able to enlist the active
aid of friendly Indians. These, together with his own industrious
activities, and the cooperation of such helpful planters as Robert
Johnson, the able Governor of the colony (1717–19 and again in
1731–35), and his Charles Town friend, Dr. Thomas Cooper, enabled Catesby to make enormous collections of flora and fauna for
his friends and patrons. As time passed, and as Catesby's returns to
England revealed their rare value, the list of friends and patrons
lengthened until Catesby was severely pressed to meet all their demands.[182] Sir Hans Sloane and William Sherard, already on poor
terms before Catesby went to Carolina, wrangled almost pettishly
over some of the collections, and Catesby's correspondence indicates
that the items sent from Carolina were distributed widely in England and on the Continent. Sloane and Sherard, who shared with
John Jacob Dillenius, a German botanist imported to assist Sherard
in the completion of his *Pinax*,[183] were the principal recipients at the
outset, but Catesby did not overlook his old friend, Samuel Dale of
Braintree. Soon Peter Collinson was promoting Catesby among his
rapidly enlarging group of friends, and receiving Carolina seeds,
roots, and the like for his garden at Peckham; and Thomas Fairchild, the nurseryman at Hoxton, Isaac Rand, Director of the
Apothecaries' garden at Chelsea, Dr. Richard Richardson, F.R.S.,
the physician, botanist, and patron of natural science at North
Bierly, Yorkshire, Dr. Richard Mead, Charles Dubois, Lord Oxford, Sir John Colleton—all these, and probably others, enriched

[182] "Mr. Peter Collinson has procured Several Subscriptions which obliges me
to send him a greater quantity than otherwise there would be occasion for."
Catesby to Sherard, Charles Town, Nov. 13, 1723, *Sherard Letters*, II, 170. The
following spring he wrote to Sherard: "I hope it cant be expected I should send
Collections to every of my Subscribers which is impracticable for me to doe."
Charles Town, April 6, 1724, *ibid.*, p. 176.

[183] In 1740 Linnaeus placed Dillenius at the head of his list of eminent botanists. See his letter to Haller, Stockholm, Sept. 15, 1740, in Sir James Edward
Smith, ed., *A Selection of the Correspondence of Linnaeus* (2 vols., London,
1821), II, 348.

their knowledge and their gardens from Catesby's Carolina collections. Even Sir Robert Walpole received items for his garden, sent at the suggestion of Governor Nicholson, who obviously had an eye for catering to persons in high places. Sloane, Dillenius, and Sherard also shared their Carolina collections with Hermann Boerhaave and Johann Friedrich Gronovius of Leiden, with Bernard de Jussieu and others at Paris, with Jakob Theodor Klein at Danzig, and with Robert Wood, of the botanic garden at Edinburgh, so that Catesby's collections made an impact upon some of the principal European scholars of the time. Moreover, Catesby enriched the gardens of Carolinians, both with new plants, shrubs, and trees introduced from the hinterlands of the colony and with seeds, bulbs, and plants from his friends in England. He also encouraged experimentation with new staples introduced from the Far East. To some extent Catesby must be credited with stimulating Carolinians in the creation of the lovely gardens for which the state has become famous.

Catesby's collections included seeds, plants, mosses, shrubs, trees, acorns, nuts, berries, cones, roots, shells (both mollusks and crustaceans), birds, animals, snakes, insects, fishes, amphibia, and a few Indian curios for Sir Hans Sloane's collections.[184] He also made notes of his observations on topography, geology, climate, agriculture, and the Indians, together with figures of birds, plants, animals, and fishes, some of them in color. In general, he sought to lay by at least one of each of his specimens for his own use and to keep his notes and figures of birds and animals, "which to distribute Seperately would wholly Frustrate that designe [i.e., to prepare his own natural history], and be of little value to those who would have so small fragments of the whole."[185] In his collecting, he attempted to visit the same area at different stages of plant growth: "My method is never to be twice at one place in the same season. For if in the spring I am in the low Countrys, in the Sum-

---

[184] Catesby to Sloane, Charles Town, Nov. 27, 1724, *Sloane 4047*, fol. 290: "an Indian Apron made of the Bark of Wild Mulberry. This kind of cloath with a kind of Basket they make with Split cane are the only Mecanick Arts worth Notice."

[185] Catesby to Sloane, Charles Town, Aug. 15, 1724, *Sloane 4047*, fol. 212.

mer I am at the heads of Rivers, and the next Summer in the low countrys, so I visit the two different parts of the Country."[186] He was constantly attentive to problems of ecology, to the height and girth of plants, shrubs, and trees observed, and to their utilitarian value in medicine, building, and joinery. Much of what he sent his friends and patrons was new; in fact, speaking of one of Catesby's earliest shipments, Sherard wrote to Richardson, "I had a letter this week from Mr. Catesby at Carolina, who sent me two quires of dry'd plants, 40 of wch were new. . . ."[187] As peace prevailed at the time, the seaways were free of armed interference, although "pyreats" despoiled at least one of Catesby's shipments. As a whole, however, they appear to have reached their destination safely and in good condition. On balance, it would appear that in terms of quantity of collections, quality of observations, and breadth of view, English patrons had had no naturalist in the colonies (save, perhaps, Dr. Sloane) who equaled Mark Catesby—although Catesby lacked contemporary training and concern for plant description and systematization comparable to that of John Banister thirty-odd years before.

In 1725, disappointed in an effort to persuade his patrons to finance a journey to Mexico with his Charles Town friend, Dr. Cooper, Catesby sailed to the Bahama Islands. He visited New Providence, Eleuthera, Andros, and Abaco, and made his customary collections, notes, drawings, and paintings. From his later published works, however, it appears that he concerned himself with marine life in the Bahamas more than he had done in Carolina. He devoted a year to work in the islands and returned to London in 1726. Excepting items received from friends and correspondents in America during the next twenty years, his days of collection were ended.

After his return to London, Catesby faced a difficult financial problem. He possessed rare materials upon the basis of which he proposed to prepare a natural history. But his former patrons' subscriptions of funds ceased after he no longer supplied them with New World collections, and he lacked the money with which to support himself, not considering the funds necessary to finance a

---

[186] Catesby to Sherard, Charles Town, Jan. 16, 1723/24, *Sherard Letters*, II, 174.
[187] Oct. 13, 1722, *Ms. Radcliffe Trust C. 4*, fol. 84.

costly book with colored illustrations such as he was planning. After vainly casting about for support, he determined to do his own work. He enlisted the direction of a French artisan, Joseph Goupy, who taught him the art of etching on copper plates so that Catesby was able to prepare his own plates for illustrations in his book and color them as well. Sir Hans Sloane gave him advice and encouragement. Peter Collinson loaned him an undisclosed amount of money without interest, and Catesby found employment with nurserymen, first with his old friend Thomas Fairchild at Hoxton, and after Fairchild's death (1729) with Christopher Gray at Fulham. The work had a double value for Catesby, for in addition to the money earned, he was able to observe many of his American flora growing in the gardens—and enhance the nurserymen's profits from the sale of such exotic seeds, roots, shrubs, and trees.

At the same time, he slowly advanced the text and the illustrations for his book. His friends and former patrons gave assistance. William Sherard, before his death in 1728, and J. J. Dillenius helped with the Latin names for his plant specimens; for birds, animals, and fishes he generally followed Ray and Willughby, with some attention to Sloane's treatment of Jamaica; for mollusks he referred to Lister's works, although in the matter of fossils he subscribed to ill-favored diluvialist views of Thomas Burnet and John Woodward, who contended that both the origin and the distribution of fossils were the consequence of the biblical Flood—an explanation, however, which recognized the organic origin of fossils and denied the ancient formed stones theory. By the spring of 1729 Catesby had completed the first portion of his book. In order to encourage subscriptions and enhance the sales by publishing it in parts at a small price, he prepared it in numbers of twenty plates each, later assembled into bound volumes. The first of these he presented to the Royal Society on May 22, 1729, as a guest of the Society. Succeeding parts were similarly presented at subsequent meetings until the fifth part was given on November 23, 1732, completing the first volume—although its publication date, as printed on the title page, was 1731.[188] Catesby was a guest at several meetings of the Society, and on February 1, 1732/33, Peter Collinson presented him for

[188] *Journal-Book*, XIV, 336, 393 (Jan. 8, 1729/30), 516 (Nov. 19, 1730); XV, 190 (Nov. 23, 1732).

membership. He was subsequently elected, and on May 3, 1733, formally admitted Fellow of the Society.[189] Already the several parts of his book had been enthusiastically reviewed at meetings of the Society, and these "free advertisements," together with the undeniable worth and beauty of the work, multiplied subscriptions for its purchase. He continued in the same manner as the second volume advanced, although he was able to employ help in coloring some of the plates. The first twenty plates of the second volume were presented to the Society on January 20, 1734/35, but it was December 15, 1743, before Catesby presented the "last part" of his volume. Actually it was April 16, 1747, before the Appendix was completed, and Peter Collinson announced in a letter to Linnaeus that "Catesby's noble work is finished."[190] In its final form, the "List of Encouragers of this Work" had grown to 166, including "several Dutch physicians & scholars of the time"—a considerable expansion of the original twelve who had patronized Catesby's collections in Carolina at the outset in 1722. They included the late Queen Caroline, the Queen of Sweden, the Prince of Liechtenstein, the Princess of Wales, the Duke of Bourbon, twenty peers of the realm, the Russian Ambassador to the Court of St. James, the Bishop of Derry (Dr. Thomas Rundle), the Royal Society, Sir Hans Sloane, Peter Collinson, Samuel Dale, and many other patrons of long standing, together with a number of colonial supporters. This last group sounded a new note, inasmuch as few American colonists had had the interest (and the funds!) to support a scientific publication. They included, besides six colonial governors and proprietors, William Byrd II, John Randolph, and Benjamin Whitaker of Virginia; Alexander Hume and Alexander Skene of South Carolina; John Bartram of Pennsylvania; and Colin Campbell of Jamaica. Catesby's twenty years of self-sacrifice and unrelenting labor produced works which ultimately attracted wide support.

And, as Collinson said, Catesby's *Natural History of Carolina, Florida, and the Bahama Islands* was a "noble work." Richard

---

[189] *Ibid.*, XV, 230, 283, 292; *Certificates, 1731–1750*, p. 41. Catesby made drawings for the *Register-Book* of the Society as payment for his bond (twenty-one pounds). *C.M.*, III, 118, 142.

[190] Presented July 2, 1747, *Journal-Book*, XX, 307. Officially, the Appendix was not published until 1748. See Frick and Stearns, *Mark Catesby*, p. 44.

Pulteney, writing later in the same century, said that it was "the most splendid of its kind that *England* had ever produced."[191] It contained more than 200 colored plates picturing 171 plants, 9 quadrupeds, 113 birds, 33 amphibia, 46 fishes, 31 insects, an undistinguished map of Carolina, Florida, and the Bahama Islands, an account of the history, topography, climate, soils, botany, zoology, and agriculture of the areas, an essay about the Indians of Carolina (heavily indebted, with credit, to John Lawson's account), and a description of "The Manner of making Tar & Pitch in Carolina"[192]— all printed in parallel columns in both French and English (the French translator is unknown). Catesby's forte lay in collection and illustration, although his descriptions of specimens were good, if uneven and occasionally too brief. His illustrations were sometimes too brightly colored, and often lacked the detail considered necessary today for scientific purposes; but on the whole they were dynamic, beautifully posed, and almost always identifiable. He early perfected the method, later employed by Wilson, Audubon, and others, of combining zoological, ornithological, and botanical specimens on a single plate, thereby depicting an ecologically correct setting for birds, trees, plants, and animals. Many of his plates were very beautiful, leading both to their removal from the original volumes to be used for decorative purposes and to their more recent reproduction for sale to the same end. Scientifically, his birds and flora excelled his fish, animals, and insects; his butterflies were beautifully done, but more decorative than scientifically useful. He contributed nothing toward the systematic classification of his specimens, although his nomenclature, especially in ornithology, persisted in modified forms even after the Linnaean binomial system, which came into use shortly after the *Natural History* had appeared, had rendered obsolete Catesby's terminologies. Catesby's finest contributions were in the field of ornithology. In this he went beyond mere illustration and description of particular birds, to speculate about migratory habits and the reasons for them. On March 5, 1746/47, he read to the Royal So-

191 Pulteney, II, 226.
192 Both the account of the Indians and the description of the making of tar and pitch had been read previously to the Royal Society. *Journal-Book*, XIX, 98–99, 101 (May 5, 1743), 106 (May 19, 1743).

ciety a paper on the subject which was published in the *Philosophical Transactions*, in a shortened form in the *Gentleman's Magazine*, and its substance included in the second volume of the *Natural History*.[193] Based upon his observations in America, Catesby rejected current theories that birds hibernated in caves, hollow trees, or beneath the waters of ponds or that birds of passage flew to their retreats above the atmosphere. Instead, Catesby contended, they flew to their new destinations in the normal manner, to seek freedom from the cold and avoid a reduced supply of food as winter approached. This commonsensical explanation was marred, however, by Catesby's insistence upon the old notion that climatic uniformity existed in equal degrees of latitude north and south of the equator and that birds of passage wintered as far south of the equator as they had summered north of it. Still, his explanation was superior to those of his contemporaries—and no less satisfactory than many advanced long afterward.[194] Ornithologists often refer to Catesby as the "founder of American ornithology," and one wrote as recently as 1929 that his *Natural History* "forms the basis of the ornithology not only of the Southern States, but of the whole of North America."[195]

After Catesby's death (1749), George Edwards (1693–1773), author of *A Natural History of Uncommon Birds* (4 vols., 1743–51), published a revised edition of the *Natural History* (1754). He used Catesby's plates and made minor changes in the text; but the edition was less valuable than the original, mostly because Edwards' coloring of the plates was, in the main, inferior to Catesby's. A third edition appeared in 1771 with Linnaean names assigned to Catesby's specimens. But the work was less satisfactory than Edwards', being done under the direction of the new publisher, Benjamin White,

[193] *Ibid.*, XX, 218–223; *Phil. Trans.*, XLIV, Pt. II, No. 483 (March–May, 1747), 435–444; *Gentleman's Magazine*, XVIII (1748), 447–448. The *Gentleman's Magazine* also published plates and excerpts from the *Natural History*. See *ibid.*, XXI (1751), 10, 11; XXII (1752), 276, 300, 364, 412, 572; XXIII (1753), 29, 128, 180, 268, 324, 512, 609.

[194] Frick and Stearns, *Mark Catesby*, pp. 63–64; W. L. McAtee, "Unorthodox Thoughts on Migration," *The Auk*, n.s., LVII (1940), 135–136.

[195] Witmer Stone, "Mark Catesby and the Nomenclature of North American Birds," *The Auk*, XLVI (1929), 447–454. See also Allen, "History of American Ornithology," pp. 463–478. For other recent articles on Catesby, see Frick and Stearns, *Mark Catesby*, p. 113.

who obviously undertook it for profit rather than for science. Still, it demonstrated the continued interest in and saleability of Catesby's work. The *Natural History* was also pirated and published on the Continent, in whole or in part, in German, Latin, Dutch, and French. It became a widely used—and abused—book of reference, even into the present century. Linnaeus, although he temporarily altered the system of classification in natural history (*Systema Naturae*, 1735, and *Genera Plantarum*, 1737) and permanently revolutionized nomenclature (*Species Plantarum*, 1753 and later), made extensive use of Catesby's works, both in botany and in zoology, and adopted many of his descriptions and binomial names of plants, birds, and fishes from Catesby's *Natural History*.[196] European scholars of the Enlightenment, and later scholars both in Europe and America, leaned heavily on the work, especially in the study of fauna. Regarding flora, Linnaeus' works, and especially Thomas Walter's *Flora Caroliniana* (1788), largely replaced Catesby before the end of the eighteenth century.

Besides an active part in the work of the Royal Society after his admission as Fellow in 1733, Catesby was also much engaged in the introduction to English gardens and parks of new species of American flora, especially trees. To this end he labored in the nurseries of Thomas Fairchild and his successor Stephen Bacon at Hoxton, and of Christopher Gray at Fulham; and he contributed to the gardens of his friend and patron Peter Collinson at Peckham, of Dr. Richard Richardson of North Bierly, of James Sherard (William's brother) at Eltham (a catalogue of which was published by Dillenius as *Hortus Elthamensis* in 1732),[197] and of many other gardens, some of them belonging to noble patrons such as Lord Perceval and Lord Oxford. Also, to further this end he prepared a second book, *Hortus Britanno-Americanus* . . . , completed shortly before his death in 1749 but not published until 1763.[198] At first glance this book—smaller and less pretentious than the *Natural History*—appears to be little more than a nursery catalogue for his friend Christopher Gray

---

196 *Ibid.*, pp. 59, 70, 78, 84–85.

197 Dillenius' *Historia Muscorum* acknowledged the contributions of Catesby, together with John Bartram of Pennsylvania and John Clayton and John Mitchell of Virginia, in supplying him specimens and information about various mosses.

198 Republished as *Hortus Europae Americanus* . . . (London, 1767).

of Fulham; but closer inspection reveals that, although the eighty-five trees and shrubs are largely illustrated by means of smaller copies of those already shown in the *Natural History*, the descriptions are wholly different, with greater detail to stimulate interest in growing American species and to supply information for their successful cultivation. The *Hortus* also corrected a few errors of the *Natural History*, and pictured some trees and shrubs not included in the earlier work. It emphasized the variety of trees and shrubs afforded by American forests and pointed out their uses in England both for their timber and for ornamental purposes. But it had been prepared with a narrower objective in view than had the *Natural History*, its author had not lived to see it through the press, and the *Hortus* leaves a suspicious impression of having been a publisher's profit-seeking promotion piece, tardily rescued from the dead author's literary remains. No wonder, then, if Catesby's *Hortus Britanno-Americanus* fell far short of his *Natural History* as a work of scientific significance.

Many of the collections, drawings, and paintings made by Catesby are still extant. The dried plants he shipped to Petiver, Sloane, and others of his correspondents in the London area are in the British Museum.[199] Others, those sent to Sherard and Dillenius especially, are in the Oxford herbaria.[200] Among the *Sloane Manuscripts* in the British Museum are unpublished drawings and paintings of American birds, plants, butterflies, and other insects made by Catesby.[201] The correspondence of and about Catesby reveals another point of significance. Even while Catesby was in the colonies, his patrons—especially Petiver, Sloane, Sherard, and Collinson—made his name known widely in England, the Netherlands, France, and Germany. Both by his associations in the colonies and by his later correspondence, Catesby himself came to know most of the mainland colonists, and a few in the West Indies, who displayed an interest and ability in natural history. After his return to England, and particularly after he was elected Fellow of the Royal Society, Catesby extended his associations with both European and American scholars.

---

[199] Dandy, pp. 110–113 and especially H.S. 212, 232.

[200] Vines and Druce, *Morisonian Herbarium*, p. 234; Clokie, pp. 77, 80, 86.

[201] *Sloane 3339*, fols. 73–73v; *4438*, fol. 32; *5267*, fols. 83, 93; *5271*, fols. 174, 315, 319–321; *5283*, fols. 8, 15, 46, 59, 63.

Although his place was secondary to that of Peter Collinson, he played a significant supporting role in the creation of an international circle of scholars and collectors embracing English, European, and colonial correspondents and friends. In England, even as his early patrons passed from the scene, he found younger replacements, such as Collinson for Petiver (d. 1718), Dillenius for Sherard (d. 1728), and many others. In Europe his correspondence was modest, though widened indirectly by that of Sloane, Collinson, and Dillenius. He himself corresponded with Gronovius at Leiden; with his English friend, Dr. John Amman, who moved to Russia, became Professor of Botany at St. Petersburg, a member of the St. Petersburg Academy, and chief promoter of the St. Petersburg botanic garden, where he experimented with American flora sent to him by Catesby and other English correspondents;[202] and with Linnaeus, although the latter at first formed an undeservedly low opinion of Catesby, and the correspondence did not mature into a wide exchange. In the American colonies, besides members of his sister's family, William Byrd, Custis, and John Randolph of Virginia, and Governor Robert Johnson and Alexander Skene of South Carolina (all of whom, with other colonials, he had known personally in the colonies), Catesby enlarged his correspondence to include John Clayton of Virginia, for whom he served as intermediary (with Collinson and others) with Gronovius, the Dutch co-author of Clayton's *Flora Virginica*;[203] with General James Edward Oglethorpe, the founder of Georgia, whom Catesby recommended for fellowship in the Royal Society; with John Bartram, the homespun Quaker genius of Pennsylvania, from whom Catesby received some of the specimens described in *Hortus Britanno-Americanus*;[204] and with Dr. John Mitchell, the physician, botanist, and cartog-

[202] Amman was author of the rare *Stirpium Rariorum in Imperio Rutheno sponte proventientium Icones et Descriptiones* (St. Petersburg, 1739).

[203] Pt. I, 1739; Pt. II, 1743. A revised edition appeared in 1762, although, as the Berkeleys have shown, Clayton himself had prepared a better revision that was not published. See Edmund and Dorothy Smith Berkeley, *John Clayton: Pioneer of American Botany* (Chapel Hill, 1963).

[204] William Darlington, ed., *Memorials of John Bartram and Humphrey Marshall* . . . (Philadelphia, 1849), *passim*. Professor Frick found further Bartram-Catesby correspondence among the *Bartram Papers* and the *Gratz Manuscripts* in the Library of the Historical Society of Pennsylvania (Philadelphia).

rapher of Urbanna, Virginia. After Dr. Mitchell returned to England in 1746, he became Catesby's close friend. Catesby took him as a guest to meetings of the Royal Society, and in 1748 was one of the Fellows who proposed Dr. Mitchell for membership (he was elected in that year). In the same year, when Peter Kalm, Linnaeus' former protégé, now a professor at Abö in Finland, visited England en route to America, Catesby met him and entertained both the Swedish visitor and Dr. Mitchell at his home one afternoon.[205] Catesby's colonial correspondents were not only part of the international brotherhood of scientists in the 1730's and 1740's, but also, by their growing familiarity with one another—a familiarity fostered by such English friends as Catesby, Collinson, and many others—the colonials were beginning to carry on a scientific correspondence and exchange among themselves. Thus began—on an ever-widening intercolonial basis—a circle of colonial scientists indigenous to America.

Georgia

According to a propaganda sheet issued by "the trustees for Establishing the Colony of Georgia in America," the original purpose of the promoters was fivefold: to relieve the poor of England and Europe by financing their establishment as colonists in America; to relieve the burden of their care and simultaneously stimulate English manufacturing and employment in the supply of "cloathes, working Tools, & other necessaryes" for the new colonists; to make "such a Barrier as will render the Southern Provinces of British North America safe from Indian & other Enemys"; to extend Christianity and the Empire; and to supply Britain with silk, wine, oil, dyestuffs, drugs, and other semitropical items which "nowadays [we] do purchase from Southern Countries."[206] One third of the

---

[205] *Journal-Book*, XX, 171 (Dec. 18, 1746). Mitchell visited the Society repeatedly before he was elected F.R.S. See *ibid.*, pp. 161, 176, 187, 191, 212, 523, 527; Peter Kalm, *Kalm's Visit to England on His Way to America in 1748*, tr. Joseph Lucas (London, 1892), pp. 17, 51–52, 118–120. Kalm went to the American colonies armed with letters of introduction from Dr. Mitchell and Peter Collinson. Adolf B. Benson, ed., *The America of 1750: Peter Kalm's Travels in North America* (2 vols., New York, 1937), I, 17.

[206] "Some Account of the Design of the Trustees for Establishing the Colony of Georgia in America," *Sloane 3986*, fols. 38–39.

original trustees were Fellows—or later became Fellows—of the Royal Society. Later trustees included other Fellows, and some of the investors in the enterprise were similarly affiliated.[207] Several of these men were patrons of Mark Catesby: General Oglethorpe, John Perceval, Earl of Egmont, Lord Petre, Earl of Derby, the Lord Bishop of Derry, Sir Hans Sloane, Charles Dubois, and Philip Miller of the Apothecaries' garden at Chelsea. Dr. Stephen Hales (1677–1761, F.R.S.), inventor, early discoverer of a method to measure blood pressure in horses, and author of *Vegetable Staticks* (1727), which inaugurated the science of plant physiology, was also a trustee. The presence of such promoters, together with the mercantilistic objects for which the enterprise was undertaken, promised an interest in natural history from the outset.[208] Indeed, in laying out the settlement of Savannah, the trustees provided that "a publick Garden was laid out, which was designed as a Nursery, in order to supply the People for their several Plantations with White Mulberry-trees, Vines, Oranges, Olives, and other necessary Plants; a Gardener was appointed for the Care of it, and to be paid by the Trustees."[209]

To oversee the management of this garden, upon the success of which the economy of the colony was expected greatly to depend, the trustees had entered into an agreement (October, 1732) with Dr. William Houstoun (1695–1733), a friend and correspondent of both Sloane and Miller. Educated at St. Andrews, in Scotland, Houstoun sailed to the West Indies as a ship surgeon in the mid-1720's but returned (1728) to Leiden, where he studied medicine with Dr. Hermann Boerhaave and won his degree in 1729. At this point he entered the service of the South Sea Company as a surgeon and traveled to the West Indies and Central and South America. There he added to his own herbarium, and made significant collections of plants for Sloane and Miller.

On July 1, 1731, President Sloane communicated to the Royal Society a letter from Dr. Houstoun, dated at Kingston, Jamaica,

---

[207] A. D. Candler, ed., *The Colonial Records of Georgia* (26 vols., Atlanta, 1904–09, cited hereafter as Candler), I, 27ff.

[208] *Ibid.*, II, 382. Oglethorpe was sending specimens from Georgia to Sloane as early as Sept. 19, 1733. *Sloane 4053*, fol. 53.

[209] Candler, II, 382.

December 9, 1730. The letter accompanied plant specimens, an account of the "way of manufacturing the Cochineal by the Indians at the town of Guaspaltepaque in the Valley of Oaxaca in Mexico," and "An Account of his inquirys concerning the nature of other plants and Drugs such as Jallop [jalap] and Contrayerva [*Dorstenia*] &c."[210] His "Account of the Contrayerva" was published in the *Philosophical Transactions*.[211] To Sloane he reported: "I met with a great many plants on the Continent which I could not possibly reduce to any Genus yet described, and therefore have made bold to characterise some of them, giveing them the names of Botanists, which is a practise now authorised by custom. But as I have but few books here to consult, it's very possible they may have been described already by some person or other; wherefor I desire to submit my Nova Genera, in that as in all other points to your better judgement."[212] In the autumn of 1731 Dr. Houstoun returned to London, was guest of Philip Miller at several meetings of the Royal Society, and on January 18, 1732/33, was elected Fellow of the Society.[213] Already, on October 3, 1732, he had contracted with the Georgia trustees to sail to Jamaica by way of Madeira, to study viniculture and wine-making at Madeira and collect vine cuttings and "any other useful plants" there for the Georgia garden, and then, with the cooperation of the South Sea Company, to proceed from Jamaica to the Spanish colonies at Cartagena, Porto Bello, Campeachy (Campeche), and Vera Cruz. He should "Stay at each of these places as long as shall be found necessary for obtaining all that can be gott from them; and upon his Return from one of them to Jamaica, to leave the Plants he shall bring over, with

---

[210] *Journal-Book*, XIV, 636–637.

[211] XXXVII, No. 421 (Oct.–Dec., 1731), 195–198. Dr. Houstoun demonstrated an excellent current knowledge of botany, worthy of one of Dr. Boerhaave's students. Experiments made by Houstoun in 1728, in conjunction with the celebrated Dr. Gerhard van Swieten, later physician to Maria Theresa in Vienna, presented to the Royal Society by Philip Miller, Feb. 10, 1736/37 (*Journal-Book*, XVIII, 40–41), were published in *Phil. Trans.*, XXXIX, No. 441 (April–June, 1736), 230–237. They dealt with the effects of injuries to the thorax in animal respiration.

[212] Quoted in Dandy, p. 140; also quoted extensively in Edith Duncan Johnston, *The Houstouns of Georgia* (Athens, 1950), pp. 12–13. Chap. 2 of this book treats of Dr. William Houstoun.

[213] *Journal-Book*, XV, 223; *Certificates, 1731–1750*, p. 34.

19. A drawing of the *Dorstenia (D. contrayerva* from Mexico) to accompany Dr. William Houstoun's "Account of the Contrayerva."

Phil. Trans., XXXVII, No. 421 (1731), facing p. 185.

20. A specimen of a Jamaican plant sent
by Dr. William Houstoun to Sir Hans Sloane about 1730.

Sloane Herbarium, H.S. 146, fol. 150. *Courtesy of the British Museum
(Natural History).*

the Person he shall judge most capable and willing to take care of them, while he goes to the remaining Places in Search of others."[214] It was estimated that these voyages would require up to two years' time, after which Dr. Houstoun was to go to Georgia. As Lord Petre, Sir Hans Sloane, the Society of Apothecaries, Charles Dubois, James Oglethorpe, the Earl of Derby, and the Duke of Richmond subscribed funds separately for Dr. Houstoun's salary (£200 per year)—and ostensibly received specimens for themselves—these voyages "for Improving Botany and Agriculture in Georgia" were intended to cost the trustees little more than bookkeeping services.[215]

Unfortunately, upon his return to Jamaica from one of his voyages to the Spanish Main, Dr. Houstoun died (August 14, 1733). Some of his collections had been shipped to Charles Town (as instructed) for the Georgia trustees; others remained in Jamaica. He left his herbarium, drawings, and "Catalogue of Plants" to his friend Philip Miller, from whom they much later passed to Sir Joseph Banks. Sir Joseph published Houstoun's "Catalogue" and drawings as *Reliquae Houstonianae, seu Plantarum in America meridionali, à Gulielmo Houston, M.D. . . .* (1781). In the meantime, the plants which Dr. Houstoun had collected in the West Indies and Central and South America for Sir Hans Sloane, Philip Miller, Lord Petre, Charles Dubois, and others were settled in the herbaria of the British Museum and Oxford, where they still remain.[216]

Dr. Robert Millar, a physician with an Edinburgh degree in medicine (d. 1742), succeeded in Dr. Houstoun's place for the

---

[214] Candler, II, 5–6.

[215] Lord Petre paid the lion's share of fifty pounds per year, although the Earl of Derby often equaled it. See *Sloane 4053*, fol. 164. The "Trustees Journal" is replete with references to this account. See Candler, I, *passim*, and later volumes. A rough addition indicates that between 1732 and 1739 the subscribers paid about £850 into this fund.

[216] Dandy, pp. 139–140; Clokie, p. 186. Philip Miller published some of Houstoun's plants in various editions of his *Gardener's Dictionary* (1731 and later) and in *Figures of Plants* (1755–60). James Britten, "Philip Miller's Plants," *Journal of Botany*, LI (1913), 132–135. When Linnaeus visited London in July, 1736, Miller gave him duplicates of Houstoun's plants. Linnaeus adopted many of Houstoun's plant names for Central and South American plants, and Gronovius named the genus *Houstonia* in his honor. See also Edith Duncan Johnston, "Dr. William Houstoun, Botanist," *Georgia Historical Quarterly*, XXV (1941), 325–339; Joseph Krafka, Jr., "Medicine in Colonial Georgia," *ibid.*, XX (1936), 326–344.

Georgia trustees. He was recommended by Sir Hans Sloane, for whom he had made collections in the Levant; but in spite of Sloane's recommendation, in which others who had subscribed funds for Dr. Houstoun concurred, there appears to have been abroad the opinion that Millar's "abilities are far short of his predecessor."[217] Nonetheless, Dr. Millar was employed (March 6, 1733/34) for the unexpired term of Dr. Houstoun's agreement (stated as two years and forty-six days), and at the end of that period (1736) his contract was extended for two years more. The trustees contributed fifteen pounds per year and refused to guarantee the payment of the subscribers' funds, which had fallen behind. However, the subscribers continued to support Millar as they had supported Houstoun, although by 1736 their enthusiasm—and their funds—showed signs of weakening.[218]

Meanwhile, Dr. Millar left England (May, 1734) for Jamaica, where he endeavored to take over the work which Dr. Houstoun's death had terminated.[219] Soon he embarked on a South Sea Company vessel for Porto Bello and Panama, where he "Collected Seeds & Specimens of all Such Plants and Trees that grow about that City"; unable to acquire "*Jesuit's Bark* and the *Balsam of Peru*," he was promised "some Seeds of both Trees remitted to me by some Gentlemen trading to the places of their Native Growth." He returned to Jamaica, "from whence [he wrote] by the first Ship which Sailed for England I sent these Seeds & Specimens to the Care of Mr. Millar at Chelsea to distribute Such part of them to the Honble The Subscribers for encouraging & improving Botany & Agriculture as he knew would be Satisfactory to them & to sow and preserve Such as should be found useful for being sent to propagate in Georgia."[220] He wrote to Sloane from Jamaica on December 10, 1734, sending seeds and plants to him and Lord Petre. The plants included "one called by the Spaniards the Flower of the Holy

---

[217] Philip Miller to Dr. R. Richardson, June 18, 1734, *Ms. Radcliffe Trust C. 9,* fol. 13. See also John Amman's statement quoted in Dandy, p. 165.

[218] Candler, II, 59–61, 174; III, 193; XXI, 38; Lord Petre to Sloane, Aug. 31, 1736, *Sloane 4054,* fols. 295–296.

[219] Millar to Sloane, Kingston, Jamaica, Aug. 8, 1734, *Sloane 4053,* fol. 250. Millar found no cooperation from Houstoun's friends, who would "neither give nor lend" anything of Houstoun's.

[220] Millar's "Narrative of my Proceedings . . ." in Candler, XXI, 191–195.

Ghost" (probably the orchid, *Peristeria elata*). He reported that he had been told in Panama that there were three sorts of Jesuits' bark from Peru, each with a different colored flower (white, red, purple), that the balsam of Peru really came from Nicaragua, and that he would make further inquiries regarding ipecacuanha.[221] The following January Dr. Millar sailed to Cartagena, where he won permission of the Spanish authorities to travel into the country. Here he found "Balsamum Capivi" (copaiba), the tolu balsam, "true Ipecacuanha," and many other things. He collected roots and seeds of the balsam trees, but as the ipecacuanha was not in seed, he boxed about 150 of the plants, some of which he transplanted in Cartagena, in the garden of the South Sea Company's physician, and the rest he took to Jamaica, where he transplanted them "into two Gardens . . . and put them under the Care of two Careful Persons." He sent a lengthy account to President Sloane of the Royal Society, describing in detail the copaiba, the "Tolu Trees, the true Ipecacuanha Plant &c."[222] Ill health and want of shipping postponed his next voyage until late November, 1735. He then embarked for Campeachy, where he found contrayerva (*Dorstenia*) and "transplanted Some hundreds of the Plants in to Boxes in Order to preserve them." He also found the trees "which produce the Gum Elemi" (probably *Elaphrium elemiferum*), but finding no seeds or transplantable young trees, he "established a Correspondence with a Spanish Priest . . . from whom I expect to be furnished with Seeds at a proper Season."[223] He then sailed from Campeachy in a Spanish vessel bound for Vera Cruz. In Mexico he hoped to collect jalap, cochineal, and "Several other useful Drugs," but the Spanish authorities there refused to permit him to land, and ultimately shipped him on a Spanish man-of-war to Havana. He might have been sent to Spain but for the civility of the Spanish captain, who

221 *Letter-Book*, XXI, 349–352; read March 27, 1735, *Journal-Book*, XVI, 117–118. The letter revealed that Millar had also visited Tobago.

222 Dated Kingston, Jamaica, July 10, 1735, *Letter-Book*, XXII, 73–77; read Nov. 20, 1735, *Journal-Book*, XVI, 196–197.

223 "I leave this evening for Campeche and La Vera Cruz. You won't hear from me as I will be voyaging about for about 6 months and I didn't want my best patron to think I was neglecting him." Millar to Sloane, Kingston, Nov. 25, 1735, *Sloane 4054*, fol. 147. He reported that he arrived at Campeachy on Dec. 10. See Millar, "Narrative," *loc. cit.* See also Candler, XXI, 68–69.

permitted him to transship to an English vessel shortly before they reached Havana, by means of which he was transported to England.[224] Subsequently he returned to Jamaica, but not before Sir Hans Sloane had provided him with a copy of James Petiver's directions for collecting and preserving "Plants, Insects, &c."[225] He tried vainly to obtain permission to collect in Mexico, and in December, 1737, he sailed to Port-au-Prince, hoping still to find means of entrance at Vera Cruz.[226] Unsuccessful, he returned to Jamaica, disappointed, weary, and in ill health. He wrote to the Georgia trustees on May 26, 1738, related his second failure to obtain admission to Mexico, and added:

> I would Directly had Proceeded to Georgia with what I have gott of the Ipecacuanna [growing in the Jamaican gardens], in my way home to England, but having some few things to do in this Island, for some of my chief Subscribers, I have choose to defer it till the Beginning of next year. And I hope before then to be favoured with a letter from you Directing me in what manner I should Proceed in Georgia Notwithstanding my time of contract will now soon be expired and I not inclining to continue longer in this way of life. Yet with a great deal of pleasure, According to my first Design, I will proceed to Georgia under your Directions, and During my short stay wch I shall be obliged to take ther, shall do any thing in my Power that shal seem to you most Proper, for the Good of the Colony, and what may tend to the General Benefitt & Inclinations of those Gentlemen who have been my subscribers. . . .[227]

Under date of August 23, 1738, the "Journal of the Earl of Egmont" (Perceval) recorded that a letter had been ordered to Millar to the effect that "it would please us if he went to Georgia with his collections, but that having been long Subscribed to him, and

---

[224] Millar, "Narrative," *loc. cit.* This was dated from London, July 7, 1736.
[225] Millar to Sloane, Kingston, Feb. 12, 1736/37, *Sloane 4053*, fol. 57. In a letter of March 1, 1736/37, he informed Sloane that he had married Mistress Mill, daughter of Richard Mill, Esq., of Jamaica, on Feb. 27, 1736/37. *Ibid.*, fol. 69.
[226] He collected birds, evidently in Jamaica, for Peter Collinson, and other items there for his patrons. See his letters to Sloane, Kingston, July 22, Dec. 6, 1737, *ibid.*, fols. 147–148, 244. In another, undated, letter, he reported shipments of seeds from Jamaica and the Leeward Islands to Collinson, Miller, Sloane, and Lord Petre. *Sloane 3059*, fol. 356.
[227] Candler, XXII, Pt. I, 150–152.

seen no fruits of the expense we had been at, but on the contrary disappointment of our expectations, we could not be at the charge of sending him to Georgia."²²⁸ So ended Dr. Robert Millar's association with both his subscribers and the Georgia trustees. He returned to England in September, 1739, and though he advised the trustees that indigo and snakeroot would do well in Georgia, he "gave us no Satisfactory Acct of the roots plants &c he was employed to collect in America for Georgia."²²⁹ Millar attended the Royal Society as a guest early in 1740, and efforts were made to secure his appointment as Professor of Botany and Anatomy at the University of Glasgow, but they came to nought; and Dr. Millar died in 1742, a disappointed and broken man. Sir Joseph Banks later purchased his herbarium, but many of his dried plants are in the British Museum and a few are at Oxford.²³⁰

In spite of numerous gifts of seeds, cuttings, and plants from well-wishers of the colony, neither the public garden at Savannah nor Georgia as a whole prospered in the 1740's. Discord arose over the trustees' system of land tenure, refusal to admit slavery in Georgia, regulation of the Indian trade, ban on the importation of strong liquors, and other matters. Finally, in 1752, the trustees resigned their patent, and Georgia became a royal colony. One of its early governors (1756–60) was Henry Ellis (1721–1806). In 1746 Ellis had been appointed by Parliament to command a voyage in search of the Northwest Passage. The effort failed, but Ellis' *Voyage to Hudson's Bay . . . for Discovering a North West Passage* (1748) attracted wide attention among scientists. He was received as Fellow of the Royal Society (February 8, 1748/49) "as a Gentleman of merit, of great curiosity, and an uncommon zeal for making of discoveries, and promoting Natural History, Geography, and Navigation."²³¹ As Governor of Georgia he sent reports to the Royal Society about Georgia weather via John Ellis, a London merchant and naturalist who had become a Fellow of the Royal

---

²²⁸ *Ibid.*, V, 65.

²²⁹ *Ibid.*, V, 229. There were lingering disputes about Millar's financial settlement. *Ibid.*, V, 351, 384.

²³⁰ *Journal-Book*, XVIII, 40, 98 (Feb. 14, 1739/40; May 8, 1740); *Sloane 4059*, fols. 354–355; Smith, ed., *Correspondence of Linnaeus*, II, 22; Dandy, p. 165; Clokie, p. 212. See also Krafka, "Medicine in Colonial Georgia," *loc. cit.*

²³¹ *Certificates, 1731–1750*, p. 408.

Society in 1754 and had won considerable reputation by *An Essay Towards a Natural History of the Corallines* . . . (1755), in which he had expanded Jean André Peyssonnel's discoveries insisting upon the animal nature of corals and corallines.[232] Governor Ellis' letters described the "Extraordinary heat" of Georgia. He had an "approved thermometer" which, as a former servant of the African Company, he had used in equatorial Africa. But in Africa his thermometer had never registered above 87° whereas in Georgia it had risen to 102°.[233] He also commented on sudden changes of temperature in Georgia, reporting that his thermometer had registered 86° on September 10, 1757, and the next day fell to 38°. In 1758–59 the two Ellises, with the cooperation of the Royal Society, experimented with various methods to preserve the fecundity of nuts, seeds, and acorns sent on long voyages in hot, humid weather. The experiments were associated with Governor Ellis' attempts to introduce in the Georgia economy new products, including chestnuts and the cork-bearing oak from Spain. John Ellis prepared seven different parcels of seeds, all wrapped differently, but upon their arrival in Georgia Governor Ellis found only one of them contained fecund seeds—one in which the seeds had been coated with beeswax and covered with brewer's loam moistened in a solution of gum arabic.[234] Because of ill health, Governor Ellis returned to Eng-

---

[232] *Certificates, 1751–1766*, p. 20. Henry Ellis and John Ellis were cousins according to Spencer Savage, "John Ellis (?1705–1776) and His Manuscripts," *Proceedings of the Linnean Society of London*, 146th Sess., Oct., 1933–May, 1934 (London, 1934), pp. 58–62. Peyssonnel's views had been presented to the Royal Society in a manuscript entitled "Traite du Corail contenent les novelle découvertes qu' on a fait sur le corail. . . ." William Watson read the substance of it in translation to the Royal Society in May, 1752 (*Journal-Book*, XXII, 99–100), and it was printed in *Phil. Trans.*, XLVII (1751–52), 445–469. Both it and John Ellis' *Essay* aroused much dispute in the Society, and William Stukeley, in particular, held out for years that polyps, corals, and the like "are the intermediate Link between the Animal and the Vegetable world, having a mixt life, which partakes of both." *Journal-Book*, XXV, 94–95 (April 16, 1761).

[233] *Ibid.*, XXIV, 184–185 (Nov. 16, 1758). Ellis' letter was dated Georgia, July 27, 1758. A similar report was sent on Feb. 14, 1759—a cold winter in Georgia, reported at 23°. *Ibid.*, p. 319; read May 17, 1759. See also *Phil. Trans.*, L, Pt. II (1758), 754–756.

[234] *Journal-Book*, XXIV, 225–226 (Jan. 18, 1759), 462–466 (Dec. 20, 1759); XXV, 139 (Dec. 11, 1760); *Phil. Trans.*, LI, Pt. I (1759), 206–215. John Ellis also sought "to ascertain the tree that yields the common Varnish used in China and

land in 1761. The following year he was appointed Governor of Nova Scotia (1761–63), but he delegated his duties to a deputy and did not go there in person. Still later he went to Naples, Italy, to engage in marine researches, and died there, having attracted scientific attention to his work as a hydrographer.[235]

In the meantime, John Ellis had entered into a scientific correspondence with Dr. Alexander Garden (c. 1730–91), an able physician and naturalist of Charles Town, South Carolina (later F.R.S.). Ellis presented to the Royal Society many of Dr. Garden's communications, including descriptions and specimens of flora, one of which, a native of the Cape of Good Hope sometimes known as Cape jasmine, Ellis named the *Gardenia* in honor of his Charles Town correspondent.[236] He also reported on the preparation of dye from cochineal, with Dr. Garden's accounts of the male and female insect as examined under a microscope.[237] In 1764 John Ellis was appointed Agent for the new English province of West Florida.[238] In this capacity he presented the Royal Society with "the skin of a Tyger from West Florida,"[239] and, more important, introduced to English gardens a new species of West Florida tree (*Illicium floridanum*, Ellis).[240] But these later activities, of course,

Japan; to promote its Propagation in our American Colonies; and to set right some Mistakes Botanists appear to have entertained concerning it." *Phil. Trans.*, XLIX, Pt. II (1756), 806–876.

235 John Nichols, ed., *Literary Anecdotes of the Eighteenth Century* (9 vols., London, 1815), IX, 533.

236 *Journal-Book*, XXIV, 118–121 (Nov. 20, 1760). See also *ibid.*, XXVII, 413 (Dec. 20, 1770), where Ellis argued that the loblolly bay was a new genus which he identified with the gardenia. *Phil. Trans.*, LI, Pt. II (1760), 929–935.

237 *Journal-Book*, XXVI, 261–263 (Dec. 23, 1762); *Phil. Trans.*, LII, Pt. II (1762), 661–666.

238 *C.O. 324/52*, p. 21 (April 2, 1764) (Public Record Office, London).

239 *Journal-Book*, XXVII, 240 (Nov. 16, 1769).

240 *Ibid.*, XXVII, 406–408 (Dec. 13, 1770). The tree, referred to as a new species of *Illicium linnai*, or star anise tree (*Illicium floridanum*, Ellis), was said to have been discovered in April, 1765, in a swamp near Pensacola by a Negro servant of William Clifton, Chief Justice of West Florida. Ellis received it in July, 1765, turned it over to William Aiton, botanic gardener to the Princess Dowager of Wales at Kew. In Jan., 1766, John Bartram, "the King's botanist for the Floridas discovered it on the banks of the river St. John in East Florida and describes it in a letter to the late Mr. Peter Collinson." Ellis, however, felt that the East and West Florida varieties differed. *Phil. Trans.*, LX (1770), 524–531. In 1774, as Agent for the island of Dominica, Ellis published and presented to

did not relate to Georgia. So far as John Ellis' connections with Georgia were concerned, they appear to have been restricted to the cooperative activities with Governor Ellis mentioned above. Indeed, whatever contributions the scientific correspondents associated with Georgia made, they appear as a whole to have pertained more to the West Indies and Central America than to Georgia herself. Strangely enough, with the exceptions of General James Ogle-thorpe and Governor Henry Ellis—neither of whose scientific contributions were significant, at least in their Georgia context—none of them appear actually to have set foot in the colony!

---

the Royal Society *An Historical Account of Coffee, with an Engraving and Botanic description of the tree* . . . (London, 1774). See *Journal-Book*, XXVIII, 304.

The West Indies

New Lines of Communication

THE LINES OF SCIENTIFIC INTERCOMMUNICATION between England and the West Indies shifted in the eighteenth century. Until about 1750, Barbados and Jamaica tended to dominate the flow of correspondence and exchange of data. The servants of the South Sea Company, augmented increasingly after the outbreak of the mid-century colonial wars[1] by military and naval officers associated with the powerful British installations in the West Indies, carried scientific exploration into Central and South America. The Bermudas and the Bahamas assumed a place in the exchange less significant than their earlier contributions had promised, and Nevis, projected on the scene near the end of the seventeenth century, attained scant notice in the eighteenth. Indeed, excepting the Reverend Mr. William Smith's *A Natural History of Nevis* (London, 1745), Nevis scarcely attracted scientific attention before the end of the Old Colonial Era. Mr. Smith was rector of St. John's Parish, Nevis, for five years after his introduction there on April 18, 1716. He traveled over the island and among the other Leeward Islands and evidently had visited New England and South Carolina, as he referred to

---

[1] The War of Austrian Succession, or "King George's War" (1740–48); the Seven Years' War, or "French and Indian War," or "Great War for the Empire" (1756–63—though it actually began in the mainland British colonies of North America in 1754). The role of the British military was evident in North America before the end of the French and Indian War, especially in Canada, the Great Lakes region, and the Ohio Valley—and it continued there during the American War for Independence.

Boston and Charles Town as if he had been there. He had kept a diary of his observations and had made some collections, particularly of shells. His *Natural History of Nevis* was written in the form of letters to his friend, the Reverend Mr. Charles Mason, Woodwardian Professor at Cambridge. Although Mr. Smith referred to Sir Hans Sloane's works about Jamaica, he revealed no familiarity with other scientific works or personages of the time. He does not appear to have been in correspondence with the Royal Society or, in fact, with other scientific men of the day. *A Natural History of Nevis* is the work of a literate amateur with only the educated layman's knowledge of natural history. As such, it holds value for the student of general colonial history but is of little significance in the study of the development of colonial science.

Neither the Temple Coffee House Botany Club nor other groups of English patrons appear to have supported scientific expeditions to the West Indies in the eighteenth century, excepting, of course, the patronage afforded Mark Catesby in the Bahamas. The Secretaries and other Fellows of the Royal Society continued to promote collections, correspondence, and studies about West Indian flora, fauna, and natural history in general. During the first half of the eighteenth century, however, scientific communication and exchange centered about the promotional activities of James Petiver and Dr. (later Sir) Hans Sloane. Before his death in 1718, Petiver was said to have had "the greatest correspondence both in East and West Indies of any man in Europe,"[2] and his promotions in the West Indies were impressive. Sloane, by virtue of his visit to the West Indies, his books about them, and his continuing interest in the islands, became the doyen of English scholars of West Indian affairs during the first half of the eighteenth century. Until Sloane's death in 1753, these two men were responsible, directly or indirectly, for the majority of the scientific intercommunications between England and the West Indies, much of which, during Petiver's lifetime at least, was brought to the attention of the Temple Coffee House Botany Club, and nearly all of which was placed before the Royal Society of London.

[2] Robert T. Gunther, ed., *The Further Correspondence of John Ray* (Oxford, 1928), p. 279.

The Bermudas

Except for a collection of shells, corals, and fish received by the Royal Society in 1708,[3] only two principal correspondents of the Bermudas contributed to English scientific endeavors in the eighteenth century before 1770. The first—and more fruitful—of these was John Dickinson. Dickinson was a correspondent of James Petiver, and like other Petiver correspondents, he was continually hounded by the latter to press on energetically to collect plants, berries, seeds, insects, shells, corals, and other natural products of the islands. In return, Petiver supplied him with instructions about collecting, preserving, packing, and shipping his specimens, together with quires of brown paper and other materials for the work, catalogues of trees, plants, and other natural history items to guide Dickinson in his selections (including early parts of Leonard Plukenet's *Phytographia* and copies of Petiver's *Musei Petiveriani*), and English newssheets relating the events of England and Europe. Further, Petiver played upon the vanity of Dickinson—as he did with nearly all of his correspondents—to induce him to collect in order to see his name in print as the "discoverer" of this or that specimen which Petiver published in his *Musei* or elsewhere.[4] Dickinson began his collections for Petiver in the early 1690's and continued into the early years of the eighteenth century.[5] It seems likely that he also contributed to the herbarium of Plukenet, for the

[3] *Journal-Book*, XI, 157 (Dec. 22, 1708). Presented by Mr. Hill and placed in the Society's Repository.

[4] For a description of Petiver's methods, see *Petiver*, pp. 257–270. In spite of his shoddy (but effective!) methods, it is to Petiver's credit that he was generous in giving credit to his donors, that he paid expenses involved, and that he sometimes paid fees for the collections themselves.

[5] Because Petiver often omitted dates in his letters, it is impossible to mark the beginning and the end of the correspondence with accuracy. In the second of Petiver's letters extant, he referred to the receipt of Dickinson's letter of April 10, 1692, and evidently the correspondence had been underway for at least a year at that time. The entire Petiver-Dickinson correspondence is in *Sloane 3332*, fols. 19v, 22v–24, 25–25v, 28–29 (March 15, 1692/93), 34v (May 11–12, 1693), 75 (June, 1694), 115–115v, 146 (Sept. 9, 1695); *3334*, fols. 62v–63 (May 16, 1701); *4063*, fol. 14 (April 9, 1700), 33 (July 6, 1700), 63 (Jan. 23, 1700/01). Those in *Sloane 4063* are Dickinson's letters to Petiver.

latter figured some of Dickinson's plants both in his *Phytographia* and in his *Almagestum Botanicum*,[6] but, although Plukenet credits Petiver with the donation of some of Dickinson's specimen's, the cool relations between the two men suggest that Petiver would not have shared his collection widely with Plukenet. Petiver recognized "my Kind Friend Mr. John Dickinson" for contributions described in the *Musei Petiveriani*,[7] and, as Dickinson's plants included some of the earliest specimens to arrive in England from the Bermudas, they excited much interest among botanical scholars in the early years of the eighteenth century. Indeed, as the references given in Dandy's *Sloane Herbarium* illustrate, they have attracted much attention of more recent botanical scholars, especially owing to the fact that sixty-one species of Bermuda plants are not known to have grown originally anywhere else in the world. As their seeds are believed to have been wafted to the Bermudas from Florida or the West Indian islands in the distant past, they may represent earlier forms in the evolution of present-day flora in the latter areas.[8] Dickinson's collections included some of the earliest specimens of these plants. As they ultimately fell into the hands of Sir Hans Sloane, Dickinson's dried plants sent to Petiver and Plukenet still lie in the natural history herbarium of the British Museum available for the scrutiny of scholars.[9]

More than half a century elapsed before further Bermudan collections were brought to the attention of the Royal Society. On April 2, 1758, James Theobald, F.R.S., read to the Society extracts of a letter from Dr. George Forbes of Bermuda, originally addressed to John Eaton Dodsworth. The letter described various marine animals found in Bermuda waters, accompanied by some drawings of them made by an unidentified "young man." One of these was described by Dr. Charles Morton, F.R.S., as an "exanguious [exsanguinous] testaceous animal, not turbinated" and identi-

---

[6] *Phytographia* (1691–96), Tables 61, 124, 248; *Almagestum Botanicum* (1696), pp. 324, 348. Plukenet also had many of Dickinson's plants in his herbarium. See Dandy, pp. 124–125.

[7] *Centuria Octava* (1700), pp. 75, 79–80.

[8] Dandy, pp. 124–125. See also W. B. Hemsley, "Bermuda Plants in the Sloane Collection," *Journal of Botany*, XXI (1883), 257–261; Nathaniel Lord Britton, *Flora of Bermuda* (New York, 1918), pp. vii–viii.

[9] *Ibid.*

fied as a limpet, a mollusk of the genus *Patella*. The description, with the drawing, was published in the *Philosophical Transactions*.[10]

The Bahama Islands

The Bahamas entered into the scientific correspondence of the eighteenth century only slightly more than they had in the seventeenth, when John Locke's lonely communication about poisonous fishes was presented to the Royal Society (1675). Captain Thomas Walker, however, sent Bahamian plants to his father, who had a garden in Westminster, where Leonard Plukenet saw them and figured some of them in his *Phytographia* and mentioned others in his *Almagestum Botanicum*. Captain Walker subsequently became Judge of the Vice-Admiralty Court in the Bahamas, and in 1701, evidently through the good offices of Robert Ellis of Charles Town, South Carolina, he undertook to collect items in the Bahamas for James Petiver. On November 13, 1701, he wrote to Petiver from New Providence Island: "By Books and directions received from your good ffriend Mr. Ellis In South Carolina I have collected Sundry trees and plants of our Countrey [although I] Know not if anything Collected may be accounted rare with you."[11] The letter was accompanied by leaves from the Brazilwood tree, the mahogany tree, a Madeira tree, and the "saffron-tree," together with several "Plants . . . of vertue." Captain Walker promised to "endeavour by all Opportunitys from hence to putt you upp Some Insects Vigitables (& other rarityes as I come by them . . .)" and begged for more brown paper in which to dry his collections. Petiver read the letter and showed the specimens to the Royal Society on December 30, 1702, whereupon Dr. Sloane declared that the "saffron-tree" was identical with the "star-apple" (genus *Chrysophyllum*) which he had observed in Jamaica.[12] Petiver also referred to Captain Walker's contributions in his *Musei Petiveriani*.[13] No further correspondence between Petiver and Captain Walker is extant until, in February, 1705, Captain Walker wrote to Petiver from South Carolina. The Spanish had captured New Providence in the course

[10] L, Pt. III (1758), 859–860. See also *Journal-Book*, XXIII, 225.
[11] *Sloane 4063*, fol. 123.
[12] *Journal-Book*, XI, 8.
[13] (1695–1703), Appendix, unpaged.

of the War of Spanish Succession, and Captain Walker had been forced to flee, "for that I have lost my Estate by the Enemy & was Stript of all down to the Shirt on my back and Barbarously and Cruelly used in my body by the Spaniards chiefly because I Executed that Commission [as Judge of the Vice-Admiralty Court], I am removed to this place out of there Way and have a great charge of Children. . . ."[14] He sought his salary for his commission as judge, had applied for help to "my Friend," Governor Francis Nicholson of Virginia, and "by the assurance Mr. Ellis gives me of your great goodness and reddyness in Doing for Persons in Distress," he hoped that Petiver would petition in his behalf to the Admiralty Board in London. He had sent a collection of trees and herbs from the Bahamas before the enemy had taken New Providence Island, and he proposed to send more if he were restored to his estates there. But at this point the correspondence ended, and Captain Walker contributed nothing further to Petiver's collections. The *Sloane Herbarium* in the British Museum (Natural History) still has some of the natural history specimens he sent to Petiver.[15]

Except for Petiver's display to the Royal Society of some crabs (May 22, 1706),[16] no further communication from the Bahamas reached the Royal Society until 1724. Then, on April 30, Dr. John Douglas printed "Some Remarks on the Island of New Providence made in the year 1723: by Mr. Lightbody Surgeon in the Service of the Bahama Company."[17] Of Mr. Lightbody nothing further is known, but Dr. Douglas was well known both in the Royal Society and in the West Indies. He had been associated with Petiver "a-Herbarizing" in England as early as 1709, and was one of four brothers all of whom became Fellows of the Royal Society. In 1712 he went as "surgeon at the Generals House" in Antigua (the "General" being his brother Walter, d. 1768, Governor of the Leeward Islands), from whence he made collections for Petiver. Before 1719 he returned to London, was elected F.R.S. in 1720,

14 Feb. 20, 1704/05, *Sloane 4064*, fol. 58.
15 Dandy, p. 229. Perhaps Captain Walker arranged for John Dudgeon, Jr., "Merchant in Bridgetown, Barbados," to send fifteen plant specimens from New Providence in 1704. See Dudgeon to Petiver, Feb. 28, 1703/04, *Sloane 3321*, fol. 134.
16 *Journal-Book*, XI, 93. The donor was not identified.
17 *Register-Book*, XI (1722–24), 517–528; *Journal-Book*, XIII, 383.

and practiced surgery in London until his death in 1743. He often presented communications to the Royal Society from correspondents in Antigua and other islands of the West Indies, and Mr. Lightbody was one of these.[18]

Mr. Lightbody ended his "Remarks" with the hope that "the imperfections of this will be excused, it being my first performance of this kind." His observations were those of an amateur, without descriptions in scientific terms. He began with commonplace remarks on the weather, complained of the lack of a good water supply, and stated that the soil was "abundantly fertile" considering that it was never more than two feet deep, with rocks cropping out at several places. He gave more extensive attention to the trees, mentioning Madeira wood, mahogany, cedar, "a little ebony," manchineel (whose beautiful apple blisters the hand and "proves mortal, when eaten"), palmetto, boxwood, Brazilwood, the *Cortex eleutheria* (cascarilla) or *Croton eluteria* ("used by the distilleries in Carolina & New England, to make their liquors more agreeable"), cabbage tree, candleberry, China orange (introduced from Lisbon), wild sweet orange, bitter orange, citron, lemon, lime, banana, and pomegranate. Pineapples, which were "reckon'd the most delicious fruit in the West Indies," were surpassed, in Lightbody's opinion, by those of Guinea. Wild figs and wild grapes grew in profusion. Watermelons, pumpkins, cotton (which Lightbody described as a shrub), sugar cane, tobacco, aloes, and indigo did well, although the lack of adequate water thwarted the development of indigo as a staple product. European garden plants generally throve. Rice and Indian corn did well, but most planters prepared their bread-stuffs from cassava root. Yams grew to enormous size and their edibility surpassed the potato, which did better in Guinea. Indeed, an abundance "of European plants and roots would grow there, were it not for the lazyness of the inhabitants, who are so devoted to their Cups and idleness [and prefer] Pyracy, which would still take place, if it were not for the Governor and Fort." The fishes of New Providence were far inferior to European varieties, and Lightbody referred only to the jewfish, kingfish, mackerel, and

[18] The four Douglas brothers were (besides Dr. John), James, physician (F.R.S., 1706); Colonel Walter (F.R.S., 1711); and George (F.R.S., 1732). See *Col. F.R.S.*, pp. 206–207.

shark, the latter of which infested the harbor. Green turtles were plentiful and good to eat, and there were other varieties, though of no value. Birds were less plentiful than in other West Indian islands. Lightbody mentioned wild pigeons, curlews, sea hawks, pelicans, hummingbirds, and flamingoes, whose feathers were used to make muffs for the ladies. Cattle, brought to the island from Carolina and Rhode Island, did not survive long, "which they impute to the Grass of the place." Swine and goats throve, and the flesh of the wild hog was considered superior to that of European pigs. Domestic fowls were good but undersized. Lightbody praised the iguana, whose flesh, he said, was like a capon, though exceedingly fat. There were "conies" like big rats with short tails; freshly killed, they tasted like pork. Various snakes existed, but Lightbody did not particularize. In time of rain, many white frogs appeared, but they were not venomous "like the Bull frog of Carolina, which they resemble"(?). The much-discussed Bahama spider was seldom seen in New Providence; it was two inches long and had six eyes. Two very troublesome insects were the "Muscato" and the "Gigger" (chigoe). However, as the rest of the island was overgrown with shrubs and "prickle bushes," Lightbody had not investigated it. His description was the best which the Royal Society had received from the Bahamas. It fell far short of accounts received from elsewhere in the West Indies—and it was the last which the Royal Society received during the Old Colonial Era.

Antigua

James Petiver began an effort to establish a scientific correspondence and collection of natural history specimens from Antigua as early as 1694. On February 5, 1693/94, he addressed a letter to Dr. Daniel Mackenning, who had recently sailed for Antigua, to seek his cooperation in the collection of trees, shrubs, and plants there. Petiver had learned of Dr. Mackenning through information imparted by a mutual friend, Mr. Michael Poultry. He sent his usual instructions, collecting books, "3 or 4 Quires of Brown Paper," "a Catalogue or 2 (which I have lately printed) of such Druggs whose Plants whence they are produced we are either altogether ignorant of or but very obscurely know." Indeed, said Petiver, "We

344

are as yet almost totally ignorant of what your Isle produces & for
that reason I would instruct you to gather 2 or 3 samples of all ye
Trees Shrubs & Herbs you Shall find, not omitting ye most common Grass, Rush, Moss, or Fern. . . ." He particularly hoped for
a specimen of a fern whose "leaf was shaped like a Halbeard (being
one of ye most beautiful Plants I ever saw) was some few years
since gathered in your Island by one Hornall (wch hath now
been figured) a Gardener & for yt Reason cal'd Hornals Antego
Halbert Fern & I doubt not but very many more may be dayly discovered by you, wch as soon as I receive I will publish to ye world
with a gratefull & just acknowledgment of you, their first De-
tector."[19]

In spite of Petiver's blandishments, Dr. Mackenning did not
respond. Petiver sent him two more pleas in 1694 and 1695 with
no reply.[20] Late in 1700, however, when Petiver wrote to Thomas
Grigg (variously addressed as "Mr.," "Dr.," and "Captain"), of
Parham Plantation, Antigua, hoping to receive "wtever Plants
Shells Insects &c as yr Island affords," and offering "wtever Re-
taliation you shall desire,"[21] he opened up a fruitful exchange that
continued until 1715. Obviously, some of the correspondence has
not been preserved, and the War of Spanish Succession interrupted
it at times.[22] Associated with Captain Grigg in collecting for
Petiver were two female relatives, the sisters "old Mrs. Rawlins &
Poor Mrs. [Rachael] Chapman." Petiver spoke of them as if he believed them indigent and called upon Captain Grigg to give them
aid, although they were able, as Petiver testified, to send out slaves

[19] *Sloane 3332*, fols. 45v–46.
[20] *Ibid.*, fols. 55, 162v.
[21] Dec. 11, 1700, *Sloane 3334*, fol. 9. Petiver may have got in touch with
Captain Grigg through the offices of Captain George Searle, a ship captain who
sometimes collected shells for Petiver in Antigua. See Petiver to Searle, March
19, 1713/14, *Sloane 3340*, fol. 14, and Petiver's dedication of Table 27 of his
Varia Opera "To his Kind Friend, Capt. George Searle."
[22] This becomes evident from the letters extant, most of which relate to the
years 1712–15. They include: Grigg to Petiver, Oct. 24, 1712; n.d., labeled "Rec'd.
July 3, 1713"; Aug. 20, 1714; Mrs. Rachael Grigg to Petiver, March 1, 1715/16, in
Sloane 4065, fols. 69, 176, 240; *3322*, fol. 25; and Petiver to Grigg, March 25,
1712; Jan. 16, 1712/13; Jan. 27, 1713/14; March 17, 1713/14; Oct. 25, 1714; Feb.
25, 1714/15; Oct. 15, 1715; Petiver to Mrs. Grigg, Jan. 17, 1712/13; Oct. 20, 1713;
Dec. 27, 1715, in *Sloane 3338*, fols. 38, 129, 129v; *3339*, fols. 81, 149; *3340*, fols.
13–13v, 59, 74v, 182, 184–185.

on collecting trips. In 1712 Captain Grigg married, and his wife, Madame Rachael Grigg, also began collecting insects, butterflies, and shellfish for Petiver and entered into correspondence with him directly, encouraged, perhaps, by Petiver's wedding gift of six pairs of gloves. Petiver supplied all of them with collecting materials, including a "muscipula," or flytrap, and he received in return several small collections of plants, shells, insects, butterflies, and a few marine animals and plants. The exchange languished when Petiver became ill in 1715, and although Madame Grigg sought to revive it in March, 1716, it had ceased permanently. But the contributions of Captain Grigg and Mesdames Grigg, Chapman, and Rawlins swelled Petiver's collections, and some of the plants they sent are still extant in the *Sloane Herbarium* of the British Museum.[23] Petiver published accounts of some of them. Mrs. Rachael Chapman's gift of shells from Antigua appeared, with a grateful acknowledgment by Petiver, in the *Philosophical Transactions;*[24] Captain and Mrs. Grigg's contributions received similar recognition in *Petiver Varia Opera.*[25]

Petiver's later correspondence with the Griggs in Antigua often was sent by way of Dr. John Douglas, Petiver's long-standing London friend who went to Antigua in 1712 to serve as "Surgeon at ye General's House," i.e., Dr. Douglas' brother, General Walter Douglas, F.R.S., Governor of the Leeward Islands.[26] Both of the brothers Douglas in Antigua contributed to Petiver's collections, although those of Dr. John were clearly the more important. A lively correspondence existed between Dr. Douglas and Petiver

[23] Dandy, p. 133. Petiver's correspondence suggests that the contributions of the Griggs ran more to insects, shells, and other marine creatures than to plants.

[24] XXIV, No. 299 (May, 1705), 1952–60.

[25] N.d. (1715?), Table 29. This Petiver publication is rare. I have used the British Museum copy, 724. K. 1 (1–5).

[26] See p. 342 above. In 1708 Petiver had endeavored to established scientific correspondence with one "Mr. Rickets, Gardiner at Mr. Gales at Falmouth in Antegoa." Rickets had been recommended to Petiver by Thomas Fairchild, the Hoxton nurseryman who was Mark Catesby's friend. But in spite of Petiver's offer to pay five shillings for every quire of paper filled with dried plants (not to exceed "4 Sprigs of a sort"), and a like sum for each wide-mouth quart bottle "filled with small Birds," fish, snakes, frogs, lizards, insects, etc., or for each oyster barrel filled with marine shells and corals, or for every 100 butterflies and moths (not to exceed "5 of each sort"), Mr. Rickets did not appear to respond. *Sloane 3336*, fols. 39v–41.

between 1712 and 1715.[27] Petiver showered his friend with collecting materials and instructions, newssheets, copies of the *Philosophical Transactions*, his own *Pterigraphia Americana* (1712), books on medicine and surgery desired by Dr. Douglas, William Derham's *Lectures* (i.e., *Physico-Theology, or a Demonstration of the Being and Attributes of God, from His Works of Creation* (London, 1713), being the Boyle Lectures for 1711–12), and jocular badgering lest "you will let my muscipula [flytrap] grow mouldy." Dr. Douglas reported on September 8, 1712, that he had "a pretty Collection" ready but, as the French threatened at the moment, he had delayed shipment. On July 20, 1713, he sent a large box of plants accompanied by his accounts of them, and later he sent shells, insects, and more plants. Many of the plants passed to Sir Hans Sloane when he purchased Petiver's collections after Petiver's death, and they rest in the British Museum (Natural History).[28] Some were published by Petiver as gifts from "My worthy Friend Mr. John Douglass Surgeon, and Brother to his Excellency the Governour of Antego."[29]

Dr. Douglas continued to serve as an intermediary between the West Indies and the Royal Society when he returned shortly after Petiver's death. On March 22, 1720/21, for example, he read "A Letter from the Revd. Mr. James Field, Rector of St. John's in Antigua to Mr. John Douglas, of a Wound of the Stomach."[30] Probably it was by the hands of the Douglas brothers that Walter Tullideph became an Antiguan correspondent of Sir Hans Sloane in the late 1720's. Tullideph had served as an amanuensis to Dr. James Douglas (1675–1742; F.R.S., 1706), anatomist, and brother to Dr. John and General Walter, before he settled in Antigua about 1726, first as a planter and then (1730) as a medical practitioner. His correspondence with Sir Hans suggests that he had known the latter before he went to Antigua.[31] Sir Hans presented him with the two volumes of his handsome *Voyage*, and Tullideph correlated

27 Extant are ten letters from Petiver to Douglas and four from Douglas to Petiver: *Sloane 3322*, fol. 54; *3337*, fol. 49v; *3338*, fols. 79, 128; *3339*, fols. 13v–14, 81v–82v; *3340*, fols. 12v, 58v–59, 64v–65v, 76; *4065*, fols. 62, 114, 121.
28 Dandy, pp. 127–128.
29 *Phil. Trans.*, XXIX, No. 346 (Nov.–Dec., 1715), 353–364.
30 Dated Windsor, Antigua, Aug. 24, 1720, *Letter-Book*, XV, 329–330.
31 Letter to Sloane, Antigua, May 7, 1728, *Sloane 4049*, fol. 160.

his Antiguan plant specimens with Sloane's descriptions of Jamaica plants. He sent Sloane at least two collections of plants, one of fourteen specimens in 1727, and another of thirty-five specimens in 1729. His notes demonstrate that he was a competent botanist, as befitted a member of the Botanical Society founded in London in 1721, and he subsequently sent Sir Hans many notes about the medicinal values of West Indian plants. His Antiguan specimens, added to Sloane's collections, are, of course, in the British Museum (Natural History).[32]

After Sir Hans Sloane's retirement (1741), only occasional scientific communications reached the Royal Society from Antigua before the end of the Old Colonial Era. William Shewington made observations on the transit of Mercury over the sun on May 16, 1753. He sent them to Benjamin Franklin, who passed them on to Peter Collinson, who, in turn, presented them to the Royal Society on November 15, 1753.[33] Two years later, the Reverend Mr. Francis Byam of Antigua sent to the Royal Society via William Fauquier, F.R.S., a chart of the rainfall in Antigua between January, 1751, and December, 1754, together with a fossil stone that had been found in a quarry on a mountainside about two miles from the sea and 300 yards above the high-water mark of the island. Split open,

[32] Dandy, pp. 221-222. It seems likely that Tullideph collected the plants in the two-volume hortus siccus which William Mathew, Governor of the Leeward Islands, presented to the Royal Society in 1750. Vere Langford Oliver, *The History of the Island of Antigua* . . . (3 vols., London, 1894-99), III, 155. Governor Mathew (d. 1752) was admitted Fellow of the Royal Society on March 10, 1719/20, shortly after a letter from him was read (Jan. 14, 1719/20) about the "Language spoken by the Caribee Indians . . . of Dominica." He appears in the minutes of the Society but once (Nov. 2, 1731)—and that regarding financial matters—until, on Nov. 1, 1750, he presented the Society with the above-mentioned hortus siccus. It was referred to William Watson, who reported to the Society (Nov. 15, 1750) that it contained "very curious and uncommon Plants, the growth of the Caribee Islands in America and many of them well preserved." The collection had been made by an unidentified "practitioner in physic" and presented to Governor Mathew. It contained 111 specimens. *Journal-Book*, XII, 400-401, 431; XXI, 401, 419-420; C.M., III, 96. In 1761 George Edwards presented to the Society a "Worm, or Insect" from Dominica given him by Governor Mathew with an engraving of it by Edwards. *Journal-Book*, XXV, 59-60 (Feb. 26, 1761).

[33] *Journal-Book*, XXII, 402. The Society referred them to James Short, F.R.S., the English astronomer.

the stone revealed "the exact portraiture of a fish (on each stone) which we call an old wife." Mr. Byam's communication was read to the Royal Society (December 11, 1755) and published, with a figure of the stone, in the *Philosophical Transactions*.[34] A second "drawing of a curious fossil" from Antigua was presented to the Society in 1759, this time of a starfish imbedded in granite. But its Antiguan donor was not named.[35] In 1759 Dr. George Macaulay read before the Society "An Account of the *Oleum Ricini* [*Ricinus communis*, or palma Christi], Commonly Called Castor Oil, and its effects as a medicine, especially in Bilious disorders by *Dr. Thomas Fraser*, of Antigua."[36] As castor oil was little known in Europe, though widely used in the West Indies, the account stirred up considerable interest. Dr. Mathew Maty, Secretary of the Society, confirmed the relation from similar information received from Guadaloupe, and Dr. Fraser's "Account" was published in the *Medical Observations and Inquiries . . .* in 1764.[37] Yet another new drug was described by Dr. James Farley, of Antigua, in 1768. On March 17 of that year, Dr. Mathew Maty read "A Letter from Dr. Donald Monro, F.R.S., to Mathew Maty, M.D., Sec. R.S. enclosing one from Mr. Farley of Antigua, on the good Effects of the Quassi Root in some Fevers." Dr. Farley referred to quassia (*Quassia amara*, L.), the Surinam root, which he had found effective in the treatment of fevers. Indeed, he argued that it was as effective as Peruvian bark (quinine), especially if mixed with *Radix serpentaria virginiana* (snakeroot). The letter was published in the *Philosophical Transactions*.[38] And the last word to the Royal Society from Antigua before the end of the Old Colonial Era was read to the Royal Society on March 11, 1773. Written by Rebecca Warner of Antigua on September 8, 1772, it had been addressed to a Miss Warner of Hatton Street, London, and was brought to the Society's attention by Joseph Warner, F.R.S. It merely described a hurricane which struck Antigua on August 31, 1772,

---

[34] XLIX, Pt. 1 (1755), 295–296.
[35] *Journal-Book*, XXIV, 306 (May 3, 1759).
[36] March 26, 1759.
[37] (3rd ed., 6 vols., London, 1763–84), II (1764), No. XVI, 235–240.
[38] LVIII (1768), 80–82.

effecting terror and great damage, although the sugar cane was not ruined and Miss Warner reported the prospect of "a tolerable good Crop."[39]

Barbados

Between 1695 and 1715, the *Sloane Manuscripts* in the British Museum record the receipt of an uncommonly large number of seeds, plants, and trees from the New World. They derived from both the mainland colonies of North America and from the West Indies, but by far the larger proportion probably came from Barbados. The records are tantalizing because of their brevity, and the immediate colonial collectors, or donors, if named at all, are difficult to identify. Joseph Gibbs, James Weir, Dr. Baskervill, a Colonel Russell, one Mistress Cranfield, and "Mr. London" are named as donors.[40] The identity of none of these persons is clear. "Mr. London" is especially troublesome because he may have been George London, the well-known gardener of the Fulham Palace gardens and later of the royal gardens—although there is no record that he went to the American colonies; or he may have been William London of Barbados, that "ingenious Drudge" who in 1680 had offered his services to John Evelyn.[41] Whoever these collectors and donors may have been, they appear to have been cooperating with members of the Temple Coffee House Botany Club, especially with William Sherard, Dr. Hans Sloane, and James Petiver, who, in turn, supplied the seeds, plants, and trees to the Chelsea Physic Garden and, more particularly, to the flourishing gardens both at Badminton, Gloucestershire, and in Chelsea, so lovingly and lavishly maintained by Mary, Duchess of Beaufort, with the advice and assistance of William Sherard. John Ray, Leonard Plukenet, Dr. Sloane, William Sherard, James Petiver, and Jacob Bobart the Younger all profited from these collections either by additions to

[39] *Journal-Book*, XXVIII, 93–95; *Add. Mss. 5169D*, fols. 225–227.

[40] See, for example, Sloane's endorsement of the "List of seeds sent mee by Mr. London August 1695 from Barbados," *Sloane 3343*, fol. 249. Others sent by "Mr. London" are listed in *ibid.*, fols. 39, 55–56, 63–66, 232, 240, 263–264, 276; for other shipments, see *1968*, fols. 182–183; *3329*, fol. 153; *4070*, fols. 65–85, 145; *4071*, fols. 110 and *passim*; *4072*, fols. 299–300.

[41] See Chapter 7 above, pp. 218–219.

their own herbaria, by the opportunity to study the Barbados exotics growing in the gardens at Badminton, Chelsea, Fulham, and elsewhere, or by both.[42] Excepting only Dr. Sloane and William Sherard, they referred to the Barbados plants and trees in one or another of their published works; Sherard included them in his incomplete and unpublished "Pinax";[43] and Sloane compared them with his Jamaican collections and often spoke with authority at meetings of the Royal Society regarding the identity, provenance, and uses of West Indian flora.

In the meantime, James Petiver sought his own correspondents and collectors in Barbados. After several vain attempts,[44] he established his only fruitful correspondence with a resident there in 1711. It seems likely that the exchange was a by-product of Petiver's associations with Captain George Searle, the shipmaster who plied his trade between the West Indies and London and who, as has been noted, contributed collections of shells to Petiver from Antigua. Captain Searle was a nephew of Captain Thomas Walduck of "Rupert's Fort in Leeward of Barbados." Little is known about Captain Walduck, but from his letters it appears that he was a military man who had been stationed in the West Indies for fourteen years. He had traveled widely among the British colonies, including the mainland colonies in North America, and he demonstrated considerable familiarity with the Spanish colonies as well. Furthermore, he was obviously widely read in the history of European colonization in America, possessed a contemplative and experimental bent of mind, together with firm—and not very favorable—opinions about the English settlements at Barbados. About 1710 Captain Walduck began a remarkable series of letters to his nephew John Searle of London, brother of Captain George Searle.

---

[42] See Dandy, pp. 209–215.

[43] At the Bodleian Library (16 vols.). References to the others' use of the Duchess of Beaufort's gardens, etc., are given in Dandy, pp. 209–215. See also *Journal-Book*, X, 211 (Feb. 19, 1700/01), when Sloane loftily comments upon Barbados specimens before the Royal Society to the effect that "all these were found by him in Jamaica." Dr. Sloane was often patronizing about West Indian matters brought before the Society.

[44] See his letter to Dr. Bane, Feb. 8, 1698/99, *Sloane 3333*, fols. 262v–263; and his letters to James and Claudius Hamilton between 1700 and 1708, *Sloane 4063*, fol. 22; *3336*, fols. 19v, 39v.

Six of these letters survive.[45] They include a description of the physical location of Barbados, its flora and fauna, its political and economic history, a survey of the benefits conferred by the New World upon the Old, a critical relation of the social customs of the Barbadians, and contemplations as to why the Ancients had not discovered the New World. In his letter of October 29, 1710, after a description of the astonishing Aztec and Inca ruins of Central and South America, and an account of the Spanish system of colonial government, Captain Walduck suggested to his nephew that the Royal Society of London should cultivate a wider correspondence in Spanish America for philosophical purposes. In a subsequent letter of March 8, 1710/11, he told his nephew: "When you design an acquaintance with some Gentleman of the Royall Society, you must not understand me that I am qualified of keeping a correspondence with those learned Gentlemen in those Noble Studies & experiments they imploy themselves in, but I shall be very glad & spare no time, expence, or labour in any usefull observation or experiment (let them prescribe ye method) they shall put me upon & lyeth within my ability."[46] At this point John Searle got in touch with James Petiver and showed him some of the letters from his uncle. Petiver made haste to write to Captain Walduck on January 1, 1710/11:

> With Mr. Searles permission I have had the ffavour to peruse most of your very curious Letters with no small satisfaction & delight, & shall thinke my selfe not a little happy to have the

[45] Copies of these letters are in *Sloane 2302*, fols. 1–24. The first two bear no date. The others, all from Barbados, are dated Oct. 29, Nov. 12, 24, March 8, 1710, respectively. Some of the originals of these, addressed to John Searle, are in *Sloane 3338*, fols. 74–77. Two others, addressed to Petiver, are in *Sloane 2302*, fols. 25–28. One is dated Sept. 17, 1712; the other bears no date. However, Petiver's recapitulation of these letters (*Sloane 3337*, fols. 135–137v) casts some doubt on these dates. Another letter of Walduck to Petiver, May 20, 1714, is in *Sloane 4065*, fols. 164–165, and in *Letter-Book*, XV, 92–94. See also Walduck's "Acct. of ye Rattlesnake," *Sloane 3339*, fols. 113–116v. Letters from Petiver to Walduck are in: *Sloane 3330*, fol. 41 (undated); *3337*, fols. 135–135v (Jan. 1, 1710/11); *3338*, fols. 3v–4 (Nov. 1, 1711), 119–120v (Dec. 25, 1712). See also *Journal-Book*, XI, 314, 401; XII, 114. The principal Walduck letters were published in the *Journal of the Barbados Museum and Historical Society* (Bridgetown), XV (1947–48), 27–51, 84–88, 137–149. I am indebted to Professor Carl Bridenbaugh of Brown University for this reference.

[46] *Sloane 2302*, fols. 20v–24.

Honour of Corresponding with a Person of your Intelligence,
Ability & Meritt.

I am Sr the more emboldened to hope this Favour since it
seems to be your desire (towards ye Conclusion of one of your
fformer Letters) to have a Correspondence with a Person of the
Royal Society of wch I have had the honour for severall Years
to be a Member, at ye same time you mention the sending of a few
Trifles or rather Curiosities for so I esteem all things that are the
productions of Nature, wch you had picked up in your perambu-
lations. . . .

I have Sr for more than 20 years dedicated a great many Leesure
hours to the Contemplations of Natures manifold Wonderful productions with an inexpressible satisfaction & delight as you will
probably see by the Contents I herewith send you. My first & early
progress began at home and after I had made some progress in ye
knowledge & Collecting all ye Plants att London especially ye
Medicall ones, that part being an absolute necessary part of my
Profession, I then proceeded to ye Animal Kingdom, ye various
classes of these from ye greater Animals as Beasts, Birds, & Fish
even to ye minutest Insect not many of wch have escaped my
Observation, so yt in a few years we have already discovered neer
2000 besides ye divers sorts of Land, River & sea Shells, ye Fabrick,
Beauty, Variety & Use of all these may justly deserve our Admiration as well as Consideration & with ye Holy Psalmist say
How manifold are thy Works O Lord & in Wisdom hast thou
made them all.

It is Sr to such Curious Persons as yr selfe yt we at this distance
must owe what your parts of the late discovered World can afford
us. Your residence there may give us great light into many things
wch we as yet but imperfectly know & others are totally ignorant of, by gathering things in all seasons & consequently in
their several states of growth or vegetation by wch we shall be
able to give better description and more accurate Figures of ym.
So that I doubt not but with your Assistance in some time we may
be able to give a tollerable Acct. of ye Natural productions of
your Island in Respect to its Animals, Vegetables, & Minerals as
also to ye politicall & Trading part wch you have already so well
begun.[47]

---

[47] *Sloane 3337*, fols. 135–135v. In the "Petiver Adversaria" (*Sloane 3330*, fol. 41)
is a copy of an undated letter to Walduck in which Petiver wrote: "Your first
letter Sr Relating to ye lawes & Government of Barbadoes being annexd to some

Petiver obviously sent with his letter some of his publications with figures of plants, and no doubt instructions and supplies for further collecting. He wrote again on May 17, 1711, and Captain Walduck received both of Petiver's letters on June 22, 1712. He replied on September 17 following. "I am heartily disposed," he said, "to oblige you in all manner of ways I am able, and As I ever took delight in things of this Nature, so now I shall doe it with more pleasure. . . ." He thanked Petiver for his "valuable Presents," and sent him "one book of plants, with as many of their names and Virtues as I could learne, their uses I have gott from our Physicians (shall I call them) nurses, old women and Negroes, and for the future, I will take care by some Experiment or other not to be imposed upon but will serve you with truth to the utmost of my capacity. . . ."[48] He also sent three bottles filled with worms, crabs, a poison lizard, and spiders, and related some experiments he had made. He had read somewhere in Francis Bacon's works that holes bored through fruit trees and then plugged would cause the trees to bear seedless fruit; but Captain Walduck only found that it caused the trees to die. He had shut up a lizard in a sealed bottle and found that the lizard died. He told of a Negress whose twins were one white and one black; of an eighty-seven-year-old woman who suckled her grandchild; and of "some supposed Diabolical Practises of the Negroes." Petiver read the letter and displayed the collection from Walduck to the Royal Society on November 20, 1712.[49] He wrote to Walduck on December 25 to give his thanks and those of the Royal Society, to bid for further collections, and to send a butterfly net.[50] Sometime during the next year Captain Walduck sent an "Acct. of ye Rattlesnake," which Petiver reported to the Royal Society on January 7, 1713/14.[51] It was founded upon hearsay evidence which Captain Walduck had gathered during visits to the mainland colonies. Although he de-

---

private business I had not ye Honour to see." Petiver hoped that later letters would treat of "pulse, fruit, Flesh, Birds & Fish as also to ye numerous insects & . . . Animals."

[48] *Sloane 2802*, fols. 25–26.
[49] *Journal-Book*, XI, 314.
[50] *Sloane 3338*, fols. 119–120v.
[51] *Journal-Book*, XI, 401; also *Sloane 3339*, fols. 113–116v.

clared that the rattlesnake was the "most pernicious creature in ye English Empire upon ye Main of America," he also said there was a smaller, even more poisonous, snake in Carolina—but he did not name it. He reported that the rattlesnake was oviparous,[52] that it sometimes killed itself with its own bite, that snakeweed was an antidote to its poison. But the Society did not take kindly to Captain Walduck's relation, and recorded that "the Capt[n] seemed to give too much credit to the ill-grounded reports of the Vulgar." However, Walduck's next communication excited more favorable comment. By experiments in Barbados to force the more rapid growth of vegetables, Captain Walduck reported that water impregnated with niter was very effective, and he suggested that niter was one of the great causes of the growth of vegetation.[53] His last communication was reported to the Society by Petiver on May 17, 1716, giving an account of "the phoenomenon in the Air the 6th of March last."[54] The records fail to disclose the nature of the phenomenon.

As a whole, Captain Walduck's letters contributed more to the political, economic, and social history of Barbados and the West Indies than to the furtherance of science. They bore considerable charm together with the stamp of a widely read, largely self-educated man, with a conscious bent toward Baconian philosophy, observant (but not highly critical), naively experimental, and genuinely eager to contribute to the new science. But Captain Walduck displayed no scientific knowledge beyond that of an interested amateur, and his methods left much to be desired, with too much credit given, as the Royal Society observed, "to the ill-grounded reports of the Vulgar." His collections held more permanent value. Petiver celebrated his contributions by dedicating Table 19 of his *Pterigraphia Americana* "to His very Curious Friend Capt Thomas Walduck at Barbados," and gave additional recognition in an article in the *Philosophical Transactions* for 1715. Walduck contributed to Petiver's herbarium which, added to the

[52] Denied at the next meeting of the Society (*Journal-Book*, XI, 402) by reference to Dr. Edward Tyson's dissections previously published in the *Philosophical Transactions*.

[53] Walduck to Petiver, May 20, 1714, *Sloane 4065*, fols. 164–165; *Letter-Book*, XV, 92–94.

[54] XXIX, No. 346 (Nov.–Dec., 1715), 357.

Sloane collections, is still extant in the British Museum (Natural History).[55]

Scientific activity in Barbados attracted little attention for several years after Petiver's death.[56] On January 7, 1730/31, Dr. Benjamin Hoadley, F.R.S., communicated from William Stevenson "an Account of an Observation of an Eclipse of the Moon, on July 29, 1729, made in Barbados by Mr. Stevenson's Brother [Thomas?]." It was remarkable principally because Mr. Stevenson's calculations of this and other eclipses at Barbados convinced him that the island was about two and a half degrees farther west than previously supposed. He was thanked for the communication, asked to continue with others, and his account was published in the *Philosophical Transactions*.[57] Both Sir Hans Sloane and Philip Miller (of the Apothecaries' garden at Chelsea) held correspondence with residents of Barbados in the 1730's. Dr. Thomas Townes kept Sir Hans informed about the health of Barbadians and the diseases prevalent, but appears to have added nothing to the Sloane collections.[58] Mr. B. Hothersal consulted with Philip Miller about coffee production in Barbados and about efforts to make wine from sugar cane. Miller had sent him vine cuttings from which he hoped to produce wine which, mixed with sugar syrup, would make a potable beverage in quantity. He also reported that he could produce 2,000 pounds of coffee in three or four years if a method of husking it by machines could be contrived.[59] Neither the wine nor the coffee projects flourished.

On November 3, 1743, President Martin Folkes presented to the Royal Society "a sort of zoophyton [zoophyte]" from Barbados.

[55] Dandy, pp. 228–229.

[56] Until the 1730's, only two notices from Barbados are recorded in the minutes of the Royal Society; one (Nov. 6, 1718, *Journal-Book*, XII, 248–249) reported receipt of two centipedes from Mr. Charles Mayne; the other (June 12, 1722, *ibid.*, XIII, 208) was of a presentation of a large map of Barbados made by the donor, one Mr. Mayo, who had prepared it "from an actual Survey."

[57] XXXVI, No. 416 (Nov.–Dec., 1730), 440–441. See also *Journal-Book*, XIV, 538–539. Mr. Stevenson's brother may have been Thomas Stevenson of Barbados. See below.

[58] Townes to Sloane, Barbados, April 30, 1732, *Sloane 4052*, fol. 105.

[59] Hothersal to Miller, Barbados, July 12, 1737, *Letter-Book*, XXV, 135–137. Miller read the letter to the Royal Society Jan. 18, 1738/39, *Journal-Book*, XVII, 352.

The specimen was the gift of the Reverend Mr. Griffith Hughes of St. Lucy's Parish, Barbados. Born in Wales about 1707, Mr. Hughes had matriculated at St. John's College, Oxford, in 1729. He left the college without a degree in 1732, and, bearing testimonials from the college officials certifying to his fitness for ordination, he was ordained by the Bishop of London and sent as a missionary of the Society for the Propagation of the Gospel to Pennsylvania, where he served among Welsh settlers at Radnor, Perquahoma, and other frontier settlements.[60] Owing to ill health, he left Pennsylvania without leave in the autumn of 1735 and went to Barbados, where he was employed by St. Lucy's Parish. He returned to Pennsylvania briefly in 1736, resigned his post there, and went back to his parish in Barbados.[61] How long he served at St. Lucy's Parish is uncertain. He was in London in 1743 when he presented the zoophyte to the Royal Society, and a few weeks later (December 22) he visited the Society in person and presented a piece of asbestos from Pennsylvania.[62] He was back in England permanently by March, 1748, when he was proposed for membership in the Royal Society (March 3, 1747/48; elected, June 9), and during the same year he proceeded both B.A. and M.A. from Oxford *in absentia*.[63] His letter to President Folkes in 1743 was published, with a figure of the zoophyte, said to resemble "the Flower of the Marigold," in the *Philosophical Transactions*,[64] and he was said to have diligently

[60] *D.N.B.*; J. Butler and G. S. Boulger, *A Biographical Index of Deceased British and Irish Botanists* (2nd ed., London, 1931), p. 158; Benjamin F. Owen, ed., "Letters of the Rev. Griffith Hughes, of St. David's Church, Radnor, Pennsylvania, 1733–1736," *Pennsylvania Magazine of History and Biography*, XXIV (1900), 139–148; *Calendar of Treasury Books, 1731–1734*, p. 343; *Pennsylvania Archives*, 3rd Ser., XXIV (1898), 423.

[61] Some of his parishioners in Pennsylvania complained to the Bishop of London that Hughes had abandoned them, but they did not mourn his loss, saying that he was a rambling sort of preacher whom they could not understand and that they preferred a man of riper years. William Wilson Manross, comp., *The Fulham Papers in the Lambeth Palace Library, American Colonial Section, Calendar and Indexes* (Oxford, 1965), p. 105.

[62] *Journal-Book*, XIX, 178.

[63] *Ibid.*, XX, 463, 527; *Certificates, 1731–1750*, p. 366; Joseph Foster, *Alumni Oxonienses . . .* (4 vols., Oxford, 1891), II, 706.

[64] XLII, No. 471 (Nov.–Dec., 1743), 590–593. On Feb. 13, 1745/46, a Mr. Bryant was present at the Royal Society meeting by request. He had been appointed Professor of Mathematics and Philosophy at Codrington College, Bar-

inquired into the natural history of Barbados and communicated other observations to the Royal Society before his election. On February 16, 1748/49, he read a paper entitled "The Murex" to the Society, an inquiry into the purple dye of the Ancients produced by a shellfish similar to that yielding a deep crimson dye found by Mr. Hughes in Barbados.[65] The paper was part of a book that Hughes was preparing on the natural history of Barbados. Additional portions of the book were presented to the Society in 1749, and on May 24, 1750, Hughes presented *The Natural History of Barbados. In Ten Books* (London, 1750) to the Society.[66]

Whatever else may be said of the book, it was a masterpiece of salesmanship in the sense that the author had rounded up a remarkable list (nearly 500) of subscribers to the undertaking. Beginning with "His Most Christian Majesty" (Louis XV) and "His Royal Highness the Prince of Wales," the list included many English peers and gentlemen (to some of whom the plates were dedicated), three colonial governors (Greenville of Barbados, Mathew of Antigua, Oglethorpe of Georgia), Peter Collinson, John Bartram, the Library Company of Philadelphia, and thirty-three subscribers from Virginia, headed by Thomas, Lord Fairfax, and including John Robinson, President of the Council, the Reverend Mr. William Dawson, President of the College of William and Mary, and other prominent Virginians.[67] Given such patronage, one might reasonably expect that *The Natural History of Barbados* would be

---

bados, and, because he was shortly to go there, "would be ready to presecute any Inquiries concerning Natural History which might be thought serviceable by any Gentlemen of the Society. Whereupon Mr. Baker proposed that some Inquiry might be made Concerning that plant Animal the account whereof was formerly communicated by the Reverend Mr. Hughes." *Journal-Book*, XX, 46. But if Mr. Bryant responded from Barbados, there is no record of it. Indeed, if Codrington College, built between 1714 and 1742 with funds from the Society for the Propagation of the Gospel and the bequests of Colonel Christopher Codrington (1668–1710), formerly (1697–1703) Governor of the Leeward Islands, gave any significant direction to scientific studies in Barbados, there was no echo of it in the Royal Society, though the Society was familiar with the college from its foundation in 1714. *Journal-Book*, XII, 16 (July 29, 1714). See also John Aubrey, *Letters Written by Eminent Persons . . . and Lives of Eminent Men* (2 vols., London, 1813), I, 128–133.

[65] *Journal-Book*, XXI, 69–71.

[66] *Ibid.*, XXI, 85–86 (March 16, 1748/49), 354.

[67] The list of subscribers fills five pages following the Preface of the book.

a scientific work comparable to Sloane's works on Jamaica, Catesby's treatment of Carolina and the Bahamas, or even Dr. Patrick Browne's *Civil and Natural History of Jamaica* published only six years later. But its scientific value fell short of all these works, primarily because Mr. Hughes, while demonstrating a wide grasp of ancient classical works, appears to have had only passing familiarity with the works of his more immediate scientific predecessors and little mastery of the techniques developed for the scientific descriptions of flora and fauna. He had a microscope, and he referred to the "ingenious and learned discoveries" of Malpighi, Grew, Ray, and Hales, but he displayed little actual use of either the microscope or the discoveries of the "ingenious and learned" scholars named. He followed John Woodward's geological views regarding the origins and distribution of fossils, although he argued that the force of the biblical Deluge was less severe in equatorial regions and that therefore the fossils were nearer the surface in those regions. He assured his readers that "This I can with Truth say, that I have not represented one single Fact, which I did not either see myself, or had from Persons of known Veracity," but the most outstanding scientific values of the book stemmed from the fact that, as the author stated at the end of the volume, the "Explanatory Notes of all Botanical and Technical Terms made use of in the foregoing Work" were done by "the *accurate* Mr. Miller *of* Chelsea," i.e., Philip Miller of the Apothecaries' garden at Chelsea. There was a competent map of Barbados made by the geographer Thomas Jeffrey, from surveys supplied by Mr. Hughes, and twenty-nine plates of uneven quality, the best ones (eight in all) made by Georg Dionysius Ehret (1708–70), the most eminent illustrator of the day both in terms of scientific accuracy and in terms of beauty of design. With one exception, however, Ehret's plates illustrated flora (probably from dried specimens), and they were not among his best works. The twenty-odd other plates, mostly of animals and fishes, were generally poor, both in scientific accuracy of detail and in artistic design.

Of the ten books of *The Natural History of Barbados*, Book I was a general description of the size, location, and peoples of the island. The author sought to demonstrate that the biblical Deluge had enveloped the island from east to west, but he referred to archeological evidence to prove that Indians once had inhabited Barbados. The population of the island, both black and white, was diminishing. Book II treated the diseases common to the people: the dry belly-ache, once so deadly and arising, in the author's opinion, from the bizarre rum and other fermented cane-juice drinks concocted by the Barbadians, was less severe than formerly; dysentery, yellow fever, smallpox, and leprosy—the last especially among Negroes—were the principal illnesses. Book III dealt with fauna—ten land animals, thirty-four birds (including thirteen birds of passage), and fifty-four insects, including a few reptiles. Domestic animals (not included in the above) were of the same breeds as those in England, though smaller in stature in Barbados. There were few venomous reptiles. The black spider, the "Forty-Leg," and the Surinam scorpion were the most poisonous insects. The number of birds and insects cited was respectable, although the descriptions were hardly scientific, and the attention to fauna in general was far from complete. Books IV to VIII, inclusive, were devoted to trees, shrubs, and plants. After a general oversight of the flora (Book IV), they were arranged into those of the "Pomiferous Kind" (Book V), of the "Bacciferous Kind" (Book VI), of the "Capsule-bearing Fruits" and of the "Pruniferous Kind" (Book VII), of the "Siliquose Kind," of the "Anamalous Kind," and these were followed in Book VIII by capillary and parasitical plants; grasses, reeds, and gramineous plants; vines; and a final section "of Gramineous Plants." Book IX turned to seaweeds, exsanguinous animals, crustaceans, shellfish, corals, "the Animal Flower" (the zoophyte shown to the Royal Society in 1743), and "The Murex," extracted from the paper of that name read to the Society in 1749. The last book, containing only fifteen pages, was devoted to "the Sea and its Inhabitants." A glance at this summary of contents will confirm the conclusion that *The Natural History of Barbados* was heavily weighted in favor of botany; actually, of 314 pages in the book, fewer than a hundred were devoted to fauna of all kinds. Perhaps this was fortunate, insofar as the botanical portions of the book were the areas in which Philip Miller was able to assist the author in achieving a more nearly up-to-date scientific arrangement and nomenclature than otherwise might have been likely. Even so, the

Reverend Griffith Hughes's *Natural History of Barbados* was not a distinguished scientific work, and though it appears to have been popular reading—as so many of its kind were in early eighteenth-century England—it was never reckoned among the great scientific contributions of the time. Nor did its author appear further in the scientific annals of the Old Colonial Era.[68]

Natural history attracted less attention in Barbados after the appearance of Mr. Hughes's book—that is, for the remainder of the Old Colonial Era. On June 30, 1759, Thomas Stevenson, possibly the same Mr. Stevenson who in 1729 had made observations on the eclipse of the moon, addressed a letter to James Bradley at Greenwich, as follows:

> Altho I am intirely a stranger to your person, I am not so to your character as an Astronomer; I shall therefore, make no apology for inclosing you a Sett of Observations made of the long Expected Comit [Halley's Comet], which altho made without any Instruments, yet, as he approached near to some remarkable known Stares & was Visible for a long time, in which he moved throw a large Trackt of the Heavens; I have some hope they may be of some service to you in order to settle his Elements & to determine the true time of his Period. I would not have presumed to have sent you these Uncorrect Observations had I not been almost certain that you could not see him in England till his return from the Southward, and that you would be glad of any helps so early as mine are which from the situation of this Island I was inabled to make. . . .[69]

Stevenson's observations, made from March 21 to May 12, 1759, were, of course, of limited value, being made without instruments. He had ordered a ten-foot telescope from London but it arrived just after the comet's disappearance. He had a good pendulum clock, he gave the location of his house in Barbados with precision, and in spite of his lack of a telescope, Stevenson's observations and

---

[68] Mr. Hughes appears to have returned to Barbados but he corresponded no more with the Royal Society. In 1756 the Council of the Society gave him notice to pay arrears in dues before Dec. 21, 1757, or face ejection (*C.M.*, IV, 184, Dec. 9, 1756), but I have found no record as to whether the Council took further action.

[69] *Bradley 95*, fols. 49–52 (Bodleian Library).

calculations indicated that he was an amateur astronomer with more than ordinary qualifications. He disagreed, however, with Halley's opinions regarding this comet:

> My chief reason was [he wrote] The difference of the times of the Periods. And I . . . still imagine that the Comets in 1607 & 1682 are not the same, notwithstanding their Elements so nearly agree; but that they are two distinct Comits, & that one of them is nearly in his Aphelion when the other is in his Perihelion. This Hypothesis will make each of their Periods agree in point of time; which I think is much more consistent with the Heavens then that the same Comet should be alternately accelerated & retarded above a year, which I must own, I think more improbable (and is almost Impossible) Than that there should be two Comits of nearly the same Elements. . . .

Perhaps, as Stevenson concluded, "I am soaring far above my Sphere, but hope you will Excuse my blunders as my intentions are good." If he was still about in 1763 and 1764, he must have been delighted when Nevil Maskelyne, F.R.S., then of Trinity College, Cambridge, but soon (1765) to become Astronomer Royal, made a series of observations of the satellites of Jupiter (November 13, 1763, to August 10, 1764) "at Willoughby Fort and at the Observatory on Constitution Hill, both adjoining Bridge Town."[70] Maskelyne went to Barbados after having observed the transit of Venus at St. Helena (1761) by appointment of the Royal Society. His reference to the "Observatory on Constitution Hill" seems to suggest a temporary location (Maskelyne used his own instruments supplied by the Royal Society) rather than a permanent installation at that time; but if the Royal Navy had a permanent observatory there it must have been one of the earliest in the West Indies.

In the meantime, Mr. Abraham Mason of Barbados presented to the Royal Society (November 6, 1760) a paper entitled "A Discovery in Magnetism."[71] The paper was enthusiastically endorsed by the Reverend Mr. T. H. Croker, Reader of the Middle Temple, who attended the Society in the name of Mason and demonstrated

[70] *Journal-Book*, XXVI, 190–191 (Dec. 20, 1764). Maskelyne's report to the Royal Society was published in *Phil. Trans.*, LIV (1764), 389–392.
[71] *Journal-Book*, XXIV, 98–99 (mispaged 95–99).

his experiments to the Society. The core of Mason's argument was that the "attraction of Magnetism is exerted to the Horizon only, and that is the power of gravity which depresses the Needle below the Horizon." Mr. Croker returned to the Society (February 2, 1761) with further "proofs" of Mason's hypothesis, and on May 7 he presented a book entitled *Experimental Magnetism, or the Truth of Mr. Mason's Discoveries Proved. . . .*[72] Abraham Mason sent two further communications to the Society but they were less controversial. One was a "sea animal" received on December 17, 1761, and later described by John Ellis in the *Philosophical Transactions* as "an Encrinus" (encrinite), which, said Ellis, "explains to what kind of Animal those Fossils belong, called Starstones, Asteriae, and Astropodia";[73] the other was a letter describing a great tidal wave at Barbados (March 31, 1761) similar to that on the day of the great earthquake at Lisbon (November 1, 1755) and followed by an epidemic of severe colds.[74]

On April 21, 1768, Dr. Thomas Lashley of Barbados was proposed for admission as a Fellow of the Royal Society. He was elected on November 24 following, and Dr. Nicholas Munkley paid his admission fees on December 8. But it was November 25, 1784, before Dr. Lashley attended the Society in person and was formally admitted Fellow.[75] He appears to have made no communications to the Society in the interim. Dr. John Fothergill, F.R.S., a London Quaker physician who maintained a widespread scientific correspondence as well as a botanical garden famous throughout Europe, communicated "An Account of a late epidemical Distemper, extracted from a Letter addressed to Gedney Clarke, Esq., by William Sandiford, M.D., of Barbados," in 1770. It was published in *Medical Observations and Inquiries . . .* the following year and dealt with Barbados cases in the summer of 1769.[76] It was the last scientific report of any significance from Barbados

---

[72] *Ibid.*, XXIV, 103.

[73] LII, Pt. I (1761), 357–365. See also *Journal-Book*, XXV, 198.

[74] *Journal-Book*, XXV, 77–78 (March 11, 1762); *Phil. Trans.*, LII, Pt. II (1762), 477–478.

[75] *Journal-Book*, XXVII, 80, 136; XXXII, 9; *Certificates*, *1767–1778*, unpaged. Dr. Lashley's supporters included Drs. Mark Akenside and William Hunter.

[76] IV (1768), No. XXIV, 305–320.

before the American War for Independence. Only two more communications from Barbados reached the Royal Society before 1800.

Jamaica

Dr. Hans Sloane was the principal scientific intermediary between Jamaica and England during the first half of the eighteenth century. He owned properties in Jamaica, had friends and relatives (by his marriage) there, and his own firsthand information about the island, together with his published works on the natural history of Jamaica, caused him to be looked upon as the foremost authority regarding Jamaican subjects. His offices as President of the Royal College of Physicians and as Secretary, member of the Council, and President of the Royal Society served further to enhance his position. Until mid-century, the majority of the most significant scientific communications from Jamaica came by the hands of Hans Sloane.

However, Dr. Sloane held no monopoly of Jamaican correspondence. The Secretaries and other Fellows of the Royal Society also received occasional letters and collections from the island.[77] James Petiver's efforts to establish fruitful exchanges with persons in Jamaica were impressive—and almost pathetically unproductive. Petiver's correspondence between about 1693 and his death in 1718 reveals that he approached at least a dozen persons in or bound for Jamaica in the hope of developing a scientific exchange.[78] The Reverend John Smyth of Port Royal was an example of one of Petiver's several failures. Dr. Smyth was known favorably by Samuel Dale of Braintree, and in 1693 Petiver sought his assistance in making collections in Jamaica. He supplied him with collecting supplies, "plain, full & easie directions," catalogues of Jamaican plants to serve as a guide, and his usual appeal to vanity with the promise that Smyth could see his name in print if he supplied specimens which Petiver could use in his publications. But Dr. Smyth's re-

---

[77] See, for example, Robert Thompson to Dr. Nehemiah Grew, Port Royal, Aug. 20, 1696, reporting the "Epidemicall feaver" in Jamaica, *Letter-Book*, XI, Pt. I, 219–222; reported July 13, 1697, *Journal-Book*, X, 43. No further letters from Thompson were reported to the Society.

[78] See his letters in *Sloane 3332, 3333, 3335–37, 4063, passim.*

sponses were few, and he sent little or nothing by way of collections. Petiver repeatedly urged him on, pointedly wrote in 1695
that "any child of 6 years is capable of doing it," and finally, in
1696, told the doctor that his promises were merely "the happy
invention of yr Mercuriall brain" and ended his letter saying
that "If I receive not a Collection of Plants from you by ye next
shipping I shall desire you will deliver ye Paper &c to a Friend" to
whom Petiver would direct him. Not surprisingly, the correspondence lapsed at this point.[79]

Petiver's earliest productive exchanges from Jamaica stemmed
from his friend, Captain Patrick Rattray, a shipmaster who himself
made collections of Jamaican shells and also delivered his own and
others' collections to Petiver in London free of charge. Between
1696 and 1699, Captain Rattray drafted into Petiver's service Mr.
Anthony Bigg and his son of the same name, Messrs. Roger and
John Fenwick, and the Reverend Mr. Henry Passmore, all of
Spanish Town.[80] These men collected Jamaican plants, shells,
small land and marine animals, and fossils for Petiver. The Fenwicks were reported to have prepared a herbal of Jamaican plants
which Petiver asked to see.[81] And all of these men were acknowledged as donors to Petiver's collections in the latter's publications.[82]
After the turn of the century he found new Jamaican contributors.
James Fraser, a ship's surgeon, sailed out from Plymouth in 1703
after promising to follow Petiver's instructions, and during the next
four years he supplied Petiver with items from Jamaica, Bermuda,

[79] The letters to Smyth are in *Sloane 3332*, fols. 33, 43, 48v, 54, 55v–56, 74v,
116v, 125v–126, 166v–167v, 172v. The quotations are from fols. 125v–126 [n.d.,
autumn, 1695?] and 191v–192 (Jan. 13, 1695/96). Other vain efforts of Petiver in
Jamaica involved Captain Wentworth of "ye Governour Brigadier Hamilton's
Regiment in Jamaica"; one Madame Carter; Allen Broderick; W. Burdon; one Mr.
Morton; William Brown, a ship's surgeon; and Dr. Archibald Stewart.

[80] See *Sloane 3333, passim*. Mr. Passmore was acquainted with Hugh Jones,
William Vernon, and Dr. David Krieg, currently collecting natural history specimens in Maryland. See *ibid.*, fols. 133v–134, 237v–238.

[81] *Ibid.*, fols. 128–128v, 231. Petiver plied them with copies of his early "Centuries," Dr. Sloane's *Prodromus*, "Darby Ale," and other gifts. But whether he
saw the Fenwicks' herbal is not clear.

[82] *Sloane 3324*, fols. 75–78 (Petiver's printed list of butterflies); *Musei
Petiveriani Secunda & Tertia* (1698), pp. 121, 291; *Quarta & Quinta* (1699), p.
19; *Octava* (1700), p. 756; *Musei Petiveriani* (1695–1703), pp. 18, 19, 72; *Gazophylacii* (1702–06), p. 40; Dandy, p. 174.

and Guinea.[83] Perhaps his greatest service was to introduce (1707) Dr. David Crawford, of Port Royal, to Petiver, and these two continued a correspondence until 1712.[84] Dr. Crawford was well informed in natural history and had studied with Tournefort in Paris, with whose *Elémens de botanique* (3 vols., Paris, 1694) he was familiar. When Petiver urged him to read Dr. Sloane's early publications about Jamaica, Dr. Crawford appears to have been unimpressed, partly, at least, because Sloane's prescription for the cure of the dry belly-ache differed from his own, which he insisted was both new and better and felt that Petiver and his friends failed properly to appreciate it.[85] Indeed, Dr. Crawford's correspondence yielded more by way of medical discussions than it did in terms of collections, and when Petiver failed to print an account of a "gally wasp" (galliwasp, a small lizard, *Celestus occiduus*) with which Crawford had supplied him, Dr. Crawford complained of his poor performance. At this point, too, the correspondence extant ceased (May 17, 1712).

Two more persons who went to Jamaica raised high hopes in James Petiver. These were James White and Dr. Thomas Hoy. White had been known to Petiver since about 1700, when he was serving as ship's surgeon for the East India Company.[86] Later he practiced medicine in Lisbon and in 1713 was physician to Lord Lexington, British Ambassador to Spain, at Madrid. He assisted Petiver in the latter's efforts to find scientific correspondents in Spain and in the Spanish colonies of the New World, and upon his return to England in 1714 he and Petiver discovered they had mutual friends at Oxford.[87] The following year White gave up his medical career and went to Jamaica, where he became rector of

[83] *Sloane 3321*, fols. 130, 215; *3335*, fols. 48v–49v.

[84] The Petiver-Crawford correspondence is scattered among *Sloane 3321*, fols. 234, 279; *3335*, fol. 49; *3336*, fols. 18, 25, 39; *3337*, fols. 71–71v, 158.

[85] Dr. Crawford's cure was hog gum, a gum or resin from various West Indian trees. He said that the cure had been discovered by one Captain Abraham in a dream. See his letters to Petiver, Sept. 25, 1708, *Sloane 3321*, fol. 234, and May 17, 1712, *ibid.*, fol. 279.

[86] See his letter to Petiver, Nov. 4, 1702, *Sloane 4063*, fol. 180.

[87] See the White-Petiver correspondence in *Sloane 4065, passim; 3337*, fols. 151–151v, 152, 157–158; *3338*, fols. 143–144v; *3340*, fol. 52; *Journal-Book*, XI, 365–366 (June 18, 1713). The mutual friends were Jacob Bobart and Richard Dyer, Fellow of Oriel College.

the Parish of Vere. Petiver delightedly addressed him in Jamaica (July 20, 1716) in the expectation of enlisting White's services as a collector there.[88] But White replied (April 2, 1717) that "I am sorry to tell you that the heat is so extreme, plantations so distant from one another, that it is almost impracticable to go a simpling on this Island."[89] White's excuses were not convincing, but he stated that "My worthy Friend Mr. [Henry] Barham who has practised physick here for many years with success & is the best Botanist that we had on the Island has promised me that you shall have some shares of anything you want & is truly a rarity. . . ." But Petiver died before he could get under way an exchange with Barham, although, as will be noted below, Barham became a contributor to Sir Hans Sloane's collections.

Dr. Thomas Hoy (1659–c. 1725), a friend of John Evelyn and described by him as "a very learned man," was an Oxford graduate (B.A., 1680; M.A. and B.M., 1684; M.D., 1689), was admitted Candidate to the Royal College of Physicians (1693), and was chosen Regius Professor of Medicine at Oxford in 1698.[90] He was elected Fellow of the Royal Society in 1707[91] and subsequently settled in Jamaica. On February 4, 1713/14, Secretary Richard Waller of the Royal Society, acting upon instructions of the Society, formally invited Dr. Hoy to undertake a philosophical correspondence with the Society.[92] Dr. Hoy replied on April 28, 1714, promised that "when any thing Observable shall offer itself within the small compasse of my Inquirys . . . I shall do my self the honour of preferring to their [i.e., the Society's] consideration,"[93] and sent some curiosities, including a bark like cinnamon, and an account of an experiment he had performed in chemistry. The bark puzzled the Society and Secretary Waller called for additional information

[88] *Sloane 3340*, fols. 242v–243.

[89] *Sloane 4065*, fols. 278–279. White complained loudly to the Bishop of London about religious irregularities in Jamaica. See Manross, comp., *Fulham Papers*, pp. 250–255.

[90] Sir Humphrey Rolleston, "The Personalities of the Oxford Medical School from 1700 to 1880," *Annals of Medical History*, n.s., VIII (1936), 280–281.

[91] *C.M.*, II, 144 (July 2, 1707); *Journal-Book*, XI, 126 (Dec. 1, 1707).

[92] *Letter-Book*, XV, 33–35; *Journal-Book*, XI, 407. Waller's letter, in the same vein as those dispatched by the Secretaries in the 1660's, indicates no alteration in the purposes and policies of the Society.

[93] *Letter-Book*, XV, 100–104; XVI, 73–76; *Journal-Book*, XII, 9 (July 8, 1714).

about it, although Dr. Sloane later reported that it had been described in his *Prodromus* as well as in a previous number of the *Philosophical Transactions*.[94] Hoy's chemical experiment baffled the Society, as the Fellows were unable to repeat it with success—which is also baffling because it was merely a precipitation of copper from "Jamaica mineral" by the use of nitric acid and a reducing agent to produce "a most elegant red color." Robert Boyle had done it years before.[95] Dr. Hoy's reply to the Royal Society's invitation led James Petiver to address him (July 20, 1716) in the hope of establishing an additional correspondent and collector in Jamaica.[96] But Dr. Hoy appears to have ignored Petiver's appeal, although it was repeated in September of the same year. Indeed, Dr. Hoy sent little, if anything, more to the Royal Society. In 1725 he wrote to William Sherard, whom he had known for years, sent some seeds of the logwood tree which Sherard had requested, reported that he was experimenting with cochineal, that he had a small collection of Jamaican plants and animals, that Jamaica was "so indolent & lazy that . . . Sugar & Indico are the only Curiositys of the Jamaica Philosophers." Dr. Hoy was "very old," was suffering "the dismal distemper of these parts they call the Bellyach," and proposed to return to England and take up residence at Bath. Perhaps he did. In any case, there is no further record of him.[97]

Dr. Hans Sloane's correspondents in Jamaica were numerous, and several of their communications were reported to the Royal Society.[98] One of the most fruitful of these was Henry Barham (c.

[94] *Letter-Book*, XV, 107–109; XVI, 82–84; *Journal-Book*, XII, 56, 57 (May 5, 1715).

[95] Information kindly supplied by Professor Douglas McKie of the University of London.

[96] *Sloane 3340*, fols. 243v, 260v, 267v.

[97] Dr. Hoy to W. Sherard, Kingston, Sept. 20, 1725, *Sherard Letters*, V, 601 (Royal Society Archives).

[98] See, for example, *Journal-Book*, XI, 79 (Jan. 2, 1705/06), an account of a severe storm at sea off the Jamaica coast, source of information not revealed; *ibid.*, XI, 324 (Jan. 22, 1712/13), two letters about a destructive hurricane of Aug. 28, 1712, authors not cited; *ibid.*, XIV, 134 (Nov. 2, 1727), extract of a letter from George Wilkinson, a supercargo, Jamaica, June 25, 1727, to the effect that bees cohabit in tropical places where there is "no winter." Other communications from Jamaica that were not laid before the Society are among the *Sloane Manuscripts*. See, for example, Francis Rose to

1650–1726). After serving an apprenticeship in surgery, Barham became a naval surgeon, traveled widely, and finally settled in Jamaica about 1680. He developed a lucrative medical practice there and was appointed Surgeon General of the military forces in the island. Obviously, Dr. Sloane had made Barham's acquaintance when he went to Jamaica, and Barham had supplied him with Jamaican curiosities.[99] Indeed, Barham himself was very knowledgeable regarding the flora and fauna of Jamaica. He studied "the Specifick qualityes of Plants" in Jamaica for many years and drew up a "Hortus Americanus Medicinalis" which he submitted to Sloane for study and correction (1711), evidently with an eye toward its publication, although it was not published during Barham's lifetime.[100] When Sloane sent him the first volume of his *Voyage* (1707), Barham studied it with care and wrote to Sloane from Spanish Town, May 10, 1712, about

> the great Benefitt I have Received thereby and I think the Whole Island ought unanimously Joyne in their Thanks to you; for the great Pains, Industry & Labor in Compiling Soe usefull a work. . . . [But some islanders were ungrateful because they had not been consulted.] Some are Dissatisfyed with putting names in your Observations of Diseases, Others that the Practis is very mean & Plaine (I am Sure it is Safe). Butt the main Objection is; that you have writt the Names and their Several Kindes of Plants in Latin wch very few understands in this Island. . . . they alsoe wish you had been larger in the Virtues of them. . . .[101]

Sloane, Jamaica, Sept. 19, 1716, about governmental changes and family matters (Rose was a kinsman of Sloane), *Sloane 4044*, fol. 222; and Peter Collinson to Sloane, London, June 2, 1734, regarding a Jamaican sloth and other animals received, *Sloane 4053*, fol. 234. It is a matter of interest to note that on Nov. 8, 1716, "Mr. Williams a black Native of Jamaica was balotted for and rejected" by the Royal Society. *C.M.*, II, 236. Who propounded Mr. Williams is not revealed. Nine members of the Council rejected him, including President Isaac Newton and Dr. Sloane.

99 Barham to Sloane, London, Oct. 21, 1717, *Sloane 4045*, fol. 55, is a brief autobiographical account; see also *D.N.B. Sloane 4072*, fols. 305–308, is "A Catalogue of Seeds, Nutts & Cones of Plants; as also some Roots, Bark, and Gums Growing in Jamaica. . . . Collected by Henry Barham at His own Cost and Charges" for Sir Hans Sloane.

100 It was published in Kingston, Jamaica, 1794, by Alexander Aikman, but erroneously attributed to Barham's son Henry Jr.

101 *Sloane 4043*, fols. 45–46v.

And he went on at considerable length to discuss his own discoveries regarding the virtues of Jamaican plants and mineral waters. He supplied Sloane with additional specimens of Jamaican flora and fauna, tested and generally confirmed many of Sloane's statements about the medicinal properties of Jamaican plants and minerals, and assisted Sloane in the completion of his two-volume *Voyage to the Islands . . . with the Natural History. . . .*[102] About 1716 Barham returned to England, settled in Chelsea, practiced medicine, and experimented with the rearing of silkworms and the manufacture of silk in England. He was evidently on very amicable terms with Sir Hans, with whom he sometimes consulted about medical cases, and he contributed Jamaican seeds to the Chelsea Physic Garden and to the royal gardens at Hampton Court.[103] Sir Hans proposed him for fellowship in the Royal Society (October 31, 1717), and he was formally admitted three weeks later. On the same day (November 21) Sir Hans read a letter describing experiments which Barham had made with the hot medicinal spring waters of Jamaica, the only mineral content of which he had found to be a sulphurous gas.[104] In December Barham read an account of a "Globe of Fire or Draco Volans" (meteorite) which he had seen fall in Jamaica about 1700 "with Remarks on the Weather, Earthquakes, etc. of that Island."[105] Early the next year, Sir Hans Sloane read Barham's description of the diseases of Negroes in Jamaica,[106] and Barham presented the Society with the skin of a galliwasp from Jamaica and the eggs from

[102] "Many extracts from Barham's Mss., relating to the medicinal properties of plants, are transcribed into Sloane's copy of his *Natural History.*" Dandy, p. 87. See also his letter to Sloane, Nov. 13, 1713, about a poisonous plant discovered by Negroes, *Sloane 4043*, fol. 208; read April 15, 22, 1714, *Journal-Book*, XI, 420–422.

[103] *Sloane 4078*, fol. 259; *4045*, fol. 29 (April 29, 1718).

[104] *Journal-Book*, XII, 186, 191, 194. Sloane's report was based upon Barham's letter of Nov. 6, 1717, describing his ingenious tests of the mineral waters. *Sloane 4045*, fols. 58–60. For other similar experiments by Barham, see *ibid.*, fols. 68–71, 77–80, 89–91.

[105] *Journal-Book*, XII, 201 (Dec. 19, 1717); *Phil. Trans.*, XXX, No. 357 (July–Sept., 1718), 837–838.

[106] *Journal-Book*, XII, 211–212 (Feb. 20, 1717/18). Based on Barham's letter of Jan. 29, 1717/18, *Sloane 4045*, fols. 90–91. Much of this merely confirms Sloane's account in his *Voyage*. Barham suggested that the yaws "was the original of the Neopolitan or French disease"; though usually incurred by copulation, it sometimes is passed on without, or so Barham insisted.

an alligator which he had dissected there.[107] In mid-April, 1718, Barham wrote to Sloane to report that he had completed the revision of his "Hortus Americanus Medicinalis" setting forth the "known vertues and experienced Qualityes as I gained them from Spaniards, Indians, and Negroes." It was intended for the popular use of planters, with English names for the vulgar Jamaicans who knew no Latin, but it was also an exercise to demonstrate to the Royal Society that "I would willingly be doeing of something and throw in my mite and not be all together a Worthless Member."[108] As further evidence, he presented the Society (May 14, 1719) with his book, *An Essay upon the Silk Worm* . . . (London, 1719), and the following February 11, 1719/20, Sir Hans Sloane read a report about the further progress of Barham's experiments in producing silk, which he argued could be as economically produced in England as in France.[109]

In the meantime, Barham, long convinced by his observations that Jamaica contained unexploited mineral deposits, won appointment as mineral superintendent of a company formed to undertake silver-mining. To oversee this enterprise, of which he had been the prime mover, he returned to Jamaica in 1720.[110] He promised Sir Hans to be diligent in making collections for him, and requested his advice about printing a history of Jamaica which he had prepared from manuscript records.[111] In the midst of delays and hopes dashed by the ultimate failure of the mining scheme, he continued to write to Sir Hans and to send him specimens and observations from Jamaica.[112] He told of earthquakes, an eclipse of the moon (June 17, 1722), a destructive hurricane (August 28, 1722),[113] a description

---

[107] *Journal-Book*, XII, 214–215 (Feb. 27, March 6, 1717/18, respectively).

[108] London, April 17, 1718, *Sloane 4045*, fols. 108–109.

[109] *Journal-Book*, XII, 314, 415–417.

[110] He wrote to Sloane soon after his arrival, bubbling with excitement about fresh symptoms of rich mines—copper, silver, and gold. He believed rumors that the Spanish had had a gold mine which they closed when the English captured Jamaica. April 17, 1720, *Sloane 4047*, fols. 337–338.

[111] Barham to Sloane, Spanish Town, Oct. 26, 1721, *Sloane 4046*, fols. 140–141.

[112] Letters of May 14, July 5, Sept. 13, 1722; Jan. 3, 1722/23; April 30, 1724; July 3, 1725, *Sloane 4046*, fols. 242–243, 260–261, 284, 325–325v; *4047*, fol. 165; *4048*, fols. 15–16.

[113] This one Sloane read to the Royal Society, Dec. 6, 1722, *Journal-Book*, XIII, 227–228.

of logwood, which he was reputed to have introduced to Jamaica, with a specimen of its branches in flower, and continued to discuss the medicinal virtues of Jamaican plants. However, when the mining venture failed, he found himself in reduced circumstances. He told Sir Hans that "I Must Now follow my Old Way of Selling Medicines to the Planters as I used to Doe & if you Know of any Apothycarry that Will Adventure that way & Will Let me Know His proposals . . . [I will send him an invoice] having Delt that Way above 20 years."[114] He was cast down further by Sloane's "admonishings on my Hortus. I know it would sell here to the planters, but if not worth printing return it to me." He said that he would be proud "to serve you in collecting specimens of plants, but feel I can do little as I am growing old and fat," and employed persons were too careless with them.[115] The following May (1726) he died. His informative history of Jamaica was never published, though the manuscript of 140 folios still lies among the *Sloane Manuscripts*[116] and some of his dried plants are still in the *Sloane Herbarium*.[117]

Dr. Rose Fuller, son of Sir Hans Sloane's stepdaughter Elizabeth Rose, went to Jamaica in 1733 and remained there until about 1756. Dr. Fuller (d. 1777) had been educated at Cambridge and Leiden and had been admitted Fellow of the Royal Society (May 4, 1732) almost a year before he had gone to Spanish Town.[118] He practiced

[114] Jan. 3, 1722/23, *Sloane 4046*, fols. 325–325v.

[115] July 3, 1725, *Sloane 4048*, fols. 15–15v.

[116] *Sloane 3918*. This is probably the copy mentioned by Henry Barham, Jr., in his letter to Sloane, May 20, 1727, *Sloane 4048*, fols. 299–300. The younger Barham, having made a fortune by medical practice and a fortunate marriage, returned to England in the 1730's. *Sloane 4053*, fol. 32.

[117] Dandy, p. 87.

[118] *Certificates, 1731–1750*, p. 25; *C.M.*, IV, 167–171. In a letter to Sloane dated at Spanish Town, March 16, 1732/33, Dr. Fuller stated that he had landed in Jamaica the first of March previously. *Sloane 4052*, fol. 299. Somehow Dr. Fuller was admitted to the Society without payment of contributions. In 1756 the Council found him in arrears to a total of £63:18:00, and voted that if he would pay forty guineas, he would be excused from further contributions, *C.M.*, IV, 170–171 (Feb. 19, 1756). By this time Dr. Fuller was probably back in England, as he was elected M.P. for New Romney in that year and served in Parliament until his death in 1777. See John and J. A. Venn, *Alumni Cantabrigiensis*, Pt. I (4 vols., Cambridge, 1927), II, 185. He had been a visitor to the Royal Society on March 16, 1726/27. *Journal-Book*, XIV, 59.

medicine in Jamaica for twenty-three years, during which time he often wrote to Sir Hans, sometimes signing himself as "your most obedient Grandson." He was acquainted with Dr. William Houstoun, the Georgia trustee-collector, whom he had probably known at Leiden.[119] He experimented with West Indian medicinal plants and occasionally sent specimens to Sir Hans, especially an ointment prepared from arum (of the family Araceae).[120] His most significant communication, however, related to astronomical observations made of a comet seen in Jamaica in January and February, 1736/37. Dr. Fuller's observations were made with the naked eye, but he promised to send others made with good instruments by one skilled in astronomy. Sir Hans reported Dr. Fuller's observations to the Royal Society (April 20, 1738), where they excited much interest because they antedated all other reports of the comet, and they were published in the *Philosophical Transactions*.[121] But those observations made with instruments in Jamaica and promised by Dr. Fuller were not forthcoming.

Dr. Fuller doubtless referred to observations made by Colin Campbell (d. 1752), of Black River, Jamaica, astronomer and friend of Edmond Halley. Halley's interest in Jamaica as a site for astronomical observations dated as far back as 1675, when his "Coadjutor," Charles Bouchar, had reported that there was "no Country better" for the purpose.[122] His interest was heightened by his voyage in 1700, when he cut a huge figure eight in the South Atlantic and saw for himself the possible advantages to be derived there for astronomy, navigation, and geography.[123] Shortly after he became Astronomer Royal (1721), a ship captain, Bartholomew Chandler, sent Halley observations of a lunar eclipse (June 17, 1722) taken in the harbor at Port Royal and they helped to correct

---

[119] He told Sloane of Houstoun's death in a letter of May 21, 1733, *Sloane 4052*, fols. 352–352v.

[120] Letter of June 4, 1734, *Sloane 4053*, fol. 227.

[121] *Journal-Book*, XVII, 241; *Phil. Trans.*, XL, No. 446 (July–Dec., 1737), 111–123. Dr. Fuller's observations were published with others from Dr. John Kearsley from Philadelphia, and from Madras, Rome, and Lisbon. All had been turned over to James Bradley for comparison and study.

[122] See Chapter 7 above, pp. 233–234.

[123] Raymond P. Stearns, "The Course of Capt. Edmond Halley in the Year 1700," *Annals of Science*, I (1936), 294–301.

knowledge of the longitude of the place.[124] Similar observations made by Joseph Harris at Vera Cruz, revised and published by Halley in the *Philosophical Transactions*,[125] corrected the position of Vera Cruz. Colin Campbell entered the picture shortly afterward, although his antecedents are unclear. On November 12, 1730, Halley proposed him for membership in the Royal Society. He was elected the following December 10 but, being in Jamaica at the time, was not formally admitted Fellow until November 7, 1734, when he visited London.[126] On that occasion he also presented the Society with a few of the natural products of Jamaica.[127] But his principal contributions related to astronomy, for which he was said to have furnished himself "with an Apparatus of Instruments not unworthy the Observatory of a Prince."[128] He was associated with Joseph Harris in some of the latter's observations in Jamaica, and Harris, who was obliged to return to England in 1732 because of ill health, reported to Professor James Bradley, the Oxford astronomer, that "Mr. Campbell hinted it some [that] if something should be said in favour of what we have done, it would be of Some sanction to him in that part of the World where astronomy is in but little Esteem."[129] The following year Professor Bradley communicated a paper to the Royal Society entitled "An Account of some Observations made in London, by Mr. George Graham and at Black River in Jamaica by Colin Campbell Esqre.

[124] *Guard-Book*, C-2, No. 50; *Letter-Book*, XVI, 315–317. Captain Chandler had a five-foot telescope and an old but reliable watch.

[125] XXXV, No. 401 (Jan.–March, 1728), 388–389. Harris gave data collected during an eclipse of the sun on March 11, 1727/28. He also communicated "An Account of some Magnetical Observations made in the Months of May, June and July, 1732, in the Atlantick or Western Ocean; as also the Description of a Water-Spout," *ibid.*, XXXVIII, No. 428 (April–June, 1733), 75–79. The "Magnetical Observations" had been made in Jamaica; the water spout was seen near the coast of Florida. George Graham, F.R.S., received the report and submitted it for the *Philosophical Transactions*.

[126] *Journal-Book*, XIV, 513, 528; XV, 343–346; XVI, 17–18; *C.M.*, III, 78. Besides Halley, Dr. James Campbell and Stephen Hales recommended Campbell to the Society. Was he a kinsman of Dr. James Campbell, F.R.S.?

[127] A hummingbird and two shells, Nov. 14, 21, 1734, *Journal-Book*, XVI, 27, 34.

[128] *Phil. Trans.*, XXXVIII, No. 432 (April–June, 1734), 303.

[129] George Graham to Bradley, London, July 15, 1732, and Joseph Harris to Bradley, London, July 3, 1732, *Bradley 44*, fols. 6–7, 12.

concerning the going of a clock in order to determine the difference between the lengths of Isochronal [isochronous] Pendulums
in those Places."[130] The clock, built by George Graham in London
and carefully tested and adjusted for the different temperature
range of Jamaica, had been sent to Colin Campbell, who tested
it at Black River. The principal object of the experiment was to
test the validity of Newton's assertion that the force of gravity
diminishes in proportion to the distance from the equator (*Principia*, Bk. III, Prop. XX). Campbell found that the greater heat of
Jamaica lengthened the pendulum and slowed the clock eight or
nine seconds per day; but beyond this there was a further retardation by 1′ 58″ unaccounted for except by the diminution of gravity
at Jamaica, a result that accorded exactly with Newton's calculations.

Campbell's astronomical endeavors in Jamaica were made in close
conjunction with Halley, the principal objectives being "to make
Observations to complete the Catalogue of the Southern Constellations." After Halley's death (1742), Campbell sold his equipment
to Alexander Macfarlane of Jamaica, who continued the work.[131]
Macfarlane (d. 1755) was proposed as a candidate for the Royal
Society (November 20, 1746) by James Short, the English astronomer, and others, and elected February 19, 1746/47.[132] Evidently
he never appeared at the Society for formal admission, as no record
of it appears. He sent to Professor Bradley observations on an
eclipse of the moon, a transit of Mercury, and "32 Days of the
Comet all wch are plain & Easy";[133] and he sent similar observations
to James Short, who laid them before the Royal Society on November 15, 1744.[134] Such observations, when compared with others
made in London, enabled John Catlin and James Short to calculate
afresh the longitudinal differences between London and Kingston,

[130] *Journal-Book*, XV, 343–346 (Dec. 6, 1733); *Phil. Trans.*, XXXVIII, No. 432
(April–June, 1734), 302–314.
[131] Was this the same Colin Campbell who was licensed as a priest for Jamaica
in 1751? See Manross, comp., *Fulham Papers*, pp. 315, 333.
[132] *Certificates, 1731–1750*, p. 341; *Journal-Book*, XX, 152–153.
[133] Macfarlane to Bradley, April 21, 1744, *Bradley 46*, fols. 11–13.
[134] *Journal-Book*, XIX, 309–310. The observations of the comet were Dec. 26–
Feb. 24, 1743/44; on the lunar eclipse, Oct. 21, 1743; and on the transit of Venus,
Oct. 25, 1743.

Jamaica, as they reported in a paper presented to the Royal Society on November 1, 1750.[135]

Dr. Patrick Browne's *Civil and Natural History of Jamaica* (London, 1756; 2nd ed., 1769) brought the natural history of Jamaica to its finest development before the end of the Old Colonial Era.[136] Dr. Browne (c. 1720–90) was born in Ireland. He visited relatives in Antigua in 1737, but ill health compelled his return to England. Subsequently, he studied botany and related subjects at Paris, and then moved to Leiden, where he won the doctor's degree in medicine in 1743. During his years of study in Europe, he became acquainted with some of the principal scholars in natural history, especially with Johann Friedrich Gronovius, the Leiden botanist, and Carolus Linnaeus (Karl von Linné), the great systematist of natural history and father of modern systematic botany, with whom he maintained a scientific correspondence until his death. He practiced medicine in London for two years and then moved to the West Indies, first to Antigua and other sugar islands, then to Jamaica, where he continued to practice medicine and studied the geology and botany of the island. In 1751 he read a paper to the Royal Society about the medicinal properties of "worm grass" or pinkroot (*Spigelia marilandica*), a vermifuge, and early the following year Peter Collinson presented to the Society another account by Dr. Browne, this time relating to the geology and mineral waters of Montserrat.[137] In 1755 he published a new map of Jamaica, and the next year the first edition of the *Civil and Natural History of Jamaica*, a work anticipated by his friends to be "much superior to that of Sloane."[138]

The *Civil and Natural History of Jamaica* was, scientifically speaking, the most up-to-date treatment of the natural history of any of the British colonies in America before the end of the Old

135 *Ibid.*, XXI, 403.

136 Omitted from the narrative are Dr. Thomas Cockburn's account of his treatment of a Negro woman stunned by lightning, Aug. 14, 1736 (read Dec. 23, 1736, *Letter-Book*, XXIII, 117–119; *Journal-Book*, XVII, 14–16), and James Traill's collections of seeds and plants at Guanaboa, Jamaica, for Philip Miller of the Chelsea Physic Garden. Traill to Miller, July 1, 1737, *Letter-Book*, XXIV, 25–27.

137 *Journal-Book*, XXI, 710–711 (Nov. 7, 1751); XXII, 19–22 (Jan. 9, 1752).

138 Dr. Peter Ascanius to Linnaeus, London, April 7, 1755, in Sir James Edward Smith, ed., *A Selection of the Correspondence of Linnaeus* (2 vols., London, 1821), II, 480.

Colonial Era. Dr. Browne was familiar with the works of nearly all of the scholars of his day and several of them—notably Linnaeus, Stephen Hales, and Gronovius—were subscribers to his book. Thus, in addition to Ray and Tournefort of the previous generation, Dr. Browne's knowledge embraced Linnaeus' new sexual system of plant classification (though not the binomial system of nomenclature, the full explication of which had not reached Dr. Browne before his own book went to press);[139] the pioneer work in plant physiology as represented by Stephen Hales's *Vegetable Staticks* (1727); Catesby's work, together with George Edwards' *A Natural History of Uncommon Birds, And of Some other Rare and Undescribed Animals* . . . (4 vols., 1743–51—a useful compilation, but not based upon firsthand field observations by the author); and Peter Artedi's new classification of fishes as published posthumously by his friend Linnaeus.[140] In the arrangement of shellfish, however, Dr. Browne, for want of fresher guidance, followed Dr. Martin Lister's work, modified by his own "New and easy Method of classifying native Fossils; in which they are disposed according to their concurring Properties."[141] And though he mentioned John Ellis, he could not accept the new conclusion that sponges and corals were animals, and continued to list them as plants. The *Civil and Natural History of Jamaica* included a map of the island, a map of Port Royal, a map of Kingston, and fifty engravings of plants, insects, fishes, and one animal. The illustrations, though not colored as Catesby's had been, were made (with the exception of five relating to fauna) from designs drawn by Georg Dionysius Ehret, previously noted as the pre-eminent botanical artist of the mid-eighteenth century, whose works were generally such a fine compromise between the demands of scientific accuracy and the artistic requirements of design that both botanists and artists have united in praise of them.[142]

[139] The second edition of Browne's work supplied Linnaean names.

[140] Dr. Browne referred to Artedi (1705–35) as "that happy *genius*" and arranged his fishes in accord with Artedi's system. *Civil and Natural History*, p. 440.

[141] *Ibid.*, p. 35.

[142] Wilfrid Blunt and William T. Stearn, *The Art of Botanical Illustration* (3rd ed., London, 1955), Chap. 12, "The Age of Ehret," pp. 143ff. The plates in neither Browne's *Jamaica* nor Hughes's *Barbados* (see above) represent the best of Ehret's work, and on balancing them against Catesby's colored plates it is difficult to say that they are better.

SCIENCE IN THE BRITISH COLONIES OF AMERICA

Dr. Browne devoted fewer than sixty of his 490 pages to the civil history of Jamaica. Part II of the book, which treated natural history, was divided into three books. The first consisted of eight sections dealing with geography and geology—very brief treatment of waters, salts, bituminous deposits, metals, soils, clays, marl, and miscellaneous minerals. Book II, of about 230 pages, was devoted to the flora of Jamaica, arranged "nearly" according to the new sexual system of classification of Linnaeus. Dr. Browne claimed that whereas Sir Hans Sloane had collected about 800 species of Jamaican plants, he had found above 1,200. He included an excellent account of the culture of sugar cane, an essay which argued that Jamaican coffee, if handled properly, could compete with Arabic sources of supply, and an informative description of the cultivation and processing of indigo. Book III, with more than 100 pages, was devoted to fauna, including the most complete treatments of insects and fishes to date, and a good, but somewhat less impressive, account of birds. As the plates used in the production of the first edition were destroyed by fire, the second edition (1769) was published without plates, with the addition, however, of an index giving Linnaean nomenclature. Dr. Browne returned to Ireland after the publication of the *Civil and Natural History of Jamaica* and he subsequently published a "Catalogue of the Birds of Ireland" classified after the Linnaean system, and a "Catalogue of Fishes observed on our Coasts, and in our lakes and rivers." At his death in 1790, he left unpublished manuscript copies of a "Catalogue of the Plants now growing in the Sugar Islands" and a "Catalogue of such Irish Plants as have been observed by the author, chiefly those of the Counties of Mayo and Galway." His extensive herbarium of Jamaican plants was purchased in 1758 by Linnaeus and is preserved among the latter's collections. It became the source of much of Linnaeus' study of West Indian specimens as well as that of his student, Daniel Carl Solander, soon to become so popular in England.[143]

After Dr. Browne, the most significant works[144] relating to the

143 Roy Anthony Rauschenberg, "Daniel Carl Solander, Naturalist on the 'Endeavour,'" *Transactions of the American Philosophical Society*, n.s., LVIII, Pt. 8 (1968), 14.

144 Dr. Anthony Robinson (d. 1768), an ingenious Jamaican physician, was a competent botanist who added to the lists of nondescript plants and made

natural history of the West Indies were those of D. Nicholaas Joseph Jacquin (1727–1817). Born in Leiden, he studied with Dr. Gerhard van Swieten (later Baron van Swieten) at Vienna, and, upon completion of his studies, was dispatched by the Emperor Francis I (husband of Maria Theresa) to the West Indies and Central and South America to procure new plants and specimens for the newly constructed gardens at Schönbrunn. He embarked on his voyage in 1753 and spent five years in the Caribbean area, visiting Jamaica, Santo Domingo, Curaçao, and the South and Central American continent. He collected a sizable herbarium, but when it was damaged by ants he supplemented his notes and descriptions of new species with watercolor drawings, in the preparation of which he was unusually able. Upon his return to Vienna, Jacquin was appointed Director of the imperial gardens at Schönbrunn and Vienna. He left in 1763 to become Professor of Chemistry at Chemnitz, near Dresden, but returned after five years to become Professor of Botany and Chemistry and Director of the botanic garden at the University of Vienna. He remained in Vienna the remainder of his life, and was made Baron Jacquin in 1806. He was a prodigious writer and illustrator of botanical works, but only two of his publications pertained to his American collections. The first was a small book entitled *Enumeratio Systematica Plantarum Quas In Insulis Caribaeis vicinaque Americes continente detexit novas, aut jam cognitas emendavit* (Leiden, 1760), which included flora hitherto nondescript and not included in Linnaeus' works. The other, *Selectarum Stirpium Americanarum Historia . . .* (Vienna, 1763), included descriptions of plants from Jamaica, Santo Domingo, Martinique, and other Caribbean areas, all classified and named after the Linnaean methods, embracing items from the works of Dr. Patrick Browne, Mark Catesby, Gronovius, Linnaeus, Petiver, Plukenet, Father Plumier, Sir Hans Sloane, Tournefort, Dillenius, and many others. These were followed by 183 folio plates illustrat-

corrections to those described by Sloane and by Browne. But his sadly prepared scraps of manuscripts were never published in his own day, and his principal contributions lay in the discovery of a new vegetable blue dye and, more important, the process of manufacturing a vegetable soap from the juice of the American aloe leaf, a soap that was equally miscible both in salt and in fresh water and therefore of great value to mariners. See Frank Cundall, "Dr. Anthony Robinson, of Jamaica," *Journal of Botany*, LX (1922), 49–52.

ing the plants—plates made by Jacquin himself and, in some copies, hand colored. This volume was reprinted in 1781 and in 1788. In 1780, however, Jacquin prepared a deluxe edition of this work, limited to twelve copies (and therefore excessively rare), with original watercolor illustrations throughout. As these were made from the original watercolor drawings Jacquin had made during his travels in the Caribbean, they were the equivalent of type specimens and accordingly possess high scientific value, though their artistic worth is relatively small. Jacquin's herbarium of American plants, though somewhat damaged, contained descriptive notes of much value, and it still lies in Vienna. Under his direction, the Schönbrunn gardens became one of the most celebrated of the late eighteenth century, with huge greenhouses to protect the plants. A French traveler related that "Tropical birds, flying among the palm-trees, bamboos and sugar-canes, gave the illusion that the visitor had been transported into the heart of America."[145] Certainly Jacquin added materially to the knowledge of Caribbean plants, and he carried the study of American flora into eastern Europe, where it had lain relatively dormant since the time of Charles de L'Escluse.

After 1760 the flow of productive scientific exchange between Jamaica and the homeland diminished markedly.[146] The colonial wars of the mid-eighteenth century, the new imperial designs for colonial government after the Treaty of Paris in 1763, and the American War for Independence led to a greater concentration of British naval and military forces in the West Indies, especially at Jamaica. With these forces went a number of military surgeons and

[145] Blunt and Stearn, *Art of Botanical Illustration*, pp. 156–158. Jacquin's principal works are listed in *Biographie Universelle* (55 vols., Paris, 1811–53), Vol. XX. See also A. H. R. Grisebach, *Flora of the British West Indies Islands* (London, 1864), pp. vii, 4, 16.

[146] Exceptions were few and of little import. The most significant were as follows. On March 10, 1763, Emanuel Mendes da Costa, Librarian of the Royal Society, reported a letter from Andrew Peter Dupont of Jamaica about a gelatinous fish called the "Pope fish," or ruff, with a drawing and descripton by Robert Long of Jamaica. *Journal-Book*, XXV, 42; *Phil. Trans.*, LIII (1763), 57–58. On March 20, 1766, John Williamson related "A New Method of treating the Opisthotonos and Tetanus" as practiced in Jamaica by Dr. M. J. Marx and certified by Dr. Lionel Chalmers of Charles Town, South Carolina. It consisted of the use of mercurial unguent and opium so as "to produce a spitting." *Journal-Book*, XXVI, 403–406.

physicians associated with the army and navy, and a few merchants, some of whom took a passing interest in Jamaican or other West Indian flora and medicinal plants. Before the end of the eighteenth century, eight of these men were elected Fellows of the Royal Society.[147] But few of them made significant contributions to scientific knowledge; they did little or nothing (with one exception) to promote science in the New World; and in most cases their election to the Royal Society reflected adversely upon the good judgment of the Society. Moreover, they were transients in the New World, and while three of them later developed into eminent scientists in England, they properly belonged to the English scene rather than the colonial, and in point of time they generally extended beyond the Old Colonial Era—a break-off point of no political significance in the West Indies, as none of the island

[147] David Riz (F.R.S., 1766), merchant, made no contributions to the Royal Society and was ejected in 1783 for nonpayment of dues. Dr. John Martin Butt (F.R.S., 1767; d. 1769) made one contribution before his election regarding a "vegetative fly" of Santo Domingo, supposed to have been half plant and half animal. *Journal-Book*, XXVI, 107–108 (May 24, 1764). Dr. Tesser Samuel Kuckhan (F.R.S., 1772) sent a box of preserved insects to the Society before he died in 1775. *Ibid.*, XXVIII, 357 (April 21, 1774). John Taylor (F.R.S., 1776), elevated to a baronetcy (1778), died (1786) with no record of any contributions to the Society. Dr. William Wright (1735–1819; F.R.S., 1778; admitted, 1780), eminent physician, botanist, and promoter of science, especially in Scotland, sent two botanical-medical papers to the Society, both of which were published. *Journal-Book*, XXIX, 337–338 (April 24, 1777), 344–345 (May 1, 1777); *Phil. Trans.*, LXVII, Pt. II (1777), 504–506. These were scientific descriptions, with the medical properties, of a species of *Cinchona caribæa*, L., and of *Geoffræa sponsa*, L. Dr. Wright also communicated a medical paper (May 17, 1781) to the effect that in Jamaica he found that pregnant women with smallpox generally miscarried. *Journal-Book*, XXX, 525–527. He prepared a hortus siccus. of Jamaican plants and contributed greatly to the development of natural science and medicine —a very worthy member of the Royal Society. A good sketch of him is in the *D.N.B.* Dr. John Hunter (1728–93; F.R.S., 1786) read two papers of observations in Jamaica (Dec. 20, 1787, May 1, 1788, *Journal-Book*, XXXIII, 86–91, 197), but his career as an eminent anatomist and surgeon pertained principally to the English scene. Dr. Gilbert Blane (1747–1834; F.R.S., 1785), later Bart. (1812), reported a hurricane of Oct. 10–11, 1780 (*Journal-Book*, XXX, 454–457, Feb. 22, 1781), but, like Hunter above, his career largely pertained to the English scene (see *D.N.B.*). Bryan Edwards (F.R.S., 1794) spent many years in Jamaica, where he made a fortune as a merchant. He contributed little to science, but was author of the useful *History of the British Colonies in the West Indies* (2 vols., London and Dublin, 1793; later eds., 1794, 1801). All of these men save Dr. John Hunter, with their certificates of election to the Royal Society, are reported at greater length in my *Col. F.R.S.*, pp. 178–246.

colonies joined the mainland colonies of North America in asserting independence from the homeland.

Indeed, with few exceptions, the scientific correspondents and
collectors in the West Indies as a whole tended to be birds of passage. Though many of them were able scientific men, they were
seldom permanent residents in the island colonies or part of the
indigenous social fabric there. Several of them were medical men
who, after practicing medicine in the West Indies for years, retired
to England; others were priests who similarly returned to their
homeland after a stint of service in the West Indies; still others
were ship captains, ship surgeons, army or navy medical men,
mobile by the nature of their occupations, and never part of the
island communities; and some were merchants and planters, anxious
to fill their coffers and return to the homeland. Such representa-
tives were ill-fitted socially to promote successfully the new science among the more settled residents of the islands—to develop
enclaves of scientifically oriented colonists with both interest and
ability in the sciences. Perhaps, as Dr. Thomas Hoy lamented,
"Sugar & Indico are the only Curiositys of the Jamaica Philosophers." But the promotions—and the promoters—of the Royal Society and other less organized agencies of science appear to have
made less headway in the West Indies than they did in the mainland colonies of North America. The West Indians still depended
upon the homeland at the end of the Old Colonial Era, in science
and its applications to human life as in the economic aspects of
their lives. They did not develop an indigenous group of scientific
and technological men, able to forge ahead without constant nourishment and new supplies of personnel from the mother country, to
the extent that the North American colonists were able to do.
Their enormous slave populations were wholly untrained in experimental science, and their white populations were too transient, or
too money-minded, or both, to create a community of scientists
among the permanent residents of the islands.

The Spanish Main

The real and fancied treasures of Central and South America had
caught the imagination of Englishmen since the days of the first

Elizabeth. As the early frenetic search for gold and silver gave way to a more measured demand for trade and trading privileges, the guarded monopoly of Spain continued to be a subject of English attack. But to the search for trade and trading privileges was added a search for knowledge—to learn the closely kept secrets of the Jesuits' bark and other drugs of the Spanish Main, to discover the nature of the dye-woods and other rare timbers of Central America, to become familiar with the geography and topography, the climate, and the prevailing winds, the ocean currents and the ocean floor, the flora and the fauna, the celestial map of the tropics and of the southern hemisphere—these became objectives too, sometimes allied with the utilitarian search for trade, sometimes desired for their own sakes. From its establishment in 1660, and throughout the Old Colonial Era, the Royal Society of London was ever alert for new data from the Spanish Main. Attempts to set up a scientific correspondence and exchange in the Spanish Main met with repeated rebuffs from Spanish authorities, so the Society was forced to be content with whatever crumbs of information it could gather. In 1668 the *Philosophical Transactions* published one of these early gleanings, an account of an Englishman who, disguised as "a Biscaner," penetrated Mexico in 1664 and spent two years prospecting and experimenting with the mineral ores of the region.[148] The Scottish attempts to establish a New Caledonia at Darien yielded the Society (May 31, 1699) "Part of Captain Penny Cooks Journall from the Madera Islands to New Caledonia in Darien."[149] James Petiver managed to get a few collections of plants and shells from Surinam and from Brazil (the latter from Edmond Halley's surgeon, George Alfrey, on Halley's voyage to the South Atlantic in 1700);[150] and in 1700 he learned of very promising sources of fur-

---

[148] *Phil. Trans.*, III, No. 41 (Nov. 16, 1668), 817–824. *Sloane 2752*, fols. 29–71, contains two accounts of English voyages into the Bay of Honduras "to Cutt Logwood," one in 1679, the other in 1682. But these do not appear to have been presented to the Royal Society. Robert Hooke's papers (*Ms. 1757*, Guildhall Library, London) contain "A Journal into ye South Sea by Basil Ringrose . . ." (to Darien, 1680), which was clearly presented to the Society about 1686.

[149] The journal ran from July 19 to Dec. 28, 1698. *Register-Book*, IX (1700–1713), 4–12; *Classified Papers, 1660–1740*, VII (1), No. 56; *Journal-Book*, X, 129. Read by Robert Hooke.

[150] *Sloane 3333*, fols. 9–13v, 58v, 67–68v (1696–97); *Musei Petiveriani Centuria Quarta & Quinta* (1699), p. 43.

ther collections from his friend at the Edinburgh Physic Garden, Professor James Sutherland. Sutherland wrote on March 25, 1700, that

> All the Surgeons and Apothecaries apprentices in this place are usually my Schollars at the Physick Garden and after they have served in a Shop five years they seek Occasions of going Surgeons in Ships to the East or West Indies, or other forrain place that Offers, and so spend their time abroad for the space of six or seven years before they return and set up a shop of their own. And I assure you I shall not be wanting for the future to oblidge every one of them to do you all the Service they are capable. . . .[151]

Professor Sutherland added a postscript to ask for "severall Coppies of your printed Instructions that I might give to any of my Scholars that have Occasion to travell abroad," and he gave Petiver the names of several. But Archibald Stewart, a surgeon who went to Darien in 1700, was the only one who actually made collections for Petiver in the New World.[152]

The organization of the English South Sea Company (1710), together with the commercial concessions wrung from the Spanish in the Treaty of Utrecht (1713), opened new opportunities for scientific penetration of the Spanish Main. Petiver soon was in correspondence with surgeons in the service of the South Sea Company: David Patton, surgeon on the *Elizabeth*, bound for Vera Cruz; Mr. Cooke, "Surgeon of ye South Sea Settlement at Panama"; Captain Dover, "Chief Agent to ye South Sea Company at Buenos Aires"; "Mr. Will. Toller, Surgeon at Buenos Aires"; and John Burnett, surgeon at Porto Bello and later to Philip V of Spain.[153]

---

[151] *Sloane 4063*, fols. 9, 22; see also fols. 17, 23, 26, 32, 43, 59, 101, 110. Some of the "Schollars" also collected for William Sherard.

[152] *Sloane 4063*, fols. 9, 22; *Musei Petiveriani* (1695–1703), p. 52. Petiver's Spanish connections turned up the inadequate description of the cinchona tree of Peru published in *Phil. Trans.*, XXIV, No. 290 (March–April, 1704), 1596; perhaps also the account of the Indians and Jesuit missionaries in lower California in *ibid.*, XXVI, No. 318 (Nov.–Dec., 1708), 232–240.

[153] Letters in *Sloane 3340*, fols. 66v–68, 161, 170v–173, 261–263, 275v–276v, 330v–331; *3322*, fol. 97; *4065*, fol. 285; *4072*, fol. 295. Petiver wrote to Toller at Buenos Aires, Nov. 19, 1716: "I can assure you that nothing can better or sooner recommend you to the South Sea Company's Favour or service than Communications of this Kind & especially of such Plants, Minerals &c. as relate to dying or an Medicinall use. One of the S.S. Company seeing ye Communications wch Mr.

Of all these men, however, only John Burnett added to Mr. Petiver's collections.[154]

Occasionally specimens reached the Royal Society from Surinam and the Spanish Main. Thomas Forster, F.R.S., presented to the Society a collection of insects, snakes, and lizards from Surinam in 1714.[155] Sir Hans Sloane presented a specimen of a Paraguayan herb, said to be used in place of tea, and Levinius Vincent's descriptions of Surinam frogs, reptiles, and other animals aroused much interest in the Society.[156] In 1730 Don John de Herrera sent "Astronomical Observations made at Lima and the Havannah from 1724 to 1730" to the Royal Society and sought a scientific correspondence with it,[157] and the following year Mr. Pym, South Sea Company Factor, sent observations on the satellites of Jupiter (1720–26) made by Father Bonaventura Suarez, a Jesuit in Paraguay. Father Suarez had the use of a sixteen-foot telescope, a five-foot quadrant, and a pendulum second clock.[158] Obviously the Royal Society, by the good offices of the South Sea Company's employees, was beginning to establish some scientific communications with Spanish residents in South America.

In 1722, after Petiver's death, John Burnett, now at Cartagena, opened scientific correspondence with Sir Hans Sloane, a correspondence that continued after Burnett became one of the physicians to Philip V of Spain, sometime before 1733. In 1722 Burnett

Burnet had made to me upon my printing of them immediately preferred him to be Surgeon at Porto Bello whither he is now gone. . . ." Petiver also urged Toller to become "intimate with the Jesuits of your parts yt are curious in Naturall History" and encourage them to correspond with Petiver. *Sloane 3340*, fols. 275v–276v.

154 Shells, butterflies, plants, fishes, an "Ostrich Egg," and an account of an abortive Negress—all in 1716–18. See Petiver, *Petiveriana, seu Collectanea Naturae* . . . (London, 1717), Nos. 476–486; Dandy, pp. 109–110.

155 *Journal-Book*, XII, 19 (Oct. 28, 1714).

156 *Ibid.*, XII, 106–107 (April 5, 1716). Vincent was an F.R.S. See his *Catalogus et Descriptio Animalium Volatilium, Reptilium, & Aquatilium* . . . (The Hague, 1726) and *Descriptio Pipae, Seu Bufonis Aquatici Surinamensis* . . . (Haarlem, 1726). William Sherard wrote to Richard Richardson, London, Nov. 22, 1722, that by means of the South Sea Company "I shall have (if it please God I live) what ever I desire from all places they trade to. . . ." *Ms. Radcliffe Trust C. 4*, fol. 86 (Bodleian Library).

157 *Letter-Book*, XXII, 409–411; read Nov. 12, 1730.

158 *Journal-Book*, XIV, 553–554 (Jan. 28, 1730/31).

had just been transferred to Cartagena, and he offered his services to Sloane. He recommended to Sloane an unnamed but "very Ingenious Gentleman a Spaniard in this town who is a Mathematician & Engineer to the King." This gentleman had made astronomical observations in Chile, Peru, "& most places of the Indies." He had sent them to Edmond Halley and had received no acknowledgment. He would send further observations if he felt that his efforts were appreciated, and he needed a better quadrant, which Burnett begged Sloane to send at his (Burnett's) expense.[159] He repeatedly renewed his pleas in behalf of the Spanish engineer, but without results; Halley did not see fit to acknowledge receipt of the astronomical observations (if, indeed, he had received them), and Sloane did not supply the quadrant. In the meantime, Burnett sent Sloane collections of butterflies, shells, earths, and medicinal roots and plants. The two physicians discussed medical cases that had come under their care, and Burnett petitioned Sir Hans to seek preferment in the South Sea Company. He hoped (vainly) to become a factor at Porto Bello, in a place which had fallen open, and was indignant because the company officials objected to him on the grounds that he had not been bred up in business:

> It does not follow [said Burnett] that a man who is born in a stable should be a horse; I have served Apprenticeship enough in their service to be able to Comprehend all the Mystery I find in this way of Merchandize; especially that of a ffactor, where there is an Invoice for our Government. . . . The Qualifications more necessary in a South Sea Comp's ffactor, than being an old Spainish Merch^t, are, a generous temper & liberall Education, knowing when to do & receive benefits.[160]

In fact, Burnett was disappointed in Cartagena, and he felt that company factors dealt unwisely with the people there. The city was run down, had never recovered from the French capture of it twenty-five years earlier, and the people, mostly Negroes and mulattoes, were poor. The company's factors, having neither the "experience nor rhetorick" for it, failed to win the personal regard

---

[159] Burnett to Sloane, Cartagena, April 6, 1722, *Sloane 4046*, fols. 227–228. Other letters by Burnett to Sloane are in *ibid.*, fol. 288; *4047*, fols. 29–30, 164, 198, 323–324, 329–330v, 333–334; *4048*, fols. 26–26v, 70, 120; *4049*, fols. 266–268, 314–314v.

[160] Aug. 6, 1723, *Sloane 4047*, fols. 29–30.

of the people there and did not make clear the *mutual* advantages of their commerce with England. He urged Sloane to arrange for the South Sea Company, in conjunction with the Royal Society, to send "me missionary (as the Jesuits do) from this to Porto Bello, Panama, Lima, Potosi, & so home by way of Buenos-Ayres, makeing what observations I am capable of both with respect to trade & commerce & with respect to the Naturall History of these Countreys. . . . Provided the Comp^y & Royall Society made it worth while, by a Sallary answerable to such an undertaking: the Late Court of Directors had such a notion and did design it. . . ."[161] But nothing came of this suggestion.

In the meantime, Burnett established a philosophical correspondence with a Spanish Franciscan missionary at Santa Fe, "a very ingenious man & somewhat of a Vertuoso who has promised me everything wch that Country affords & which I shall transmitt to you. . . ." He continued to send Sloane occasional collections, and Sloane responded with medicines, medical news from England (especially news about the success of inoculation for smallpox and a new method of operating for the stone), and all the new English books and pamphlets "worth reading." In October, 1727, Burnett wrote from Jamaica, having been ordered out of Cartagena when the Spanish king threatened war because of alleged misdoings of the South Sea Company, contrary to the commercial treaties between Spain and England. He turned up in Paris in 1728 "in pursuit of an affair of consequence to me that is pending at the Court of Madrid."[162] He requested a few letters of recommendation to Sloane's Paris friends. Early the next year, he was in Jamaica in a state of uncertainty. His next letter, in April, 1733, was from Seville, where he now served as one of the physicians of "his C: [atholic] Majestys bed-chamber." In this new capacity Burnett served as a lead into the affairs of the Spanish court. In 1734 he sent Sloane "a Copy of the Institutions of a New Accadamy lately Established in Madrid, in Imitation of that of London & Paris, of which I have the honor to be a Member & doubt not by your favour & protection, to convince them, that they are Severall years behind hand with the learned part of the World, not for want of Genius but

---

[161] March 17, 1724/25, *ibid.*, fol. 329.
[162] *Sloane 4050*, fol. 21.

want of application & practice."[163] Sloane sent the new Spanish Academy packets of the *Philosophical Transactions* and, at Burnett's suggestion, Ephraim Chamber's new *Cyclopedia* . . . (2 vols., 1728), and other works.[164] And Burnett countered with the works of the enlightened Spanish monk, Benito Jeronimo Feijoo y Montenegro, and other Spanish authors. Perhaps his most treasured gift to Sloane, however, was a hortus siccus "of all the Herbs in Spain" collected under the direction of the new Spanish Academy.[165]

Other communications from the Spanish Main reached the Royal Society more directly. Late in 1731, "Mr. John Gray of the Navy Office, Author of a treatise on Gunnery," was recommended for election to the Society. Gray was elected and admitted early in 1732,[166] and in 1736, "now residing at Carthagena in the Spanish West Indies," he sent the Society "An Account of the Peruvian or Jesuits Bark . . . extracted from some Papers given him by Mr. William Arrot, a Scotch Surgeon, who had gather'd it at the Place where it grows in Peru."[167] Arrot had learned of four kinds of Peruvian bark: *colorado* (red), *amarylla* (yellow), *crespilla* (curling), and *blanca* (white); but he had actually found only two, the yellow and the white. He believed the other two were variations produced by differences in soil and climate. He reported that the best bark grew high in the mountains near the city of Loxa (now Loja, in Ecuador), about 110 leagues southeast of Guayaquil. It was best when cut between September and November and transported at once to lower altitudes to dry. He expected it to become

---

[163] Madrid, Dec. 10, 1734, *Sloane 4053*, fols. 363–364.

[164] July 2, Oct. 10, 1736 (N.S.), *Sloane 4054*, fols. 266–267, 314–315v. Burnett wrote that "all books must be Inspected into by the Inquisition, which occasions Some Small delay."

[165] *Ibid.*, fols. 266–267.

[166] *Certificates, 1731–1750*, p. 21; elected March 16, 1731/32, admitted March 22 following.

[167] *Journal-Book*, XVI, 366–369 (Oct. 28, 1736); *Phil. Trans.*, XL, No. 446 (July–Dec., 1737), 81–86. Communicated by Philip Miller. Arrot's information, as reported by Gray, closely approximated in point of time and in point of fact that gathered by the French botanist, Joseph de Jussieu, in Peru. But Jussieu was the better botanist and presented a more complete botanical account based upon direct observations of the cinchona tree. See M. L. Duran-Reyals, *The Fever Bark Tree* (Garden City, N.Y., 1946), and L. Suppan, "Three Centuries of Chinchona," *Proceedings of the Celebration of the 300th Anniversary of the First Recognized Use of Chinchona* (St. Louis, 1931).

increasingly scarce, because it was harvested only by Indians, who, by the cruelties and hard usage of the Spaniards, were approaching extinction. Gray further reported: "He [Arrot] could not tell me by what Artifice or Stratagem the *Jesuits* have got this Bark to be called after them, if not that they carried it first into *Europe*, and gave themselves out as the first Discoverers of its Virtues: But . . . its Qualities and Use were known by the Indians before any Spaniard came among them. . . ." Later in the same year Gray reported, via George Graham, experiments he had made with a thermometer and Graham's isochronous clock in Cartagena—a repeat of the experiment made by Colin Campbell in Jamaica in 1733 to test Newton's theory regarding the diminution of the force of gravity near the equator and with similar results, though Gray found the retardation of the clock to be 2' 11" at Cartagena as opposed to 1' 58" at Jamaica.[168] Gray also enclosed meteorological observations made at Cartagena between October 27 and November 12, 1734. He lamented that Halley had promised him astronomical information and, as this had not been forthcoming, Gray then "expected to have got some from Mr. Campbell at Jamaica, as also the Loan of some Instruments. When I was there, I wrote him twice, but had no answer; so that I came here without either Telescope, Meridan-Instrument, or the places of any Stars: by which means I have not been able to observe any of the Eclipses of Jupiter's Satellites, as I promised Mr. Bradley, for which I am not a little concerned." A year later (July 29, 1737) Gray wrote to Philip Miller. He was returning Graham's clock, sent further meteorological observations from Cartagena, and sent money to purchase two thermometers, promising to make regular observations at Cartagena with one, while a friend sixty leagues away would do the same with the other. He also promised Miller to collect some seeds and plants of the ipecacuanha which Miller had requested.[169]

Captain William Walker of Shenston, near Lichfield, Staffordshire, was elected Fellow of the Society on February 9, 1737/38,

[168] Gray to Graham, Cartagena, July 22, 1736, *Letter-Book*, XXIII, 21–26; read Nov. 11, 1736, *Journal-Book*, XVI, 379–380.

[169] *Letter-Book*, XXIV, 412–414; read June 8, 1738, *Journal-Book*, XVII, 281–282. This was Gray's last recorded letter, but in 1756 the Council of the Society ordered copies of the *Philosophical Transactions* sent to him as they had not yet been delivered. *C.M.*, IV, 180 (Nov. 18, 1756).

after having recently presented to the Society "a large parcell" of natural curiosities from the Spanish Main, together with "a curious piece of Hydrography being a plan of the Bay of Honduras." Captain Walker, who died in 1743, was said to be a gentleman "well versed in Mathematics, Astronomy, Mechanicks, especially the art of Ship-building, and in Natural Knowledge." He had been at sea for thirty years, during which time he had "made several curious observations in the Art of Navigation, and in Hydrography, and hath been a dilligent collector of Natural Curiosities."[170] And on January 21, 1741/42, Dr. Joseph Atwell communicated from Captain Edward Legge (1710–47), F.R.S., the latter's observations of a lunar eclipse (December 2, 1740) at St. Catherine's Isle off the coast of Brazil, with some corrections regarding the longitude of places along the east coast of South America.[171] Captain Legge, of H.M.S. *Severn*, who had been a Fellow of the Society since February 26, 1735/36, reported that the position of St. Catherine's Isle, as shown in John Senex's map of the South Atlantic, was in grave error and that these mistakes led to some of the misfortunes he and other British seamen had experienced in navigating about Cape Horn. Captain Legge accompanied George, Lord Anson, in his voyage to the Pacific in 1740–42 and sent the Royal Society a journal of his observations with variations of the compass and meteorological records.[172] He was later Commander in Chief at the Leeward Islands.

Early in 1746 the Royal Society had the good fortune to obtain Don Antonio de Ulloa's "Journal of the Observations made in Peru by the French Astronomers and himself." This was the lengthy account of the observations made by a joint French and Spanish expedition to Peru commonly known as the Condamine Expedition, after the French commander, Charles Marie de la Condamine (1701–74), a mathematical geographer.[173] It was part of a widespread effort on

[170] *Certificates, 1731–1750*, p. 139; *C.M.*, III, 310, 351; *Journal-Book*, XVII, 132.

[171] *Journal-Book*, XVIII, 324; *Phil. Trans.*, XLII, No. 462 (Jan.–Feb., 1741/42), 18–19.

[172] *Journal-Book*, XVIII, 433 (June 24, 1742).

[173] Robert Finn Erickson, "The French Academy of Science Expedition to Spanish America, 1735–1744" (unpublished Ph.D. thesis, University of Illinois, 1955), *passim*. A good popular account of the expedition, placed in its historical setting, is Tom B. Jones, *The Figure of the Earth* (Lawrence, Kan., 1967).

the part of the French Academy to discover the true shape of the world by comparing the lengths of a degree of the meridian in the arctic region, in the temperate zone, and near the equator. A French-Swedish expedition made measurements in Lapland, the French made measurements in France, and the Condamine Expedition made measurements in Peru, near the equator. As Peru was under Spanish authority, the French sought Spanish permission, the result being a joint operation: a French party with Condamine, six members of the French Academy (including Joseph de Jussieu, the botanist), and various menials to carry their instruments and supplies; and a Spanish party, with Don Antonio de Ulloa (1716–95) and Don Jorge Juan y Santacilla (1712–73), both young naval men, together with menial assistants. Philip V of Spain formally authorized the expedition in 1734. The French and Spanish parties sailed separately in 1735, met at Cartagena, and from thence went together to Porto Bello, crossed the Isthmus of Panama, and made their laborious way to Quito, where they arrived in 1737. Using Quito as a base of operations, they mapped the region, established base lines, and, measuring by triangulations, they struggled until 1744 before the work was completed. The Spanish team was recalled to Lima temporarily in 1740, when the War of Austrian Succession began, and did not return until January, 1744. After the work was completed, various members returned to Europe in their own way, some of the French staying on at Quito awaiting safer conditions, for the war still raged. Don Antonio de Ulloa sailed around Cape Horn in a French frigate. His ship was damaged by an English attack in the Atlantic, put in at Louisburg for repairs, was captured by English ships, his papers were seized, and he was taken to England, where he arrived in December, 1745. Martin Folkes, President of the Royal Society, reported to the Society on May 29, 1746, as follows:

Upon the representation of the Case to his Excellency the Right Honourable the Earl of Harrington, Secretary of State, and an Honourable Member of this Society, He, out of his great Humanity and regard to knowledge, caused the Author to be put in such a Course as to have his Papers refer'd to the President of the Royal Society, that upon a proper report of their being as represented they might be regularly restored back to him, which being accordingly done, the President at the same time procured leave to oblige the Society with the Notes he took about them.[174]

Accordingly, President Folkes "obliged" the Society with a full report on Don Antonio's notes of the Condamine Expedition. In the meantime, too, Don Antonio was proposed (May 15, 1746) as a Fellow of the Society. On December 11 following he was elected.[175] Already he had been amicably permitted to return to Spain. Thus ended, for the time being, an astonishing sequence of events during which a subject of a foreign power with which England was at war had been captured at sea, his papers seized, their contents expropriated by his captors and, before they had been tendered to his own government which had authorized the undertaking, presented to his captors' foremost scientific society, while in the meantime the enemy prisoner was freed and, in an atmosphere of apparent sincerity and mutual cordiality, elected to fellowship in the said society, and freely permitted to return to his homeland, an enemy state!

Of course the Royal Society had heard about the Condamine Expedition, and Fellows of the Society had long viewed it with a jealous eye, as if the French Academy had put one over on them in gaining such wide access to the Spanish Main—as in fact it had. The Royal Society Fellows knew something of the progress of the expedition,[176] but they itched to learn the secrets of natural history, topography, geography, meteorology, astronomy, and all else that the Condamine Expedition was likely to reveal. Don Antonio de Ulloa's journal gave them considerable insight into the places visited in South America, the painstaking methods used to measure the length of a degree of the meridian in equatorial Peru, the instruments employed, and the results of the calculations.[177] The person

---

[174] *Journal-Book*, XX, 108–120.

[175] *Certificates, 1731–1750*, p. 329.

[176] See, for example, Ludovik Godin, Astronomer Royal at Paris, to the Society, read Dec. 19, 1734, *Journal-Book*, XVI, 48ff.; XVII, 209–210 (March 9, 1738/39).

[177] The Condamine Expedition found the first degree of latitude from the equator to be 56,750 toises; measurements of a degree between Paris and Amiens (1740) revealed 57,075 toises; that of Lapland (1739) was 57,422 toises. These calculations appeared to confirm the theories of Hooke, Newton, and Huygens that the earth was flattened toward the poles. Errors in the French calculations

of Don Antonio de Ulloa was secondary to the contents of his journal, and once the latter had become available, the English were quite willing to set him free, especially since he had proved cooperative, tractable, and even personally attractive to his captors. Indeed, as subsequent events revealed, Don Antonio became a loyal Fellow of the Royal Society and was soon sending to it communications and specimens.[178]

Whether or not there was a causal relationship, the mutual regard established between Don Antonio de Ulloa and the Royal Society was soon followed by an increase in communications received by the Royal Society from the Spanish Main. On April 10, 1746, Dr. Jacob de Castro-Sarmento, F.R.S., presented the Society with a Spanish account of Paraguay and was asked to digest its contents and report them at a later meeting.[179] The following year (May 14, 1747), Dr. de Castro-Sarmento read observations made between April 12 and May 17, 1744, on a comet appearing at the town of St. Ignazio in Paraguay. The observations had been made by Father Bonaventura Suarez and sent to the Society by Dr. Matthew Sa-

tended to cancel each other so that Condamine estimated the difference in the earth's axes at 1/303.6, a figure remarkably similar to that of a later calculation in 1891, which estimated the difference at between 1/290 and 1/300. See Charles Marie de la Condamine, "Extrait des Operations Trigonométriques et des Observations Astronomiques, Fastes pour la Mesure des Degré du Méridien aux Environs de l'Equateur," *Histoire de l'Academie Royale des Sciences*, LI (Paris, 1751); "Journal du Voyage Fait par Ordre du Roi à l'Equateur . . . à la Mesure des Trois Premiers Degrées du Méridien," *ibid.*, *Supplement*, XI (Paris, 1751); Antonio de Ulloa, *A Voyage to South America* (2 vols., London, 1760). By the seizure of Ulloa's journal, the Royal Society got these basic results several years ahead of their official publication, together with the results of several astronomical observations and other data. Two informative accounts of Ulloa are: Arthur P. Whitaker, "Antonio de Ulloa," *Hispanic American Historical Review*, XV (1935), 155–194, republished in Lewis Hanke, ed., *Readings in Latin American History* (2 vols., New York, 1966), I, 257–276; and Whitaker, "Antonio de Ulloa, the *Delivrance*, and the Royal Society," *Hispanic American Historical Review*, XLVI (1966), 357–370.

178 As early as 1746, Ulloa sent a letter from Seville (dated Sept. 14) with a sample of a South American poison and an account of it. *Journal-Book*, XX, 173 (Dec. 18, 1746).

179 *Ibid.*, XX, 79. The author was not identified. Jacob de Castro-Sarmento (1691–1762) was a Portuguese who had migrated to England. He had published an article about Brazilian diamonds in *Phil. Trans.*, XXXVII, No. 421 (Oct.–Dec., 1731), 199–201.

rayva of Rio de Janeiro.[180] Dr. Sarayva also sent an account of a Brazilian balsam, together with a sample and a description of its production and uses.[181] Father Bonaventura Suarez sent further astronomical observations from Paraguay made in 1746, of a comet seen near the end of 1748, of two lunar eclipses, and other astronomical data, all of which were referred to James Bradley for further study.[182] Dr. Sarayva sent a historical account of Paraguay written in 1744 by Father Pedro Luzano, a Jesuit missionary at Cordova.[183] On October 27, 1748, Don Pedro de Maldonado, of Quito, Peru, Gentleman of the Bedchamber to the King of Spain and Governor of the province of Las Esmeraldas, Peru, being in London, was proposed as a candidate for fellowship in the Royal Society, but the gentleman died before the balloting day had arrived.[184] Don Jorge Juan and Don Antonio de Ulloa, the two Spaniards officially appointed to accompany the Condamine Expedition, presented the Society (December 15, 1748) with the first three (of a total of five) volumes of their work entitled *Relación Histórica del Viaje a la América Meridional* . . . (Madrid, 1748),[185] and the following April Don Jorge was nominated for fellowship in the Royal Society. He was elected Fellow on November 9, 1749.[186]

These evidences of a more relaxed relationship between the Royal Society and Spain came to an abrupt halt with the outbreak of the Seven Years' War. A decade after the war was over, however, the Secretary of the Royal Society, acting upon orders from the Council and with the approval of the Society at large, addressed to the Spanish Ambassador in London, Prince Masseran, himself a Fellow of the Society, the following letter:

> I am directed by the Royal Society to apply to your Excellence for some of the more uncommon Natural productions of South

[180] *Journal-Book*, XX, 267.

[181] *Ibid.*, XX, 300 (June 25, 1747).

[182] Reported Jan. 28, 1747/48, Jan. 19, 1748/49, Jan. 10, 1750/51, *ibid.*, XX, 436; XXI, 49, 486.

[183] Reported May 5, 1748, *ibid.*, XX, 508. The account treated of the Platte River area.

[184] *Certificates, 1731–1750*, p. 379; *Journal-Book*, XX, 552.

[185] *Ibid.*, XXI, 14.

[186] *Certificates, 1731–1750*, p. 397.

America, and Mexico, as it happens that many of them have not yet been accurately described by any Naturalist whatsoever.

It is entirely submitted to your Excellence Whether this request of the Royal Society should be laid before his Catholick Majesty, if it should, we can presume to do it only through your Excellence.

We take the Liberty to transmit herewith some short directions translated into Spanish and French, as possibly they may be of some use to those who may be employed in collecting Natural productions.

If we may hope that your Excellence will consider this our request as deserving the notice of his Catholick Majesty, we would more particularly desire the fish which may be caught near the Island of Iuan-Fernandez, On the coasts of Chili, of Aquapulca, and of California as also of the lakes in the neighbourhood of Mexico. We could likewise request the rarer insects throughout the whole Continent of South America, as they may be packed in a small compass.

As for the larger Animals we would rather request their Skeletons together with some of their hair.

We cannot however omit the desiring the Condor (or Great American Vultur) &c. &c. &c.

Signed, C.[harles] Morton F.R.S.[187]

This letter was delivered to the Spanish Ambassador, and on January 28, 1773, the Ambassador replied "that he would not fail to represent the same to his Royal Master; and to do his utmost to show his regard for, and to forward the purpose of the Society in every respect."[188] By July 29, 1773, the Society heard a translation of a letter from the Ambassador, dated from Aranjuez, April 30, reporting that he had laid the Society's request before His Catholic Majesty and his ministers, who replied as follows:

the Royal Society should receive from His Majesty's dominions in

[187] *Journal-Book*, XXVIII, 34–35 (Dec. 23, 1772). This "Rough Draught of a Letter from the Royal Society to the Spanish Ambassador, F.R.S." was formally approved both by the Council and by the Society at large. See also *Add. Mss. 5169D*, fols. 178, 184. Prince Masseran had sent to the Society "A short Account of the Observations of the late Transit of Venus, made in California, by Order of his Catholic Majesty . . . ," which was printed in *Phil. Trans.*, LX (1770), 549–550.

[188] *Journal-Book*, XXVIII, 55.

America any of those singular productions which they may request, and can be spared; provided that we may be permitted in
like manner to procure the Natural Curiosities of His Britannic
Majesty's Dominions, all this being very proper in order to maintain that good harmony and Correspondence so necessary to that
important end, the general instruction of the Public, by the knowledge of Natural History. The Utility of this will result to all
equally, at the same time that such mutual endeavours will contribute to adorn and compleat both Cabinets of Collections. To
effectuate this, you will be pleased to keep a correspondence
with Don Pedro Davila, director of the King my Master's
Cabinet. . . .[189]

This letter, in turn, was transmitted to the Earl of Dartmouth,
English Secretary for the Colonies, in order to seek royal approval
of the proposed exchanges with the Spanish government. The Earl
laid it before George III, who gave his assent to the plan. At this
point the Royal Society, spurred forward by its recently appointed
Committee on Natural History (originally created to take care of
the extraordinary gifts presented by the Hudson's Bay Company[190]), began to envision a broad new plan whereby collectors
chosen by the Society—and including taxidermists—would make
natural history collections in the American colonies at royal expense (estimated at £150 to £200 per year) and gather specimens
in triplicate, one each for the Spanish "Cabinet," for the recently
established British Museum, and for the Royal Society. In this
fashion it was anticipated that, by means of the exchange program
proposed by His Catholic Majesty of Spain, a considerable harvest
of the natural productions of the West Indies and of both North
and South America would flow into England.[191] Unfortunately,
however, these ambitious plans failed to materialize. The outbreak
of the American War for Independence foiled them completely.[192]

[189] C.M., VI, 179–181.

[190] See Chapter 7 above, pp. 254–256.

[191] C.M., VI, 182, 184, 186–187 (July 26, Sept. 9, Sept. 16, 1773).

[192] Don Antonio de Ulloa sent to the Society "An Observation of the Solar
Eclipse of the 24 June, 1778, on board the Spanish Ship called the *Spain*, by Don
Antonio De Ulloa, F.R.S., Commander of the Spanish Squadron from Vera Cruz."
Reported Dec. 24, 1778, *Journal-Book*, XXIX, 645–648; *Phil. Trans.*, LXIX, Pt. I
(1779), 105–119. The value of the observations was marred by uncertainty of the
ship's position, the ship's motion, and inadequate instruments.

The Royal Society had only a fleeting prospect of the realization of their broad-based plan. Whether the English government would have underwritten the expense involved never really came to the point of decision, although the Earl of Dartmouth had assured the Society (March 25, 1773) that "I am commanded to assure them [the Society] that his Majesty's Governors in the respective Provinces [of North America] shall be directed to give every assistance in the power to the person or persons who shall be entrusted with that Commission [i.e., to make natural history collections]."[193] But even though the American Revolution prevented the development of these plans, the Royal Society had succeeded, momentarily at least, and in an indirect manner, in breaching the barriers to the Spanish Main. However, had the proposed plan developed into a more permanent exchange with the Spanish Main, the Society would have been dependent upon Spanish agents, who, with the exception of some of the missionaries and the professors in the universities of Spanish America, would have been, like most of the Society's correspondents in the British West Indies, transients in the New World, unlikely to create a trained clientele of scientists among the permanent residents either of the British or the Spanish colonies.

[193] Reported April 1, 1773, *Journal-Book*, XXVIII, 108. Already the Society had arranged for a representative to go to the island of St. Vincent to collect natural productions there.

TEN

The Northern Mainland Colonies

New England

IT IS WELL KNOWN THAT THE PEOPLE of New England suffered severe spiritual trials and political upheavals during the last quarter of the seventeenth century. To a deep concern about their religious health were added Indian wars, war with the French, the machinations of Edward Randolph, attacks upon the Bay Colony charter (which was annulled in 1684), the regime of Sir Edmund Andros, political troubles in the homeland culminating in the "Glorious Revolution" of 1688, currency problems, and adjustments to the new charter of the Province of Massachusetts Bay of 1691. These events jarred New Englanders, excited deep fears, and provoked animosities which gave rise to new factions among the colonists. The Boston Philosophical Society which Increase Mather had fostered in the early 1680's suffered "a fatal and total Interruption" in consequence of the "Calamity of the Times."[1] During these same years, the Royal Society of London fell into decline, scientific communications were reduced to a low ebb, and cooperative efforts to advance experimental philosophy were temporarily immobilized.

Scientific prospects brightened, however, during the 1690's. The Royal Society recovered, thanks to Dr. Hans Sloane and others of a rising new generation of Fellows. And, although the Boston Philosophical Society was not revived, some of its former participants continued the pursuit of scientific learning, while the *com-*

---

[1] Cotton Mather, *Parentator. Memoirs of Remarkables in the Life and Death of . . . Dr. Increase Mather . . .* (Boston, 1724), p. 86. Cf. also Chapter 5 above, pp. 150ff. The intellectual impact of these troubles is best described in Perry Miller, *The New England Mind: From Colony to Province* (Cambridge, Mass., 1953).

munity of scientific interests which the Boston Philosophical Society had imparted to New Englanders was resumed and transmitted to the eighteenth century. Indeed, this community aspect of the New England scene, evident only at Boston among all the British colonies of America in the seventeenth century, was soon paralleled in Philadelphia and elsewhere, until, shortly after mid-century, it included a number of interlocking "circles" of colonial scientists.[2] These men exchanged scientific specimens, observations, information, and opinions not only with the Royal Society and other European groups and individual scientists but also—of prime significance for the ultimate development of American science—with each other. The isolation of the colonial scientist was breaking down as communities of scholars arose, giving rise to fruitful exchange and mutual encouragement. Before the end of the Old Colonial Era, these interlocking communities, interested and knowledgeable in scientific observations and experiments, spread from New England to South Carolina, with occasional tendrils of inquiry reaching into Georgia, Florida, the trans-Appalachian West, and the West Indies. Out of these movements the first American scientific societies developed, more or less in imitation of the Royal Society of London and similar overseas organizations. Thus, Henry Oldenburg's admonitions to John Winthrop, Jr., followed by the latter's donation of his telescope to Harvard College, bore fruit in American society.

Members of the newly organized Temple Coffee House Botany Club in London stimulated the re-establishment of scientific commerce between New England and the homeland. Dr. Sloane's "Catalogues of Plants" recorded receipt of forty "Seeds Rec'd from New England, 1696/7," which George London cultivated in the Fulham Palace gardens.[3] In 1696 Benjamin Bullivant, a London apothecary who ran afoul of the regulations of the Society of Apothecaries, migrated to Boston. James Petiver, his brother apothecary, saw to it that Bullivant went equipped with collector's supplies and full directions to collect, preserve, and return all pos-

2 Probably the New England town system of early settlement contributed to this community aspect of early scientific endeavor, and Harvard College supplied a focal point. This community aspect arose somewhat later in other colonies as the urban movement grew in the eighteenth century.

3 *Sloane 3343*, fol. 55.

sible plants, shrubs, insects, fishes, reptiles, and the like.[4] Before the
year had passed, Bullivant had joined forces with the pious, un-
stable, but well-connected speculator in New England mines, Mr.
Hezekiah Usher, whom he recommended to Petiver as an additional New England correspondent.[5] Together they sent Mr.
Petiver six New England butterflies (October 20, 1696), and Bullivant accompanied the collection with a letter to describe the specimens. Petiver presented both the butterflies and the letter to the
Royal Society on December 2, 1696, and was ordered "to give ye
Gentleman thanks for ye pains he had been at."[6] Petiver obeyed,
writing both to Bullivant and to Usher, January 12, 1696/97. He
urged them to send further collections of New England flora and
fauna, sent a colored drawing of one of the butterflies they had
sent, promised publication of their contributions, and sent a copy
of John Josselyn's *New-England's Rarities* as a guide for their
further collections.[7] Usher's death in 1697 terminated his part in
these enterprises,[8] but Bullivant sent a collection to Petiver, including fireflies (described as "a Glow-worm volant"), grasshoppers,
other insects, and a "Hum-bird." He included a letter (dated at
Boston, January 15, 1697/98) with an account of "some Naturall
Observations he had made in those parts."[9] Petiver showed the
letter to the Royal Society (May 25, 1698) and it was promptly
published in the *Philosophical Transactions*.[10] Bullivant's letter dis-

[4] Petiver "to Mr. Bullivant when he went for Boston," May 26, 1696, *Sloane
3332*, fols. 211–212. Cf. also *Petiver*, pp. 321ff.

[5] Usher was the brother of John Usher, Lieutenant Governor of New Hamp-
shire (1692–97 and 1704–15). He had previously (1680's) corresponded with Robert
Boyle. *Boyle Papers*, Bl. VII, No. 36 (Royal Society Archives). His revealing will
was published in *The Historical Magazine*, 2nd Ser., IV (1868), 120–126. See also
The Diary of Samuel Sewall, 5 *Coll. M.H.S.*, Vols. V–VII (1878–82), V, 456;
VI, 31.

[6] *Journal-Book*, X, 8; *Sloane 3341*, fols. 23–24.

[7] *Sloane 3332*, fols. 226v, 243–245 (Usher), 234–234v (Bullivant). Usher appears
to have made a collection of New England insects. See also *Sloane 3324*, fols.
4–12.

[8] Petiver, as yet unaware of Usher's death, wrote to him, Aug. 29, 1697, *Sloane
3333*, fols. 57–58. He also addressed Bullivant the next day, *ibid.*, fols. 31v–32v.

[9] *Letter-Book*, XI, Pt. II, 61–62; *Guard-Book*, B-2, No. 46. Bullivant described
Josselyn's book as "short but nervous."

[10] *Journal-Book*, X, 70; *Phil. Trans.*, XX, No. 240 (May, 1698), 167–168.

cussed Indian remedies, butterfly eggs "near as big as a Wrens," tortoises, grasshoppers, clams, the "Hum-bird" (which he had kept in captivity for a time, feeding it with honey), and observations on the fireflies. Petiver discussed Bullivant's observations in a letter of May 30, 1698, and sent a copy of the *Philosophical Transactions* with Bullivant's letter, together with some newssheets and a further plea for New England specimens.[11] At this point, however, the correspondence inexplicably ended.[12]

Further scientific exchange between New England and the homeland was scanty until after the War of Spanish Succession, which ended in 1713. "Mr. [Henry?] Newman of New England was permitted to be present" at the Royal Society's meeting of May 7, 1707,[13] and a few communications were subsequently received.[14] However, Thomas Brattle (1658–1713), who had associated with members of the Boston Philosophical Society, contributed astronomical observations used by Newton in his *Principia*, and visited John Flamsteed, the Astronomer Royal, at

[11] *Sloane 3333*, fols. 134v–136.

[12] Did Bullivant return to England? On April 19, 1711, Petiver reported receipt of a letter from a Mr. Bullivant in Northamptonshire. *Journal-Book*, XI, 216.

[13] *Ibid.*, XI, 116.

[14] On Feb. 1, 1709/10, Richard Waller, soon (1710) to be elected Secretary of the Royal Society, presented from an unnamed donor "a branch of New-England Tea, where it was called partridge berry and used as Tea in that Country." *Journal-Book*, XI, 184. Thomas Banister (1684–1716) sent a letter to John Chamberlayne, F.R.S., dated at Boston, Jan. 4, 1711/12, to describe a remarkable "Ball of Fire, in appearance near as big as the full moon" observed "to rise westward of Albany" and pass over New Haven. The meteorite, if such it was, had been observed in the previous summer (1711), and was said to have passed from Albany to New Haven and become lost over the Atlantic horizon in a space of about two minutes. Chamberlayne read the letter to the Society on April 24, 1712. *Letter-Book*, XIV, 322, 383–384; *Journal-Book*, XI, 287. Subsequently, Banister sent an account of the "strange effects of the Late great droughts in New-England" in 1714. *Journal-Book*, XII, 47; read March 3, 1714/15. Dr. Sloane communicated a letter from Mr. Thomas Newton of Boston to Sir Matthew Dudley relating to strange weather phenomena reported from Newcastle, Pennsylvania, on Oct. 3–17, 1715. *Ibid.*, p. 113; read May 10, 1716. One Captain Candeler submitted a report of an eclipse of the moon, March 15, 1717/18, made at Cambridge, but the observer was not identified. *Ibid.*, p. 282; read Feb. 19, 1718/19. And Governor Francis Nicholson presented to the Society a copy of Herman Moll's map of New England and the other English plantations, July 7, 1720. *Ibid.*, XIII, 33.

Greenwich in 1689, had entered into correspondence with Flamsteed.[15] Brattle had sent observations of solar eclipses made in 1694 and 1703 and of lunar eclipses made in 1700 and 1703, all of which had been reported in the *Philosophical Transactions*.[16] Further, in a letter dated October 31, 1711, Brattle wrote to John Chamberlayne to describe the incidence and mortality of smallpox in Boston.[17] Then, about eighteen months afterward (May, 1713), Thomas Brattle died. When the Royal Society learned of his death,

> Mr. [William] Derham proposed [October 29, 1713] that Mr. Brattle's papers lately dead in New-England might be endeavoured to be procured for the Society which was accordingly ordered.
>
> Mr. Chamberlayne having a correspondence with Mr. Brattle's Brother [William (1662–1717), a clergyman and sometime tutor at Harvard College] was desired to use his Interests to that end which he was pleased to promise.[18]

Chamberlayne then added a sweetener, for on December 3, 1713, at the Society,

> The President [Isaac Newton] was pleased to communicate . . . a Letter he had lately received from Mr. Chamberlayne in which he proposed that the Reverend Mr. William Brattle of Cambridge in New-England (Brother to Mr. Brattle deceased, whose Papers and Manuscripts relating to Philosophical Enquiries and observations the Society had desired in whose hands they now are) might be proposed as a Candidate to be chosen into this Society which

---

[15] *Phil. Trans.*, XXIV, No. 292 (July–Aug., 1704), 1630–38; Frederick G. Kilgour, "The Rise of Scientific Activity in Colonial New England," *Yale Journal of Biology and Medicine*, XXII (1949), 125–126. In view of his associations in New England, Brattle's complaints about his isolation and lack of encouragement are not wholly convincing. However, his charge against the state of mathematical instruction in New England was well taken. The statement (p. 126) that Brattle was "the only man in the Boston area who was engaged in science" appears too strong, although he may have been the only one actively engaged in astronomy at the moment.

[16] *Journal-Book*, XI, 121; *Phil. Trans.*, XXIV, No. 312 (Oct.–Dec., 1707), 2471–72.

[17] *Journal-Book*, XI, 275; read March 6, 1711/12. Brattle stated that smallpox epidemics had occurred in Boston in 1666, 1678, 1690, and 1702, and he predicted another in 1714—a twelve-year cycle that the Society believed "may deserve farther Inquiry." The epidemics were said to be "as mortal as the Plague here in England."

[18] *Ibid.*, XI, 381.

was referred to the consideration of the next Council. The same letter also proposed Collonel John Leveret Esqr President of Harvard College of Cambridge . . . as a Candidate which was also referred to the next Council.[19]

The Council took affirmative action on both candidates on December 7, 1713;[20] and on March 11, 1713/14, both William Brattle and President Leverett were elected Fellows of the Royal Society.[21] President Leverett (1662–1724) returned his thanks "for the respects of the Society in electing him"—although he appears to have contributed nothing to it directly. Indirectly, however, he contributed much toward the development of the scientific spirit in the New World. For as Professor Morison has related,[22] the "great Leverett," as President of Harvard (1708–24), both fostered science at the college and successfully pursued policies which endued Harvard with a liberal spirit and possibly saved it from becoming a mere sectarian institution devoted to the training of ministers in the tradition of the old, narrow, New England orthodoxy. William Brattle, however, wrote to John Chamberlayne "declining [to] accept of the Title of member of the Society as unqualified. . . ."[23] There is no evidence that the Royal Society received Thomas Brattle's philosophical papers.

Cotton Mather (1663–1728), F.R.S.

In the meantime, one of the most significant figures in early New England science had become a contributor to the Royal Society and a promoter of science in the colonies. This man was Cotton Mather, son of Increase Mather and associated with his father in the Boston Philosophical Society in the 1680's. Much has been written about Cotton Mather in recent years, although a full-scale, balanced, reasonably objective evaluation of the man still remains to be done—understandably, in view of the great mass of

[19] *Ibid.*, XI, 390.

[20] *C.M.*, II, 215.

[21] *Journal-Book*, XI, 414.

[22] Samuel Eliot Morison, *Three Centuries of Harvard, 1636–1936* (Cambridge, Mass., 1936), Chap. 4.

[23] Both Leverett's acceptance and Brattle's declination were reported to the Society on Aug. 4, 1715. *Journal-Book*, XII, 76.

materials to be digested and related to the full context of the times. Still, enough has been done to expose the enormous complexity of Cotton Mather and to demonstrate that he was neither the innocent, misunderstood, much maligned minister of God that he so frequently represented himself, nor the ignorant, hard-nosed, witchhunting bigot that his detractors would have posterity believe. Few informed persons would now deny Cotton Mather's deep, wideroaming scholarship, understanding, industry, and public-spiritedness. But his equally remarkable powers of self-delusion, self-conceit, arrogance, and passion for public recognition are distasteful. One can admire, respect, appreciate—even pity—Cotton Mather; but it is very difficult to like him or to place unalloyed trust in his motives. These complexities were further enhanced by the obvious fact that Mather demonstrated a considerable capacity for growth during his moderately long life. He was not the same man in the 1720's that he had been in the 1680's. And one of the marked aspects of this growth lay in his development as a natural philosopher.

Mather betrayed an early interest in science. Plagued by a speech defect in early youth, he questioned his fitness for the ministry and devoted much study to medicine. Later, when he had overcome his speech difficulty and had taken up the ministry, he retained a lifelong interest in the new science. No doubt this interest was sustained both by the fact that it was customary at that time for ministers to consider the cure of souls and the cure of bodies to have a vital interrelationship (and, as has been noted, many ministers practiced medicine in their flocks), and by the fact that Puritans welcomed the new science as a handmaiden of theology, as a new instrument with which to discover the mind of God. The interest stimulated in New England by the comets of 1680 and 1682 clearly excited Cotton Mather's further interest in science, and the activities of both Cotton Mather and his father during the witchcraft episodes of the next decade or so betray more than a suspicion (whatever else may be said of them) that both father and son experimented with some of the alleged victims along the lines of what would today be termed psychosomatic medicine.

Over the years Cotton Mather collected a sizable library, including many of the latest works in experimental philosophy, and when, during the controversies over inoculation for smallpox in the 1720's, he boasted that he had read more books in physic than had the physicians, he probably had. Indeed, his *Christian Philosopher* (1721) was a convincing testimonial to his wide-ranging knowledge, if not complete understanding, of the new science. Certainly, when in 1711 he began to collect accounts of American curiosities with the intent of presenting them to the Royal Society of London, he was no neophyte as a man of science.

On November 12, 1712, Mather addressed to the notorious Dr. John Woodward[24] of London the first of a long series of letters to describe American scientific curiosities. The letter was intended for presentation to the Royal Society, and between November 17 and 29 he composed thirteen such letters, seven to Woodward and six (beginning November 24) to Richard Waller, Secretary of the Royal Society.[25] These were accompanied by a covering letter, dated at Boston, December 1, 1712, in which Mather addressed Waller as follows:

> For some very good Reasons, which yet there is no need of mentioning, I have addressed certain Letters to you; and joyn'd you with my Excellent friend Dr. John Woodward in a small Collection of *Curiosa Americana*, to the making whereof I devoted near a fortnight a Little before the sailing of the fleet from hence, as far as my many other Employments would give me Leave.
>
> I could have been glad I could have provided a Richer Entertainment for you; But your Candour will accept of these mean things because they were the best, that I could in so little time as I could allow myself to be master of. I despair not that if I live, I may annually entertain you with something or other from these parts of the world that a Secretary of the Royal Society may judge it proper to take some notice of. And tho' I cannot presume

[24] Mather had had previous correspondence with Woodward, and internal evidence suggests that Woodward had solicited Mather for geological data and specimens from the New World.

[25] All of this first series are in *Guard-Book*, M-2, Nos. 21–33. Mather planned to "write one letter per Day, till I has [*sic*] passed thro' a Fortnight." He wrote none on Sunday, Nov. 23, two on Nov. 24. Cf. Frank Alexander Tredinnick, Jr., "Cotton Mather, Puritan Scientist: A Study of His *Curiosa Americana*" (unpublished M.A. thesis, Columbia University, 1947).

so much upon my own merits as to dream of being thought worthy to be admitted a member of that more than illustrious Body

Society [sic] but rather to belong to the order of the *fratres Ignorantiae*, yet I will here venture to subscribe the Promise which your Candidates make at their admission; That I will Endeavour to promote the good of the Royal Society of London for the Improvement of Natural Knowledge.

I have Entreated my Dear & Honoured Dr. [Woodward] that he would obtain an Interview with you, and there at a perusal of some Testimonies with how much Regard, I am Sir

Yours to Command,
Cotton Mather.[26]

Thus began Mather's *Curiosa Americana*. In the course of the next twelve years, additional communications swelled the total to at least eighty-two letters, most of them sent in multiples, like the first series, of four to a dozen letters at a time.[27]

---

26 *Letter-Book*, XV, 44–45.

27 Not all of the letters which made up Mather's *Curiosa Americana* are preserved. Some of them are known only by title, and, indeed, there is no agreement as to how many there were. Tredinnick depended heavily upon G. L. Kittredge, "Cotton Mather's Scientific Communications to the Royal Society," *Proceedings of the American Antiquarian Society*, n.s., XXVI (1916), 18–57, and Kittredge's other articles, "Some Lost Works of Cotton Mather," *Proc. M.H.S.*, XLV (1912), 418–479, and "Further Notes on Cotton Mather and the Royal Society," Publications of the Colonial Society of Massachusetts, *Transactions*, *1911–1913*, XIV (1913), 28–92. Holographs of Mather's letters are in *Guard-Book*, M-2, Nos. 21–56; *Classified Papers, 1660–1740*, XXIII (2) ("Inoculations"), No. 31; *Misc. Mss. M.M. 1* (Royal Society Archives); and *Sloane 4065*, fols. 255–255v. Copies of many of these, with others for which the originals are lost, are in *Letter-Book*, XV, 44–47, 383–400; XVI, 417–422, 434–436, 459–460, 495–498; XVIII, 157–159; XIX, 43–72. Transcriptions of most of these are among the Gay transcriptions in the Library of the Massachusetts Historical Society (Boston), together with Mather's own (incomplete) *Curiosa Americana* (*Ms. C. 61.2.6*), the unpublished *Biblia Americana* (6 vols.), from which Mather culled materials for the *Curiosa*, and other Mather materials. The original manuscript of *The Angel of Bethesda* is in the Library of the American Antiquarian Society (Worcester, Mass.), together with a considerable number of the original volumes from Mather's library. Still other Mather items are in the Houghton Library of Harvard University. Mather's own list of the letters in the *Curiosa* is in Mather's letter to Jurin, May 21, 1723 (*Letter-Book*, XVI, 417–422), and in Petiver's commonplace book, *Sloane 3340*, fols. 277ff. Still further letters in the *Curiosa* series are noted in the minutes of the meetings of the Royal Society and will be cited in appropriate places below.

About six months passed before Mather's earliest communications were placed before the Royal Society. Waller produced the covering letter on July 16, 1713, the other letters were cursorily examined, and the Society ordered Waller to cultivate Mather's correspondence, and "to draw up a Letter in answer to his against the next meeting."[28] Waller's letter, dated July 22, was read and approved by the Society on July 23, and it explained the delay in recognizing Mather's communications of the previous November and December:

It is with all the appearance of Justice that you may think I ought to be ashamed to own the Receipt of your obliging Letter dated December ye 1st 1712, After so long an Omission of returning you a suitable answer, and [I] must beg the favour of your Accepting for my Excuse an Account of the Truth of the matter that caused the Delay. Sr, your letter to me mentioning the Curiosa Americana, I apprehended at first that they were some Collections of Natural Raritys of New England designed for the Royal Society's Repository till our common friend Dr. Woodward inform'd me that the Curiosa Americana were several Observations you had made and communicated in diverse Letters you had bin pleased to write partly to himself and the rest to me. The Manuscript of which I had not the happiness of receiving from him 'till the 15th Instant. When to my Surprize I found myself obliged in so great Debt that at present I must own myself unable to make a tolerable satisfaction, and like other unhappy insolvents, desire time for Repayment. I have not since I received the Manuscript MSS had time and Leisure to read over all your Letters nor consider duely of all that curious Collection of Discoverys and Observations you have therein communicated. Some few of them I read at the next meeting of the Royal Society who were very well pleased and Entertained with them, and I was ordered by the first oportunity to return the Thanks of the Society to you for your Information and for the offer you are pleased to make of future obligations of the like nature: which as they will be extreamly acceptable to the Society; so I shall in particular think myself happy in cultivating a Philosophical Correspondence with so candid and Learned a person. Sr, I tooke the same opportunity of proposing you to be a member of the

28 *Journal-Book*, XI, 374.

Society which was received to the satisfaction of the members present. But there being a formality requisite of being allowed of by the Council of the Society and of being chosen in full assembly of the Fellows which at this time of the year can hardly be expected, I doubt [not?] your Election will be deferred till after Michaelmas, when I need not question but I shall be able to send you an account of your being chosen: At which Time I shall not fail (God willing) to send you a Letter which I hope to render more Acceptable by some Informations of several matters of a philosophicall Nature, which have lately bin brought before the Royal Society and shall from time to time communicate such Discoverys or Observations. Esteeming my self happy in being any ways instrumental in promoting an Intercourse of such Intelligence between the old and new world and for the present desire you would please to believe me Sir

Your most obliged and humble Servant

Ric. Waller R.S. Secr.[29]

Actually the Council met on July 27, 1713, and approved Mather's candidacy as a Fellow of the Society.[30] But because of the customary "long vacation," the Society did not meet again until after Michaelmas. It was more than a year before Mather's first series of *Curiosa Americana* was presented to the Society. Secretary Waller prepared extracts of the letters, which were read to the Society on December 3 and 10, 1713, and the members of the Society raised queries and made observations of their own regarding them.[31] On December 23 Waller addressed a letter to Mather including "Several Remarks on a Manuscript of Letters sent by Dr. Mather." It was read to and approved by the Society on January 7, 1713/14.[32] Waller assured Mather that he had been elected Fellow of the Society, although

> the Ceremony of an actual Admission is wanting which you being beyond the Sea cannot be performed, and you being a natural

[29] *Guard-Book*, W-3, No. 75; *Letter-Book*, XV, 45–47; XVI, 34–35; *Journal-Book*, XI, 377 (July 23, 1713).

[30] *C.M.*, II, 211.

[31] "An Extract of several Letters from Mr. Cotton Mather to John Woodward M.D. and Richard Waller esq. S.R. Secr.," *Letter-Book*, XV, 47–59; *Journal-Book*, XI, 392, 395.

[32] *Guard-Book*, W-3, No. 76; *Letter-Book*, XV, 59–68; XVI, 46–47; *Journal-Book*, XI, 401.

408

born subject cannot be inserted among the Gentlemen of other Nations in the printed List, which is the reason that your name is not printed in the List herewith sent you. The Chief reason of printing these Lists is for ye convenience of the Fellows present in choosing the Council, President and officers for the ensuing year [and Mather, not having formal admission, which required his presence at the Society, had no vote].[33]

Waller's "Extract of several Letters from Cotton Mather, D.D. to John Woodward, M.D. and Richard Waller, Esq; S.R. Secr." was subsequently published in the *Philosophical Transactions*.[34]

Mather already had resumed communications on December 1, 1713. This "second series" of the *Curiosa Americana*, completed during December and January, 1713–14, was followed by "a Description of the Moose-Deer," dated June 21, 1714, and written in response to Secretary Waller's "particular Enquiries after the Moose in our Country."[35] In 1715 Mather sent Waller the manuscript of a volume entitled "The Christian Virtuoso," but as Waller had died before its arrival the Royal Society turned it over to the Reverend Mr. Jean Théophile Desaguliers "to look over it & give an Account of it which he promised to do."[36] This was the volume later published under the title *The Christian Philosopher: A Collection of the Best Discoveries in Nature, with Religious Improvements* (London, 1721), certainly one of Mather's major testimonies as a man of science. A third series of the *Curiosa Americana*, written to Woodward between July 2 and 13, 1716, followed, augmented by two more letters in September, a singularly important one to James Petiver (September 24) regarding a botanical experi-

---

[33] See Chapter 4 above, pp. 110–111.

[34] XXIX, No. 339 (April–June, 1714), 62–71.

[35] *Letter-Book*, XV, 122–124. The "Description" was largely based upon the accounts of others, especially that of Josselyn. Read to the Society Oct. 29, 1714. The balance of Mather's "second series" were read Jan. 28, Feb. 6, 1713/14, and Oct. 28, Nov. 4, 1714. *Journal-Book*, XI, 405–407; XII, 20, 22. Part of Mather's letter of Dec. 1, 1713, was published in William Derham, ed., *Philosophical Experiments and Observations of . . . Dr. Robert Hooke . . .* (London, 1726), p. 386.

[36] *Journal-Book*, XII, 62 (May 19, 1715). A cryptic note among Sloane's papers suggests that he sent Mather a variety of scientific publications on April 15, 1715, including John Ray's "English Herball of 600 plants." *Sloane 3331*, fol. 25.

ment performed at Roxbury, accompanied by a volume of dried plant specimens, and one to Dr. Woodward regarding "A Singular Limestone."[37] Between July 25 and December 14, 1717, Mather sent four more letters, together with a collection of fossils, to Dr. Woodward. Then, in spite of Mather's promise to continue his communications "annually," the series was interrupted until 1720. In that year Mather, having heard a false rumor that Woodward was dead, addressed a dozen letters to John Chamberlayne, who turned them over to Newton, who in turn gave them to Secretary Edmond Halley for a report to the Society. Twelve more communications to Chamberlayne arrived in 1721, and William Derham presented to the Society two pamphlets describing an aurora seen in New England on December 11, 1719, one by Mather, the other by Thomas Robie (1689–1729), a tutor at Harvard College who later turned to medicine.[38] The next two years were largely devoted to the controversy about inoculation for smallpox in Boston. Mather sought to justify his part in the promotion of inoculation (which will be discussed below) and to proclaim the successes of the bold experiment. He wrote a letter to Dr. Woodward (March 10, 1722/23) on the subject and ten letters to Dr. James Jurin (1684–1750), successor to Edmond Halley as Secretary of the Royal Society and an enthusiastic supporter of smallpox inoculation and careful student of its effects.[39] Only the first of Mather's letters to Jurin (May 21, 1723) referred to the smallpox controversy, a petulant cry against those who had attacked him. The others (June 3–August 3) continued the *Curiosa Americana* with three letters of introduction interspersed. Of the latter, the first (June 7) praised the work of the inventive and persistent Reverend Mr. Joseph Morgan (1671–c. 1745), then at Freehold, New Jersey, and urged the Royal Society to encourage his plans for an "Engine to Sail on

[37] Reported to the Society Dec. 20, 1716, Jan. 10, 17, 1716/17. *Journal-Book*, XII, 141–143. On March 10, 1715/16, Mather wrote a fulsome letter to Waller proposing to dedicate his unpublished *Biblia Americana* to him. *Misc. Mss. M.M. 1*.

[38] *Journal-Book*, XIII, 69 (March 9, 1720/21). Both pamphlets had been printed at Boston. Robie's was entitled *A Letter to a Certain Gentleman . . . of a Wonderful Meteor that Appeared in New-England, on Decemb. 11, 1719 . . .* (Boston, 1719). Further attention will be given to Robie below.

[39] Raymond P. Stearns, "Remarks upon the Introduction of Inoculation for Smallpox in England," *Bulletin of the History of Medicine*, XXIV (1950), 103–122.

the Land."[40] The second introduced to Jurin the bearer, Mr. Isaac Greenwood (d. 1745), who subsequently studied in London with Jurin and others before his appointment as the first Hollis Professor of Mathematics and Natural Philosophy at Harvard College.[41] And the third (August 3) presented to Jurin "Mr. John Perkins . . . a Physician of great learning & Wisdom & Success, and Inferior to none that we have in this Country," who had gone to London to receive a bequest that had befallen him.[42] Three meetings of the Royal Society (December 5, 12, 19, 1723) were almost wholly taken up with consideration of Mather's communications of 1723.[43] During 1724 Mather sent twelve more letters, addressed alternately to Drs. Woodward and Jurin, in September, early October, and mid-December.[44] In one of these (October 5), Mather told Jurin that he had ready for press another book, *The Angel of Bethesda: An Essay Upon the Common Maladies of Mankind . . .* , and hoped that Jurin would wish him well in its publication—a vain hope, as events turned out, for *The Angel*, Mather's last substantial scientific work, remains unpublished today, although an expert and appreciative evaluation, together with portions of the manuscript, were published in 1954.[45] In the last of these letters (December 15), Mather wrote to Dr. Jurin to introduce Zabdiel Boylston, the Boston inoculator and Mather's especially dear friend: "Yea [wrote Mather] perhaps the prince and princesse themselves if informed of such a one coming to London may not be unwilling to take some cognisance of a person so distinguished by an operation of so much consequence [i.e., Boylston's successful inoculations for smallpox

[40] For a full account of this busy, somewhat pathetic, dabbler in science, see Whitfield J. Bell, Jr., "The Reverend Mr. Joseph Morgan, an American Correspondent of the Royal Society, 1732–1739," *Proceedings of the American Philosophical Society*, XCV (1951), 254–261, and pp. 497–499 below. For Mather's letter, see *Guard-Book*, M-2, No. 41 (June 7, 1723).

[41] Mather spoke warmly of Greenwood, a Harvard graduate and supporter of inoculation for smallpox in Boston. *Ibid.*, No. 44 (June 10, 1723).

[42] *Ibid.*, No. 45 (Aug. 3, 1723). Perkins was Mather's next-door neighbor in Boston.

[43] *Journal-Book*, XIII, 327–338.

[44] Read to the Society March 25, 1725/26, April 22, 26, 29, May 6, 1726. *Ibid.*, XIII, 464, 474, 476, 480.

[45] Otho T. Beall, Jr., and Richard Shryock, *Cotton Mather: First Significant Figure in American Medicine* (Baltimore, 1954).

in Boston]."[46] Thus ended the *Curiosa Americana*.[47] There was merit, if not dignity, in Mather's rhetorical question to Dr. Jurin (October 25, 1724): "I pray tell me: Have you many Correspondents whose Remittances are more than this your poor American's?"

Professor Otho T. Beall, Jr., has demonstrated that many of the subjects treated in Mather's *Curiosa Americana*, especially those of the first series (1712), were culled from the collections of the Boston Philosophical Society, dating from the 1680's.[48] Others were taken from his unpublished *Biblia Americana*, a huge commentary on the Bible, in which Mather sought to reconcile revelation with the new science and suggest "scientific" explanations of biblical accounts. This work, upon which Mather labored for upwards of twenty years, was on the point of completion when he wrote the first letter of the *Curiosa* to Dr. Woodward in 1712. In fact, he used it as an introduction to his first subject, the discovery near Albany, New York, in 1705, of the bones of an antediluvian "giant." Thus, although some of the phenomena described in the *Curiosa* had been observed directly by Mather himself, others he had learned about from his friends and correspondents, and some of the material was several years old before Mather communicated it to the Royal Society. The sources of the data illustrate the community aspect of the scientific scene in New England as well as the fact that Mather, in his vast output of correspondence and other literary works, often used and reused the same materials.[49] He consciously patterned his letters in the *Curiosa* after those in the *Miscellanea Curiosa Medico-Physica Academiae Naturae Curiosorum, sive Ephemeridum Medico-Physicarum Germanicarum Curiosarum*, the yearbook of the Collegium Naturae Curiosum, a German philosophical society whose publication he had seen in the Harvard Col-

[46] *Letter-Book*, XVIII, 157–159.

[47] I omit consideration of a letter to Jurin, Nov. 27, 1727, with which Mather enclosed three copies of his remarks on the earthquake which shook New England, Oct. 29, 1727. *Journal-Book*, XIV, 274 (Dec. 19, 1728).

[48] "Cotton Mather's Early 'Curiosa Americana' and the Boston Philosophical Society of 1683," *Wm. & Mary Qty.*, 3rd Ser., XVIII (1961), 360–372.

[49] Mather's bibliographer lists 478 published works. These do not include a considerable array of unpublished books and correspondence. See Thomas J. Holmes, *Cotton Mather: A Bibliography of His Works* (3 vols., Cambridge, Mass., 1940).

lege Library.[50] With care for his style, often mingling wit and
fulsome remarks to the addressee, he usually stated his subject in
theoretical terms with a learned discussion of the background and
previous opinions, then plunged into the specific information about
the subject and abruptly concluded. He clung to this pattern
throughout, although the later letters tended to become somewhat
shorter than the earlier ones.

The *Curiosa Americana* treated a wide range of subjects. The
modern fields of anthropology, astronomy, biology, entomology,
geology, mathematics, meteorology, ornithology, and zoology were
represented, but the most common fields were medicine and its
allies, botany, psychology, and philosophy. To be sure, Mather included accounts of monsters and other things which demonstrated
the credulity and superstition still common in his day. Modern
commentators have often scornfully dismissed such materials as
"unscientific," thereby tossing out the baby with the bath water;
for they overlook the long struggles with credulity and superstition
out of which modern science has developed. More recently, wiser
students have pointed out that the subjects with which Mather
concerned himself were in keeping with those of the overwhelming
majority of his contemporaries, certainly with those being presented to the Royal Society and often published in the *Philosophical Transactions*. However, this is not to say that the Fellows of
the Royal Society uniformly accepted such items without criticism
—or even appreciated fully all the contributions which Mather
offered to the new science. The reactions of the Society to Mather's
first series of communications in 1712 was clearly a mixed response.
Mather's account of meteors observed in New England was most
acceptable, and the Society solicited further astronomical data. His
description of a medicine extracted from the gall bladder of a rat-

[50] Originally founded (1652) as the Gesellschaft naturforschenter Aerzte, at
Schweinfurt, the Collegium published *Miscellanea Curiosa Medico-Physica Academiae Naturae Curiosorum, sive Ephemeridum Medico-Physicarum Germanicarum Curiosarum* (Leipzig, 1670–80); it was continued as *Miscellanea Curiosa,
sive ephemeridum medico-physicarum Germanicum Academiae . . .* (Nuremberg,
1682–1706); and as *Academiae Caesaro-Leopoldinae (Carolinae) Naturae Curiosorum Ephemerides . . .* (Frankfurt and Leipzig, 1712–22). The *Nova Acta
Leopoldina* ultimately succeeded (Halle, 1934——).

tlesnake was "perfectly new" and much appreciated. His brief account of American medicinal plants, derived largely from Indian medicinal practices, was highly acceptable, although the Society indicated a preference for a more systematic catalogue and description of plants unknown in England, and James Petiver was directed to instruct Mather in the performance of this work. Mather's account of the flights of wild pigeons and his speculations (borrowed from Charles Morton) that migratory fowl flew to an unknown satellite of the world were greeted with much criticism. Although the flights of pigeons were called "incredible," the Society preferred descriptions of the birds, a knowledge of their species, and a figure of one of them (in fact, figures of American birds were much desired, especially that of the wild turkey), and the notion that migratory fowl go to some undiscovered satellite of the earth was dismissed as a "particular fancy." With regard to the "giant" unearthed near Albany, the Society doubted that the bones and teeth found were of human origin and criticized Mather for his failure to send "an exact figure" of them. It is significant, however, that neither Mather nor the Society even suggested that they might be formed stones. As to monstrous births, the Society expressed skepticism, "begging pardon of the Reverend Gentleman that attested it" —although Secretary Waller went on to relate examples of monstrous births in England without offering any further explanation. To Mather's descriptions of the power of certain "antipathies" or imaginations to affect the health of a person or, as in the case of a pregnant woman, to leave a physical mark upon her child, the Society stated that these things more likely proceeded from the imagination of others! Indeed, the Society was most unreceptive to Mather's ventures into psychosomatic medicine and dismissed them airily with the statement that "these accounts relate little to Natural Philosophy"—an attitude exhibited later toward Mather's *Angel of Bethesda*.[51]

[51] Mather had referred to the problems of "what operations of the Invisible World, there have been in communicating the *Knowledge of Medicine* unto us," and pointed out that Galen himself had learned from a dream how to cure a malady upon himself. The Royal Society's reactions are in "Some Notes and Querys drawn up by Mr. Waller upon the Reverend Mr. Cotton Mather's Letter wrote to Dr. Woodward and Mr. Waller." *Guard-Book*, W-3, No. 77; *Letter-Book*, XV, 61–68.

Until 1723 few of Mather's communications to the Royal Society excited much comment, including some that have won wide praise from twentieth-century commentators, which suggests that occasionally Mather was either failing adequately to communicate his thoughts to his English contemporaries or that the full import of some of his observations fell on fallow ground in the Royal Society.[52] In general, the letters of the *Curiosa Americana* were received with polite thanks and usually noted in the minutes of the Society's meetings. But few of them, even those which more recently have been considered significant contributions to science, were published in the *Philosophical Transactions*. When in 1717 Mather described a monster brought forth by a cow, the Society's comment was that "the draught thereof annexed seemed not sufficiently to answer the description." In the same year the Society noted that Mr. Mather, after "giving a Collection of all relations about Tritons or Maremen which he had met with in his reading," and "concluding with an Account of what had been lately seen on the Coast of New England by three credible witnesses," the description "seem'd such that it might well enough be a great seal which at a distance might deceive their sight."[53] Again, in 1723, the Society remarked dryly that Dr. Mather, after consulting "a Multitude of Authorities from Antient Writers, Historians, Poets, and Naturalists" with reference to the existence of monstrous reptiles, introduced an account of "a Supposed Snake of Monstrous size," the skeleton of which recently had been found by Virginian workmen digging a trench near the James River.[54] This implied criticism was noted at the first of three consecutive meetings of the Society devoted almost exclusively to Mather's communications of 1723, evidently prompted principally by the consideration of smallpox

[52] There is evidence to suggest both that some Fellows of the Society tended to reject scientific comments originating from an overseas Dissenter parson unknown to them personally and that the Society as a whole was beginning to receive communications from the colonies with a certain arrogant disdain, an attitude understandable, perhaps, in view of the frequent disappointments the Fellows had experienced in their long promotion of the new science in the colonies. Mather's self-conceit shone through in some of his letters to the Society, which may have had an effect upon their reception.

[53] *Journal-Book*, XII, 141, 143 (Jan. 10, 17, 1716/17).

[54] *Ibid.*, XIII, 327–330 (Dec. 5, 1723).

inoculation and reports of its remarkable success in New England.

More recent commentators have found greater merit in some of Mather's communications to the Royal Society. Mather himself denigrated his abilities in ornithology, but recent scholars in this field find that two of his descriptions of pigeons, especially his account of the flights of passenger pigeons (1712) and his relation of "The Nidification of Pigeons" (1716), though not fully appreciated at the time, contained observations of considerable scientific value.[55] Again, though Mather stated that he had had little leisure for the study of botany, he said, "I honour the Study," and in his attempts to identify plants mentioned in the Bible, he had become familiar with the works not only of the Ancients but also those of Gerard, Bauhin, Johnson, Parkinson, Tournefort, and Ray—an impressive list! Moreover, in his efforts to identify American plants, particularly those which the Indians had shown to possess "truly stupendous" medicinal virtues, he was led to oversee botanical experiments at Roxbury which pioneered in the description of plant hybridization. Professor Conway Zirkle, in his book *The Beginnings of Plant Hybridization* (Philadelphia, 1935), published Mather's letter to James Petiver, dated September 24, 1716, as "the earliest account of plant hybridization yet found."[56] Mather was familiar with Nehemiah Grew's hypothesis that flowering plants reproduce sexually and the recognition of the role of pollen in plant reproduction. His account to Petiver of the experiments ran as follows:

> In a Field not far from the City of *Boston,* there were lately made these Two Experiments.
>
> First: my Friend planted a Row of *Indian Corn* that was Coloured Red and Blue; the rest of the Field being planted with corn of the yellow, which is the most usual colour. To the Windward side, this Red and Blue Row, so infected Three or Four whole Rows, as to communicate the same Colour unto them; and part of ye Fifth, and some of ye Sixth. But to the Lee-ward Side,

[55] Arlie W. Schorger, "Unpublished Manuscripts by Cotton Mather on the Passenger Pigeon," *The Auk,* n.s., LV (1938), 471–477. F. T. Lewis, "The Passenger Pigeon as Observed by Cotton Mather," *ibid.,* n.s., LXII (1945), 306–307. See also Mather, *The Christian Philosopher,* Essay XXX, pp. 180ff.

[56] Pp. 104–107. The original of Mather's letter is in *Sloane 4065,* fols. 255–255v. He published a more formal account of the same observations in *The Christian Philosopher,* Essay XXVI, pp. 124–125.

no less than Seven or Eight Rows, had ye same Colour communicated unto them; and some small Impressions were made on those that were yet further off.

Secondly: The same Friend had his garden ever now and then Robbed of the Squashes, which were growing there. To inflict a pretty little punishment on the Theeves, he planted some *Gourds* among the Squashes (which are in aspect very like 'em) at certain places which he distinguished with a private mark, that he might not be himself imposed upon. By this method, the Thieves were deceived, & discovered, & ridiculed. But yet the honest man saved himself no squashes by ye Trick; for they were so infected and Embittered by the Guords, that there was no eating of them.

This account of the hybridization of Indian corn and squash went unnoticed until Professor Zirkle brought it to the attention of geneticists in the 1930's.

Last, in reviewing Cotton Mather's contributions to science, attention must be directed to his medical work. His role in the bold experiments made in Boston in 1721–22 to try the merits of inoculation for smallpox has been so frequently discussed that it calls for no more than a short summary in this place.[57] Inoculation for smallpox had been a folk practice in the Orient and Africa, and perhaps elsewhere, for centuries. Dr. Martin Lister and others of the Royal Society had knowledge of it from China by 1700. Cotton Mather heard about it from the Negro slave, Onesimus, who had been presented to him by his parishoners in 1707. Onesimus had been inoculated in Africa before being sold into slavery. His account aroused Mather's interest, and he questioned other blacks in Boston, several of whom confirmed the widespread practice of inoculation in Africa. The matter was discussed by the Royal Society in October, 1713, and on May 27, 1714, Dr. John Woodward communicated to the Society a Latin letter from Dr. Emanuel Timonius, F.R.S., dated from Constantinople, December, 1713, with a full account of the medical technique of inoculation and a favorable estimate of its

---

[57] Probably the best account is that of John B. Blake, "The Inoculation Controversy in Boston: 1721–1722," *New England Quarterly*, XXV (1952), 489–506. See also Solon S. Bernstein, "Smallpox and Variolation: Their Historical Significance in the American Colonies," *Journal of the Mount Sinai Hospital, New York*, XVIII (1951), 228–245; and Stearns, "Remarks upon the Introduction of Inoculation for Smallpox in England," *loc. cit.*

results. This was published in the *Philosophical Transactions* in the early summer of 1714, followed by a confirmation from Jacobus Pylinarus from Venice, published in the same journal early in 1717. In the meantime, William Sherard, now Consul at Aleppo in Smyrna, one in whom the Fellows of the Royal Society held full confidence, having been questioned by Secretary Waller, wrote that "a man that should doubt of the truth thereof at Smyrna would pass for a Grand Heretick." Mather saw the articles in the *Philosophical Transactions*, and he wrote to Dr. Woodward in 1716 that "if I should live to see *Small-Pox* again enter our City, I would immediately procure a Consult of our Physicians, to introduce a Practice [of inoculation] which may be of so very happy a Tendency."[58] In April, 1721, when a smallpox epidemic struck Boston, this was exactly what he did. As a whole, however, the physicians did not respond. After a delay of some weeks, Zabdiel Boylston, persuaded by Mather with the backing of several other ministers (including his aged father, Increase Mather), undertook to try the experiment. On June 26 Boylston inoculated his six-year-old son and two of his slaves. When these cases appeared to prove successful, he inoculated others in July. Boylston's acts raised up "an horrid Clamour." Popular opposition, backed by the opinions of the majority of Boston physicians, was immediately aroused: what could be more dangerous than deliberately to infect healthy persons with such a serious

[58] Quoted in Kittredge, "Some Lost Works," p. 422. Cf. also Reginald H. Fitz, "Zabdiel Boylston, Inoculator, and the Epidemic of Smallpox in Boston in 1721," *Bulletin of the Johns Hopkins Hospital*, XXII (1911), 315–327. Dr. William Douglass stated in 1721 and again in 1722 that he had lent Mather the *Philosophical Transactions* articles about inoculation. Because of Mather's previous relations with Fellows of the Royal Society, and his statement to Secretary Waller in 1716, two years before Douglass' arrival in Boston, I doubt that Dr. Douglass was the *original* source of Mather's familiarity with the articles of Timonius and Pylinarus. Mather had ample access to them before Dr. Douglass ever saw Boston, and his commonplace books in the American Antiquarian Society Library show evidence that he had read the *Philosophical Transactions* with care. However, the matter is arguable. Dr. Douglass, with all his early prejudices about Boston and "a certain credulous Preacher of this place called Mather," was an able and estimable man. See Douglass to Alexander Stuart, Sept. 25, 1721, *Guard-Book*, D-2, No. 2; and to Cadwallader Colden, May 1, 1722, in *The Letters and Papers of Cadwallader Colden* (*Collections of the New York Historical Society*, Vols. L–LVI, 1917–23; Vols. LXVII–LXVIII, 1934–37; cited hereafter as *Colden Papers*), I, 141–144.

disease? It was considered an offense against God and man, and, as the physicians rightly pointed out, the practice might spread the infection. Bitter charges were hurled back and forth, scurrilous pamphlets were published in defense of both sides of the controversy, the newssheets entered the fray, the Boston Selectmen ordered Boylston to desist (July 21)—an order which Boylston defied —and the entire community was thrown into turmoil. Mather and Boylston, of course, had their sympathizers, and a few other physicians, notably Thomas Robie, also inoculated persons with smallpox. But the majority of the populace, led by the majority of the local physicians, violently opposed the practice, and Mather and Boylston were inevitably the objects of their fury. By November the opposition had grown so violent that someone hurled a lighted bomb into Cotton Mather's house with a message saying, "Cotton Mather, You Dog, Dam you: I'll inoculate you with this, with a Pox to you," but the bomb failed to explode. Dr. William Douglass, a Scottish newcomer to Boston (he had arrived in 1718) and the only Boston physician with an earned academic degree in medicine, emerged as the leader of the opposition. Subsequently Douglass became persuaded of the merits of inoculation and practiced it himself, but for the moment, though he had been on friendly terms with Mather, he became one of the latter's most vigorous and knowledgeable opponents. Like Mather, Douglass was able and filled with self-pride, and in addition he had a British imperialist's low opinion of the state of colonial culture and a disdain for colonial parsons, who in his opinion meddled in affairs outside their proper, ministerial spheres. Obviously he underrated Cotton Mather.

Unfortunately, the controversy quickly degenerated on both sides to the level of name-calling and personal attacks, with less and less attention to the merits or demerits of inoculation for smallpox. Unfortunately, too, these attacks, as Perry Miller has shown,[59] left deep scars upon some of the protagonists, driving a wedge between Mather and Douglass, two of New England's ablest men of science. Both factions called for support from their friends and correspondents in neighboring colonies and in the Old World, thus helping to spread the controversy beyond New England. In the course of these sorry developments, John Checkley (1680–1753), a Boston apothe-

[59] *New England Mind*, pp. 345–366.

cary, having already raised Mather's ire by his support of Anglicanism and having temporarily won Mather's nephew, Thomas Walter, to the Church of England, espoused the cause of the anti-inoculators. Noting that Mather's name had been omitted from the printed lists of the Royal Society, although he had repeatedly written "F.R.S." after his name, Checkley called into question Mather's right to pass himself off as a Fellow, and wrote to the Society to ascertain the facts.[60] Meanwhile, it was whispered about among Mather's opponents in New England that Mather had long posed improperly as a Fellow of the Royal Society. The charge cut Mather deeply, and he fired off a long, almost childish, letter to Dr. Jurin (May 21, 1723), reviewing the events surrounding his election in 1713, his many contributions to the Society, the "more than three hundred" publications to his name, the honors bestowed upon him by European universities (including a doctorate in theology), and ending sorrowfully with

> if it be the pleasure of those Honourable Persons who compose or govern the Royal Society, that I should lay aside my pretensions to be at all related unto that Illustrious Body, upon the least signification of it, by your Pen, it shall be dutifully complied withall. I will only continue to take the leave of still communicating Annually to you (as long as I live) what Curiosa Americana I can become the Possessor of . . . it is not the Title, but the Service, that is the Heighth, & indeed the whole, of my Ambition.[61]

[60] June 15, 1721, *Journal-Book*, XIII, 112: "A Letter from Dr. John Checkley of Boston in New-England to the Secretary [Jurin] desiring a certificate under hand, of the either being or not being of Mr. Cotton Mather of the same place, a Member of this Society. This being occasioned by the Omission of Mr. Mathers name in the Printed Lists, Mr. Thomas who takes care of those Lists answered that he had order not to print in the Lists the name of such Englishmen as had not actually been admitted in person." The case was not without parallel. Cf. *Col. F.R.S.*, pp. 199–203.

[61] Mather to Jurin, May 21, 1723, *Letter-Book*, XVI, 417–422. There is something anomalous in the record, however. Woodward is reported to have said (*Journal-Book*, XIII, 280) that although Mather had been recommended for fellowship, and approved by the Council "in order to be balloted for in the Society, which was never done; he therefore desired that the said Gentleman might be now balloted for, which being granted Dr. Cotton Mather was Elected a Fellow." Earlier records of the Society, cited above, appear to show conclusively that Mather had been recommended, approved by the Council, *and*

Mather had been vindicated by the Society before he wrote this letter. On April 11, 1723, Dr. John Woodward had advised the President of Mather's election "many years ago" and it was confirmed by ballot on this date.[62] In 1727 the Society drafted a new statute to clarify election proceedings and obviate the possibility of such misunderstanding as had arisen about Mather's election.

The inoculation controversy had passed the point of its highest intensity before this time. The Boston epidemic had tapered off in the winter of 1721–22 and disappeared by the spring of 1722. During its incidence nearly half of the people of Boston, 5,889 persons, had suffered the smallpox "in the natural way." Of these, 844 had died, a mortality rate of about 15 percent. During the same time, Boylston had inoculated 242 persons, of whom six died—a mortality rate of about 2.5 percent. In other words, as Boylston showed, the danger of death by smallpox was only about one-sixth as great in inoculation as in natural infection. Thus, on the basis of this pragmatic demonstration, Cotton Mather and Zabdiel Boylston were wholly justified in their bold experiment.[63] Still, their enemies were not entirely silenced. Dr. William Douglass was too good a physician to overlook the results, and he adopted the practice of inoculation for the future. But there *had* been danger in inoculating healthy persons without isolating them from other people who were not known to be immune to the disease until the patient was fully

balloted for in 1713–14. Waller's letter of Dec. 23, 1712, approved by the Society Jan. 7, 1712/13, stated flatly that Mather had been fully elected, although "the Ceremony of an actual Admission is wanting" until Mather could appear in person at a meeting of the Society. James Petiver gave Mather the same information, April 6, 1715. *Sloane 3340*, fols. 155-157. I have found no specific mention in the minutes of 1712-13 for the final ballot—but this would not be the first error or omission found in the minutes. I incline to accept the word of Waller and Petiver that a final ballot had been taken. Certainly Mather had every right to consider himself a Fellow on the basis of information received. The entire matter was reviewed from Checkley's standpoint in "Was Cotton Mather a Fellow of the Royal Society?" *The Nation*, LIV, No. 1390 (Feb. 18, 1892), 127-128; and more soundly considered by G. L. Kittredge, "Cotton Mather's Election into the Royal Society," *Publications of the Colonial Society of Massachusetts*, XIV (1912), 81-114.

[62] *Col. F.R.S.*, pp. 184-185.

[63] In the main, I follow Beall and Shryock, *Cotton Mather*, in this recapitulation. For another account, written from a wholly different point of view, see Miller, *New England Mind*, Chap. 21.

recovered from the effects of artificial infection. This precaution was learned during the controversy, and public law made it mandatory in 1722. Religious objections—the argument that inoculation interfered with the will of God—were not stilled, however, and continue to the present day against the newer, safer technique of vaccination.

Both Mather and Boylston reported to England the results of the inoculation experience in New England. The reports made a favorable impression in many circles, especially in the Royal Society, where Dr. Jurin, the Secretary, began a systematic collection of the results in statistical form. The Boston experience proved to be an influential pioneering effort, going far to sustain and extend the practice in England (from whence, for the most part, it spread to the Continent of Europe) and in other colonies. In time the practice began to prevent smallpox by inoculations in the absence of an epidemic. Even Edward Jenner's safer "vaccination," discovered in 1798, grew out of the earlier inoculation procedure. In the meantime, throughout the eighteenth century, the Boston experience was cited over and over as evidence in support of the practice. If inoculation did not wipe out smallpox in the eighteenth century, it was the first major step in its successful control, and went far to calm the terror which smallpox epidemics had so often aroused in the peoples of Europe and America.

It has been suggested that, in view of the dangers involved and the widespread ignorance of the causes and nature of smallpox at the time, Cotton Mather was lucky in the ultimate outcome of the Boston experiment in inoculation. In a sense this is true. Still, he had not promoted the experiment in blind ignorance, for besides the materials published in the *Philosophical Transactions*, he was familiar with the success of the practice in Africa and the Orient. Knowing of these things, and unlike Dr. Douglass and the majority of the Boston physicians, Mather was open-minded enough to take a calculated risk for the public welfare. As such, his promotion of smallpox inoculation, far from being the act of an ignorant, credulous minister intent upon the maintenance of the status of the aristocracy of the cloth, was rather inspired both by *medical* evidence and by a far-reaching humanitarianism.

Beall and Shryock, in their book *Cotton Mather: First Significant*

Figure in American Medicine, have made the first expert analysis of Mather's medical knowledge and opinions. As regards the smallpox, they point out that as early as 1721 Mather had begun to suspect that the external cause of the disease would be found in pathogenic animalculae which could be seen in the pustules by means of a microscope. These microscopic "animals," he believed, entered the body through the respiratory passages, and so won direct access to the blood and the vital organs.[64] To this advanced view of the nature of the disease and its spread, Mather also set forth his answer to the query, "Why Inoculation giveth a more kindly Small-Pox than common Infection." His hypothesis—the only American one set forth at the time—was that by inoculation the animalculae attacked only the "Out-Works" of the body (usually the arm), "and at a considerable Distance from the Center of it." This enabled the body to marshal its full defenses before the disease could seize upon vital organs and "oblige him [the "Invader"] to march out the same Way he came in, and are sure of never being troubled with him any more."[65] The insistence that the virus consisted of living organisms, together with a plausible explanation of the success of artificial immunization, established Mather, in the words of Beall and Shryock, as "certainly the outstanding medical thinker as well as the moral leader throughout the epidemic of 1721."[66]

Mather advanced the animalculae theory as the cause of diseases other than smallpox. In a burst of uncommonly good prose in *The Angel of Bethesda,* he wrote:

> Every part of Matter is peopled. Every green Leaf swarms with Inhabitants. The Surfaces of Animals are covered with other Animals. Yea, the most solid Bodies, even Marble itself, have innumerable Cells, which are crouded with imperceptible Inmates. As there are infinite Numbers of these, which the Microscopes bring to our View, so there may be inconceivable Myriads yett smaller than these, which no Glasses have yett reach'd unto. The Animals that are much more than thousands of times less than the finest Grain of Sand, have their Motions; and so, their Muscles, their Tendons, their Fibres, their Blood, and the Eggs wherein

[64] *Cotton Mather,* p. 113.
[65] *Ibid.,* p. 182.
[66] *Ibid.,* p. 117.

their Propagation is carried on. The Eggs of these Insects (and why not the living Insects too!) may insinuate themselves by the Air, and with our Ailments, yea, thro' the Pores of our skin; and soon gett into the Juices of our Bodies. They may be convey'd into our Fluids, with the Nourishment which we received, even before we were born; and may ly dormant until the Vessels are grown more capable of bringing them into their Figures and Vigour for Operations. . . . It may be so, that one species of these Animals may offend in one Way, another in another, and the various Parts may be variously offended: from whence may flow a variety of Diseases. . . .[67]

This hypothesis, together with Mather's clinical approach to medicine (occasionally reminiscent of the opinions of Descartes and the works of the great Sydenham, both of whom Mather cited), his interest in genetics and the problems of generation as evidenced in his account of the pollination of plants, his pioneering concern for psychiatry and the humane treatment of mental cases, his discriminating accounts of pigeons and their nesting habits, his tendency to reject astrology, both in medicine and astronomy,[68] and his unabashed promotion of the new science to both the clergy and the laity, placed Mather in the vanguard of the colonial scientists of his day. The familiarity which he exhibited with the works of the new scientists of the Old World was unequaled by any colonial. His book *The Christian Philosopher* was a veritable encyclopedia of "the best Discoveries in Nature," demonstrating Mather's knowledge and acceptance of the world of nature as delineated by Newton ("from whom 'tis a difficult thing to dissent in any thing that belongs to Philosophy"[69]), Halley, Ray, Willughby, Boyle, Grew, and scores of lesser figures. It is obvious that by the 1720's Mather

[67] *Ibid.*, pp. 149–150.

[68] In 1712, after describing various celestial phenomena, Mather stated that although there had been much speculation about the ominous things expected to follow, he believed it was only "a *Storm* now a breeding, which, I think alwayes happened"—a considerable shift in view from that of Increase Mather in the 1680's. *Guard-Book*, M-2, No. 29. Beall and Shryock (p. 50) comment on Mather's change from theological to scientific explanations. However, I incline to the opinion that Mather's appeal to a natural explanation was only a momentary lapse—or, rather, was meaningless in that he saw no essential differences between the natural and the preternatural.

[69] P. 41.

had advanced his knowledge and acceptance of science far beyond
that of his early days in the Boston Philosophical Society. Moreover, he had become recognized as a natural philosopher in many
parts of the western world and had widely advertised the growing
scientific community of New England.[70]

Still, he remained a cleric at heart, and he clung to some of the
superstitions of the previous age. When, for example, in the course
of the *Curiosa Americana*, he described the powers of rattlesnakes
to charm their victims and the virtues of "Poor Robin's Plantain"
as an antidote for snake-bite, he evoked the image of his contemporary in Virginia, William Byrd II. Again, his account of a rain of
frogs, of a two-headed snake at Newbury, and of other "monsters"
bespoke a credulity hardly conformable to the truly scientific
mind, even in his own time. If occasionally he wavered in his opinion, he appeared, in the end, to settle with the age-old theological
opinion. All this points to the conclusion that Cotton Mather was a
transitional figure in colonial life, poised between the old theological interpretation of natural phenomena and the new scientific explanation. He retained to the end his basic clerical responses, his
inability to distinguish finally between the preternatural and the
natural, as both flowed from a personal God. In his *Diary*, where
he often recorded his innermost thoughts, he repeatedly represented his communications to the world of science as exercises to
"do good," exercises which sprang, at base, from a religious motivation, not a concern, per se, with the advancement of secular learning.[71] His two great works, the unpublished *Biblia Americana* and
the published *Christian Philosopher*, were, as Beall and Shryock
have said, two sides of the same coin: the former an effort to recon-

[70] It should be emphasized, although there is no space here for detail, that
Mather's correspondence reached beyond the Royal Society to scholars in
England, Scotland, Germany, Switzerland, the United Netherlands, and British
colonies outside New England. See Kuno Francke, "Further Documents Concerning Cotton Mather and August Hermann Francke," *Americana Germanica*, I
(1897), 55–56; Harold J. Jantz, "German Thought and Literature in New
England, 1620–1820," *Journal of English and Germanic Philology*, XLI (1942),
5ff. An example of Sir Hans Sloane's role in distributing Mather materials to
continental correspondents is in *Sloane 4055*, fols. 108–109.

[71] Worthington C. Ford, ed., *Diary of Cotton Mather*, 7 *Coll. M.H.S.*, Vols.
VII–VIII (1911–12; republished, *American Classics*, 2 vols., New York, n.d.), II,
245–246, 265–266, 282, 548, 582, 628.

cile biblical revelation with the new science, the latter a survey of the new science from the perspective of the Christian religion.[72] His vital interest in medicine sprang from the religious conviction that all human illness derived from original sin—that "Sickness of the Spirit" will naturally and inevitably lead to a "Sickness of the Body." Thus, his Christian ministry encouraged his interest in science and in social welfare (both ways of doing good), and molded him, unconsciously, into becoming a herald of the Enlightenment and of the Great Awakening in America. Notwithstanding his fundamental religious motivation, Cotton Mather was the first native-born American colonial to advance beyond the status of a mere field agent for European scientists in the New World and to demonstrate a genuine *philosophical* approach to science, with scientific *ideas* and *hypotheses* of his own, in addition to the contribution of specimens and observations of natural phenomena. These achievements marked a new level of performance by a colonial contributor to the Royal Society—and, again, heralded a new generation of colonial scientists. But if becoming a scientist requires that one admits the sciences to be *autonomous* systems of thinking, Cotton Mather was no scientist. To him, science was no more than a useful handmaiden of theology with which to probe the mind and glory of God.

Thomas Robie (1689–1729), F.R.S.

In addition to Cotton Mather, there arose among the leaders of community life in the Boston area an impressive number of contributors to science. The group embraced men of widely differing origins, but they held in common a concern for Harvard College and an interest in science. In varying degrees they were familiar with one another, were in some cases related by marriage, and as time passed even their personal antagonisms cooled. Among these leaders was Thomas Robie.

Robie was born in Boston, graduated from Harvard College (A.B., 1708; A.M., 1711), became Harvard's "Library-Keeper" in 1712 and tutor, 1713–23. In the latter year he left Harvard, married, and engaged in the practice of medicine at Salem until his

[72] P. 50.

death in 1729, a practice in which he was self-taught and in which he had become increasingly involved before he left the college.[73] He early demonstrated scientific interests and engaged in the publication (1708) of a series of annual almanacs entitled *An Ephemeris of the Coelestial Motions. . . .* Some of these included essays for the scientific education of the public; that of 1716, for example, contained an explanation, based upon William Derham's *Astro-Theology* (1715), "Why it is Hottest in Summer when the Sun is vastly furthest from us; And Coldest in Winter when it is nearest." This was followed by a discussion of thunder, lightning, and hail, based in part upon Cotton Mather's *Christian Philosopher* (which, obviously, he had seen in manuscript form) and in part upon John Wallis' earlier treatment of the subjects in the *Philosophical Transactions.*[74] Robie suggested that lightning resulted from an explosion of nitrous and sulphurous vapors in the air, although he did not envisage the cause of the explosion to be an electric spark, as Franklin did later. His commonplace book included copious notes from the *Philosophical Transactions,* and he appears to have been fired especially by Edmond Halley's articles about methods to ascertain the sun's parallax and to calculate the size of the universe. These studies only confirmed him in his acceptance of "the truth of the Newtonian Philosophy." His *Almanack* for 1720 included "an account of the Solar System, according to Copernicus and the Modern Astronomers," with tables of "The Middle Distances of the Planets from the Sun," and of "The Diameter of the Sun and Planets with the Moon," all "calculated from the latest Observations by Sir Isaac Newton's Rules." Such articles could scarcely fail to promote and to popularize Newtonian physics.

About the time of his appointment to the Harvard tutorship (1713), Robie began a scientific correspondence, chiefly with both Matthew Wright of Crew (to whom he was introduced by Edward Holyoke) and the Reverend Mr. William Derham, F.R.S., Boyle Lecturer (1711, 1712), author of *Physico-Theology* (1713),

[73] The best accounts of Robie are: C. K. Shipton's continuation of J. L. Sibley, *Biographical Sketches of Graduates of Harvard University* (cited hereafter as *Sibley's Harvard Graduates*), V (Boston, 1937), 452; Frederick G. Kilgour, "Thomas Robie (1689–1729), Colonial Scientist and Physician," *Isis*, XXX (1939), 473–490.

[74] XIX (1695–97), 653–658.

of *Astro-Theology* (1715), and prebend of Windsor. Much of his correspondence with Derham has survived, thanks to the fact that Derham communicated it to the Royal Society. On February 24, 1714/15, Derham presented to the Society a letter from "Mr. Robie of Harvard College" dated December 7, 1714. It included meteorological observations made at Cambridge, together with astronomical data collected during "an Immertion [immersion] of one of Jupiter's Satellites in August last on the 20th day 9 h 35′ 44″ P.M."[75] Early in 1716 Derham reported another communication from Robie but its contents were not divulged.[76] Again, in 1718, Derham sent in an account of two eclipses observed (1717) at Cambridge in New England (but not observable in England).[77] About eighteen months later Derham presented to the Society "two small Pamphlets printed at Boston in New England one written by Dr. Mather the other by Mr. Robie about a Meteor commonly called the Aurora Borealis which was Seen in those parts on the 11th day of December 1719."[78] The description which Robie gave was not that of a meteor but, as he recognized, of a gaudy display of the aurora borealis. His account was remarkable in that he wholly eschewed all effort to explain the event by recourse to divinity or the old philosophy. Instead, he regarded it as a natural phenomenon and sought to explain it by reference to John Wallis' nitrous-sulphureous theories printed in the *Philosophical Transactions* in 1697. As the preceding days had been hot and sultry, Robie concluded that the combustible particles had vaporized from the earth and in some way ignited to cause the fiery exhibition. He regarded it as a natural phenomenon, and expressed an abhorrence for any prognostications drawn from

[75] *Journal-Book*, XII, 45. On the following July 14 the Society received "a Construction and Observation of the Eclipse of the Sunn on the 6th of December 1713 made at Boston in New England by Mr. Holyoke in the presence of his Excellency General Nicholson." *Ibid.*, p. 73. No doubt this was Edward Holyoke (1689–1769), Harvard graduate (1705) and later President of Harvard (1737–69). Like Leverett, President Holyoke fostered scientific studies at Harvard. Cf. Morison, *Three Centuries of Harvard*, pp. 83–94.

[76] *Journal-Book*, XII, 99 (March 1, 1715/16).

[77] *Ibid.*, XII, 246 (Oct. 30, 1718). One was a lunar eclipse, March 15, 1716/17, the other a solar eclipse, no date given.

[78] Robie's pamphlet was *A Letter to a Certain Gentleman* . . . (see note 38 above).

the event. Although his explanation would be untenable today, Robie's attitude toward it was entirely scientific in spirit.

On the same day (March 9, 1720/21), Dr. Jurin read a letter from Derham with "An Account of Mr. Robie's Observations in New England of some *Northern Streamings* there, & an old Oak, that by burning, yielded a large quantity of Salt." The "*Northern Streamings*" were another description of "*Lumina borealis*," this time observed on November 24, 1720: ". . . only one thing [reported Derham] he takes notice of yt deserves observation, viz, that he *was informed by an honest & ingenious man, yt at Casco-Bay, it reached to the Zenith; & by another that at Barnstable, 80 miles South of Cambridge, it was very low in the Horizon; & so (saith he) I have had from others in different places, an account of different Heights, always lower in the Horizon to places South, higher to the Northward.*"[79] Obviously Robie had scientific correspondents in New England as well as overseas—further evidence of the rise of a scientific community in the Boston area.

Derham's letter went on to quote Robie's description of "*A white Oak tree*" in Cambridge, about one third of which "seemed really to be rotten wood, yet the decayed part, in burning, would turn almost wholly in good white Alkali, & it would run down into white lumps white & clean." One of the lumps had been brought to the College "to know what it was." Robie gave the following account of its analysis, betraying a considerable knowledge of chemistry:

> We tasted it,[80] & found it to be salt, & very strong. We dissolved it in clean water, & upon Decantation & Evaporation, without any filtration, we produced a very clean, white Salt, exceeding in strength, & whiteness, any to be bought at the Shops. We tried it many of the ways of proving the goodness of an *Alkali*. Now although *Alkalis* may be extracted from common Ashes, yet what was peculiar in this is,
>
> 1. That while it was Burning, The Wood it self would melt,

[79] Derham to Jurin, Windsor Castle, Nov. 27, 1722, *Guard-Book*, D-1, No. 67. Robie's letter to Derham was said to be dated "the latter end of the year, 1720."

[80] Robie was assisted by Samuel Danforth, Harvard graduate (1715), later judge and Councillor who was forcibly removed from his latter position during disturbances over the Boston Port Bill in 1774.

& run down into Lumps of Salt; & none of the Wood yt was sound, but only yt wch was decayed, & what was most decayed, would yield ye greatest Quantity of Salt. And

2. Whereas all other *Alkalis* of Wood made thus by Incineration, are blackish at first, & a Lixivium made of them, although often filtered, will yet be tinged with a Brown colour, occasioned from a kind of Coal, or Ashes so inclosed, or closely united to the *Alkali* in burning, as not easily to be separated by filtration, though often repeated; yet this *Alkali* was very White, even before Solution, & when dissolved, The Lixivium was not in the least tinged, but Clear like pure Water, only a very small quantity of Ashes subsided to the bottom of the Vessel, in wch the Solution was made. The *Lixivium* thence decanted needed no filtration, but when boiled up to Dryness, the Salt remained fine & white. And

3. That in Burning of this Wood, as the heat of the fire grew more intense, the Wood did, as it were, melt & clodder together in great Lumps, & did visibly bubble, & boil, with an hissing noise, like the frying of fat in a pan.

4. That whereas the Weight of the Alkali-Salt produced from other Wood, in the Common way of Incineration, is very inconsiderable, in proportion to the weight of the wood producing it before burnt, yet this Salt nearly equalled in weight, the Wood from whence it was taken.

5. Whereas the Ashes of other Wood is never so replete with Salt, as yt Salt can be seen, or in the least cause the Ashes to lump or Clodder together, yet this, the whole of it, would gather into hard & solid Lumps of white Salt, as easily to be distinguished from Ashes (tho' White) as the purest Salt of Tartar made with Nitre would be.

6. That although from other *Rotten Wood* much less of an Alkali can be produced than from *Sound Wood*, yet here it is quite contrary; the Decayed part of this Tree yielding in quantity, as aforesaid; ye other or Sound part, yeilding no more than other Wood.

After this "true & full Account of this strange & unusual Production," Robie went on to "give you our thoughts respecting the Solution of it, which we would not attempt, but being on the Spot, we have examined the Tree, & considered what (by the marks found on it) hath, in all probability, happened to it. . . . All wch we do

with humility & modesty submit to your ·Censure." Robie found
that the tree bore evidence of having been struck by lightning, and
conjectured "yt the Lightning having penetrated ye Wood, had so
altered & disposed ye parts & pores of ye wood . . . to attract, receive in, & restrain the Nitrous Salt of the Air, which through so
long a space of time, could not but be in great abundance. Even as
Salt of Tartar, or other alkali being exposed to the Air for some
considerable space of time, will be wholly reduced to a Nitrous Salt
(as *Glauber* says) & its Quantity also increased & very Considerably. . . ." Robie did not hold that the lightning had reduced the
wood to "a perfect Salt," but rather that it had "calcined it in such
measure, as to give it a like propriety or disposition of Attracting
the Nitrous Salt of the Air." If it were objected that the niter, being volatile, "would flee away in the burning of the Wood,"

> we answer, That although Nitre can't be fixed, & reduced to an
> Alkali-Salt, by calcining it per se, yet it may be so by the addi-
> tion of the powder of Charcoal (as chymists teach us) And here
> we suppose the Wood so altered by Lightning, in which this
> Nitrous Salt was lodged, as served instead of Coal in the Burning
> of it.
>
> Thus far the curious & ingenious *Mr. Robie* [wrote Derham],
> who was joyned with another ingenious man in his latter Observa-
> tions, *Mr. Sam¹ Danforth;* who submit their conjectures to the
> judg^mt of this illustrious Society.

Evidently the Royal Society raised no objections to Robie's explanation of the above phenomenon, and Robie's "Account of a large
Quantity of Alcalious Salt produced by burning rotten Wood" was
published in the *Philosophical Transactions*,[81] the first detailed relation on record of a chemical analysis by a colonial.

Robie's scientific work at Harvard led to the acquisition of so
much apparatus that the college set aside a special chamber for it
in Massachusetts Hall, the roof of which was used as an informal
astronomical observatory.[82] Here he made ready for his next major
investigation in astronomy, occasioned by the annular solar eclipse
on November 27, 1722. He prepared the public for the approaching phenomenon with a newspaper appeal entitled, "For the Enter-

[81] XXXI, No. 366 (Sept.–Dec., 1720), 121–124.
[82] Morison, *Three Centuries of Harvard*, p. 58.

tainment of the Country and the Promoting of Knowledge."[83] He assured his readers that scholars no longer were ignorant of the causes and times of such eclipses. They could predict them accurately, and careful observations of them, especially with regard to the exact times of their stages of development, gave assistance to the more accurate calculation of longitude. The eclipse would occur during the early morning of November 27, and Robie gave the late Astronomer Royal's (Flamsteed's) directions for its observation, hoping for responses from the public after the eclipse had taken place. How widely the public responded is not recorded, but on the morning of November 27, the college people turned out in force, watching from the rooftops of college buildings and dwellings nearby. Besides his own observations, Robie reported those by "Mr. [Samuel] Danforth in a room just by me"; by Mr. Nathaniel Appleton, the new Cambridge minister who succeeded the late William Brattle, at his house; by "Mr. Owen Harris an Ingenious Schoolmr. in Boston"; and by an unidentified observer at "Yale College in Connecticut Colony." He gave Yale's position as about 8′ or 10′ longitude west from Cambridge and latitude about 41½° North; but he felt that the Yale result was "very much mistaken," because the New Haven calculation estimated the extent of the eclipse to be "about 8 digits" whereas Robie found it to be "above 11" digits at Cambridge. Reports from Cape Cod indicated that there the eclipse was almost complete, and other responses to his public appeal aided him partially to confirm his calculations regarding the path of the shadow. All this information Robie transmitted to William Derham, who laid it before the Royal Society on February 20, 1723/24, and, with other items noted below, it was published in the *Philosophical Transactions*.[84]

Robie's letter to Derham was delayed, possibly because, having been caught up in Harvard quarrels, he left the college in 1723 and moved to Salem to practice medicine. Actually he had begun the practice of medicine before he left Harvard. Indeed, the above letter to Derham related the case of Nathaniel Ware, of Needham,

---

[83] Carried in the *Boston News-Letter*, No. 978 (Oct. 22–29, 1722), and *The Boston Gazette*, No. 155 (Nov. 5–12, 1722).

[84] Robie to Derham, Salem, June 4, 1723, *Letter-Book*, XVI, 429–432; *Journal-Book*, XIII, 339–340; *Phil. Trans.*, XXXIII, No. 382 (March–April, 1724), 67–70.

whom Robie had treated for a spider-bite on September 13, 1722. Further, when the controversy about inoculation for smallpox had arisen in Boston, Robie had written to Derham to inquire: "(1) What Credit the Operation hath in London? (2) Wt Effects & Consequents it produceth, & whether Persons inoculated were ever known to have the S. Pox again in the Common way? (3) Whether it is practised in England? *Or in short,* Whether it be a safe & justifiable practice?"[85] Derham passed the queries on to Sir Hans Sloane, "as no man is better able to answer than your self." Sloane's reply is not extant, but it is well known that he approved of inoculation and himself inoculated patients. In any case, Robie adopted the practice and devoted two weeks during the summer of 1722 caring for inoculated persons at the isolation hospital on Spectacle Island in Boston Harbor.[86] He also wrote to Derham (June 4, 1723) to defend the practice of inoculation for smallpox in New England. He said that he had observed no "ill Effects of Inoculation" and argued that those persons who had died after being inoculated suffered "some previous disposition to such distempers," i.e., distempers other than smallpox.[87]

Cotton Mather submitted to the Royal Society (January 9, 1723/24) an account of observations made at the "Transit of Mercury over the Suns Disk," on October 29, 1723, by Robie assisted by Mr. Owen Harris, the Boston schoolmaster.[88] Mather stated that the observations were made at Cambridge, but Robie himself transmitted these observations, together with those made of the eclipse in 1722, to Cadwallader Colden in New York, stating that the observations made on October 29, 1723, were made at Salem "thro a 9 feet Telescope."[89] The Boston scientific community was widening its tentacles.

[85] Derham to Sloane, quoting Robie, Windsor Castle, Feb. 13, 1721/22, *Sloane 4046*, fol. 201.

[86] Kilgour, "Thomas Robie," pp. 484–488.

[87] *Letter-Book*, XVI, 429–432; *Journal-Book*, XIII, 356 (Feb. 20, 1723/24).

[88] *Journal-Book*, XIII, 339–340.

[89] Nov. 9, 1723, *Colden Papers*, I, 157–159. Kilgour suggests that this letter was originally addressed to Governor William Burnet of New York and fell into Colden's hands. Robie stated that having heard from a Boston friend, Jacob Wendal, of the addressee's great knowledge and interest in astronomy, he sent the information as a matter of common interest. As the addressee was not named, it might have been either Burnet or Colden. Both had interests in astronomy.

On January 14, 1724/25, Thomas Robie was proposed as a candidate for fellowship in the Royal Society by William Derham. The Council approved, and on April 15, 1725, Robie was elected a Fellow of the Society.[90] Shortly after (June 7, 1725), Robie addressed a letter to Dr. James Jurin, Secretary of the Society, as follows:

> Tho' unknown to you, yet at the motion that was made me by Mr. [Isaac] Greenwood, I take the freedom of writing to you at this time which if acceptable and you see good, may be the beginning of a Correspondence with you if an American can be any ways worth your time to promote. Mr. Greenwood wrote me word that you would be glad of my Weather Observations I have therefore sent 'em long after I designed 'em which I suppose you will receive by Mr. Greenwood before you do this. I have sent 'em just as I wrote 'em every day and there be some little Notes of things which you must not laugh at but forgive the Simplicity of them; if they can be of any use to you I shall be glad if not I pray I may have 'em again. . . .[91]

Robie continued to ask Dr. Jurin's opinions, as a physician, regarding the treatment of "Hysterick Disorders." He cited Sydenham's observations and treatment of such cases, but had not found them answerable in all instances. Jurin's reply, if made, does not appear to have survived, and, strangely, there appear to be no further letters between the two men extant in the Royal Society Archives. Because he had already sailed for New England, Greenwood, who had been Robie's student at Harvard, did not deliver the meteorological journal to Jurin at once. But he subsequently sent it to Derham. Years later, in 1731, when the latter drew up abstracts of several "Meteorological Diaries" for submission to the *Philosophical Transactions*, he included Robie's and thus it was published,[92] nearly three years after Robie's death in 1729. It was not Robie's best work, having been made between the years 1715 through

---

[90] *Journal-Book*, XIII, 437, 471; *C.M.*, III, 237. *Journal-Book*, XIII, 471, records that Robie was present on April 15, 1725, but I have found no evidence of his going to England. Moreover, the Council minutes note that he was not formally admitted F.R.S.

[91] *Letter-Book*, XVIII, 211–212; dated from Salem.

[92] Derham to Sloane, Upminster, Aug. 4, 1731, *Sloane 4051*, fol. 294; *Phil. Trans.*, XXXVII, No. 423 (April–June, 1732), 261–273.

1722, while he had been at Harvard—and, in those days, without either a barometer or a thermometer with which to register atmospheric pressures and temperatures accurately.

Thomas Robie's contributions to science were not, in themselves, outstanding. But he took an active part in promoting science in the Boston community, and by his activities helped to popularize science in the public mind. He trained up no truly great colonial scientists, although Isaac Greenwood, the first Hollis Professor of Mathematics and Natural Philosophy at Harvard, was his pupil; and he may have awakened scientific interests in Thomas Clap, who graduated from Harvard in 1722 and later, as President of Yale (1740–66), initiated the earliest vigorous, continuing study of science. Harvard President Leverett recorded in his unpublished "Book Relating to College Affairs" that "It ought to be Remembered, that Mr. Robie was no small honour to Harvard College, by his Mathematical Performances and by his Correspondence thereupon with Mr. Derham & other learned persons in those Studies abroad."[93] But the most significant feature of Thomas Robie's career was neither his analytical mind nor his adherence to Newtonian physics, but his clear adoption of the scientific method, unfettered by superstition and theological concern. In this regard he far outshone Cotton Mather. For whereas Mather could never completely rise above the superstitious tendency to see dire omens in such natural phenomena as comets, heavy storms, and earthquakes, Robie calmly stated, with reference to the so-called meteor of 1719, that "I don't mean that the Sight was not surprising to me, for I have said it was before, but I only mean that no Man should fright himself by supposing that dreadful things will follow. . . ."[94] Despite the fact that he had served as a minister in New England, Thomas Robie was one of New England's first true scientists.

Zabdiel Boylston (1680–1766), F.R.S.

Zabdiel Boylston, who shared with Cotton Mather the widespread obloquy dispensed during the Boston controversy over

[93] Quoted in Kilgour, "Thomas Robie," p. 484.
[94] Cited in *ibid.*, p. 490.

smallpox inoculation, was born in Brookline, Massachusetts. He was trained in medicine by his father, Dr. Thomas Boylston, and by John Cutter, a Boston physician. For many years before the inoculation disputes arose, Boylston operated an apothecary shop in Boston and, like many others at the time, practiced medicine as well.[95] He also exhibited a lively interest in natural history and made a considerable private collection of plants and animals. Before being persuaded to undertake the experiment of inoculation for smallpox, he examined both the evidence from the pages of the *Philosophical Transactions* as supplied by Cotton Mather and also made his own survey of evidence which he gathered from interviews with Boston blacks who had been inoculated in Africa before having been sold into slavery. Fortified with this information, so widely scorned by his opponents, Boylston assumed a dignified stance during the public uproar, content to allow the facts to speak for themselves. On July 15, 1721, he inserted in the *Boston Gazette* a quiet "Advertisement" in self-defense, stating that

> I have patiently born with abundance of Clamour and Ralary, for beginning a new Practice here, (for the Good of the Publick) which comes well Recommended, from Gentlemen of Figure & Learning, and which well agrees to Reason, when try'd & duly considered, *viz.* Artificially giving the Small Pocks, by Inoculation, to One of my Children, and Two of my Slaves, in order to prevent the hazard of Life, which is often endanger'd by that Distemper in the common way of Infection; And as the thing was new, & for fear of erring in doing, I left it wholly to Nature, which needed no help in my Negro Man, who was taken ill a day or two before the other two, in which time the Symptoms abating, caus'd me to hope for the same in the others . . . and they never took one grain or drop of Medicine since, & are perfectly well. And for Encouragement, no one need fear, in this way, of having many Pustules, of being scarr'd in their face, or of ever having the Small-Pocks again. This is fully cleared by those Gentlemen, *viz* Doctor Emanuel Timonius of Constantinople and Fellow of the Royal Society, and Jacobus Pylarinius (a Physician as appears by his Writings) the Venetian Consul at Smyrna, who have try'd it,

[95] See his advertisements in the *Boston News-Letter*, No. 467 (March 23–30, 1713), and No. 804 (Sept. 7–14, 1719).

and known it try'd, upon Thousands & with good Success, as they have inform'd us; And in the three, of whom I have had by Experience, I found their Account just and true. And in a few Weeks more, I hope to give you some further proof of their just and reasonable Account.

Saturday, July 15th, 1721.   Zabdiel Boylston[96]

In the tumultuous months that followed, Boylston maintained this calm, reasonable attitude, though privately "affrighted" by the rage of the people who opposed his work.[97] His booklet, *Some Account of what is said of Inoculation or Transplanting the Small-Pox. By the Learned Doctor Emanuel Timonius and Jacobus Pylarinius. With some remarks thereon. To Which are added, a Few Quaeries in Answer to the Scruples of many about the Lawfulness of this Method*, issued in Boston in August, 1721, preserved the same quiet tone—although the *Boston News-Letter* accused Boylston of being "*illiterate*," incapable of understanding the "Writings of those Foreign Gentlemen," and undertaking a practice which was "deemed *Wicked* and *Fellonius*" by the learned men in England.[98] And his later summation of his experience, *An Historical Account of the Small-Pox Inoculated in New England* (London, 1726), retained the same professional calm, marked further by remarkably accurate and complete clinical observations of his cases and a pioneering use of medical statistics.

Early in August, 1723, the *Boston Gazette* printed an announcement: "These are to desire all persons indebted unto Doctor Zabdiel Boylston to (send or come and) pay their Debts. And likewise to desire all persons that have any Demands on said Boyl-

---

[96] *Boston Gazette*, No. 85 (July 10–17, 1721). No. 88 (July 27–31) contained a letter addressed to "the Author of the Boston News-Letter" protesting against the "unhandsome treatment" accorded Boylston, signed by both Mathers, Benjamin Colman, Thomas Prince, John Webb, and William Cooper. Both Colman and Cooper published defenses of Boylston in 1721. See John T. Barrett, "The Inoculation Controversy in Puritan New England," *Bulletin of the History of Medicine*, XII (1942), 169–190.

[97] When the public feared further spread of the epidemic by inoculations, Boylston inserted a letter in the *Boston Gazette*, No. 112 (Jan. 8–15, 1721/22), to quiet them.

[98] *Boston News-Letter*, No. 912 (July 17–24, 1721). James Franklin's new paper, the *New-England Courant*, soon entered the fray to stir up public opinion heatedly against Boylston and the Mathers.

ston to bring in their Accompts and receive their Money, he designing in a short time a Voyage for London."[99] Boylston went to London late in 1724, designing to learn the new lateral method of cutting for the stone (lithotomy), recently developed in France. Both Cotton Mather and Paul Dudley sent letters to recommend him to Dr. James Jurin, Secretary of the Royal Society. Mather, as has been noted, suggested that Boylston's presence in England might fittingly deserve court recognition. Dudley stated that Boylston was held "in high esteem in his own Countrey for his great skill long experience and happy success in surgery," was "an Ingenious and worthy man," and "God made him a singular Blessing in our late terrible small pox" as "the happy instrument of saving many scores of lives by the new method of Inoculation."[100] The Royal Society was well prepared to receive him with exceptional honor and courtesy, for it had received numerous relations of his success in inoculation for smallpox, and Dr. Jurin had played a leading role in the study and promotion of inoculation in England.[101] Moreover, Boylston had recently sent to the Society an account of ambergris found in bull spermacetic whales by Nantucket whalers—an account published in the *Philosophical Transactions* shortly before his arrival in England.[102] When he first visited the Society (February 11, 1724/25) he presented it with a "large lump" of ambergris and remarked upon its medicinal virtues.[103] A few weeks later (May 19, 1726) Boylston again attended the Society and presented a copy of his recently published *Historical Account of the Small-Pox*.[104] The following week (May 26), Dr. Johann Georg Steiger-

[99] No. 193 (July 29–Aug. 5, 1723). The same paper (No. 221, Feb. 10–17, 1724/25) carried an advertisement by Boylston to let his two-acre garden in Boston and sell the stock in his shop.

[100] Dudley to Jurin, Boston, Dec. 19, 1724, *Guard-Book*, D-1, No. 83; *Letter-Book*, XVIII, 159–160.

[101] See Stearns, "Remarks upon the Introduction of Inoculation for Smallpox in England," pp. 116–122.

[102] XXXIII, No. 385 (Oct.–Dec., 1724), 193. The account drew attention because the origin of ambergris had long been in dispute. Boylston reported its discovery in a "cyst or bag" near the genitals of some of the bulls captured, although not in all of them. Whether it was "naturally or accidentally produced in that Fish," said Boylston, "I leave to the Learned to determine."

[103] *Journal-Book*, XIII, 448.

[104] *Ibid.*, XIII, 584. Dr. Jurin commented on the book, and concluded that the mortality in smallpox inoculation in New England was about the same as that

thal proposed Boylston as a Fellow of the Royal Society. The Council acted affirmatively, and on July 7, 1726, Boylston was elected. Being present, he signed the obligation and was formally admitted on the same day.[105]

Subsequently, Boylston presented lectures about his observations in inoculation for smallpox both to the Royal Society and to the Royal College of Physicians, lectures which the historian of medicine in Massachusetts has characterized as a "masterly clinical presentation."[106] In the course of the summer of 1726, Boylston made the acquaintance of many of the Fellows of the Royal Society, especially Sir Hans Sloane (who succeeded Sir Isaac Newton as President of the Society in 1727) and Dr. Jurin, to whom Boylston wrote after his return to Boston to thank him for "Your handsome & kind Treatment of me from time to time."[107] Boylston arrived in Boston about the first of December, 1726. On the same day that he wrote to Dr. Jurin he also addressed a warm letter of thanks to Sloane for "the Honours done me," and dispatched a five-pound stone which had been removed from a gelding in Boston on December 26, 1724, "just after my departure for London."[108] On February 1, 1726/27, he addressed Jurin to give a further description of the stone and the case of the gelding from which it had been removed, with a further comment on the affinity of the diseases which afflict both man and beast. Jurin presented the stone and Boylston's comments to the Society on April 13, 1727.[109]

Boylston's communications to the Royal Society were thereafter infrequent. He suffered asthma during the last forty years of his life, gradually relaxed his medical practice, and after about 1740 retired from it completely. On April 25, 1735, in response to an inquiry from Dr. Cromwell Mortimer, Secretary of the Royal Society,

in England. At an earlier date Boylston had consulted Sloane about the dedication of his book to Princess Caroline. Boylston to Sloane, London, n.d., *Sloane 4048*, fol. 241.

[105] *Journal-Book*, XIII, 585, 604; *C.M.*, II, 296.

[106] Henry R. Viets, "Some Features of the History of Medicine in Massachusetts during the Colonial Period," *Isis*, XXIII (1935), 389–405.

[107] Boylston to Jurin, Boston, Dec. 14, 1726, *Letter-Book*, XVIII, 384–385.

[108] Boylston to Sloane, Dec. 14, 1726, *Sloane 4048*, fol. 238.

[109] Boylston to Jurin, Boston, Feb. 1, 1726/27, *Letter-Book*, XVIII, 412–414; *Journal-Book*, XIV, 68–69.

about the fate of Thomas More, whom a group of English sub-scribers had sent (1722) to collect specimens of the New England flora,[110] Boylston wrote that More had not been seen in New Eng-land since 1724, when he sailed for London; and he went on to describe briefly experiments he had made with rattlesnakes "suc-cessively for three or four years past." He had procured several of them each spring:

> One Summer I had thirteen and try'd many ways to feed them, and reconcile them to Confinement; in order to send a number of them to the Royal Society: that the Curious that way might have opportunity of making Experiments as to the effect of their Bite, &c., particularly of their Power of charming their Prey to their mouth &c. But I am yet at a loss, they never feeding on any thing sufficient to carry them thro' the Winter; but the longest Liver died from Winter to Winter before February was half out, and the most of them before January began. . . .[111]

Evidently Boylston knew nothing of Sir Hans Sloane's experiments (1730) or of his conclusions regarding the myth about rattlesnakes charming their victims.[112] His account was read to the Society on June 26, 1735, without comment.[113] The following January Boyl-ston wrote to Dr. Mortimer that "I have not yet wholly got over the fascinating or charming Influence or Power of the Rattle Snake; but intend to make, next Summer, some further Experiments, and send them with the Antidote or Cure for such venemous Bites."[114] He also sent "a little Phial of natural Balsam taken from the blisters found in the Bark of our white Cedar." He said it was "new to me," but from "the fine Smell" he expected to find "some-thing extraordinary in its Virtue." He sent a second phial of the balsam to "the great Sir Hans Sloane for his approbation." The Royal Society ignored Boylston's comments about rattlesnakes, but Sir Hans Sloane judged of the balsam, as Boylston had, that because of "the fineness of its odour" its virtues would be found very use-

---

[110] See below for further treatment of More.
[111] Boylston to Mortimer, Boston, April 25, 1735, *Letter-Book*, XXI, 467–469.
[112] See p. 290 above.
[113] *Journal-Book*, XVI, 171–172.
[114] Boylston to Mortimer, Boston, Jan. 15, 1736/37, *Letter-Book*, XXIII, 183–184.

ful. One Mr. Stevens, a surgeon who had tried the balsam, asserted that it was "every whit as good as the Balsam of Capivi and somewhat less nauseous to the Stomach."[115] Dr. Mortimer inquired further about the snake-bite cure and asked for more of the balsam. His inquiries brought forth Boylston's last communication to the Royal Society, on December 17, 1737:

> I am sorry that I can't at present give you any tolerable account of the Antidote or Cure of the Rattle Snake bite: for tho' I have had at times so many living Serpents, yet thro' hurry of business, and want of health, I have made but few Experiments that way. But if it may please Providence, that my Life may be continued untill my son the Bearer, who comes for further improvement in the healing Art, returns to take my business, and I can get a convenient house built on a little Farm I have five miles from Boston, where I purpose to make my retreat: I hope to send that learned and august Society some communications that may in some measure compensate for my long Silence.
>
> I send you herewith by my Son all the Cedar Balsam I have, or at present can get. It is very scarce and dear. I have never met with any before nor since: but believe I shall in a short time. As for the account given by that intelligent Surgeon before the Royal Society; that it answers in all cases, where Balsam Capivi is proper, I something doubt: for as I take it, this Balsam is hotter, and more styptic; and as it has a much greater consistence, it should be in less quantity, and well mix'd with some proper Substance to open the Body, and render it less subject to adhere to the Coats of the Stomach and Bowels.
>
> I also send you part of a Buck, or Hart Bezoar, which to some may be thought extraordinary or curious. It was found in the Stomach of a large Buck kill'd by our Hunters, and sent to Town for Sale, expecting a great price. It being very solid and heavy, I had a mind to see the inside: and upon breaking it, found it with the Shot in the middle, as you see it. And altho' these Stones have generally something like such a beginning, yet it may be admired, that the wound in the Stomach, which may be supposed made by the Shot, should not have destroyed the Animal. Pray spare me, if my Reasoning shou'd prove trifling. . . .[116]

[115] *Journal-Book*, XVII, 48 (Feb. 24, 1736/37).

[116] Boylston to Mortimer, Boston, Dec. 17, 1737, *Letter-Book*, XXIV, 180–181; read Feb. 16, 1737/38, *Journal-Book*, XVII, 191.

Two days later, Boylston addressed his last letter to Sir Hans Sloane. The bearer, he said, was

> my youngest son of two who comes to spend a year in London in order to acquire further knowledge in ye Healing art. He is the first fruits of Inoculation in ye American World, viz. in 1721, when my Practice therein made such a Clamour as was never made here before or since, & the fatigues therein cost me my Health, viz. brot on me a convulsive Asthma with Tubercles in my Left Lobe together with such a Cartarrhical distillation upon my whole breast that I have not had for ye last twelve years twelve hours ease in twenty-four. . . .[117]

Shortly afterward Boylston retired from medical practice, and though he lived for more than a quarter-century thereafter, devoted his time to gardening and horse-breeding. If he performed any further experiments with rattlesnakes, the cedar balsam, or anything else, they were not communicated to the Royal Society. His principal claim to scientific fame rests upon his pioneering work in the inoculation for smallpox, in which his moral leadership and his professional skill gave him a just claim to recognition in his own time and gratitude of succeeding generations. He was the first American colonial-born physician to be celebrated in England.

Benjamin Colman (1673–1747)

The Reverend Mr. Benjamin Colman was no scientist, but he swelled the Boston community of scientific amateurs in the early eighteenth century and exerted considerable influence on behalf of the new learning. Born in Boston of well-to-do parents, he graduated from Harvard College in 1692, a favorite student of tutor John Leverett. He returned to Harvard to take a master's degree in 1695. He then toyed with the notion of joining with the group of New England missionaries that included Joseph Lord, Petiver's later correspondent from Dorchester, South Carolina, but changed his mind and sailed for England instead. Unfortunately, his ship was taken by a French privateer, and Colman was stripped of nearly all his possessions. When he subsequently made his way to

[117] Boylston to Sloane, Boston, Dec. 19, 1737, *Sloane 4055*, fol. 248.

England he was almost destitute. However, he soon made the acquaintance of a number of distinguished men in England, some of whom became frequent correspondents after his return to New England. Among these was Henry Newman, a graduate of Harvard in 1687, later a member of the legal profession associated with the Middle Temple who transmitted several scientific communications from New England to Secretaries of the Royal Society. In 1699 Colman accepted an invitation to become minister of the newly organized Brattle Street Church at Boston, and he returned to New England. He devoted the remainder of his life to that charge, rising in the estimation of the community until he was recognized as a moderate, highly learned man with a graceful, even elegant, style of preaching and writing. Many of his sermons were published. By his learning and his pleasing personality, Colman was able to bridge the chasm dividing some of the factions of the Boston community, especially between the Mathers and their adversaries. He declined the presidency of Harvard when he was chosen as successor to John Leverett in 1724.

Already he had served as the principal agent to persuade the London merchant, Thomas Hollis (1659–1731), to contribute to Harvard (1719) a parcel of books and about £300 for poor scholars, gifts which were followed (1720) by nearly £700 for scholarships for students in divinity and (1721) an annual endowment of £40 for a chair subsequently known as the Hollis Professorship in Divinity. All this prepared the way for Isaac Greenwood (see below) to touch Hollis for an endowed chair at Harvard in mathematics and natural philosophy, which was established in 1727 with Greenwood as the first incumbent. To the £1,200 as endowment for the Hollis Professorship in Mathematics and Natural Philosophy, the generous benefactor added a new telescope and other scientific apparatus for Harvard. Hollis donated additional scientific equipment before his death in 1731, and members of his family subsequently provided more.[118] Indirectly, Benjamin Col-

---

118 I. Bernard Cohen, *Some Early Tools of American Science* . . . (Cambridge, Mass., 1950), pp. 6, 12, 133ff. Probably the best sketch of Colman is in Shipton, *Sibley's Harvard Graduates*, IV (Cambridge, Mass., 1933), 120–127. See also Theodore Hornberger, *Scientific Thought in the American Colleges, 1638–1800* (Austin, 1945), pp. 26–27; and Michael Kraus, *The Atlantic Civilization: Eighteenth-Century Origins* (Ithaca, 1949), pp. 76–78.

man was responsible for the promotion of scientific work in New England to a very substantial extent.

More directly, Colman lent active support to the Mathers and Boylston during the controversy about inoculation for smallpox in Boston. He published *Some Observations on the New Method of Receiving the Small-pox, by Ingrafting or Inoculation* (Boston, 1721), in which he praised Boylston's regimen with those whom he inoculated—a regimen which Colman observed "from my house (which faces into the Doctor's [Boylston's] yard"—and he suggested the hypothesis, developed further by Mather's *Angel of Bethesda*, that smallpox was transmitted by the inhalation of "Animalcules," swarms of which "Animated Atoms" had been discovered by the microscope in every pustule of the smallpox. Colman's book was a modest, well-written defense of Boylston's work. Paul Dudley sent a copy to Henry Newman in London, and Newman sent it to Dr. Jurin, who presented it to the Royal Society (April 12, 1722) with Newman's comment that "Mr. Colman is my old acquaintance, and has always had the esteem of a person of great veracity, and is not only Justly reckon'd among the first Rank of Divines, but among the best Philosophers in that Country. . . ."[119] A year later (March 8, 1722/23), Colman sent Newman a letter which was similarly transmitted to the Society (May 16, 1723) with a complete statistical summary of the results of the smallpox in the Boston area, as follows:

> Some honest & appointed men have gone from house to house thro' the Town, & made the following report
>
> Had the Small Pox in Boston in the Natural Way . . 5742
> Of these died 841
> Inoculated in the Town 147
> Of these Died 3
> Inoculated in the neighboring Towns
> By Dr. Boylston 98
> Of these died 4
> By Dr. Thompson . . . . about 30
> By Dr. Roby . . . . 7 or 8
> So that seven were all that died after inoculation in Town or Country. Of five of these he gives much ye same account as had

119 Newman to Jurin, London, April 10, 1722, *Letter-Book*, XVI, 253–254.

been given by Capt. Osborne, but was not acquainted with the particular circumstances of the other two.[120]

With the ice broken by his friend Henry Newman, Colman wrote directly to Dr. Jurin (May 15, 1723) to describe a "small cock humming bird" and "a small fish called Hypocampus" (*Hippocampus*, a sea horse), which he presented to the Society.[121] Another example of his interest in science was a letter to White Kennett, Bishop of Peterborough (September 5, 1728), to describe the earthquake in New England on October 29, 1727. Bishop Kennett had requested Colman to send his meteorological observations, and Colman sent the account of the earthquake, saying that although he did not know what New England Fellows of the Royal Society had reported about it, "my desire is that what relates to ye Earthquake may be communicated to Dr. Jurin or some Gentlemen of ye Royal Society."[122] The Bishop did so, and the letter was presented to the Royal Society on April 17, 1729, and published in the *Philosophical Transactions*.[123] It corroborated other reports to the effect that the center of the disturbance had been at Newbury, that

[120] "Extract of a Letter from the Revd Mr. Colman of Boston in New England to Henry Newman Esq of the Middle Temple, dated March 8th 1722/3," *Classified Papers, 1660–1740*, XXIII (1) ("Inoculations"), No. 35. On the left margin are Dr. Jurin's computations of the odds: "$147 + 98 + 30 + 7 = 282$. $5742:841::6:8:1$; $282:7::40+:1$." See also *Journal-Book*, XIII, 287–288. Captain Osborne's account was Captain John Osborne, "A brief Account relating to the Inoculation of the Small Pox in New England viz: what number came under that operation, how many died that underwent it; and likewise the number of those who died of the small Pox in the natural way, and of those who recover'd." Dated London, Jan. 30, 1722/23, and presented to the Royal Society Jan. 31, 1722/23, *Letter-Book*, XV, 360–363. Both Osborne and his wife, living in Roxbury at the time, had been successfully inoculated by Mather's nephew, Thomas Walter. Osborne concluded his account with the observation that "whenever it shall please God to visit a people with any sore judgement or affliction, and at the same time to open or discover a way whereby the violence of it may be appeased & mitigated; whether it is not their duty thankfully to embrace it." Dr. Jurin added: "Verily it is."

[121] *Journal-Book*, XIII, 337; presented Dec. 19, 1723.

[122] Colman to Bishop Kennett, Boston, Sept. 5, 1728, *Guard-Book*, C-2, No. 78. Colman also commented favorably about Governor William Burnet (1688–1729), F.R.S., of Massachusetts, and his famous father, Bishop Gilbert Burnet, who had befriended Increase Mather in London during the negotiations for the Massachusetts charter of 1691.

[123] *Journal-Book*, XIV, 318–320; *Phil. Trans.*, XXXVI, No. 409 (May–June, 1729), 124–127.

the quake threw up "many cart-loads of a fine Sand and Ashes" smelling of sulphur, and was succeeded by lesser quakes in the days following. But Colman did not speculate about the nature and cause of earthquakes. This letter and his earlier communications illustrated, however, his interest in science and his association with others in the Boston community of scientifically oriented men.

Isaac Greenwood (1702–45)

Born in Boston, Isaac Greenwood graduated from Harvard College in 1721, a worthy pupil of his tutor Thomas Robie. Early caught up in the controversy over smallpox inoculation, he wrote a satirical defense of Cotton Mather and Zabdiel Boylston entitled *A Friendly Debate; or, A Dialogue Between Academicus* [Boylston] *and Sawny* [Dr. William Douglass] *& Mundungus* [John Williams], *Two Eminent Physicians, About some of their Late Performances* (Boston, 1722). A youthful indiscretion, the satire identified Greenwood's sympathies in the controversy, but also heightened the personal nature of the attacks, especially as Greenwood represented Dr. Douglass as a semiliterate libeler with "a liberal Education at Billingsgate." Greenwood studied for the ministry with Cotton Mather for a time, but a desire to see more of the world led him to England in 1723, armed with letters of introduction to Dr. James Jurin from both Cotton Mather and Thomas Robie. In London his early activities are uncertain. He may have studied medicine; it is recorded that he preached to Dissenter congregations and in this manner became acquainted with Thomas Hollis, the benefactor who had endowed Harvard's professorship of divinity. But his greater interest in mathematics and natural philosophy led him to consort more and more with English men of science. Probably his introduction to Dr. Jurin smoothed the way, and he came to know William Derham, visited the Royal Society, conversed with Sir Isaac Newton, and met several Fellows of the Society, some of whom later became his correspondents. He became closely associated with Dr. Jean Théophile Desaguliers (1683–1744), F.R.S., Curator of Experiments at the Royal Society, inventor of the planetarium, and a popular lecturer and writer about Newtonian physics, astronomy, and mechanics, all of which he demonstrated

446

21. Cotton Mather (1663-1728), whose Christian ministry impelled him to embrace such a concern for and knowledge of the new science that he became a transitional figure in colonial thought and a herald of the maturation of colonial science.

Courtesy of the American Antiquarian Society.

22. Isaac Greenwood (1702-45), nurtured by Cotton Mather, went beyond his mentor to become the first Hollis Professor of Mathematics and Natural Philosophy at Harvard College. Although his career was foreshortened by intemperance, he promoted science widely and helped to make it a vital force in New England culture.

Courtesy of the New-York Historical Society.

by means of a variety of cleverly contrived pieces of apparatus. Greenwood attended Desaguliers' lectures, became his assistant, and it was said that on occasion Greenwood himself gave the lectures and demonstrations when Desaguliers was indisposed or otherwise unable to be present. By midsummer, 1725, Greenwood had determined to acquire apparatus of his own and return to Boston to teach mathematics and natural philosophy after the manner of Desaguliers. He consulted with Hollis about this design, and learned that Hollis was planning to establish, by his will, a professorship of mathematics and natural philosophy at Harvard College. Greenwood persuaded Hollis to set up the chair at once, and to designate him as the first incumbent. Hollis agreed, although he was dismayed when Greenwood suddenly departed from London (July, 1726), leaving behind him unpaid debts of about £300, including board and room owed Dr. Desaguliers and such fripperies as "3 pair perle colour silk stockings."[124]

Greenwood arrived in Boston fired by the example of his London mentor, Desaguliers, and anxious to impress the Harvard authorities in order to win the appointment there. He brought with him apparatus for scientific demonstrations, and soon (January, 1727) launched a series of sixteen public lectures, the first of its kind in New England and a model of such lecture series, which soon were to become widespread in all of the American colonies. Greenwood advertised his lectures widely, including the publication of a kind of "outline series" as *An Experimental Course in Mechanical Philosophy* . . . (Boston, 1726). According to this, the course enabled subscribers (at four shillings a head) to acquire "a competent Skill in Natural Knowledge" by means of "various Instruments, and Machines, with which there are above Three Hundred Curious, and useful Experiments performed," the whole intended to make "such Persons as are desirous thereof . . . better acquainted with the Principles of Nature, and the wonderful Discoveries of the incomparable Sir Isaac Newton. . . ." The performances persuaded the Harvard Corporation to appoint Greenwood, in spite of Hollis'

[124] Shipton, *Sibley's Harvard Graduates*, VI (Boston, 1942), 471–480. See also Leo G. Simons, "Isaac Greenwood, First Hollis Professor," *Scripta Mathematica*, II (1934), 117–124; Morison, *Three Centuries of Harvard*, pp. 79–80; Hornberger, *Scientific Thought*, p. 44.

hesitation, as the first Hollis Professor of Mathematics and Natural Philosophy. He assumed the chair in 1727 and was formally installed in February, 1728, one of the first men in the colonies able to make his living by science.[125]

In the same year Greenwood undertook a scientific correspondence with Dr. James Jurin, Secretary of the Royal Society. His first letter, dated May 7, 1727, was written in behalf of Cotton Mather, who, but for a temporary indisposition, "would otherwise have wrote Himself." Mather had long endeavored "to engage his Friends in making Meteorological Observations" in response to Jurin's invitation, but "the unskillfulness of Some, & the Business of others have I think hitherto prevented his Design." At last, however, Mather had "prevailed upon an ingenious Tradesman of this Town Mr. [Grafton] Feveryear by Name, by presenting Him with a Barometer, & some other things as to obtain the Inclosed Observations." Evidently this was the first barometer used for meteorological observations by New Englanders, as Robie's reports, which Greenwood had forwarded to William Derham after his return from London in October, 1726, had been made without instruments, and Mather wanted to assure Jurin of the accuracy of Feveryear's work. Greenwood also volunteered on his own behalf to send the Royal Society "an annual Account of the State of the Weather &c. in these Parts of the World, especially at Cambridge," where he had recently been "chosen by the College their Hollisian Professor of the Mathematics & Experimental Philosophy, wch place, has some peculiar Advantages for Observation, above most of the same Nature in the World," by virtue of its "very large Apparatus of Glasses, and other Instruments" (many recently furnished by the largess of Thomas Hollis), together with "10 Pensionary Scholars of the 2 upper Classes who will always be ready to continue on the Observations in case of Sickness, absence, or any other Accident." Further, Greenwood promised astronomical reports "of Such Eclipses &c as may occur," as the college provided

[125] The College of William and Mary had appointed one Mr. Le Fevre as Professor of Natural Philosophy in 1711, but he was discharged in 1712 for consorting with a "London hussy." The Reverend Hugh Jones succeeded him, introduced algebra to William and Mary students (1717–22), but did not become the popularizer of science that Greenwood became a few years later. Hornberger, *Scientific Thought*, pp. 25–41.

"very good Instruments for that Design," including "the same Quadrant that Dr. Halley had to observe on the Southern Constellations at St. Hellena, besides several good Telescopes."[126]

Three days after writing the above letter in Mather's behalf, Greenwood addressed Dr. Jurin to present "A New Method of Composing the Natural History of Meteors." He was inspired to compose his thoughts by Jurin's "Invitatio ad Observationes Meteorologicas" mentioned in the letter above. He believed that his "Imperfect Thoughts" were "new, and uncommon," and if they were likely to be "conducive to the Promotion of Physical Knowledge, I doubt not in the least, but that you will improve them in such a Manner as that they become most useful." They required "Care & Application," but "the Advantage that may accrue from thence, to Posterity, [would be] a sufficient Ballance to the little Inconveniences there may be therein to the present Age." Briefly, Greenwood's "New Method" proposed that the Royal Society of London, in cooperation with the French Academy of Sciences, serve to collect standardized reports (a form was set forth) on winds and weather at sea from seamen sailing to all parts of the world. The Royal Navy already kept such journals, and the cooperation of commercial ship captains should be enlisted. Greenwood felt that many years would pass before similar weather observations could be systematically made on land, even in the capitals and university towns of Europe—and less could be expected from Africa, Asia, and the Americas. But the possibilities of obtaining reports from the great numbers of vessels plying the seas of the world was great, "if I may judge from the Trade of the little Town where this Letter is dated, there must be many Thousands; for there are seldom less than 8. or 9. Hundred voyages made to and from this Port in a year. . . . I need not say of how much Importance it would be to the Trading part of the World, were We able to define the more frequent & reigning Winds of every Climate [so that sailors might plan voyages in order to arrive] in the most probable Dispatch to his Port." Greenwood envisaged that, in time, charts could be prepared to show the limits of the winds at sea, their direction and velocities in different latitudes and in the various sea-

126 Greenwood to Jurin, Boston, May 7, 1727, *Guard-Book*, G, Pt. II, No. 6; *Letter-Book*, XVIII, 428–430.

sons of the year, the direction and speed of ocean currents in the various seas, and the variations of the compass. Perhaps, too, hurricanes could be accurately forecast, unusual events, such as tricks of lightning, might profitably be noted, "any Extraordinary Rendezvous of Fish &c that are used in the affairs of Life" might be spotted, and the location of ports could be perfected.[127] Thus, well over a century before Matthew Fontaine Maury issued his *Wind and Current Chart of the North Atlantic* (1847), Greenwood anticipated a similar work for all of the oceans of the world, to be compiled by the joint efforts of the two foremost scientific organizations of the time. It was something new for a colonial to outline major useful tasks for the Royal Society to undertake, and though the Society collected many meteorological bits and pieces, it published nothing comparable to Greenwood's vision.

Dr. William Rutty succeeded James Jurin as Secretary of the Royal Society (1727), and Greenwood addressed Dr. Rutty on May 10, 1729, to thank the Society for publishing his "New Method of Composing a History of the Weather" and for the Society's "Compliment" regarding it. He also offered to make further observations upon natural phenomena and begged for instructions.[128] Nine days later he sent Dr. Rutty "A brief Account of Some of the Effects and Properties of Damps." This was an account of damps recently found in two old Boston wells, both of which had lain unused for a time, had been covered, and had become reservoirs "of all the dregs of the neighbouring streets." Two workmen had lost their lives upon descending in one, and a third had been hauled up in a stupefied condition. Greenwood had been called in to investigate them. He lowered lighted candles and lanterns, all of which were extinguished at a depth of about six feet. Both a kitten and a bird died upon being lowered. He tested the "elasticity" of the damps with a small bell and found it to conform to that of ordinary air, and he found that the moisture of the damps was also conformable to that of the air. After leaving the wells un-

---

[127] Greenwood to Jurin, Boston, May 10, 1727, *Guard-Book*, G, Pt. II, No. 7; *Letter-Book*, XVIII, 430–434; *Phil. Trans.*, XXXV, No. 401 (Jan.–March, 1728/29), 390–402.

[128] Greenwood to Rutty, Cambridge, May 10, 1729, *Guard-Book*, G, Pt. II, No. 8; *Letter-Book*, XIX, 393–394.

covered for a few days, the damps were dissipated, and workmen could safely continue their tasks. All this Greenwood related in detail, appending a query to the Society to seek an explanation of these transient subterranean exhalations.[129] Dr. Rutty acknowledged receipt of Greenwood's account in a letter of February 18, 1729/30. He had communicated it to the Royal Society (January 15, 1729/30) and published it in the *Philosophical Transactions*, a copy of which "I would convey to you, did I but know a safe method." He would send it, together with other recent copies of the *Philosophical Transactions*, "as soon as I hear from you." As regards further communications from Greenwood, Dr. Rutty depended upon his judgment, "being so well qualified in this part of Learning." However, Dr. Rutty observed that "The part of Philosophy least known, and where there is a great deal worth knowing is rational chymistry, divested of its Rosicrucian Jargon: there being a multiplicity of now latent properties in Substances which by a thorough Scrutiny might thereby appear, and prove to be of great Use. This we are now prosecuting sedulously, and Experiments made in parts at so great a distance would be much the more valuable, as they may be tried upon Bodies as yet unknown in our Climate. . . ."[130]

If Greenwood tried his hand at "rational chymistry," he did not communicate the results. But on October 24, 1730, he addressed Dr. Rutty from Cambridge, thanked him for the copies of the *Philosophical Transactions*, which had been received, and enclosed a twelve-page "Account of an Aurora Borealis seen in New-England on the 22ᵈ of October, 1730. . . ."[131] Dr. Rutty had died (June 11, 1730) before Greenwood had written; but his successor as Secretary of the Society, Dr. Cromwell Mortimer, noted that Greenwood's "Account" was "an accurate and minute description." Greenwood had included eight drawings of the phenomenon to

---

[129] *Journal-Book*, XIV, 400–402; read Jan. 15, 1729/30; *Phil. Trans.*, XXXVI, No. 411 (Oct.–Dec., 1729), 184–191.

[130] Rutty to Greenwood, London, Feb. 18, 1729/30, *Letter-Book*, XIX, 480–481.

[131] Greenwood to Rutty, Cambridge, Oct. 24, 1730, *Guard-Book*, G, Pt. II, No. 10; *Letter-Book*, XIX, 416–417; presented March 4, 1730/31. The letter was delivered by one of Greenwood's former students at Harvard, and Greenwood begged Rutty "to direct Him how He may see the celebrated Musaeum &c of the Royal Society."

illustrate the aurora from different angles and at different times, all carefully tied in with the account. Mortimer presented it to the Royal Society on March 4, 1730/31, and the "Account" was published in the *Philosophical Transactions*.[132]

In the meantime, inquiries had come to the Royal Society which led to Greenwood's last communication. On March 13, 1725 (N.S.), John Villa, a Northamptonshire clergyman who had gone to Berlin, in Prussia, to teach English to the Princess Royal, addressed "A Letter to Dr. Jurin relating to an Inscription upon a Rock in America." Villa had seen an account of the inscription on Dighton Rock in the *Philosophical Transactions* for 1714 (supplied by Cotton Mather). He had showed it to one Mr. De la Croze of Berlin, who was skilled in reading ancient Oriental inscriptions. The latter had suspected that the inscription was Japanese, but he found it unskillfully reproduced and asked for a more exact copy. Accordingly, Villa sought Jurin's aid in obtaining another reproduction of the inscription "by some Judicious and skillful hand." Jurin applied to Cotton Mather, but received no reply. Whereupon Jurin consulted with John Eames, F.R.S., a knowledgeable tutor of science, who applied to Isaac Greenwood.[133] Greenwood sent two figures, one of the entire face of the rock, and one with a detail of the inscription, on December 8, 1730. He stated that he had been assisted by a Reverend Mr. Fisher and others, presumably from near the site of Dighton Rock in Taunton River.[134] But unfortunately the figures miscarried. Eames again applied to Greenwood, who, on April 28, 1732, sent new copies of the inscription, "one taken by himself in September, 1730, the other formerly by the Revd Mr. [John] Danforth in the year 1680."[135] He also furnished a description of the stone, its position in the river, and an Indian tradition

[132] XXXVII, No. 418 (March–May, 1731), 55–69. Mortimer wrote to Greenwood, Oct. 25, 1731, to tell him of Rutty's death, to send a copy of the *Philosophical Transactions* with Greenwood's "Account," and to beg his continued correspondence. *Letter-Book*, XIX, 494–495. The same number of the *Phil. Trans.* (pp. 69–72) included a description of the aurora as seen in Maryland by Richard Lewis of Annapolis, communicated by Peter Collinson.

[133] See Villa's letters in *Letter-Book*, XVIII, 168–169; XIX, 354–355.

[134] Greenwood to Eames, Cambridge, Dec. 8, 1730, *Sloane 4432*, fols. 185–190.

[135] *Letter-Book*, XIX, 438; read June 15, 1732, *Journal-Book*, XV, 149–151. The inscriptions supplied by Greenwood, including the original by Danforth in 1680, are in *Register-Book*, XVII (1731–32), 214–219.

given by Danforth, possibly relating to the meaning of the inscription to the effect "that there came a wooden house, and men of another country in it, swimming up the River of Asoonet, that fought the Indians, and Slew their Sachem &c." In comparing the two copies of the inscription supplied by Greenwood, the Society recorded that "it manifestly appears that the Characters are much impaired by time, and the traces grown much fainter than they were formerly." But no one hazarded a guess as to the meaning of the inscription—nor was it reported what, if anything, Mr. De la Croze made of it. Indeed, the curious inscription on Dighton Rock is a puzzle to the present day, and the descriptions and copies of it supplied by Greenwood are still useful for its study.[136]

Besides his communications to the Royal Society, Greenwood continued to promote Newtonian science in New England. His *Arithmatick Vulgar and Decimal . . .* (Boston, 1729) was the first American publication on the subject, and it improved the teaching of mathematics throughout New England.[137] His lecture on the death of Thomas Hollis (April 7, 1731), published as *A Philosophical Discourse Concerning the Mutability and Changes of the Material World* (Boston, 1731), expounded Newtonianism further and went far to demonstrate that Greenwood had forsaken the apocalyptic view of the world and subscribed to a purely mechanistic explanation. When Hollis' nephew of the same name sent over additional philosophical apparatus in 1732, including a double microscope, a large armillary sphere, and "a very costly Orrery"— the first in America—Greenwood gave public lectures (1734) on the orrery that attracted attention as far away as Philadelphia. A Boston versifier celebrated the lectures—published as *Explanatory Lectures on the Orrery . . .* (Boston, 1734)—in a sixty-four-line outburst, including the following:

> Greenwood, with what Delight we hear you prove
> The hidden Laws by which those Bodies move,

136 For a thorough review of the evidence regarding Dighton Rock, see Edmund Burke Delabarre, "Early Interest in Dighton Rock," Publications of the Colonial Society of Massachusetts, *Transactions, 1915–1916*, XVIII (1917), 235–299. The evidence above supplements that of Professor Delabarre but does not conflict with it.

137 Leo G. Simons, "Isaac Greenwood's Arithmetic," *Scripta Mathematica*, I (1933), 262–264.

Describe the Rings that shape their rapid Course,
And bring to light Attraction's wondrous Force.
Now kind'ling Meteors through the Heavens may rise,
And blazing Comets sweep along the Skies:
No more we'll gaze with superstitious Fear,
While you the Secret Laws of Nature clear. . . .
Long may you stay below to clear our Sight,
Inform our Mind, and set our judgment right:
And when your Soul shall upward take its way,
To join great NEWTON in the Realms of Day,
Nature unveil'd, you clearly shall behold,
And find her such as you before had told,
Grasp all her Motions at one single View,
And be well pleas'd to find your Doctrine true.[138]

During summer holidays Greenwood often gave series of public lectures and sometimes taught private classes in mathematics. In the midst of these activities, however, he struggled constantly with the demons encaged in alcohol, and by 1737 "confessed the Charge of intemperance." The Harvard Overseers let him off with an admonition and his promise of reformation. But within a year he had tried their patience too far, was removed from his chair (July 13, 1738), and his apparatus (which he attempted to claim as his own) was retrieved for the college. He then "set up a School of experimental Philosophy" in Boston, advertising mathematics, both theoretical and practical, and various branches of natural philosophy, "as Mechanics, Optics, Astronomy, &c."[139] He was reputed to be an excellent teacher, but the demand for his services fell short of his needs. By 1740 he had sold off most of his properties and tried his luck lecturing in Philadelphia, where Benjamin Franklin arranged the loan of apparatus from the equipment owned by the Library Company. But he had no better success there, returned to Boston (1742), and, falling further and further into drunkenness, went to

[138] *Pennsylvania Gazette*, No. 322 (Jan. 28–Feb. 4, 1734/35), p. 2. Dated at Boston, Dec. 16, 1734, and addressed "To Mr. Greenwood, Hollisian Professor of Mathematicks and Astronomy at Cambridge, occasioned by the late astronomical lectures," the lines were unsigned. They were probably reprinted from the *Boston Gazette* of Dec. 16, 1734, but that particular issue appears to be no longer extant.

[139] *Boston Weekly News-Letter*, No. 1828 (March 30–April 5, 1739).

sea, dying on a voyage to South Carolina (October 12, 1745).

Still, in spite of his regrettable last years, Isaac Greenwood contributed greatly to making science a living part of New England culture. Although an accurate observer of scientific phenomena and a rational philosopher in the tradition of the eighteenth-century Enlightenment, he contributed comparatively little to advance science as a whole. He kept well abreast of the newest scientific developments in England and cited in his works materials from the latest copies of the *Philosophical Transactions*, from Stephen Hales's *Vegetable Staticks* (London, 1727), a pioneering work in plant physiology, and other contemporary works. He went far to improve mathematical studies at Harvard and elsewhere in New England. His popular lectures and visual demonstrations with scientific apparatus after the manner of his former master, Dr. Desaguliers, were imitated in every colonial college, and were soon the model of many traveling public lecturers in cities and towns throughout the mainland colonies—their success an index of the growing popular interest in science. Scientific observations, experiments, and demonstrations, as well as books, were becoming a widespread method employed in teaching natural philosophy. And in addition to these things, Isaac Greenwood, the pupil of Thomas Robie, whom he excelled, raised up, in turn, a pupil of his own who was destined to become one of the greatest colonial scientists. This man was John Winthrop (1714–79), great-grandnephew of the John Winthrop who, as the first colonial Fellow of the Royal Society, began promotion of the new science in New England in the 1660's. Professor Winthrop, as he became known (for he succeeded Greenwood as Hollis Professor at Harvard and held the chair until his death in 1779), became a productive scholar, and he carried forward mathematics, astronomy, and experimental physics at Harvard, bringing them to reputable heights during the last generation of colonials in the Old Colonial Era.

Paul Dudley (1675–1751), F.R.S.

Paul Dudley belonged to a haughty, aristocratic Massachusetts family which reached back to the Bay Colony's beginnings and was connected by marriage, in Paul's time, with the Winthrops and the

Sewalls. His father, Joseph Dudley, was prominent in politics, serving as Commissioner of the New England Confederation (1677–81), President of the Dominion of New England (1686–89), Chief of the Council of New York (1690–92), and Governor of Massachusetts (1702–15), where his autocratic management of affairs made him unpopular, especially with the Mathers. Paul attended Harvard College (A.B., 1690; A.M., 1693), studied law in Boston and at the Inner Temple in London, was called to the bar, and was appointed Attorney General of Massachusetts in 1702. He subsequently had a successful career in law and politics, was appointed Judge of the Superior Court of Massachusetts in 1718 and Chief Justice in 1745. During these years he overcame the handicap of his father's unpopularity, became highly esteemed and honored as a great lawyer, and even Cotton Mather came to accept him kindly.[140]

Both Joseph and Paul Dudley exhibited an interest in natural history, and both, by their sojourns in London, had come to know Fellows of the Royal Society and entered into correspondence with them from New England. About 1701, Joseph Dudley submitted to John Chamberlayne two papers which Chamberlayne presented to the Royal Society (February 11 and March 11, 1701/02), and urged their publication in the *Philosophical Transactions*.[141] The second, which appears to have been lost, set forth an alleged "Solution of that nice question how Women in the State of Innocency [i.e., before the Fall of Adam] could have been Freed of the Pains of Child-birth &c"; the first, which is preserved, was a statement of Dudley's opinions about "the Circulation of the Several Juices in Fruit Trees" with an ingenious explanation of the alteration of fruit by grafting. Joseph Dudley held that

[140] Dean Dudley, *The Dudley Genealogies and Family Records* (Boston, 1848), pp. 73–74, *passim;* Dudley, *History of the Dudley Family* (2 vols., 2 supplements, Wakefield, Mass., 1886–1901), I, 521–526; *D.A.B.;* Shipton, *Sibley's Harvard Graduates*, IV, 46–54.

[141] *Journal-Book*, IX, 295, 301. See also Chamberlayne's letters to Sloane in 1702–03 appealing for the return of Dudley's communications, for "as we thought the Royall Society would certainly publish them, we kept no copies thereof," and Chamberlayne sought to return them to Dudley. *Sloane 4038*, fol. 349; *4039*, fols. 20, 44, 51, 73, 87. Evidently Sloane found and returned the account of fruit trees, as it is in *Guard-Book*, C-2, No. 38.

in Animalls, as in fruits, every one originally was distinct & perfect suo genere, & had no inclination nor borrowing from another, & that all alterations & mixtures are the Luxuriant Device (though not unlawfull) of men, & that he that first grafted a Tree was brother to Anak that found mules in ye Wilderness . . .[142] the Curse & miserable alterations of the Seasons since the fflood . . . [led to] the true Reason, Use, & Benefit Originally of Grafting . . . to Carry Fruits more Northward or Southward, as farr as they will bear the Alteration of the Sun, which by experience is found to be done much further by a cyon [cion] set upon a root in its own Country, then by the Removing of the whole plant; the Nurse being at home in her own Soil, though the Bratt be Forreign. . . . the Constituent parts of all Fruits being the Rind, the Flesh, & the seed or heart Answers perfectly to those of the Tree, Bark, Body & Pith, & accordingly the Nourishment of each of them passes through the Several Parts of the Tree into the Fruit. . . .

The Cause of the alteration of Fruit upon Grafting must needs be principally, the Filtration of the Juice through the Sieve or fine Webb made at the setting on of the Cyon; it cannot be the Juice from the Earth, that's the same as with a Crabb; it cannot be only the formation & Tincture of the Cyon, because the same Cyons differ much upon a thorn, upon a Crabb, upon a Large Wilding; but above all the Demonstration of the Value & Influence of ye Filtre appears in doubling, & repeating it, which you may do as often as you please, a Pippin grafted upon a Crabb is a good Pippin, Cutt it off, & sett it upon itself, it is a Renate [rennet], Larger, better digested, thinner skinn'd then before, & so it will bear a much higher advance in the repetition. . . .

There are by the abovesaid Methods of Grafting . . . Infinite Confusions & mixtures brought into all sorts of Fruits of Different species, which make so great Alterations that pusle all Philosophers to solve, saving alwaies you may observe, that Vines & Shrubbs, stoned Fruits & seed fruits, will keep Distinct, & not to be Jumbled. . . .

From wt is above you see the Reason why a Tree propagated from a seed of ye best grafted pippin or Renate will be again a wild apple, because the Nourishment that made the seed passed by the Pith wch knows nothing of ye fine sieve where ye Juice past nearest the Bark that made the flesh of the former apple, &

142 Genesis 36:24.

it bringing nothing with it but the Juice & Pith of the Centre of the Tree wch has nothing to do with the flesh of the Apple. . . .

From hence also ye Vulgar opinion of the Descent of ye sapp of trees annually in Autumn & return in the Spring is demonstrated to be in error, for if there were so great a flux & reflux of Digested Juice it would come to pass in a few years . . . yt the Pipes & Conveyances of ye stock below would be so tinctured with that better Juice of ye graft above that it would lose all it's originall ill nature, the contrary whereof is every day seen . . . it is apparent no Juice can possibly return downwards being spent wholly in Leaves, fruit, & additionall Substance of the tree which demand a great Quantity of sapp to be condensed into such sollid substance. . . .

For one who knew nothing of callus tissue, and cites no experimentation or microscopic observations, Dudley's explanations were arresting.

Paul Dudley may have imbibed his interest in natural history from his father, and surely he inherited some of his father's personal connections with Fellows of the Royal Society.[143] When he sent his first communication to the Royal Society in 1719, he dispatched it to his father's friend, John Chamberlayne, and accompanied it with a letter expressing warm sentiments of personal regard to Chamberlayne, with "my Duty to the Royal Society," "my Dame Chamberlayne," and "Mr. [Henry] Newman."[144] The communication described the making of maple sugar, "our Physitians look upon it not only as good for the Common Use as the West Indie sugar But to Exceed all other for its medicinall Vertue." And he proposed to send additional communications and curiosities from New England, especially "with respect to Locusts (I mean the Insect)." His account of maple sugar was presented to the Royal Society (October 29, 1719) and published in the *Philosophical Transactions,* although Sir Hans Sloane rightly objected that it "was no new discovery but had been practiced in Arcandia [Acadia] in

---

[143] *Journal-Book,* IX, 57 (Nov. 15, 1704), records that "Mr. Dudley" was permitted to visit the Royal Society. Which Mr. Dudley—Joseph or Paul—is not revealed.

[144] Paul Dudley to Chamberlayne, Boston, June 20, 1719, *Guard-Book,* D-1, No. 72; *Letter-Book,* XVI, 167–169.

New France & the same method was communicated to the Society many years ago."[145]

On March 16, 1720/21, John Chamberlayne sent to Sir Hans Sloane a letter "from the Ingenious Mr. Dudley," dated from Boston the previous January 7, giving "An Account of the Poyson-Wood-Tree in New England."[146] Dudley's description of the tree, probably poison sumac, one of several varieties of *Rhus*, was supplemented by another from William Sherard which he had received from Thomas More, the elusive "Pilgrim Botanist," who was currently collecting botanical specimens in New England.[147] Both accounts were published in the *Philosophical Transactions*, and specimens of the tree, with seeds, were received from Dudley, More, and Catesby.[148] The following June 20, the Society heard another Dudley paper about "a Method lately found out in New-England for Discovering where Bees Hive in the Woods, in order to get their Honey." Although often attributed to Paul Dudley, the manuscript sources suggest that it may have originated from his father, Joseph Dudley.[149] The paper, published in the *Philosophical Transactions*, described how to prepare a chart with compass and ruler to set the courses of wild bees returning to their "hives" and locating the "hive" by intersecting lines—all postulated upon the belief that bees always returned to their hives in a straight line. It also observed that honeybees had been introduced to America from England "near a hundred years ago," that none was indigenous to New England, as witnessed by the fact that the Indian tongues had no word for honeybee.[150]

[145] *Journal-Book*, XII, 363–364, 369; *Phil. Trans.*, XXXI, No. 364 (Jan.–April, 1720), 27–28.

[146] *Journal-Book*, XIII, 76–77; Chamberlayne to Sloane, March 14, 1720/21, *Sloane 4046*, fol. 74.

[147] See George Frederick Frick and Raymond P. Stearns, *Mark Catesby: The Colonial Audubon* (Urbana, Ill., 1961), Appendix, pp. 114–127.

[148] *Phil. Trans.*, XXXI, No. 367 (Jan.–April, 1721), 145–148.

[149] The *Journal-Book*, XXXI, 29–31, refers to it as from "Col. Dudley, Gov'r of N.Y. communicated by Mr [Thomas] Bates" (June 30, 1721). This compounds confusion, but it seems unlikely that the Society's Secretary, who knew both Dudleys, would have referred to Paul as "Col. Dudley" and have associated him with New York. Moreover, Paul Dudley, up to this time, had sent all his communications for the Society via John Chamberlayne.

[150] *Phil. Trans.*, XXXI, No. 367 (Jan.–April, 1721), 148–150.

Paul Dudley wrote to Chamberlayne (March 15, 1720/21), commenting jocularly that "I had sent you some further Curiosities from New England by this Conveyance but that I am affraid of falling under your Censure & Dr. Mather for sending them by Douzens."[151] Indeed, the number of Dudley's communications to the Royal Society in the early 1720's threatened momentarily to rival the copiousness of Cotton Mather's *Curiosa Americana*. He sent "A Description of the Moose-Deer in America," an effort to improve upon John Josselyn's earlier account, and to distinguish between the light grey moose, said to herd together, and the larger black moose, which was less gregarious, only four or five commonly being seen together. It appears likely that Dudley had consulted with Massachusetts Governor Samuel Shute, who sent a moose to Samuel Dale of Braintree about this time, and there was much speculation in England, based upon the works of John Josselyn, John Clayton, Daniel Neale's recent *History of New England* (2 vols., London, 1720), Robert Beverley's *History and Present State of Virginia* (London, 1705; 2nd ed., 1722), Mark Catesby's specimens of deer horns, a specimen of antlers sent to Dale by Catesby's sister Elizabeth Cocke, of Virginia, Dudley's account, and Governor Shute's moose—all in an effort to distinguish clearly among American deer, elk, and moose, though without clear results.[152]

On the same date (March 15, 1720/21) Dudley sent a lengthy communication to Chamberlayne which, possibly for reasons of political expediency, was not published in the *Philosophical Transactions*. Of "A Short Account of a Trading Voyage performed by Joseph Kellug An English Man of New England in Company with Six French Men from Canada to Mississippi in two Cannos made of Birch Bark, with some general remarks made by the Said Kellug," Dudley wrote that "The Journall of Mr. Kellug's Trading Voyage from Canada to Missasipi is what I took from his own mouth and then digested into the method you see, and tho' he be not a man of

151 *Guard-Book*, D-1, No. 73; *Letter-Book*, XVI, 195–196.

152 See Dudley's "Description" in *Phil. Trans.*, XXXI, No. 368 (May–Aug., 1721), 165–168; Samuel Dale in *ibid.*, XXXIX, No. 444 (Nov.–Dec., 1736), 384–389; Dale to Sloane, Braintree, March 26, Sept. 26, 1732, *Sloane 4025*, fols. 96, 98v; *Letter-Book*, XX, 90–96; *Journal-Book*, XIII, 91–93. Dale concluded that Dudley's "Description" was the best account of moose at hand.

Letters, yet has so much probity and ingenuity that you may depend upon the truth of what he says. . . .”[153] Joseph Kellogg (1691–1756) had been carried into Canadian captivity in 1704, when the French and Indian attack destroyed the frontier Massachusetts village of Deerfield. He remained a prisoner in New France until 1714, when he was released to return to Massachusetts. During these years he demonstrated unusual abilities in the Indian tongues and accompanied a French trading voyage as interpreter in 1710, possibly the first Englishman to penetrate the Illinois country.[154] Dudley found that Kellogg’s firsthand information about the geography of the Great Lakes region made it possible for him to correct errors in John Senex’s map of “North America, Corrected from the Observations Communicated to the Royal Society at London, and the Royal Academy at Paris” (London, 1710), and other North American maps later included in Senex’s *A New General Atlas . . . of the World* (London, 1721).[155] With this in mind, he prepared an account of Kellogg’s observations, obviously made with a copy of Senex’s map of 1710 spread out before the two men.

> I have sometimes thought [he wrote Chamberlayne] it might not be improper to communicate his corrections and observations to Mr. [Anthony] Hammond sometime of the Navy to whome Mr. Senex’s Map of North-America is dedicated or to Mr. Halley of the Royal Society to whom that of South America is addressed and who indeed corrected it. But it is yours and I leave it wholly to your disposal. If you should send it to either of those Gentlemen I must pray you to do me the Hon^r of mentioning my name with Respect to them; without doubt Mr. Senex will be thankfull of a sight of it. . . .

Chamberlayne sent Kellogg’s “Journall” to Sir Hans Sloane, who presented it to the Royal Society (May 11, 1721), where it provoked a careful study of Senex’s map, with revisions of the locations

---

153 See note 151 above.

154 For more complete details, and Kellogg’s journal in full as reported by Dudley, see Raymond P. Stearns, “Joseph Kellogg’s Observations on Senex’s Map of North America (1710),” *Mississippi Valley Historical Review*, XXIII (1936), 345–354.

155 Kellogg also described flora and fauna of the Illinois country.

formerly accorded Lake Nipissing, Lake Superior, Lake Michigan, and Fort Crevecoeur, with correction of the name of "Michigan," "misplaced for another name of Lake Huron whereas it ought to have been placed to the Lake Illinois for another name of that Lake."[156] At the same meeting, Sloane proposed Dudley for fellowship in the Royal Society. The Council approved (October 26, 1721), and Paul Dudley was elected Fellow on November 2 following.[157]

While the Royal Society was considering Dudley's candidacy for fellowship, Dudley himself was off to Albany as an agent to visit the Indians, the result of which journey was summarized in his report entitled "A Short Account of the Names, Situations Numbers &c. of the Five Nations of Indians in Alliance with New York" (1721). To the Royal Society, however, he sent "An Account of the Cataract or Falls of Niagra, taken at Albany October the 10th 1721 by P. Dudley from Monsieur Borrassaw a French Native of Canada." The identity of "Monsieur Borrassaw," whose name was surely misspelled, is unknown. The account stated that "a late measurement" of the falls showed it to be only twenty-six fathoms high, whereas Father Louis Hennepin's *Description de la Louisiane* (1683–84) had described it as 100 fathoms. Dudley also had visited the falls of the Schenectady River in New York, and confirmed reports of large pike and trout found in the lakes of western New York.[158] Shortly afterward, Dudley wrote to Henry Newman of the Middle Temple to enclose various accounts of smallpox inoculation in Boston and "a Commission . . . to buy for him the Philosophical Transactions for the last Twenty years."[159] On April 14, 1722, Dudley wrote to Chamberlayne that "Doctor Mather acquaints me that he sent you the Narrative of the New Island, that Rose up among the Azores in the month of November,

156 *Journal-Book*, XIII, 93–94.

157 *C.M.*, II, 257; *Journal-Book*, XIII, 129. Chamberlayne had recommended Dudley for fellowship to Sloane on May 16, 1721. *Ibid.*

158 *Guard-Book*, D-1, No. 74; read March 29, 1722, *Journal-Book*, XIII, 178–179; *Phil. Trans.*, XXXII, No. 371 (April–May, 1722), 69–72; Dudley, *History of the Dudley Family*, I, 524.

159 Newman to Jurin, April 10, 1722, *Letter-Book*, XVI, 253–256 (April 12, 1722).

1720. . . ."[160] He enclosed a copy of Mather's *A Surprising Relation of a new burning Island lately raised out of the Sea near
Tercera* [Terceira] (Boston, 1721), together with "Six different
draughts of the said Island made by Capt. Robertson, representing
how it appeared to him in so many different views, as he sailed
round it in January and December, 1720 in order to survey it."
Enclosed also was "A Small piece of the Cinder or Pumice Stone
thrown out by the Eruption and taken up by Capt. Robertson, near
the burning Island. It is very light and porous, of a dark sad colour,
insipid, very friable and grows harder by Soaking in water."[161] This
was succeeded (October 25, 1722) by "An Account of a new sort
of Molasses made of Apples; and of the degenerating of Smelts." The
molasses, said to be a good substitute for cane molasses, had been
developed by one J. Chandler of Woodstock, Massachusetts.[162] A
few days later (November 3, 1722) Dudley sent to Dr. Jurin two
letters with various specimens of New England flora and fauna and
"An Account of the Rattlesnake" based upon his own observations
and "from those which he can rely upon." One letter was of espe-
cial interest because Dudley spoke of the activities of Thomas
More, the collector currently supported in New England by Sir
Hans Sloane and other English subscribers. "The diligent and ingenious Mr. More," reported Dudley, "has lost no time in his
searches after the productions of nature in this Countrey since his
arrivall and it has been a great pleasure to me to have given him
some assistance therein."[163] The specimens included two bottles of
the molasses made of apple juice, several plants, many of which
were said to grow larger in New England than in old England, a
hummingbird's nest with young birds in it, a swift, and some smelts
in a bottle of spirits. By their markings Dudley distinguished among
three kinds of rattlesnakes, described the habits of rattlesnakes in
some detail, expressed his belief in their power to charm their victims and in the virtue of bloodroot as an antidote for rattlesnake-

---

[160] *Guard-Book*, D-1, No. 75; *Letter-Book*, XVI, 254–255.
[161] Read May 24, 1722, *Journal-Book*, XIII, 197–198.
[162] *Phil. Trans.*, XXXII, No. 374 (Nov.–Dec., 1722), 231–232.
[163] *Guard-Book*, D-1, Nos. 76, 77; *Letter-Book*, XVI, 340–342; read Jan. 10,
1722/23, *Journal-Book*, XIII, 238–243.

bite, and added his observations on the development of their young based upon dissections of several female rattlesnakes at intervals before they had given birth.[164]

The frequency of Paul Dudley's communications to the Royal Society was reduced after 1722. He was ill during the summer of 1723, and sent (November 13) only one report, "An Account of an Extraordinary Cure by Sweating in Hot Turff, with a Description of the Indian Hot houses."[165] It related the case of seventy-four-year-old Peter Coffin of Exeter, who in 1704 had been cured of a painful "Surfeit" by sweating in hot turf saturated with oil. The treatment had been found useful for the cure of "colds, Surfeits, Sciatica's, and Pains fixed in the Limbs." The Indian "Hot houses" referred particularly to those of Nantucket and were similar to earlier reports. On the same day (November 13) Dudley addressed Dr. Jurin to state that he was "preparing a small box of some further Curiosities from our poor Countrey," but was uncertain as to when he would be able to send it.[166] The contents of the box, and its fate, are not disclosed. The next year, however, Dudley sent one of his most significant letters to the Royal Society. Addressed to Dr. Jurin, October 3, 1724, it was reported to the Society on February 4, 1724/25, and subsequently published in the *Philosophical Transactions* as "Observations on some of the Plants in New-England, with remarkable Instances of the Nature and Power of Vegetation."[167] The paper reflected Dudley's long absorption with his avocations, gardening and horticulture, and included some observations whose importance escaped the notice of natural scientists for many years. Much of the paper was devoted to an account of the superiority of New England garden vegetables and fruits (especially apples, pears, and peaches) in their size,

---

[164] *Phil. Trans.*, XXXII, No. 376 (March–April, 1723), 292–295. On Jan. 24, 1722/23, a report on vipers in Italy confirmed Dudley's account of the charming powers of snakes and vipers. *Journal-Book*, XIII, 248.

[165] *Classified Papers, 1660–1740*, XIV (2) ("Physic"), No. 16; reported Jan. 23, 1723/24, *Journal-Book*, XIII, 348–350; *Phil. Trans.*, XXIII, No. 384 (July–Sept., 1724), 129–132.

[166] *Guard-Book*, D-1, No. 79; *Letter-Book*, XVI, 488–489.

[167] *Guard-Book*, D-1, No. 80; *Letter-Book*, XVIII, 143; the original copy of Dudley's "Observations" is among *Classified Papers, 1660–1740*, X (2) ("Botany &c."), No. 3; *Journal-Book*, XIII, 444ff.; *Phil. Trans.*, XXXIII, No. 385 (Oct.–Dec., 1724), 194–200.

quality, and productivity to the same varieties grown in England, with a few instances of experimentation in grafting fruit trees in New England. But at one point he presented observations about the hybridization of Indian corn that improved upon those given by Cotton Mather in 1716:

Our Indian Corn is the most Prolifick grain that we have and commonly Produces twelve hundred and often two thousand grains from one, but the fairest Computation is thus, Six quarts of this grain will plant an Acre of Ground, & it is not unusual for an Acre of good ground to produce fifty bushels of Corn. The mention of *Indian* Corn obliges me to take notice of an extraordinary Phoenomenon in the Vegetation of that Grain, viz., the Interchanging or mixing of Colours after the corn is planted. For your better understanding this Matter I must observe, That our Indian Corn is of Several Colours, as blue, white, red and yellow; and if they are planted Separately, or by themselves, So that no other Sort be near them, they will keep to their own Colour, i.e., the blue will produce blue, the white, white, &c; But if in the same Field, you plant the blue Corn in one Row of Hills (as we term them) and the white, or yellow, in the next Row, they will mix and Interchange their Colours; that is, Some of the Ears of Corn in the Blue Corn Rows shall be white, or yellow; and Some again, in the white or yellow Rows, shall be of a blue Colour. Our Hills of Indian Corn are generally about four Foot asunder, and so continued in a streight line, as far as the field will allow, & then a second Line, or row of hills, and so on; and yet this mixing and Interchanging has been observed when the distance between the Rows of hills has been several yards; and a worthy Clergyman, of an Island in this Province (The Reverend Mr. [Experience] Mayhew, of Martha's Vineyard), assures me that the Blue Corn has thus communicated or Exchanged even at the distance of four or five rods; and, particularly in one place, where there was a broad ditch of water betwixt them. Some of our People, but especially the Aborigines have been of opinion, that this Commixtion, and Interchange, was owing to the roots, and small Fibers reaching to and Communicating with one another, but this must certainly be a mistake, Considering the great distance of the Communication, especially at some times and Cross a Canal of water; for the smallest Fibres of the Roots of our Indian Corn, cannot extend above four or five foot. I am therefore humbly of Opinion,

That the Stamina or Principles of this wonderful Copulation or mixing of Colours are carried thro the Air by the wind, and that the Time, or Season of it is when the Corn is in the Earing, & while the milk is in the grain for at that Time, the Corn is in a sort of Estuation and emits a strong scent. One Thing, which confirms the Airs being the medium of this Communication of Colours in the Corn, is an Observation of one of my neighbours that a close high board fence, between two fields of Corn that were of a different Colour, entirely prevented any mixture or alteration of colour from that they were planted with. . . .

Professor Conway Zirkle has pointed out that Dudley's comments formed the basis of all of Philip Miller's descriptions of the variety of crosses in Indian corn which appeared in the various editions of the latter's *Gardener's Dictionary* (London, 1731, and later).[168] Dudley himself wrote to Jurin (December 11, 1724) that "since my letter of plants & Vegetation I have happily seen Mr. [Richard] Bradley's fine treatise of gardening wherein I find my phenomenon of our Corn mixing Colours thoroly accounted for. If I had met with that book before writing my letter I would not have troubled the Society with it; however, I have the satisfaction to find my own notion of the Aer's being the medium of the interchange or mixture, confirmed. . . ."[169] Like Mather's earlier account of plant hybridization, both Bradley's and Dudley's escaped the notice of geneticists until the twentieth century.

In a letter to Dr. Jurin (October 10, 1724), Dudley stated "that I have done my self the honor to present the Society . . . with a Cane made out of the jaw of a Spermaceti Whale, and also a tooth taken out of the same fish of which creature I hope in my next to give the society a very particular acct."[170] Actually, the account was not sent until April 5, 1725.[171] Entitled "An Essay upon the Natural History of Whales with a particular Account of the

---

[168] Zirkle, *The Beginnings of Plant Hybridization*, pp. 129–131.

[169] *Guard-Book*, D-1, No. 82; *Letter-Book*, XVIII, 156–157. Dudley referred to Richard Bradley's *New Improvement of Planting and Gardening, both Philosophical and Practical* (London, 1717). For Bradley, see Pulteney, II, 129–133.

[170] *Guard-Book*, D-1, No. 81; *Letter-Book*, XVIII, 148–149; received May 6, 1725, *Journal-Book*, XIII, 477.

[171] Dudley to Jurin, April 2, 5, 1725, *Guard-Book*, D-1, Nos. 84, 85; *Letter-Book*, XVIII, 171, 172–186.

Amber-greese, found in the Sperma Caeti Whale," it was based upon information Dudley had gathered from men engaged in whaling, especially one Mr. Atkins of Boston, the Reverend Mr. Daniel Greenleaf of Yarmouth, and Mr. Isaac Coffin of Nantucket, both Yarmouth and Nantucket being "places Famous for the Whale-Fishing." The account described the varieties of whales found in New England waters, their habits and utility, and in particular the source of ambergris, which, as Atkins said, was an animal production found only in male spermaceti whales, especially old ones, in little bags just above the testicles. Dudley likened it to the musk found in musk deer and muskrats. The Royal Society received the account with enthusiasm (June 3, 1725) and it was published in the *Philosophical Transactions*.[172]

Dudley complained of ill health in 1725 and 1726, but on January 21, 1726/27, he sent to Dr. Jurin seven brief communications after the pattern of Mather's *Curiosa Americana*.[173] Presented to the Royal Society at various times during 1727,[174] they consisted of "An Account of a Stone taken out of a Horse at Boston in New England, in the Year 1724" (the same case reported by Zabdiel Boylston in the previous February);[175] accounts of three human cases of "uncommon Cures and Observations," one of which had taken place more than a decade earlier; "A Short Account of the Aurora Borealis that appear'd on the 27th of March 1726 at night in Roxbury, & many other Places in New England"; "A Surprizing Instance of ye Nature and Effect of Lightning, June 24th, 1724"; and "Salt Water, or the Spray of ye Sea carried many Miles into the Countrey &c." Later in the same year (November 13) Dudley sent "An Account of the several Earthquakes which have

---

[172] *Journal-Book*, XIII, 491–493; *Phil. Trans.*, XXXIII, No. 387 (March–April, 1725), 256–269. Dudley promised to send later a specimen of "the Yellow Liquor in ye Ambergrease Whale," but was unable to supply it until 1736, when he sent a bottle of it to the Society by Zabdiel Boylston's son. *Guard-Book*, D-1, No. 89; *Letter-Book*, XVIII, 408–409; *Guard-Book*, D-2, No. 68; *Letter-Book*, XXIII, 184–185; *Journal-Book*, XVII, 49 (Feb. 24, 1736/37). Dudley was unhappy about errors in the printed version of his account of whales and sent corrections to Jurin, Dec. 21, 1725. *Guard-Book*, D-1, No. 88; *Letter-Book*, XVIII, 277–278.

[173] *Guard-Book*, D-1, Nos. 89–91; *Letter-Book*, XVIII, 402–409; XIX, 170–177.

[174] *Journal-Book*, XIV, 71–72 (April 20), 105 (June 15), 148–150 (Dec. 7).

[175] See p. 439 above. Dudley's account was published in *Phil. Trans.*, XXXIV, No. 398 (April–June, 1727), 261–262.

happen'd in New-England, since the first Settlements of the English in that Country, especially of the last, which happen'd on Octob. 29, 1727." The relation briefly described eight earthquakes, ending with that of 1727, for which the Royal Society had already received (and published) an account by Benjamin Colman.[176] But Dudley, referring to Gilbertus Jacchæus' *Institutiones physicæ* (Amsterdam, 1644) and other ancient works, sought to classify the intensity of the New England quake of 1727. The Society heard Dudley's "Account" (January 25, 1727/28), and Dudley sent an additional letter (February 2, 1727/28) to point out that the quake of 1727 had been felt at Barbados and Martinique. But as the Society did not publish the "Account," Dudley sent a second copy (April 17, 1735), suggesting that after a "late Terrible Earthquake at Portsmouth in England, which the public news papers make mention of will render the publishing this from New England the more seasonable. . . ." And so it was published in the *Philosophical Transactions.*[177]

Dudley was further engaged at this time with meteorology. Dr. Jurin's appeal for meteorological data in the *Philosophical Transactions* (1723), fortified, perhaps, by Isaac Greenwood's grandiose plan of 1727, induced Dudley to undertake weather observations. He mentioned his intent in a letter to Jurin (November 9, 1725), but at that time he possessed no barometer and feared that "my Riding the [court] Circuit and frequent absence from home will render it impracticable for me to be constant in my remarks."[178] By 1729, however, he began weather records, and on April 12, 1733, forwarded to the Royal Society via Cromwell Mortimer "a Journal of the winds and weather" for four years.[179] Later in 1733 he sent a lengthy "Account of the Locusts in New England," including

176 See note 123 above.

177 XXXIX, No. 437 (April–June, 1735), 63–73. Both copies of Dudley's "Account" (not quite identical) are in *Guard-Book*, D-1, Nos. 92, 94; *Letter-Book*, XVIII, 471–482. Cf. *Journal-Book*, XIX, 167–168, 182; Dudley to William Derham, April 17, 1735, *Guard-Book*, D-1, No. 93.

178 *Guard-Book*, D-1, No. 87; *Letter-Book*, XVIII, 269–270.

179 *Guard-Book*, D-1, No. 95; *Letter-Book*, XX, 245. The "Journal" was presented to the Society Oct. 25, 1733, and referred to Dr. Derham. *Journal-Book*, XV, 321.

"some of the Locusts preserved in Rum, and a small hand-box with a few of the Carcases and Cases of the Locusts dried," with a couple of twigs "where you will see how they open the Twigs, and deposit their Eggs, though the Sticks are very much dried."[180] Dudley believed that the New England locust was identical to those of Egypt as mentioned in the Old Testament. He described the creature and the manner of its generation, with much attention to its alleged periodicity of appearance at seventeen-year intervals. Indeed, he said that he had begun his essay on locusts in 1716, "but was willing to wait one period more before I offer'd it to the Society, for fear it would not answer: but I am now well satisfied it may be relied upon." He had observed three seventeen-year intervals, in 1699, 1716, and 1733, and he was convinced that the life cycle of the New England locust required seventeen years to complete. Accordingly, he was taken aback when his paper was read to the Royal Society (June 13, 1734), and President Sloane "took notice that Mr. Dudley had, under the name of a Locust, described the Cicada; and by confounding one Insect with the other had created to himself a difficulty to reconcile the accounts which he had read of the one with the Observations he had made of the other."[181] Sloane also produced from his collections specimens of the cicada to prove his point, and an Italian cicada was sent to Dudley for his edification. Dudley, however, was not convinced. On April 18, 1735, he addressed a long letter to Dr. Mortimer "containing Explanations of, and additions to his account of the periodical Locusts in New-England."[182] He insisted that the New England creature was a locust and, with Isaac Greenwood's cooperation, had confirmed this to their mutual satisfaction by reference to a variety of "the best Authors." Further, he had both the word of the Reverend Mr. Weld of Attleboro, where the "locusts" abounded, and

---

[180] Dudley to Mortimer, Dec. 18, 1733, *Add. Mss. 4433*, fols. 4–11; *Letter-Book*, XXI, 99–114.

[181] *Journal-Book*, XV, 446–448.

[182] *Letter-Book*, XXII, 241–246; read March 4, 1735/36, *Journal-Book*, XVI, 274–275. The Reverend Mr. Punderson of Graunton confirmed Dudley's account of the "locusts." *Letter-Book*, XXI, 319–327; read March 20, 1734/35, *Journal-Book*, XVI, 113.

the evidence of New England almanacs, that the "locusts" demonstrated a life cycle of seventeen years:

> I am not a little surprised [he wrote] that you should imagine the great appearance of our Locusts every seventeen years should be owing to any accidental cause. I cannot reconcile it to my poor Philosophy, or the Laws of Nature, that so regular an effect should be produced by accident. And you go upon a mistake, if you think that in the intervening sixteen years there are Locusts to be seen: for there was no such thing as a single Locust to be seen or heard of the last year. I do not wonder that the learned World are surprized at the remarkable Cycle of this Animal: because it is indeed very extraordinary, and I suppose never heard of but in this country. . . .

He denied that the New England "locust" underwent any metamorphoses, holding that "a true proper Locust is immediately produced from the Egg," whereas, as he understood it, the Italian cicada undergoes a series of metamorphoses during the course of a year. He described experiments made with New England locusts' eggs, in conjunction with Weld of Attleboro, which purported to show that the New England locust was hatched from the egg and assumed mature form in a matter of minutes. This rebuttal was presented to the Royal Society on March 4, 1735/36. To settle the matter, the Society sent him a specimen of an Egyptian locust, which he had insisted was identical to the New England variety. Dudley wrote to Mortimer (January 25, 1736/37):

> Having now a little more leisure (though not health) I am willing to employ it by telling you once more how much you have obliged me in the great Curiosity of an Egyptian Locust; and thereby effectually convincing me that I was in the wrong to call our Cicadae by the name of Locusts: however I doubt not you will be so good as to excuse my mistake. But after all, what name we shall give the Cicadae in the English tongue I am at a loss; and as to our common people, let the learned world say or do what they please, they will continue to call them Locusts still.
>
> I am sensible my Essay must undergo a great Reform, by reason of this mistake: and I shall endeavour to do it as soon as possible, and send it a second time. . . .[183]

183 *Letter-Book*, XXIII, 286–287; read April 28, 1737, *Journal-Book*, XVII, 90.

There is no record, however, that Dudley submitted a revised essay about the New England cicada.

Indeed, he was already concerned with another—his last—communication to the Royal Society. This was a dissertation concerning the different sorts of evergreens growing in New England, a study for which Dudley had collected materials for many years. Sent with a specimen of the "bayberry tree" (probably wax myrtle, *Myrica cerifera,* a plant sometimes confused in England with the Jamaican bayberry tree, *Eugenia acris*). It was read to the Royal Society in two parts (April 1, 8, 1736), and described, with their respective uses, sixteen varieties: white pine, pitch pine, sapin or popple pine (*Sapium?*), apple pine, hemlock, spruce (distinguished into white, black, and red), white cedar, red cedar, savin, juniper, holly, and ivy.[184] As the original of Dudley's dissertation does not appear to be among the Archives of the Royal Society, it is possible that Peter Collinson, F.R.S., got hold of it. Collinson (1694–1768), a wealthy merchant, naturalist, and antiquary, who after 1730 became a prime promoter of the sciences in the colonies, wrote to his friend and protégé, John Bartram of Pennsylvania, on June 7, 1736:

> I have now a very curious account before me, sent by Paul Dudley, from his house in Roxbury, New England, October 24, 1735; who very ingeniously describes the Evergreens of New England in two sheets of paper. . . .
>
> My kind friend, Dr. Witt, sent me some years agone, several small plants that he called Spruce; but, by the very particular description of P. Dudley, they prove to be the Hemlock; for I have two fine plants, in my garden, which agree exactly with his description of the Hemlock; and, to confirm me that P. Dudley is right, I had this year, come from Newfoundland, two fine Spruce trees, which both grow, and prove very different plants from what the Doctor sent me; but agree exactly with P. Dudley's description of the Spruce. This I send by way of information, and to put thee on observing what you have, of these kinds, growing near you.[185]

184 *Journal-Book,* XVI, 296, 302–303.
185 William Darlington, ed., *Memorials of John Bartram and Humphrey Marshall* . . . (Philadelphia, 1849), p. 79. As I have seen no evidence of correspondence directly between Collinson and Dudley, perhaps Collinson got it, or a copy of it, from the Royal Society.

It was fitting that Paul Dudley's long series of communications to the Royal Society should end with the account of evergreens growing in New England, and particularly that he was able to make a valued distinction between the spruce and the hemlock. He had made scientific communications to the Society for nearly eighteen years, the longest continuous series of any colonial up to his time. Naturally the contributions varied widely in value, but Dudley had proved to be an honest, critical observer throughout. His error regarding the cicada was an honest mistake which he retracted without temper when the evidence was placed before his eyes. His contributions contained several items of lasting scientific value: Kellogg's "Journall" and the correction of the height of Niagara Falls added to more accurate geographical knowledge; the history of whales solved the age-old problem of the origin of ambergris; the "Observations on some of the Plants in New-England" provided useful botanical and horticultural details, especially the explanation of the function of winds in the hybridization of Indian corn; and his dissertation about New England evergreen trees was the most complete yet received in England, with a valuable distinction drawn between the spruce and the hemlock. Other, lesser, contributions were valuable principally in their corroborative details. Dudley was strictly an amateur scientist, retaining some of the superstitions of the past. He prepared no broad synthesis of the natural history of New England comparable to that of John Josselyn; but he was better informed than Josselyn, a more accurate and critical observer. Moreover, his works frequently portrayed the fact that he consulted with others in the rising New England community of scientists. His letters clearly indicate that on many scientific questions he had discussed the problems with Cotton Mather, Thomas Robie, Isaac Greenwood, Thomas More, Zabdiel Boylston, Benjamin Colman, and others over a wide area of New England. His observations and opinions, therefore, were tempered and fortified, so that in some cases they represented corporate views of an informed body of scholars. Among those of his group, Paul Dudley may well rate as one of New England's best students of natural history before the end of the Old Colonial Era.

Thomas More (fl. 1670–1724)

Thomas More was a visitor to New England during the years 1722–24. There appear to be no records by which one can establish his origin, date of birth, education, or year of death. Still, he flitted in and out of English scientific correspondence from about 1670 until 1724. He was associated with Jacob Bobart and Edward Lhwyd at Oxford in the 1690's, probably as a menial.[186] Early in the eighteenth century, More collected for Lhwyd in southwestern England, where he impressed William Musgrave, F.R.S., of Exeter, so favorably that the latter wrote to Dr. Hans Sloane to recommend him to the attention of the Royal Society. In 1704 More turned up in London, visited the Society (July 12), and amused the Fellows with "some of his Tables for Reducing Nature under several heads." Evidently More was trying his hand at system, and Edward Lhwyd wrote to James Petiver later in the same year referring to "the Bearer, Mr. More, the great Philosopher, who promises to give us an account of the whole creation from an Angell to an Attome." This note of humorous sarcasm recurred frequently in references to More by members of the English circle of naturalists. Still, they appear to have appreciated his energy and enthusiasm as a collector, and Petiver, Sloane, Lhwyd, and others employed the services of the "Pilgrim Botanist," as they called More.

In 1709–10, when five Mohawk chiefs were transported to England, entertained with much pomp, and even received by Queen Anne, Thomas More somehow met the chiefs, won their friendship, and was invited to visit them in the New World. Thereafter, he wanted to visit New England and New York to renew his friendship with them. An opportunity arose in 1721.

William Sherard had recently returned to England from his consulate in Aleppo and was making plans to raise subscriptions with which to support Mark Catesby on a voyage of scientific discovery and collection in Carolina. As these plans were maturing, Sherard

---

186 I have discovered little more about Thomas More than appears in the sketch in the Appendix of Frick and Stearns, *Mark Catesby*. I refer the reader to that work for most sources of information.

met the "Pilgrim Botanist." Shortly afterward (December 7, 1721), Sherard wrote to Richard Richardson: "The Pilgrim Botanist, Mr. More, whom we heard of in Wales, is desirous of going to New England and the rest of our Colonies in North America. He is an excellent collector of all parts of Natural History, and desires no more than a poor subsistence. . . . It is a pity he is not younger, and I am sorry I did not know him sooner; he would have done more service than all that have been sent abroad." At this point Sherard enlarged his design to send Catesby to Carolina by soliciting subscriptions from his friends to send Thomas More to New England as well. During the course of 1722, Sherard, Richardson, Sloane, Charles Dubois, John Bellers, Dr. Richard Mead, and Dr. Hermann Boerhaave, the famous Leiden professor of medicine, botany, and chemistry, each made an annual subscription of a pound (or, in the case of Richardson, Mead, and Boerhaave, a guinea) for the support of More's expedition. Some of the subscribers were the same men who supported Catesby, but there were fewer of them, and the total sum subscribed was less lucrative than Catesby commanded—although it appears likely that Sherard, or others, contributed additional funds to pay the costs of ocean passage. The "Pilgrim Botanist" set out with a supply of seeds and a letter of introduction from Dr. James Jurin, Secretary of the Royal Society, to Paul Dudley, and he arrived in New England in September, 1722. According to a memorandum among Sherard's papers, More was to go to all the colonies north of Virginia "to seek for and discover all sorts of Land and Sea Plants, Seeds, Fruits, Barks, Metalls, Stones, Sparrs, Ores, Marchasites, Earths, Serpents, Insects, Fishes, Beasts, Birds, and all such other Naturall Bodys frequenting those Countrys as yet unknown to Us. . . ."

More's arrival in New England was too late, in Sherard's opinion, "to perform any great matters this year." Moreover, New England had had a hot, dry summer, and More found most of the "trees & herbs quite parcht dryed & burnt up." But he set to work quickly, hired an Indian servant, and collected a box of specimens to send to his patrons. Paul Dudley embraced him warmly, introduced him to Governor Shute and other "Grandees of the Country," all of whom "highly approved of my design & made me very welcome." More had a reunion with one of his Indian friends, "The

Prince of the Mohawks," who had come to Boston on a diplomatic mission to help "mediate a peace between us and the ffrench Indians now at war with us." And he attended the ceremonies at Harvard College (October 23, 1722) when Edward Wigglesworth was formally installed as the first Hollis Professor of Divinity. A few days later (October 27) he addressed a letter[187] to Sherard to describe his experiences and to give a brief account of twenty-six plant specimens he had ready to ship. Both the box of specimens and the letter greatly disappointed William Sherard. He complained to Richardson (February 23, 1722/23) that although More had been very diligent, "most of what he gathered were common, and spoiled in coming over by his fault; for he put in the same box with dried plants, Fruits, and seeds, Limes, Gourds, and such like trash, and to fill it up, put sea-weeds atop. I took him for a great philosopher, and shall give him orders how to pack for the future. . . ." The letter demonstrated that the "Pilgrim Botanist" was not abreast of the botanical knowledge of the day. His language was quaint, with hardly a trace of description that could be called scientific, and his mind was filled with little more than superstitious folklore regarding the specimens he had gathered. Moreover, while he purred with delight at the reception given by the Bay Colony "Grandees," he was severely critical of the colonists as a whole, whom he found to be "Idle, Ignorant, and Poor and proud and which is worse don't care much for being better." He was shocked by their wastefulness, especially by their use of fine timber (walnut, oak, etc.) for firewood, "for Lazyness to bring in worser fewel." Also, he had hatched a secret plan to save His Majesty great sums of money, supply New England pitch, tar, and hemp for the Royal Navy, and better regulate New England—if his patrons would intercede with the King and obtain supporters for the project. But he would not reveal the details— would "not open my Pack till I can sell some of my goods"—and so he betrayed dreams of grandeur which might possibly wean him away from the work for which his patrons supported him in New England.

William Sherard sent his new "orders" to More on March 20,

---

[187] Published in full in *ibid.*, pp. 120–124. The quotations above are from this letter.

1722/23, but it was many months before he heard from the "Pilgrim Botanist" and a year before More's next shipment reached England. In the following July, Paul Dudley reported that "our friend Mr. More is gone to the neighbouring colonys with all the recommendations we could possibly give him from hence." In August Sherard reported that he had heard from More, who was sending "specimens and seeds which I expect in 2 months." But More had gone "to visit his old acquaintances the Indian Kings that were in England," and it was December before he shipped his second collection of specimens to England. In a letter to accompany it, dated from Boston, December 12, 1723,[188] he commented briefly on eight seed specimens sent (mostly trees) and promised more to come; but his comments were of greater utilitarian than of scientific value, and they were curiously intermingled with personal anecdotes and bitter comments about New Englanders, the most arresting of which ran as follows: "I have travel'd thro Six Countrys of New England this year and Doctor Dilenius was all wrong when he represented this as a hospitable Country, for here they bite like sharks and in some things outdoe the very Hollanders themselves. . . ." In fact, months before (May 16, 1723) he had written to Lord Carteret, Secretary of State, to complain even more widely against New England "order, oeconomy, Government, and husbandry": the "Ship of this Government has sprung so many leaks, your Lordp., as a Master Carpenter ought to stop 'em, and I as your mate have made good pluggs for you to drive in, etc. . . ." After this presumptuous beginning, More offered fourteen "pluggs" for the reform of civil government and economy in New England—a program reminiscent of that of the late despised Governor Joseph Dudley—and then he reached more personal concerns. He had been unable to acquire land on which to experiment with plants, "unless," he said, "I bought it." He wanted a patent for a tin mine he claimed to have discovered, and as he was especially bitter against the New Englanders' flagrant disregard of markings on trees reserved for H.M. Navy, he sought an appointment as H.M. "Forest Ranger," i.e., Surveyor General of the Woods, an office currently vacant. In this way More apparently drove himself

[188] Part of this letter is missing. The remaining fragment is published in *ibid.*, pp. 124–127.

into a state of mind scarcely conducive to good relations with the colonists and likely to lead to the neglect of his patrons' interests. Indeed, there is no record of anything further he collected to send to the latter, and in the winter of 1723–24 he returned to England, evidently to press his claims at court in person. Sherard noted on April 25, 1724, that "He [More] is come over & will return next month. I hope to get a place of one of the King's rangers for preserving timber in that country, which will give him a power & opportunity of making his subscribers amends." But in the end, the "Pilgrim Botanist" received none of the favors he sought. Nor did he return to New England. In fact, he passed out of English botanical circles as anonymously as he had entered them.

Thomas More's patrons gained very little for their subscriptions, especially when the results were compared with the remarkable fruits of Mark Catesby's labors. The *Sloane Herbarium* lists one specimen he had gathered in Cornwall, but nothing from the New World.[189] In 1730 Philip Miller published a catalogue listing a fir tree and "The New England Ash," which he said had been grown in the garden at Chelsea from seeds supplied by Thomas More. The fir tree was a reintroduction to England from the New World (it had formerly been growing in Bishop Compton's garden at Fulham Palace), but the ash was new, "rais'd from Seeds sent from *New England* by Mr. More, Anno 1723." These two trees appear to have been the only lasting recorded fruits of More's expedition to New England.

Dr. William Douglass (c. 1691–1752)

Dr. William Douglass was Scottish born, educated at Edinburgh, Leiden, and Paris, but from which institution he took his medical degree is uncertain. He had a sound background in the humanities, especially languages, in which he had a command of Greek, Latin, French, Dutch, and English. At Paris he learned to revere Tournefort, to whom he later referred as "the prince of botanists," although, as Tournefort died in 1708, it is unlikely that Douglass had studied under his direction. In medicine he was said to have been greatly influenced by Archibald Pitcairne (1652–1713), the cele-

---

[189] Dandy, p. 168.

brated Scottish physician, poet, and satirist, from whom Douglass may have imbibed the irreverence toward clergymen and religion that disturbed his New England associates. It is noteworthy, in any case, that by virtue of his Scottish birth and youthful associations in Scotland and on the Continent, Douglass was a "foreigner" in New England. With his close correspondent, Cadwallader Colden of New York, another Scot, he belonged to a small but rising group of educated Scots who were destined to exercise an influence on science out of proportion to their numbers in the colonies and with an intellectual orientation turned toward Scotland and the European Continent more than toward England and the Royal Society of London. Douglass himself might well have exerted a much wider influence than he did but for his stubborn independence of mind, his patronizing attitude toward New Englanders (to whom he obviously felt himself superior), his indiscreet assertion of unorthodox religious views, and his unpleasant tendency to adopt partisan causes with uncompromising ardor and an occasional suggestion of malignancy in his strong expressions of partisanship. Dr. Douglass was a strong hater, and although he became President of the Scottish Charitable Society in Boston (1736–52), he appears to have kept aloof from his more "Americanized" neighbors—or they kept aloof from him. Thus, although he arrived in Boston in 1718, after a brief sojourn in the British West Indies, with letters of introduction to both of the Mathers and to Benjamin Colman, he soon became critical of his new associates and never quite "belonged" to the New England community of scholars. He spent the remainder of his life in New England, but while he won their respect, he failed to win the wholehearted approval of New Englanders.[190]

Still, Dr. Douglass was an uncommonly able man, and at the outset of his career in New England he was the only physician in the

---

[190] Douglass deserves more complete study. The principal sources for his life, other than his letters and published works, are: *D.A.B.*; George H. Weaver, "The Life and Writings of William Douglass, M.D.," *Bulletin of the Society of Medical History of Chicago*, XI (1921), 229–259; Charles J. Bullock, ed., "A Discourse Concerning the Currencies of the British Plantations," *Economic Studies*, II (1897), 265–375; Dr. John B. Beck, *Medicine in the American Colonies* (Albany, 1850; reprinted with Foreword by Dr. Charles F. Fishback, Albuquerque, 1966), *passim.*

Boston area with an earned medical degree—which may have been another source of trouble, for he evidently advertised the fact and was fully as jealous of his status as a medical practitioner as Cotton Mather was of his cloth. But the two men did not cross swords until about three years after Douglass' arrival, for the doctor's broad interests and abilities in scientific matters must have won Mather's early respect. The immediate occasion of the antipathy between the two men was the controversy over smallpox inoculation in Boston. Dr. Douglass, feeling that his superior medical education should naturally elevate him to the position of arbiter in medical disputes, resented the fact that inoculation for smallpox was inaugurated by Zabdiel Boylston, a medical practitioner who had no medical degree, and that it was publicly promoted by Cotton Mather, a parson meddling in affairs outside his competence and proper sphere of activity. When James Franklin launched his new paper, the *New-England Courant*, in August, 1721, Dr. Douglass was a frequent contributor of articles attacking the practice of inoculation as "rash and dubious," being "the blind Conduct of *Empiricks and Montebanks*" "profoundly ignorant of the Matter," and he suggested both that the inoculators perpetrated a felony and that the ministerial promoters were blind instruments of Satan.[191] Dr. Douglass was instrumental in organizing the Boston physicians against the practice and he was early at the center of the vicious attacks upon Boylston and his ministerial supporters. He solicited Cadwallader Colden's advice from New York, and gathered anti-inoculation opinions from Europe to support his case.[192] He also wrote several pamphlets attacking both the inoculators and the practice of inoculation, one of which was communicated by "Dr. Steward" (Alexander Stuart, F.R.S.) to the Royal Society of London.[193]

191 *New-England Courant*, Nos. 1–5, 14, 16, 17, 19, 20, 22, 24, 26, 32, 42 (Aug. 7, 1721–May 14–21, 1722).

192 Jared Sparks, ed., "Letters from Dr. William Douglass to Cadwallader Colden of New York," *4 Coll. M.H.S.*, II (1854), 164–189, Nos. 3, 4; *Colden Papers*, I, 141–145; Blake, "The Inoculation Controversy in Boston," *loc. cit.*

193 Douglass to Stuart, Sept. 25, 1721, *Guard-Book*, D-2, No. 2; *Journal-Book*, XII, 163 (Nov. 16, 1721). Dr. Douglass was not a *direct* correspondent of the Royal Society. His publications are listed in Weaver, "Life and Writings of William Douglass."

Dr. Douglass, of course, had a case. He pointed out that persons inoculated for smallpox might, if not kept in isolation until the infection had run its course, communicate the disease to others, and that proof of the alleged immunization from future attacks was as yet inconclusive. But he also charged that inoculation led to the plague and other vile diseases, and even published a tongue-in-cheek, grisly article entitled "A Project for reducing the Eastern Indians by Inoculation."[194] These wild swings not only alienated Cotton Mather and the supporters of inoculation but also inflamed popular opinion. In the long run, too, after Boylston's successes became apparent, they weakened Douglass' case. Indeed, Douglass was himself too good a physician to ignore the success of inoculation for smallpox. He confessed to Colden (May 1, 1722) that smallpox "seems to be somewhat more favourably received by inoculation than received in the natural way," and though he still held it to be "a sin against Society to propogate infection by this means," he gradually relaxed his opposition.[195] By 1730, when he published *A Practical Essay Concerning the Small Pox* (London), he adopted a milder tone, and soon he adopted the practice himself. Indeed, in his rambling book, *A Summary, Historical and Political, of the First Planting . . . and Present State of the British Settlements in North America* (2 vols., Boston and London, 1755), he wrote that inoculation for smallpox "is a very considerable and most beneficial improvement in that article of medical practice."[196] But there is no evidence that he ever apologized to Mather, Boylston, or the other promoters of inoculation whom he had so wantonly attacked.

Like other European-trained physicians, Dr. Douglass held a low opinion of medical practices in the colonies. He wrote disparagingly of it to Colden in 1721, and his more considered opinion in *A Summary* was that the "practice of physic is so perniciously bad, that excepting in surgery, and some very acute cases, it is better to let nature under a proper regime take her course . . . than to trust to the honesty and sagacity of the practitioner."[197] However,

194 *New-England Courant*, No. 2 (Aug. 7–14, 1721).

195 Sparks, ed., "Letters," No. 3.

196 II, 406. He still argued, however, that the first promoters of inoculation had been too extravagant in their recommendations of it.

197 II, 351–352; Sparks, ed., "Letters," No. 1.

he developed a lucrative practice himself and lived handsomely upon his fees. Moreover, he made significant contributions to American medicine. When the serious "throat disease" (scarlet fever) first attacked New Englanders in the mid-1730's, Dr. Douglass published *The Practical History of A New Epidemical Eruptive Military Fever, with an Angina Ulcusculosa, which prevailed in Boston, New England, in the Years 1735 and 1736* (Boston, 1736), which medical historians have praised as the first —and the best—clinical description of the disease in English. Douglass described the course of the disease in detail, noted the greater susceptibility of children to it, and performed autopsies on persons who had died of it. Moreover, his treatment of cases by the use of calomel was one of the most successful at the time.[198] Indeed, Dr. Douglass pioneered in the use of mercuric preparations in colonial medicine, a practice in which colonial physicians were advanced beyond the European practitioners of the day.

Dr. Douglass was versatile, accumulated a large library, and devoted himself to a wide range of scientific interests. He was especially knowledgeable in natural history and cited in this connection the works of Bauhin, L'Escluse, Plukenet, Morison, Tournefort, Ray, Cornuti, Sloane, and Catesby. He complained that New England had no "botanick writers" comparable to the early French Recollets in Canada or Father Plumier in the French West Indies.[199] He wrote to Cadwallader Colden in 1721 that he had already collected "above seven hundred Plants within . . . four or five miles from Boston," and he planned to extend his searches. The Boston epidemic of smallpox interrupted his botanical reasearches, but he later resumed them and reported that his collections totaled 1,100 plants.[200] He exchanged plant specimens and botanical information with Colden, but apparently did not send collections or descriptions to Europe, although "a particular description of the way of manufacturing and making Tar & Pitch and extracting Turpentine from the Firs with an Account of the Several Species of Firs in

---

[198] Weaver, "Life and Writings of William Douglass," pp. 237–238; Beck, *Medicine in the American Colonies*, pp. 29–32, 42ff.

[199] *A Summary*, I, 121; II, 216–217n.

[200] Sparks, ed., "Letters," pp. 164–165; *Colden Papers*, I, 308; *A Summary*, II, 57–58n, 216–217n.

that Country [New England]" was communicated to the Royal Society by Dr. James Douglas, F.R.S., on January 11, 1727/28.[201] His later publication, *A Summary*, contains much botanical data, especially regarding New England trees, and it is evident that Dr. Douglass was a competent botanist.[202]

He also demonstrated more than a passing interest in other scientific endeavors. Although he had no instruments when he first arrived in New England (and knew of none in the Boston area except the astronomical gear at Harvard College), he prepared a journal of the weather for 1720, sent a copy to Colden, and later exchanged meteorological data with Colden regarding the weather in New York and Philadelphia. He planned to borrow Harvard's astronomical instruments to observe an eclipse of Jupiter's first satellite in order to establish the geographical position of Boston (which Thomas Robie subsequently did), and later he exchanged astronomical data with Colden concerning these questions.[203] Associated with this interest in the correct latitude and longitude of places in the New World was Douglass' concern for geography and the improvement of maps in America. He traveled widely over New England to measure distances and to prepare local surveys for the preparation of a map, but, as he confessed to Colden, he was unable to reduce his materials to a satisfactory map.[204]

Dr. Douglass adopted a strong partisan position against paper currencies in the colonies, especially in Massachusetts. His letters to Colden repeatedly refer to this subject. He gathered information about the paper currency emitted in New York, New Jersey, and

201 *Journal-Book*, XIV, 156. Whether the two Drs. Douglas[s] were kinsmen is unknown. No copies of Douglass' papers read to the Society appear in the Royal Society Archives. At the same meeting Dr. Douglass communicated an account of the New England earthquake of Oct. 29, 1727, prepared by one "Mr. Wats" of Boston and dated Nov. 8, 1727. The inference is that Wats's paper had been transmitted by Dr. William Douglass, who also corresponded with Colden about this and other earthquakes. *Colden Papers*, I, 234–239, 244–248.

202 See especially II, 52–59, 65–66, 203–209. Douglass' notes about whaling and fishing (I, 56–62, 287–304) are commercial rather than scientific in treatment.

203 Douglass to Colden, Sept. 14, 1724, *Colden Papers*, I, 164; Sparks, ed., "Letters," No. 10.

204 *Ibid.*, Nos. 8, 10; *A Summary*, I, 452–464, 466–471; II, 21–22. Douglass was familiar with Joseph Kellogg's journal sent by Dudley to the Royal Society. *Ibid.*, I, 181–182. He said that Dudley was planning a map of New England but was "very jealous" of his information. *Colden Papers*, I, 164.

other colonies, and in 1739 stated that he was engaged "in reducing all our Paper currency to a regular obvious Scheme."[205] He published an anonymous discourse on the subject in London and in 1740 published it in Boston under his own name. Entitled *A Discourse Concerning the Currencies of the British Plantations in America*, the treatise is a valuable source for the historical study of the subject, in spite of Dr. Douglass' ranting tone and his subsequent railings against Governor William Shirley.[206] Dr. Douglass had lost money in Massachusetts real estate transactions, and he opposed the "cheap money" policies of his day with all the bitterness of a burnt creditor.

Dr. Douglass' final book, *A Summary*, was published posthumously and never completed as he had planned it. Though badly organized, it illustrates the broad interests of the man, some of his able qualities, and some of his violent prejudices. Its digressions supply useful hints about many of his opinions and intellectual operations which are frequently left undeveloped. One of these, referring to North American trees, suggests that Dr. Douglass was either impatient with the usual treatments of natural history—the "useless virtuoso part of natural history concerning figured stones, curious marcasites . . . unusual petrefactions, shells of all sorts" and "stiff scholastick enumerations"[207]—or that he sought to rise above the minutiae of taxonomy and get on with the more fundamental scientific matters of plant physiology, plant chemistry, and plant generation. If the latter be true, his mind had a glimmering of the new age beginning in natural history with the works of such men as Stephen Hales, Albrecht von Haller, Pieter Camper, and others in Europe who were laying the foundations of modern plant physiology, morphology, comparative anatomy, and the like. But the hint is tantalizingly insufficient for such a conclusion, and one is led to the disappointing opinion, in view of Dr. Douglass' personality, that it merely demonstrated his impatience with the multitudinous detail which consumed the attention of so many of his fellow naturalists.

[205] Douglass to Colden, Nov. 12, 1739, *ibid.*, II, 196–200.
[206] The *Discourse* was reprinted, Boston, 1751, and edited and reprinted by Bullock in "A Discourse Concerning Currencies," pp. 265–375.
[207] *A Summary*, II, 52–53n.

Still, Dr. Douglass contributed to "the wave of the future" in a different sense. On February 17, 1735/36, when the "throat distemper" was afflicting Bostonians, he wrote to Cadwallader Colden to remind him, "you may remember that some years ago you proposed the forming of a sort of virtuoso society, or rather correspondence" among colonials with scientific interests. Douglass referred to Colden's previous suggestion (c. 1728) that "a certain number of Men would enter into a Voluntary Society for the advancing of Knowledge." Colden had proposed that Boston would be the proper center of it, as it had "the greatest number of proper persons." Each member would submit a paper semi-annually, and the governors of the colonies would be called upon to aid by granting the society a franking privilege. Douglass had rejected the suggestion in 1729 on the grounds that there were not enough persons interested to support it.[208] Now, in 1736, Douglass organized such a society among the medical practitioners of the Boston area, one of the earliest of its kind in the colonies. He outlined five projects the society planned to undertake and stated that it would publish medical memoirs from time to time. One of these, published in 1741 "for the Comfort and Benefit of the unhappy and miserable Sufferers, by the excruciating Pain, occasioned by a Stone in the Bladder," described an operation in the presence of the Boston Medical Society after the manner of "Mr. Cheselden's late improvement of the *lateral* way" by Mr. Gardner of Boston, "who had some Part of his Education in the Hospital and Infirmaries of *London* and *Paris*."[209] Thus, the way was being prepared for American societies of scientists and other learned men to supplement, and eventually largely to replace, the role which the Royal Society of London had played since the 1660's.

The Boston Scientific Community

It is obvious from the foregoing pages that during the last fifteen years of Cotton Mather's life there arose in and about Boston a considerable group of persons concerned, in varying degrees, with scientific interests. Some of them, like Cotton Mather and his

[208] Douglass to Colden, March 31, 1729, *Colden Papers*, VIII, 190–193; Colden to Douglass, n.d., *ibid.*, I, 271–273; II, 146–147.
[209] *Boston Gazette*, No. 1029 (Nov. 10, 1741).

father, were "holdovers" from the earlier group associated in the Boston Philosophical Society.[210] Others were younger men, including newcomers to New England, with subsequent members who extended the life of the group and broadened its significance to the end of the Old Colonial Era and, indeed, to the present day. The group embraced men of widely differing origins, factions, and employment, but they held in common an interest in science —and personal differences among scientists often stimulate scientific development fully as much as their amiable intercourse. In Cotton Mather's day the group numbered upwards of twenty men,[211] seven of whom were elected Fellows of the Royal Society.[212] To them may be added a less identifiable "fringe group," including such participants in the inoculation controversy as John Williams, Lawrence Dalhonde, John Perkins, the Reverend Mr. John Wise, and, more important (in terms of the popularization of science in New England), a small group of ephemeral contributors to the Royal Society, two of whose communications were published in the *Philosophical Transactions*.[213] And to these

[210] The Mathers, Samuel Sewall, Thomas and William Brattle, and possibly Joseph Dudley.

[211] Besides those listed above were: Thomas Robie, Samuel Danforth, John Leverett, Edward Holyoke, Isaac Greenwood, Benjamin Colman, Thomas Prince, Ralph Walter, Paul Dudley, Dr. William Douglass, the transient Thomas More, and Governors Samuel Shute, William Burnet, and Jonathan Belcher. To these may be added John Winthrop (1681–1747), grandson of John Winthrop, the first colonial F.R.S., and brother-in-law of Paul Dudley. Although in and out of New England and of little merit as a scientist, Winthrop cashed in on his family reputation and was himself elected F.R.S. See Raymond P. Stearns, "John Winthrop (1681–1747) and His Gifts to the Royal Society," Publications of the Colonial Society of Massachusetts, *Transactions, 1952–1956*, XLII (1964), 206–232.

[212] C. Mather, William Brattle, John Leverett, T. Robie, P. Dudley, Governor Burnet, and John Winthrop. Although his influence in New England was indirect and his interest in science minimal, Elihu Yale (1649–1721) was enlisted by Cotton Mather as the benefactor of the Collegiate School at Saybrook, Connecticut, which later became Yale College, and Yale was elected F.R.S.

[213] John James, the Reverend Mr. Andrew le Mercier, and Robert Harris, of Boston; Mrs. Martha Gerisk and John Harrison of Cambridge; William Walker and Theodore Coker of Rhode Island; Thomas Jenner of Charlestown; and Matthias Plant of Newbury. Plant and Harrison made the *Phil. Trans.* See XLI, No. 462 (Jan.–Feb., 1741/42), 33–42; XLVIII (1751–52), 184–187. Similar ephemeral communications to the Society continued until the end of the Old Colonial Era, and the movement spread to many of the other colonies—indicative, no doubt, of the quickening of scientific interests among the people at large.

may be added at least three men who were the products of this generation of New Englanders and whose future influence on colonial science is well recognized: Thomas Clap (1703–67), Harvard graduate in 1722 and later President of Yale College (1745–66), where he became a contributor to the Royal Society and was instrumental in reorganizing the curriculum with heavy emphasis upon science and in building up the scientific apparatus of the college to a high level of usefulness;[214] John Winthrop (1714–79), Harvard graduate in 1739, student of Isaac Greenwood, whom he succeeded as Hollis Professor of Mathematics and Natural Philosophy, frequent contributor to the Royal Society, to which he was elected Fellow, and the colonies' most renowned astronomer and teacher of mathematics and natural philosophy;[215] and Benjamin Franklin, destined to become, among other things, America's first scientist and promoter of scientific learning. Though only a fifteen-year-old printer's apprentice when his brother's newssheet, the *New-England Courant,* opposed inoculation for smallpox, this clever, calculating, bookish youth obviously absorbed something of the current scientific interests of the Boston community. What else could have induced his strong desire, as expressed in his *Autobiography,* to see Sir Isaac Newton and to cultivate the acquaintance of Sir Hans Sloane when he visited London in 1725?[216]

It appears evident that this early eighteenth-century generation of New England scientists reached new levels of accomplishment and sophistication in their scientific achievements. In their communications with the Royal Society they were no longer content to serve merely as field agents for their masters in the homeland. To the specimens, descriptions, and observations which they sent to the Royal Society they added philosophical speculations, hypotheses, and scientific *ideas,* some of them founded upon some scientific experimentation and measurement. To be sure, these accomplishments were few and generally proffered with a humility still due

---

[214] Shipton, *Sibley's Harvard Graduates,* VII (Boston, 1945), 27–50; *Journal-Book,* XIX, 261–262, 292; XXVI, 119–120, 123, 133–138; author of *Conjectures upon the Nature and Motion of Meteors which are above the Atmosphere* (Boston, 1744; republished, Norwich, Conn., 1781).

[215] Brooke Hindle, *The Pursuit of Science in Revolutionary America, 1735–1789* (Chapel Hill, 1956), p. 88 and *passim.*

[216] See his letter to Sloane, June 2, 1725, *Sloane 4047,* fol. 347.

to the more advanced scientific achievements of the homeland. But they marked a growing independence of mind on the part of colonials, and they heralded a day, no longer far distant, when colonial scientists would be able to muster such a degree of self-reliance in scientific matters that they would dare to pit their knowledge against that of the mother country and, in a few instances, to teach the masters. Moreover, the colonial scientist no longer operated in isolation, without books and instruments, on a raw frontier where, as John Winthrop had said in 1668, there were "all thinges to doe, as in the beginninge of the world." If Thomas Brattle felt in 1703 that "I am here all alone by myself, without a meet help in my Studies . . . & that little insight wch I have into these matters, I have gotten it, as they say, proprio Marte, there never having been any one person that could give me ye least Instruction, Assistance, or Satisfaction in these Studies,"[217] certainly such complaints could not be truthfully voiced after 1730. By that time Mather, Robie, Dudley, Greenwood, Douglass, and others had formed a small community of scientifically-minded men, armed with books and apparatus, able to confer with one another, to exchange data and opinions, and to give mutual encouragement, support, and competition. Their situation hardly rivaled that of the Fellows of the Royal Society, but the latter must have been conscious of their enlarged potential in scientific work, for about 1736 the Secretary of the Society, Dr. Cromwell Mortimer, invited Paul Dudley to set up in Boston "a Company here subservient to the Royal Society" and cooperating with it.[218] Dudley felt compelled to beg off because of ill health and the pressure of his work. But Dr. William Douglass' Boston Medical Society, organized about the same time, offered a substitute of a sort—and an entering wedge for further scientific organizations in the colonies.

Isaac Greenwood's scientific lectures and demonstrations broadened the popular base of scientific interest in the Boston community while his successful appeals to the generosity of Thomas Hollis enabled Harvard College to obtain the most splendid "Philosophi-

217 Brattle to Flamsteed, Dec. 15, 1703. Brattle enlarged upon his complaint in 1705. Both letters are quoted in Kilgour, "The Rise of Scientific Activity," pp. 125–126. Cf. note 15 above.

218 Dudley to Mortimer, Jan. 25, 1736/37, *Letter-Book*, XXIII, 286–287.

cal Apparatus" in the colonies.[219] The literature of the times further demonstrated a growing interest in science. The late Perry Miller pointed out the quantitative immensity of New England literature during the years 1700–1730, and that by the 1720's scientific knowledge was being widely diffused.[220] Books, pamphlets, broadsides, even sermons directed public attention to scientific matters. The yearly almanacs frequently instructed their readers in the new science. Edward Holyoke's *Almanacks* for 1712 and 1713 republished from the *Philosophical Transactions* extracts from Edmond Halley's account of trade winds and from William Petty's *Political Arithmetic.* Daniel Travis' *Almanack . . . For . . . 1723* reprinted Cotton Mather's account of "Thunder and Lightning" from *The Christian Philosopher.* Nathaniel Ames, Jr., wrote in 1734 that the Copernican hypothesis of the universe "is now by infallible Methods of Reasoning from *Geometrical* Principles render'd indisputable . . . every Objection against it has been fully answered, and no Man of sense pretends to dispute it. . . ." However, as these mathematical principles are "above the capacity of the generality of men," he set out to explain the proofs of Copernicanism, hopefully within the capacities of his readers. Two years later, Ames described "minute Animalculae, each made compleat and perfect in their Kind, and want no parts either convenient or necessary for them; [and he went on to relate that] The Learned have sufficiently exploded the Notion of . . . spontaneous Generation of Animals, and will not allow the vilest and most contemptible Insect to be generated without Parents, Male and Female, of the same Kind." To illustrate the point further, Ames pointed to the life cycle of botflies in horses, easily observed by anyone at the time, and he quoted Cotton Mather to prove the falsity of ancient ideas about spontaneous generation. In 1737 Ames discussed astronomy. Borrowing from William Derham's accounts of the planets (based principally upon Christian Huygens' observations), he discussed the sizes of the planets in relation to that of the earth, their relative distances from the sun, and the possibilities of

[219] See Cohen, *Some Early Tools of American Science,* especially Chap. 2 and Appendix I, with illustrations.
[220] *New England Mind,* pp. 395, 444.

their being inhabited "in like Manner as this Earth is." In 1740 Ames presented an essay on Newton's law of gravity, the "Cement (as 'twere) of the whole Creation."[221] In the meantime, Nathan Bowen of Marblehead, in *The New-England Diary: Or, Almanack . . . 1723*, presented a remarkable diagram to illustrate a lunar eclipse. In 1731 Bowen printed a popular exposition of "*The Three Laws of Motion laid down* by Sir Isaac Newton, by which we think that every Thing, that relates to Motion may be explained." In 1733 he explained why the Gregorian calendar has an extra day in leap year. In 1735, borrowing from Sir William Petty and Edmond Halley, he explained the construction and use of mortality tables in actuarial practices. In 1736, again relying upon Halley, he sought to explain "why it is so much hotter with us in *Summer*, than in *Winter*, tho' the *Sun* is vastly nearer to us in *Winter*, than in *Summer*." And in 1737 he described "some Discoveries in the [solar] System of the World, made by the great Sir Isaac Newton," with definitions of the primary and secondary planets and conjectures about other nebulae which might be discovered when more powerful telescopes became available.[222] By and large, these brief, popular scientific essays were couched in simple terms, with diagrams (woodcuts), so that almost every literate person could comprehend their contents.

The newssheets likewise devoted space to scientific subjects from time to time. On February 8, 1717/18, the *Boston News-Letter* printed "for the benefit of the Publick" an account from "one of the Publick Prints of London" about how "to prevent wheat from being Smutty," by steeping the seed in a brine of salt and alum. This appears to have been the first of a series appearing in colonial newspapers during the remainder of the Old Colonial Era to offer up-to-date aids to agriculture, horticulture, and animal husbandry —although colonial farmers often responded negatively to such advices, and before many years had passed, newspapers were pub-

221 Nathaniel Ames, Jr., *An Astronomical Diary, Or, An Almanack . . . 1734* [*1736, 1737, 1740*] (Boston, 1734, 1736, 1737, 1740). Later editions (1746, 1750, 1753) continued astronomical articles, including (1746) "A Dialogue *between* a Scholar *and a* Clown" to prove the Copernican theory of the universe.

222 Nathan Bowen, *The New-England Diary: Or, Almanack . . . 1723* [*1731, 1733, 1735, 1736, 1737*] (Boston, 1723, 1731, 1733, 1735, 1736, 1737).

lishing articles to dispel farmers' prejudices about the "Folly of farming by Book."[223] In the issue of February 15–22, 1720/21, the *Boston News-Letter* printed a detailed description of "The Method of preparing Tar in Russia, with Remarks." In the number for October 1–8, 1722, it noted discussions of geological and anthropological items at a recent meeting of the Royal Society in London, directing its readers to future numbers of the *Philosophical Transactions* for further details. Number 997 (February 28–March 7, 1722/23) deplored the divisions in New England arising out of smallpox inoculation, declared the paper's neutrality in the quarrels, proposed to print "The History of Nature among us, as well as of Political and Foreign affairs," and urged "Ingenious Gentlemen" to contribute materials "so that this Paper may in some Degree, serve for the *Philosophical Transactions* of *New-England*. . . ." Three weeks later was published a response, "Concerning Cause of the late Extraordinary Inundation" by an unusually high tide along the New England coast on February 24, 1722/23. However, public responses to the plea of the editor (B. Green) were infrequent and concerned meteorological phenomena mostly.[224] The *Boston Gazette* (June 1–8, 1724) published a London dispatch describing a recent series of surgical operations at St. Bartholomew's Hospital in which the surgeons employed "the new Way, above the *Os Pubis*" to cut for the stone. Samplings of other New England newspapers demonstrate a continuing concern for scientific subjects during the 1720's and 1730's, with a marked increase in the output after 1740. How widely the public mind was drawn from the old, apocalyptic interpretation of natural phenomena to the rising, rational, mechanistic view is impossible to measure with any accuracy. But the differences clearly demonstrated between the basic outlook of Cotton Mather and of Thomas Robie were symptomatic. Few New England scientists after Mather gave weight to the old religious views; rather, they subscribed to the new, mechanistic concept of the universe that was characteristic of the Enlightenment. Given the

[223] See, for example, the *Newport Mercury*, Nos. 27 (Dec. 19, 1758), 29 (Jan. 2, 1759), 47 (May 8, 1757).

[224] *Boston News-Letter*, Nos. 670 (Feb. 18, 1717/18), 827 (Feb. 15–22, 1720/21), 557 (Oct. 1–8, 1722), 997 (Feb. 18–March 7, 1722/23), 1000 (March 21–28, 1723). Later issues, to the mid-1730's, revealed little more.

agencies for popularization which New England possessed, surely enlightened public opinion was not far behind.

New York, New Jersey, and Pennsylvania

Evidence of scientific interest in New York is scanty before the eighteenth century. Among the *Classified Papers, 1660–1740* in the Royal Society Archives is a "Description of ye Towne of Mannadoux [New Amsterdam] in New Netherland, as it was in Sept. 1661,"[225] but there is no evidence about the authorship, and the "Description" offers historical data of little scientific consequence. No other reference to New York occurs until December 7, 1692, when Robert Hooke read to the Society "a Description of the County of New York in North America. . . . The whole being a compendious Naturall History of that Countrey."[226] The "Description" was from the pen of Mr. Charles Lodwik, nephew of Mr. Francis Lodwik, F.R.S., and had been solicited by the latter and Robert Hooke some years earlier. Charles Lodwik addressed his uncle and Hooke from New York, May 20, 1692, to apologize for his delay, it having been "full 4 years since I rec'd your Commands to give you what Acct. I was capable of ye Constitution of this Country. . . ." The events of Leisler's Rebellion had occasioned the delay so that "for Almost 3 years we had enough to do to exercise all our brains to secure our persons & that little we had from ye Cruelty & Tyranny of an ungovernable Mobb. . . ." Young Lodwik had collected "ye Opinions of our greatest Sages here, where my young experience would not lett me conclude," but he felt that the "Description" fell short of what it ought to be, "for where Mrs. of Ships are ye chiefest Mathematicians & ye Natives Geographers, with such tools you must not expect a good Fabrick. . . ." The "Description" itself, covering about eleven folio pages, described the geography and location of New York

225 VII (1) ("Travel, etc."), No. 8.
226 *Journal-Book*, VIII, 141. Dr. Waller, Secretary of the Society, later found the original of the "Description," with its covering letter to Mr. Francis Lodwik, F.R.S., and Dr. Hooke, among Hooke's papers, and he presented them to the Society again, Nov. 26, 1713. *Ibid.*, XI, 388–389. Both are preserved in *Sloane 3339*, fols. 93–98v.

City, its diversity of peoples and religions, its climate and soil, its economy, its natural animals and birds, the Indians and stories of Indian medicine. But it is a traveler's account, by one who demonstrated no scientific knowledge, informative for the general historian but of small import for the history of science. Of greater specific value were the specimens of plants, seeds, fishes, and articles of Indian manufacture which young Lodwik presented to the Royal Society upon his return from New York in the autumn of 1693.[227]

About 1692 James Petiver attempted to establish a scientific correspondence with the Reverend Dr. Alexander Innis at Boscobell House near Hempstead, Long Island, but when Dr. Innis failed to respond, Petiver addressed a sharp letter to him (November 6, 1695) for failing to keep "ye promise of a clergyman" and terminated the effort.[228] Edward Southwell presented the Royal Society "a very curious and Entire Malucco Crabshell taken up at New York" and a comb said to be made of a man's skull from New York in 1698,[229] and Dr. Hans Sloane sought a scientific exchange with one P. Gordon going to reside in New York in 1702, but it appears to have yielded nothing.[230] In 1705 Edward Hyde, Viscount Cornbury, Governor of New York and New Jersey (1702–08), sent to his kinsman Henry Hyde, second Earl of Clarendon, a tooth and some bones of the "gyant" unearthed near Albany to which Cotton Mather later referred, and they were presented to the Royal Society on March 6, 1705/06.[231] One Robert Gamble wrote from New York, November 22, 1713, to describe his voyage and the shortage of specie in New York, but when Petiver invited him to engage in a philosophical correspondence, Gamble did not reply.[232] In 1722 William Burnet, F.R.S., Governor of New York and New Jersey (1720–28), communicated to the Royal Society observations of a solar eclipse made at Fort St. George in New York, and the Society (February 14, 1722/23) turned them over to Edmond

227 *Journal-Book*, VIII, 196, 209 (Oct. 18, Dec. 6).

228 *Sloane 3332*, fols. 19, 38, 53v, 125, 168.

229 *Journal-Book*, X, 93 (Dec. 28).

230 Gordon to Sloane, Portsmouth, April 17, 1702, *Sloane 4038*, fol. 330.

231 *Journal-Book*, XI, 84. See also *Sloane 4064*, fols. 86, 93.

232 Gamble to Samuel Sutton, *Sloane 3339*, fol. 110; Petiver to Gamble, n.d., *ibid.*, fol. 147.

Halley for study.[233] Two years later, Governor Burnet sent observations of four eclipses of the first satellite of Jupiter, "made by his order for determining the Longitude of the fort of New York, to which was added an account of the present state of the Variation of the Magnetic Needle at that place."[234] The eclipses had been observed on August 9, 25, September 10, 1723, and June 26, 1724. Burnet had been assisted in the work by Cadwallader Colden and James Alexander. Colden calculated the latitude of Fort St. George at 40° 40′ N and, on the basis of the observations, determined the longitude at 74° 57′ 30″ W. The observations and calculations were published in the *Philosophical Transactions.*[235]

The cooperation of Colden and Alexander with Governor Burnet in 1723 suggests the nucleus of a scientific community forming in New York, although it did not develop into significant proportions until after the mid-century. James Alexander (1691–1756), a Scottish-born Jacobite, who fled to New York after the vain efforts to seat the Stuart pretender upon the English throne at the accession of George I in 1714, became a prominent lawyer who served on the councils of both New York and New Jersey, as Attorney General of New Jersey, and as Surveyor General of New Jersey. He was closely associated with his fellow Scot, Cadwallader Colden (1688–1776), in various scientific activities. Colden, who has already been noted as a scientific correspondent of Dr. William Douglass of Boston, was born in Ireland of Scottish parentage, graduated from the University of Edinburgh (1705), studied medicine in London, and migrated to Philadelphia in 1710, where he practiced medicine and engaged in mercantile enterprises. Prior to his arrival in Philadelphia, Dr. Charles Preston, Superintendent of the Edinburgh Physic Garden, recommended Colden to James Petiver so that by "your advice & assistance in his studies you can show him ye way of making Specimens of dry plants & of collecting other natural curiosities; he Designs in ye spring for Virginia where he will be a good Correspondent for us."[236] Colden agreed

[233] *Journal-Book*, XIII, 258.
[234] *Ibid.*, XIII, 454–455 (March 11, 1724/25).
[235] XXXIII, No. 385 (Oct.–Dec., 1724), 162–165.
[236] Sept. 24, 1708, *Sloane 4064*, fol. 178. Whether Colden originally had planned to go to Virginia or whether Preston spoke loosely is impossible to say. Cf. p. 246n above.

to serve Mr. Petiver, who wrote to him (October 27, 1709) urging him to take ample supplies of brown paper, wide-mouth bottles, and other necessaries for collecting natural history specimens in Pennsylvania, for "That part of the Continent I have as yet never seen any thing from so that whatever you send will be New to me, As well the Insects, Shells, & Fossills as the Plants. And be assured Sir as soon as ever I shall be so happy as to receive anything from you with directions how to write You shall hear largely from me with my Remarks on what you shall send as also some Tables of Plants, Insects &c wch I have lately engraven relating to Carolina Virginia &c. . . ."[237] Shortly before his departure, Colden addressed Petiver from Oxnam (October 17, 1709):

> The curious world is bound to you for the toil & cost you put yourself to in serving them; and I would be glad to show how much I think my duty to give you all the Assistance in my power. You may assure yourself I will omit no opertunities of doeing it.
>
> In Obedience to your Commands I give you the trouble of this that you may know that I design to go for Pensylvania from a Port near this within three weeks or a Month at furthest. Please to lay your Commands on me as soon as possible before I go . . . to be left for the Post to Jedburgh at the Post-office in Edinburgh.[238]

However, in spite of his assurances to Petiver, there is no evidence that Colden sent the latter anything from America. Indeed, in 1711 Petiver had had no word from him and wrote to Dr. Preston to "pray let me know where he is & when you heard from him."[239] To Petiver, Cadwallader Colden was another of many promising collectors who failed to perform.

In Philadelphia, Colden made the acquaintance, among others, of James Logan (1674–1751) and Dr. John Kearsley (1684–1772). Logan, a Scot and a self-made man, had arrived in Philadelphia in 1699 as a mere secretary in the service of William Penn; but he remained to become the Proprietor's principal American represen-

---

[237] *Sloane 3337*, fol. 63v.

[238] *Sloane 4064*, fol. 213. This letter is mutilated, and the signature is torn off. But I believe that there is no doubt about the identity of the author. See also Petiver to Preston, March 22, 1710/11, *Sloane 3337*, fols. 122v–123v.

[239] *Ibid.*

tative in business affairs. He enriched himself by land speculation and the fur trade, became an influential statesman, acquired one of the best private libraries in the colonies, and was "Philadelphia's First Scientist."[240] Kearsley, English born, came to America in 1711 and settled in Philadelphia in 1717. A versatile man, he was the architect who designed Christ Church in Philadelphia, taught so many young men the art of medicine that his medical office was sometimes referred to as the "first college" of Pennsylvania, and communicated to Peter Collinson several papers and plant specimens which were presented to the Royal Society of London. These dealt with the enchanting powers of rattlesnakes, which Dr. Kearsley rejected, having made experiments with results in agreement with those of Sir Hans Sloane's observations; with the value of various herbs in the cure of snake-bite, which, in some instances, he had observed Indians to use successfully, although he felt that most of them were of no value; with the "locusts" of Pennsylvania, about which he agreed in part with Paul Dudley, although he believed that there was more than one variety; with the medicinal values of American ginseng, which he had found to be of little efficacy; with inoculation for smallpox, which he reported had been very successful in Philadelphia; with medical cases he had encountered in Philadelphia; and with observations of a comet which appeared over Philadelphia on January 27, 1736/37, and of a solar eclipse on February 18 following. He also sent specimens of various herbs used for snake-bite; of Pennsylvania ginseng, scammony, gentian, and ipecacuanha; of a "locust"; of a shell of a "land tortoise"; and of maple sugar.[241] Had Cadwallader Colden remained in Philadelphia, doubtless he would have become an active member of the city's community of scientists, which by mid-century had

[240] Frederick B. Tolles, *James Logan and the Culture of Provincial America* (Boston, 1957); Tolles, "Philadelphia's First Scientist: James Logan," *Isis*, XLVII (1956), 20–30; *Colden Papers*, I, 102–103.

[241] Letters from Kearsley to Collinson: Nov. 18, 1735, *Letter-Book*, XXII, 324–340; *Journal-Book*, XVI, 336–338; March 19, 1736/37, *Letter-Book*, XXV, 115–116; March 21, 1736/37, *ibid.*, XXIV, 319–325; *Journal-Book*, XVII, 96–97; May 12, 1737, *ibid.*, pp. 251–253; May 4, 1738, *ibid.*, p. 327; July 26, 1739, *ibid.*, pp. 480–481. Dr. Kearsley stated (March 21, 1736/37) that there was only one astronomer who would undertake to make astronomical observations and that he mistrusted the results "for want of good instruments and a serene air." He did not identify the observer. Was it Thomas Godfrey?

become the leader in the development of colonial science. However, though he presently moved to New York, he was a close friend and correspondent of these and other members of the Philadelphia circle.

In 1715 Colden returned to Scotland, married a Scottish girl, and, through the offices of an unidentified friend, transmitted to the Royal Society a paper "containing some Propositions . . . relating to the Animal Oeconomy [physiology] and the Nature and cure of fevers," read to the Society on January 12, 1715/16.[242] Two years after his return to Philadelphia, Colden visited New York, where he was persuaded to stay by promises of preferment from the popular Governor, Robert Hunter, F.R.S.[243] In 1720 Colden was appointed Surveyor General of the Province of New York and was launched upon a career in politics. During the next year he was elevated to the Council, on which he served for forty years, and subsequently became Lieutenant Governor (1761) and Acting Governor on several occasions before his death in 1776. In the meantime he practiced medicine, which he gave up after a few years, took advantage of his position as Surveyor General to enrich himself by land speculation, and gathered a library rich in scientific materials.[244] He also became active in scientific affairs, became internationally known (though not always favorably) for his scientific endeavors, and engaged in a wide scientific correspondence both with other colonial scientists and with eminent scientists in Europe. He early became interested in Indian affairs and published *The History of the Five Indian Nations of Canada* (London, 1727; enlarged ed., 2 vols., 1745, 1755). The book was designed to correct French accounts of the Iroquois, to upgrade English opinion of the civilization of the Six Nations, for whom Colden held a warm sympathy, and to emphasize the strategic importance of Iroquois friendship with the English. Based upon French accounts, minutes of the English Commissioners for Indian Affairs, Indian treaties, and personal observations, Colden's book brought him

[242] *Journal-Book*, XII, 89. There is no evidence that Colden was present at the meeting.

[243] After his retirement as Governor of New York, Hunter urged Colden to prepare a natural history of the province. *Colden Papers*, VIII, 157.

[244] See his order to London booksellers, *ibid.*, I, 41–43, *passim*.

plaudits from both sides of the Atlantic, and the treatise still remains a valuable source for the history of the Iroquois—though less useful for anthropological research.[245] In 1728 Colden built a new country home near Newburgh, New York, named Coldengham, where he lived the life of a country gentleman and indulged himself frequently in scientific endeavors. He lived there until 1761, when he became Lieutenant Governor of New York, after which he moved to a country seat called Spring Hill, near Flushing, Long Island. In the next chapter below, Colden will enter the colonial scientific scene repeatedly, for he lived to the very end of the Old Colonial Era.[246]

Except James Alexander, noted above, who held high offices both in New York and New Jersey (sometimes simultaneously) and might, therefore, be cited with the latter province, there was little scientific activity in New Jersey. The most frequent contributor to the Royal Society was Joseph Morgan (1671–c. 1745), who sent (1699) a paper "concerning the dressing of Leather in West Jersey in America, by the help of their [i.e., the animals'] brains."[247] Morgan was born in Connecticut, received the bachelor of arts degree at Yale (1702), and an honorary master's degree (1719) from the same institution. He had been licensed to preach (1697), ordained (1700), and had ministered to various congregations in Connecticut and New York before he moved to Freehold, New Jersey (1709), and later to Maidenhead and Hopewell, near Trenton (1728). Cotton Mather had already brought his name to the attention of the Royal Society,[248] Francis Nicholson trans-

245 Peter Collinson saw the enlarged edition of 1745 through the London press. *Ibid.*, II, 250–251, 271–272. There have been more recent American reprints.

246 No satisfactory life of Colden has been written. The *Colden Papers* provide an immense reservoir of materials from which some useful fragments of Colden's life have been published. See especially A. J. Wall, "Cadwallader Colden and His Homestead at Spring Hill, Flushing, Long Island," *New York Historical Society Quarterly*, VIII (1924), 11–20; Louis Leonard Gitin, "Cadwallader Colden as Scientist and Philosopher," *ibid.*, XVI (1935), 169–177; Jacob Judd, "Dr. Colden's Cure," *ibid.*, XLV (1961), 251–253; Brooke Hindle, "A Colonial Governor's Family: The Coldens of Coldengham," *ibid.*, pp. 233–250; Hindle, "Cadwallader Colden's Extension of Newtonian Principles," *Wm. & Mary Qty.*, 3rd Ser., XIII (1956), 459–475; *D.A.B.*

247 *Journal-Book*, X, 156 (Dec. 13, 1699). There is no evidence as to who communicated this paper.

248 See p. 410 above.

mitted one of his papers to the Society in 1720,[249] and soon Benjamin Franklin began to note Morgan's proposals in the *Pennsylvania Gazette*, to print some of his materials, and to transmit some of his "scientific" papers to the Royal Society. But the Reverend Joseph Morgan was a rustic, ill-educated eccentric.[250] His published works include a remarkably good Utopian novel,[251] theological tracts, and sermons with a strict Calvinistic accent. His communications to the Royal Society, especially between 1732 and 1739, included an eighty-eight-page paper about "The Original of all Nations," in which the author argued that internecine wars among Noah's descendants drove some of them into Tartary and to America; "Some further Improvement of ye Astronomical Philosophy of Sir Isaac Newton & others," setting forth an atomic view of matter and arguing "That Heat is nothing else but ye agitation of Fireparticles (*Friction* will excite you to agitation . . ."); "The Times Prophesied, Known by Scripture: without ye Help of Humane History"; a proposition to settle the Mississippi Valley ahead of the French together with a way to keep the American colonies forever loyal to Great Britain by dividing America into small, compact settlements, each with a chartered government by the King; and a variety of lesser papers. But Morgan's passion was his "invention" of a light, flat-bottomed boat, propelled by a complicated system of oars and sails, able to navigate shallow streams and even to pass "with Ease & pleasure on Smooth Ice." This was exhibited in New York as early as 1714, pressed upon the English Board of Trade, repeatedly presented to the Royal Society, and, as late as 1745, urged upon the British Navy. Morgan felt that his boat would be a valuable asset in the British settlement and defense of North America; he offered it for the good of mankind; and he hoped to win fame and fortune from it. But no one responded favorably to his repetitious but rather fuzzy expositions, and the plans he so

[249] *Journal-Book*, XIII, 19 (June 2, 1720). Morgan supposed that a piece of wrought iron found in a Jersey salt marsh was evidence that the descendants of Cain had traveled to America before the Flood.

[250] Dr. Alexander Hamilton presented an amusing picture of Morgan in 1744. See Carl Bridenbaugh, ed., *Gentleman's Progress: The Itinerarium of Dr. Alexander Hamilton, 1744* (Chapel Hill, 1948), pp. 35–36.

[251] *The Kingdom of Basaruh*, ed. with an informative Introduction by Richard Schlatter (Cambridge, Mass., 1946).

urgently presented brought him only disappointment and heart-break. Moreover, the Royal Society generally ignored his communications. His early papers were turned over to Captain Fayer Hall, F.R.S., who reported to the Society (June 7, 1733) that "he had perused the Miscellaneous papers of Mr. Morgan of Maidenhead in New-Jersey, relating to various Subjects in Navigation, Mechanics, Husbandry, Government, &c; but finding the author throughout in pursuit of vain subjects, contrived, as should seem, rather for the amusement than instruction of the Reader, he was of opinion that it would be needless to trouble the Society with any Extract of the Contents."[252] When Morgan subsequently pressed for a more specific response to his communications and made the "remarkable Observation" that "Julius Caesar by neglecting to read a Letter lost his Life, and did not know the cause," the Royal Society replied tartly (and not altogether honestly) that "it is not their custom to pass judgment either in approving or in disapproving any Proposal, unless it comes from one of their own Members."[253] Subsequent communications from Morgan were "not read nor registered but an answer sent" by the Royal Society. These manifold slights upon Morgan's proposals ultimately drove the man to drink. He lost his church and faded into obscurity. Still, in spite of the doubtful scientific value of his work, and the confusion and absurdity of some of his proposals, Morgan demonstrated, as Whitfield J. Bell, Jr., has observed, "some minor significance": "A country clergyman of little education who was drawn to science and scientific ideas; an amateur immured in Maidenhead, New Jersey, who corresponded with the Royal Society; a curious and pathetic figure, Joseph Morgan is a sort of case study in trans-Atlantic cultural ties, an index of the extent and influence of scientific ideas and institutions in eighteenth-century colonial America."[254] He did not represent colonial America at its scientific best; but he signified the transmission of a scientific spirit, in an understandably muddled form, to the lower echelons of society.

252 *Journal-Book*, XV, 308–309.
253 *Ibid.*, XVI, 23 (Nov. 14, 1734).
254 Bell, "The Reverend Mr. Joseph Morgan," p. 261. Morgan's communications to the Royal Society are in: *Guard-Book*, M-3, Nos. 35–40, 48, 66–67; *Letter-Book*, XX, 200; XXI, 167–168; XXIV, 383–384; *Journal-Book*, XVI, 23; XVIII, 31.

On October 22, 1755, Governor Jonathan Belcher's lady wrote from Elizabethtown, New Jersey, to Dr. Stephen Hales to report a strange phenomenon in Lake Ontario, whose waters recently "rose and fell five feet and a half three Severall times in the Space of half an hour." The Society speculated that the event was somehow associated with an earthquake felt in Europe on November 1 following.[255] The following July 26, 1756, Lady Belcher communicated to Dr. Hales another account of two men who, working in copper mines some fifty miles apart at depths of 80 and 140 feet below the surface respectively, did not feel an earthquake which terrified persons on the surface of the earth in New Jersey (November 18, 1755).[256] The Society felt that this evidence supported the contention that the cause of earthquakes originated in the air. Nearly ten years later, the Reverend Mr. Samuel Finley, President of the College of New Jersey, communicated to the Reverend Mr. Samuel Chandler, F.R.S., a letter regarding the migration of birds (May 1, 1765). Though he believed that swallows were migratory, Mr. Finley reported that he had found some that did not migrate in hollow trees, caves, and along the mud banks of rivers.[257] Thus ended scientific communications from New Jersey to the Royal Society during the Old Colonial Era.[258]

William Penn was the second American colonial (if he may be so described) to become a Fellow of the Royal Society of London.[259] On the day of Penn's election (November 9, 1681), John Houghton "presented to the Society from Mr. Penn his Map of Pennsylvania for which the Society returned their thanks."[260] Thus the new proprietary grant was early drawn to the attention of the Society. But no further scientific returns from Pennsylvania reached the Royal Society for many years. In June, 1694, at a discussion in

[255] *Journal-Book*, XXIII, 283 (Feb. 19, 1756).

[256] *Ibid.*, XXIII, 587–588 (March 31, 1757).

[257] Read Nov. 21, 1765, *ibid.*, XXVI, 313–314.

[258] Soon after the American Revolution, André Michaux and Pierre Paul Saunier set up gardens under French auspices in New Jersey and South Carolina as a source of supply of botanical specimens and seeds for the Paris Jardin des Plantes. See William J. Robbins and Mary C. Howson, "André Michaux's New Jersey Garden and Pierre Paul Saunier, Journeyman Gardener," *Proceedings of the American Philosophical Society*, CII (1955), 351ff.

[259] *Col. F.R.S.*, pp. 196–197.

[260] *Journal-Book*, VII, 32; *Egerton 2381*, fol. 226-b (British Museum).

the Society about the geography of North America, Penn was said to have told Robert Hooke that both the St. Lawrence and Delaware rivers ran far inland, and several of the early accounts of Pennsylvania, some of them promotion pamphlets, treated fleetingly of the geography and natural productions of the province.[261] But none of them qualified as a scientific treatise, and none appear to have attracted attention in the scientific circles of England or elsewhere. Not until 1717, when Governor William Keith (1680–1749) sent to the Royal Society observations of an eclipse of the sun which he had made at Philadelphia on September 23 of that year, was Pennsylvania again represented in the flow of scientific communications from the American colonies to the Royal Society.[262] By that time James Logan and Dr. John Kearsley, both of whom have been noted above, had won attention on the colonial scientific scene, but it was several years afterward that Philadelphia began to exhibit a scientific community to rival that of Boston. Indeed, by around 1740 a number of intersecting circles of colonial scientists were emerging. Strengthened by new and more able scientific personalities, stimulated by enlarged intercolonial scientific communications, encouraged by new and wider contacts with the European world of science, and supported by an ever-widening popular interest in a rapidly growing colonial society with maturing cultural institutions, this last generation of colonial scientists brought scientific developments in the mainland British colonies of North America to the point of independence from the homeland—to the end of the Old Colonial Era. With their achievements the next chapter is principally concerned.

261 *Journal-Book*, IX, 163 (June 13, 1694); Albert Cook Myers, ed., *Narratives of Early Pennsylvania, West New Jersey and Delaware, 1630–1707*, in J. F. Jameson, ed., *Original Narratives of Early American History* (19 vols., New York, 1906–17; reprinted, 1952), *passim;* Francis Daniel Pastorius, *A Particular Geographical Description of the lately discovered Province of Pennsylvania . . .* , tr. from the original German by Lewis H. Weiss in *Memoirs of the Historical Society of Pennsylvania*, IV, Pt. II (1850); Frederick B. Tolles, *Quakers and the Atlantic Culture* (New York, 1960), especially Chap. 4.

262 *Journal-Book*, XII, 192; presented to the Society by Edmond Halley, Nov. 14, 1717.

ELEVEN

The Emergence of American Science

Societal Growth and Urbanization

SCIENCE BLOSSOMED IN THE BRITISH AMERICAN colonies during the last generation of the Old Colonial Era.[1] The number of persons engaged in scientific pursuits increased materially, and their scientific accomplishments, in some instances, compared favorably with those of European scholars for the first time. These developments appear as a culmination of all the promotional efforts of the Royal Society of London and of other friends and patrons of the colonists in the Old World in conjunction with a sudden and remarkable growth of colonial society itself.

At base, this growth is most readily observed in the rapid expansion of colonial population. Between 1740 and 1770 the total population of the British colonies on the mainland of North America rose from 940,563 to 2,148,076.[2] It more than doubled during a generation which was involved in two bitter colonial wars, "King George's War" and its almost consecutive successor, the French and Indian War (or "Great War for the Empire," as the latter more recently has been called).

This population growth was unevenly spread over the colonies.

---

[1] The scientific developments in the American colonies after about 1740, unlike those of the earlier years, have been treated by a variety of scholars, especially by Brooke Hindle, *The Pursuit of Science in Revolutionary America, 1735–1789* (Chapel Hill, 1956). Accordingly, this chapter rests heavily upon these earlier, secondary works, supplemented from time to time from primary sources which have escaped the attention of previous writers or upon which I place a somewhat different emphasis.

[2] The statistics include Negroes but not Indians. See *The Statistical History of the United States from Colonial Times to the Present* (2 vols. in 1, Stamford, Conn., 1965), p. 756.

The population of New England and of the southern colonies almost doubled while that of the middle colonies nearly tripled. Pennsylvania alone rose from a population of 85,637 in 1740 to 240,057 in 1770; that of New York from 63,665 to 162,920. Virginia was the most populous colony in 1770 (447,016), but the *rate* of population growth had been most rapid in the middle colonies, especially in Pennsylvania.

Although rural areas appear to have expanded in population as rapidly if not somewhat more rapidly than towns, urbanization was a matter of prime importance, especially with regard to cultural development. Carl Bridenbaugh has pointed out that the principal colonial towns which had arisen by 1740 (Boston, Newport, New York, Philadelphia, and Charles Town) had developed into cities by 1770. Boston, which had been the cultural capital of the colonies during the first forty years of the eighteenth century, was overshadowed by Philadelphia before 1770.[3] Indeed, each of the five original colonial towns with which Professor Bridenbaugh deals had approximated the status of a cultural metropolis by 1770, with both Philadelphia and Boston clearly in that category. In addition, by 1770 there had arisen a host of secondary urban communities, such as Salem, Hartford, New Haven, New London, Marblehead, Middletown, Springfield, Providence, Norwich, and Portsmouth in New England; Albany and Lancaster in the middle colonies; and Williamsburg, Annapolis, Baltimore, Norfolk, and Savannah in the southern colonies.[4] If colonial society was still predominantly rural in 1770, nonetheless it contained sizable urban centers of wealth, commercial significance, and cultural leadership.

Both the population expansion and the relative urbanization were reflected in almost every facet of colonial experience. Communications, in terms of roads and postal services, improved rapidly, par-

[3] See Carl Bridenbaugh, *Cities in the Wilderness* (New York, 1938); *Cities in Revolt* (New York, 1955; paperback ed., 1964); and, with Jessica Bridenbaugh, *Rebels and Gentlemen: Philadelphia in the Age of Franklin* (New York, 1942; paperback ed., 1942).

[4] See Bridenbaugh, *Cities in Revolt*, p. 217; Bridenbaugh, *Seat of Empire: The Political Role of Eighteenth-Century Williamsburg* (Williamsburg, 1950). Charles O. Paullin, *Atlas of Historical Geography of the United States*, ed. John K. Wright (Washington, 1932), p. 42, shows an increase of *all* colonial towns from 205 in 1700 to 730 by 1775.

ticularly in the northern and middle colonies, where the greater urbanization was taking place. Printing establishments for books, pamphlets, and newspapers multiplied rapidly. Libraries, both public and private, grew in number and quality. The famous Library Company of Philadelphia, founded by Benjamin Franklin and his friends in 1731, came to serve not only the usual functions of a library but also as a repository for scientific apparatus and as a museum for scientific specimens.[5] A Library Society was established at Charles Town in 1748,[6] and the New York Corporation Library, founded in 1730, was opened to the public in 1746.[7] To these must be added the Redwood Library founded at Newport in 1747, various lending libraries, and a considerable number of private libraries scattered widely over the colonies, the owners of which frequently lent books to their friends.[8] The libraries of the colonial colleges further enlarged these facilities, and the number of colleges was enlarged between 1740 and 1770. To the earlier colleges—Harvard, Yale, and William and Mary—were added the College of New Jersey (Princeton, 1741), the Philadelphia Academy (which soon became the College of Philadelphia, 1755, later

[5] Leonard W. Labaree, ed., *The Autobiography of Benjamin Franklin* (New Haven, 1964), pp. 130ff.; Austin K. Gray, *Benjamin Franklin's Library: A Short Account of the Library Company of Philadelphia, 1731-1931* (Philadelphia, 1936), *passim;* Dorothy F. Grimm, "Franklin's Scientific Institution," *Pennsylvania History*, XXIII (1956), 437–462; Bridenbaugh and Bridenbaugh, *Rebels and Gentlemen*, pp. 86–96.

[6] Frederick P. Bowes, *The Culture of Early Charleston* (Chapel Hill, 1942), *passim.*

[7] Louis Leonard Gitin, "Cadwallader Colden as Scientist and Philosopher," *New York Historical Society Quarterly*, XVI (1935), 177; Bridenbaugh, *Cities in Revolt*, pp. 181, 380–385.

[8] James Logan of Philadelphia had the finest private scientific library in the middle colonies, and it was given to the City of Philadelphia, complete with building, upon Logan's death in 1751; but those of Cadwallader Colden in New York, Dr. William Douglass in Boston, Dr. John Mitchell, William Byrd II, John Clayton, and Isham Randolph, all of Virginia, and Drs. William Bull and Alexander Garden of Charles Town, were sizable. Specialized libraries, especially in medicine, similar to those described by Bridenbaugh for Philadelphia, also arose in Boston, New York, Charles Town, and elsewhere. See *Rebels and Gentlemen*, pp. 91ff. The influence of the Library Company of Philadelphia and the multiplication of colonial libraries is well described in Margaret Barton Korty, "Benjamin Franklin and Eighteenth-Century Libraries," *Transactions of the American Philosophical Society*, n.s., LV, Pt. 9 (1965).

the University of Pennsylvania), King's College (Columbia, 1754), the College of Rhode Island (Brown, 1764), Queen's College (Rutgers, 1766), and Dartmouth College (1769). In varying degrees, the newer colleges acquired libraries and scientific apparatus and served, with the older ones, as centers of scientific study and promotion. Books, especially books relating to science, which had been a scarce commodity in the early colonies, had become more readily available by 1770, and in each of the major cities there were booksellers from whose hands almost any book in print could be purchased.

Newspapers and almanacs, which previously had been confined largely to New England, multiplied rapidly during the second third of the eighteenth century. And, as they had done in the Boston area, they continued to serve as occasional media for the dissemination of scientific information. Many of the newspapers launched proved to be ephemeral, but several continued publication to the end of the Old Colonial Era—and a few beyond. The *Boston News-Letter*, founded in 1704, continued to 1776; the *Boston Gazette* ran from 1719 to 1798; and the *Boston Evening Post* from 1735 to 1775. In Philadelphia the *Pennsylvania Gazette* (Franklin's paper) ran from 1728 to 1815; the *Pennsylvania Journal* from 1742 to 1793. In New York the *New York Gazette*, launched in 1725, altered its title to the *New-York Gazette or Weekly Post-Boy* in 1744 and continued until 1773; and these were paralleled by the *New-York Weekly Journal* from 1733 to 1751. In Charles Town the *South Carolina Gazette* ran from 1732 to 1775. In Williamsburg the *Virginia Gazette* continued from 1736 to 1780. In Annapolis the *Maryland Gazette*, beginning in 1745, ran to 1777. And at Newport the *Newport Mercury* was published from 1758 to 1775. These papers, as well as those with a shorter life span, varied in nature and quality. Franklin's *Pennsylvania Gazette* generally excelled its competitors both in the quality of materials published and in its circulation. Moreover, by his financial interests in other papers and in paper manufacture, and by training in his print shop apprentices who went forth to set up other print shops and other newspapers, Franklin exerted over colonial newspapers an influence which reached far beyond Philadelphia. The *South Carolina Gazette*, the *New York Gazette* (after 1742), and others as far away as the West Indies were established and published by printers under Franklin's patronage.[9] The newspapers often published news of a scientific nature—"Science in the *Virginia Gazette, 1736–1780*" has been the subject of a recent study[10]—and the diffusion of the new science, undertaken first by the Boston newspapers, spread over all the colonies.[11] Similarly, the role played by the early New England almanacs was continued. Nathaniel Ames's *An Astronomical Diary, Or, An Almanack* was published by his son of the same name and appeared yearly, with varying titles, until 1775. Daniel Leeds, Andrew Bradford, and Thomas Godfrey published almanacs in Philadelphia, and a host of others appeared in other colonial towns, although many of them were short-lived. The best known was, of course, Franklin's *Poor Richard's Almanac* (1732–67) which, though it published less scientific matter per se than several of its competitors, constantly cultivated a scientific state of mind by its occasional articles reflecting Franklin's scientific interests and by its gentle, though devastating, ridicule of superstition from the pen of the pretended astrologer, Richard Saunders ("Poor Richard").[12] A thorough study of science in the colonial almanacs is wanting, but a liberal sampling of those published between 1740 and 1770 indicates that they continued to perform a role in the popularization of science similar to that earlier in the century demonstrated above.

More specifically in the realm of science and its diffusion, series of lectures illustrated with scientific apparatus, similar to those inaugurated by Isaac Greenwood at Boston in 1727, became suf-

---

[9] A complete list of colonial newspapers, with essential facts about them, is in Clarence S. Brigham, *History and Bibliography of American Newspapers, 1690–1820* (2 vols., Worcester, Mass., 1947). See also Carl Van Doren, *Benjamin Franklin* (New York, 1937), pp. 95–101, 119–123.

[10] By Richard A. Overfield in *Emporia State Research Studies*, XVI, No. 3 (Emporia, Kan., 1968).

[11] Magazines fared badly in the colonies. Although several were founded, the only one to survive—and that only from 1743 to 1746—was the *American Magazine and Historical Chronicle*, edited at Boston by Jeremiah Gridley. The *American Magazine, or Monthly Chronicle for the British Colonies* (Philadelphia) and the *New American Magazine* (New York), both later, failed within a year or two. All of them, while they lasted, published scientific articles of significance, mostly materials reprinted from English journals.

[12] Bridenbaugh and Bridenbaugh, *Rebels and Gentlemen*, pp. 332–333.

ficiently well received to become a lucrative source of income for lecturers in the colonies. Records of them depend heavily upon newspaper advertisements and occasional diaries and letters, and it is probable that they are far from complete. The number of repeat performances given by the known lecturers is uncertain, especially those which took place in secondary towns without a newspaper, and it appears likely that some lectures by obscure speakers escaped newspaper notice. However, besides Greenwood, who, as has been noted, lectured in Philadelphia (1740) and Charles Town (1744–45) as well as in Boston, several of these lecturers are identifiable. A few of them were, like Greenwood, scientists of note, and many of them lectured widely through the colonies over a period of several years. After Greenwood, one of the early important lecturers was Dr. Archibald Spencer, a Scot from Edinburgh who lectured along the Atlantic seaboard between 1743 and 1751, when he was ordained and accepted a living in Maryland. Benjamin Franklin heard him in Boston in 1743, and Spencer was one of the sources from which Franklin acquired an interest in electrical experiments.[13] Indeed, before Dr. Spencer died in 1760, Franklin purchased his apparatus.[14] In the meantime, however, Dr. Spencer had given his course of lectures at Newport, Philadelphia, Annapolis, Williamsburg, and possibly elsewhere.[15] Ebenezer Kinnersley (1711–78) was, after Greenwood, the most able scientist to give lectures in the colonies, and he was said to possess the best apparatus for scientific demonstrations. English born, his parents took him to Pennsylvania when he was less than three years old and he grew up in Baptist surroundings and became a Baptist preacher. However, his opposition to religious enthusiasm during the Great Awakening led him into difficulties with churchmen, and

13 I. Bernard Cohen, *Benjamin Franklin's Experiments: A New Edition of Franklin's Experiments and Observations on Electricity* (Cambridge, Mass., 1941), pp. 19ff. There has been confusion about Spencer's names. See J. A. Leo Lemay, "Franklin's 'Dr. Spence': The Reverend Archibald Spencer (1698?–1760) M.D.," *Maryland Historical Magazine*, LIX (1959), 199–216. Spencer's lectures dealt with science in general, not with electricity alone.

14 J. A. Leo Lemay, *Ebenezer Kinnersley, Franklin's Friend* (Philadelphia, 1964), p. 52.

15 *Boston Evening Post*, May 30, Aug. 1, 1743; *Pennsylvania Gazette*, April 26, May 3, July 26, 1744; *Virginia Gazette*, Jan. 9, 1745/46; *Maryland Gazette*, Sept. 26, 1750; Bridenbaugh, *Cities in Revolt*, pp. 206–207.

his independent, rational views placed him closer to the Enlightenment than to evangelical Christianity. In Philadelphia Kinnersley became associated with Benjamin Franklin in the latter's electrical experiments about 1745. He was an apt student, and he and Franklin became close friends. As Kinnersley lacked employment, Franklin encouraged him

> to undertake shewing the experiments for money, and drew up for him two lectures, in which the experiments were rang'd in such order, and accompanied with such explanations in such method, as that the foregoing should assist in comprehending the following. He procur'd an elegant apparatus for the purpose, in which all the little machines that I had roughly made for myself were nicely form'd by instrument-makers. His lectures were well attended and gave great Satisfaction; and after some time he went thro' the Colonies exhibiting them in every capital Town, and pick'd up some Money.[16]

Thus, with Franklin's help, Kinnersley took to the road as a lecturer in 1749. If, at the outset, the lectures were largely Franklin's work, Kinnersley made them his own. He constantly revised, rephrased, and enlarged them, and, as he was "a Master of words as well as of Experiments"—as James Alexander told Cadwallader Colden—his lectures were immensely popular, being both entertaining and informative. Together with demonstrations, such as "The Force of the Electrical Spark, making a fair hole thro' a Quire of Paper. Small Animals killed by it Instantaneously. . . . Spirits of Wine also kindled by a Spark after it has passed thro' ten feet of Water. . . . The Salute repuls'd by the Ladies Fire; [or Fire darting from a Lady's Lips. so that she may defy any Person to salute (i.e. kiss) her.] A Battery of eleven Guns discharged by Fire issuing out of a mans Finger . . . ,"[17] Kinnersley also expounded Franklin's explanations of electricity, the identity of lightning and electricity, and the utility of lightning rods (demonstrated by min-

[16] Quoted in Lemay, *Ebenezer Kinnersley*, p. 63.
[17] From Kinnersley's advertisement in the *Maryland Gazette* (May, 1749) quoted in *ibid.*, pp. 68–69; also reprinted from the *Boston Evening Post*, Oct. 7, 1751, in William Northrop Morse, "Lectures on Electricity in Colonial Times," *New England Quarterly*, VII (1934), 367–369. The portion in square brackets is from Morse's reprint.

iature houses and ships struck by a stroke of electricity; those with lightning rods were unhurt; those without went up in flame). He was judicious, and he forestalled criticism both by his own "highest Christian Character," as a Newport writer observed, and by his care to point out to his hearers that such experiments exalted "our ideas of the great Author and God of Nature."

Kinnersley toured the colonies from early in 1749 to the summer of 1753. He delivered his lectures in Annapolis, Williamsburg, Charles Town, Philadelphia, Boston, Newport, New York, and St. John's (Antigua), spending five months in Boston, two in Newport, and almost three in New York. Everywhere he made the acquaintance of scientifically-minded men, performed experiments with some of them, and became widely known. In January, 1750, the *Gentleman's Magazine* of London published an article entitled "By a Number of Experiments, lately made in Philadelphia, several of the Principal Properties of the Electrical Fire were demonstrated, and its effects shewn." Long believed to have been from the pen of Franklin, this article has recently been identified as an outline of Kinnersley's lectures as published nine months before in the *Maryland Gazette*.[18] After his return from Antigua in 1753, Kinnersley accepted a position, offered through the good offices of Franklin, as Master of the English School of the Philadelphia Academy. Two years later, when the College of Philadelphia opened, he was chosen "Professor of the English Tongue and of Oratory." He remained in these positions, except for a visit to Barbados in the winter of 1772–73, until his health failed in 1775, when he retired to the country. But he did not give up his lectures on electricity and other aspects of science. Revised repeatedly in the light of newer research, the lectures were given several times a year at the Philadelphia Academy up to and including the early months of 1774. In the meantime, too, he continued experiments in electricity.[19] On March 12, 1762, he dispatched to Franklin, then in London, an il-

[18] Lemay, *Ebenezer Kinnersley*, pp. 62–63. Kinnersley's syllabus was published in 1764. *Ibid.*, pp. 103, 124.

[19] See Franklin to Cadwallader Colden, April 18, 1754, in *The Letters and Papers of Cadwallader Colden* (*Collections of the New York Historical Society*, Vols. L–LVI, 1917–23; Vols. LXVII–LXVIII, 1934–37; cited hereafter as *Colden Papers*), IV, 440–441. See also Lemay, *Ebenezer Kinnersley*, pp. 96–97.

lustrated account of twelve experiments designed to clarify Franklin's theory of electrical repulsion; to test whether the amount of electricity in the air is as great "at the height of two or three hundred yards, as near the surface of the earth" (he found it to be greater in the higher altitudes); to describe his "electrical air thermometer," with which he proved, contrary to Franklin's view, that "lightning does not melt by a cold fusion" but by a hot one, that is, that electricity produces heat. Franklin took the letter to the Royal Society, where it was read (November 18, 1762; March 24, April 14, 1763), and subsequently published, with the illustrations, in the *Philosophical Transactions*.[20] Again, on October 13, 1770, Kinnersley sent to Franklin a letter to describe damages inflicted by lightning on Philadelphia buildings and to relate a series of experiments he had performed to test the electrical conductivity of charcoal made of various kinds of wood. Franklin read the letter to the Royal Society on December 10, 1772, and it was published in the *Philosophical Transactions* the following year.[21] Excepting Professor Isaac Greenwood, none of the other public lecturers in science in the colonies appear to have contributed to the meetings of the Royal Society.

Electricity excited wide interest in the colonies as it did in Europe around the mid-eighteenth century.[22] A variety of lecturers, anticipating lucrative profits, imitated Greenwood, Spencer, and Kinnersley: William Claggett, the Newport clock-maker, aided by Captain John Williams (Newport, Boston, and possibly elsewhere, 1746–47); Daniel King of Salem (Salem and Boston, 1747); Richard Brickell (New York, 1748); the peripatetic and elusive Transylvanian, Samuel Dömjén, who traveled from Philadelphia

[20] *Journal-Book*, XXV (1762), 216–217; (1763), 52–53, 56–61; *Add. Mss. 5166* (Charles Morton's notes), fols. 113–114; *Phil. Trans.*, LXIII (1763), 84–97.

[21] *Journal-Book*, XXVIII, 6–7; *Add. Mss. 5169D* (Morton's notes), fols. 153–154; *Phil. Trans.*, LXIII, Pt. I (1773), 38–39.

[22] One of the sources of this interest was the publication in the *Gentleman's Magazine*, XV (1745), 193–197, of "An historical account of the wonderful discoveries, made in Germany, &c. concerning Electricity," reprinted in the *American Magazine and Historical Chronicle*, II (1745), 530–537. To this stimulus, however, must be added a variety of contemporaneous communications to colonial men of science from the Royal Society and especially from Peter Collinson. Cf. Bridenbaugh, *Cities in Revolt*, pp. 206–207; Lemay, *Ebenezer Kinnersley*, pp. 54–62.

23. Jean Théophile Desaguliers (1683-1744), Curator of the Royal Society,
thrice Copley Medalist, inventor of the planetarium,
and popular lecturer on and demonstrator of experimental philosophy,
from whom Greenwood, Kinnersley, and others introduced the art
to the colonies.

Duncan Campbell Lee, *Desaguliers of No. 4 and His Services to Free-Masonry*
(London, 1932), frontispiece.

24. A design of the "thunder-house" similar to a device employed by
colonial lecturers to demonstrate the identity of lightning and electricity.
The church on the left is burning from a bolt of lightning while the toy
model on the right is burning from an electric spark set off from
a Leiden jar by the demonstrator.

Courtesy of Mr. D. P. Wheatland, Historical Scientific Instrument Collection,
Harvard University.

through Maryland, Virginia, North Carolina, and South Carolina, living "eight hundred miles upon electricity" (1748); David James Dove, the brilliant but somewhat choleric Philadelphia schoolmaster (Philadelphia, 1750); Lewis Evans, the Philadelphia surveyor and cartographer, who purchased Dove's apparatus and prepared a course of thirteen lectures about natural science and electricity at the College of New Jersey (1751) and then moved on to New York, Newark, Philadelphia, and as far south as Charles Town; Robert Skiddy, who presented a course about electricity (Charles Town, 1754); Joseph Hiller, a Boston jeweler (1756); William Johnson, the Irish Quaker, who rivaled Kinnersley as a popular lecturer on electricity and toured widely over the colonies between 1763 and his death in 1768; David Mason and William Jones (Boston, 1765); Dr. John Shippen, recently returned from graduate study at Rheims, who gave illustrated lectures on fossils (Philadelphia, 1770); Christopher Colles, who offered lectures on geography, natural philosophy, and physics (Philadelphia, 1772–73); Benjamin Rush, who delivered lectures on chemistry (Philadelphia, 1774–75); Dr. Abraham Chovet, who presented a series of lectures on human anatomy and physiology, demonstrated by "his elegant Anatomical Wax-Work Figures" (Philadelphia, 1774–76); and Arthur John O'Neile, who attracted large audiences up to the eve of the American Revolution. Not all of the lecturers, as has been noted, concentrated on electricity, and in addition to those mentioned above, Professor John Winthrop of Harvard gave occasional public lectures of astronomy and meteorology, and a variety of anatomical and medical lectures were given in New York, Philadelphia, Charles Town, and possibly elsewhere from time to time. After the mid-century, lecture-going became a popular source of colonial entertainment in the urban communities. How many of the "curious" men and women who attended them were persuaded to adopt a scientific frame of mind is impossible to say. But there is evidence that some adopted a rational explanation for natural phenomena, and the lecturers at least exposed thousands of colonials to the experimental method.[23]

[23] Although most of the evidence for the lecturers appeared in the colonial newspapers, many of these have been cited in secondary materials. See Morse, "Lectures on Electricity in Colonial Times," pp. 364-374; Bridenbaugh and

All of these developments demonstrate that by the end of the Old Colonial Era colonial society, or urban parts of it, had become remarkably mature. Lawrence Henry Gipson has cited this relative maturity as a factor in the events leading up to the American Revolution, which brought the Old Colonial Era to a close.[24] And just as Americans were becoming increasingly self-confident in their capacity to manage their own affairs in matters relating to politics and economics, without further direction from the mother country, so they were becoming self-confident in their abilities in cultural and intellectual life, including science. They had developed a society with relatively sophisticated urban communities, containing a number of highly cultivated, wealthy people with education, leisure time for study and experimentation, and the ability to speak with and understand one another in intellectual affairs. These persons had libraries at hand; coffeehouses, inns, lecture halls, and the like in which to gather together; print shops with which to publish and disseminate their findings by means of books, pamphlets, newspapers, and almanacs; a growing capacity to intercommunicate among themselves on an intercolonial, even continental, stage; and they were developing, besides a continued interchange with the Royal Society of London, wider and more firm scientific relations with Scottish, Dutch, French, German, Italian, and Swedish men of science in the Old World—a trend already noted in connection with the latter years of Mark Catesby's career, and one which may have decreased further the colonists' sense of dependence upon the Royal Society and the mother country at large.[25] Moreover, their colonial "public"—their intellectual hinterland—was becoming more and more receptive to the secular, rational, scientific approach to knowledge, including knowledge pertaining to the world of na-

---

Bridenbaugh, *Rebels and Gentlemen*, pp. 65, 330–333, 354–358; Bridenbaugh, *Cities in Revolt*, pp. 200–201, 206–208, 330, 380, 409; Lemay, *Ebenezer Kinnersley*, pp. 48–81, 88–92, 96–102; Lawrence Henry Gipson, *Lewis Evans* (Philadelphia, 1939), pp. 8–14; Bowes, *Culture of Early Charleston*, pp. 84ff.; Cohen, *Benjamin Franklin's Experiments*, Appendix I, pp. 401–408; Louis B. Wright, *The Cultural Life of the American Colonies* (New York, 1957; paperback ed., 1962), pp. 187, 235–236.

[24] See especially Gipson's *The Triumphant Empire: Thunder Clouds Gather in the West, 1763–1766* (*The British Empire before the American Revolution*, Vol. X, New York, 1961), pp. 1–25.

[25] See Chapter 8 above, pp. 324–326.

ture. To be sure, not every American was free from the superstitions of the past, nor are they to this day; but as the Enlightenment advanced in America more and more Americans had become receptive to the proposition that the enigma of man's existence could be unraveled rationally without the aid of revelation or theology. Even if the urban masses had not become wholly secular, they appear to have become able to compartmentalize comfortably the "how" and the "why" of natural phenomena in their minds as many do today.

Still, when all these things regarding the relative maturity of colonial society have been said, one must guard against presenting too favorable a picture of the impact of science. Certainly science and scientists enjoyed societal advantages in 1770 that had not existed in 1700, and the people at large, particularly in urban areas, appear to have become more receptive to scientific explanation. But the colonies as a whole were still largely agrarian, shading off on the north, the west, and the south into remote frontier districts as raw, except for their greater proximity to the seaboard communities, as New England had been in the younger John Winthrop's day. Thousands of colonists had little opportunity for formal education, seldom, if ever, saw (or could have read) a book, a newspaper, or an almanac, and never witnessed an illustrated lecture on science. The bulk of the colonial men of science were still amateurs, and the total number who became scientists of international distinction was still small, even by the end of the Old Colonial Era.

Moreover, disregarding the inward qualities of colonial scientists, in one outward respect they still lagged behind their European counterparts. They were still dependent upon the Old World for fine precision instruments for observation and measurement in scientific endeavor. At the outset, of course, colonials had been wholly dependent upon the work of the Old World craftsmen for all scientific instruments. As the seventeenth century wore on, they became skillful in fashioning instruments of wood, especially for surveying and navigation. In the eighteenth century, European craftsmen migrated to the colonies to some extent, and colonials became increasingly skillful in metal-working, particularly in the manufacture of clocks and guns. Before the end of the Old Colonial Era, a few colonial craftsmen were able to cast type and even to produce such remarkable mechanical contraptions as orreries.[26] Colonial shipwrights, wheelwrights, gunsmiths, clock-makers, furniture-makers, ironmongers, silversmiths, and others in the manufacture of daily necessities, including a few luxuries, reached a high level of performance. Thomas Godfrey (1704–49), a Philadelphia glazier with a mechanical and mathematical bent of mind, invented, with the patronage of James Logan and other Philadelphians, an improved quadrant for ascertaining latitude at sea (1730) which almost took precedence over that of John Hadley (1682–1744), an Englishman for whom the instrument was named.[27] Indeed, by 1770, American instrument-makers were able to supply most of the scientific instruments necessary for teaching and demonstrating the more simple scientific experiments. But for the finer precision instruments required in scientific observation and measurement, especially microscopes, telescopes, barometers, and thermometers, Americans were still dependent upon Old World craftsmen, a situation which persisted long after political independence had been won.

[26] Silvio A. Bedini, *Early American Scientific Instruments and Their Makers* (U.S. National Museum, *Bulletin*, No. 231, Washington, 1964), *passim* (lists instrument-makers, pp. 155–171); I. Bernard Cohen, *Some Early Tools of American Science* . . . (Cambridge, Mass., 1950), *passim;* Carl Bridenbaugh, *The Colonial Craftsman* (Chicago, 1961), *passim;* Theodore Hornberger, *Scientific Thought in the American Colleges, 1638–1800* (Austin, 1945), *passim;* William M. and Mabel Sarah Coon Smallwood, *Natural History and the American Mind* (New York, 1941), pp. 197–202.

[27] Godfrey's quadrant was perfected in Nov., 1730, and tested satisfactorily on voyages to Jamaica and Newfoundland. Logan sent an account of it to the Royal Society (May, 1732), which turned it over to Halley for a report. Halley dawdled somewhat, perhaps because he found Godfrey's instrument essentially identical to one invented in England by Hadley in May, 1730, and reported to the Society in May, 1731; and Halley's report to the Society on Godfrey's instrument was not made until January 31, 1733/34. Logan had believed that Godfrey's invention would be "wholly new" to Halley, but Halley suspected trickery and made uncomplimentary remarks about Logan, who then complained of Halley's "unhandsome conduct." In 1734 the British Admiralty approved of the Hadley instrument, and Hadley properly received the credit for the invention—although Godfrey had been only a few months behind. The Royal Society Archives contain much material about the dispute. See *Journal-Book*, XV, 374–378, 379–380; XVI, 64–67; *Letter-Book*, XX, 225–235, 415–435; XXI, 225–235; *Phil. Trans.*, XXXVIII, No. 433 (Dec., 1734), 441–450; S. P. Rigaud, ed., *Correspondence of Scientific Men of the Seventeenth Century* (2 vols., Oxford, 1841), I, 282–288, 313–319.

New Sources of Promotion and Inspiration

The Royal Society of London continued to serve as a clearing house for scientific information communicated by colonials, and its promotion of science in the colonies was unabated. However, the Society's prestige waned in the middle years of the eighteenth century, partly because of its tendency to elect to fellowship many dilettantes of little or no scientific accomplishment.[28] The retirement (1741) and death (1753) of Sir Hans Sloane removed an active collector and promoter of natural history in the colonies. Mark Catesby, who was well known in the southern colonies and who had befriended colonials in their scientific interests, died in 1749. Still, these losses were compensated for by the contemporary rise of a new generation of English promoters of science in the colonies. Certainly the most helpful of these was Peter Collinson (1694–1768), previously noted as a benefactor of Mark Catesby. Collinson was a well-to-do London cloth merchant of Quaker ancestry who adopted science as his avocation, became a knowledgeable gardener and amateur botanist, and developed an enormous scientific correspondence and exchange among colonials and European scientists between about 1730 and his death in 1768. Friendly, generous, and considerate of the needs of his correspondents, Collinson excelled James Petiver and Sir Hans Sloane as a promoter of scientific activity, and indeed rivaled that of the Royal Society, to which he was elected in 1728. Collinson created an extraordinary garden, first at Peckham and after 1749 at Mill Hill, and exchanged seeds, plants, trees, shrubs, and other natural history specimens with his correspondents, together with expert advice about the nurture of plants. His enthusiasm for gardening was infectious. He aroused the interest of others and became influential in the development of gardens in England and elsewhere. He also collected for others, especially for his friend Philip Miller, already noted as developer of the Chelsea Physic Garden and author of the widely admired *Gardener's Dictionary*, for the estates of English noblemen, such as the Prince of

[28] E. N. da C. Andrade, *A Brief History of the Royal Society* (London, 1960), pp. 7–8, 25; Raymond P. Stearns, "The Royal Society of London: Retailer in Experimental Philosophy, 1660–1800," *University of New Mexico Publications in History*, IV (Albuquerque, 1952), 45–46.

Wales, the Dukes of Marlborough, Richmond, Norfolk, and Bedford, the Earls of Lincoln, Bute, and Leicester, and for wealthy garden fanciers such as Horace Walpole. These men paid for their collections, and by the encouragement and good services of Collinson, who ultimately found as many as fifty-seven subscribers for him, John Bartram (1699–1777) of Philadelphia, with the further assistance of James Logan and other Philadelphians, was enabled to rise from an obscure, ill-educated Quaker farmer with a green thumb to become the foremost botanical collector and gardener in the American colonies. Collinson served as a purchasing agent for the Library Company of Philadelphia, to which he made frequent gifts. In 1745 his gifts of books and electrical apparatus were vital factors in stimulating Benjamin Franklin to enter upon his famous experiments in electricity. Similarly, Collinson acted as a principal intermediary to introduce to the Royal Society of London the scientific works of a host of colonial men of science, and it was he who, by his correspondence, introduced several of them both to one another and to Continental European scientists.

In fact, during the middle third of the eighteenth century, no one man did as much to stimulate science and to establish intercommunication among scientists as Peter Collinson. Although his principal interests ran to natural history, he promoted scientific work in many areas. His avocation as a gardener, horticulturalist, and collector led him to become something of a botanist in his own right, and the *Hortus Collinsonianus*, later published by Lewis Weston Dillwyn (Swansea, 1843), indicated that Collinson introduced to English gardens about 180 new species of plants, trees, and shrubs. But his wide and stimulating correspondence with scientific men, which embraced nearly every colonial of significance as well as English and European scientists of the first rank, performed a unique service in the promotion of scientific endeavors. Scientific circles, especially those of the New World, would have been far less tightly knit together without the pen and the generous assistance of Peter Collinson.[29]

[29] Peter Collinson, "An Account of the Introduction of American Seeds into Great Britain," reprinted in *Journal of Botany, British and Foreign*, LXIII (1925), 163–165. Earl G. Swem, "Brothers of the Spade," *Proceedings of the American Antiquarian Society*, n.s., LVIII, Pt. I (1948), 17–190, gives a long list of Collin-

In addition to Collinson, John Ellis, Dr. John Fothergill, and Dr. John Coakley Lettsom were helpful patrons of colonial science in the mid-eighteenth century. Ellis has been noted above with reference to his association with Georgia and his appointments as Agent for West Florida (1764) and later for Dominica (1770), but his promotion of colonial science extended beyond these activities.[30] Ellis' correspondence reveals that his interest in the introduction of plants of commercial value into successful colonial cultivation, such as tea, coffee, cinchona, and the Oriental "varnish tree," was closely related to the objects of the English Royal Society of Arts (founded in 1754), with which Ellis was prominently associated; that he was an enthusiastic defender of Jean André Peyssonnel's discovery of the animal nature of corals and corallines which, in addition to repeated defenses of Peyssonnel's views in the Royal Society, led to his publication of *An Essay towards a Natural History of the Corallines . . .* (London, 1755); and that he sought to widen the acceptance of Linnaeus' new system of classification and nomenclature in natural history. In pursuit of these objectives, Ellis engaged in wide correspondence, both with colonials and with European scientists, thereby performing a role similar to, though less extensive than, that of Peter Collinson.[31] Both Dr. John Fothergill (1712–80) and Dr. John Coakley Lettsom (1744–1815) were successful London physicians with many professional honors, including membership in the Royal Society. Despite the

son's colonial correspondents (pp. 154–156); Ernest Earnest, *John and William Bartram, Botanists and Explorers* (Philadelphia, 1940), well demonstrates the Bartrams' debt to Collinson; Dr. John Fothergill, *Some Account of the Late Peter Collinson . . .* (London, 1770), includes a testimonial by Franklin; Pulteney, II, 275–277, includes a warm appreciation; N. B. Brett-James, *Life of Peter Collinson* (London, 1925), is inadequate. Professor George Frederick Frick, of the University of Delaware, is currently occupied with a fresh treatment of Collinson's life.

30 See Chapter 8 above, pp. 333–336.

31 In 1758 Ellis urged the Royal Society of Arts to set up botanical gardens in the American colonies. Much of Ellis' correspondence is in Sir James Edward Smith, ed., *A Selection of the Correspondence of Linnaeus* (2 vols., London, 1821, cited hereafter as Smith), I, 82ff., 284ff., and in Spencer Savage, ed., *Catalogue of the Manuscripts in the Library of the Linnean Society of London. Part IV. Calendar of the Ellis Manuscripts* (London, 1948). Cf. also Savage, "John Ellis (?1705–1776) and His Manuscripts," *Proceedings of the Linnean Society of London,* 146th Sess., Oct., 1933–May, 1934 (London, 1934), pp. 58–62.

disparity in their ages, they were close friends (Dr. Lettsom edited Fothergill's works after the latter's death), and their circle of friends in London embraced Peter Collinson, John Ellis, and Benjamin Franklin. Both of them were Quakers, with wide communications among the Society of Friends, and both became physicians of high repute with unusual achievements in their profession.

Dr. Fothergill was the first physician to recognize and describe diphtheria in England, but his scientific interests led him into active work in natural history as well. With Collinson's assistance, he developed one of the finest private botanical gardens in Europe at Upton, Essex, and he made extensive collections of insects, shells, birds, and animals. Dr. Lettsom was born at Little Jost Van Dyke Island in the Virgin Islands of the British West Indies, and he always considered himself to be an American. As a medical man he was especially well known for his work in the treatment of fevers, especially malaria and yellow fever, but like Fothergill he was interested in natural history. He published *The Natural History of the Tea Tree* (London, 1772; 2nd ed., 1799), and prepared a handbook for collectors entitled *The Naturalists and Traveller's Companion* (London, 1772; 2nd ed., 1774; 3rd ed., 1799). In the 1790's he presented a collection of 700 mineral specimens to Harvard College. Both men were widely and favorably known among colonial men of science, and with Collinson and Ellis they contributed largely to the promotion of scientific interests in the American colonies.[32]

All of these men—Peter Collinson, Philip Miller, John Ellis, John Fothergill, and John Coakley Lettsom—enlisted the interest and cooperation of garden fanciers and publishers and collectors of scientific prints and drawings, thereby enlarging and strengthening scientific circles on both sides of the Atlantic. Landscape gardening, previously the avocation of crowned heads and wealthy noblemen, more and more attracted the attention of well-to-do commoners as the eighteenth century advanced. By the end of the Old Colonial Era, it had become a popular pastime of hundreds of "gentlemen" in England and on the Continent of Europe. Few of these men were

32 The lives of both Fothergill and Lettsom are treated in the *D.N.B.* See also James Johnson Abraham, *Lettsom: His Life, Times, Friends and Descendants* (London, 1933).

scientists, but many of them became informed amateurs in botany. They often employed gardeners of skill and even greater knowledge of botany, and, of greater significance in the development of science, they patronized collectors in far-off places and the more or less scientific authors of a number of garden books prepared to guide them in the acquisition and cultivation of exotic trees, shrubs, and decorative plants from various parts of the world. This popular fancy infected many colonial governors, planters, wealthy merchants, and professional men, and gardening became a widespread interest of well-to-do colonials both in the West Indies and on the mainland of North America. To some extent, these men experimented in developing new methods of cultivation, heated glasshouses (greenhouses), grafting, and cross-fertilization of plants, thereby contributing occasionally both to scientific techniques and scientific knowledge. Although their interest centered upon exotic trees, shrubs, and flowers, they did not overlook new additions to their orchards and kitchen gardens, so that they promoted the knowledge of many species of flora. By their frequent communications with expert gardeners, nurserymen, and collectors—communications often extended by the catalytic effects of such men as Peter Collinson—they sometimes became closely associated with the foremost botanists and other scientists of the day. Several colonials, some of them scientists, others mere amateur gardeners, were drawn into this international circle of garden fanciers. Among others, these included Jared Eliot of Killingworth, Connecticut; Cadwallader Colden of New York; James Logan, Dr. John Kearsley, and Joseph Breintnall of Philadelphia; Dr. John Mitchell, John Clayton, Sir John Randolph, Isham Randolph, Colonel John Custis, William Custis, the brothers William and Francis Fauquier, all of Virginia; Dr. Alexander Garden, Henry Laurens, Dr. William Bull, John Gregg, and Dr. Thomas Dale, of Charles Town; and Governor Henry Ellis of Georgia.[33] Among these, too, were John Bartram and his son William, of Philadelphia, who were enabled, largely by means of the patronage of garden fanciers and lovers of natural history both in the colonies and overseas, to emerge as collectors

---

[33] The list is incomplete; it emphasizes colonials whose correspondence appeared in scientific circles. For further information about colonial gardens, see Bridenbaugh, *Cities in Revolt*, pp. 143–146, 337–340.

and naturalists with international recognition and to develop the most famous botanical garden in the New World.

The publication of drawings, especially colored prints of flora and fauna, became a popular if not always profitable enterprise in the eighteenth century. As copper engravings replaced the older woodcut illustrations, artists and craftsmen found employment both in the preparation of illustrations for scientific works in botany and zoology and in selling "nature books," with colored pictures, to nature lovers. Eleazar Albin (fl. 1713–59) was one of the first of these in eighteenth-century England: "Teaching to Draw, and Paint in Water-Colours, being my Profession, first led me to the observing of *Flowers* and *Insects*, with whose various Forms and beautiful Colours I was very much delighted. . . ."[34] At first, he said, he painted them for his "own Pleasure," but soon James Petiver, Sir Hans Sloane, and Mary, the Dowager Duchess of Beaufort, encouraged him to do more. Soon he had "a good Family" of patrons, including William Byrd II, of Westover, Virginia, and with the support of such subscribers he published 100 hand-colored plates of caterpillars, moths, butterflies, and other insects, with a one-page description of each, as *A Natural History of English Insects* . . . (London, 1720). Four years later, a second and finer edition appeared, with "Large Notes And Many Curious Observations by W[illiam] Derham, Fellow of the Royal Society." The success of these ventures led Albin to publish *A Natural History of Birds* . . . (3 vols., London, 1731–38; 2nd ed., 1738–40) and subsequent volumes on spiders, English song birds, and edible fish. As far as possible, Albin used live specimens for his models, and his search for specimens, live, stuffed, or dried, led him to communicate with collectors from the New World, especially with Mark Catesby and his friends.[35]

A contemporary of Albin was Georg Dionysius Ehret (1708–70), already noted as the most eminent portrayer of botanical art in the mid-eighteenth century. German born and self-trained, Ehret

---

[34] Albin, *A Natural History of English Insects* . . . (London, 1720), Preface, dated Jan., 1713/14.

[35] *Ms. Radcliffe Trust C.3*, fols. 85, 95 (Bodleian Library); John Nichols, ed., *Illustrations of the Literary History of the Eighteenth Century* (8 vols., London, 1817–58), I, 370–371.

was a gardener uncommonly able in botany as well as an artist. He wandered widely, supporting himself by his twin skills, meeting many of the leading botanists and gardeners of Europe, and finally (1736) settled in England. He married Philip Miller's daughter and became a popular figure in London society both as an art instructor and as a botanical draughtsman and painter. His colored prints sold rapidly—Dr. Richard Mead, the Royal Physician, purchased 200 of them at a guinea apiece—and Ehret stated that "If I could have divided myself into twenty parts, I could still have had my hands full."[36] Ehret associated with the most prominent botanists and patrons of natural history in England and western Europe, and although he appears to have had few direct communications with American colonists, he made drawings and prints of colonial plants and was familiar with American botanical scholars. He prepared drawings for some of the plates in the Reverend Griffith Hughes's *Natural History of Barbados* (1750), for John Ellis' *Essay towards a Natural History of the Corallines* (1755), and for Dr. Patrick Browne's *Civil and Natural History of Jamaica* (1756). Peter Collinson planned (1755) for Ehret to do the plates for the second edition of Johann Friedrich Gronovius' *Flora Virginica*, based upon plant specimens and information supplied by John Clayton of Virginia, but the project was, instead, completed by Gronovius' son, Laurenz Théodor, and published (1762) without illustrations.[37] Still, by the wide popularity of his prints, drawings, and paintings, Ehret added popular interest to natural history and opened additional sources of patronage for colonial collectors and scientists.

Among the publishers of popular prints and illustrated books in natural history, no one cast his nets more widely in the collection of rare specimens than George Edwards (1694–1773). A self-made artist and naturalist, Edwards acquired a grammar-school education in Essex and was apprenticed to a cultivated tradesman who left his

[36] Wilfrid Blunt and William T. Stearn, *The Art of Botanical Illustration* (3rd ed., London, 1955), p. 146.

[37] Edmund and Dorothy Smith Berkeley, *John Clayton: Pioneer of American Botany* (Chapel Hill, 1963), pp. 128–140. See also Ehret, "An Account of a Species of Ophris [*Guinandra diandre*, L.] . . . ," *Phil. Trans.*, LIII (1763), 81–83. Sent by John Bartram, the plant was grown in Collinson's garden, drawn and described by Ehret.

small collection of scientific books to his apprentice. The books quickened Edwards' interest in science and he determined to abandon a business career and spend his life in travel and study. He traveled widely in Holland, Norway, and France, visited many of the principal scientists, viewed their collections, studied natural history, "applying himself to drawing and Colouring such animals as fell under his notice," and made his way by selling his pictures.[38] He returned to England in 1733, acquired the sponsorship of James Theobald, F.R.S., a Lambeth businessman zealous for the promotion of science, and attracted the attention of Sir Hans Sloane, who provided him with a sinecure as Librarian of the Royal College of Physicians. Soon Edwards was making scientific illustrations for Sloane, the Royal Society of London, and a wide number of English scientists and lovers of natural history. He pulled together his business acquaintances, whose interests extended widely over England's vast colonial possessions; his scientific friends, who included Linnaeus, Gronovius, other Dutch scholars, and many English scientists; and a host of amateur naturalists and gardeners of England, who purchased his paintings and subscribed to his publications. He was a close friend of Mark Catesby, from whom he borrowed specimens, imitated techniques, and, after Catesby's death, revised and republished Catesby's *Natural History of Carolina, Florida, and the Bahama Islands.* He was also closely associated with Peter Collinson, John Ellis, and Dr. John Fothergill, and his colonial friends and correspondents included Dr. John Mitchell of Virginia, the Reverend Griffith Hughes of Barbados, Dr. Patrick Browne of Jamaica, Governor Henry Ellis of Georgia and Nova Scotia, John Drayton of South Carolina, Benjamin Franklin, both John and William Bartram of Philadelphia, and various other correspondents in Maryland, the British West Indies, and Hudson's Bay. In addition, Edwards frequently presented specimens to the Royal Society, was awarded the Copley Medal by the Society (1750) for his book, *A Natural History of Uncommon Birds, And of Some other Rare and Undescribed Animals* . . . (4 vols., London, 1743–51), and was elected Fellow of the Society in 1757.[39] Few men with so little

[38] *Some Memoirs of The Life and Works of George Edwards, Fellow of the Royal and Antiquarian Societies* (London, 1776), pp. 6–7.
[39] *Journal-Book*, XX, 3, 6, 7, 145, 175, 353; XXIII, 526, 611, 620–621.

claim to solid scientific accomplishment managed to tap so many lines of scientific communication.

Edwards was a talented artist, exact in his portrayal of birds and mammals, and careful in his descriptions.[40] But unlike Mark Catesby, he did not visit the colonies and view his specimens alive and in their native habitat. Some of them he saw alive in English aviaries and parks or as live specimens sent by his overseas correspondents. But the majority of them he saw only stuffed, skinned, or dried, often in poor condition, for the taxidermy of the day was crude and imperfect. Before he completed his *Natural History of Uncommon Birds*, Edwards saw, probably by the hand of Collinson, the manuscript journal of John Bartram's journey, with Conrad Weiser and Lewis Evans, to Onondaga in 1743.[41] Soon he was in direct communication with Bartram, and his second major work, *Gleanings of Natural History, Exhibiting Figures of Quadrupeds, Birds, Insects, Plants, &c. . . .* (3 pts., London, 1758–64), was heavily indebted to both John and William Bartram for specimens, descriptions, and drawings. His description of the ruffed grouse was based upon a specimen sent to Collinson by John Bartram in 1750 together with additional data gathered by correspondence with Bartram, upon the basis of which Edwards had presented a paper to the Royal Society on January 17, 1754.[42] Similarly, in his description of the Maryland Yellowthroat, Edwards wrote that his original drawing was based on a specimen preserved in spirits, but "Since the writing of the above, I have received the Yellow-Throat, together with a drawing of it, very neatly and exactly done, by Mr. William Bartram, of Pensylvania, who hath enabled me to give a further account of this bird; for he says, it frequents thickets and low bushes by runs (of water, I suppose, he means) and low grounds; it leaves Pensylvania at the approach of winter, and is supposed to go to a warmer climate."[43] Indeed, the second part of Edwards' *Gleanings* is largely

[40] Elsa G. Allen, "The History of American Ornithology before Audubon," *Transactions of the American Philosophical Society*, n.s., XLI, Pt. 3 (1951), 480–486.

[41] Edwards, *A Natural History of Uncommon Birds*, IV, 222–225.

[42] Edwards, *Gleanings*, I, 79–83; Allen, "History of American Ornithology," p. 484. Edwards republished the account of the ruffed grouse in *Essays Upon Natural History, And other Miscellaneous Subjects* (London, 1770), pp. 188–190.

[43] *Gleanings*, I, 54–58.

drawn from specimens and other materials supplied by the younger Bartram. Near the beginning, Edwards noted that "These Birds, with many others, were shot near Philadelphia, in Pensylvania, by my friend Mr. William Bartram, who sent them to London, for me to publish the figures and natural history of them."[44] And, after a long list of specimens attributed to him, Edwards acknowledged receipt of many unknown birds from William Bartram, "who obliged me at one time with fourteen American birds, mostly non-descripts, with some short accounts and observations concerning them, in a letter dated Pensylvania, June, 1756. All which are figured in this Second Part of my Gleanings, &c."[45] Both of Edwards' books mentioned above were published in French translations, and both German and Dutch translators published pirated editions of Edwards' revision of Mark Catesby's *Natural History of Carolina*.[46] After Edwards' death, Carl Linnaeus, the Swedish botanist, who had subscribed to the *Gleanings of Natural History*, published *A Catalogue of the Birds, Beasts, Fishes, Insects, Plants, &c. Contained in Edwards's Natural History In Seven Volumes, With Their Latin Names* (London, 1776). Thus Edwards'—and the Bartrams'—works were given Continental currency in Europe, and the manifold lines of colonial scientific communication once again were extended.

The promotional efforts of these men, often working in conjunction with the Royal Society, combined with the societal growth of the American colonies and the increasing mobility of colonial scientists, tended to shorten the time lag which hitherto had existed between European scientific discoveries and colonial knowledge of them. This was not universally true. In chemistry, some aspects of physics (especially in studies of heat and light), and higher mathematics and the mathematical sciences generally, the colonials demonstrated neither the interest nor the ability of such men as Hermann Boerhaave, Stephen Hales, Henry Cavendish, Joseph Black, Karl

[44] *Ibid.*, II, 139–142. William Bartram was also one of the subscribers to Pt. II.

[45] *Ibid.*, II, 192. Other specimens from William Bartram are described in *ibid.*, I, 91–100; II, 144–145, 165–166, 173–174, 185–200; III, 300, 319. Other contributors from the American colonies included Mr. Elliott, a Carolina merchant; Mr. Brooks, a surgeon in Maryland; and Dr. Patrick Browne of Jamaica.

[46] See George Frederick Frick and Raymond P. Stearns, *Mark Catesby: The Colonial Audubon* (Urbana, Ill., 1961), p. 110.

Wilhelm Scheele, Joseph Priestley, the Bernoullis, Leonhard Euler, and a number of other Old World scientists of the time—although several Americans were familiar with Boerhaave's *Elementa Chemiæ* (1724; English ed., London, 1735), Hales's *Vegetable Staticks* (London, 1727), Hales's contributions to animal physiology in his *Statical Essays* (London, 1733), and several others. In applied sciences, such as medicine and building construction,[47] the colonists made a better showing, both because of the migration to the colonies of European artisans, physicians, and surgeons, and because of the increasing numbers of colonials who were trained abroad in medicine and surgery.[48] At the very close of the Old Colonial Era,

[47] Washington felt the lack of qualified engineers severely during the War for Independence, and he pleaded for the establishment of an engineering school. After the war, when the first permanent bridge was constructed across the Schuylkill River near Philadelphia, no American engineer was capable of the task and an Englishman was imported for it. The United States Military Academy, established at West Point in 1802, was the first engineering school in the nation. Cf. J. L. Ringwalt, *Development of Transportation in the United States* (Philadelphia, 1966), p. 35.

[48] Dr. William Bull, of Charles Town, was the first native-born North American colonial to obtain the M.D. degree abroad, at Leiden. Between 1744 and 1770 thirty-one American colonials graduated M.D. from the University of Edinburgh. Of these, at least the following set up practice in the mainland colonies: Dr. John Moultrie (1749), Charles Town; Dr. Thomas Clayton (1758), Virginia; Dr. William Shippen (1761), Philadelphia; Dr. William Smibert (1763), Boston; Dr. John Morgan (1763), Philadelphia; Dr. Arthur Lee (1764), Virginia; Dr. Samuel Bard the Younger (1765), New York; Dr. Adam Kuhn (1767), Philadelphia; Dr. Benjamin Rush (1768), Philadelphia; Drs. Peter Fayssoux and Thomas Caw (1769), Charles Town; Dr. William Logan (1770), Philadelphia. Of the above, Drs. Bard, Shippen, Morgan, and Kuhn also studied at London with Drs. William and John Hunter; Dr. Morgan also studied in Paris. Dr. Thomas Dale, Charles Town, and Dr. James Lloyd, Boston, were trained wholly in London, and Dr. John Jones, New York, held a medical degree from Rheims, although he also studied at Leiden, Edinburgh, and Paris; Dr. Thomas Cadwalader, Philadelphia, after an apprenticeship with his uncle, Dr. Thomas Evans, completed his medical training in France (Rheims?) and also studied with Dr. William Cheselden in London. Dr. Thomas Bond, Philadelphia, was trained at Paris; his brother, Dr. Phineas Bond, Maryland, studied at Leiden, Paris, Edinburgh, and London. The foregoing list excludes European-born physicians who migrated to the colonies after receiving the M.D. abroad: Drs. William Moultrie, Lionel Chalmers, John Lining, and Alexander Garden, all of Charles Town; Drs. Cadwallader Colden and Peter Middleton, of New York; Dr. William Hunter, of Newport; and Dr. William Douglass, of Boston—all of them Scots. Doubtless there were other less well-known physicians who had won their M.D. degrees abroad. But of about 3,500 physicians in the mainland colonies at the time of the American Revolution,

too, medical schools and hospitals were being set up in the colonies to reduce American dependence on European facilities for the training of physicians and surgeons. But, as will be demonstrated further below, in natural history (especially botany), geology, geography (particularly cartography), observational astronomy, and that area of physics devoted to the new study of electricity, Americans ran closely abreast and in some instances slightly in advance of their European contemporaries.[49]

The Impact of Carl Linnaeus

Before the American Revolution, no aspect of scientific knowledge in the eighteenth century underwent as great a revolutionary change in Europe as natural history, particularly botany. The change was effected principally by the work of the Swedish scholar, Carolus (or Carl) Linnaeus (1707–78).[50] Born into a poor Lutheran pastor's family, Linnaeus was educated at the universities of Lund and Uppsala. Early fascinated with botanical studies, he read most of the works of early botanical scholars, mastered Tournefort's method of classifying plants according to their flower structure, became familiar with Sébastien Vaillant's criticisms of Tournefort, and through Vaillant appears to have been introduced to the problem, so widely

---

Shryock estimates that less than 5 percent held degrees of any sort. Richard H. Shryock, *Medicine and Society in America, 1660–1860* (New York, 1960), p. 9. The foregoing list has been compiled from: *List of Graduates in Medicine in the University of Edinburgh from* MDCCV *to* MDCCLXVI (Edinburgh, 1867); Robert William Innes Smith, *English-Speaking Students of Medicine at the University of Leyden* (Edinburgh, 1932); Francisco Guerra, ed., *American Medical Bibliography 1639–1783* (New York, 1962); and James Thatcher, *American Medical Biography, or Memoirs of Eminent Physicians Who Have Flourished in America* (2 vols., Boston, 1828; reprinted in one vol., New York, 1967).

[49] For European developments, see Allan Ferguson, ed., "Natural Philosophy through the Eighteenth Century and Allied Topics," *Philosophical Magazine*, 150th Anniversary No. (London, 1948), pp. 10–164; and A[braham] Wolf, *A History of Science, Technology, and Philosophy in the 18th Century* (New York, 1939), *passim*.

[50] Carolus Linnaeus was the Latinized form of the name, usually partially Anglicized to Carl Linnaeus. In 1762 Linnaeus was ennobled by the King of Sweden and adopted the name of Karl von Linné. The surname originated from the Swedish word for linden tree.

discussed in the early eighteenth century, of sexuality in plants. Indeed, one of Linnaeus' earliest works was an essay on this subject, and he evidently pondered over the matter for several years. Linnaeus advanced rapidly at Uppsala, and in 1735 went to Holland, where he acquired a medical degree in short order at the University of Hardwijk. In Holland he made the acquaintance of Dr. Hermann Boerhaave, the famous (but aging) Professor of Medicine, Botany, and Chemistry at Leiden; Johann Friedrich Gronovius, the Dutch botanist soon to play a prominent role in the development of American botany; Isaac Lawson, Physician General of the British Army in Flanders and a patron of botanical studies; and George Clifford, a wealthy banker with a remarkable botanical garden at Hartecamp, near Haarlem. These men were so impressed with a little manuscript that Linnaeus showed them that Gronovius and Lawson paid the cost of its publication. Thus came into being the first edition of one of Linnaeus' most revolutionary works, entitled *Systema Naturae* (Amsterdam, 1735). George Clifford also persuaded Linnaeus to remain in Holland as his private physician and superintendent of his garden at Hartecamp. Linnaeus lived at Hartecamp for about three years, during which time he worked closely with Gronovius, became familiar with many American plants which the latter received from his American correspondents, especially John Clayton of Virginia, and found time to publish several more botanical works, including a descriptive catalogue of Clifford's garden, entitled *Hortus Cliffortianus* (Amsterdam, 1737). When his friend Peter Artedi accidentally drowned in an Amsterdam canal, Linnaeus edited his *Ichthylogia* (1738), an innovative study of fishes, the first to advance beyond the much earlier work of John Ray. Linnaeus also made short visits outside the Netherlands. In England (July, 1736) he met Sir Hans Sloane, Peter Collinson, John Jacob Dillenius, Sherard Professor of Botany at Oxford, Philip Miller (who gave him duplicates of William Houstoun's plants from the Caribbean area), and others. In 1738 he returned to Sweden after a short visit to Paris, where he met the great successors to Tournefort, Antoine de Jussieu and his younger brother Bernard, members of an extraordinary family of French botanists whose work in the classification of plants was destined to supplant that of Linnaeus.[51] In Sweden Linnaeus practiced medicine in Stockholm, founded, with kingly cooperation and approval, the Swedish Royal Academy of Science, and in 1741 returned to Uppsala as Professor of Medicine and Natural History. He enlarged and enriched the botanical garden at Uppsala until it vied with the other principal gardens in Europe, and he remained active at the university until illness forced him to retire in 1766.[52]

Linnaeus was a great teacher, and students flocked to his classes in great numbers. He trained his students thoroughly, including field trips to collect botanical specimens. He also managed funds from the Swedish Academy of Science to send several of his students to collect in distant lands—Africa, China, Asiatic Russia, and elsewhere. One of his students, Peter Kalm, visited North America during the years 1748–51. Kalm went by way of England, where he visited the Royal Society (April–May, 1748),[53] and, as has been noted, spent an afternoon with Mark Catesby and Dr. John Mitchell, recently returned from Virginia.[54] Collinson provided Kalm with letters of introduction to several of his colonial correspondents, and Kalm arrived at Philadelphia in September, 1748. He was cordially received by Franklin and John Bartram, and he spent the autumn of 1748 collecting specimens in Pennsylvania and New Jersey. In the spring of 1749 Kalm traveled to New York

[51] Antoine de Jussieu (1686–1758) edited many of Tournefort's works, and succeeded him as Professor at the Jardin des Plantes; Bernard (1698–1777) was Demonstrator at the Jardin des Plantes, established the garden at the Trianon Palace for Louis XV, and laid foundations for a natural system of plant classification. Another brother, Joseph (1704–79), died in South America, where he had gone as a botanist with the Condamine Expedition. Their nephew, Antoine Laurent de Jussieu (1748–1836), became Professor at the Jardin des Plantes and author of *Genera Plantarum* (1789), which elaborated upon the system of classification founded by Bernard and served as the basis for the present-day natural classification of plants. His son, Adrien (1797–1853), also succeeded to the professorship of botany at the Jardin des Plantes.

[52] Knut Hagberg, *Carl Linnaeus*, tr. Alan Blair (New York, 1951); Norah Gourlie, *The Prince of Botanists: Carl Linnaeus* (London, 1953); Alice Dickinson, *Carl Linnaeus, Pioneer of Modern Botany* (New York, 1967). William T. Stearn gives a splendid summary of Linnaeus' life in the Introduction to *Species Plantarum, a Facsimile of the First Edition, 1753* (London, 1957)—a very useful work.

[53] *Journal-Book*, XX, 490, 516, 523 (April 21, May 19, May 26, as guest of Collinson, Mortimer, and Watson respectively).

[54] See Chapter 8 above, p. 326.

and New France, spent the summer in the St. Lawrence Valley, and returned to Philadelphia for the winter of 1749–50. In the spring of 1750 he collected in New Jersey and western Pennsylvania, then returned to Canada by way of the eastern Great Lakes, Niagara Falls, and Oswego. In October he returned to Philadelphia for the winter, sailed for England in February, 1751, and, after a month in England, arrived in Stockholm in June. Kalm had spent nearly two and a half years in North America, and the account he wrote of his travels presents useful information about the colonies at the mid-century.[55] His extensive collections, both from New France and the English colonies, greatly enlarged Europeans' knowledge of the flora of the St. Lawrence Valley and of the middle English colonies. Linnaeus had known of only nine species of American plants culti-vated in Swedish gardens before he visited Holland; he learned of many more in Holland, England, and France and listed about 170 North American species in the *Hortus Cliffortianus;* his *Species Plantarum* (1753) mentioned about 700 North American species, of which about ninety had been supplied by Kalm, including sixty new ones. Kalm later supplied Linnaeus with seventy-five more American species and in 1758 Linnaeus purchased Patrick Browne's Jamaican herbarium, so that in all, Linnaeus had become familiar with about 780 American plant species.[56] Some of these Linnaeus had received directly from colonial correspondents or indirectly by the hands of Collinson, Miller, or Gronovius. Indeed, several of the specimens from Kalm had been supplied to him by John Bartram or Cadwallader Colden. Kalm's visit to North America enabled him

---

[55] First published as *En Resa til Norra America* (3 vols., Stockholm, 1753–61); English ed. (London, 1770–71), by John Reinhold Forster; American ed., by Adolph B. Benson, *The America of 1750: Peter Kalm's Travels in North America* (2 vols., New York, 1937). Kalm published an account of Niagara Falls in the *Pennsylvania Gazette*, Sept. 20, 1750, which Collinson read to the Royal Society Dec. 6, 1750. *Journal-Book*, XXI, 440–446. Kalm reported the height of the Falls, as measured by a French engineer, as 137 feet. Kalm's *Travels*, however, provoked surprise and indignation among the colonists. Franklin commented dryly: "It is dangerous conversing with these Strangers that keep Journals." *Colden Papers*, VII, 185–186.

[56] C. Skottesberg, "Linné, Kalm et l'Études de la Flore Nord-Americaine au xviii Siècle," *Les Botanistes Français en Amérique du Nord avant 1850* (*Colloques Internationaux du Centre National de la Recherche Scientifique, Paris, 11–14 Sept., 1956*, Paris, 1957), pp. 179–187.

to become closely acquainted with many colonial men of science, especially in the Philadelphia community. His presence stimulated further colonial interest in botanical studies and went far to acquaint colonials with Linnaeus' revolutionary work in natural history. Cadwallader Colden wrote to Linnaeus (February 9, 1748/49) that though he had become discouraged in his early botanical attempts in America, he was so inspired by the "new lights" which Linnaeus' work had cast upon the subject that he had resumed his efforts—nearly forty years after his early discouragements had impelled him to ignore James Petiver's pleas for assistance.[57] Colden may have exaggerated the case somewhat—for his correspondence demonstrates that Peter Collinson, John Bartram (who visited Colden in 1742), Gronovius, and others had reinvigorated his interest in natural history about five years earlier.[58] But he mastered the Linnaean system of plant classification, and for a time entered vigorously into the exchange of plants, seeds, plant descriptions, and other items that engaged the attention of the natural history circle. He entered into scientific correspondence and exchange with Linnaeus (among others) and helped to widen colonial familiarity with Linnaeus' work. Perhaps the latter's fine reputation, spread further among the colonists during Kalm's visit, impelled young Adam Kuhn (1741–1817), later Professor of Materia Medica and Botany at the College of Philadelphia, to attend the University of Uppsala, where he studied with Linnaeus between 1761 and 1764, the only colonial to have sat at the feet of the master.

Linnaeus' works were manifold, but their total effects centered about the twin problems of system and nomenclature. Linnaeus was a born encyclopedist and a masterly Latinist. He struggled to find the key to a natural classification of flora. Failing in this effort, he set forth an attractive, relatively simple, "artificial" system based upon the sexual characteristics exhibited by plants, i.e., the number of male organs (stamens) and female organs (pistils) observed in (or found to be absent in) the flowers of plants. The fundamentals of the plan were expounded in his *Systema Naturae* in 1735 and

[57] See Chapter 10 above, pp. 493–494; *Colden Papers*, IV, 95–99.
[58] *Ibid.*, II, 277–283, 451–458; III, 10–11, 31–33, 42–45, 50–52, 54–58, 60–61, 68–70, 83–92, 96–98, 125–126; V, 215–217.

enlarged upon in the *Genera Plantarum* (Amsterdam, 1737) and in his later works. He grouped the genera of plants into twenty-four classes arithmetically, according to the number of stamens (or their apparent absence). Within each class, the genera were arranged into smaller groups, or "orders," according to the number of pistils. Linnaeus' work also went far to clarify the principles for the accurate definition of genera and species. His system was "artificial" because it took account of only a few marked characteristics of plants and failed to unite them by their natural affinities. In effect, it retained the principle of the fixity of species, although Linnaeus became aware of the hybridization of plants, which he explained as a product of time whereas the natural orders themselves had been the work of the Creator. Moreover, Linnaeus himself came to view his method only as a steppingstone to a natural system, but the facility with which his plan of classification could be applied recommended it, especially to the American colonists, and his sexual system prevailed widely from about 1760 until the early nineteenth century, when it was gradually supplanted by a natural system, the nucleus of which was first set forth by Antoine Laurent de Jussieu's *Genera Plantarum* (Paris, 1789).

Still, the Linnaean system of classification was not embraced everywhere with enthusiasm. French scholars generally ignored it, and English botanists resisted it at the outset. English and Russian scientists initially protested that the system was "lewd" and "licentious," and that "such loathsome harlotry as several males to one female would never have been permitted in the vegetable kingdom by the Creator."[59] In time, however, most of the critics accepted the fact that in the world of plants males and females were "made to act like husbands and wives in unconcerned freedom." Fortunately, the Linnaean system had been abandoned before the Victorian Age, although as recently as the mid-1920's Alfred Noyes warned,

> . . . "Beware of old Linnaeus
> The Man of the Linden-tree."
> So beautiful, bright and early
> He brushed away the dews,

[59] Johann G. Siegesbeck, St. Petersburg academician, quoted in Linnaeus, *Species Plantarum, a Facsimile*, Introduction, pp. 23–24.

531

He found the wicked wild-flowers
All courting there in twos. . . .[60]

However, there appears to have been no outcry against the alleged "lewdness" of Linnaeus' system in the American colonies.

Linnaeus' contribution to nomenclature in natural science had more permanent influence. The core of it was the binomial method of naming plants whereby two names were assigned to a specimen, the first "an unchangeable generic name" and the second the name of the species. The plan was not wholly original, but Linnaeus was the first to apply it rigidly without variation. He experimented with it in some of his early works, but his *Species Plantarum* (2 vols., Stockholm, 1753) employed it throughout and the tenth edition of his *Systema Naturae* (1759) applied it to zoological nomenclature as well. The method led to an unfortunate discontinuity in nomenclature, but its utility and simplicity made it attractive to scientists, and Linnaeus' masterly use of Latin in his generic descriptions provided botanists with a useful new technical vocabulary. Certainly Linnaeus had contributed greatly to the solution of the problem of nomenclature, a fact that was recognized in 1905, when, at an international conference held in Vienna, botanists agreed to adopt the Linnaean botanical names employed in the first edition of the *Species Plantarum* as the starting point for modern botanical nomenclature, abandoning all earlier names unless Linnaeus, too, had adopted them. The development of the binomial system in the naming of new plants is carefully guided today by rules drawn in the *International Code of Botanical Nomenclature*, adopted in 1959. Thus, Linnaeus' binomial nomenclature lives on.[61] In the eighteenth century, however, English scholars, generally revering the works of John Ray, grumbled loudly against the Linnaean innovations. Just as the new system of classification was looked upon as lewd, so the new nomenclature jarred them because it involved discontinuity and because they often resented the new names employed by Lin-

---

[60] *The Torch-Bearers II: The Book of Earth*. For the Linnaean system, see *Species Plantarum, a Facsimile*, Introduction, pp. 24–34. A somewhat simpler description of the system, well designed for the "general reader," is in Dickinson, *Carl Linnaeus*, pp. 113–127.

[61] A very illuminating treatment is William T. Stearn, *Botanical Latin* (London, 1966).

naeus.[62] Patrick Browne was the first English scholar to adopt the Linnaean system of classification (though not of nomenclature) in his *Civil and Natural History of Jamaica* (1756). Philip Miller grudgingly adopted the system of classification in the *Gardener's Dictionary* in 1759, but he resisted the nomenclature until 1768. The English reaction, however, was not uniform. John Hill's *British Flora* . . . (London, 1759–60) and William Hudson's *Flora Anglica* . . . (London, 1762) used both of the Linnaean innovations, and by the end of the Old Colonial Era the Linnaean systems had captured English scientists as a whole. Colonial scientists were more receptive to both the Linnaean classification and nomenclature. In this instance, at least, they did not lag behind their contemporaries in the Old World. If, as has been suggested, their enthusiastic acceptance of Linnaeus' innovations was furthered by their propensities for descriptive botany, they also advanced beyond to give attention to morphology, ecology, and experiments in cross-breeding.

Scientific Communities in the Colonies

In his excellent account of *John and William Bartram, Botanists and Explorers*, Ernest Earnest observed that "The eighteenth-century world seems strangely small. Almost all the important men knew each other."[63] The literary remains of the eighteenth-century men of science, both in Europe and in the American colonies, amply demonstrate the truth of this statement. By virtue of occasional visits with one another, and more especially by means of the exchange of letters, scientific specimens, and scientific data, colonial men of science developed a net of intercommunications and mutual acquaintanceships which enveloped nearly all of the colonial sci-

---

[62] Collinson protested to Linnaeus (April 10, 1755) against the latter's "making *new names*, and altering old and good ones for such hard names that convey no idea of the plant. . . ." But the Earl of Bute, who Collinson said (1755) was in the first rank of English botanists, best expressed the early British attitude: "I cannot forgive him [Linnaeus] the number of barbarous Swedish names, for the sake of which he flings away all those fabricated in this country. . . . I own I am surprized to see all Europe suffer these impertinences. In a few years more the Linnaean Botany will be a good Dictionary of Swedish proper names." Smith, I, 33, 35, 36–37.

[63] P. 68.

entists in America and reached out to include many of the principal scientists of Europe as well. To some extent their intercommunications were marked by the beginnings of scientific specialization. Botanists, gardeners, and persons concerned with natural history in general tended to consult together and to form a "natural history circle." Others, too, tended to correspond principally with their colleagues in astronomy, meteorology, medicine, and so on. But the lines of scientific communication remained flexible. Specialization in the sciences had not yet taken place. Even the most "specialized" colonials, such as Benjamin Franklin, Cadwallader Colden, John Winthrop, John Clayton, James Logan, and Dr. Alexander Garden, consulted with their contemporaries about a variety of subjects which, in the present-day spectrum of the sciences, would include physics, botany, zoology, meteorology, astronomy, mathematics, and many others. To most scientists of the latter years of the Old Colonial Era, "science" still included all aspects of "natural philosophy." Specific scientific disciplines had not yet clearly emerged, and their lines of demarcation were still blurred and indistinct.

Moreover, even at the close of the Old Colonial Era, relatively few colonial men of science were professional scientists. Only the medical men—physicians, surgeons, and pharmacists—a few professors of medicine or natural philosophy, and the itinerant popular lecturers made their livings in scientific endeavors. The majority of colonial scientists were amateurs whose livelihoods were won in the marketplace, on the plantation, in law, in the pulpit, in the classroom, or in the public service. Indeed, at this time, a large proportion of European scientists were amateurs by the same token.

Still, by their mutual devotions to scientific learning, these men gravitated toward one another. In the principal population centers —in New England, New York, Philadelphia, Williamsburg, and Charles Town—communities of colonial scientists arose, with close, personal intercommunications and fellowship.[64] Among these communities, by letters, visits, and the exchange of specimens and scientific information—all fostered by the stimulating promotional activities of the Royal Society of London, the eager correspondence of such men as Peter Collinson, the widening public interest of the

[64] The scientific personnel in each of these communities is partially revealed below.

colonists themselves, and the societal growth of colonial institutions —there arose, by the 1750's, an intercolonial network of scientific communication which was both continental and intercontinental in scope. The scientific correspondence of Professor John Winthrop of Harvard reached out over New England to Philadelphia and to various astronomers in England; that of Cadwallader Colden of New York extended to New England, Philadelphia, Charles Town, Scotland, England, the United Netherlands, Sweden, and elsewhere; that of Benjamin Franklin embraced nearly all of the British colonies in America, England, Scotland, Ireland, France, Germany, and Italy; that of John Clayton of Virginia went to Philadelphia, Charles Town, England, the United Netherlands, and Sweden; and that of Dr. Alexander Garden of Charles Town spread to Georgia, Virginia, Philadelphia, New York, England, Sweden, and the United Netherlands. Other colonials wrote less extensively, but they multiplied these scientific communications many times, both among the colonists themselves and with Europeans. No longer were colonial scientists isolated from one another in a wilderness. They had become active participants in the scientific activities of the entire Western world. And the time was not far distant when they would create their own American model of the Royal Society of London.

American Scientists: James Logan (1674–1751)

Practically all colonial men of science exhibited an interest in natural history, and most of the following were primarily naturalists. Some of the physicians devoted their efforts primarily to medical problems; Cadwallader Colden made an unhappy pass at speculative physics; Benjamin Franklin, interested in almost everything, made his greatest discoveries in physics; and Professor John Winthrop, to whom attention will be turned at greater length in a later section of this chapter, was primarily an astronomer.

James Logan, already mentioned as one of the early scientists of Philadelphia,[65] was largely a self-educated man. As he prospered in land speculation and the fur trade, and rose to a high station in the politics and social life of Philadelphia, he acquired a vast library which, after his death, was presented to the city, with its own build-

---

[65] See Chapter 10 above, pp. 494-495.

ing, as a public trust. He was an enthusiastic scholar with a broad-ranging mind. He cultivated languages, both ancient and modern, and was especially fascinated by mathematics. His library contained probably the best collection of classical works in the colonies, including many rare scientific volumes. He may have been the first in the colonies to import Sir Isaac Newton's *Principia Mathematica*. His fidelity to the Proprietor led him into serious political difficulties from time to time, and his enemies felt that he was arrogant. Crippled by a fall in 1728, and partially paralyzed by a stroke in 1740, his temper was not improved, and he tended to withdraw from the society of Philadelphia scientists in his later years. But his mind was not affected, and it ranged widely through numismatics, philology, mathematics, astronomy, optics, botany, and moral philosophy. He engaged in philosophical correspondence with the Royal Society of London, Sir Hans Sloane (to whom he sent a rare Hebrew coin),[66] Peter Collinson, Carl Linnaeus, and others. At the same time that he sent the coin (September 20, 1735), Logan dispatched two letters to Sloane which subsequently were published in the *Philosophical Transactions*. One, "Concerning the crooked and angular Appearance of the Streaks, or Darts of Lightening in Thunder-Storms," suggested that as the rays of lightning must pass through different densities, owing to clouds, they "suffer very great Refractions."[67] The other, with "Some Thoughts concerning the Sun and Moon, when near the Horizon, appearing larger than when near the Zenith," confirmed John Wallis' previous observations to this effect and ascribed it to vapors that cause the sun and moon, when near the horizon, to appear spheroid in form by refracting light unevenly.[68] He was much interested in mathematical optics, appears to have mastered Newton's works, and published two books on the subject. The first, *Canonem pro Invendis Refractionum . . .* (Leiden, 1739), proposed variant theories regarding the motion of the moon; and the second, *Demonstrationes de radiorum lucis . . .* (Leiden, 1741), claimed to prove that, Christian Huygens to the contrary notwithstanding, the laws of spherical aberration could be

---

[66] *Journal-Book*, XVI, 255–258 (Feb. 12, 1735/36). See also *Sloane 4058*, fol. 162; Rigaud, ed., *Correspondence of Scientific Men*, I, 287–292, 323–341.

[67] *Phil. Trans.*, XXXIX, No. 441 (April–June, 1736), 240.

[68] *Ibid.*, XXXIX, No. 444 (Nov.–Dec., 1736), 404–405.

worked out with absolute mathematical rigor, and he proposed improvements in Huygens' methods of treating lenses.[69]

But Logan's greatest contribution to science was in botany. In a letter to Collinson of November 12, 1734, he described some experiments he had made with maize "concerning the Generation of Plants and Animals."[70] He found that materials cited in Miller's *Gardener's Dictionary* from Claude Joseph Geoffroy and Dr. Patrick Blair were at variance with his own observations. Geoffroy had argued that seeds would grow to full size and appear perfect without impregnation by the *"farina foecundans"* (pollen), but Logan's carefully described experiments with maize proved the contrary. Dr. Blair had written a confused, rambling account to describe how pollen reached the seed, but Logan cited his own observations on various flowers, fruits, and vegetables to prove that it reaches the seed by way of apertures in the husk, which are covered by a liquid that conveys the impregnating substance to the seed. Logan concluded, therefore, "that from the Experiments and Observations that have been made, it appears a reasonable Opinion, and which may well enough be embraced, till the contrary is discovered: That Seeds in general must be impregnated by the Farina before they can come to their proper growth; that this Farina is produced by the Apices; and that the Styles are the only conveyance by which it passes into the Seed-bed, and there impregnates the grains that are thereby brought to maturity." However, Logan was still unsatisfied: "Whence is this Animalcule [*farina*], or where, or by what means is it form'd?" He found all natural bodies, both animal and vegetable, "are no other than a complication . . . of Tubes or Canals, some larger, others smaller, some firmer, others laxer; all serving more or less to convey or contain certain fluids of various kinds. . . ." Still, in spite of all recent researches, "we know nothing of what ought properly to be term'd Generation." He had studied various books on the subject sent to him by Linnaeus. He was puzzled by the fact that pollen floats freely and idly in the air. Is the act of impregnation in plants purely adventitious? Why are there so many more particles of pollen than are required? These

[69] Frederick B. Tolles, "Philadelphia's First Scientist, James Logan," *Isis*, XLVII (1956), 20–30.

[70] *Letter-Book*, XXI, 241–266; read Jan. 23, 1734/35, *Journal-Book*, XVI, 70–74.

matters Logan found incompatible with the law of nature, "to do everything by the shortest ways, as well as to do nothing in vain." He looked in vain for any "special reason," and he pointed out the lack of satisfactory knowledge in these matters.

From the minutes of the meeting, it appears that the Royal Society missed the point of Logan's queries. The minutes record that Logan had argued that the "Animalculum or Embryos of either Plants or Animals were not contained within themselves . . . but in the Air. And then [he] proceeds to show how by a certain sort of Attraction they are gathered by the Farina or seed of the one sex, and carried into the Ovary of the other."[71]

When Logan learned of this he protested that he had sent the letter to Collinson to be showed only to proper judges, that it had not been intended for a public hearing, and that it was "to my very great Surprise" that Collinson had presented an abstract of it to the Royal Society.[72] Whereupon he prepared a second account of his experiments with Indian corn, omitting his speculative queries that had led to such misinterpretation, and fortifying himself further by adding that his conclusions with regard to maize could not be applied to other plants "without a great variety of Experiments on different Subjects." Collinson presented this to the Royal Society and it appeared in the *Philosophical Transactions* in 1736 as "Some Experiments concerning the Impregnation of the Seeds of Plants."[73] An enlarged Latin version was subsequently published under the care of Gronovius as *Experimenta et Meletemata de Plantarum generatione* (Leiden, 1739), and in 1749 Dr. John Fothergill published an English translation of this in London. Actually, although sexuality in plants had long been recognized and both Cotton Mather and Paul Dudley had previously described plant hybridization, Logan's account was the first to demonstrate experimentally the function of the various plant organs in the sexual reproduction of maize. Linnaeus honored Logan by naming a genus of plant after him, and Richard Pulteney, noting that "Several ingenious gentlemen in *America* pursued botanical investigations with great success about

71 *Ibid.*, XVI, 70–74.
72 Logan to Sloane, Sept. 20, 1735, *Letter-Book*, XXII, 178–186; read Feb. 12, 1735/36.
73 XXXIX, No. 440 (Jan.–March, 1736), 192–195.

this period," affirmed that Logan's experiments were "considered, and appealed to, as among the most decisive in establishing the doctrine they were intended to illustrate and confirm," i.e., the sexuality of plants.[74]

Dr. John Mitchell (1690?–1768), F.R.S.

The high quality of much of John Mitchell's scientific work has provoked several competent studies of him, but many of the details of his personal life are still unknown.[75] Whether he was American born or not is a matter of dispute, although scholars have tried to claim him as a native Virginian—a rather ironic claim in view of Mitchell's low opinion of the colonies and most of the colonists. Whether he possessed a medical degree is uncertain, although he studied at Edinburgh and possibly at Leiden, and the certificate signed by six worthy Fellows of the Royal Society when Mitchell was proposed for fellowship in 1748 referred to him as "Doctor of Physic."[76] Probably he was born and educated in Great Britain. Between 1735 and 1746 he was in Virginia, principally at Urbanna, and because of ill health he went to England, where he lived, mostly in London, until his death in 1768. His scientific correspondence embraced scientists in Scotland, England, the United Netherlands, and Sweden, including Dr. Charles Alston, Professor of Botany at Edinburgh, whose pupil he had been; Peter Collinson, Mark Catesby, and a number of Fellows of the Royal Society whom he came to know personally; J. J. Dillenius at Oxford; J. F. Gronovius of Leiden; and Carl Linnaeus of Uppsala, who named the partridgeberry, *Mitchella repens*, after him. In America, he

[74] Pulteney, II, 277–278.

[75] The best studies are: Theodore Hornberger, "The Scientific Ideas of John Mitchell," *Huntington Library Quarterly*, X (1947), 277–296; Lyman Carrier, "Dr. John Mitchell, Naturalist, Cartographer, and Historian," *Annual Report of the American Historical Association*, I (1918), 201–219; Herbert Thatcher, "John Mitchell, M.D., F.R.S., of Virginia," *Va. Mag. Hist. & Biog.*, XXXIX (1931), 126–135, 206–220; XL, 40–62, 268–279, 335–346; XLI, 59–70, 144–152; *D.A.B.* Other materials appear in *Va. Mag. Hist. & Biog.*, XIV (1905–07), 240; XXX (1922), 206; *Wm. & Mary Qty.*, 2nd Ser., VI (1926), 316–317; *Tyler's Historical and Genealogical Magazine*, VIII (1926–27), 286.

[76] *Certificates, 1731–1750*, p. 372. Signed by Cromwell Mortimer, William Watson, Thomas Birch, Richard Graham, Mark Catesby, and Martin Folkes.

visited Philadelphia in 1744 and met John Bartram and Benjamin Franklin, with whom he also corresponded, and he communicated with Cadwallader Colden and Peter Kalm. He was a close friend of his fellow Virginian, John Clayton (1694–1773), Clerk of Gloucester County, whose interest in botany rivaled that of his distant neighbor. Obviously Mitchell swam in the main currents of the colonial natural history stream.

Mitchell's scientific interests included botany, zoology, physiology, medicine, cartography, climatology, agriculture, chemistry, and electricity. The advertisement for the sale of his property at Urbanna, when he prepared to go to England in 1745, included "a large garden containing many curious plants and herbs . . . the furniture and proper utensils of an Apothecary's shop and a small Chemical Laboratory (which is mostly new) with sundry drugs and medicines; likewise a choice collection of Books, both ancient and modern, in several languages, chiefly consisting of the most approved authors in the several branches of medicine, natural history, and Philosophy."[77] He made large collections of plants with the triple intent of finding new, nondescript specimens, improving upon plant classification, and discovering new medicinal flora of use to mankind. His garden, his library, his laboratory, and his correspondents were employed to these ends, and he used them all with diligence. Some of the earliest evidence of his botanical work is in a letter he sent (October 4, 1738), with forty-two specimens and seeds of Virginia flora, to Dr. Charles Alston, his former teacher at Edinburgh, with promise of more to come.

> I have endeavoured to reduce these plants to their proper genera [he wrote], with their specific difference, which if I am not mistaken in, nothing would be more acceptable than to admonish me of my error; I have likewise offered at the synnonymous names of their first discoverers which I know would be preferable to any description I could give of them; but am sensible how subject I shall be to err in this point, for want of that proper aid from books, that is not to be had here (not to mention ye conversation of the learned, which has descouraged me not a little in my

[77] *Virginia Gazette*, Nov. 14, 1745. Published in *Tyler's Historical and Genealogical Magazine*, VIII, 286.

Botanical enquiries [in 1738 Mitchell had not yet met Bartram or corresponded with Colden and other colonial botanists]. . . .[78]

Mitchell's citations indicate that he had at hand, or had extensive notes from, a remarkable number of botanical authors, including John Banister's catalogue of Virginia plants, published in the first volume of John Ray's *Historia Plantarum*, together with the works of Tournefort, Parkinson, Morison, Oviedo, Cornuti, Boerhaave, Hermann, Bobart, and his contemporaries, Catesby, Linnaeus, and Dillenius. One of the plants he had previously sent to Dillenius, with "some hundred more," and he was preparing to ship still others. He had observed several "that I think new genera of plants," and he described his efforts to test the medicinal virtues of Virginia flora, and sought advice for further work of this kind.

During this same year (1738), Mitchell sent to Collinson for presentation to Sir Hans Sloane a Latin manuscript entitled "Dissertatio brevis de principiis botanicorum et zoologicorum." Collinson endeavored to publish it, but before a publisher had been found, Mitchell asked him to delay it until he could complete a second part. This, entitled "Nova genera plantarem Virginiensum," was completed in 1741 and sent to Collinson (March 11), although Mitchell apparently delayed its publication until he could revise it.[79] Ultimately Collinson arranged for the publication of both in the *Acta Physico-Medica Academiae . . . Leopoldina . . . Ephemerides . . .* (Nuremberg, 1748, Vol. VIII, Appendix, pp. 178–202).[80] With the "Nova genera," sent to Collinson on March 11, 1741/42, Mitchell sent a collection of 560 plants, twenty-five of which he believed to be new. In four of these Mitchell is now be-

[78] Published by Herbert Thatcher in *Va. Mag. Hist. & Biog.*, XL, 50–57, from the original in the University of Edinburgh Library.

[79] Collinson wrote to Linnaeus (Jan. 18, 1743/44) that Mitchell "has made many discoveries in the vegetable world" and that one of his Latin essays was ready for the press, "but he intends to add another, so the printing of the first is deferred." Yet Collinson had received the copy of the second sent in 1741, or at least the collection of plants sent with it; hence, I conclude that Mitchell held up the second copy (or possibly the earlier "Dissertatio") for revision.

[80] The papers were republished at Nuremberg in 1769. Thatcher has published translations of the dedicatory letters to Sloane and Collinson in *Va. Mag. Hist. & Biog.*, XL, 98–103.

lieved to have been mistaken, and eleven more were described by other botanists before Mitchell's essays were printed. He is credited, then, with the discovery of ten new genera at the time of his publication in 1748.[81]

Of far greater significance, however, was Mitchell's attitude toward the nature and problems of classification set forth in the "Dissertatio." The first important American discussion of taxonomy, the essay was set forth as an improvement upon the "natural" system of John Ray. Mitchell discussed both plants and animals, and he made a most logical effort to place taxonomy on a genetic basis. It is likely that if his suggestions had been followed, the concept of species would today be more precise and classifications would be on an objective basis, without "splinters" and "lumpers." The gist of his argument was simply that, as regards both plants and animals, if a male and female are able to produce offspring, they are of related species; mutually cross-fertile individuals that produce prolific offspring are of the same species. Those of different species but which are recognized to belong to the same genus because of their resemblance in many primary attributes cross-fertilize but produce "mules" which cannot propagate the species. "This seems to me from the above," Mitchell wrote, "that the generic relationships . . . may be discovered by [breeding] experiments. . . ." Moreover, in the classification of plants Mitchell leaned toward a "natural" system based upon the attributes of the whole plant, rather than upon the organs of regeneration alone, as Linnaeus had done:

> In determining new genera, when nature seemed to allow it, I noted the differences, closely taking into account the whole plant, as well as any more outstanding part: but the generic differences I only sought from the fertilizing parts which I made generic ones (even if to some they are less important) where this seemed to be pointed to by the rest of the plant.
>
> On the other hand, however, where the structure pointed the other way, through an assemblage, as it were, of very many attributes, I made the differentiating character of fertilization specific rather than generic. Some later writers would seem to have

[81] See the conclusions of Dr. T. A. Sprague, of the Royal Botanic Garden at Kew, cited in Hornberger, *Scientific Thought*, p. 281.

partly determined their genera in a different way from us, ac-
knowledging hardly any other differentia, but the very marked
one of fertilization, whereas Ray, Tournefort, Vaillant, Boerhaave,
Dillenius, great names in whose footsteps I strive to tread . . .
have marked out many genera accurately by giving heed both
to the whole plant and to the fertilizing element as well. . . .[82]

Thus Mitchell accepted the Linnaean system of classification with
reservations, although later on he came to regard his ideas to be in
substantial agreement with those of Linnaeus; "if I mistake not," he
wrote the Swedish naturalist in 1748, "our systems support each
other."[83] His initial reaction to the Linnaean system of classification
was conservative, harking back to the "great names" of the past;
but in another sense he was forward-looking, concerned with a
more natural system which considered the attributes of the whole
plant, such as the Jussieus later worked out in France, and which,
in turn, replaced the Linnaean system. Of the later developments he
could not be conscious, of course, and any claim for him as a
"purer" scientist than Linnaeus must be balanced against Mitchell's
utilitarian interest in the medicinal uses of plants, which led him to
refuse to group any "wholesome" plants in the same genera as
"poisonous, harmful ones," evidently without concern for other
attributes of the plants which might suggest relationships. This
irrational note suggests, as Hornberger has said, that Mitchell's
botany was shaped largely by utilitarian ends. No wonder, then,
that Linnaeus, who was little concerned with botanical medicine,
wrote of Mitchell's "Nova genera" that "All these things were very
difficult for me to make out."[84]

Mitchell continued to collect plants until he sailed for England
in 1746, when he embarked with "more than a thousand specimens." Unfortunately, his ship, which also carried plants sent by
Cadwallader Colden to Linnaeus, was plundered by a Spanish
privateer, and Mitchell arrived in London in a destitute condition.[85]
He subsequently recovered his plant collection, but it was mostly

[82] Quoted from the "Dissertatio" as translated in Conrad Zirkle, *The Begin-nings of Plant Hybridization* (Philadelphia, 1935), p. 143.

[83] Smith, II, 448.

[84] *Ibid.*, II, 429.

[85] *Ibid.*, II, 391; Carrier, "Dr. John Mitchell," p. 203.

ruined. His interest in American botany continued for some years. He ordered specimens from John Bartram and sent American specimens to Linnaeus. In 1749–50 he made "a long and laborious journey with the Duke of Argyle [Archibald Campbell, third Duke of Argyll] to the uttermost parts of Scotland, over the mountains and wilds of that country as well as through various counties of England."[86] He collected many northern plants previously found by Linnaeus in Sweden and Lapland, but he discovered nothing new. After his journey Mitchell's concern for botany gave way to other interests. What influence he may have exerted over colonial botany is impossible to say with exactitude. John Bartram held him in high esteem and reported that he was "an excellent Physition & Botanist."[87] Colden and Clayton also thought highly of him. If he contributed anything beyond minutiae to his colonial colleagues, it may well have been his unusual concern for and abilities in taxonomy.

Early during his residence in Virginia, Mitchell acquired a pair of opossums; by observing them carefully and dissecting them, he sought to improve upon the earlier accounts of their generation as set forth previously in the folk tales of Robert Beverley, John Lawson, and William Byrd II, as well as in the more informed description of Dr. Edward Tyson. In 1741 he sent Collinson "An Account of the Male and Female Opossum," which Collinson presented to the Royal Society on February 10, 1742/43.[88] The article contained "several curious Observations he had made on the anatomy of one of each sex, especially with regard to the organs and parts of generation; the production of the Foetus from the mammae of the female without placenta, umbilical string or navel, and the use of the False Belly." Fellows of the Society were still unsatisfied, and at the request of the President two Fellows sent off additional inquiries to Mitchell, whose reply, "Further Observations on the Opossum," dated July 8, 1745, was read to the Society on March 20, 1745/46.[89] He gave a full account of the opossum's

86 Smith, II, 449–451; *Colden Papers*, IX, 87–91.
87 *Ibid.*, III, 79.
88 *Journal-Book*, XIX, 32.
89 *Ibid.*, XX, 73–76. Thatcher printed the entire paper in *Va. Mag. Hist. & Biog.*, XL, 338–346.

anatomy, disagreed with Dr. Tyson's description of the structure and function of the female organs, and gave a "brief account of what I have been able to discover hitherto, in relation to the structure of the organs of generation and gestation of the opossums; the manner of gestation in utero, and an ocular view of the manner of delivery out of it." He laid to rest the old story, perpetuated by Lawson and Byrd (among others), that the fetus grew from the mammae; rather, he found that it was delivered from the womb by a passage different from that of other animals. He described in detail the attachment to the teats, peculiar to the marsupials; and, though his account varies in some points from present-day understanding of the details, he displayed throughout both meticulous care and patient, intelligent observation.[90] Few, if any, zoological papers originating in the colonies before the end of the Old Colonial Era equaled Mitchell's observations on the male and female opossums.

About 1743 Mitchell prepared a long paper entitled "An Essay upon the Causes of the different Colours of People in different Climates," thereby directing his attention to a problem, noted above, which had perplexed scholars for many years both in Europe and in the British American colonies. The immediate occasion was the offer of a prize by the Bordeaux Academy for an essay upon the causes of the color of Negroes. Unhappily, Mitchell's paper was too late to enter the competition, but he sent it to Collinson, who presented it to the Royal Society, and it was read serially at several meetings between May 3 and June 14, 1744.[91] Mitchell stated that in the course of his investigations he had made several experiments on living subjects, but the essay was more notable for the author's marshaling of authorities and close reasoning than for his experimentation. Basically, his authorities were William Cowper's *Anatomy of Humane Bodies* (London, 1698), Marcello Malpighi's *Epistolae Anatomicae* (Amsterdam, 1662), and Newton's *Optics* (London, 1704). He flatly denied Malpighi's hypothesis that there was black fluid in the skin of Negroes and argued that color differences in human beings arise from climatic effects,

---

[90] Hornberger, *Scientific Thought*, p. 285.

[91] *Journal-Book*, XIX, 243, 249, 256, 274-277. Mitchell's "Essay" was published in *Phil. Trans.*, XLIII, No. 474 (June–Dec., 1744), 102-150.

differences in peoples' way of life, differences in the structures of their skins, and differences in the amount of light refraction from the skins of various colored peoples owing to their varied skin structures. The essay consisted largely of seven propositions: 1. That the color of whites proceeds from the color of parts under the epidermis, for the epidermis is a transparent membrane and has no color of its own. 2. That the skins of blacks are thicker and more dense and no color is transmitted through them; therefore they appear black. 3. That the part of the skin of Negroes which appears black is the *corpus reticulare cutis* and the external lamella of the epidermis; all other parts are of the same color as those of white people except the fibers which pass between those two parts. But the *corpus reticulare Malpighi* is black in Negroes. 4. That the color of Negroes does not proceed from any black humors in their skins, for there is none in any part of their bodies more than in white people. 5. That the epidermis, especially the external lamella, is divided into two parts by its pores and its scales, 200 times less than the particles of bodies on which their colors depend, as has been proved by the observations of Newton and Anton van Leeuwenhoek. 6. That to determine and explain the proximate cause of the color of Negroes, Indians, white people, etc., is to determine the extent to which their skins intercept light; darker persons' skins intercept light more, and when the rays of light are wholly intercepted, the body appears black. 7. That the influence of the sun in hot countries, and the ways of life of the inhabitants of those countries, are the remote causes of the colors of Negroes, Indians, etc.; and the ways of living practiced by white people made their color whiter than they were originally, or would be naturally. The sun makes the skin thicker and more dense, and the relative nakedness of African natives increases it. White spots on blacks arise from the same causes as red spots on whites, i.e., a distension, dilation, and rarity or pellucidity of the vascula of the epidermis. Mitchell stated, in conclusion, that all people may have naturally descended from one and the same parents; their different environments over the years led to alterations in their skins which explain their differences in color.

Mitchell's arguments are, of course, outmoded. He knew nothing about cell structure, pigmentation, genes, Mendel's law, or Darwinism. But his logic was impressive, and his appeal to the authority of Sir Isaac Newton was bound to impress an eighteenth-century audience, especially the Royal Society of London.

When Mitchell went to Philadelphia in 1744 he took along a manuscript account of the epidemics which struck Virginians in 1737, 1741, and 1742, supposed to have been yellow fever. Franklin saw the paper, had a copy made for Cadwallader Colden to see,[92] and proposed to print it. But Mitchell demurred. His reasons for objecting were couched in evasive and contradictory terms. They were revealed at great length to Colden, who opened a correspondence with Mitchell after he had seen the paper.[93] The paper was not ready for publication, Mitchell said, and "was never intended to be," having been prepared originally for Edinburgh friends who were skilled in medical arts; it omitted explanations, especially with regard to the dissections he had made of the bodies of victims of the disease, which would be necessary for less knowledgeable American readers; he felt that "a fuller account of other things relating to the Disease" was desirable, but he had had no opportunity to do it; and he deplored the "itch" to print which infected the world. But at the same time, Mitchell revealed that he had previously sent the paper to Alexander Munro (1697–1767), Professor of Anatomy at Edinburgh, and to his friend John Clark (d. 1757), in the expectation that it would be printed in the *Medical Essays and Observations*, published at Edinburgh between 1731 and 1735. But the *Medical Essays*, unbeknown to Mitchell, had been "discontinued before this was wrote."[94] The burden of the whole piece was, very apparently, that Mitchell held a low opinion of colonial medical practitioners and of the intelligence of colonials in general, and he simply did not want to have his paper published under such lowbrow auspices. Recognition at Edinburgh counted for much more.

All this appears ironic in view of the situation at the time, and even more so in view of the subsequent history of Mitchell's paper. Mitchell's low opinion of current colonial medical practice

92 *Colden Papers*, III, 77–78 (Oct. 25, 1744).

93 *Ibid.*, VIII, 314–318 (Sept. 10, 1745); III, 151–154 (Sept. 12, 1745); VIII, 328–338 (Nov. 7, 1745).

94 *Ibid.*, VIII, 321.

was widely justified and shared by many others. His own five autopsies on the victims of the epidemic in Virginia, accompanied by careful clinical observations, were exceptional at the time and place. His treatment of the disease, however, was to "thin" the blood of persons infected by judicious bleeding and follow with purges to rid the system of excess choler—a treatment entirely unexceptional at the time and wholly in keeping with medical theories prevalent in the colonies. Still, the paper which Mitchell was so reluctant to see in print ultimately became his most famous work, "a classic among early American essays," as one medical historian has called it.[95] Franklin gave a copy to Dr. Benjamin Rush, who, following some of its suggestions during the yellow fever epidemic in Philadelphia in 1793, believed it helped him to save 6,000 lives. Twice printed in the early nineteenth century,[96] it added greatly to Mitchell's fame and is still his best-known writing. The crowning irony, however, is the more recent opinion of medical writers that the disease described in Mitchell's paper was not yellow fever at all. It took place, as Mitchell stated, in "the Winter & Spring Season" when the mosquito now known to transmit it is not normally present in Virginia. Moreover, an obscure contemporary colonial physician noted at Philadelphia that "the yellow fever in Virginia Described by Dr. John Mitchell Differs from that which appeared in Pensilvania in the Same Period of Time."[97] Indeed, Mitchell's description of the symptoms of the disease in Virginia significantly omits some of the principal clinical symptoms which attracted the attention of other colonial writers. Duffy, in his *Epidemics in Colonial America*, suggests that the Virginia epidemics were outbreaks of dengue, an infection closely resembling mild yellow fever

[95] Wyndham B. Blandon, *Medicine in Virginia in the Eighteenth Century* (Richmond, 1931), p. 54.

[96] In *Philadelphia Medical Museum*, I (1805), 1–20, and in *American Medical and Philosophical Register*, IV (Oct., 1813), 181–215. The latter also published letters on yellow fever from Colden to Mitchell and from Mitchell to Franklin (IV, Jan., 1814, 378–387). See also Francis Packard, "Dr. John Mitchell's Account of the Yellow Fever in Virginia in 1741-42, Written in 1748," *Annals of Medical History*, n.s., VI (1934), 91–92; Packard, *History of Medicine in the United States* (2 vols., New York, 1931), I, 113-114; N. G. Goodman, *Benjamin Rush, Physician and Citizen* (Philadelphia, 1934), pp. 164-195.

[97] John Duffy, *Epidemics in Colonial America* (Baton Rouge, 1953), p. 151.

in its early stages. But Dr. Gordon W. Jones, who also insists that the Virginia epidemics were not yellow fever, believes that the fever may have been a "louse-borne relapsing fever, with a seasoning of typhus, or the less likely Weil's disease."[98] Whatever it was that Dr. Mitchell described in his famous paper, evidently it was not yellow fever.

In the late summer of 1745, Mitchell's letters indicated that he suffered ill health. He was unable to practice medicine and he felt that he must soon return to England or lose his life.[99] Accordingly, he sold his Virginia property and embarked for England early in 1746. The vicissitudes of his voyage have already been noted, together with his close association in London with Mark Catesby, Peter Kalm (then on his way to America), and others of the Royal Society. He was a frequent visitor of the Society late in 1746, in 1747, and in 1748, as the guest of various Fellows.[100] On June 9, 1748, he was nominated for fellowship in the Society, and on November 17 and 24 following, before his election had taken place, he presented a paper to the Society entitled "An Account of the Preparation and Uses of the various kinds of Potash."[101] Though based in part upon his observations in America, the paper surveyed the methods and various kinds of wood employed in the production of potash in Germany, Russia, Sweden, France, Spain, and England, presented a chemical analysis of potash, and related its uses in manufacturing. Mitchell held that a Spanish potash, made from herbs, was superior to all others, and he recommended that a trial should be made to produce these plants (barilla and saltwort) in the colonies of North America, as "we seem to want nothing more than some proper Production," he stated acidly, "for the vast Tracts of Land we are possessed of there." On December 15 following, he was elected Fellow of the Royal Society. A week later he was formally admitted.[102]

98 *Ibid.*, p. 152; Jones, "Dr. John Mitchell's Yellow Fever Epidemics," *Va. Mag. Hist. & Biog.*, LXX (1962), 43–48.

99 Mitchell to Franklin, Sept. 12, 1745, *Colden Papers*, III, 151–154; also published by Thatcher in *Va. Mag. Hist. & Biog.*, XLI, 144–147.

100 *Journal-Book*, XX, 166, 171, 176, 187, 191, 212, 344, 523, 538, 576, 580.

101 *Ibid.*, XX, 579–581.

102 *Ibid.*, XXI, 17.

Except during his botanizing journey to Scotland with the Duke of Argyll in 1749–50—apparently Mitchell's last active fling at botany—he appears to have spent most of his time in London. He gave up the practice of medicine, possibly because of continued ill health, and turned his attention to history and politics, evidently using his personal familiarity with the American scene to cash in on the lively booksellers' market for materials about the American colonies. It seems strange that, having forbidden the publication of his work on "yellow fever" in America, he should have turned into a literary hack in London, but possibly necessity and bad health left him little choice. In any case, in the next several years he is believed to have written part of a section on the English in America in the second volume of the revised edition (1748) of John Harris' *Voyages and Travels* (2 vols., London, 1744–48), and four books published anonymously: *A New and Complete History of the British Empire in America* (3 vols., London, 1756); *The Contest in America between Great Britain and France by an Impartial Hand* (London, 1757); *The Present State of Great Britain and North America* (London, 1767); and *American Husbandry* (2 vols., London, 1775).[103] The preparation of these works, together with the excitement engendered by "the contest in America between Great Britain and France" for the control of the interior of North America, led Mitchell to study the rival territorial claims in the New World between the contestants. The result was publication of a *Map of the British and French Settlements in North America* (London, 1755), a major contribution to American cartography in spite of its political motivation.

Collinson informed Linnaeus in a letter (April 10, 1755) that "Dr. Mitchell has left Botany for some time, and has wholly employed himself in making a map, or chart, of all North America, which is now published in eight large sheets for a guinea, and coloured for a guinea and a half. It is the most perfect of any published, and is universally approved. He will get a good sum of money by it, which he deserves, for the immense labour and pains

---

[103] Carrier, "Dr. John Mitchell," pp. 205–218, reviews the evidence in favor of ascribing these works to Mitchell. Cf. Harry J. Carman, "Authorship of American Husbandry," in Carman, ed., *American Husbandry* (New York, 1939), pp. xxxix–lxi.

he has taken to perfect it."[104] Mitchell had collected materials from his American correspondents, read the accounts of the English colonial exploration and discovery in the colonies, and consulted French authorities.[105] His map was dedicated to the Earl of Halifax and the other members of the Board of Trade and Plantations, and it was certified by John Pownall, Secretary of the Board, as having been prepared at the request of the Board of Trade and made chiefly from "Draughts, charts, and Actual Surveys" of the colonies, many of which had been made specifically for the purpose. Mitchell's map, then, bore the stamp of high authority, and it quickly became the most important map of North America, used as the principal basis for British territorial claims in the St. Lawrence and Mississippi valleys. Indeed, as it was employed to settle boundary disputes from the eighteenth century on to as late as 1932, it has come to be regarded as "the Most Important Map in American History."[106]

But Mitchell's map was imperialistic in tone and content. He sought to extend British dominion in North America far beyond the actual English infiltration at the outset of the French and Indian War. All was English east of the Mississippi and Ohio rivers and south of the St. Lawrence saving the Florida peninsula, Michigan's northern peninsula, and a small triangle north of Lake Champlain (but including a portion of the province of Ontario north of a line drawn nearly due east from Sault Ste. Marie). Moreover, as the map illustrates, Mitchell believed that the future of the British colonies in North America rested upon their extension into the Ohio Valley, "A Fine Level and Fertile Country of Great Extent, by Accounts of the Indians and Our People." As Hornberger has stated, "The effect upon British ambitions in America is suggested by a copy which appeared in the *Gentleman's Magazine* for July,

---

104 Smith, I, 34.

105 See *Colden Papers*, IX, 98–107 and *passim; Va. Mag. Hist. & Biog.*, XXX (1922), 206; "Remarks on the Journal of Batts and Fallam in their Discovery of the Western Parts of Virginia in 1671. By John Mitchell, M.D., F.R.S.," *Sloane 4432*, fols. 6–9; Hornberger, *Scientific Thought*, pp. 290–291.

106 Colonel Lawrence Martin, Chief of the Division of Maps, Library of Congress, and former Geographer of the Department of State, in Hunter Miller, ed., *Treaties and Other International Acts of the United States* . . . , III (Washington, 1933), 328–351. Cf. also Martin's article about Mitchell in the *D.A.B.*

1755, with observations indicating plainly that the path of conquest was marked out."[107] With the map, Mitchell presented to the Board of Trade a report, later published as *The Contest in America between Great Britain and France by an Impartial Hand*. In view of the evidence, one may properly question the impartiality of the hand. But undeniably Mitchell's map, however political in its motivation, presented a knowledge of the geography of the North American continent superior to that of any of its predecessors.

Mitchell's studies appear to have led him into becoming, in spite of his protestations of having "an Impartial Hand," a thoroughgoing British expansionist imbued with the tenets of English mercantilism. *The Present State of Great Britain and North America* (1767) approximated his intent, announced to Franklin in a letter of 1745,[108] to prepare "a Natural & Medical History of my countrey," for which he said he had kept a journal of observations for many years. In this book he advanced the thesis that though America contained enough English colonists, they were located in the wrong places. The climate along the Atlantic seaboard being what it was, the colonists ought to be in the southern Mississippi Valley. Located as they were, they were unable to produce the staples— silk, wine, cotton, dyestuffs, and oil—that Great Britain's economy needed, and they would be forced, as New England had been already, into commerce and manufacturing which merely competed with the mother country. The area east of the mountains in America was too small, its soil too poor, and its climate too unhealthful, to supply Great Britain with her needs. Mitchell presented a very bleak picture of the American colonies, whose cold and unhealthful climate he had found distasteful and whose soils were generally unfit for the type of agricultural production for which he felt they should properly be employed.[109] Evidently he remembered America as a cold place, much colder than England,

---

[107] *Scientific Thought*, p. 291; see *Gentleman's Magazine*, XXV, 296, and insert.

[108] *Colden Papers*, III, 151–154.

[109] Mitchell supported his condemnation of the seaboard colonies by citing Catesby, Douglass, Jared Eliot, Bartram's journal of his travels in East Florida, meteorological observations kept by Dr. John Lining at Charles Town, and records kept at King's College in New York of the temperatures for Jan., 1765. Cf. Hornberger, *Scientific Thought*, p. 293.

and he sought to explain the extreme cold in America in terms of prevailing northwest winds sweeping over a land mass that extended to the North Pole, embracing in the north high mountains constantly covered with ice and snow. He ridiculed the notion, advanced by some, that the temperature would moderate after the forests had been cleared, for he argued that that part of the continent which is coldest—the Hudson's Bay region—"is so barren, that it does not bear a tree or a bush." To obviate these colonial defects, Mitchell advanced drastic new colonial policies. The solution, as he saw it, was to move the center of the colonial population into the lower Mississippi Valley (which he had never seen!) where soil and climate furnished a prospect, he believed, to construct a colonial agricultural society which would correspond to his mercantilistic a priori concept of what the colonies should be. Similar policies were set forth in *American Husbandry* (1775), in the preparation of which Mitchell may have had a hand, for it devoted much space to the "defects" of colonial farming methods and found the best hope for the future in the Mississippi Valley, where, as Mitchell had said in *The Present State*, "Colonies should live *merely by their Agriculture*, without either Manufactures or Trade, but what is confined to their Mother Country." Still, the useful proposals for the improvement of colonial agriculture, by crop rotation, diversification, replenishment of soils by fertilization, and agricultural experimentation, were not, as Hornberger has pointed out, clearly foreshadowed by Mitchell's earlier works.[110]

Mitchell dabbled in chemistry from time to time, but he produced nothing of significance. He also experimented with electricity and presented one of Franklin's electrical experiments to the Royal Society in December, 1759.[111] Despite his remarkable map of North America, nearly all of his scientific work after his return to England was tainted by his political prejudices or marred by his unhappy personal memories of the American colonies, especially of the American climate, upon which he apparently placed the blame for the break in his health. His scientific work in the colonies,

110 *Ibid.*, p. 295.

111 His experiments with electricity made "of fine Silk Stockings," were presented to the Royal Society at the same time. *Journal-Book*, XXIV, 454–455 (Dec. 20, 1759); Carrier, "Dr. John Mitchell," pp. 204–205.

however, had been fresher and more objective. In botany, zoology, physiology, and medicine, his work was generally imaginative, marked by meticulous care, close observation, and a considerable measure of clinical skill. Probably his best work was done in taxonomy, especially in the treatises published in Germany, in which he momentarily approached scientific greatness in his ideas about arriving at concepts of species and genera and in his discussions of plant classification. However, even in the latter his utilitarian concern for medicinal botany limited his scientific accomplishment and hardly marked him as a seeker after knowledge for its own sake—that fugitive ideal of the philosophers of science. Like most scientists he sought recognition, and like most colonials he especially treasured European recognition. Both his personality and his career are somewhat reminiscent of his contemporary at Boston, Dr. William Douglass, although Mitchell displayed greater breadth of interest and skill. Neither man held colonial society in high esteem, but Mitchell appears to have been the more dissatisfied and disappointed. In addition, he was plagued by ill health and had the misfortune to be captured by a privateer, his personal effects and vast botanical collection lost, and no doubt his life endangered. In spite of all, he ranks high among the scientists in the colonies of his day.

John Clayton (1694–1773)

Dr. John Mitchell appears to have had little association with the Virginia community of scientific men except with his friend John Clayton. Indeed, the men in the Virginia community, probably because of their lack of a bustling urban center, were less closely knit together personally than those of other colonial scientific groups. The colonial capital at Williamsburg, with its governor's residence, the periodical meetings of the House of Burgesses, the court sessions, and the College of William and Mary, was the principal social, cultural, and intellectual center of the Old Dominion, but it hardly could compete with the larger commercial centers such as Philadelphia, Boston, and Charles Town—also the seats of colonial government in their respective colonies. The College of William and Mary, especially between 1758 and 1764, when Dr.

William Small, a Glasgow graduate in medicine, served as professor, first of mathematics and then of philosophy, was a lively scientific center to which the Lieutenant Governor, Francis Fauquier, F.R.S., lent prestige and support. Dr. Small led in the formation of a society founded at Williamsburg in 1759, modeled upon the Royal Society of Arts founded in London five years before, to encourage scientific experimentation and new discoveries, arts, and manufactures. Some investigations were undertaken with niter and medicinal properties of Virginia plants, but probably the greatest accomplishment of the group was to quicken the scientific interests of Thomas Jefferson, then a student at the College of William and Mary. Initially, Jefferson's interest in science may have been implanted by his father, Peter Jefferson, a surveyor and cartographer; but his scientific bent was strengthened by Dr. Small and the Williamsburg society. Dr. Small returned to England in 1764. Two years later Dr. Arthur Lee, F.R.S., a native Virginian with a medical degree from Edinburgh, settled at Williamsburg to practice medicine. Just prior to his arrival in Williamsburg, Dr. Lee had presented to the Royal Society of London (May 1, 1766) a report on twenty-five well-conceived "Experiments on the Peruvian Bark," designed to "confirm the pharmaceutic treatment by the Bark where it is just; to correct it where it is erroneous; & to improve what may be defective therein."[112] Dr. Lee was elected Fellow of the Society shortly afterward (May 29, 1766), but his presence in Williamsburg did not compensate for the departure of Dr. Small, and after two years Dr. Lee forsook medicine for law

---

112 *Journal-Book*, XXVI, 422–423. On Williamsburg society, see Bridenbaugh, *Seat of Empire*, and Richard M. Jellison, "Scientific Enquiry in Eighteenth-Century Virginia," *The Historian*, XXV (1963), 292–311. Benjamin Franklin signed the bond for Arthur Lee's contributions and paid his admission fee to the Royal Society, for which he was reimbursed about six months later. Lee was formally admitted to the Society after his return to England, on Nov. 10, 1768. *Journal-Book*, XVIII, 119; *Certificates, 1751–1766*, unpaged. From then on until early in 1776, Lee was a frequent attendant at meetings of the Society and presented (Jan. 28, 1773) a paper entitled "Extracts of Some letters from Sir William Johnson, Bart., to Arthur Lee, F.R.S., on the Customs, Manners, and Language of the Northern Indians of America." *Journal-Book*, XXVIII, 58–61. Later, Lee declared "that from the time of the establishment of American Independence he had considered himself as no longer a fellow" of the Society, and he was formally dropped on Jan. 17, 1788. *C.M.*, VI, 291–292.

and politics and returned to London for legal training. However, about Williamsburg were the plantations of the Carters, the Randolphs, the Byrds, the Lees, the Custises, and others with an interest in gardening, and some of these men were in touch with John Bartram, Peter Collinson, various nurserymen and botanists in England, and occasionally with the Royal Society.[113] Among these men too, at Gloucester Court House, lived the prosperous and knowledgeable Clerk of Gloucester County, John Clayton.

John Clayton was a distant relative of that scientifically-minded minister of the same name who had served Jamestown between 1684 and 1686 and had sent scientific communications regarding Virginia to the Royal Society.[114] English born and educated (probably in law), John Clayton went to Virginia about 1720 to join his father, who since about 1705 had been in Virginia, first as secretary to the Lieutenant Governor and later as Attorney General of the colony.[115] The younger Clayton acquired land, became a well-to-do planter, and served as Clerk of Gloucester County, Virginia, until his death in 1773. His interest in natural history may have been inspired by Mark Catesby, with whom he formed a close friendship while Catesby was in Virginia, and soon Clayton was entering into scientific correspondence on his own. He developed a botanical garden and he communicated with his fellow Virginian, William Byrd II. Catesby introduced him to Gronovius and probably to Collinson; Collinson put him in touch with John Bartram; and soon Clayton was well known to members of the natural history circle on both sides of the Atlantic. As early as 1729 he sent to one Mr. Pole, a London woolen draper, a "Box of Natural Rarities," which Pole passed on to the Royal Society.[116] He sent many collections, mostly of Virginia plants, to Catesby, after the latter's return to England, and Catesby shared them widely with others, especially with Gronovius in Holland, by whose hand Linnaeus, who was in Holland until 1736, became familiar with them. In the late 1730's Clayton prepared "A Catalogue of Plants,

[113] Swem, "Brothers of the Spade," pp. 39, 166; *Journal-Book*, XVI, 381; XXI, 44; XXIV, 178–179.

[114] See pp. 183ff. above.

[115] Details of Clayton's life are drawn largely from the excellent recent study of him by the Berkeleys, *John Clayton: Pioneer of American Botany*.

[116] *Journal-Book*, XIV, 383 (Dec. 11, 1729).

Fruits, and Trees Native to Virginia" and sent it to Gronovius. Gronovius, with the assistance of Linnaeus, revised it in accordance with the Linnaean method of classification (Clayton had followed John Ray's system) and published it in 1739 as *Flora Virginica*. Gronovius appears to have taken rather questionable liberties with Clayton's manuscript, having revised and published it under his own name without Clayton's consent—although he acknowledged Clayton's contributions in the Introduction to the book. However, it spurred Clayton on to become familiar with the Linnaean system and he continued to send Virginia collections, mostly by the hand of Catesby, to Gronovius. By 1740 Clayton had mastered the Linnaean system to such an extent that he was able occasionally to offer corrections to the master in the latter's description of New World flora, and Linnaeus, in recognition of Clayton's contributions, named the charming American early woods wildflower, the spring beauty, as it is popularly known, the *Claytonia*. In 1743 Gronovius published a second part of *Flora Virginica* from additional materials supplied by Clayton. By this time Clayton had become a widely respected member of the natural history circle. When Franklin, Bartram, and other friends launched the feeble, short-lived American Philosophical Society in 1743 at Philadelphia, Clayton, Mitchell, Dr. Arthur Lee, and others of the Virginia community were invited to become corresponding members.

Clayton continued his studies and collections in natural history until the end of his life. In 1748 he made a journey west of the mountains into the Great Valley to collect materials, and in 1752 he sent to Collinson "an extract from a Manuscript . . . concerning the Smoaky Weather in Virginia and a remark concerning an uncommon kind of humming bird," which Collinson presented to the Royal Society.[117] Collinson said that the "Smoaky Weather," usually ascribed to fires set by the Indians in hunting and to the tar kilns in southern Virginia and North Carolina, arose from rising vapors in a calm period following moist weather, but he was unclear as to whether this explanation arose from his own or Clayton's thinking about it. In 1755 Clayton joined John Bartram to publish in the *Gentleman's Magazine* of London "Some Remarks Made on Dr. Alston's Dissertation on the Sexes of Plants by two celebrated

[117] *Ibid.*, XXII, 47–48 (Feb. 13, 1752).

botanists of North America, both dated June 10, 1755."[118] Dr. Alston, Professor of Botany at Edinburgh and Dr. John Mitchell's former mentor, had argued against the bisexuality of plants, and the two colonial botanists presented smashing evidence in support of the contention. In the meantime, too, Clayton sought to persuade Gronovius to prepare a revised edition of the *Flora Virginica*. He himself worked at it and in 1757 notified Collinson that he had completed it, although he had no publisher.[119]

Collinson undertook to supply the want. Clayton sent him the manuscript, John Ellis read it with approval, and Georg Ehret was engaged to prepare plates for the proposed volume. But before the manuscript was published, Gronovius' son, Laurenz Théodor, completed the revision over which his father had dawdled for several years, and published a revised edition of *Flora Virginica* in Leiden in 1762. In consequence, Collinson's plan to publish Clayton's revision was abandoned. The younger Gronovius' edition, however, was distinctly inferior to Clayton's work. It lacked many items, including plants from west of the mountains in Virginia, which Clayton had intended to include. It had no illustrations, whereas Ehret's plates for Clayton's volume promised, in view of Ehret's reputation, to be superior prints. And the younger Gronovius did not employ the new Linnaean nomenclature, which Clayton had supplied. Indeed, Benjamin Smith Barton wrote in 1805 that Clayton's work supplied the "best foundation of our knowledge of the plants of a considerable part of the tract of country now called the United States."[120] John Clayton had prepared sound scientific principles for the study of botany in the United States according to the most up-to-date knowledge of his day. But circumstances beyond his control robbed him of the public recognition that was his due. His collections and manuscripts were destroyed by fire during the War for Independence, although many of his collections, including a copy of his revised *Flora Virginica*, are preserved in European

---

[118] *Gentleman's Magazine*, XXV (1755), 407–408. Collinson submitted the letters for publication. See William Darlington, ed., *Memorials of John Bartram and Humphrey Marshall* . . . (Philadelphia, 1849, cited hereafter as Darlington, *Mems.*), p. 203.

[119] Collinson to Linnaeus, Dec. 25–27, 1757, Smith, I, 42.

[120] Quoted in the Berkeleys' *John Clayton*, p. 178.

herbaria at the British Museum, Oxford, Cambridge, and Stock-holm.

Cadwallader Colden (1688–1776)

If until recently John Clayton of Virginia failed to win the public recognition as a scientist which was his due, Cadwallader Colden may have been accorded a reputation as a scientist beyond his true merits. The reader was introduced to Colden in the previous chapter,[121] when he first appeared on the scene at Philadelphia and subsequently established himself in New York. But a more detailed treatment of his career was postponed because Colden lived until the end of the Old Colonial Era and acquired a prominent place among the colonial men of science who constituted the last generation of colonial scientists. Like most of his contemporaries, his interests were broad, encompassing a study of the Indians of upper New York, natural history, botany, medicine, geology, meteorology, mathematics, theoretical physics, astronomy, cartography, and moral philosophy. He was active in the promotion of scientific affairs in the colonies, early suggested the creation of a scientific society, took a leading part in the founding of the New York Society Library (1754), encouraged the establishment of The Society of The Hospital of the City of New York in America (1767), and engaged widely in scientific correspondence both with his fellow colonials and with scientists in Europe. Obviously, most of his colonial contemporaries held Colden in high esteem; Peter Collinson and J. F. Gronovius promoted his efforts in Europe; and Carl Linnaeus, in a moment of extravagance, referred to Colden as "Summus Perfectus." Some twentieth-century writers have extolled Colden's merits similarly. Louis Leonard Gitin, in an evaluation of "Cadwallader Colden as Scientist and Philosopher" in 1935, placed Colden as the man who, "next to Franklin, was the most eminent scientist and philosopher in America," and he continued with the doubtful assertion that "Practically no progress had been made in botanical science in America prior to the time of Colden."[122] Brooke Hindle, in his *Pursuit of Science in Revolutionary America*,

121 See pp. 493–496 above.
122 *New York Historical Society Quarterly*, XVI, 166–177.

1735–1789 (1956), and in his long article about "Cadwallader Colden's Extension of the Newtonian Principles" (1956), has given a far more restrained and judicious account of Colden's scientific work.[123] Hindle's treatments are, on the whole, excellent; but a full-length, scholarly biography of Colden, with a balanced perspective in the assessment of his scientific accomplishments, is wanting. However, the *Colden Papers*, published by the New York Historical Society,[124] supply the bulk of Colden's correspondence, and, with his published works and the letters of his contemporaries, furnish a large mass of materials upon which to formulate a judgment of the man.

Indubitably Cadwallader Colden was an intelligent, well-educated man with a thirst for knowledge. He fashioned his life in such a manner as to acquire political sinecures whereby to obtain wealth as a means of leisure for study, and he frequently applied to his friends for aid in obtaining yet more political appointments whereby to enlarge his material well-being—and his capacity for leisure.[125] As a colonial he exhibited an unusually contemplative bent of mind with a creative, but not always well-disciplined, imagination. His ideal appears to have been that of the cultured English gentleman, free from other worldly cares to sit in "my Sweet & Calm old Mansion," as he wrote to Collinson, looking down "on the Busie Vain World below."[126] His approach to learning was to read and think "by turns," filling in the intervals with learned conversation.[127] However, excepting infrequent visitors and

---

[123] Hindle, *The Pursuit of Science*, pp. 39–48 and *passim;* Hindle, "Cadwallader Colden's Extension of Newtonian Principles," *Wm. & Mary Qty.*, 3rd Ser., XIII (1956), 459–475.

[124] *The Letters and Papers of Cadwallader Colden* (*Collections of the New York Historical Society*, Vols. L–LVI, 1917–23; Vols. LXVII–LXVIII, 1934–37; reprinted as *The Letters and Papers . . . 1711–1775*, 9 vols. in 4, New York, 1968). Additional manuscript materials repose in the New York Historical Society Library. Cited are the *Collections of the New York Historical Society.*

[125] Several such letters exist among the *Colden Papers.* See, for example, Colden's letters to Collinson in *ibid.*, IV, 312–314; IX, 95–97, 116–119.

[126] *Ibid.*, VI, 288–291. He also referred fondly to his garden, with plants from all quarters of the world.

[127] From Colden's "Introduction to the Study of Natural Philosophy," unpublished manuscript in the New York Historical Society Library, cited by Hindle,

occasional journeys to New York, Boston, or elsewhere, his conversations were limited to the members of his family, for his "mansion" at Coldengham, to which he removed in 1728,[128] was far distant from the conversational exchanges possible in an urban community. To some extent, of course, Colden's wide scientific correspondence compensated for his lack of mature, informed conversationalists. He told Peter Collinson in May, 1742, that it was one of the happy incidents of his life to have the good fortune to correspond with him and "to communicate some thoughts in natural philosophy which have remained many years with me undigested for we scarcely have a man in this country that takes any pleasure in such kinds of speculations."[129] To the shortcomings of his chosen way of life was added the fact that as he read, contemplated, and conversed, he seldom engaged in any scientific experimentation. He was capable of close observation, as his botanical works demonstrate, but he was opposed to excessive industry lest it blunt the imagination. He appeared to believe that such great syntheses of knowledge as Sir Isaac Newton's *Principia* were the product primarily of creative imagination, and he ignored the laborious experimentation often required to test the validity of hypotheses born of creative imagination. In short, Colden failed to comprehend the full import of the *experimental* method in science; and many of his works were faulted by this failure. In consequence, although he despised the "mere Scholar" who acquired all his learning from

The Pursuit of Science, p. 42. Colden's children were educated at home by means of "conversations." See Brooke Hindle, "A Colonial Governor's Family: The Coldens of Coldengham," *New York Historical Society Quarterly*, XLV (1961), 233–250.

128 Colden's removal to his country estate was made by stages. He built Coldengham in 1728 and moved his family there, making occasional trips to New York and elsewhere to attend to his business interests. In 1739 he left the oversight of his farm largely to his wife and children and retired to his study. In 1750 he retired from public life to "philosophical amusement." In 1762, after he was chosen Lieutenant Governor of New York, he left Coldengham in charge of his son, Cadwallader Jr., and purchased an estate at Spring Hill, near Flushing, Long Island, still outside New York City. He died at Spring Hill in 1776. See A. J. Wall, "Cadwallader Colden and His Homestead at Spring Hill, Flushing, Long Island," *New York Historical Society Quarterly*, VIII (1924), 11–20.

129 *Colden Papers*, II, 257. Colden's mood led him to add an autobiographical sketch.

books, and yearned to make a mark for himself in the world of science, his works tended to be exercises in rational thinking rather than experimental or clinical in nature.

Colden was trained as a physician, and although he made vain efforts in Philadelphia to set up a medical lectureship, he abandoned the practice of medicine soon after his removal to New York, and confessed in 1753 that he had "entirely laid aside the practice of physic upwards of twenty years."[130] Still, his inactivity in the profession did not deter him from setting forth medical treatises. He published an *Essay on the Iliac Passion . . .* (Philadelphia, 1741), the only evidence of his interest in intestinal disorders. In *An Abstract from Dr. Berkeley's Treatise on Tar-Water with Some Reflexions Thereon, Adapted to Diseases Frequent in America* (New York, 1745) he expanded Bishop George Berkeley's metaphysical claims for the curative properties of tar water, already proclaimed a panacea by the Bishop.[131] In the same year he published *An Essay on Yellow Fever* (New York, 1745), although when Dr. John Mitchell objected to Colden's views, the latter admitted that he had never seen a case of yellow fever (nor, as would appear from the account above, had Dr. Mitchell!). In 1751 Colden published in the *Gentleman's Magazine* an article on the value of pokeweed (*Phytolacca*, probably *P. decandra*) as a cure for cancer.[132] Based upon reports of the successful use of the juice of pokeweed in the cure of cancers by the Anglican "reverend Dr. *Johnson*" of Stratford, Connecticut, Colden sounded a note of caution, for hitherto he had known of no certain method to cure "genuine cancers." But he passed on the reports, described pokeweed, and suggested that "there seems to be some kind of analogy between cancers and the tumours made by some insects laying their eggs in

[130] Colden to Fothergill, Dec. 24, 1753, in *Medical Observations and Inquiries . . .* (3rd ed., 6 vols., London, 1763–84), I (1763), 211–229.

[131] A letter by James Alexander identified Colden as the author. *Colden Papers*, III, 102. Colden referred to the Bishop of Cloyne's *Chain of Philosophical Reflections concerning the Virtues of Tar-Water* (London, 1744).

[132] XXIV (1751), 305–308; dated from New York, Dec. 15, 1750. Colden received the information about pokeweed from Samuel Johnson of Stratford, Connecticut. *Colden Papers*, III, 148–149. Richard Brooke, a surgeon formerly in Maryland, confirmed the virtues of pokeweed in a letter to Dr. John Pringle, read to the Royal Society March 8, 1753. *Journal-Book*, XXII, 285–286.

leaves, on the bark or fruit of vegetables, and in the flesh of animals." But he had not tested the hypothesis and only proposed it "to the curious and learned, for further inquiry." Two years later he sent a letter to Dr. Fothergill "Concerning the Throat-Distemper," subsequently published in the London *Medical Observations and Inquiries*, although he confessed that the information was "What I chiefly learned . . . from the late Dr. [William] Douglas[s] of Boston."[133] And finally, he published a more substantial work entitled *Treatise on Wounds and Fevers* (New York? 1765) which, according to Francis Packard's *History of Medicine in the United States*, "was long regarded as a standard authority in this country . . . referred to and quoted by medical authors."[134] Colden's correspondents included several physicians with whom medical cases were sometimes discussed, and he was a close friend of John Bard, a New York physician. Bard had been a medical apprentice of Dr. John Kearsley of Philadelphia, and he later (1788) served as first President of the New York Medical Society. Colden recommended him to the Royal Society and interceded with Dr. John Fothergill to persuade the latter to publish Bard's article, "A Case of an extra-uterine foetus," in the London *Medical Observations and Inquiries*. Bard's son, Dr. Samuel Bard, was reported to have learned botany by the hand of Colden's accomplished daughter, Jane.[135]

After Colden left New York City in 1739 to settle into a life of study and contemplation at Coldengham, his letters leave the impression that he was restless, casting about for a subject to occupy his mind. He toyed with a project to design an accurate portable quadrant for surveyors and cartographers. His old friend, James Alexander, submitted it to Peter Collinson for English consideration, but George Graham (1673–1751), the English mechanician, found it faulty and Colden abandoned the idea. However, Alexander had introduced Colden to a correspondence with Collinson which proved to be mutually enriching. Colden devised a plan to

[133] I (1763), 211–229.

[134] I, 505. I have been unable to find copies of the *Essay on the Iliac Passion, An Essay on Yellow Fever,* and the *Treatise on Wounds and Fevers* and am forced to rely on secondary accounts.

[135] *Journal-Book,* XXI, 303–305; *Medical Observations and Inquiries,* II (1764), 369–372; Thatcher, *American Medical Biography,* pp. 97–143; *Colden Papers,* II, 207–208.

print from plates, a kind of stereotyping whereby toilsome type-setting could be obviated. Franklin was pleased with it, but when Collinson took it to William Strahan, a London printer, Strahan reported that it had been tried before and found impractical. At Collinson's suggestion, Colden set out to enlarge his previous work, *The History of the Five Indian Nations of Canada* (1727). He worked at it during the winter of 1741–42, carried the account forward to 1697, and sent the manuscript to Collinson, who arranged for its publication in London in 1745. Another printing was made a decade later. Meanwhile, he discussed with Collinson the latter's opinion that among "noxious animals" there was born an excess of males over females, but they arrived at no firm conclusion for want of scientific data.[136]

In the meantime, too, Colden hit upon a subject that caught his fancy. About 1742, he picked up a copy of Linnaeus' *Genera Plantarum* (1737), set about to master the Linnaean system of classification, and soon was launched on a study of botany. At the suggestion of Collinson, John Bartram visited Colden in 1742. The two men found each other mutually agreeable, Bartram fired Colden's enthusiasm for botany to new heights, and they entered upon a scientific correspondence and exchange that continued for many years. Colden began to collect and study the flora in the vicinity of Coldengham, to exchange plant specimens and descriptions with Collinson as well as Bartram and other colonials, and to cultivate exotic plants in his garden. Collinson was delighted with Colden's progress, especially in his mastery of the Linnaean system. He wrote to Linnaeus (January 18, 1743/44) that "Your system, I can tell you, obtains much in America"; Colden, Clayton, and Mitchell were "complete Professors" of it.[137] With Collinson's aid and encouragement, Colden was soon engaged in a scientific exchange with J. F. Gronovius and with Linnaeus himself. He sent plant specimens and descriptions to Gronovius, who found them to be valuable additions to the Virginia materials received from John Clayton, and he drew up a catalogue, using the Linnaean system of classification, of the New York flora found in the vicinity of Col-

---

[136] For these varied activities, see *ibid.*, II, 208–211, 257, 277–283; III, 10–14, 27–29; *Journal-Book*, XIX, 159.

[137] Smith, I, 8–9; see also *ibid.*, III, 42–45.

dengham. Gronovius sent it on to Linnaeus, who was so favorably impressed with it that he published it as "Plantae Coldenhamiae in provincia Novaboracensi Americanes sponte Crescentes" in the *Acta Societates Regiae Scientiarum Upsaliensis* (IV (1743), 81–135; V (1744–50), 47–82). Moreover, when Linnaeus published his *Species Plantarum* (1753), he gave a prominent place to "C. Colden" as a source of his knowledge of New World plants, and he gave the name *Coldenia* to a plant in the *Flora Zeylanica* (Stockholm, 1747).

Still, Colden was not content with his botanical studies. In a letter to Gronovius in December, 1744, he made clear that he perceived some of the faults of Linnaeus' sexual system of plant classification, and he expressed opinions on the subject which evidently pointed toward a "natural" system.[138] But he pursued the matter no further, thereby, perhaps, passing by the opportunity to rival Dr. John Mitchell, whose "Dissertatio" had set forth novel thoughts regarding problems of classification. Colden continued, somewhat desultorily, to collect and exchange flora and some other natural history specimens with his correspondents.[139] But botany had lost much of its charm, and Colden tired of the endless round of collecting and exchanging plants, seeds, and the like with members of the natural history circle. Moreover, as his eyes lost their sharpness with advancing age, he found it difficult to examine plant specimens accurately. Accordingly, he sought to turn over his botanical activities to his daughter Jane. On October 1, 1755, he wrote to Gronovius:

> I often thought that Botany is an amusement which may be made greater to the Ladies who are often at a loss to fill up their time. . . . Their natural curiosity & the pleasure they take in beauty & variety of dress seems to fit them for it far more than men. The chief reason that few or none of them have hitherto applied themselves to this study I believe is because all the books of any value are wrote in Latin & so filled with technical words that the obtaining the necessary previous knowledge is so tiresome and disagreeable that they are discouraged at the first set out &

138 *Colden Papers*, III, 83–92.
139 He sent collections of fossils to Collinson, and in 1748–49 investigated the cicada briefly. *Ibid.*, III, 50–52, 68–70, 96–98, 125–126; IV, 131–132; V, 80–83, 116–119, 215–217; *Journal-Book*, XXI, 77–78, 104.

give it over before they can receive any pleasure in the pursuit.

I have a daughter who has an inclination toward reading and a curiosity for natural philosophy and natural history, and a sufficient capacity for attaining a competent knowledge. I took the pains to explain Linnaeus's system, and to put it in an English form for her use. . . . She is now grown very fond of the study, and has made such a progress in it as I believe would please you if you saw her performance . . . she now understands in some degree Linnaeus's characters, notwithstanding that she does not understand Latin. She has already a pretty large volume in writing, of the description of plants. She was shown a method of taking the impression of the leaves on paper with printer's ink, by a simple kind of rolling press, which is of use in distinguishing the species. . . . She has the impressions of three hundred plants in the manner you'll see by the samples. That you may have some conception of her performance and her manner of describing, I propose to enclose some samples in her own writing, some of which I think are new genus's. . . . If you think, sir, that she can be of any use to you, she will be extremely pleased at being employed by you, either in sending descriptions, or any seeds you shall desire, or dried specimens of any particular plant you should mention to me. She has time to apply herself to gratify your curiosity more than I ever had; and now when I have time, the infirmities of age disable me. . . .[140]

And so Colden transferred his botanical correspondence as much as possible to his daughter. Jane was industrious and surprisingly competent. She exchanged specimens and seeds with John Bartram (she met both John and William Bartram in 1753), with Alexander Garden of Charles Town, with Collinson, and with Gronovius. She prepared a "Flora of New York" with some 340 illustrations which was ultimately purchased by Sir Joseph Banks and deposited in the British Museum (Natural History). Her plant descriptions were often good, and she displayed a housewifely concern for the uses of plants in cookery and as household remedies for sickness. At least one of her plant descriptions was published in the Scottish *Essays and Observations* (Vol. II, Edinburgh, 1757). Her work excited considerable comment in the natural history circle, more, it would

140 *Colden Papers*, V, 29–32.

appear, because of her sex than because of her abilities. A female botanist was a rare thing to contemplate, "being the only Lady that I have yett heard of," wrote Collinson, "that is a professor of the Linnaean System. . . ."[141] But no one appears to have accepted her work on the same level as that of her father and of other male collectors in the colonies. However able Jane Colden may have become as a botanist, her unconventional activities as a woman relegated her to the outer rim of the natural history circle.[142]

Cadwallader Colden's coolness toward botany began years before his daughter had become skilled in the subject—and years before "infirmities of age" had made field trips more difficult to carry out. The fact was that Colden had fastened upon a new subject for study and contemplation, one that he believed to be of prime scientific significance, and one that could elevate him above the company of Gronovius and Linnaeus to the exalted ranks of Newton, Galileo, and the other founders of the new science. He was coy about announcing it. He wrote to Collinson as early as June, 1744, that his botanical activities might be interrupted because his thoughts had turned to another subject "so bold" that he dared not even mention it until after he had consulted further with friends in America.[143] Actually, his correspondence shows that his thoughts had been moving in a new direction since early in 1743, when he began a scheme of applying fluxions to physics with the notion that fluxions were founded in nature. He had consulted with Captain John Rutherford, a cultivated Scot who commanded the garrison at

141 *Ibid.*, V, 139–141.

142 Examples from Miss Colden's "Flora of New York" are published in H. W. Rickett and Elizabeth Hall, eds., *Botanic Manuscript of Jane Colden, 1724–1766* (Garden Club of Orange and Dutchess counties of New York, 1963). See also Hindle, "A Colonial Governor's Family," *loc. cit.*; James Britten, "Jane Colden and the Flora of New York," *Journal of Botany, British and Foreign*, XXXIII (1895), 12–15; Anna Murray Vail, "Jane Colden, an Early New York Botanist," *Contributions from the New York Botanical Garden*, IV (1966–67), 21–34. Jane Colden's "Flora Nov-Eboracensis" is in the British Museum (Natural History) Catalogue, No. 26.e.19. It contains 341 plant descriptions, in English, and figures of many of them at the end. The figures are neat, but as they lack flowers and root structures, as well as sufficient detail of stems and leaves for present-day botanical use, they are more curious than valuable.

143 *Colden Papers*, III, 60–61.

Albany, and Rutherford had expressed some doubts about the matter but found that it helped the imagination prodigiously.[144] In 1744 Colden circulated among some of his other colonial friends—including James Alexander of New York and James Logan and Benjamin Franklin of Philadelphia—a rough draft of his ideas, and his communication sent his friends scurrying to consult whatever books they could find pertinent to the subject in an effort to comprehend Colden's meaning. In 1745 Colden published in New York a small, incomplete edition of his manuscript so that other scholars could have access to it, under the title of *An Explication of the First Causes of Action in Matter; and the Cause of Gravitation*, and dedicated it to James Alexander. Although Colden was diffident in setting forth new principles in physics, "different from those of all writers before me," and "to attempt to explain the cause of gravitation after all the great men in philosophy have failed, and after Sir Isaac Newton stopt short," nonetheless he pretended to have discovered the cause of gravitation.

As Brooke Hindle has said, "No more audacious claim to intellectual eminence was ever made in Colonial America. . . ."[145] Colden's approach to this fundamental problem in theoretical physics was rational, nonexperimental, and ill-informed. He sought to explain the mystery of gravitation without "pomp" or a vast array of mathematical demonstrations. Apparently he was totally unfamiliar with the works of the greatest mathematical physicist of his day, Leonhard Euler (1707–83); and he demonstrated a like ignorance of the remarkable Swiss family of eight mathematicians and scientists, the Bernoullis, as well as of the pertinent materials to be found in the works of Leibnitz, Huygens, Hooke, and others. He had studied Newton's *Principia* and *Optics*, although it is apparent that he had not mastered Newton's concept of intertia, his laws of motion, or the remarkable balance of forces exerted upon the planets. Indeed, in keeping with his high regard for the human

[144] Rutherford to Colden, March 2, 1742/43, *ibid.*, III, 6–10. Other letters from Rutherford are in *ibid.*, *passim*.

[145] "Cadwallader Colden's Extension of Newtonian Principles," *loc. cit.* This article is the best exposition available of Colden's book, giving his arguments in greater detail than space allows here. See also Hindle, *The Pursuit of Science*, pp. 43–47.

25. Cadwallader Colden (1688-1776) of New York.

Courtesy of the New York Chamber of Commerce.

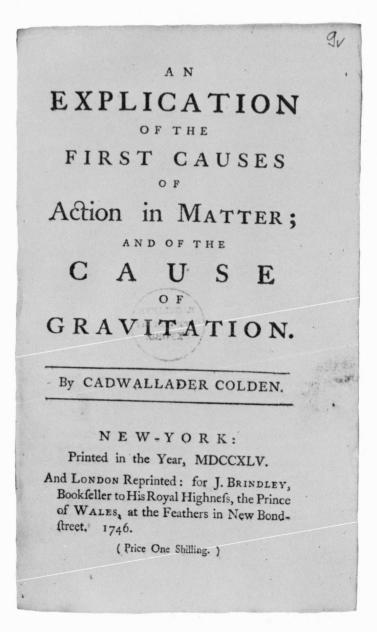

A N

EXPLICATION

OF THE

FIRST CAUSES

O F

Action in Matter;

AND OF THE

CAUSE

O F

GRAVITATION.

By CADWALLADER COLDEN.

NEW-YORK:

Printed in the Year, MDCCXLV.

And LONDON Reprinted: for J. BRINDLEY,
Bookſeller to His Royal Highneſs, the Prince
of WALES, at the Feathers in New Bond-
ſtreet. 1746.

(Price One Shilling.)

26. The title page of Colden's controversial work, later enlarged and
published in London as *The Principles of Action in Matter* (1751).

Courtesy of the American Philosophical Society Library.

faculty of creative imagination, Colden appears to have given greater attention to Newton's "General Scholium" in the *Principia* and "Queries" at the end of the *Optics*, where Newton relaxed and indulged himself in daring speculations and hypotheses, than to the main body of Newton's work, where he held his imagination in check and set forth his theses by rigid reasoning, carefully designed experiments, and mathematical demonstrations. It was largely from clues drawn from the "General Scholium" and the "Queries," without a clear understanding of some of the fundamentals of Newton's works, aided neither by further experimentation and observation of the phenomena nor by consultation with the pertinent works of other scientists, that Colden set out to explain "the causes of Gravitation."

Colden was an ardent disciple of Newton and considered his own work, as Hindle has said, as an "extension" of Newtonian principles. To explain gravity, he began by dividing the material of the world into three substances: *ether*, a subtle substance similar to that postulated by Newton, although Colden's explanation of its function in relation to gravitation differed from Newton's hypothesis; *resisting matter*, i.e., bodies having mass and occupying space; and *moving matter*, which Colden essentially equated with light, i.e., corpuscular light, like that of Newton. Gravity, according to Colden, was the force exerted by ether upon all planets and stars. As there were fewer ether particles between the sun and each of the planets than there were surrounding them, Colden assumed that there would be less force exerted by the ether between them while the superior force exerted around them would cause each planet to move toward the sun. Simultaneously, however, the force exerted by light particles emanating from the sun prevented collisions, while light also imparted a tangential motion to the planets in some mysterious, undefined fashion. Indeed, Colden ascribed to light the sole power of movement found in the universe, and it was responsible for impelling the planets in their orbits, for imparting to them velocities proportional to their respective distances from the sun, and even for the rotations of planets on their axes. Thus the movements of the planets toward the sun occasioned by gravity (ether) were counterbalanced by the force of light moving away from the sun, so that each planet took the course of least resistance by moving in a direction at right angles to the rays of light reaching it from the sun.

Colden's colonial friends were bewildered by the *Explication*.[146] They turned to the writings of scientists they believed to be related to Colden's speculations, calling them to Colden's attention in the hope that his work might be strengthened and improved. Captain Rutherford summarized the ideas of Descartes, Malebranche, and Boerhaave, and sent Colden one of the works of Leibnitz; James Alexander, his best friend to whom the *Explication* was dedicated, sent him Flamsteed's *Historia Coelestis Britannica* (1725) and referred him to some pertinent materials in the *Philosophical Transactions;* James Logan called Colden's attention to a "particularly appropriate work," the *Dissertatio de Gravitate Aetheris* (1683) by Jacob Bernoulli;[147] Samuel Johnson (1696–1772), Anglican leader in New England and later (1754–63) first President of King's College in New York, sent Colden the works of Bishop George Berkeley (1685–1753) in the hope that he might be won over to the Bishop's metaphysical point of view. Franklin, Lewis Evans, John Bartram, and Thomas Hutchinson (1711–80), the learned Massachusetts merchant and legislator, all saw Colden's work and none of them could comprehend it.[148] Colden was disappointed by his colonial friends' reaction but not, evidently, seriously concerned by it. Perhaps their lack of understanding only magnified his own. At any rate, he gave little attention to the books they recommended to him. And, for some reason, none of his friends appear to have been led to Newton's *Principia* and *Optics*, where they would have found the keys to Colden's baffling work.

The initial reaction from overseas was similar. Colden sent Collinson nine copies of the *Explication* (June 20, 1745). He believed, as he wrote to Collinson, that he had opened a new method for improving astronomy, navigation, and geography. He proposed to expand the work in order to encompass "an entire Theory of the Earth's motion," and he suggested that he might be entitled to a part

146 Franklin to Colden, Oct. 25, 1744, *Colden Papers*, III, 77.

147 *Ibid.*, III, 274.

148 Most of Colden's colonial friends' responses to the *Explication* are in *ibid.*, *passim.*

of the "great reward" offered by the English Parliament for the discovery of a method to determine longitude at sea.[149] An English printer quickly pirated the New York edition and published the *Explication* in London (1746). Collinson spread the treatise far and wide both in England and on the Continent. He presented a copy to the Royal Society (December 11, 1746), and after he had sent a copy to Linnaeus he wrote (October 26, 1747) that Dr. Colden presented "a new system, which he desires may be thoroughly examined." He recommended it to the attention of the Swedish Royal Society, and lamented that there was no Latin edition so that more scholars could read it.[150] However, comments on Colden's treatise were slow to materialize. European scholars appear to have been as baffled as the colonials had been. Collinson wrote (August 3, 1747) that he had received little response, although "a Certain great Mathematician amongst us" charged that Colden's *Explication* was a plagiarism which had originated in Europe; but this man, added Collinson, "is a Little touched in his pericranium." Still, Collinson held to the conviction that "some one or other will at Last Saye Something for or against It."[151] Meanwhile, translations of the *Explication* were published in German (Hamburg, 1748) and in French (Paris, 1751); and Collinson arranged for the publication of a revised and enlarged English edition under a new title, *The Principles of Action in Matter* . . . (London, 1751).

By this time, too, criticisms were beginning to be heard. One of the earliest was from Samuel Johnson of Connecticut, who pointed out to Colden (January 12, 1746/47) that two bodies in ether would have as much pressure exerted between them as from the other sides.[152] Collinson reported (August 3, 1747) that one English reader had declared that "Mr. Colden is mistaken in every part of his Conjectures."[153] German readers, said Collinson, approved of the treatise as far as page 34, but beyond that were at a loss to comprehend it; and Abraham Gotthelf Kästner (1719–1800), who made the German translation, felt that Colden's definitions of the

---

149 *Ibid.*, III, 117–120.
150 *Journal-Book*, XX, 167; Smith, I, 19–20; *Colden Papers*, III, 367–369, 371–372.
151 *Ibid.*, III, 410–412.
152 *Ibid.*, III, 331.
153 *Ibid.*, III, 412.

three forms of matter were inadequate and that his mathematical understanding was deficient.[154] Dr. John Bevis (1693–1771), F.R.S., the most prestigious English astronomer to comment on Colden's work, read it twice and consulted with a friend before he finally found so many objections that he was loath to recite them to Colden in full, and instead wrote a mild letter merely expressing his general disapproval.[155] The most devastating criticism, however, was by Leonhard Euler, the great Swiss mathematician and physicist then at the Berlin Academy. Euler wrote to the Reverend Mr. Caspar Wetstein, Chaplain to the Dowager Princess of Wales, on November 21, 1752, a letter read to the Royal Society of London (January 11, 1753), a copy of which Collinson forwarded to Colden. Euler stated that Colden's book (he had read the revised London edition) contained "many ingennious reflections," but that the author

> has acquitted himself but ill in the explanation he has undertaken to give of it; and that he discovers [i.e., betrays] his ignorance of Mechanics in his attempts to attack the best established propositions of the late Sir Isaac Newton . . . [he] shows but little knowledge of the principles of Motion and entirely disqualifies the author from Establishing the true Forces requisite to the Motion of the Planets, from whatever Course He may attempt to Derive them. Besides his explication founded on the Elasticity of Matter, is so ill managed, that it is absolutely contrary to the first principles of Hydrostaticks. What an absurdity it is to assert, that the Ether between Two of the Coelestial Bodies, has not the same Spring with that of the Rest. . . .[156]

Cadwallader Colden was deeply hurt by such criticism. He had struggled hard to revise his manuscript and never shook off the conviction that somehow he had not made himself properly understood. He protested that he had intended no attack upon Newton, was "confident that I have asserted nothing contradictory to what Sir Isaac has demonstrated but surely it may be hoped that some im-

154 Collinson to Colden, June 20, 1748, *ibid.*, IV, 67; Hindle, *The Pursuit of Science*, p. 46.

155 See Collinson's letters, March 13 and June 3, 1755, and Bevis to Colden, Aug. 10, 1755, *Colden Papers*, V, 6–7, 12–15, 22–24.

156 *Journal-Book*, XXII, 223–224; *Colden Papers*, IV, 355–357.

provement may be made in knowledge besides what he has done."[157]
He prepared a reply to Euler which Collinson laid before the Royal
Society (January 10, 1754).[158] But the Society heard the title only
and referred it to Dr. William Brakenridge for a report. Dr.
Brakenridge was a Doctor of Divinity and his report to the Society,
if ever made, was not likely to possess much weight. The Society's
action indicated the Fellows' intent to wash their hands of the
matter. But Colden persisted in his belief that his arguments were
sound, and he began to prepare a new revision for publication.
When Collinson was unable to find a publisher in England, he
turned to Edinburgh, evidently hoping that correspondents at his
alma mater would receive the manuscript with favor and find a
printer for it. Dr. Robert Whytt (1714–66), Professor of the
Theory of Medicine at Edinburgh, and his Edinburgh correspondent, disagreed with Colden's conception of the nature of light, felt
that various aspects of his hypothesis needed to be confirmed by
experiment, and reported that Dr. Matthew Stuart (1717–85), Professor of Mathematics at the University of Edinburgh, believed the
phenomena described by Colden could be explained on Newton's
principles alone.[159] At this point Colden nearly gave way to despair
and thought of burning his papers.[160] But he recovered his confidence, complained that "the English affect too much to despise all
Theory," that the spirit of the times was destructive to new learning,
and arranged with Dr. Robert Whytt and Adam Ferguson (1723–
1816), Professor of Philosophy at Edinburgh, to deposit his final
revision of *The Principles of Action in Matter*, with related papers,
in the Library of the University of Edinburgh.[161] There they remain to this day.

Colden's momentary despair in 1760 soon gave way to a renewed
confidence in his principles, and he sought to expand them to embrace the entire material world. He planned a new theory of the
motion of the moon, and persuaded his crippled son, David, to

157 Colden to Collinson, June 3, 1755, *ibid.*, V, 12–15.

158 *Journal-Book*, XXII, 452. See also Collinson's letter, Sept. 1, 1753, *Colden Papers*, IV, 404–406.

159 Whytt to Colden, Oct. 27, 1758, *ibid.*, V, 261–263.

160 Whytt to Colden, Oct. 20, 1760, *ibid.*, V, 349–358.

161 Whytt to Colden, May 16, 1763, *ibid.*, VI, 217–219; see also *ibid.*, V, 349–358; VI, 217–219, 272–274; Hindle, *The Pursuit of Science*, pp. 46–47.

write "A Supplement to the Principles of Action in Matter," whereby an attempt was made to apply his father's explanation of the cause of gravity to the cause of electrical attraction and other electrical phenomena.[162] Colden also encompassed the field of natural history and sought to explain "The Animal OEconomy . . . Mechanically according to the Laws of Matter in Motion." He prepared "An Inquiry into the Principles of Vital Motion" and brought his theory to bear upon the motion of microscopic life in fermentation, putrefaction, and sperm.[163] Most of these efforts rested on speculation, without experimentation, but Colden was better informed about natural history and medicine than he was about theoretical physics, and his explanations, though unscientific, were more reasonable than his attempt to explain gravitation.

The Principles of Action in Matter was Colden's passion until the end of his life. Indeed, he undertook but little work of a scientific nature unrelated to it. With James Alexander, he lent assistance to Lewis Evans in the preparation of the latter's *Map of Pensilvania, New-Jersey, New-York, And the Three Delaware Counties* (Philadelphia, 1749);[164] he questioned the widespread belief that waterspouts were the means by which water was drawn into clouds and correctly asserted that they were created by tornadoes;[165] he sent Collinson an account of an earthquake which shook his house badly on November 18, 1755, which Collinson presented to the Royal Society (March 18, 1756) and which was subsequently published in the *Philosophical Transactions*;[166] he corresponded with Franklin regarding the latter's experiments with electricity, although he confessed to Franklin that his notions about electricity were "confused & indigested";[167] and he continued to send plants

---

[162] Collinson reported, May 7, 1761, that John Canton (1718–72), F.R.S., the electrician, was unable to verify the electrical experiments because he could not understand Colden's principles. *Colden Papers*, VI, 31; see also *ibid.*, V, 220–227.

[163] See Whytt to Colden, March 17, 1761, *ibid.*, VI, 15–17; Hindle, *The Pursuit of Science*, p. 47.

[164] Evans to Colden, March 3, 1748/49, *Colden Papers*, IV, 107–108; Gipson, *Lewis Evans*, Chap. 2, especially pp. 18, 21.

[165] Collinson to Colden, March 10, 1754, *Colden Papers*, IV, 377–379; Colden to Collinson, May 28, 1754, *ibid.*, IV, 445–448.

[166] *Journal-Book*, XXIII, 320–321; *Phil. Trans.*, XLIX, Pt. I (1755–56), 443.

[167] Colden to Franklin, Oct. 28, 1757, *Colden Papers*, V, 207–208; see also *ibid.*, III, 236–237; IV, 314–317, 319–323, 325–328, 351–353.

and other natural history specimens to Collinson and others from time to time, although most of the work of collecting was done by his daughter Jane or by his son David.[168] In retrospect, it appears that Cadwallader Colden's most significant contributions to science lay in the field of botany, although even in this area his work scarcely equaled that of James Logan, Dr. John Mitchell, John Clayton, and Dr. Alexander Garden; and Patrick Browne's work on Jamaica excelled them all.[169] Colden's vast scientific correspondence helped to tighten the ties of the natural history circle, particularly among the scientific communities of the mainland American colonies, with all of which he was in communication. His manifold failures, especially in theoretical physics, although partially explicable in terms of his inadequate preparation and the cultural poverty of his isolated environment, were more largely attributable to his peculiar bent of mind. If he understood, he seldom applied in his work the methods of experimental science. The exalted value he placed upon creative imagination led him to speculate upon matters for which he lacked requisite preparation and to equate his speculations, too often buttressed neither by experiments of his own nor by pertinent discoveries of other scholars (with whose works he was singularly unacquainted), with the scientific syntheses of a Newton, a Galileo, or an Aristotle. Perhaps he was lazy, unwilling to undertake the tedious hours of studious preparation and experimental trial and error which scholarship necessitates. Certainly he was ambitious, anxious to obtain recognition and to have his name inscribed among those of the greatest scholars of mankind. It is easier to forgive his human thirst for fame—for without a measure of it a scholar's work is buried from the sight of his contemporaries and is essentially lost upon the world—than to overlook his arrant audacity.

John Bartram (1699–1777)

Thanks to William Darlington, who edited and published the *Memorials of John Bartram and Humphrey Marshall* . . . (Phila-

---

[168] David Colden engaged in electricity experiments and in 1768 sent Collinson an account of his observations regarding the supposed power of snakes to charm their prey. Collinson to Colden, July 2, 1768, *Colden Papers*, VII, 142–144.

[169] See Chapter 9 above, pp. 376–378.

delphia, 1849), the life and letters of John Bartram and his cousin, Humphrey Marshall, have been known to scholars for well over a century. In consequence, the place of John Bartram in the saga of early American science has become better known than that of several of his contemporaries, some of whom made greater contributions to science in the colonial era than Bartram himself. Still, Bartram exhibited a rustic charm, an integrity, and a scientific insight remarkable for a man of limited education; and because of his ability, his agreeable personality, and his wide travels in the colonies, he came to be personally acquainted with a larger number of his colonial contemporaries with scientific interests than anyone of his generation. His scientific correspondence, both in the colonies and in Europe, compared favorably with that of most of his colonial contemporaries, and his travels in the mainland colonies of North America, in the interests of botany and natural history, were excelled by none.

Born of humble Quaker parentage at Darby, Pennsylvania, John Bartram was orphaned at an early age by the death of his mother and the migration of his father to North Carolina. He was reared by his grandmother and his uncle, Isaac Bartram, from whom he inherited a small farm at Darby. He rose to some prominence in the Quaker Meeting at Darby, married, had two sons, one of whom died in infancy, and his wife died in 1727. The next year Bartram purchased a farm along the Schuylkill River, about three miles from Philadelphia, enlarged a stone house on it with his own hands, remarried (in 1729), and lived there for the remainder of his life. Here, too, he laid out a botanic garden of about five acres which in time became the most famous of its kind in America.

John Bartram prospered as a farmer, enlarged his lands by further purchases, and considered himself a farmer all his life, with the bucolic pride and a touch of inverse snobbery that often infects the tiller of the soil. His interest in botany may have sprung from an early concern for medicine and medicinal plants. The traditional story is that, as he was resting from the plow one day, he idly plucked a daisy, grew interested in its intricate structure, and resolved to undertake the study of flowers and plants. He purchased some botanical books and a Latin grammar and began his study. Soon his dedicated interest stirred Philadelphians to come to his assistance. Dr. Christopher Witt (1675–1765), a skillful physician and a learned man—though he held theosophic views and a reputation as a conjurer—had a fine garden, a botanical exchange with Peter Collinson, and a generous impulse to share his knowledge, his books, and his garden with Bartram. Isaac Norris (1671–1735) and his son of the same name (1701–66), merchants and prominent Quaker politicians, had an estate called Fair Hill with many unusual plants with which Bartram became familiar. But in the early botanical education of John Bartram, James Logan and Joseph Breintnall were of greater importance. Logan, already noted as a scholar, bibliophile, and botanist, had a garden at Stenton, and he extended generous aid. He lent Bartram books from his library, including such early works as those of Nicholas Culpeper, William Turner, and others, and in 1729 he presented Bartram as a gift a copy of John Parkinson's *Paradisi in Sole Paradisus Terrestris*.[170] Logan also assisted Bartram in reading the Latin of Linnaeus and instructed him in the use of a microscope. Joseph Breintnall sometimes had business dealings with Peter Collinson, and when Collinson implored Breintnall to collect and exchange plants, seeds, and botanical information, Breintnall recommended Bartram for the purpose, thus opening a rich and lasting relationship between Bartram and the English promoter. It was probably Breintnall, too, who, as a member of Benjamin Franklin's early junto, introduced Bartram to the rising printer and publisher who later became the colonies' greatest scientist. Obviously, by the early 1730's John Bartram was beginning to take an active part in the intellectual life of Philadelphia. He read widely, and though his spelling and grammar were occasionally imperfect, he developed an English literary style which reflected pseudo-classical elements derived from his love of literature and was especially vivid in matters of description. He never really mastered Latin, in spite of Logan's tutelage. His interest in secular learning gradually altered his religious outlook. Like his contemporary, the Baptist preacher Ebenezer Kinnersley, Bartram rejected evangelical Christianity and adopted the rational outlook of the Enlightenment. Nominally he remained a Quaker all his life, retaining the Quaker forms of per-

---

170 For Culpeper, Turner, and Parkinson, see above, pp. 16–17, 58–60, 64–65, respectively.

sonal address, the aversion to extravagance in dress and manner of living; but in 1757 the Quaker Meeting at Darby disowned him for "disbelieving in Christ as the Son of God." Bartram's God had become the God of nature, his religion the natural religion: "It is through the telescope I see God in his glory," he wrote. Quaker orthodoxy had given way to deism; John Bartram the Quaker had become John Bartram the rational scientist.

If Bartram's early botanical education owed much to the scientific community of Philadelphia, his career as a botanical collector and gardener developed under the patronage of Peter Collinson. The two began correspondence in the early 1730's and continued until Collinson's death in 1768. Though they never met personally, their correspondence assumed an intimacy which bespoke rare mutual respect, trust, and friendship. Collinson, in particular, came to adopt an almost paternalistic attitude toward Bartram, to whom he appears to have attributed both a degree of poverty and of childlike innocence which the prosperous farmer and shrewd collector scarcely deserved. Besides supplies for collecting seeds, plants, insects, and other items of natural history, books for his library, a pocket compass for his travels, a microscope for his observations, and seeds and plants for his garden, Collinson also sent Bartram clothing for himself, "a Calico gown for thy wife and some odd little things that may be of use amongst the children and family," occasional advice about means better to please his English patrons, and introductions to members of the international natural history circle on both sides of the Atlantic. When Bartram proposed to make a botanical expedition into Virginia in 1738, Collinson gave him detailed instructions to call upon Colonel John Custis, William Byrd II, and Isham Randolph, sent letters to each of them to introduce Bartram, and added the following personal advice:

> One thing I must desire of thee, and do insist that thee oblige me therein: that thou make up that drugget clothes, to go to Virginia in, and not appear to disgrace thyself or me; for though I should not esteem thee the less, to come to me in what dress thou will,— yet these Virginians are a very gentle, well-dressed people—and look, perhaps, more at a man's outside than his inside. For these and other reasons, pray go very clean, neat, and handsomely

dressed, to Virginia. Never mind thy clothes: I will send more
another year.[171]

In this, as well as in other references to Bartram, Collinson betrays
an image of his colonial friend that smacks of the effete European's
view of the "natural man" in the American wilderness. If Bartram
ever resented Collinson's attitude, he never mentioned it. He accepted the gifts with humble thanks, performed his tasks as a collector with industry and skill, grew in knowledge under his patron's
generous and considerate care—and profited from the subscriptions
which Collinson collected to support his efforts.

Bartram began making collections for Collinson in the early
1730's. Almost from the outset, it was a two-way scientific exchange, for Collinson sent seeds and plants from his own garden,
from Philip Miller, and from other English and Continental
sources, thereby enriching Bartram's garden with many exotic
trees, shrubs, and plants. Collinson also enriched Bartram's botanical
knowledge. For "thy own improvement in the knowledge of plants
[he told Bartram on January 25, 1735/36], thou shalt send me
another quire of duplicates of the same specimens; I will get them
named by our most knowing botanists, and then return them again,
which will improve thee more than books; for it is impossible for
any one author to give a general history of plants."[172] Moreover,
Collinson distributed Bartram's collections to Lord Petre, Sir Hans
Sloane, and others who paid Bartram for his collections. In 1736 it
was proposed that Lord Petre make Bartram an "annual allowance"
of ten guineas and Collinson set about to enlist other English
subscribers. He wrote to Bartram (March 12, 1735/36):

This, we think, will enable thee to set apart a month, two, or
three, to make an excursion on the banks of the Schuylkill, to
trace it to its fountain. But as so great an undertaking may require

---

[171] Collinson to Bartram, Feb. 7, 1737/38, in Darlington, *Mems.*, pp. 88–89.
Collinson repeatedly sent clothing and gifts to Bartram and his family. See *ibid.*,
pp. 63, 69, 72–73, 95, 147–150, 150–153, 170–171. In 1755 Collinson warned Bartram:
"Please give nobody a hint, how thee or thy wife come by the suit of clothes.
There may be some, with you, may think they deserve something of that nature."
Ibid., p. 70.
[172] *Ibid.*, pp. 63–64.

two or three years, and as many journeys, to effect it, so we must leave that wholly to thee. But we do expect, that after harvest, and when the season is that all the seeds of trees and shrubs are ripe, thou will set out; and them that happen not to be ripe when thou goes, they may have attained maturity when thou comes back. We shall send thee paper for specimens and writing, and a pocket compass,—expect the'll keep a regular journal of what occurs every day, and an exact observation of the course of the river, which, with a compass, thee may easily do. . . .[173]

Thus was John Bartram set upon his travels. Collinson found many different subscribers as years passed, collected the money (ultimately set at five guineas a box), and forwarded it to Bartram. In this manner John Bartram made natural history become a source of income. The demands of his patrons sometimes became exacting, and they were almost immediately enlarged to embrace fossils, birds, snakes, insects, animals, and so on until they included the whole gamut of natural history.[174]

Indeed, although John Bartram's reputation came to rest upon his botanical work, his earliest publication referred to his dissection of a rattlesnake found near Germantown. He wrote to Collinson that in dissecting the rattlesnake

I met with what has not been observed before by any that I can remember: that is a Cluster of Teeth on each Side of the upper Jaw, at the root of the great Fangs, thro' which the Poison is ejected. I observed in the same case that the two main Teeth were sheath'd, in which lay four others at the root of each Tooth in a cluster together, of the same Shape and Figure with the great ones; and I am apt to think for the same use and purposes: if by accident the main Teeth happen to be broken; as was the fellow to this I send you. . . . I am not certain whether this is an uncommon case. Perhaps others have not dissected the head of this Animal with the same care that I have done. I wish you would make Inquiry about it. . . .[175]

[173] *Ibid.*, pp. 72–73. In a letter of Nov. 3, 1735, Bartram had "modestly" proposed that he be paid for his "pains and trouble" in collecting. Not all of the patrons, however, subscribed as much as ten guineas a year.

[174] See *ibid.*, pp. 67–68, 73, 76–77, 81–82.

[175] *Letter-Book*, XXII, 94–95; read Dec. 18, 1735, *Journal-Book*, XVI, 219–220.

Collinson presented the above letter to the Royal Society, and it appeared in the *Philosophical Transactions* later on.[176] Bartram's dissection of the rattlesnake impelled Collinson to ask Bartram's opinion about the alleged capacities of rattlesnakes to charm their prey.[177] Collinson referred to Sir Hans Sloane's recent experiments which appeared to deny the popular myth,[178] but William Byrd II had insisted upon the reality of the snake's powers to charm, and Bartram's friends, Joseph Breintnall and Christopher Witt, supported Byrd's contention.[179] On the other hand, Dr. John Kearsley of Philadelphia upheld Sloane's view, and reported that an Indian named Taughtahanah, "said to be a fellow of good Sense," also denied that rattlesnakes possessed the power to fascinate their prey.[180] Bartram upheld the popular view espoused by Byrd, Breintnall, and Witt.[181] Obviously, opinion on this subject was divided in the Philadelphia community. The discussion also embraced the question of the value of snakeroot in the treatment of snake-bites, and the Philadelphians divided along similar lines. Breintnall and Witt praised the virtues of snakeroot, and Kearsley denied its usefulness; but Bartram offered no opinion.[182]

[176] XLI, Pt. I, No. 456 (Jan.–June, 1740), 358–359.

[177] Collinson to Bartram, Sept. 20, 1736, in Darlington, *Mems.*, pp. 81–82.

[178] See p. 291 above.

[179] See Christopher Witt's letters to Collinson, Nov. 11, 1735, Sept. 27, 1736, *Letter-Book*, XXII, 187–194; XXIII, 181–183; *Journal-Book*, XVI, 259–260; XVII, 47–48; Breintnall to Collinson, Nov. 3, 1735, *ibid.*, XVII, 44–46.

[180] Kearsley to Collinson, March 21, 1736/37, *Letter-Book*, XXIV, 319–325.

[181] Bartram to Collinson, Feb. 27, 1737/38, *ibid.*, XXV, 118–121; *Journal-Book*, XVII, 339–340.

[182] Witt to Collinson, March 24, 1738/39, *Letter-Book*, XXVI, 18–21; *Journal-Book*, XVII, 474–475. Breintnall was less positive, and when he was bitten by a rattlesnake in 1746, he did not resort to snakeroot for the cure. *Ibid.*, XX, 79–81; Kearsley to Collinson, March 21, 1736/37, *Letter-Book*, XXIV, 319–325; *Journal-Book*, XVII, 251–252, 480–481. Kearsley also questioned the medicinal values of American ginseng, but he praised the virtues of scammony and gentian. These men also sent other scientific data to Collinson about this time, all of which Collinson took to the Royal Society. Witt described the severe winter of 1733–34: *Letter-Book*, XXI, 307–308; *Journal-Book*, XVI, 107–108; Kearsley sent observations of a comet (Jan. 27–Feb. 21, 1736/37), and a solar eclipse (Feb. 18, 1736/37) witnessed at Philadelphia and observed "by a reflecting telescope of Mr. Hadley's make": *Letter-Book*, XXIII, 307–309; *Journal-Book*, XVII, 96–97; Brientnall sent specimens of ginseng, insisting that it was identical to that from China, reported observations on an aurora borealis observed at Philadelphia (Dec. 29, 1736), a

As his interest was diverted from botanical matters to other aspects of natural history, Bartram reported the annual appearance in Pennsylvania of vast numbers of caterpillars (tent caterpillars) which moved from west to east to "devour whole Woods before them," of great flocks of pigeons (carrier pigeons), and of bears— all indications to Bartram that there must be "very great Forests and a fertile Country to the Westward, that can maintain & support so many Millions of Pigeons (besides other Animals)."[183] He also noted the periodicity of the appearance of locusts in Pennsylvania, sent specimens to Collinson, and agreed with Paul Dudley's description of them. Similar accounts received from Christopher Witt and Dr. John Kearsley led Collinson to prepare "Some Observations on the Cicada of North America," later published in the *Philosophical Transactions*.[184] Bartram also became interested in wasps. In 1737 he sent Collinson "An Account of some very curious Wasps Nests made of clay in Pensilvania," with specimens of the nests and a description of the generation of the wasps; the whole was published, with drawings, in the *Philosophical Transactions* for 1745.[185] Bartram sent further specimens of the mud wasps in 1745, a large black wasp in 1749, a dragonfly in 1749, and an "Account of May-

---

meteor (Nov. 19, 1737), and two earthquake shocks (Dec. 7, 1737): *Letter-Book*, XXV, 111–112; *Journal-Book*, XVII, 332–334. The astronomical observations were published in *Phil. Trans.*, XLI, Pt. I, No. 456 (Jan.–June, 1740), 359–360. Breintnall also reported some experiments he had made to show that the sun's rays were hotter in summer than in winter and that the sun's rays penetrate colored more than white materials. See Stephen Bloore, "Joseph Breintnall, First Secretary of the Library Company," *Pennsylvania Magazine of History and Biography*, LIX (1935), 42–56. How much Bartram was associated with these observations and experiments is unclear, but it appears likely that he was acquainted with them and with the cross-currents of scientific opinion in Philadelphia.

183 Bartram to Collinson, April 26, 1737, *Letter-Book*, XXV, 148–151; *Journal-Book*, XVII, 370.

184 LIV (1764), 65–68, with illustrations. See letters of Kearsley (March 19, 1736/37), Bartram (April 26, 1737), and Witt (Sept. 16, 1737) to Collinson, *Letter-Book*, XXV, 112–115; *Journal-Book*, XVII, 332–334; XXVI, 51–53. Later accounts from Bartram and Cadwallader Colden, in 1748, with specimens, led Collinson to distinguish between the locust and the cicada. *Ibid.*, XXI, 77, 348– 349. Bartram stated that the locusts were no plague upon the country, but rather a benefit, as they afforded food for fowls, beasts, and Indians, "who boil them, after they have plucked off their wings."

185 XLIII, No. 476 (April–July, 1745), 363–366. See also *Journal-Book*, XIX, 400–401; *Sloane 4055*, fol. 259.

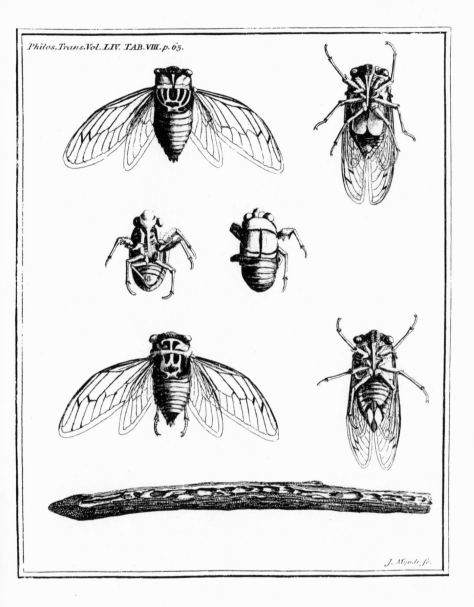

J. Mynde. sc.

27. Cicadas, with twig showing punctures made by the females
in depositing their eggs, supplied by John Bartram
to Peter Collinson for the latter's article,
"Some Observations on the Cicada of North America."

Phil. Trans., LIV (1764), 65.

28. George Edwards' picture of the ruffed grouse supplied
to him by John Bartram in 1750 (see p. 523 above).

Phil. Trans., XLVIII (1754), 499.

flies in Pennsylvania" in 1750, all of which were reported to the Royal Society by Collinson.[186] Also, on February 24, 1763, Collinson read to the Royal Society "Observations made by Mr. John Bartram of Pensulvania, on the yellowish Wasp of that Country," a paper published in the *Philosophical Transactions* in the same year.[187] In the meantime, on July 4, 1742, Bartram had sent to Collinson a variety of observations regarding the function of the "fibrous roots" of salt-marsh mussels, the nature of oysters, and an account of a hummingbird he had tamed so that it would eat from his hand—all of which, omitting the hummingbird, was published, with figures, in the *Philosophical Transactions* for 1744.[188]

But Bartram's principal interest lay in botany. He made a brief expedition, probably in the spring of 1736, to the Rattlesnake Mountains, and in the autumn of the same year he traced the course of the Schuylkill River, as proposed by his English patrons. He also kept a journal of his observations and prepared a map, as he had been instructed. Collinson was delighted with both: "now I can read and travel at the same time," he said; and Lord Petre was equally well satisfied.[189] The map, "very prettily done," as Collinson said, was probably the earliest one to depict the region beyond the Blue (Kittatinny) Mountains, and Bartram made a side trip to Crystal Cove, near Reading, and probably visited Conrad Weiser, James Logan's friend. In the course of the next thirty years John Bartram made similar botanical excursions through the North American colonies. Repeatedly he visited nearby counties in Pennsylvania, New Jersey, Delaware, and Maryland. In 1738 he traveled about 1,100 miles into Virginia, stopped over at Williamsburg, visited William Byrd II, Colonel John Custis, Isham Randolph, and called on John Clayton, who unfortunately was not at home at the time, and returned by way of the Virginia back country. In 1742 he traveled through East and West Jersey, up the Hudson River to the falls in the Mohawk River, climbed the Cats-

---

[186] *Journal-Book*, XIX, 400; XXI, 213–214, 232, 305–306.

[187] LIII (1763), 37–38. See also *Journal-Book*, XXV, 30–31.

[188] LXIII, No. 474 (June–Dec., 1744), 157–159; *Journal-Book*, XIX, 292–293. Bartram stated that the salt-marsh mussel partakes of a vegetable nature in that it has roots in the soil, possibly for nourishment and to keep it from being carried away by tides, yet it is an animal.

[189] Collinson to Bartram, Dec. 14, 1737, in Darlington, *Mems.*, p. 104.

kills, and visited Cadwallader Colden at Coldengham, where the two men found each other mutually stimulating and attractive. The next year he joined Conrad Weiser and Lewis Evans in an expedition to Onondaga in New York, making a side trip to Fort Oswego on Lake Ontario. His journal of this expedition (one of three, for both Weiser and Evans left journals as well) was subsequently sent to Collinson, who published it in London as *Observations on the Inhabitants, Climate, Soil, Rivers, Productions, Animals, and Other Matters Worthy of Note, Made by Mr. John Bartram in His Travels from Pensilvania to Onandaga, Oswego and Lake Ontario in Canada* (1751). He returned to New York in 1744 and made a trip up the Susquehanna River in 1745. But the Indian War made travel hazardous for the next few years. In 1754 he took his four-teen-year-old son, William, on a trip to New York. They visited the Coldens, where they met Dr. Alexander Garden, recently from Scotland, who subsequently settled at Charles Town, where he soon became a naturalist almost as well known as Bartram himself. In 1755 Bartram took "Billy" on a trip to Connecticut to visit Jared Eliot. Five years later he went to South Carolina, visiting John Clayton in Virginia en route. At Charles Town he renewed his acquaintance with Dr. Alexander Garden, met Mrs. Martha Logan, author of the first American book on gardening, praised her garden, and began a mutual exchange of seeds. In 1761 he accompanied Colonel Henry Bouquet, who had distinguished himself under General John Forbes in the capture of Fort Duquesne from the French (1759), to Pittsburgh, and made a three-day voyage down the Ohio River. The next year he explored the interior of South Carolina, returning by way of the Shenandoah Valley. By this time he had collected seeds, plants, fossils, and some fauna, as he said, from Nova Scotia to Georgia. But he still longed for new regions to explore.

After the Peace of Paris in 1763, at which Great Britain acquired East and West Florida from Spain, Bartram wrote to Collinson that "I should be exceedingly pleased if I could afford it, to make a thorough search, not only at Pensacola, but the coast of Florida, Alabama, Georgia, and the banks of the Mississippi."[190] Bartram

[190] March 4, 1764, *ibid.*, pp. 261–262.

hoped for a royal appointment as King's Botanist for America.[191] Collinson brought forth all his powers of persuasion at court, but William Young, a relatively unknown German from Philadelphia, by means of some devious negotiations, won the title of Botanist to the King and Queen in 1764, at £300 per year, thereby throwing Bartram and his friends into momentary consternation. But Franklin and Collinson renewed the plea for Bartram, and in 1765 Bartram was appointed Royal Botanist—but with the miserly stipend of fifty pounds per year.[192] In June, despite his sixty-four years, Bartram set off for Florida. His son William, who had been in business at Cape Fear, North Carolina, joined him in Charles Town, and together they made their way to St. Augustine in Florida. From St. Augustine they traced the course of the St. John's River southward to its sources, and prepared a map of the stream with its principal tributaries. John Bartram kept a journal which, published as a *Description of East Florida, with a Journal kept by John Bartram of Philadelphia, Botanist to His Majesty for the Floridas upon a journey from St. Augustine up the River St. John's, as far as the Lakes*, passed through three London editions between 1766 and 1769.[193] However, this was John Bartram's last botanical expedition. Shortly after his return to Philadelphia, his old patron and friend, Peter Collinson, died (1768), and Bartram's health was declining. In 1771, as his sight failed, he turned over most of his business affairs to his son John, and in 1777 he died. He left his estate and the care of his garden to John, Jr., although William joined his brother in its oversight in 1778, and together they made over the famous botanic garden into a nursery.[194]

191 Bartram to Collinson, Oct. 15, 1764, *ibid.*, pp. 266–267.

192 Collinson to Bartram, April 9, 1765, *ibid.*, pp. 268–269. Bartram still could count upon subscriptions from private patrons. Collinson said, "At the same time that thou art collecting seeds for the King, where thou finds plenty, thou may think on me and thy other correspondents."

193 Collinson was present when Bartram's first box of plants was opened for the King. Collinson to Bartram, Feb. 17, 1768, in Darlington, *Mems.*, p. 298. But Collinson died soon afterward and, at Bartram's request, Franklin took charge of the presentation of Bartram's next box. After the American Revolution began, Franklin requested Bartram to send his collections, if any, to Paris, for the French gardens. See Franklin to Bartram, May 27, 1777, *ibid.*, p. 406. But Bartram died (Sept. 22) before any shipments could be made.

194 I omit further attention to William Bartram (1739–1823) because his

John Bartram was peculiarly the product of scientific promotion, although his own personal capacities for growth led him to become a widely respected figure in the international natural history circle. Members of the Philadelphia scientific community launched him upon an extraordinary career of self-education, and Peter Collinson enabled him to expand his vision by an equally extraordinary series of expeditions of exploration, discovery, and collection, and by scientific correspondence and exchange abroad. To the early assistance of James Logan and Joseph Breintnall in Philadelphia must be added the somewhat later encouragement of Benjamin Franklin. In March, 1742, Franklin published in his *Pennsylvania Gazette* "A Copy of the Subscription Paper for the Encouragement of Mr. John Bartram," to make "an annual Contribution for his Encouragement" in the collection of "the Production of Nature in general." Already, the item stated on March 10, 1741/42, "*Near 20 £ a Year is already subscribed,*" at the post office in Philadelphia.[195] As Bartram's principal English patron, Lord Petre, died in the following July, this local financial assistance helped greatly to support Bartram's travels until Collinson could find new English subscrib-

---

mature career largely took place after the Old Colonial Era. His remarkable talents as an illustrator of natural history, which led Collinson (May 12, 1756) to compare him to Ehret because "his performances are so elegant" (Smith, I, 39), soon attracted the attention of George Edwards. His contributions to Edwards' work have been noted above. But William Bartram was a dreamer and a poet, and when his father sought to establish him in a business that would insure a livelihood, he failed repeatedly. After John Bartram returned from East Florida to Philadelphia in 1766, William stayed behind, determined to try his hand as an indigo planter. Again he failed. Then Dr. John Fothergill, who had admired William's illustrations of natural subjects, employed him to collect plants in Florida and to make drawings of the flora and fauna there. Between 1773 and 1777, William traveled widely in East and West Florida, Carolina, and Georgia, collecting, making drawings, and preparing the materials out of which he later produced his minor American classic, *Travels through North & South Carolina, Georgia, East & West Florida . . .* (Philadelphia, 1791). There are many later editions, both in Europe and America. Some of his illustrations for Dr. Fothergill have recently been published by Joseph Ewan, ed., *William Bartram: Botanical and Zoological Drawings, 1756–1788. Reproduced from the Fothergill Album in the British Museum* (Memoirs of the American Philosophical Society, Vol. LXXIV, Philadelphia, 1968). See also Nathan Bryllion Fagin, *William Bartram: Interpreter of the American Landscape* (Baltimore, 1933).

[195] Quoted in Earnest, *John and William Bartram*, p. 25.

ers.[196] Again, in 1743, Franklin persuaded the directors of the Library Company of Philadelphia to present Bartram with free access to the books in the library—a gift amounting to the cost of a share in the company (forty shillings) and annual dues (ten shillings). At about the same time, Bartram himself helped to establish a similar library at Darby, his old home, and persuaded Collinson to act as its purchasing agent. Later still, after Franklin had become Deputy Postmaster General (1751), he extended franking privileges to Bartram for his correspondence.[197] Indeed, over the long run, Bartram had no more helpful colonial friend than Benjamin Franklin.

Collinson's services to Bartram, in addition to those already noted, added further to Bartram's opportunity for scientific growth. He sent books, information regarding new scientific developments abroad, and introduced him not only to the Royal Society of London, but also to other scientists in the colonies and in Europe. Bartram's Philadelphia associates, especially Logan and Franklin, extended his colonial friendships somewhat—to Conrad Weiser and Jared Eliot, for instance. But it was Collinson who recommended him to the Virginia community of scientists and gardeners whom Bartram visited in 1738 and later.[198] Likewise, it was Collinson who urged Bartram to visit Cadwallader Colden, and at Coldengham he met Captain Rutherford and Dr. Alexander Garden. The latter subsequently extended Bartram's exchange to Charles Town, where he also corresponded with Dr. Lionel Chalmers and others. These acquaintanceships ripened, and Bartram entertained at his farm from time to time, besides many of his Philadelphia friends, Colden and his sons, Captain Rutherford, Dr. John Mitchell, Dr. Alexander Garden, and John Clayton, all of whom admired his botanical garden and exchanged scientific data and plant specimens. When Peter Kalm visited Philadelphia, James Logan complained that he saw no one but Franklin and Bartram.

196 See Collinson to Bartram, July 3, 1742, in Darlington, *Mems.*, pp. 157–158. "*All our schemes are broke*," cried Collinson; "he that gave motion is motionless, —all is at an end."

197 *Ibid.*, pp. 179–180.

198 See "Letters of John Clayton, John Bartram, Peter Collinson, William Byrd, Isham Randolph," *Wm. & Mary Qty.*, 2nd Ser., VI (1926), 303–325.

Similarly, thanks to Collinson's promotions of Bartram's work in Europe, Bartram's scientific correspondence and exchange was enlarged to include Mark Catesby, J. J. Dillenius at Oxford and his successor, Dr. Humphrey Sibthorp, Sir Hans Sloane, Philip Miller, Daniel Carl Solander, Dr. John Fothergill, Dr. John Hope, Professor of Medicine and Botany at Edinburgh, J. F. Gronovius at Leiden, and Carl Linnaeus at Uppsala. With Linnaeus, Bartram received little in exchange, for, as Collinson said, "It is a general complaint that Dr. Linnaeus receives all, and returns nothing."[199] But he won the high respect of Linnaeus, who was said to have referred to Bartram as "the greatest natural botanist in the world." In 1739 Dr. Thomas Bond of Philadelphia presented some of Bartram's dried plants to Antoine de Jussieu, Professor at the Jardin des Plantes in Paris, and Bartram's work became known both to Jussieu and to the Director of the Paris garden, Buffon. Franklin later introduced Bartram to T.-F. D'Alibard, a scientist at Paris, and during the War of Austrian Succession and the Seven Years' War, French scholars received some of Bartram's collections which had been captured at sea. Bartram took the broad view of the philosopher of mankind, common to the Enlightenment, that he did not mind if his collections fell into the hands of so-called enemies so long as they made scientific use of them and did not heave them into the sea. Accordingly, during the wars, he sometimes addressed shipments intended for Gronovius or Collinson to Jussieu, D'Alibard, or Buffon, with instructions to forward them to Gronovius, who, from his location in Leiden, might be able, if called upon, to deliver them in England. Actually, too, the French did send some captured collections to Gronovius, but they also retained some for themselves.[200]

Bartram received several tokens of appreciation from his European correspondents. Mark Catesby, in gratitude for Bartram's collection of specimens needed to complete his *Natural History of Carolina*, presented Bartram with sections of his book as they appeared serially from the press.[201] Dillenius sent several gifts, includ-

[199] Collinson to Linnaeus, March 27, 1747/48, in Smith, I, 17–18.

[200] Darlington, *Mems.*, pp. 169–170, 316, 352–354.

[201] *Ibid.*, pp. 128–130, 319–320; *Bartram Papers*, I, 21, 24, 97 (Pennsylvania Historical Society Library, Philadelphia); *Gratz Mss.* (H.S.P.), Case 12, Box 6

ing Philip Miller's *Gardener's Dictionary* and his own *Hortus Elthamensis* . . . (Oxford, 1732), which Bartram praised as "the completest of that kind that ever was wrote."[202] Sir Hans Sloane, in return for Indian artifacts, insects, and dried plants, sent Bartram his two-volume natural history of Jamaica, the works of James Petiver, and five guineas with which Bartram asked Collinson to purchase a silver cup, "which I or mine may keep to entertain our friends withal, in remembrance of my noble benefactor."[203] J. F. Gronovius, who carried on a lively exchange of fossils and insects with Bartram, sent Bartram his *Flora Virginica* and several of Linnaeus' works, including the *Systema Naturae*. Their literary exchange, however, was sometimes difficult for Bartram, who often sent Gronovius' Latin descriptions to Collinson for translation, and begged Gronovius to write in English, "which I can understand much better than *Latin*, which is troublesome to us to understand your sentiments."[204] Bartram's friends remembered him in other ways as well: in Stockholm they elected him to the Royal Academy of Science in 1769, and the Edinburgh Society of Arts and Sciences sent him a gold medal in 1772.[205]

John Bartram was a superb collector, naturalist, and explorer, and his mind came to encompass an almost encyclopedic knowledge of the natural history of America.[206] But his lack of concern for system in the study of natural history disqualified him from becoming either a botanist or a zoologist in the strict meaning of the terms. Both his letters and the journals kept of his travels demonstrate a remarkable capacity for close observation of details with reference to plants and animals, fossils, geology, soils, topography, and Indians. Despite his lack of formal education, he developed a literary style of charm and clarity, granting his occasional slip in orthography and grammatical construction: "Good grammar and

---

(*idem*). I am indebted for these manuscript references to Professor George Frederick Frick, of the University of Delaware.

[202] Darlington, *Mems.*, pp. 160–161.

[203] *Ibid.*, pp. 163, 302–305. The cup was engraved, "The Gift of Sr Hans Sloane Bart. To his Frd John Bartram Anno 1742." See also *Sloane 4019*, fols. 130–131; *4057*, fols. 56, 157, 174; *4069*, fol. 90; *Colden Papers*, III, 189–190.

[204] Darlington, *Mems.*, pp. 352–354; *Colden Papers*, III, 209–211.

[205] Darlington, *Mems.*, pp. 405, 444–446.

[206] See Peter Kalm's comments quoted in Earnest, *John and William Bartram*, pp. 70–71.

good spelling [he wrote to Collinson], may please those that are more taken with a fine superficial flourish than real truth; but my chief aim was to inform my readers of the true, real, distinguishing characters of each genus, and where, and how, each species differed from one another, of the same genus."[207] He held a rational, highly practical, no-nonsense approach to the study of nature, generally devoid—excepting the matter of rattlesnakes' power to charm their victims—of superstition. Like Catesby and Collinson, he dismissed the widespread belief that swallows and perhaps some other birds hibernate in caves or hollow trees or beneath the waters of ponds. He disputed the point with both Peter Kalm and Linnaeus, and with Collinson argued that if the theory were correct, swallows must possess special organs to enable them to breathe under water.[208] He prepared a Preface and notes to Franklin and David Hall's edition of Thomas Short's *Medicina Britannica* which amounted to an early American pharmacopoeia, and he prepared "An Essay for the improvement of estates, by raising a durable timber for fencing and other uses" which was published in Franklin's *Poor Richard Improved* for 1749. He appears to have done little by way of experimentation, although in 1739, evidently at the request of Gronovius, he followed up the experiments in the hybridization of plants formerly reported by Cotton Mather, Paul Dudley, James Logan, and others. He employed a microscope, gave a careful description of the sex organs of plants, and attempted to determine the precise time when female flowers were "in heat." In a letter to William Byrd II he described his experiments as follows:

> I have this spring made several microscopical observations upon the malle and femall parts in vegetables to oblige some ingenious botanists in Leyden, who requested that favour of mee which I have performed to their satisfaction and as a mechanicall demonstration of the certainty of this hypothesis of the different sex in all plants that hath come under my notice. I can't find that any vegetable hath power to produce perfect seed able to propogate without the conjunction of malle seed any more than animals and by a good microscope the malle and femall organs is plainly dis-

[207] Nov. 3, 1754, in Darlington, *Mems.*, p. 196.
[208] See Frick and Stearns, *Mark Catesby*, pp. 63–64; Earnest, *John and William Bartram*, p. 76.

covered. I have made several successful experiments of joyning several species of the same genus whereby I have obtained curious mixed Colours in flowers never known before but this requires an accurate observation and judgment to know the precise time when the femall organs is disposed to receive the masculine seed and likewise when it is by the masculin organs fully perfected for ejection. I hope by these practical observations to open a gate into a very large field of experimental knowledge which if judiciously improved may be a considerable addition to the beauty of the florists garden.[209]

Bartram produced a hybrid, or "mule," a flesh-colored lychnis (rose campion or "ragged robin"). But his account is disappointing in its lack of details, and in spite of the "very large field" for further experimentation, there are no records that Bartram continued in it.[210]

Although Bartram never published his thoughts in any organized fashion, he engaged in philosophical reflection about the mysteries of nature. He was careful to observe the ecology of plants and animals and, like many others of the eighteenth century, subscribed to the "great chain of being" as an a priori concept of nature study.[211] His letters repeatedly reveal his awareness of the interconnectedness of things in nature, and of "a universal balance mentained tho matter seems never at rest but always in A state of contraction or expantion."[212] He speculated with Lewis Evans and others about geological matters, postulated that there may "be vast chains of mountains in the sea, as well as on land," the tops of which may be sand banks or islands, and if soundings were made to discover them, perhaps profitable new fishing banks might be found.[213] He held, contrary to the notions of his day, that limestone and marble were formed in the earth by a mixture of slime, or mud, and the nitrous or marine salts therein. And in 1756 he proposed to Dr. Alexander Garden an idea for the construction of a geological map substantially the same as that set forth by Dr. Martin Lister in

[209] Quoted in *ibid.*, pp. 81–82.

[210] Zirkle, *The Beginnings of Plant Hybridization*, pp. 144–146.

[211] See Ralph C. Ritterbush, *Overtures to Biology: The Speculations of Eighteenth-Century Naturalists* (New Haven, 1964), pp. 11–12 and *passim*.

[212] *Bartram Papers*, I, 42; Earnest, *John and William Bartram*, p. 64.

[213] Bartram to Gronovius, Nov. 30, 1752, in Darlington, *Mems.*, pp. 359–360.

1684—and still nearly fifty years before the first geological map was made in Europe.[214] He set forth a practical plan for mounting the bones of a mastodon found by Christopher Gist and George Croghan at Big Bone Lick in Kentucky and shipped (1767) to Benjamin Franklin and the Earl of Shelburne in London. These "elephant bones" excited much controversy in European scientific circles for several years, centering about Peter Collinson's initial contention (backed by William Hunter) that the bones were the bones of elephants, although the grinders (teeth) were not. The dispute raged beyond the American War for Independence, but some of the problems might have been simplified if John Bartram's advice had been followed in 1768.[215]

However, John Bartram's principal contributions to science lay in his collections, his garden, and his scientific correspondence. By means of his wide correspondence he became a great broadcaster of scientific views in the colonies while his garden and constant exchanges of seeds and plants enriched both the gardens and the knowledge of other colonials. He was the field agent par excellence of the entire colonial era, and his letters and shipments abroad similarly enriched the gardens and knowledge of Europeans. His collections inspired other colonists to emulate his example, and before his death the Proprietor's gardener, James Alexander (not to be confused with Cadwallader Colden's friend of the same name, who died in 1756), Thomas Lees, William Young (who was appointed Botanist to the King and Queen in 1764), and Bartram's younger cousin, Humphrey Marshall, long a close correspondent with Dr. John Fothergill, all had become collectors of colonial flora and fauna. Alexander, of course, had the fine garden of the Penns to work, and Marshall created an excellent one at Marshallton in Chester County nearby. No longer could Bartram complain, as he had to Collinson and Catesby years before, that "Our Americans have very little taste for these amusements." Yet with all his abilities to enlist the interests of others, with all his powers of close observa-

[214] See pp. 167–168 above; Earnest, *John and William Bartram*, pp. 82–83.
[215] Darlington, *Mems.*, pp. 238–239, 294; *Colden Papers*, VII, 132–133; Collinson and Hunter in *Phil. Trans.*, LVII, Pt. II (1767), 464–469; LVIII (1768), 34–35. For a full account of this matter, see Whitfield J. Bell, Jr., "A Box of Old Bones: A Note on the Identification of the Mastodon, 1766–1806," *Proceedings of the American Philosophical Society*, XCIII (1949), 169–177.

tion and discernment, with all his capacities for rational speculation, and with his enlightened views of natural philosophy as a philosophy for all mankind, it is probably just as well that John Bartram did not succumb to the importunities of Cadwallader Colden and others to prepare a natural history of the American colonies. Dr. Alexander Garden, who had corresponded with Bartram for years and considered him to be a "worthy man," was surprised when Bartram was appointed Royal Botanist in 1765. After Bartram had been his guest for nine days in Charles Town and had departed for East Florida, Dr. Garden wrote to John Ellis (July 15, 1765) that Bartram

> knows nothing of the generic characters of plants, and can neither class them nor describe them . . . his knowledge is rude, inaccurate, indistinct, and confused, seldom determining well between species and varieties. He is however alert, active, industrious, and indefatigable in his pursuits, and will collect many rare specimens. . . . He is well acquainted with soils and timber. . . . He appears to me not very credulous, which is another matter . . . to give the title of King's Botanist to a man who can scarcely spell, much less make out the characters of any one genus of plants, appears rather hyperbolical.[216]

Evidently Bartram was aware of his deficiencies. The urge to collect and the lure of the unknown wilderness drove him to activities for which he was far better qualified. Moreover, as he had remarked to Colden years before, his European correspondents afforded him "the secret pleasure of modestly informing them of some of their mistakes."[217]

The Charles Town Scientific Community: Dr. Alexander Garden (1730–91), F.R.S.[218]

Charles Town was relatively isolated throughout the colonial era. The system of roads that developed in the northern and middle colonies did not extend to Charles Town, and communications

---

[216] Smith, I, 537–538.

[217] Quoted in Hindle, *The Pursuit of Science*, p. 21.

[218] Students of the history of South Carolina must distinguish between Dr. Alexander Garden, who flourished between 1752 and 1782, and the Reverend Alexander Garden, who went to South Carolina in 1720 as a Society for the

were largely by sea. Postal services were poor. It was often possible to send a letter via the West Indies or England to Philadelphia, New York, or Boston with greater dispatch than to wait for ships sailing directly to the northern communities. In consequence, Charles Town's ties with the northern communities were slow to develop, and its communications tended to follow the shipping lanes. Nonetheless, a scientific community arose in Charles Town, and after 1730 its ties with the northern colonial communities strengthened rapidly. Its personnel consisted primarily of physicians, many of whom drifted into politics, into the planter class, or both. But a surprising number of them were well-educated men, with earned degrees in medicine from European universities. Dr. William Bull (1710–91), of whom mention has been made above, was the first native-born American to receive the M.D. degree (at Leiden), but he early forsook medicine for politics and served as Acting Governor of South Carolina for some years before the American Revolution. Dr. Thomas Dale (1699–1750), a nephew of Samuel Dale, physician and botanist of Braintree, Essex, was said to hold a medical degree from Leiden. He arrived in Charles Town about 1725, entered politics, made considerable botanical collections, some of which he sent to his uncle and to William Sherard, and in his will directed that his entire natural history collection be sent to his friend, J. F. Gronovius of Leiden. But when a smallpox epidemic struck Charles Town in 1738, Dr. Hale opposed inoculation and raised a controversy somewhat similar to that in Boston seventeen years earlier.[219] James Kilpatrick (d. 1770), a Scot who settled in Charles Town about 1717, played the role of Zabdiel Boylston in this controversy, and inoculated about 800 Charlestonians with only eight deaths before the anti-inoculation forces man-

---

Propagation of the Gospel missionary, served St. Philip's Church in Charles Town for several years after 1720, was appointed Commissary for the Bishop of London in 1729, and remained in South Carolina until the early 1740's. See William Wilson Manross, comp., *The Fulham Papers in the Lambeth Palace Library, American Colonial Section, Calendar and Indexes* (Oxford, 1965), pp. 133, 136–138, 142, 147, 148, 149; *Boston Evening Post*, No. 346 (March 22, 1742/43).

[219] Miller Christy, "Samuel Dale (1659?–1739), of Braintree, Botanist, and the Dale Family: Some Genealogy and Some Portraits," *Essex Naturalist*, XIX (1919), 53–59; Robert E. Seibels, "Thomas Dale, M.D., of Charleston, S.C.," *Annals of Medical History*, n.s., III (1931), 50–57.

aged to outlaw the practice. Kilpatrick subsequently went to England where he changed his name to Kirkpatrick, took a medical degree, and won an enviable reputation as a famous inoculator.[220]

Another physician who belonged to the Charles Town scientific community was Dr. John Moultrie (d. 1773?), a Scot with a medical degree from Edinburgh.[221] He became a locally popular obstetrician, and in 1755 organized and became the first President of the Charles Town Faculty of Physic, a local medical society designed for the "better Support of the Dignity, the Privileges, and Emoluments of their Humane Art"—but the emphasis appears to have rested on the prompt collection of emoluments, which fomented criticism in the local press.[222] As a man of science, however, Dr. Moultrie was far outclassed by Dr. John Lining (1708–60), a third Scot with his medical degree from Edinburgh. Dr. Lining settled at Charles Town about 1730. Like many others, he took notice of the unhealthful climate of the community and the prevalence of much sickness and high mortality rates. In 1741 he began to keep "Meteoro-statical Tables" of the weather, two of which he sent to Dr. James Jurin of the Royal Society of London with a notice of his intent "to discover the causes of the regular returns of the Epidemical Diseases which prevail in these parts."[223] His tables included a record of daily temperatures, barometric pressures, rainfall, and winds. But Dr. Lining also added details of a painstaking series of metabolism records he had made on himself integrated with his meteorological data. These included records of his weight and pulse rate, taken twice daily, morning and evening; the quantity of urine and amount of perspiration excreted; the number and quantity of stools; and his food and liquid intake daily. He re-

220 Joseph I. Waring, "James Killpatrick and Smallpox Inoculation in Charleston," *ibid.*, X (1938), 301–308; Duffy, *Epidemics*, pp. 26–36. A letter from the Reverend Mr. S. Garden of Charles Town (March 8, 1738/39) was reported to the Royal Society (May 17, 1739), stating that of 850 persons inoculated (450 whites, 400 blacks), seven died (4 whites, 3 blacks); that of 2,190 persons naturally infected with smallpox (700 whites, 1,490 blacks), 480 died (180 whites, 300 blacks). *Letter-Book*, XXV, 288–290; *Journal-Book*, XVII, 438.

221 His son, also Dr. John Moultrie (1729–98), took his M.D. at Edinburgh in 1749, and later became Lieutenant Governor of East Florida.

222 Joseph I. Waring, "An Incident in Early South Carolina Medicine," *Annals of Medical History*, n.s., I (1929), 608.

223 Read May 19, 1743, *Journal-Book*, XIX, 104–105.

gretted the lack of proper instruments to analyze his blood and urine periodically. The tables were complete for an entire year; and Dr. Lining held that such data helped in "illustrating the Nature and predisponent Causes of Epidemic Diseases." The materials were sent in two letters to Dr. Jurin, dated January 22, 1740/41, and April 11, 1741, respectively. Dr. Jurin presented them to the Royal Society on May 19, 1743, and they were published in the *Philosophical Transactions*.[224] On January 29, 1743/44, Dr. Lining sent additional material of the same nature, also published in the *Philosophical Transactions*.[225] Again, on September 30, 1746, and on April 9, 1753, Dr. Lining sent meteorological tables of the weather in Charles Town, the latter including a table of the rainfall over a period of fifteen years (1738–52 inclusive), with particular note of a thunderstorm (June 30, 1750) during which more than five inches of rain fell in two hours and of a hurricane (September 16, 1751) in which nearly three inches of rain fell. Dr. Lining was hoping to test the theory, scorned by Dr. John Mitchell, that "clearing the land of its woods" changed the climate, but he offered no conclusions.[226] In 1753 Charles Pinckney, then in London, sent Dr. Lining several queries regarding electrical experiments with a kite, probably stemming from reports of Franklin's famous experiment to draw electricity from thunderclouds in 1752. Dr. Lining replied on January 1, 1754. Evidently he had planned to repeat Franklin's experiment, but, as he said, he was hindered from completing it by an attack of gout. However, he explained his preparations for the experiment and included an anxious inquiry regarding the death of Professor Georg Wilhelm Richman, reputedly killed by lightning at St. Petersburg, Russia, during the performance of a kite experiment similar to Franklin's. Lining's letter was read to the Royal Society (May 9, 1754) and referred to William Watson, F.R.S., for reply. Watson confirmed the account of Richman's death (July 26, 1753) and warned Dr. Lining to ground his apparatus

---

[224] LXII, No. 470 (April–June, 1743), 491–509; *Journal-Book*, XIX, 104–105.

[225] XLIII, No. 475 (Jan.–March, 1745), 318–330; read March 21, 1744/45, *Journal-Book*, XIX, 385.

[226] *Ibid.*, XX, 523–524 (May 26, 1748); XXII, 388–389 (July 12, 1753); *Phil. Trans.*, XLVIII, Pt. I (1753), 284–285.

to the earth in making the kite experiment.[227] Lining corresponded with Franklin about electrical matters, but his activities appear better to illustrate the widespread interest in the subject than any significant contribution he made to the knowledge of it.[228] His treatise, "A Description of the American Yellow Fever," written in a letter to Dr. Robert Whytt at Edinburgh, was of far greater importance. Probably the first American account of the true yellow fever, it was a competent clinical account of the progress of the disease tied in with a detailed weather report for comparative uses. It was published posthumously in the Edinburgh *Essays and Observations*.[229] Dr. Lining's contributions to medicine, especially with regard to relationships between climatic conditions and health, were imaginative, accurately done, and useful, the first fruits of the scientific community taking shape in Charles Town.

Dr. Lining retired from medical practice about 1755 to nurse his gout and oversee an indigo plantation he had acquired. He made a few experiments with vegetable dyes, but Dr. Alexander Garden reported in 1757 that he was "taken with trifling observations, and does not pursue them . . . as he has turned planter now altogether, and quite done with practice."[230] His studies regarding the interrelationships of climate and disease were in some measure appropriated, enlarged upon, and ultimately published by his erstwhile partner in medical practice, Dr. Lionel Chalmers (1715–77). Chalmers, however, was a competent physician and scientist in his own right. A Scot by birth, he appears to have had a medical education at St. Andrews, and later (1756) was awarded the medical degree at Edinburgh through the kind intercession of Dr. Robert Whytt.[231] He settled at Charles Town about 1737, married there, and ultimately became associated with Dr. Lining, both as a partner

---

227 *Journal-Book*, XXII, 541–542, 591–594. I assume that the Charles Pinckney who made the queries was the husband of Elizabeth Lucas Pinckney, the indigo specialist. Dr. Lining's letter was published in *Phil. Trans.*, XLVIII, Pt. II (1754), 757–764.

228 Cohen, *Benjamin Franklin's Experiments*, pp. 331–345.

229 II (1771), 404–432.

230 Garden to Ellis, May 6, 1757, in Smith, I, 409.

231 Joseph I. Waring, "Lionel Chalmers, Medical Author," *Bulletin of the History of Medicine*, XXXIII (1958), 349–355.

in medicine and as a student of meteorology and the relation of weather to human illness. In 1755 Dr. John Huxham (1692–1768), F.R.S., presented to the Royal Society a letter from Chalmers with a meteorological record of Charles Town from June, 1750, to January, 1755, a detailed list of the diseases present monthly at Charles Town during the same period, from all of which "Dr. Chalmers had deduced a Considerable number of Aphorisms concerning the effects of the Weather on the Human body which are subjoined to the Account."[232] Unlike Dr. Lining, however, Chalmers gave no attention to barometric readings or humidity, which he believed really show "how far the air is pure and Elastic" rather than the weight of a column of air, and he tentatively concluded that changes in the temperature of the air were far more significant to health than pressure or humidity. He continued to enlarge upon these studies, however. In 1768 he sent several essays on the relations of climate to disease to the American Society held at Philadelphia for Promoting and Propagating Useful Knowledge and finally he crowned his researches with the publication of the impressive work, *An Account of the Weather and Diseases of South-Carolina* (2 vols., London, 1776).[233] In the meantime, too, Dr. Chalmers had sent (1753) to Dr. John Fothergill an article highly valued in English medical circles at the time entitled "Of the Opisthotonos and Tetanus," which Fothergill published in the London *Medical Observations and Inquiries*.[234] Chalmers also published *An Essay on Fevers* (Charles Town, 1767; republished, London, 1768; German ed., *Ein zersuch uber die fieber* . . . , Riga, 1773). His contributions to eighteenth-century medicine were impressive, and when he died (1777) he left scientific apparatus valued at £500.[235]

[232] *Journal-Book*, XXIII, 158–159, 162–164 (June 12, 19, 1755).

[233] Hindle, *The Pursuit of Science*, pp. 133, 182; Michael Kraus, *The Atlantic Civilization: Eighteenth-Century Origins* (Ithaca, 1949), pp. 202, 205.

[234] I (1763), 87–110. John Wilkinson, F.R.S., quoted widely from Chalmers in a paper read to the Royal Society March 20, 1766, on "A New Method of treating the Opisthotonos and Tetanus, as practiced by a Gentleman in the West India Islands with General Success." *Journal-Book*, XXVI, 403–406. The West Indian physician was Dr. M. J. Marx, but where in the West Indies he practiced is not divulged. The treatment was by use of "Mercurial Unctions."

[235] I have been unable to locate *Liber de Febribus*, a tract attributed to Chalmers as of 1765. Cf. Waring, "Lionel Chalmers," *loc. cit.*

Probably the ablest of the Charles Town community of scientists before the close of the Old Colonial Era was Dr. Alexander Garden (1730–91). Born in a scholarly clerical family near Aberdeen, Scotland, Garden had an austere childhood. But at Marischal College, Aberdeen, he acquired an excellent classical education, and as an apprentice to Dr. James Gordon, Professor of Medicine in the college, he was trained in medicine and botany. Forced to make his own way, Garden went to London (1746), qualified as a ship's surgeon's mate at sixteen years of age, and served in the British Navy for about four years. In 1750 he left the navy to resume his medical studies at Edinburgh, where he studied under the direction of Dr. Alexander Munro (1697–1767), Professor of Anatomy, Dr. John Rutherford (1695–1779), Professor of the Practice of Medicine, and Dr. Charles Alston (1683–1760), Professor of Botany and Materia Medica, with whom Dr. John Mitchell had earlier studied. Garden's love of botany, as he wrote later, was planted by Gordon at Marischal College and cultivated by Alston at Edinburgh, where he was taught the botany of Ray and Tournefort, for Alston strongly resisted Linnaean principles and rejected the notion of bisexuality in plants.[236] Garden completed the work at Edinburgh for a medical degree, but being unable to pay the fees required, departed without it and made his way to Charles Town, where he arrived in April, 1752.[237]

In Charles Town, Garden became medical assistant to Dr. William Rose, of Rose Hill, Prince William Parish, quickly became enamored of the botanical riches of the area, and sent Dr. Alston and Dr. Rutherford accounts of the much-described pinkroot (*Spigelia marilandica*, L.), an anthelmintic still in use.[238] But Gar-

---

[236] See Garden's letters to John Ellis and Linnaeus in Smith, I, 284–289, 379. The ready acceptance in America of Linnaean methods by both Drs. Mitchell and Garden, after Alston's training at Edinburgh, is noteworthy. See Pulteney, II, 9–17.

[237] For details in the life of Garden, see Edmund and Dorothy Smith Berkeley, *Doctor Alexander Garden of Charles Town* (Chapel Hill, 1969). This excellent new biography supplies materials about Garden's early life hitherto unknown.

[238] Dr. Patrick Browne of Jamaica had also described it in 1751 (see p. 376 above), and Dr. John Lining sent a description of it to Dr. Robert Whytt. *Essays and Observations*, I (1754), 386–389. Garden sent a later account to Dr. Whytt which was published, with a figure of the plant, in *ibid.*, III (1771), 145–153.

den's early impressions of Charles Town were not favorable. He found few persons knowledgeable in botany, his own prime interest. The blacks appeared to know more about natural history than their masters, who devoted themselves very widely to the pursuit of idle pleasures. Dr. Bull was an exception, and he had a natural history collection in his library from which Garden borrowed significant books, but Dr. Bull was not active in the study of natural history. Other Charles Town physicians shared mutual acquaintances in Scotland, especially the younger Dr. Moultrie, who had won his medical degree at Edinburgh only two years earlier. Dr. Lining was becoming old, afflicted with gout, and on the point of retirement. Garden had a scientific interest in botany, he sought to expand his scientific correspondence with others knowledgeable in the subject, and he displayed a singular forwardness in initiating correspondence with such persons. He communicated with Dr. John Huxham, F.R.S., of Plymouth, England, sent him the account of the Carolina pinkroot, which Dr. Huxham presented to the Royal Society of London (February 27, 1755), where it was widely discussed;[239] corresponded with Stephen Hales, the plant physiologist and inventor; and got in touch with William Shipley (1714–1803). Shipley, with Hales, was the originator of the Royal Society of Arts (1754), soon to become active in the colonies, to bestow rewards "for such productions, inventions, or improvements, as shall tend to Employing of the Poor, the Increase of Trade, and the Riches and Honour of the Kingdom."[240] Garden offered enthusiastic support and was elected (March 19, 1755), that society's first colonial correspondent. In this connection he took an interest in silk culture, in viniculture, and in Eliza Lucas Pinckney's current promotion of the cultivation of indigo in South Carolina. But the hot summers of Charles Town impaired his health, and after a brief botanical excursion along the coast of South Carolina in the

[239] *Journal-Book*, XXIII, 70–71. Philip Miller, who cultivated pinkroot in the Chelsea garden, may have acquired the seeds from Garden.

[240] "The American Correspondence of the Royal Society of Arts, London, 1755–1840 . . . ," Introduction by Dr. G. C. Allan, Curator-Librarian, Royal Society of Arts, 1963 (Ser. B), *Records Relating to America in Microfilm*, gen. ed., Walter Minchinton, Swansea, Film 606 Rol 2a Reel 1. See also Derek Hudson and Kenneth W. Luckhurst, *The Royal Society of Arts, 1754–1954* (London, 1954), *passim*.

spring of 1754, he sailed to New York to avoid the heat and repair his ailing constitution.

Alexander Garden's trip in the summer of 1754 was a momentous event in his life, and to a degree in the future of American colonial science. He visited Dr. Cadwallader Colden at Coldengham during the time John Bartram and his son William were there. "Not only the doctor himself is a great botanist," he reported, "but his lovely daughter is a great master of the Linnaean method, and cultivates it with great assiduity."[241] He found John Bartram to be "a plain quaker, but a most accurate observer of nature." He read letters Colden had received from Peter Collinson, John Clayton, J. F. Gronovius, and others. He saw Colden's library, and especially was impressed by Linnaeus' *Genera Plantarum* and the *Critica Botanica* that he saw there. He had seen Linnaean works in South Carolina, in the library of Dr. Bull, but those in the library of Colden gave him a new lease on life, for having been instructed by Dr. Alston in Tournefort's system of plant classification, he had become so bewildered that he was about to give up trying.[242] He collected plants and minerals in the Catskills with Colden and Bartram and observed the utility of the Linnaean system. He visited Philadelphia, saw Bartram and his botanical garden, met Benjamin Franklin and others of the Philadelphia community, advertised the new Royal Society of Arts, in which Franklin, too, soon took an active interest, and somewhere in his journeys met John Clayton of Virginia. He returned to Charles Town (December 14, 1754), having regained his health, made the personal acquaintance of several of the principal men of science in the mainland colonies, spread the word about the Royal Society of Arts in fruitful places, enlarged his scientific correspondence for the future, and became an enthusiastic convert to the systems of Carl Linnaeus. Also, he had paid his fees and was awarded his medical degree by Marischal College on November 2.

Upon his return to Charles Town, Dr. Garden settled in medical partnership with Dr. David Olyphant, a Scot formerly in partnership with Dr. John Lining, who chose to retire from medical practice at this time. Within a year, however, Olyphant moved to

[241] Garden to Ellis, March 5, 1755, in Smith, I, 342–354.
[242] Garden to Linnaeus (via Gronovius), March 15, 1755, *ibid.*, I, 284–289.

Dorchester, and Dr. Garden succeeded to Dr. Lining's practice.[243] He immediately became very busy. He received a letter from Dr. Huxham, who argued that botanists should concentrate on the study of the medicinal qualities of plants, an old herbalist notion with which Garden disagreed.[244] He felt that botany had become a science in its own right and should no longer be a handmaiden of medicine and pharmacy. He wrote several letters to Colden, enclosing Carolina seeds, and in one of them he sent Jane Colden some African seeds that he had received from "Dr. Mounsey, Chief Physician to the Army & Physician to the Prince Royal of Russia."[245] He wrote to John Ellis, who came to play a role in Garden's career similar to that played by Peter Collinson in the life of John Bartram. He told Ellis of his trip to New York and Philadelphia, inquired about how one got elected to the Royal Society, sent him seeds and fossils, and asked him to investigate the Carolina jessamine, which Garden suspected was a nondescript "unless what Mr. Catesby has done, which besides his print is just nothing."[246] Later he persuaded Linnaeus that the Carolina jessamine (yellow jasmine, *Gelsemium sempervirens*) did not belong to the genus *Bignonia*. He opened a correspondence with Henry Baker, F.R.S., also a prominent member of the Royal Society of Arts. He addressed a letter to Linnaeus, sending it to Gronovius (to whom Ellis had introduced him), hoping to strike up a direct correspondence with the great Swede. He thanked Gronovius for a copy of the *Flora Virginica* and for copies of Linnaeus' *Fundamenta Botanica* and *Classes Plantarum*, and asked for more of Linnaeus' works together with any other important new works in physic, botany, and natural history. Dr. Garden was building up a scientific library of his own. He urged South Carolina planters to follow some of the recommendations of the Royal Society of Arts and try horse-powered threshers for rice in place of slave labor, experiment with sesame (*Sesamum*

[243] It was reported that Garden had been offered a position at King's College in New York. If he declined such an offer, it may have been in the knowledge that Dr. Lining was about to retire and that he might succeed to Lining's practice. See Pierre Gautier Jenkins, "Alexander Garden, M.D., F.R.S. (1728–1791)," *Annals of Medical History*, X (1928), 149–158.

[244] Garden to Colden, Jan. 14, 1755, *Colden Papers*, V, 1–4.

[245] *Ibid.*, V, 4–5.

[246] March 25, 1755, in Smith, I, 342–357.

indicum) as a source of oil, *Gossypium* (cotton), cochineal, and a variety of other products, but he found the planters generally uninterested. In May, 1755, he wrote Colden to report that he was on the point of accompanying Governor James Glen on an expedition into the Cherokee country which, in addition to botanical pleasures, offered "a Guinea a day beside other practice for the two troops of horse that attends him."[247] He referred to a diplomatic journey to Saluda Old Town made by the Governor. Dr. Garden kept a journal of the expedition, with an account of the plants and minerals he observed, which he sent to Dr. Huxham in the hope that it might be presented to the Royal Society. Unfortunately, however, it was lost, and no copy of it has been discovered.[248] In November, after his return, he apologized to Colden for neglect of his correspondence, as "an affair of Love quite engrossed my thoughts for a season." He referred to his impending marriage (December 25) to a sixteen-year-old Huguenot girl, Elizabeth Perroneau, daughter of a wealthy and socially well-connected Charles Town merchant. But evidently Garden did not allow the affair to paralyze his pen entirely. On the eve of his wedding he wrote a long letter to John Ellis to thank him for a copy of Linnaeus' *Species Plantarum*, for "freeing my mind from the error it laboured under in believing the Corrallines to be vegetables," with an account of his journey to Saluda, a plea for help to widen his correspondence with Bernard de Jussieu and others, and a statement that as he was about to be "Matrimonialized," "I must hurry away to meet the parson and my dear girl. . . ."[249]

After John Ellis explained to Garden the regulations regarding election to fellowship in the Royal Society, Garden balked at paying the fees. He suggested that as he lived overseas, he might be admitted as a member on the foreign list. When Ellis pointed out that the bylaws of the Society would not admit such an irregular proceeding, Garden exploded. He would not pay money to be admitted to any society under the sun, he wrote, "as I always think

[247] May 23, 1755, *Colden Papers*, V, 10–12.

[248] Garden to Ellis, March 22, 1756, in Smith, I, 371–380. In 1759 Garden told Ellis that he hoped his *Iter Saludiensis* had been burned, as he thought ill of it. *Ibid.*, I, 443–448.

[249] *Ibid.*, I, 357–360; *Colden Papers*, V, 41–43.

that these things should be a matter of Choice in the Society, not of any pecuniary award. There is no body of learned men in the world that I have a greater regard for . . . but if they do not think that I merit a place as a foreigner, when I certainly am one to all intents and purposes, I think that I have no reason to mind them so much as my private friends."[250] Evidently Dr. Garden had no clear understanding of the financial foundations upon which the Royal Society rested. But the matter festered in his mind, and he grumbled about it for years.[251] Seventeen years later he recanted, asked Ellis to nominate him, and was elected to fellowship in the usual manner. Whereupon he paid the fees and became a Fellow of the Royal Society "upon the domestic list."[252]

Garden refused to become a paid collector of seeds and natural history specimens as John Bartram had done and was doing. He would collect only for his friends, and nothing he collected was for sale to others.[253] He valued his scientific correspondence highly, and admitted to an "avaricious desire after new correspondents." He told Ellis that "Every letter which I receive, not only revives the little botanic spark in my breast, but even increases its quantity and flaming force." Such communications "fire the breast; and make life, *life* indeed."[254] As time passed, his scientific correspondents included, besides Colden, Franklin, Clayton, Bartram, and other American colonials, John Ellis, Peter Collinson, J. F. Gronovius, Carl Linnaeus, Bernard de Jussieu, Daniel Carl Solander, William Shipley, Stephen Hales, Dr. John Huxham, Henry Baker, Thomas Pennant, and Scottish friends at Aberdeen and Edinburgh.

---

[250] Garden to Ellis, March 22, 1756, in Smith, I, 377.

[251] Garden's irritation with the Royal Society showed when he sought to cheer Cadwallader Colden after the latter became despondent over the ill reception accorded his treatise on gravity: ". . . you have not been the first whose works have been Denied the Countenance of the English Society," wrote Garden; "They appear to me to be either too Lazy and indolent to examine or too conceited to receive any new thoughts from any but from a F.R.S." Garden to Colden, March 14, 1758, *Colden Papers*, V, 227–231.

[252] May 15, 1773, in Smith, I, 595. Ellis had posted his nomination on the previous March 11. *Certificates, 1767–1778*, unpaged; *Journal-Book*, XXVIII, 88–89, 148, 159. After he went to London in Dec., 1782, Garden was formally admitted, May 15, 1783. *Journal-Book*, XXXI, 373–374.

[253] See his letter to Ellis, July 6, 1757, in Smith, I, 414–415.

[254] *Ibid.*, I, 362–371.

He also collected a remarkably up-to-date library of scientific works, if one may judge from the books mentioned in his letters. What scientific instruments he possessed, besides medical instruments for his profession, is unclear. John Ellis sent him a microscope and Garden told Colden in 1756 that he had acquired one of "Mr. [John] Cuff's Microscopes for viewing water Animals." It is not clear whether this is the same instrument supplied by Ellis. He once ordered several thermometers, but reported that they were all broken in transit; presumably the loss was repaired. Evidently he also had a barometer at hand. He reported several astronomical phenomena, but there was no mention of a telescope. He was a careful workman, and was frequently praised by his foreign correspondents for the precision of his observations and the fine manner in which he prepared his shipments of collections. As a scientist, he was above all a systematist and an enthusiastic supporter of the Linnaean system. He liked system, he said, because it aids the memory, and he told Linnaeus that he had learned more in one year using the Linnaean system than he had learned in three years using Tournefort's. And he soothed Linnaeus for having failed to discover a natural system of classification, saying that "The man who gives the natural system must be a second *Adam*, seeing intuitively the essential differences of things."[255]

Garden was independent-minded and would not easily bow to the adverse opinions of his correspondents. He was also impatient with the faults of others, and sometimes gave harsh judgments of his predecessors, some of whom had not enjoyed all the more advanced techniques and information of the Linnaean age. After Patrick Browne's *Civil and Natural History of Jamaica* appeared in 1756, Garden wrote to Colden that Browne's work was much better than the one prepared by "that Most pompous, confused & illiterate Botanist Sir Hans Sloane."[256] He was equally severe in his judgment of Mark Catesby's *Natural History of Carolina*. After his trip with Governor Glen to the Saluda River, he complained that Catesby "never was 30 miles back from the Coast during his stay here." He was especially severe in his criticism of Catesby's fishes. On January 2, 1760, he wrote to Linnaeus: "It is sufficiently evident that his sole

255 To Ellis, Jan. 13, 1756, *ibid.*, I, 368–369.
256 *Colden Papers*, V, 89–92.

object was to make showy figures of the productions of Nature, rather than to give correct and accurate representations. This is rather to invent than to describe." Eleven days later he wrote to John Ellis: "I never before looked to our fishes attentively, but I am struck with astonishment at Catesby's blunders. Sometimes he forgets whole fins &c. In a word, there is nothing can possibly recommend him but the specious beauty of the colouring of the plates." And on April 12, 1761, he again wrote Linnaeus: "I have also consulted Catesby, as it seemed proper to do so, but never without disgust and indignation. I cannot endure to see the perfect works of the Most High, so miserably tortured and mutilated, and so vilely represented. His whole work, but especially his 2d volume, is so imperfect, and so grossly faulty, that to correct its errors, and supply its deficiencies, would be no less laborious, than it is necessary."[257] Garden was partially justified in his criticisms of Catesby's fishes, for Catesby was careless and inaccurate in the portrayal of some of them. He was on less firm ground in criticizing Catesby's colorings. Catesby had experienced obvious difficulties in reproducing in watercolor some of the bright colorations of the tropical and subtropical fishes, but he also had been handicapped by the fact that many of the fish changed color after their removal from the water. Garden's comments on Catesby's fishes were not uniformly justified, and his condemnation of Catesby's book as a whole, including the work on ornithology, was unduly censorious. But Dr. Garden's temper was sometimes as prickly as the *Opuntia* with which he experimented for the Royal Society of Arts.[258]

Until about 1760 Dr. Garden's scientific work, aside from his medical practice, was divided between the promotion of the objectives of the Royal Society of Arts and the collection and study of natural history specimens, especially in botany. He sent large collections of seeds and dried plants to John Ellis and Henry Baker during these years; both of them were active in the work of the Royal Society of Arts.[259] On April 5, 1756, he addressed a letter to Henry Baker with "an Account of Such of the productions of the

[257] Smith, I, 300–301, 469, 307.
[258] See Frick and Stearns, *Mark Catesby*, pp. 76–77; Berkeley and Berkeley, *Doctor Alexander Garden*, pp. 328–330.
[259] Smith, I, 345–357, 360–362, 380–382, 392–415.

province of South Carolina which are likely to be of Service both to them and Great Britain." Garden listed items which he believed could be produced in South Carolina to a much greater advantage than they were. He included vines for wine-making, sesame, cotton ("no more is planted," he said, "then what Serves to employ some old Superannuated Negro Women"), mulberry trees for silk production, cochineal (which he said was abundant on the wild *Opuntia*, or prickly pear, but the people lacked the proper methods to kill the insects so as to preserve the dye), hemp, flax, and potash. He also noted "the great want they stand in in that province of Mechanical machines to facilitate labour."[260] In the summer of 1756 he accompanied Governor Glen on a projected expedition to Mississippi, but with the arrival of a new Governor (William Henry Lyttleton), the expedition was recalled, much to Garden's disappointment, for, as he wrote, he had hoped to glut "my very soul with the view of the Southern parts of the Great Apalachees."[261] He did make some drawings of plants which were sent to Ellis, but the outbreak of the Seven Years' War led to their capture by the French, and as Garden knew nothing of the war at the time, he had made no duplicates. Indeed, the French captured nineteen of the twenty-one vessels that sailed from Charles Town in January and February, 1767, entailing the loss, among other things, of two large collections shipped to John Ellis.[262] In conjunction with Henry Baker and others of the Royal Society of Arts, Garden undertook experiments with various vegetable dyes. He found that the prickly pear would color urine of children "a very lively red colour," and likewise it made the milk of a Negro wench reddish; he found that cows pastured in an indigo field had both their urine and their milk tinctured blue. All this he reported to Henry Baker, who read it to the Royal Society (June 23, 1757), and the report was published

---

[260] Read May 25, 1758, *Journal-Book*, XXIV, 153–155. Philip Carteret Webb pointed out that the method of preparing cochineal insects was well known in England and "a Draught of the whole process" was available at the South Sea House. See also *Egerton 2381*, fol. 56v (British Museum).

[261] Garden to Charles Whitworth, April 27, 1757, in Smith, I, 393. Garden stated that the expedition left Charles Town on May 19, 1756, and had advanced 260 miles west before it was recalled.

[262] Garden to Ellis, May 6, July 6, 1757, *ibid.*, I, 392–415.

in the *Philosophical Transactions*.[263] Charles Whitworth (1714?–78), M.P., of the Royal Society of Arts, proposed that trials of different vegetables be made to see if other sources of dyes might be found, and Garden was urged to press the matter in South Carolina. Accordingly he placed a notice in the *South Carolina Gazette* in the spring of 1757 to invite the planters of South Carolina to cooperate in the effort.[264] However, in spite of the "particular directions" given to the planters, Garden reported in disgust to Ellis on August 11, 1758, that he had not received a single response.[265] But he would continue his efforts in this direction, "for its utility is visible."

In his search for new plants, Garden discovered a shrub believed to be a new genus. He sent a drawing of it, with a description and a dried specimen, to Linnaeus via John Ellis on November 30 1758. He also urged Linnaeus to name it *Ellisia* after Ellis. Ellis laid it before the Royal Society (April 26, 1759) before he forwarded it to Linnaeus. After Linnaeus had studied the materials, he classified the new genus as a *Swertia*, of the gentian family. With this, however, Garden disagreed, insisting that the new shrub was a *Duranta*, of the vervain family. He persisted in his opinion and ultimately was found to be correct—probably the first time an American colonial had dared to disagree with Linnaeus and won his case.[266]

Garden had recommended to Baker and Ellis that the Royal Society of Arts seek to exploit the cochineal of South Carolina as a source of dyestuffs. Both Philip Miller and Emanuel Mendes da Costa of the Royal Society of London asserted that both the South Carolinian cochineal and the prickly pear upon which it fed were of a "bastard kind" and would not produce dye. But John Ellis was not convinced. He asked Garden for specimens and descriptions of the cochineal, and Garden sent both in 1759. He examined cochineal "flies" under his microscope, prepared drawings and descriptions of

[263] *Journal-Book*, XXIII, 594–595; *Phil. Trans.*, L, Pt. I (1757), 296–297. The Royal Society's minutes recorded that "This Effect of the Opuntia is very common & long known."

[264] Garden to Whitworth, April 27, 1757, in Smith, I, 382–392.

[265] *Ibid.*, I, 418–428.

[266] *Ibid.*, I, 290–302; *Journal-Book*, XXIV, 299. According to Birch's notes of the Royal Society meeting, Garden's account of the *Ellisia* was "withdrawn" from articles destined for publication in the *Philosophical Transactions*. No reasons were given. *Add. Mss. 4446*, fol. 149v.

608

both the male and the female, and forwarded them, with specimens, to Ellis. It was on the basis of these materials that Ellis subsequently published in the *Philosophical Transactions* "An Account of the Male and Female Cochineal Insects, that breed on the Catus Opuntia, or Indian Fig, in South Carolina and Georgia."[267] When the Society of Arts put forth the idea of establishing a series of botanical gardens in the colonies in which to experiment with American plants, Garden endorsed the notion with enthusiasm. Indeed, he believed that the South Carolina provincial government might set up such a garden at its own expense, but the Assembly could not be persuaded to raise the funds, and the idea fell on infertile soil. Garden cooperated with John Ellis in the latter's experiments to preserve the fecundity of seeds transported over long distances by covering them with wax, and he also kept meteorological records at Charles Town to check with Governor Henry Ellis' accounts of the extreme heat of Georgia.[268] He enlisted the aid of George Roupel, an official in the customs service at Charles Town, to draw plants for him, and Ellis praised Roupel's work highly.[269] Garden contributed to John Clayton's revised and expanded *Flora Virginica*. Ellis urged Garden to send specimens of the "Palmetto Royale," which Ellis insisted was a yucca. But Garden argued that it was not a yucca, and if the plant in question was the *Sabal palmetto* (Walt.) Todd, Garden was right, for the latter belongs to a genus, family, and order different from the yucca. So far as Clayton's book was concerned, however, it hardly mattered, as his fine revision of the *Flora Virginica* was never published. Late in 1759 Garden sent Henry Baker an account of Halley's Comet which he had viewed over Charles Town on the previous April 29 and May 9. There was no mention of any astronomical instruments employed, but Garden related the comet's declination and path.[270] Thus, the year

267 LII, Pt. II (1762), 661–666, with figures. See also *Journal-Book*, XXV, 261–264; *Egerton 2381*, fol. 57v.

268 See pp. 334–335 above. Garden to Ellis, Feb. 17, 1759, and Ellis to Garden, Aug. 25, 1759, in Smith, I, 431, 438–442, 459–464, 501–506.

269 Garden to Ellis, May 11, 1759, and Ellis to Garden, Aug. 25, 1759, *ibid.*, I, 446, 459–464. Garden promoted Roupel for the office of Surveyor General of Customs for South Carolina without success. *Ibid.*, I, 485–488, 500–501.

270 Garden to Baker, May 10, 1759; read Nov. 22, 1759, *Journal-Book*, XXIII, 572–573.

1759 was a crowded one for Dr. Garden, and he complained of his want of time for collecting because, as he said, his medical practice occupied him daily from seven in the morning until nine at night and he was on call twenty-four hours a day.[271]

The next year was worse. Governor Lyttleton returned from his ill-conceived punitive expedition against the Cherokees, "as we then thought," wrote Garden, "crowned with laurels; but, alas, bringing the pestilence [smallpox] along with him, and having the war at his heels."[272] Soon Dr. Garden was worn out with inoculations, having inoculated upwards of 2,400 persons in less than two weeks. Garden developed a treatment with mercury and antimony for smallpox victims, adapted from "my learned and ingenious friend," Dr. Adam Thompson of Philadelphia, a fellow Scot from Edinburgh and author of *A Discourse on the Preparation of the Body for Smallpox* (Philadelphia, 1750). He thought he might have found a specific for the disease, but further trials proved him wrong. The epidemic raged at Charles Town from January until June; about 6,000 of the 8,000 people in Charles Town were ill, of whom 730 died. In the midst of the epidemic John Bartram visited Garden—a delightful interlude lasting nearly three weeks, but, as it happened, at a very inconvenient time. The pressures of his practice, too, led Garden to long for retirement. The collection and study of natural history specimens were more attractive, but in spite of these allurements, Dr. Garden was unable to free himself of his medical practice for many years to come. He was much pleased, however, when in the course of 1760, John Ellis, with Linnaeus' approval, named the Cape jasmine the *Gardenia*. It was reported to the Royal Society on November 20 that Philip Carteret Webb, F.R.S., had "a great variety of American and other plants" in his garden. One of these, the snowdrop or silver-bell tree, originating in North America, had first been described by Dr. Garden and John Ellis had named it *Halesia* to honor Stephen Hales. Another, "known here by the name of Cape Jasmin, Mr. Ellis, with the approbation of Linnaeus, calls, Gardenia, from his friend Dr. Alexander Garden. It was brought to England from the Cape of Good-Hope, by Capt.

[271] Letter to Ellis, July 14, 1759, in Smith, I, 449–458.
[272] Letter to Ellis, March 13, 1760, *ibid.*, I, 473–476.

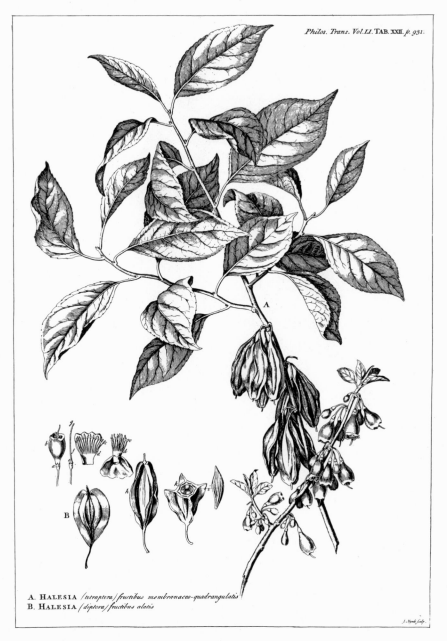

A. HALESIA *(tetraptera) fructibus membranaceo-quadrangulatis*
B. HALESIA *(diptora) fructibus alatis*

29. A drawing of the *Halesia* (the snowdrop or silver-bell tree),
named after Stephen Hales by John Ellis, to accompany
Ellis' "Account of the Plants Halesia and Gardenia."

Phil. Trans., LI, Pt. II (1760), 931.

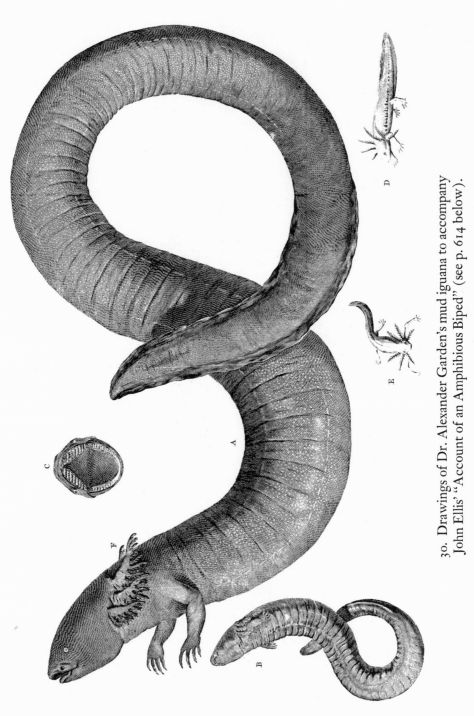

30. Drawings of Dr. Alexander Garden's mud iguana to accompany John Ellis' "Account of an Amphibious Biped" (see p. 614 below). *Phil. Trans.*, LVI (1766), 189.

Hutchinson, and presented to Mr. [Richard] Warner; but first effectually cultivated, to his own great advantage, by Mr. [James] Gordon [the nurseryman] of Mile End; and is nothing less than a Jasmine, as appears by its characters. . . ."[273] In fact, the gardenia was very popular. Ellis told Garden that James Gordon made £500 from four cuttings of it in less than three years; but, unhappily, specimens sent to Charles Town failed to survive, and Dr. Garden was unable to cultivate it in his Broad Street garden.

After 1760, prompted by Linnaeus, Garden's interest in natural history was enlarged to embrace fauna as well as flora. To his first letter to Linnaeus, in 1755, he had received no reply, and Garden felt snubbed. But John Ellis repeatedly praised Garden's abilities to the great Swede, sent him items collected by Garden, and urged him to write to Garden directly.[274] Finally, on May 30, 1759, Linnaeus wrote to Garden, who received the letter with much delight. Linnaeus, however, urged Garden to give attention to insects, fishes, and serpents. Garden obviously felt ill-prepared for such work, but he undertook it with a will and soon was engrossed in the collection and study of these items.[275] He sent fishes to Linnaeus at first, dried specimens of skins still said to be in a good state of preservation, with such remarkably careful observations that Linnaeus soon learned to depend heavily upon Garden's descriptions. Early in 1761 Garden sent a large shipment of fishes, snakes, lizards, alligators, and insects. The collection included fourteen varieties of snakes, five of which were nondescript, and a large number of fishes including several new genera, some of which were not identified until a century later. "Pray be not sparing of your criticisms," he begged Linnaeus with uncharacteristic humility, "for by their means my faltering steps may be encouraged and confirmed."[276] At

[273] *Journal-Book*, XXIV, 118–121. The Cape jasmine was not a true jasmine, although the term was applied to plants of genera other than that of the *Jasminum* about this time.

[274] See Smith, I, 85–87, 90–96, 129–133. The first letter Linnaeus sent to Garden miscarried.

[275] Garden to Linnaeus, Jan. 2, 1760, *ibid.*, I, 297–302. Garden's first letter sent directly to Linnaeus included his objections to Linnaeus' classification of the *Swertia*. Linnaeus' direction of Garden's attention to fishes and the like also led Garden to consult Catesby, with the loud outcries against him noted above.

[276] April 12, 1761, *ibid.*, I, 303–308.

the same time he sent a shipment to John Ellis, including a picture of a new plant by George Roupel. The shipment was sent by the hand of John Gregg, a merchant of Charles Town, who also collected for Ellis. Garden spoke warmly of Gregg, "who is a worthy, good lad."[277] Gregg was in and out of Charles Town until 1768, when he fell out with his business partners, went to the West Indies, settled at Dominica, and in 1772 was elected Fellow of the Royal Society.[278] Garden continued to send collections to Linnaeus, Ellis, Solander, Gronovius, and Dr. John Hope (1725–86), Professor of Medicine and Botany at Edinburgh, after whom, at Garden's request, Linnaeus named the genus *Hopea*. The specimens sent to Linnaeus appear to have been the most numerous, with many new plants, snakes, fishes, and insects, all of which Linnaeus made profitable use of in the twelfth edition of his *Systema Naturae* (3 vols., Stockholm, 1766–68). In 1769 Ellis introduced Garden to a correspondence with the popular naturalist and traveler, Thomas Pennant (1726–98), F.R.S. Pennant was author of *British Zoology* (London, 1766) and *Arctic Zoology* (2 vols., London, 1784–85), and for the preparation of the latter Garden supplied him with collections of American fishes and birds.[279] It was probably by the hand of Pennant that Garden became familiar with the works of Buffon and of Peter Simon Pallas (1741–1811), the German naturalist and collector in Russia and Siberia.[280] In 1763 another small-pox epidemic occurred in Charles Town, and Garden reported that he inoculated about 800 persons with the loss of only two children. He planned to publish a book about his smallpox treatment, but evidently never completed it.[281] At this point, too, Garden began to consider retirement, and even spoke of moving to London. Ellis wrote to Linnaeus (September 10, 1765) that Garden planned to withdraw from his medical practice to take care of his health, and he "promises to be a very good correspondent for the future."[282] In

[277] Garden to Ellis, April 26, 1761, *ibid.*, I, 508–511.

[278] *Ibid.*, I, 564–567; *Col. F.R.S.*, pp. 232–233. Gregg sent John Ellis the polyp described in *Phil. Trans.*, LIII (1763), 419–435. See also Ellis to Linnaeus, Oct. 21, 1766, in Smith, I, 189.

[279] *Ibid.*, I, 567–575, 583–589.

[280] Garden to Ellis, n.d. [1768], *ibid.*, I, 564–567.

[281] Garden to Linnaeus, June 2, 1763, *ibid.*, I, 309–317.

[282] *Ibid.*, I, 174.

1768 he planned to turn over his practice to his half-brother Francis (M.D., Edinburgh, 1768), but Francis died (September, 1770) less than two years after his arrival, and Garden was forced to carry on, though obviously at a reduced tempo. In 1771 he purchased from John Moultrie a plantation of 1,689 acres in St. James Parish, Goose Creek, about fifteen miles from Charles Town,[283] and, as his health was deteriorating, went into semi-retirement. But he continued to collect natural history specimens for Ellis, Gronovius, Pennant, Linnaeus, and Dr. Hope at Edinburgh. In 1770 he sent a "black servant" to Providence Island to collect fishes and insects. Some of the collection was lost in a storm on the return voyage, but Garden had fourteen fishes and some insects to send Linnaeus.[284] He continued to take an active part in the Royal Society of Arts, and in 1773 planted some upland rice and other Oriental seeds sent by John Ellis and distributed seed to other planters in South Carolina for trial.[285] It appears that his promotion of the Royal Society of Arts brought some response from other South Carolinians. In 1757 Moses Lindo observed that the excrement of a mockingbird that had consumed the berries of a weed called "Pouck" (pokeweed, *Phytolacca*) contained a crimson dye. Whereupon Lindo found that he could extract the dye from the berries and in 1763 sent an account of it to the Royal Society of London. The Society published it in the *Philosophical Transactions*.[286] Whether Garden's promotions reached Dr. George Brownrigg, of Edenton, North Carolina, is less certain. Brownrigg was the brother of Dr. William Brownrigg (1711–1800), F.R.S., and in 1769 he sent William Watson an account of peanut oil, which he recommended as an inexpensive substitute for olive oil and might reduce the need to import the latter. Watson read the account to the Royal Society (December 14, 1769) and it was published in the *Philosophical Transactions*. The Royal Society of Arts took note and imported thirty gallons of peanut oil from Samuel Bowen of Georgia in 1772. Either the Society's trials proved unsatisfactory or the War for Indepen-

---

283 Henry A. M. Smith, "Goose Creek," *South Carolina Historical and Genealogical Magazine*, XXIX (1928), 22–23.

284 Garden to Linnaeus, June 20, 1771, in Smith, I, 330–337.

285 Garden to Ellis, May 15, 1773, *ibid.*, I, 594–600.

286 LIII (1763), 238–239; *Journal-Book*, XXV, 140–142 (Nov. 10, 1763).

dence interrupted the experiments, for peanut oil did not become a staple of commerce until many years later.[287]

Dr. Garden's collections and observations included, in addition to many new plants, three significant zoological items. On June 5, 1766, John Ellis read to the Royal Society a paper describing a strange biped which Garden had sent him in 1765. Known in South Carolina as the "mud Iguana," three specimens had been sent by Garden, with a careful description, and Dr. John Hunter had dissected the creature before Ellis read his account. Garden had also sent a specimen to Linnaeus (May, 1765), with whom Ellis had consulted, and the specimens were tentatively considered to be "a new and very distinct genus between the Eel and the Lizard." Linnaeus believed that it was "the larva of some kind of lacerta" and urged Garden to see if it did not develop four feet at a more mature stage. But Garden insisted that it had only two feet in its maturity, and that it was a wholly new genus. Garden was correct, and ultimately Linnaeus was persuaded to create a new class of amphibia (Sirenidae) to accommodate the mud iguana (*Siren lacertina*), much to Garden's fame in Europe.[288]

In December, 1770, Garden sent Ellis and Pennant a specimen of a dried fresh-water soft-shelled turtle, which, as he said, Catesby had "amazingly" missed. He had received a specimen first from Lieutenant Lachlin McIntosh of Darien, Georgia, had studied it with care, and when it died after about three months, he had dissected it. Other specimens were found in the Savannah and Altamaha rivers, and Garden reported that the turtle grew to a large size, was deliciously edible, and was apparently nondescript, as he had found nothing in Linnaeus like it. Pennant presented Garden's specimen and description to the Royal Society (May 2, 1771) and a summary of the description was published in the *Philosophical Transactions*. It also appeared in Pennant's *Arctic Zoology*, and Linnaeus included it in the twelfth edition of his *Systema Naturae*.[289]

---

287 *Ibid.*, XXVII, 263–264; *Phil. Trans.*, LIX (1769), 379–383; "The American Correspondence of the Royal Society of Arts" (microfilm). Peanut oil had been mentioned in the first volume of Sloane's *Voyage to the Islands . . .* (1707).

288 *Journal-Book*, XXVI, 440–441; *Phil. Trans.*, LVI (1766), 189–192; Smith, I, 174, 185.

289 *Journal-Book*, XXVII, 489–492; *Phil. Trans.*, LXI, Pt. I (1771), 266–271;

Garden's last important communication from Charles Town to the Royal Society was "An Account of the Gymnotus Electricus or Electrical Eel," presented by John Ellis to the Royal Society on February 23, 1775. The electric eel had excited much attention both in Europe and America since the early 1770's. A fish that could transmit an electric shock to anything nearby quickly attracted students of electricity as well as naturalists, anatomists, and physicians. It was a fresh-water fish first found in Surinam, and ship captains transported live specimens both to the British North American colonies and to Europe, where they were placed on display and drew large crowds of onlookers willing to pay to see this curiosity of nature. The Royal Society of London published in the *Philosophical Transactions* several papers about it by John Walsh, the student of electricity, John Hunter, the surgeon and anatomist, Henry Cavendish, the brilliant chemist, and others.[290] In July, 1773, the new American Philosophical Society carried out a series of experiments with an electric eel recently brought to Philadelphia. Ebenezer Kinnersley, David Rittenhouse, Isaac Bartram, and others experimented to test the nature and the severity of the electric shock, bringing into play pith balls and the other apparatus of the electrical buffs of the time. A year later, Captain George Baker, a mariner en route to England, brought several electric eels to Charles Town. On August 14, 1774, Dr. Garden addressed his "Account of the Gymnotus Electricus or Electrical Eel" to Ellis. He had observed five of the creatures and described the largest one—although he complained that he had difficulty in examining the living specimen because of the danger of electrical shock, and he had no dead one to dissect. His specimen was three feet, eight inches long, and about ten to fourteen inches in girth at the thickest part; he was told that some found in Surinam were over twenty feet long (a gross exaggeration). Dr. Garden reported that it had a flat head and a large mouth without teeth; he said that it could swim either backward or forward, could shorten or extend its body like an angle-

Smith, I, 260, 582–599; Pennant, *Arctic Zoology*, II, Supplement, pp. 78–79; Linnaeus, *Systema Naturae* (12th ed.), I, Pt. III, 1039–40.

[290] See *Phil. Trans.*, LXIII, Pt. II (1773), 461–480; LXIV (1774), 464–473; *Journal-Book*, XXVIII, 164–175. John Walsh was given the Copley Medal in 1774 for his experiments on the electric eel. *Ibid.*, XXVIII, 517–520.

worm; and he believed that it had lungs and was amphibious, for he observed that it rose to the surface to breathe every four or five minutes. Ellis referred to Garden's account as "extremely accurate" and published it in the *Philosophical Transactions;* but Garden later referred to it as confused and inferior owing to the fact that he had scarcely recovered from fever when he did it. Evidently he had been anxious to justify his recent election into the Royal Society.[291] Perhaps he realized that he should not have characterized the electric eel as an amphibious animal without dissection.

Ill health and the approach of the War for Independence caused Garden's scientific correspondence to diminish rapidly in the 1770's. He won another argument with Linnaeus in 1772 regarding the classification of the sago palm. Linnaeus, to whom Garden had sent a specimen, had classified it as a fern, but Garden insisted that it was a *Zamia*—and so it was, the only genus of the cycad family native to North America. The death of John Ellis in 1776, "My Dear, my first, my chief Botanical Friend," as Garden had addressed him in 1770, and of Linnaeus in 1778, removed two of his closest correspondents. He appears to have had early premonitions of American independence, for he wrote to Ellis after the Stamp Act in 1765 that "The die is thrown for the sovereignty of America." He was sympathetic to many of the colonial grievances against England, and one of his "very particular friends" was Henry Laurens, the leader of the revolutionary movement in the South. But when the war began, Garden could not bring himself to take up arms against the King. He adopted a neutral stance, emphasized by his daughter's marriage to a British officer in 1781 and by his son's service as aide-de-camp to General Nathanael Greene. But after Cornwallis' defeat, the enthusiasms of patriotic Charlestonians could no longer be contained, and Dr. Garden was banished and his properties confiscated.[292]

---

[291] *Ibid.*, XXVIII, 535–537; *Phil. Trans.*, LXV, Pt. I (1775), 102–110; Garden to Ellis, March 12, 1775, in Smith, I, 603–605; Hindle, *The Pursuit of Science,* pp. 188–189. Actually, Garden's observations added little to those of Hugh Williamson at Philadelphia in Sept., 1773. See *Phil. Trans.*, LXV, Pt. I (1775), 94–101.

[292] Dr. Garden had conveyed his plantation, Otranto, to trustees for his son

Dr. Garden took up residence in London in January, 1783, bringing with him a collection of seeds from South Carolina. He attended the Royal Society on May 15, signed the Obligation, and was formally admitted to the Society.[293] Thereafter he was frequently present at the meetings of the Society, usually bringing visitors, including John Moultrie, Governor William Moultrie of South Carolina, his brother-in-law, Mr. Perroneau of Charles Town, George Roupel, Dr. John Morgan of Philadelphia, and others of his former colonial friends. On November 20, 1783, he was named one of the auditors of the treasury accounts for the year, and he was elected to the Council of the Royal Society for 1784 and 1785.[294] But he appears to have offered no papers to the Society and taken no active part in the discussions. He toured the Continent in the summer of 1788, and visited Scotland in the following year. In June, 1789, he moved to Ramsgate, where he died on June 15, 1791.

In addition to his election to the Royal Society of London and his corresponding membership in the Royal Society of Arts, Dr. Garden was honored by election to the Edinburgh Society (1760), the Swedish Royal Society (1763), and corresponding membership in the American Society for Promoting and Propagating Useful Knowledge (1768), and in the newly reorganized American Philosophical Society at Philadelphia (1769). His work as a scientist is difficult to appraise. Considering his activities in and promotion of scientific work, he published very little, seemingly content to allow his efforts to be largely appropriated by others. His enthusiastic efforts in behalf of the Royal Society of Arts betrayed a strong utilitarian bent to his thinking, yet his letters appear to demonstrate that he was more concerned with system—and more able to contribute to it, both in botany and in zoology—than any other colonial, save, perhaps, Dr. Patrick Browne of Jamaica or John Clayton of Virginia. He was at home with the scientific method, was a close and accurate observer, critical of the conclusions of others, scornful of the errors

---

Alexander, then a minor, in 1778. Alexander Jr., who became a major in the American forces, sold part of it in 1785, the remainder in 1798.

293 *Journal-Book*, XXXI, 373-374.

294 *Ibid.*, XXXI, 426, *passim*; XXXII, 32, *passim*; XXXIII, 2, *passim*; C.M., VI, 151, *passim*.

and omissions of his predecessors. He exhibited an interest, common to practically all of the scientists of his time, in many aspects of scientific knowledge—botany, zoology, medicine, surgery, oceanography, meteorology, geography, astronomy, and mineralogy—but his chief claims to competence lay in medicine, botany, and zoology. He appears to have been a respected physician, although he made no outstanding contributions to the science and art of medicine. As a botanist and zoologist he made many new discoveries useful to others, although the new plants and animals which he discovered and described with accuracy and skill were mostly published by John Ellis in the *Philosophical Transactions*, John Pennant in his *Arctic Zoology*, and Linnaeus in the twelfth edition of his *Systema Naturae*.[295] Some of the fruits of his collection still survive in the repositories at Oxford, Uppsala, and the British Museum. Still, although Garden's influence must be sought principally in the works of others, no other colonial, in describing new genera and species of flora and fauna, described them with such informed confidence in his own observations as to stand up staunchly for his opinions against such formidable authorities as Linnaeus—and be correct in his views. Cadwallader Colden sought to challenge European authorities in his work on gravity, and he was not only crushed by Leonhard Euler, but also failed to win any appreciable support in informed circles. Alexander Garden, however, convinced Linnaeus and others of the informed circles of his day (and since) that the Carolina jessamine was not of the genus *Bignonia*, that the *Ellisia* was not a *Swertia*, that the Florida cycad, *Zamia*, was not a fern, that the palmetto was not a *Yucca*, and that the "mud iguana" belonged to an entirely new genus which Linnaeus ultimately accom-

---

[295] The twelfth edition of Linnaeus' *Systema Naturae* was published at Stockholm, 1766–68. It was the last edition revised by Linnaeus himself. Additional pages appeared in 1780, edited by John Hope and labeled "Ex editione decima tertia," but the bibliographers assert that these actually belonged with the twelfth edition. A thirteenth edition, with corrections and additions, was edited by Johann Friedrich Gmelin and published at Leipzig, 1788–93. See *A Catalogue of the Works of Linnaeus . . . of the British Museum* (2nd ed., London, 1933), pp. 11–15. Laurenz Théodor Gronovius received many specimens of flora and fauna from Garden, but he appears to have published no acknowledgments of them, either in his *Museum Gronovianum . . .* (Leiden, 1777) or in his *Zoophylacium Gronovianum . . .* (Leiden, 1781).

modated in his scheme of classification. In these respects, at least, Dr. Alexander Garden was unique among colonial scientists.

The Master Electrician:
Benjamin Franklin (*1706–90*), F.R.S.

On November 9, 1749, William Watson (later Sir William, 1715–87), apothecary, physician, naturalist, and one of the few Fellows of the Royal Society who had specialized in the study of electricity, read to the Society "Part of a Paper Containing a new Theory of Thunder Gusts." As the paper was long, the remainder of it was postponed until the following week. At that time (November 16) Watson completed the paper, the full title of which was "Observations and suppositions towards forming a new hypothesis for explaining the several phaenomena of thunder gusts." Written by Benjamin Franklin of Philadelphia, the paper had been transmitted to Watson by Dr. John Mitchell, the Virginia physician who had retired to England for his health in 1746. In the minutes of the meeting, the Society's Secretary, showing unusual respect for the author of the paper, recorded that Franklin "endeavours to account for the several phaenomena of Thunder and Lightning from Electricity; but as his Paper Contains a long chain of deductions, it is impracticable to give any tolerable account of it without running out into too great a length, or doing injury to his arguments."[296] About a month later (December 4) another paper from Franklin, dated at Philadelphia, April 29, 1749, and entitled "Further Experiments and Observations in Electricity, made in Philadelphia, 1748," this time transmitted to the Society by Peter Collinson, was referred to Watson, who reported (December 21) "that he thought the paper worthy of being read at the Society" and proceeded to do so. In the course of the next two years, Collinson transmitted other letters from Franklin describing electrical experiments, some of which he had turned over to Edward Cave, the editor, who published them in the *Gentleman's Magazine*.[297] In 1751 Cave published all

[296] *Journal-Book*, XXI, 166–167, 171–172.
[297] For full details regarding the publication of Franklin's accounts of his experiments, see Cohen, *Benjamin Franklin's Experiments*, p. 84 and *passim*.

the letters, with an Introduction by Dr. John Fothergill, in a pamphlet entitled *Experiments and Observations on Electricity, Made at Philadelphia in America* (London). Collinson presented it to the Society on May 9, 1751, and the Society called upon Watson for a report of it. Watson responded on June 6 following to the effect that as the letters had already been presented severally to the Society, he would confine his report to his estimate of Franklin's opinions, which were recorded into the minutes as follows: "Upon the whole, Mr. Watson says, that Mr. Franklin appears to be a very able and ingenious Man, with regard both to Theory and practice. And altho' there are some few Opinions in this Work in wch Mr. Watson cannot agree with him, He thinks scarce anyone is better acquainted with the Subject of Electricity than himself."[298] By unanimous agreement in the Society, both Franklin's letters and Watson's comments were published in the *Philosophical Transactions*.[299] In such a manner the Royal Society became familiar with the first fruits of Franklin's experiments and observations on electricity.

Franklin's interest in science dated from his youth, but a conjunction of circumstances appears to have turned his attention to electricity in the mid-1740's. In 1743, "being at Boston, I met there with a Dr. Spence [Archibald Spencer, lecturer], who was lately arrived from Scotland, and show'd me some electric Experiments. They were imperfectly performed, as he was not very expert; but being on a Subject quite new to me, they equally surpriz'd and pleas'd me."[300] Three years afterward, in 1746, Peter Collinson presented to the Library Company of Philadelphia a glass tube commonly used to generate electricity (by rubbing), together with, as Franklin said, "some account of the Use of it in making such Experiments." It seems likely that this "account of the Use of it" was "An historical account of the wonderful discoveries, made in Ger-

---

[298] *Journal-Book*, XXI, 649–650. For the previous reports of Franklin's letters, see *ibid.*, XXI, 475–476 (Dec. 20, 1750), 491 (Jan. 17, 1750/51), 620–621 (May 9, 1751).

[299] *Phil. Trans.*, XLVII (1751–52), xxxi, 202–211. For the unanimity of the Society, see Secretary Birch's minutes, *Add. Mss. 4495*, fol. 5. Publication was delayed by the reorganization of the controls of the *Philosophical Transactions*. See p. 105 above.

[300] Labaree, ed., *The Autobiography of Benjamin Franklin*, pp. 140–241.

many, &c concerning Electricity" which had appeared in the *Gentleman's Magazine* for April, 1745.[301] Armed with this article, which not only described experiments made in Europe but also suggested possibilities for further electrical experimentation, Franklin

> eagerly seized the Opportunity of repeating what I had seen at Boston, and by much Practice acquired great Readiness in performing those also which we had an Account of from England, adding a Number of new Ones. I say much Practice, for my House was continually full for some time, with People who came to see these new Wonders. To divide a little this Incumbrance among my Friends, I caused a Number of similar Tubes to be blown at our Glass House, with which they furnish'd themselves, so that we had at length several Performers.[302]

The study of electricity was a relatively new scientific interest. William Gilbert (1540–1603), an English physician, had introduced the subject in his book, *De Magnete* (1600); Otto von Guericke (1602–86), Burgomaster of Magdeburg, better known for his invention of the air pump, constructed the first rotating frictional electrical generator in 1660 and described early experiments in the transmission of electrical power in his book, *Experimenta Nova . . . Magdeburgica . . .* (1672); and Robert Boyle (1627–91) published the first work in English on the subject, in his *Experiments and Notes about the Mechanical Origins or Production of Electricity* (1675). But electrical science essentially began in the early eighteenth century when Francis Hauksbee (d. c. 1713), F.R.S. and Curator of the Royal Society, published *Physico-Mechanicall Experiments on Various Subjects, Containing an Account of several Surprizing Phenomena touching Light and Electricity* (1709), in which he demonstrated the relationship between light and electricity. Stephen Gray (d. 1736), a pensioner of the Charterhouse who was elected to the Royal Society in 1732, conducted experiments by which he distinguished between some conductors and nonconductors of electricity. Charles François du Fay (1698–1739), a French chemist, confirmed Gray's discoveries and discovered (1734) that there are two kinds of electricity, *vitreous*

---

[301] See Lemay, *Ebenezer Kinnersley*, pp. 49–59.
[302] Labaree, ed., *The Autobiography of Benjamin Franklin*, pp. 240–241.

(generated by rubbing glass with silk) and *resinous* (generated by rubbing resin with wool or fur), and that unlike kinds attract each other but like kinds repel each other. In 1741 Jean Théophile Desaguliers, Curator of the Royal Society, with whom Isaac Greenwood had worked in London in the 1720's, won the Copley Medal (his third) from the Royal Society for extraordinary experiments in electricity. He had also discovered that the subject was highly attractive in his philosophical lectures; indeed, by this time, such lectures had won wide popular appeal in Europe and had become a fad in scientific and pseudo-scientific circles, including those of the royal courts. In 1745 two men, E. G. von Kleist, a clergyman of Pomerania, and Pieter van Musschenbroek (1692–1761), a Dutch mathematician and physicist at Leiden, almost simultaneously discovered the "Leiden jar," a primitive battery with which electrical charges could be "stored" and discharged at will. About this time, too, Franklin and his friends entered the field. Ebenezer Kinnersley, with Franklin's assistance, began his series of popular lectures and demonstrations in 1749, and by the 1750's electricity was a widely popular subject in the American colonies as it had become in the Old World.[303]

The article in the *Gentleman's Magazine*, "An historical account of the wonderful discoveries, made in Germany, &c concerning Electricity," was reprinted in the *American Magazine* (December, 1745), and it excited interest in several colonies. William Claggett, an itinerant clock-maker, fashioned his own apparatus and gave electrical demonstrations at Boston and Newport. John Winthrop, the Hollis Professor of Mathematics and Natural Philosophy at Harvard, devoted to electricity his twenty-sixth lecture in a course of philosophical lectures (May 10, 1746). Cadwallader Colden wrote to Franklin (August 3, 1747), stating that some gentlemen of New York were desirous of doing electrical experiments and, as he understood that Franklin had the apparatus, he would be glad to purchase like pieces if Philadelphia artisans could duplicate them. Franklin replied (August 6) that the apparatus could be duplicated

---

[303] B. Dibner, *Early Electrical Machines: The Experiments and Apparatus of the Two Enquiring Centuries (1600 to 1800) That Led to the Triumphs of the Electrical Age* (Burndy Library Publications, No. 14, Norwalk, Conn., 1957).

and that he would supervise its construction.[304] Already, on June 5, 1747, Franklin had sent Colden the first part of his "Electrical Journal" with a glass tube and directions for rubbing it to generate electricity. And already Franklin was beginning to note interesting distinctions, having proved to his own satisfaction that the "Electrical Fire" is an element of itself, not *created* by friction but *collected* by it.[305]

By 1747 Franklin was "totally engrossed" in the study of electricity. The next year he retired from business and planned to devote himself wholly to scientific pursuits. Several members of the Philadelphia community, especially Philip Syng (1703–89), a silversmith, Thomas Hopkinson (1709–51), a lawyer, and Ebenezer Kinnersley, the preacher about to turn lecturer, assisted him in his work, and he consulted with Cadwallader Colden and Colden's son David, who did creative electrical experimentation, as well as with James Bowdoin, Professor John Winthrop, Dr. John Lining, and other colonial men of science. He kept detailed minutes of the experiments he made, with occasional memoranda of others to be made later. Adopting William Watson's theory of electricity as a *single* elastic fluid which is transferred from a body with a greater charge to a body with a lesser charge until equilibrium is established (1746), Franklin developed it further, renamed Du Fay's "vitreous" electricity "positive" and his "resinous" electricity "negative," applied this theory to explain the Leiden jar, and pointed out that electricity, like Newton's magnetism, was capable of exerting force at a distance. He also became convinced of the identity of lightning and electricity and sought experimental means to prove it. In a letter to Dr. John Lining (March 18, 1755), Franklin quoted from the minutes of his experiments of November 7, 1749:

Electrical fluid agrees with lightning in these particulars: 1. Giving light. 2. Colour of the light. 3. Crooked direction. 4. Swift motion. 5. Being conducted by metals. 6. Crack or noise on exploding. 7. Subsisting in water or ice. 8. Rending bodies it passes through. 9. Destroying animals. 10. Melting metals. 11. Firing inflammable substances. 12. Sulphureous smell. The electric fluid

304 *Colden Papers*, III, 409–410, 414–415.
305 *Ibid.*, III, 396–398.

is attracted by points. We do not know whether this property is in lightning. But since they agree in all the particulars wherein we can already compare them, is it not probable they agree likewise in this? Let the experiment be made.[306]

The fact that "the electric fluid is attracted by points," said to have first been brought to Franklin's attention by his friend and coworker, Thomas Hopkinson, led Franklin to speculate that pointed rods might draw lightning from thunderclouds, thereby suggesting both an experiment to prove the identity of lightning and electricity and the application of the theory to the construction of lightning rods. In his letters to Collinson, published first in London by Edward Cave (1751) and later in the *Philosophical Transactions* (1753), Franklin proposed an experiment to prove the identity of lightning and electricity by placing pointed rods on the top of "some high tower or steeple" from the lower end of which a man, standing on an insulator, might draw sparks from thunderclouds during a storm. But Franklin did not immediately act upon his suggestion, and as Cave's edition of his *Experiments and Observations on Electricity* was translated into French and published at Paris early in 1752, three French scientists, Buffon, T.-F. D'Alibard, and one M. de Lor, with the enthusiastic support of Louis XV, tried the experiment at Marly in May, 1752, with complete success. Indeed, during the summer of 1752, the experiment was repeated many times—at Paris and elsewhere in France, in Belgium, Berlin, Dublin, and in London—by Watson and John Canton (1718–72), F.R.S., another electrical scholar of the Royal Society.[307] Franklin had become famous in Europe before he knew of it in America.

In the meantime, however, Franklin had conceived of another, simpler, way of testing his theory. This was the famous kite experiment, which he performed in June, 1752, a month after the French electricians had first verified his theory. But for reasons which are obscure, Franklin did not report the success of his kite experiment until the following October 1, following up a letter

[306] Leonard W. Labaree, ed., *The Papers of Benjamin Franklin* (in progress, New Haven and London, 1959———), V, 524.

[307] All of these were reported to the Royal Society. *Journal-Book*, XXII, 214, *passim*.

to Collinson on that date with a full account of the "Electrical Kite" in the *Pennsylvania Gazette* (October 19), and a description of "How to Secure Houses, etc., from Lightning" in *Poor Richard's Almanac* for 1753 a short time afterward. Possibly, as a biographer suggests, Franklin, having learned of the success of the European experiments before he was ready to report the success of his kite experiment, felt that there was no need to verify an experiment which had already been verified satisfactorily by others.[308] Franklin's letter to Collinson was read to the Royal Society on December 21, 1752. He reported "making a Machine or Kite" by which, when raised into the air during a thunderstorm, "The Electric Fire may be drawn from the Clouds to such a degree as to charge a Phail, kindle Spirits, and perform all the other Experiments which are usually done by rubbing a Glass Globe or Tube, and thereby the Sameness of the Electric matter with that of Lightning may be demonstrated."[309]

The magnitude of this basic contribution to science, together with the international reputation accorded Franklin because of it, moved the Royal Society to take unprecedented action. The Copley Medal, originating (1709) in a legacy from Sir Godfrey Copley, Bart., F.R.S., was regarded as the highest scientific distinction that the Royal Society could bestow. Since the recent death of Sir Hans Sloane, who had been the last surviving trustee of the Copley estate, the decision regarding the bestowal of the medal rested with the President and Council of the Society. On November 30, 1753, the President, George, Earl of Macclesfield, announced to the Society the decision of the Council to award the Copley Medal to "Benjamin Franklin Esquire of Philadelphia in America on Account of his Curious Experiments and Observations on Electricity." Franklin was honored, continued the President,

for though he be not a Fellow of this Society nor an Inhabitant of this Island, is a Subject of the Crown of Great Britain and must

308 Van Doren, *Benjamin Franklin*, pp. 164–168.
309 *Journal-Book*, XXII, 216–217. This letter was published in *Phil. Trans.*, XLVII (1751–52), xcv, 565–567. This volume also printed (pp. 289–291) a letter from Franklin to Collinson, dated June 20, 1751, read to the Society Nov. 14, 1751, about a ship whose masts had drawn lightning "as by points, from the cloud." Franklin recommended lightning rods on ships' masts.

be acknowledged to have deserved well of the Philosophical World and of this learned Body in particular. . . . [Indeed, the Council had not thought it fit] to confine this Benefaction within the narrow limits of any particular Country much less to the Society itself. For they were of opinion that learned men and Philosophers of all Nations ought to entertain more enlarged notions; that they should consider themselves and each other constituent parts and Fellow members of one and the same illustrious Republic, and look upon it to be beneath Persons of their character to betray a fond partiality for this or that particular district where it happened to their own lot either to be born or reside. But their benefactions should be universally diffused and as extensive as the knowledge they profess to pursue, and should be sensibly felt by all who in their respective stations contributed their proportion to the Common Stock of the whole by their endeavours to promote & advance Science and usefull knowledge wherein alone the true intent and welfare of such a Republic consist.[310]

Franklin's reaction to this idealistic, stuffy, and somewhat patronizing speech would have been delightful to know. But he was not there. When he heard of it, he dryly referred to it as "a very handsome speech." The Society gave the medal to Peter Collinson, who conveyed it to Franklin by the hand of Captain William Denny, successor to Robert Hunter Morris as Governor in Pennsylvania in August, 1756.

By this time the Royal Society had received several more communications from Franklin, through the good offices of Peter Collinson. Two letters, dated respectively September, 1753, and April 18, 1754, and read to the Society June 27 and July 4, 1754, modified some of Franklin's earlier conclusions: the luminosity in sea water was not the effect of electricity after all, and the electricity in clouds was not always negative electricity, as Franklin had previously reported. On December 12, 1754, Collinson presented a book by Franklin listed as "New Experiments on Electricity part 3d."[311] The following year (December 18, 1755), Collinson presented two more letters; the first, dated at Philadelphia, March 14, 1755, contained "Electrical Experiments, made in pursuance of those of Mr. John Canton, dated Decemb. 3, 1753; with Explana-

[310] Journal-Book, XXII, 410–414.
[311] Ibid., XXII, 587–591; XXIII, 8.

tions by Mr. Benjamin Franklin," and demonstrated that electrical atmospheres flowing around nonelectric bodies repel each other and will not mix or unite, i.e., negative charges repel negative charges, and positive charges repel positive charges; the second was a letter by Franklin to T.-F. D'Alibard of Paris, June 29, 1755, to explain how lightning rods function, to insist that the rods should be pointed and grounded, and to describe the success of such rods on the steeple of the Newbury church in Massachusetts, which Franklin had observed in November, 1754. Both letters were published in the *Philosophical Transactions.*[312]

On the following January 29, there was posted at the Royal Society the following certificate:

> Benjamin Franklin Esqr of Philadelphia, a gentleman who has very eminently distinguished himself by various discoveries in Natural Philosophy, & who first suggested the experiments to prove the analogy between lightning and Electricity, being desirous of being elected a fellow of the royal Society, is recommended by us in Consideration of his great Merit & of his many Communications as highly deserving the honour he desires.
> London, 29 January 1756.
>
> > Macclesfield
> > [Rev. Mr. William] Parker
> > [Hugh, Lord] Willoughby of Parnham
> > P. Collinson
> > W. Watson
> > Tho. Birch
> > James Parsons
> > Jno. Canton[313]

On April 29 following, Franklin was elected a Fellow of the Royal Society. Before Franklin could have acknowledged this honor, the Council of the Society, at a meeting held on July 15, 1756, upon the motion of William Watson, voted "in the Affirmative nemine contradicente" that "the Name of Mr. Benjamin Franklin who had deserved so highly of the Society, and whose affairs oblige him to reside at Philadelphia, be inserted in the Lists before his Admission and without any Fee, or other payment to the Society. And that

---

[312] XLIX, Pt. I (1755), 300–309. See also *Journal-Book*, XXIII, 213–216.
[313] *Certificates, 1751–1766,* unpaged.

such Name be continued in the Lists, So long as he shall continue to reside Abroad."[314]

In his *Autobiography*, Franklin later indicated that he had been slighted by the Royal Society at the outset; only after the Society had verification of the success of "the experiment of procuring lightning from the clouds by a pointed rod" (summer, 1752), had it "made me more than amends for the slight with which they had before treated me" and elected him to fellowship "Without my having made any application for that honor." A review of the evidence from the minutes of the Royal Society's meetings between 1749 and 1752 hardly bears out Franklin's charge. Actually, Franklin's presentation of the sequence of events appears to have been in error. He appears to have believed that Watson's report on his letters (June 6, 1751) was *after* both the French experiments and those of Watson and Canton had verified the proposed experiment with pointed rods. He also stated that when his first letter (to Dr. Mitchell) had been read to the Society (November 9, 1749), Dr. Mitchell reported that one of "the connoisseurs" in the Society laughed at Franklin's suggested identification of lightning and electricity (a suggestion already familiar to Europeans, though not yet demonstrated). Perhaps this was true (by no means all the Fellows of the Society were "connoisseurs"); but the official action of the Society as a whole was both entirely respectful and in keeping with its usual reception of little known and untried correspondents. Moreover, Watson's report on the first edition of the published *Experiments and Observations on Electricity* (given in June, 1751, *not* in 1752) was generous in word and tone, and the Society immediately and unanimously voted to publish all of Franklin's communications in the *Philosophical Transactions*. It is true that the publication was delayed until 1753, which may have exacerbated Franklin's dissatisfaction. But the delay was the result of the decision of the Council to place the responsibility for the editing of the *Philosophical Transactions* in the hands of a committee of the Society rather than leave it wholly in the hands of the Society's Secretary, as it had been since the time of its founder, Henry Oldenburg—a decision which had nothing to do with Frank-

[314] *C.M.*, IV, 176; *Journal-Book*, XXIII, 347.

lin's letters and certainly was no slight upon him.[315] It has also been suggested—and the sequence of events in Franklin's *Autobiography* may lend credence to the notion—that Franklin's election to the Society occurred as a consequence of political pressures rather than a candid recognition of his merits as a natural philosopher.[316] If, as Franklin said, he was elected "Without my having made any application for the honor," it is possible that Collinson or other Fellows initiated the action without an explicit expression of desire from Franklin himself. But in view of the Society's usual stout resistance to outside pressures in such matters, it would seem more likely that the Society acted on its own initiative in recognition of Franklin's merits, without (or in spite of) political pressures. Evidence of the latter is wholly circumstantial at best, and judgment of the matter cannot be decisive. If, as it appears in his *Autobiography*, Franklin was miffed at the early treatment accorded him by the Royal Society, it would appear to have been the result of his misunderstanding of the sequence of events, for the Society's record of events is clear, without any hint of an intention to "slight" Franklin. Even Franklin came to feel that the Society had "made me more than amends" for the supposed slight, and his relations with the Society, after he went to London in 1757, were cordial. Indeed, as Van Doren has said, he used the Society almost as if it were his club.

Franklin thanked Collinson "for the Honour they have done him in Electing him into their body" in a letter from Philadelphia dated September 23, 1756.[317] Already, in 1753, he had communicated to the Society (by way of Collinson) "A Letter from Mr. William Shewington . . . dated at Antigua, 20th of June, 1753" with an account of the transit of Mercury over the sun on May 6, 1753. He also sent to Collinson "a course of Correspondence on Philosophical Subjects between Benjamin Franklin Esquire of Philadelphia and Several of his Friends in North America." This included fifteen letters between Franklin, James Bowdoin of Boston, and Cadwallader Colden and his son David of New York.[318] Bowdoin

---

315 For an account of this transfer, see C. R. Weld, *A History of the Royal Society* . . . (2 vols., London, 1848), I, 518–523.

316 Van Doren, *Benjamin Franklin*, pp. 255–257.

317 *Journal-Book*, XXIII, 489. Extract read to the Society Feb. 24, 1757.

318 *Ibid.*, XXIII, 434–443. Collinson presented all of these letters to the Society

hypothesized that as magnetic needles are affected by lightning, perhaps lightning, electricity, and magnetism are of "the same element." He also questioned Franklin's notion that lightning was gathered into clouds over the sea, and suggested that the crookedness of lightning flashes was the result of varying densities of the air. Franklin approved of Bowdoin's explanation of the crookedness of lightning, questioned his own hypothesis about lightning being gathered into clouds over the sea, and suggested, instead, that it might be generated in the air by friction created by winds. Bowdoin approved the latter hypothesis, but pointed to the lack of proof. The discussion then moved about a paper by Franklin entitled "Physical and Meteorological Observations, Conjectures, and Suppositions," in which Franklin sought to explain rain, fog, and winds from the mutual "repulsion of the particles of air, from the different degrees of heat and cold in the upper and lower air, and from the motion of the Earth 'round its axis."[319] He also suggested that waterspouts resulted from whirlwinds. Bowdoin, and later the elder Colden, insisted that waterspouts were descending rather than ascending, but Franklin clung to his interpretation—and members of the Royal Society disputed the matter until the Secretary recorded that "It seems yet undetermined which of the two opinions is the best supported." Cadwallader Colden challenged Franklin's hypothesis regarding the origin of winds but Franklin was unmoved by Colden's objections and added

> an ingenious Conjecture . . . concerning a Region of Electrical Fire above our atmosphere, prevented by our air, and its own too great distance from joining our Earth. Perhaps some of this fluid may be low enough to attach itself to our highest clouds, and thence they becoming electrified, may be attracted by, and descend towards, the Earth and discharge their watery contents together with that Ethereal fire, & perhaps the Aurora Boreales are currents of this fluid in its own region above our atmosphere becoming by their own motion visible.

---

at a series of meetings culminating on Dec. 23, 1756. See *ibid.*, XXIII, 378, 383, 402, 405, 414, 434–443. Shewington's report on the transit of Mercury was published in *Phil. Trans.*, XLVIII, Pt. I (1753), 318–319.

[319] Later published in *ibid.*, LV (1765), 182–192.

Bowdoin suggested further that the luminous appearance of sea water may be attributed to little animals floating on the sea, like tiny fireflies. David Colden contributed a letter to defend Franklin against the French Abbé Nollet, who had attacked Franklin's theories, and the younger Colden also proposed a method to measure the speed of electric sparks moving across a given distance. Franklin raised objections to the latter on the grounds that, because of numerous variables, the experiment would not demonstrate what it was intended to demonstrate. The last letter in the series was by Franklin, a number of mathematical speculations about his well-worn magical squares. It was the only one of the series not read to the Royal Society. Taken altogether, however, the letters well justified the observation of the Secretary of the Royal Society with reference to the dispute over waterspouts, that the discussions were "carried on with great Candor on both sides, with a truly Philosophical Spirit of enquiry."·

Already, however, Franklin had become embroiled in politics. His most creative scientific work was done between the years 1743 and 1752. During these years, too, he had extended his scientific correspondence to worldwide proportions, the consequence of his scientific accomplishments and the manifold honors which were conferred upon him. And in spite of the political obligations which he assumed, he kept up, and even extended further, his scientific correspondence, clearly hoping that a day would come when he could escape from political affairs and resume his scientific work. But that day never arrived. Pennsylvania politics, colonial problems arising from the "Great War for the Empire," critical relations between the colonies and the mother country with the coming of the American Revolution, delicate diplomatic matters in France, and, finally, the Constitutional Convention, kept Franklin abroad, first in England (1757–62, 1764–75) and then in France (1776–85), by which time he had become so old and plagued with physical infirmities that a return to scientific endeavors was no longer feasible. He died in 1790.

However, although Franklin's political affairs consumed the bulk of his time and energies during these years, he retained science as his principal avocation. His reputation in Europe as a scientist was greater than he himself realized, and it aided him materially in his political affairs both in England and in France. After his arrival in England in 1757, he visited the Royal Society (November 24), signed the Obligation, and was formally admitted.[320] Thereafter he attended the meetings of the Society frequently, became personally acquainted with Fellows whom he had known before only by correspondence or by reputation, and took an active part in the discussions. On January 12, 1758, his friend Dr. John Pringle (later, in 1766, Sir John) read to the Society "An Account of the Effects of Electricity in paralytic cases. In a Letter . . . from Benjamin Franklin, Esq., F.R.S." Franklin reported that he had experimented with electric shock treatments years before—in conjunction with Dr. Cadwalader Evans of Philadelphia[321]—and he had found the effects to be only of temporary value, with no permanent cure.[322] In 1760 Franklin presented to the Society extracts from letters about experiments with electricity from Giovanni Battista Beccaria (1716–81), his Italian friend and disciple, and he read a paper (February 14) to explain some of the latter's experiments.[323] On November 20, 1760, Franklin was chosen as one of the auditors of the Society's accounts for the year, and on the following December 1 he was elected to the Council of the Society for 1760–61.[324] In March, 1761, he was appointed to the committee to select papers to be printed in the *Philosophical Transactions*.[325] In 1763 he presented to the Society (February 4) letters by himself and by John Canton to describe the latter's experiments "as they were performed in the presence of Dr. Franklyn, to account for the singularities observed by Mr. Delaval, in his electrical experiments upon Portland stone, Tobacco pipes, Wood &c. under different degrees of heat."[326] On March 18 he presented extracts from another letter by Beccaria regarding the refraction of light in

[320] *Journal-Book*, XXIII, 627–628.

[321] See Evans, "A Relation of a Cure performed by Electricity . . . Communicated Oct. 21, 1754," *Medical Observations and Inquiries*, I (1763), No. XI, 83–86. The case was treated by Franklin in Sept., 1752.

[322] Franklin's letter appeared in *Phil. Trans.*, L, Pt. II (1758), 481–483. See also *Journal-Book*, XXIV, 27–28.

[323] *Ibid.*, XXIV, 503–509.

[324] *Ibid.*, XXIV, 113 (mispaged). The Council minutes for 1760–61 show that Franklin was present at six meetings. *C.M.*, IV, 227, 279, 283, 291, 303, 305.

[325] *Add. Mss. 4446*, fols. 181, 189, 192, 199 (Birch minutes).

[326] *Journal-Book*, XXV, 34–37; *Phil. Trans.*, LII, Pt. II (1762), 456–461.

prisms; and on May 13 following he communicated a letter from Samuel Evatt, of Ashford, Derbyshire, describing a "circular sepulchral Monument" uncovered in road-building nearby.[327] He would have been gratified when, on December 16, Watson read a paper by the Honorable George, Lord Anson, Admiral of the Fleet, on "Suggestions concerning the preventing of mischiefs to ships, and their Masts by Lightning," by the use of Franklin's lightning rods.[328] But by this time Franklin had returned to Philadelphia.

After his return to England in 1764, Franklin continued his attendance at the meetings of the Royal Society. He was re-elected to the Council for 1766–67, 1767–68, and 1772–73, and was diligent in attendance at the Council's meetings.[329] On March 19, 1767, he presented to the Society a remarkable paper from Professor John Winthrop, F.R.S., of Harvard College, entitled *Cogitata de Cometis*. Winthrop had formerly (April, 1760) sent this paper to James Bradley (1693–1762), F.R.S., the Astronomer Royal, but Bradley had apparently ignored it.[330] Now, in 1767, Franklin presented to the Royal Society a revised version of the paper and read an English "Abstract of the Latin Memoir." The minutes of the meeting recorded that the paper was a very excellent piece on gravitation and the gravitational balance of the universe, suggesting that by observations and reasoning it is possible "to determine the quantity of matter and the density of the particular Comets," a wholly new approach that had not been attempted previously. The entire paper, probably Winthrop's best effort in speculative astronomy, was published in the *Philosophical Transactions*.[331]

---

[327] *Journal-Book*, XXV, 81–85, 131–132; *Phil. Trans.*, LII, Pt. II (1762), 544–546.

[328] *Journal-Book*, XXV, 242–246.

[329] *C.M.*, V, 156–157, *passim*, 204, *passim*; VI, *passim*. He attended eight Council meetings in 1766–67, fourteen in 1767–68, and ten in 1772–73.

[330] *Bradley 46*, fols. 2–10 (Bodleian Library): a bound manuscript entitled "Cogitata quaedam de Cometis; Viro Reverendo Jacobo Bradleio S.T.D. Astron: Reg: Grenov. Astron: apud Oxon: Prof: Savil. et S.R.S. Lond. et Par. quam humillime inscripta a Johanne Winthrop," dated "Cantabr. Nov. Angl. 18 Aprilis 1760," includes four figures, and is endorsed in another hand: "Winthrop. See *Phil. Trans.*, Vol. 57, p. 132. This varies from the printed copy. S.P.R."

[331] LVII, Pt. I (1767), 132–154. The revision is dated May 7, 1766. Franklin received it Jan. 14, 1767. See also *Journal-Book*, XXVI, 566–570.

Franklin also presented to the Society several letters from his English correspondents. One in particular, by Messrs. Lane and Read, presented an account of an electrometer designed by Read, a mathematical instrument-maker of Knightsbridge, to measure quantities of electricity, although the unit of measurement is not clear.[332] In March, 1769, Franklin was appointed to a committee of the Royal Society to consider an appeal from the Dean and Chapter of St. Paul's for "instructions in what manner the Cathedral of St. Paul's would be secured from damage by lightning." Watson read the committee's report (June 8), which advised construction of "a sufficient quantity of metalline conductors from the summit of the great cupola & those of the two smaller ones & downwards to the Earth." The Society resolved to continue the committee until "the business be completed."[333]

Further matters involving lightning rods attracted Franklin's attention. On March 29, 1770, he presented to the Society a letter from a ship captain, J. L. Winn, who had experimented with a lightning rod from the top of his mast into the sea and found it undeniably protective. It was hoped that the account would remove "the present backwardness of the Gentlemen of the Navy to fixing conductors on the masts of ships."[334] Again, on July 9, 1772, Franklin was appointed to a committee to advise the Board of Ordnance, whose Secretary sought expert aid in affixing lightning rods on the powder magazine at Purfleet. When the report was ready in December, Benjamin Wilson, F.R.S., a member of the committee, issued a dissenting voice, vigorously protesting the use of pointed rods, which, he said, invited lightning and increased the likelihood of a strike. He favored the use of blunted rods, which, he believed, would decrease the danger. Franklin insisted upon pointed rods, and as Cavendish, Watson, and Robertson, the other members of the committee, supported him, the advice of the committee, with the support of the Royal Society, was to equip all five of the buildings of the powder magazine with pointed lightning rods.[335]

[332] *Ibid.*, XXVI, 657–659 (Nov. 27, 1767). See also *passim.*
[333] *Add. Mss. 5169B*, fols. 176, 203 (Morton minutes); *Journal-Book*, XXVII, 189, 226–227. Others on the committee were: Edward Delavall, W. Watson, Benjamin Wilson, John Canton.
[334] *Ibid.*, XXVII, 331–332; *Phil. Trans.*, LX (1770), 188–191.
[335] *Journal-Book*, XXVII, 656; XXVIII, 3–6.

A year later (December 16, 1773), William Henley, F.R.S., presented a paper, citing evidence from Professor John Winthrop of Harvard College, further supporting the use of pointed rods.[336] But there was a humorous sequel. In 1777, after the American Revolution was under way, one of the buildings at Purfleet was struck by lightning. The damage was not extensive, but when the matter was referred back to the Royal Society for an opinion, "a fellow of the Society stood up [Benjamin Wilson] & read a paper expressing his disapprobation of the pointed conductors." But "the question being put as before was affirmed by a very great Majority [in favor of pointed rods]."[337] However, King George III, incensed at the colonials, and inclined to place heavy blame for the American Revolution upon Franklin's shoulders, was convinced that Franklin (and his pointed rods) had been responsible for the damage at Purfleet. In a fit of political frenzy, George III identified all those who favored pointed lightning rods with the revolutionary colonists. The King ordered blunt rods placed on the Purfleet magazine and upon his own palace. He also sought to persuade the Royal Society to reverse its position, without success. Sir John Pringle, Franklin's close friend, President of the Royal Society (1772–78) as well as Physician to His Majesty, told the King: "I cannot reverse the laws and operations of nature," a statement which led to his prompt dismissal from the King's service. Benjamin Wilson, of course, became a royal favorite. Some wag wrote an epigram on the absurd affair:

> While you, George, for knowledge hunt,
> And sharp conductors change for blunt,
> The Nation's out of joint:
> Franklin a wiser course pursues,
> And all your thunder useless views,
> By keeping to the point.[338]

In the meantime, Franklin had presented to the Royal Society several papers largely relating to colonial astronomical observations. The first, read December 21, 1769, was "Observations of the late

[336] *Ibid.*, XXVIII, 246.

[337] *Ibid.*, XXIX, 383 (June 19, 1777); *Add. Mss. 5169D*, fols. 149–153.

[338] Weld, *Royal Society*, II, 95–102; Dorothea Waley Singer, "Sir John Pringle and His Circle—Part I—Life," *Annals of Science*, VI (1919), 168–180.

Transit of Venus . . . made by Mr. Owen Biddle and Mr. Joel Bayley, at Lewiston, Pensylvania"; the second, read April 5, 1770, was an extract of a letter from Professor Winthrop to inquire "whether the phases of the transit of Venus be accelerated or retarded by the abberation of light"; the third, presented December 13, 1770, was an "Account of the transit of Mercury observed at Noriton in Pensylvania Novr 9, 1769, agreeable to an appointment of the American Philosophical Society held at Philadelphia for promoting useful knowledge; by Willm Smith, Provost of the College of Philadelphia, John Lukens, Esqr Surveyor Gen'l of Pensylvania, David Rittenhouse, M.A., and Mr. Owen Biddle"; the fourth, read January 10, 1771, communicated Professor Winthrop's observations on the same phenomenon; and the last of this series presented Professor Winthrop's defense of Sir Isaac Newton against "a Passage in Castillione's Life of Sir Isaac Newton." All of these items were published in the *Philosophical Transactions*.[339]

With the exception of the last paper presented by Franklin to the Royal Society before his return to America in 1775, all of his subsequent communications were supplied by his correspondents. On April 9, 1772, he presented "An Account of the Effects of lightning on the Chappel in Tottenham Court Road 22nd March, 1772" as observed by William Henley, Edward Naire, and William Jones. Henley had sent the account, and he originated three more in the course of the following year: observations on the effects of atmospheric electricity, presented April 30, 1772; an electrometer by Henley, described by Joseph Priestley, presented by Franklin on May 28, 1772; and a paper, "Account of some Experiments in Electricity Made in the Years 1771 and 1772. With the Description of a new Prime Conductor for those purposes, to which are added a few Observations on the Electricity of the Atmosphere," communicated April 1, 1773.[340] Henley was elected Fellow of the Society the following month. In June Franklin read to the Society an extract of a letter from Dr. J. M. Nooth (also elected F.R.S. in March, 1774) describing improvements in electrical machines to

---

[339] *Journal-Book*, XXVII, 335, 405–406, 419–420; XXVIII, 273–274; *Phil. Trans.*, LIX (1769), 414–421; LX (1770), 358–362, 504–507; LXI, Pt. I (1771), 51–52; LXIV, Pt. I (1774), 153–157.

[340] *Journal-Book*, XXVII, 616–617, 620, 641–642; XXVIII, 108–114.

lessen absorption.[341] On July 1 Franklin presented three letters from John Walsh, F.R.S., to describe experiments and observations, including some from France made in 1772, regarding the electric eel.[342] Humphrey Marshall, of West Bradford, Pennsylvania, John Bartram's kinsman, sent Franklin (May 3, 1773) a long manuscript to describe more than two years' work in observing sunspots. It embraced about 300 observations, with miniature pencil drawings of sunspots "done with great neatness and precision," together with some lesser observations on the satellites of Jupiter and some meteorological records. Franklin made an extract of the account which he presented to the Royal Society on February 3, 1774, and the extract was published in the *Philosophical Transactions*.[343] The last communication made by Franklin to the Society before his return to America in April, 1775, was presented (June 2, 1774) by Dr. William Brownrigg, F.R.S. It consisted of several letters by Brownrigg, Franklin, and the Reverend Mr. Farish regarding experiments made in the English Lake Country (Derwentwater) of the effects of oil on water, i.e., to see if oil would quiet troubled waters. This was a subject about which Franklin had speculated since 1757, having read about the ancient practice in Pliny. With a party from the Royal Society, he had conducted experiments in Portsmouth Harbor in October, 1773. The results had appeared inconclusive at the time, but Franklin was still confident that, given proper techniques (and sufficient quantities of oil!), the method had merit.[344]

Since his early days in Philadelphia, Franklin had promoted both science and scientists in many ways. In London he continued these promotions, serving as a pipeline to the Royal Society for both American and European scientists. He recommended numerous colonials and Europeans for membership in the Royal Society, and, conversely, he recommended prominent European scientists to the new American Philosophical Society, which he had helped to organize and of which he was President for many years. In the

[341] *Ibid.*, XXVIII, 164–167.
[342] *Ibid.*, XXVIII, 169–175; *Phil. Trans.*, LXIII, Pt. II (1773), 461–480.
[343] LXIV, Pt. I (1774), 194–195. See also *Journal-Book*, XXVIII, 284–286.
[344] *Ibid.*, XXVIII, 399–406; Van Doren, *Benjamin Franklin*, pp. 419, 433–436. The letters were published in *Phil. Trans.*, LXIV, Pt. II (1774), 445–460.

latter promotions he was materially aided by his presence in Europe, where he met personally many of the foremost European scientists. While in England he had traveled on the Continent: to Belgium and Holland in 1761, where he met the co-inventor of the Leiden jar, Pieter van Musschenbroek; to Hanover, with his friend Sir John Pringle, in 1766; to France (again with Sir John) in 1767 and 1769, where he was enthusiastically received by French scientists (he was made a foreign member of the Académie Royale des Sciences, elected to some of the French provincial academies, and both T.-F. D'Alibard and Mme. D'Alibard were singularly hospitable). All this was, of course, to his advantage when in 1776 he returned to Paris to represent the Second Continental Congress and ultimately to assist in the negotiation of the Treaty of Paris in 1783, which sealed American independence.

But in the midst of his political activities, his interest in science never wavered. As has been noted, he strove to direct American scientific communications to Paris after the war had interfered with connections to London and the Royal Society. Both in London and in Paris he consulted with a host of European scientists, exchanging views on a wide number of subjects, introducing Europeans to the fruits of his fertile mind, and, by his associations with them, enriching and deepening his own scientific information. When time permitted, he engaged in scientific experimentation and writing. After he had witnessed some of the balloon ascents in Paris (1783), he quickly wrote to President Joseph Banks of the Royal Society of London to describe the phenomena and to speculate about the future of this earliest conquest of the air by man.[345] And on his return voyage from France in the late summer of 1785, he continued researches on the Gulf Stream, which he had begun at least as early as 1775, writing his last significant scientific paper en route. He had been informed that American ship captains made quicker passages across the Atlantic than English mail packets made, and upon consultation with a Nantucket whaling captain, he concluded that it was because the Americans, understanding the Gulf Stream better, rode with the stream eastbound as much as possible, and cut across it as squarely as possible westbound in order to avoid sailing against

[345] Franklin to Sir Joseph Banks, Passy, Aug. 30, Oct. 8, 1783, *Journal-Book*, XXXI, 413; read Nov. 6, 1783.

its current any more than necessary. He sought to identify and to chart the course of the Gulf Stream, and he urged shipmasters to make frequent tests of the temperature of the sea water in order to learn when they entered and when they left the Gulf Stream. These recommendations, together with evidence which he had gathered and other fruitful suggestions for the improvement of safety at sea, he combined in "Maritime Observations," which he dispatched in a letter to Julien-David Le Roy (1724–1803), a French architect and naval buff, brother of Jean-Baptiste Le Roy, a French electrician who had been Franklin's neighbor and friend at Passy. Franklin subsequently read this and other papers prepared on the voyage to the American Philosophical Society, which printed them in the second volume of the Society's *Transactions* (Philadelphia, 1786).[346]

Franklin's scientific works have been subjected to the most searching studies accorded any of the American colonial scientists. In a number of publications, of which three are of prime significance, Professor I. Bernard Cohen has reviewed and evaluated Franklin's work with informed care. The first, *Benjamin Franklin's Experiments: A New Edition of Franklin's Experiments and Observations on Electricity*,[347] is the first American edition of Franklin's work, although it had passed through eleven editions in Europe before the American Revolution—five in English, three in French, one in Italian, one in German, and one in Latin: a revealing index of Franklin's reputation and influence abroad! Professor Cohen's second work, *Franklin and Newton: An Inquiry into Speculative Newtonian Experimental Science and Franklin's Work in Electricity as an Example Thereof*[348] is enlightening with regard to the non-mathematical works of many scientists, including Franklin and other

[346] The letter to Le Roy is in Albert Henry Smyth, ed., *The Writings of Benjamin Franklin* (10 vols., New York, 1905–07), IX, 372–413. For further information, see Franklin Bache, "Where Is Franklin's First Chart of the Gulf Stream?" *Proceedings of the American Philosophical Society*, LXXVI (1936), 731–741 plus maps. Bache concludes (p. 732) that Franklin "had the first chart of the Stream engraved, made the first scientific observations of it and was the first to ascribe correctly its causes."

[347] Cambridge, Mass., 1941.

[348] *Memoirs of the American Philosophical Society*, Vol. XLIII (Philadelphia, 1956).

colonials, who, accepting—though not always fully comprehending —Newton's answers to fundamental questions, sought to explore paths of research which Newton had postulated for future study. Among these followers of Newton, Professor Cohen finds that Franklin was a model disciple. The third work, an article entitled "The New World as a Source of Science for Europe,"[349] gives a brief overview of the work of some of the principal colonial scientists and concludes that Franklin was the only one whose work was "of such major importance that it is worthy of being recorded in every general history of scientific thought."

Professor Cohen's estimate is correct, although, as will be noted in the next chapter, by his definitions he tends to downgrade the accomplishments of several other colonials. But Professor Cohen also maintains that Franklin properly belonged to a European school, rather than an American one, and that his science was peculiarly a product of European scientific culture. It is difficult to understand why Franklin, especially the Franklin who performed his most important scientific work *before* he had spent any extended time in Europe,[350] should be considered as any more the product of European scientific culture than Cotton Mather or Professor John Winthrop, all of whom were native-born colonists, all of Boston birth. Indeed, of these three, Franklin had the least formal education. Largely self-educated, he may well have been more exposed to colonial folkways than his college-bred contemporaries. It appears possible that Professor Cohen confuses the *acceptance* of Franklin's scientific work, which he rightly points out was more celebrated in Europe than in America, with the *production* of this work, which was wholly colonial insofar as that of any colonial scientist was colonial, and subject to cultural influences identical to those of other colonials. It is true, as Professor Cohen maintains, that Franklin founded no school of science in America, left no disciples (other than Ebenezer Kinnersley), and established no tradition of

[349] *Actes du IXᵉ Congrès International d'Histoire des Sciences* (*Collections de Travaux de l'Académie Internationale d'Histoire des Sciences*, No. 12 (UNESCO), Barcelona and Paris, 1960), I, 95–130.

[350] Franklin's youthful stay in England (1724–26) was of scarcely more than eighteen months' duration, and his time was devoted largely to the perfection of his knowledge and skill in the printing trade.

continuing research bearing his name or stemming directly from his influence. However, this was equally true of all the colonial scientists. Their work was personal, and it never became institutionalized (even in the colleges) to such an extent that they established any specific "schools" of continuing scientific research traditionally associated with their names and continuing beyond their individual spans of life. In fact, no such "schools" arose in America until well into the nineteenth century, unless the medical faculty at Philadelphia, which graduated its first class in 1768, near the close of the Old Colonial Era, could be so characterized.

However, Franklin's work on electricity clearly marks him as the premier colonial scientist. He brought to his researches fresh insights, unencumbered by great mathematical learning and thorough exposure to the ideas and traditions of European speculative philosophy. Like several other colonials, he worked in areas outlined by Isaac Newton for future research, without complete mastery of Newtonian principles. Still, he had mastered the scientific method, despite a confessed weakness for framing hypotheses, and both the construction of his experiments and his careful conclusions founded upon them—all meticulously set down in his journals and well described in his letters to Collinson and others—were, as Professor Cohen has said, "a model of the experimenter's art and of literary expression in science."[351] He began his work in a relatively new field, and he quickly caught up with, and soon surpassed, the level of electrical understanding of his European contemporaries. He discovered new electrical phenomena and set forth sound principles upon which to explain, correlate, and predict electrical events. He enlarged the boundaries of Newtonian natural philosophy, and his fresh insights led to new points of view with regard to the constitution of matter, the composition of bodies in relation to their physical properties, and the behavior of various kinds of matter under controlled laboratory conditions. He entered upon a field in which only scattered bits of evidence were known, and he left it, after a scant decade of experimental study, an organized science, framed in new general principles that not only accorded with the Newtonian world view but also enlarged the contemporary view

[351] Cohen, *Franklin and Newton*, p. 73.

regarding the possibility of a complete mechanical explanation of the universe.[352]

Still, it is difficult to avoid the conclusion that Franklin was lucky. Equipped with meager knowledge for advanced study in the physical sciences, he was, at the outset, so ignorant of the work of his predecessors that he could not determine whether his discoveries actually were new. Indeed, there may be weight to the contention that the more Franklin read the less original his scientific works appeared to be. Perhaps, as Daniel J. Boorstin has said, Franklin's achievements illustrated "the triumph of naïveté over learning."[353] His naïveté prompted him to present his thoughts about electricity without fear of attack or contradiction, in language as fresh as the ideas themselves, interlarded with wit, humor, and unusual clarity. But his ideas were so original that they represented the triumph of genius over initial naïveté. And it was the quality of this genius, which included a thorough understanding of the experimental method in science, that enabled Benjamin Franklin to avoid the pitfalls into which Cadwallader Colden had fallen.

John Winthrop (1714–79), F.R.S., and Colonial Astronomy

John Winthrop was the son of Chief Justice Adam Winthrop of Massachusetts Bay Colony and the great-grandnephew of the first colonial Fellow of the Royal Society. He graduated A.B. from Harvard College in 1732 and A.M. in 1735. As noted above,[354] he had been a student of Isaac Greenwood, the first Hollis Professor of Mathematics and Natural Philosophy at Harvard, although it is clear from his commonplace book that his initial interest in science was kindled by reading Cotton Mather's *Christian Philosopher*. When, in 1738, Professor Greenwood was forced to resign because of intemperance, Winthrop, at the age of twenty-three, was appointed to the chair and publicly installed (January 2, 1738/39) as

[352] *Ibid.*, p. 285.
[353] Boorstin, *The Americans: The Colonial Experience* (New York, 1964), p. 252.
[354] P. 455.

Harvard's second Hollis Professor of Mathematics and Natural Philosophy.[355]

It was a grave responsibility—and offered great opportunities—for one so young and untried. But Winthrop rose to the occasion magnificently and retained the chair until his death in 1779. He proved to be a careful, exact, and systematic scientist who, like his predecessor, went beyond the classroom to inform the public at large about scientific phenomena with public lectures and demonstrations, newspaper accounts, and other publications. Within a few years, he emerged as the leading scientist in the Boston community. He collected one of the best private libraries of scientific books in the colonies and added to the scientific apparatus which Thomas Robie and Isaac Greenwood had collected at Harvard wherewith to lecture and demonstrate to students—with occasional public lectures as well—the laws of mechanics, light, heat, and the movements of celestial bodies in accordance with Newtonian doctrines. As Winthrop also employed the apparatus for his own scientific researches, it may well be said that he established at Harvard the first institutional laboratory of experimental physics. He introduced to Harvard's mathematical curriculum the elements of fluxions (i.e., differential and integral calculus), and he stoutly endorsed Franklin's theories of electricity with lectures and demonstrations of electrical experiments. After Harvard Hall was destroyed by fire early in 1764, with all of the scientific apparatus which Winthrop had accumulated (including, of course, that which had been inherited from Greenwood), Winthrop appealed to the public for aid. He collected donations from about 150 persons and, with the advice and assistance of Franklin from England, together with that of other able scientists and some of the finest instrument-makers in England, he assembled at Harvard a larger, more up-to-date set of scientific apparatus than the college had ever possessed before—better, probably, than that of any other colonial institution.

As a teacher, Winthrop introduced the new science and its methods to four decades of Harvard students, and he imparted the first impulse to at least one whose subsequent scientific achievements outshone his own. This was Benjamin Thompson (1753–

---

[355] *Boston Gazette*, No. 990 (Jan. 1–8, 1738/39).

1814), of Concord, New Hampshire. But Thompson did not imbibe Winthrop's patriotic fervor during the American Revolution. A Loyalist, he went to England and was elected Fellow of the Royal Society of London only a few weeks before Winthrop's death. Knighted in 1784, he subsequently entered the service of the Elector of Bavaria, who created him a Count of the Holy Roman Empire (1791). Known thereafter as Count Rumford, he won high distinctions as a physicist and adventurer, but he appears to have treasured his early days as a student of Professor Winthrop, to whom he referred as "an excellent and happy teacher." The Harvard Overseers twice sought to persuade Winthrop to accept the Harvard presidency. But while he accepted temporary appointments, he refused the position on a permanent basis, evidently preferring to pursue the scientific endeavors which won him high honors as a colonial scientist and an international reputation—the first productive scholar of the Harvard College faculty.[356]

The late Mr. Frederick E. Brasch, who pioneered in the study of American colonial science, projected a full study of Winthrop's scientific career. Unfortunately, he died before it was published—although a sketch by Mr. Brasch appeared in the *Dictionary of American Biography*.[357] Winthrop's principal works were published in the *Philosophical Transactions;* however, some were published separately, and manuscript notes of his scientific observations are preserved in the Harvard College Library. Mr. Frederick G. Kilgour has published from these a facsimile, with commentary, of Winthrop's earliest recorded scientific observations.[358] These appear to constitute the first colonial attention to sunspots, made more than thirty years before those of Humphrey Marshall in Pennsylvania. On April 19, 1739, during a period of high sunspot activity (as was later calculated), Winthrop was walking on the Boston

[356] Hornberger, *Scientific Thought*, pp. 57–61; David P. Wheatland and Barbara Carson, *The Apparatus of Science at Harvard, 1765–1800* (Cambridge, Mass., 1968), *passim;* Samuel Eliot Morison, *Three Centuries of Harvard, 1636–1936* (Cambridge, Mass., 1936), pp. 92–93; *D.A.B.*

[357] Mr. Brasch also published "Newton's First Critical Disciple in the American Colonies—John Winthrop," in *Sir Isaac Newton, 1727–1927: A Bicentenary Evaluation of His Work* (Baltimore, 1928), pp. 301–338.

[358] "Professor John Winthrop's Notes on Sun Spot Observations (1739)," *Isis*, XXIX (1938), 355–361.

Common shortly before sunset, "the air being so hazy that I was able to look on the sun, I plainly saw with *my naked eye* a very large and remarkable spot." He observed it until sundown and found his observations to be confirmed by others. The next day he was in Cambridge by six in the morning to observe this phenomenon with Harvard's eight-foot telescope (probably the same instrument which Robie had used in 1722), discovering "not only the same spot which I saw before but several others in his [the sun's] disk." That night, he added, "a considerable aurora borealis" was seen. The next day was cloudy, but on April 22 Winthrop again observed the spots with the telescope, recording that "several other persons in the country saw them likewise with their naked eye, particularly some at Medford and the ferrymen at the Charlestown ferry." On both April 20 and 22 Winthrop made rough drawings of the sunspots, but the drawings are not sufficiently accurate to determine the latitude of the spots in relation to subsequent studies. A tantalizing feature of Winthrop's notes relates to his cryptic comment under date of April 20 regarding "a considerable aurora borealis." Did Winthrop intend to imply a relationship between sunspot activity and the aurora? Such a relationship was not definitely established until after the middle of the nineteenth century, but John Eames, F.R.S., had reviewed J. J. D. de Mairan's *Traité Physique et Historique de L'Aurore Boréale* (Paris, 1731) for the Royal Society in which a possible relationship between "zodiacal light" and sunspots was suggested. The review had been published in the *Philosophical Transactions* for 1734,[359] and Winthrop may have read it there. If his note on the appearance of the aurora borealis—the only part of his account not bearing directly upon the sunspots—implied association between the incidence of sunspots and the appearance of the aurora, Winthrop's thinking was well in advance of that of other astronomers of his time. However, as he failed to develop the idea further, the note on the aurora appears to have been only coincidental, with no further implications.

The following year (1740) Winthrop observed the transit of Mercury over the sun (April 21) and an eclipse of the moon (December 21). He reported his observations in a letter (Decem-

[359] XXXVIII, No. 431 (Jan.–March, 1734), 243–257.

ber 30) to Cromwell Mortimer, Secretary of the Royal Society, and it was tardily reported to the Society on November 3, 1743, and promptly published in the *Philosophical Transactions*.[360] Winthrop described the instruments, which, as he said, had descended to him from Isaac Greenwood and Thomas Robie. They included a twenty-four-foot telescope, an eight-foot telescope with a brass quadrant of two-foot radius, telescopic sights, and cross-hairs fitted in the focus of the lens. He also had two good pocket watches, one of them with a second hand. He stated that no one had previously reported observations of a transit of Mercury in the latter's descending node, and he estimated that the transit had not been observable in Europe. His account of the equipment, the manner in which he employed it, and his record of findings and calculations added up to a very commendable proceeding, worthy of a professional astronomer. Three years later Winthrop made similar observations of the transit of Mercury over the sun (October 25, 1743), although he did not transmit a report of his results until after his correspondent, Nathaniel Bliss (1700–1764), F.R.S., Professor of Geometry at Oxford, had become Astronomer Royal (1762–64). The report was sent in a letter from Cambridge, June 20, 1763, read to the Royal Society on November 10 following, and published in the *Philosophical Transactions* for 1769.[361] This report also revealed Winthrop's further objectives in making these observations of the transits of Mercury; namely, to determine the exact longitude of Cambridge in relation to London, a matter of importance in cartography, navigation, and future astronomical observations at Cambridge.

In 1743 Winthrop began a "Meteorological Diary of the Weather, and the Heights of the Barometer and Thermometer for every day from the Latter end of the year 1743 to the latter end of the year 1747." He sent a copy of this to Cromwell Mortimer (April 15, 1748) for communication to the Royal Society. It was presented to the Society on June 16, 1748, and the following year Winthrop sent a meteorological account of the weather at Cam-

---

[360] *Journal-Book*, XIX, 140–142; *Phil. Trans.*, XLII, No. 471 (Nov.–Dec., 1743), 572–578.
[361] LIX (1769), 505–506.

bridge from 1743 to the end of 1748, together with a paper of observations on electricity. These were presented to the Society on May 25, 1749, and the paper on electricity, as it raised some objections to assertions of William Watson, was referred to Watson for a reply. On June 1 Watson read his answer, which, he said, "fully obviates all Mr. Winthrop's objections," and Winthrop, who was on the point of becoming Franklin's loyal disciple in matters pertaining to the new electrical science, let the matter drop.[362]

In view of Winthrop's observations of the transits of Mercury in the 1740's, it is strange that he took no active part in similar observations of the transit of Mercury in 1753. Elsewhere in the colonies, especially in New York and Philadelphia, the phenomenon attracted considerable attention, although the actual observations reported were disappointing. The middle colonies appear to have become stirred up about it when French instructions to observers in Quebec were routed by way of New York as time was short and that was the quickest way.[363] The instructions, prepared by Joseph-Nicolas Delisle of the Académie des Sciences, fell into the hands of James Alexander, who obtained permission to publish them and sent translations both to Cadwallader Colden and Benjamin Franklin, who already knew of the approach of the transit of Mercury from instructions issued by the Royal Society for its observation, printed copies of Alexander's translations of the French instructions (superior to those of the Royal Society, which erred in advising that the beginning as well as the end of the transit would be visible in America) and distributed them over the colonies in the hope of stimulating observations. At least two further objectives motivated Alexander and Franklin. They felt that the colonial colleges should be encouraged to provide "a proper Apparatus for making such Observations," and they believed that the experience won by colonials in observing the transit of Mercury in 1753 would

---

362 *Journal-Book*, XX, 538; XXI, 131–132, 134–135.

363 *Colden Papers*, IV, 346–351, 363–372. For fuller treatment, see Hindle, *The Pursuit of Science*, pp. 82ff.; I. Bernard Cohen, "Benjamin Franklin and the Transit of Mercury in 1753," *Proceedings of the American Philosophical Society*, XCIV (1950), 222–232; Harry Woolf, *The Transits of Venus: A Study of Eighteenth-Century Science* (Princeton, 1959), pp. 44–71.

help to prepare colonial observers for the more important task of observing the transit of Venus over the sun in 1761.[364] For it was confidently believed that calculations resulting from observations of this phenomenon in several different parts of the world would yield more exact data regarding the distance between the sun and the earth, hitherto known only in broad relative terms. Colonial observers in colleges and elsewhere should use the observations of the transit of Mercury in 1753 to ascertain their respective latitudes and longitudes accurately, and should acquire and become familiar with the use of astronomical instruments and the like "that they may be expert at taking Observations of that Kind before the Transit [of 1761] happens."

Unfortunately, on the day of the transit of Mercury (May 6, 1753), Alexander and the New York group could do nothing because of cloudy skies. Evidently similar conditions balked the efforts of Franklin and his friends in Philadelphia. And it may have been the same in Boston, although there is no evidence that Winthrop had prepared elaborately for the event. Perhaps, having twice observed the transits of Mercury, he felt no pressing need to make further remote preparations for the transit of Venus in 1761, of the importance of which he was fully aware. The net result of all the promotion in the colonies was the observations of the transit of Mercury by Captain Richard Tyrrel in Antigua, where copies of Franklin's "circulatory letters" had been sent to his nephew, Benjamin Mecom. Captain Tyrrel's observations were done with a "valuable" collection of instruments made by John Bird of London. They were sent by William Shewington to Franklin, who communicated them to Peter Collinson, who in turn transmitted them to the Royal Society (November 15, 1753). They were published in the *Philosophical Transactions* soon afterward.[365]

The devastating earthquake which destroyed Lisbon, Portugal, on November 1, 1755, created wide interest in earthquakes in scientific circles. John Winthrop could not have heard of it when,

---

[364] [James Alexander], *Letters Relating to the Transit of Mercury over the Sun, which is to Happen May 6, 1753* (Philadelphia, 1753).

[365] XLVIII (1753), 318–319. Captain Tyrrel's calculations for the longitude of Antigua were 61° 45' W. James Short calculated it at 61° 22' 30" W. See also *Colden Papers*, IV, 388, 393–394.

on November 18 following, a lesser quake shook New England. His account of the New England phenomenon, which he sent (January 10, 1756) to Thomas Birch, Secretary of the Royal Society, was still timely when it was read to the Society on January 13, 1757.[366] The earthquake consisted of four successive shocks extending over a period of nearly eighteen hours. It was felt as far away as Annapolis Royal in Nova Scotia on the northeast, Lake George on the west, Maryland on the south, and an unknown distance eastward into the Atlantic. Its course ran from the northwest toward the southeast over an area about 550 miles wide and 800 miles long. Winthrop pointed out that its course was similar to some of the other quakes reported in New England (which he briefly reviewed) and that the center of them appeared to be near the Merrimack River. It broke china and glass, wrecked chimneys, injured roofs and brick buildings, shook down stone fences, and left chasms in the earth. It came upon New England like a long, rolling, swelling sea, with a roaring noise as it approached. Winthrop accompanied his account with a meteorological record for the year in New England, but he avoided any suggestion that weather conditions had anything to do with the cause of the earthquake. The Royal Society found that Winthrop's relation agreed with accounts it had received from New York, Philadelphia, and the West Indies.

The earthquake terrified many persons in New England, and Winthrop hastened to quiet their fears by *A Lecture on Earthquakes: Read in the Chapel of Harvard College . . . November 26 1755. On Occasion of the Great Earthquake which Shook New-England the Week Before. . . .*[367] He took pains to ascribe earthquakes to purely physical causes, and vigorously denied the contentions of those who sought to explain the quake as a direct intervention of the finger of God in earthly affairs. But the aging Reverend Mr. Thomas Prince (1687–1758), who, at the time of the New England earthquake in 1727, had published a sermon entitled *Earthquakes the Works of God and Tokens of his just Displeasure*, took occasion to reprint the sermon with an appendix to suggest that a secondary cause of earthquakes might be electrical in nature,

[366] *Journal-Book*, XXIII, 445–449; *Phil. Trans.*, L, Pt. I (1757), 1–18.
[367] Published in Boston, 1755.

perhaps a consequence of the erection of so many lightning rods in New England. Winthrop then added an appendix to his *Lecture on Earthquakes* before it was published and specifically denied Mr. Prince's contentions. When the latter sent a protest to the *Boston Gazette*, Winthrop firmly reiterated his stand in *A Letter to the Publishers of the Boston Gazette, etc. Containing an Answer to the Rev. Mr. Prince's Letter, Inserted in Said Gazette, on the 26th of January, 1756.*[368] Winthrop softened the blow somewhat by stating that all "natural effects" refer to the agency of God in true "philosophy," but probably many readers agreed with the verdict of Jared Eliot, who wrote to Ezra Stiles (March 24, 1756) that Winthrop had "laid Mr. Prince flat on [his] back."[369] This personal triumph, however, was insignificant compared with Winthrop's fundamental understanding of the nature of earthquakes. He described the shock as a "kind of undulatory motion," a kind of "wave of earth" like a wave of water, terms repeated, in essence, in his account sent to the Royal Society.[370] This recognition of the wave-like nature of earthquakes antedated by nearly five years a somewhat similar concept set forth by the Reverend John Mitchell, of whose essay (1760) Professor Wolf has stated that it "must be regarded as the beginning of scientific seismology."[371]

Comets, like earthquakes, aroused the fears of superstitious folk. The comet of 1758 was no exception, although its return, previously predicted by Edmond Halley, delighted colonial scientists with an interest in astronomy. It was observed by Winthrop at Cambridge, and by Ezra Stiles in Newport, Thomas Clap in New Haven, and Theophilus Grew in Philadelphia. But only Winthrop sought to quiet the fears of the superstitious and to enlighten the more rational segment of the public with *Two Lectures on Comets, Read in the Chapel of Harvard College . . . in April, 1759 On Occasion of the Comet Which appear'd in that Month. With an Appendix, Concerning the Revolutions of that Comet, and of some*

---

[368] Published separately, Boston, 1756.
[369] Quoted in Hindle, *The Pursuit of Science*, p. 95.
[370] Winthrop, *A Lecture on Earthquakes*, p. 6.
[371] Wolf, *A History of Science, Technology, and Philosophy in the 18th Century*, p. 398. Mitchell, "Conjectures concerning the Cause, and Observations upon the Phaenomena of Earthquakes . . . ," *Phil. Trans.*, LI, Pt. II (1760), 566–634.

Others. . . .[372] This exercise evidently led to further thought and calculation which provided data for his Latin memoir, *Cogitata de Cometis*, which, as has been observed above, he sent to James Bradley, the Astronomer Royal, on April 18, 1760.[373] When Bradley failed to give it publicity, Winthrop revised the account and sent it to Franklin (May 7, 1766), who communicated it to the Royal Society (March 19, 1767), and it was published in the *Philosophical Transactions* for 1767. As the Royal Society recorded, *Cogitata de Cometis* offered novel means whereby "to determine the quantity of matter and the density of particular Comets" by a method never before set forth. It was Winthrop's most original contribution to speculative astronomy.

Meteors likewise attracted the attention of colonial scientists. On July 30, 1760, Winthrop addressed to Thomas Birch, Secretary of the Royal Society, "An Account of a Meteor seen in New England, and of a Whirlwind felt in that Country. . . ."[374] The meteor had appeared on May 10, in calm, fair weather. It transcribed a circle about eighty miles in diameter with Bridgewater at the center, moving from northeast to southwest, accompanied by loud explosions, and appearing as a ball of fire four or five inches in diameter with "a tail of light behind it seen in bright sunshine." The whirlwind occurred on July 10 at Leicester, moving from southwest to northeast with enough volume to uproot trees and move stones of 150 pounds. It cut a swath of about forty rods wide, leveled crops and fences, and destroyed the home of David Lynde. Winthrop had speculated on its cause but found no explanation. "Mr. Winthrop," recorded the Royal Society Secretary, "is very particular in describing all the effects of this whirlwind, but excuses himself from meddling in any wise with the cause." But when Winthrop inquired of the Royal Society for an explanation, the Society offered none.[375]

Little more than a year afterward (November 17, 1761), Win-

---

[372] Published in Boston, 1759; republished as *Two Lectures . . . also, an Essay on Comets, by A. Oliver, Jun. Esq. with Sketches of the Lives of Professor Winthrop and Mr. Oliver* [by John Davis] (Boston, 1811).

[373] See note 330 above.

[374] *Phil. Trans.*, LII, Pt. I (1761), 6–16.

[375] *Journal-Book*, XXV, 14.

throp sent to Dr. John Pringle "An Account of several fiery Meteors seen in North America. . . ." He had received from Franklin an account of a meteor reported in England, and he gave relations of three others seen in New England. The first had been seen at Cambridge on June 3, 1739, moving from south to north, very large, bright, and noisy, and was heard for eighty miles around. The second had been reported by Thomas Clap (1703–67), President of Yale College, on November 24, 1742. It too was very bright and noisy, but only four to six inches in diameter. It moved from southwest to northeast for about 200 miles before it exploded in the air and disappeared. The third, viewed by a shipmaster at St. John's, Newfoundland, on May 4, 1760, moved from northeast to southwest and disappeared in the sea. After reference to a discussion recently published in the *Philosophical Transactions*, wherein Dr. Pringle had described the figure, magnitude, height, path, and velocity of meteors, Winthrop asked "What centre these bodies may most probably be supposed to revolve round?" He considered the possibility of the meteors revolving about either the earth or the sun, and concluded, after a number of mathematical calculations, that they revolved about the sun.[376]

In this matter Winthrop disagreed with Thomas Clap, the colonies' most serious student of meteors. Clap's interest appears to have been aroused by the meteor of 1742, to which Winthrop referred above. During the years that followed, Clap gathered further information, corresponded with Winthrop, Ezra Stiles, and others on the subject, and drew up a paper. This paper, entitled "Conjectures upon the Nature and Motion of those Meteors which are above the Atmosphere," was circulated among his colonial correspondents in 1756. Later he sent it (March 1, 1763) to Dr. John Pringle, who was himself something of an authority on meteors, and Pringle presented it to the Royal Society. Together with some letters by Clap to Pringle, the paper was read at meetings of the Society held on June 7, 21, and July 5, 1764. Clap argued that, besides those meteors which fell to earth, there were three meteors ("terrestrial comets") which revolve about the earth constantly, just as Edmond Halley had contended that comets move about the

---

[376] *Ibid.*, XXVI, 117–119 (June 7, 1764); *Phil. Trans.*, LIV (1764), 185–192.

sun. He estimated that these meteors were at least one-half mile in diameter, were as solid as iron, weighed about two billion tons, and moved at a velocity of 500 miles per hour. He concluded that they revolve around the earth "in long ellipses, their least distance [from the earth] being about 20 or thirty miles; that by their friction upon the atmosphere they make a constant rumbling noise, & collect electrical fire; and when they come nearest to the earth, or a little after, they make an explosion as loud as [that] of a large cannon [a discharge calculated to be 583 billions of "electrical snaps"]."[377] Clap also suggested that these meteors might serve to purify the air and make it more salubrious to mankind. The paper was not published in the *Philosophical Transactions*, probably because many of its concepts had appeared earlier in England. Fantastic as it may appear today, the notions were taken seriously by many scientists in the eighteenth century on both sides of the Atlantic, and Ezra Stiles published Clap's *Conjectures* (Norwich, Conn., 1781), years after the Yale President's death.

In the meantime, extraordinary preparations had been made, both in the colonies and by the Royal Society, to observe the transit of Venus across the sun on June 6, 1761. The importance of this observation had been emphasized in astronomical circles for many years. The object was to determine the solar parallax, i.e., the angle subtended by the radius of the earth at the sun. To determine this angle, a method had been devised by Edmond Halley in 1716, to use observations of the time in which Venus was seen to cross the sun's disk. It was hoped that, by comparing the results of several such observations, made at several widely separated locations on the earth, the solar parallax could be established. From this, astronomers could compute the distance from the earth to the sun and the distances of the planets from the sun. Hitherto, astronomers had been able to express the distances of the planets from the sun only in relative terms, i.e., not in actual miles but rather in terms of their distance from the sun compared with the earth's distance from the sun. Moreover, knowledge of the actual distances in miles between the sun, the earth, and other planets would make possible a more accurate concept of the size of the solar universe. But all this de-

[377] *Add. Mss. 5167*, fols. 156v–157v, 161v, 177v–182v (Morton minutes).

pended upon observations of the transit of Venus across the sun, an event of very infrequent occurrence. The transit of Venus occurs only four times in 243 years, at irregular intervals of 8, 121½, 8, 105½, 8, 121½ years, in sequence. The last opportunity had been in 1639, when two remarkable young astronomers, Jeremiah Horrocks (1617?–41) and William Crabtree (1616–44?), had made observations in England. But their observations were inadequate for the purpose, and the next transit was 121½ years later, or June 6, 1761, with another opportunity eight years afterward, on June 3, 1769. To seize upon these rare opportunities and settle the problem of the solar parallax was the hope of the astronomers. Inevitably, it was accompanied by an atmosphere of urgency, excitement, and a sense of competition among the various observers.

The necessary preparations involved, besides the astronomical apparatus, exact knowledge of the latitude and longitude of the observation points in order to determine the distance between them. The duration of the transit, that is, the time of contact of Venus with the sun either upon ingress or egress (preferably both), required a fine timepiece whose rate of "going" had been determined precisely. Moreover, the observers had to be skillful in the use of all the instruments employed and thoroughly familiar with the "drill," or the exact order in which every move took place, in order to execute them efficiently and without delay. All these things had been made reasonably clear both by the *Philosophical Transactions* and by promotions emanating from the French astronomer, Joseph-Nicolas Delisle, and several colonials were familiar with them. But when the time arrived, the only adequate telescope in Philadelphia was temporarily in London for repairs, and no observations were made there. Ezra Stiles was ready at Newport on the appointed day, but when the sun rose he was unable to see a trace of Venus. John Winthrop was the only British colonial in North America ready and able to observe the phenomenon.

Winthrop's preparations were extraordinary. Anxious to make the observations in Newfoundland, where the end of the transit would be visible, he conspired with his friend, James Bowdoin, a member of the provincial council, to persuade Francis Bernard (1711?–79), recently appointed Governor of Massachusetts Bay, to find means to finance an expedition. Governor Bernard recommended the undertaking to the Massachusetts General Court, wisely emphasizing the utilitarian benefits to navigation expected from the enterprise, and the legislature responded handsomely—one of the first instances of significant public support of scientific endeavors in the colonies. The "Province-Sloop" was assigned to carry Winthrop and his assistants, with their apparatus, to Newfoundland, and the Harvard Corporation permitted Winthrop to take any of the college instruments he required, provided they were insured against loss or damage.

Winthrop sailed from Boston on May 9, 1761, taking with him, in addition to the apparatus, two Harvard students as assistants and the college carpenter. His apparatus, as later described to the Royal Society, consisted of "an Excellent Clock, Hadley's quadrant with Nonius's divisions, a refracting telescope with wires at half-right angles for taking differences of right ascension and declination, and a Nice reflecting Telescope adjusted by Cross levels, and having vertical and horizontal Wires for taking Corresponding Altitudes, or differences of Altitudes and of Azimuths."[378] The party arrived at St. John's, Newfoundland, on May 22, and the carpenter, under Winthrop's direction, mounted the apparatus on an "eminence" nearby. There was some difficulty in establishing the longitude of the place, and it was not until June 11 that Winthrop was able to make observations whereby to calculate it with respect to Greenwich. Meanwhile, also, the party was plagued by "infinite swarms of insects" and "their venemous stings." When the transit took place, Winthrop made the observations while one of his assistants recorded the observations and the other "counted" the clock. Winthrop observed the egress of Venus from the sun as well as making other observations which enabled him to trace the path of the planet across the sun. Taken altogether, it was a very successful effort, and Winthrop soon prepared reports of his work in great detail.

The Reverend Jonathan Mayhew (1720–60), pastor of the West Church, Boston, appraised Thomas Hollis, F.R.S., of Winthrop's voyage in a letter read to the Royal Society on November 5,

[378] *Journal-Book*, XXVI, 160–163 (Nov. 15, 1764).

1761, reporting that Winthrop "had a very favourable observation of the Transit of Venus, and would speedily transmit the particulars."[379] Winthrop sent an account of his observation to James Bradley on September 21, 1761.[380] This was probably a copy of his book, *Relation of a Voyage from Boston to Newfoundland, for the Observation of the Transit of Venus, June 6, 1761* . . . (Boston, 1761), which was also received by the Royal Society (February 18, 1762) and "referred to James Short, to give some account of it."[381] Another report of Winthrop's observations reached the Society on November 10, 1763, this time from Professor Nathaniel Bliss of Oxford, Bradley's assistant at the Royal Observatory and later (1762–64) himself Astronomer Royal.[382] This, too, was delivered to James Short for a report to the Society. Indeed, James Short (1710–68), F.R.S., served the Society to sift all the reports received from observers of the transit of Venus of 1761, and from them he calculated an authoritative solar parallax. In the meantime, Short entered into correspondence with Winthrop to raise questions about the longitude of St. John's and the equation of time as found by equal attitudes. Winthrop's replies were read to the Society on November 15, 1764, and both his "Observation of the Transit of Venus" and an "Extract of a Letter . . . to James Short" were published in the *Philosophical Transactions* for 1764.[383] Winthrop had expressed the hope that his observations would be published in the *Philosophical Transactions*, for as an American he had an obvious pride in his achievement and he wanted to "do honour to the government who employed me." He added, further, "Perhaps, too, posterity may be glad to see, and make use of, the only observation of this rare phænomenon that was made in America."

Winthrop had additional cause for pride because the observations made under the auspices of the Royal Society were not impressive. The Society started to lay its plans too late, and was plagued by

379 *Ibid.*, XXIV, 168.
380 Winthrop to Bradley, Cambridge, Mass., Sept. 21, 1761, *Bradley 44*, fol. 147.
381 *Journal-Book*, XXV, 51.
382 *Ibid.*, XXV, 140.
383 *Phil. Trans.*, LIV (1764), 277–283; *Journal-Book*, XXVI, 160–163.

further delays and confusion. Possibly the Seven Years' War explained the tardy planning and it certainly explained at least one of the delays. It was January 31, 1760, before the Society gave attention to the matter, and late June of that year before the Council, in conjunction with James Bradley, began to consider where the observations should take place and who should make them. It decided to have observations taken at St. Helena and at Benkoelen in Sumatra, and it applied to the King for financial aid. George II granted £800, with the promise of a like sum if observers made it to Benkoelen. Bradley and Nevil Maskelyne (1732–1811), F.R.S. and later (1765–1811) Astronomer Royal, were chosen to go to St. Helena, but Bradley's health forced him to back out (he died in 1762), and Charles Mason, Bradley's assistant, was at first chosen in Bradley's stead. Later, it was determined to send Mason, with Jeremiah Dixon, to Benkoelen, and to send Robert Waddington, a mathematical instrument-maker, with Maskelyne to St. Helena. But when Mason and Dixon (the same pair who later established the boundary between Pennsylvania and Maryland) set out, their vessel was crippled by a French man-of-war and forced to return to port. Before they could set forth again, they protested that time was too short to arrive in India in time to observe the transit. But the Royal Society sternly ordered them to make the attempt, although when they subsequently arrived at the Cape of Good Hope they decided that they could never get to Benkoelen in time, and set up their apparatus at the Cape of Good Hope, where, contrary to their instructions, they made their observations. The Society was forced to rent instruments for the observers, and their expense, together with the money necessary for payment of the observers and their diet, etc., exhausted the Society's grant from the King, and as George II died before the second part of his promised grant was received (assuming he might have awarded it at all in the circumstances), the Society found itself in serious financial difficulties. Moreover, the observers at St. Helena were unsuccessful in their observation of the transit of Venus—although Maskelyne made some other useful observations. And, out of all this effort and expense, the Society received only the observations of Mason and Dixon at the Cape of Good Hope, observations which, fortunately, proved worthy of consideration.[384] John Canton made observations of the transit in London.

When James Short compared Winthrop's observations with those of Mason and Dixon, he found they yielded a value of 8.25" for the solar parallax. After he had compared this further with other data at hand, he estimated that the mean horizontal parallax was 8.56" (it is now internationally accepted as 8.80"). Winthrop's were the only American observations that Short took into account, and he stated that they had been done "with great care, and as much exactness as the low situation of the sun at that time would permit."[385] Short's computation of the solar parallax was challenged and never universally accepted, but another opportunity was offered to make new observations of the transit of Venus and compute the solar parallax in 1769. In the meantime, upon the recommendation of Benjamin Franklin (June 27, 1765), and a certificate signed by Franklin, James Short, Thomas Birch, John Canton, and Charles Morton, John Winthrop was elected Fellow of the Royal Society on February 20, 1765. A note at the bottom of his certificate recorded that "Benj. Franklin signed a Bond for him for his Contributions & paid his admission fee in November, 1767."[386]

Preparations for the observation of the transit of Venus on June 3, 1769, were far more extensive on both sides of the Atlantic than they had been in 1761. Debates over the results of the earlier attempts served to advertise the event. Newspapers and almanacs discussed it repeatedly, published directions for the general public, and the attention of the entire Western world was focused upon the matter.[387] The inconclusive results of the efforts made in 1761, together with the knowledge that no further opportunity would be offered until 1874, created a new sense of urgency, with Great

[384] For further details about the Royal Society's difficulties, see *ibid.*, XXIV, 492ff.; *C.M.*, IV, 241–257; *Add. Mss. 4444*, fols. 6–7 (Birch, "Letters Relating to the Royal Society").

[385] Quoted from the *Philosophical Transactions*, in Hindle, *The Pursuit of Science*, p. 100; see also Woolf, *The Transits of Venus*, pp. 44–96.

[386] *Journal-Book*, XXVI, 290, 374, 375; *Certificates, 1751–1766*, unpaged.

[387] As excellent secondary accounts of the transits, especially for that of 1769, are available, this treatment is largely drawn from them and presents less detail. See Hindle, *The Pursuit of Science*, Chap. 8; Woolf, *The Transits of Venus*, pp. 150ff.; Brooke Hindle, *David Rittenhouse* (Princeton, 1964), Chaps. 3–5.

Britain, France, Russia, and some other states contending with one another in the hope of discovering an acceptable solution to the problem involved. American colonial scientists became infected with the spirit of competition, anxious to prove their spurs both to the homeland and to the world at large. The Royal Society of London, determined not to be caught napping a second time, resolved as early as June 5, 1766, "that one or more Astronomical Observers be now sought out and engaged to observe the next transit of Venus in 1769."[388] No war threatened the implementation of the designs, and led by the young, able, and vigorous Nevil Maskelyne, who had become Astronomer Royal in 1765, careful plans were drawn well in advance in order to enable observers to be at the places determined upon, fully prepared by the time of the phenomenon to make observations. George III gave £4,000 to cover the expenses, and the Board of Admiralty, the Hudson's Bay Company, the East India Company, and others were successfully enlisted to lend cooperation and assistance.[389] Ultimately the Society received reports of observations made in Hudson's Bay, India, northern Sweden, Quebec, West Florida, California (made in conjunction with the King of Spain), Tahiti (where the famous Captain James Cook (1728–79), under orders from the British Admiralty, was on his first voyage to the South Seas, assisted by Charles Green, in the employ of the Royal Society), Scotland, Ireland, Gibraltar, and elsewhere—not including those reports received from the colonists of North America.[390]

In the colonies of North America, public interest in the transit of 1769 exceeded that of England and western Europe, and the number of observations made of the event reflected this concern. There was, however, one major disappointment. Nevil Maskelyne sought to encourage a colonial expedition to Lake Superior, where the entire duration of the transit of Venus could be observed. When he spoke with Franklin about this, Franklin recommended Professor John Winthrop, for he believed that Massachusetts was the only colony "likely to have a spirit for such an undertaking," and

---

[388] *C.M.*, IV, 223–225. For details of later propositions, see *ibid.*, *passim*; *Journal-Book*, XXV, *passim*.

[389] See pp. 252–253 above.

[390] *Journal-Book*, XXVII, 235–240, 245, 256, 503; *C.M.*, IV, 281–282.

possessed of "a person and instruments suitable" for it.[391] But Winthrop's health had deteriorated, and he did not feel able to undertake the trip himself. He recommended, instead, that Thomas Danforth, a Harvard tutor, be supported for the expedition. The question of funds to support the expedition was more difficult. The Massachusetts legislative body which had supplied money to finance the expedition to Newfoundland in 1761 was no longer in existence in 1768, when the problem arose. Thanks to the Circular Letter which the General Court had issued to urge other colonies to oppose the Townshend Acts, the General Court had been dissolved by action of the English government. Winthrop took up the question with his old friend, James Bowdoin, who might be able to enlist support from Governor Bernard again, or, possibly, find that General Thomas Gage, commander of the English troops recently sent to Boston, could "make the expence of it a contingency within his own department." But neither the Governor nor the General, although both were friendly toward the undertaking, could find the funds for its support. In the end, the expedition to Lake Superior was abandoned—a victim of the contest brewing between Great Britain and the American colonies which had deprived Massachusetts of its General Court. Winthrop and his friends were forced to be content with observations made at Cambridge. This, however, necessitated little by way of preparation, inasmuch as Winthrop could lay his hands upon all the apparatus needful. With typical concern, however, for the interest and instruction of his students and of the public, Winthrop clearly explained the methods and the purposes of the observations in *Two Lectures on the Parallax and Distance of the Sun, as Deductible from the Transit of Venus*, soon published at the request of his students (Boston, 1769).

When Franklin told Maskelyne that Massachusetts was the only colony "likely to have a spirit for such an undertaking" as an expedition to Lake Superior, he was unaware of the spirit which had infected his home town. For in Philadelphia the scene had changed materially since 1761. A considerable array of excellent new telescopes and other apparatus was being assembled. A variety of capable observers were at hand. One of them, the young David Rit-

---

[391] Quoted in Hindle, *The Pursuit of Science*, p. 147.

tenhouse (1732–96), was already on the path that led him to become, after Franklin's death, the most celebrated leader of science in the New Republic and the second citizen of the United States, after James Bowdoin of Boston, to become a Fellow of the Royal Society of London on the foreign list (1795).[392] Rittenhouse had assisted Charles Mason and Jeremiah Dixon in surveying the "Mason and Dixon Line" between Maryland and Pennsylvania (1763–65), and he and other colonists in Pennsylvania and Maryland had perfected their techniques by observing and associating with these two seasoned experts from England in various other scientific works in Pennsylvania.[393] In 1767 Rittenhouse began the design of an orrery which, when completed in 1771, was easily the most excellent device of its kind constructed by a colonial, and it spread Rittenhouse's fame far and wide.[394] The following year, Rittenhouse constructed an observatory for use in observing the transit of Venus in 1769, and built other apparatus, including a 144-power refracting telescope.

By 1769 the Philadelphia scientific community was well prepared to make observations of the transit of Venus. Several persons had purchased, or borrowed, the necessary apparatus, and amid an unbecoming spate of personal competition and jealousies, the way was clear for a multiple series of observations in and near Philadelphia. The most important source of bickering centered about the rivalries between the members of the newly revitalized (1767) American Philosophical Society and the American Society held at Philadelphia for Promoting and Propagating Useful Knowledge, which, originating in a young junto beginning in 1750, had found

---

[392] Bowdoin was elected in 1788. Cf. *Col. F.R.S.*, pp. 242–244.

[393] In addition to their survey of the boundary, Mason and Dixon, with the cooperation and support of Thomas Penn and Lord Baltimore, and with additional apparatus supplied by the Royal Society in 1765, measured a degree of latitude in Pennsylvania (1766–67), made observations to determine the difference in gravity between the Royal Observatory at Greenwich and "the place set up in Pennsylvania," and made other astronomical observations, including an eclipse of the moon and the immersion of Jupiter's first satellite. See *C.M.*, V, 33, 44, 110–128, 163–164; *Journal-Book*, XXVI, 40–42, 575–576; XXVII, 124–125, 143–144, 348; *Add. Mss. 5167*, fols. 37–39v.

[394] For details, see Hindle, *David Rittenhouse, passim*. Actually, Rittenhouse constructed two orreries; the first was sold to the College of New Jersey, the second to the College of Philadelphia.

new life in 1767. Both societies existed in Philadelphia, and both aspired to become general scientific societies, but each was dominated by a rival political faction in the city: the Philosophical Society by the Proprietary party leaders, the American Society by the liberal Quaker faction; and, although their objects were similar and some of their members acknowledged membership in both societies, the Philosophical Society tended to emphasize basic science somewhat more than the American Society, whose interests tended more toward utilitarian (especially agricultural) objects. Their rivalry was at its peak in the two years immediately preceding the transit of Venus in 1769, but about six months before that event took place, the two societies agreed to unite as one, to be known as the American Philosophical Society, held at Philadelphia, for Promoting Useful Knowledge (December 20, 1768). On January 2, 1769, they met together for the first time and elected officers. As President, they agreed on the only man whose scientific reputation transcended party considerations: Benjamin Franklin. But Franklin was away in England at the time, and leadership in the new combined society rested largely upon the shoulders of Dr. Thomas Bond, the Vice-President, who had been the principal reactivating force in the American Philosophical Society, and the rivalries and antagonisms between the two groups had not been completely stilled by the time of the transit of Venus.

In September, 1768, before the union of the two societies, James Dickinson, an obscure member of the Philosophical Society, unaware of the Royal Society's intention to make observations at Hudson's Bay, had presented a plan to observe the transit at James Bay, an arm of Hudson's Bay. Dickinson also proposed that the Philosophical Society appeal to the public, to the King, and to the Pennsylvania Assembly for financial support. Very likely the Assembly would have responded favorably, but the petition submitted to it by the Philosophical Society did not make clear the advantages of James Bay, where the entire transit could be observed; rather, it merely pointed out that the solution of the problem of the solar parallax depended upon a multiplicity of observations, such as might be made at Pittsburgh, Fort Chartres, a place on the Mississippi, or Hudson's Bay, compared with those made at Philadelphia, New York, Boston, and other parts of the world. In consequence, as no one pressed for observations at James Bay or other more distant places, and as there were advantages in concentrating the Philosophical Society's limited resources upon a number of observations made near Philadelphia, even though only the ingress contacts of the transit could be viewed, plans to journey to more distant places were abandoned. However, the Pennsylvania legislature responded promptly with an appropriation not to exceed £100 for the purchase of a telescope. In February, 1768, the united American Philosophical Society learned that the telescope was indeed on its way, and it speedily requested the Assembly for permission to erect an observatory in State House Square and asked for additional funds to pay for its construction. The Assembly both granted the permission and appropriated another £100 for the observatory. Within the American Philosophical Society, a committee was set up to plan the observations. Three separate sites were selected. At Norriton, David Rittenhouse's farm some twenty miles northwest of Philadelphia, the Reverend William Smith, Provost of the College of Philadelphia, led a party consisting, besides himself, of Rittenhouse, John Lukens, Surveyor General of Pennsylvania, and John Sellers, a representative in the Pennsylvania Assembly for Chester County. Rittenhouse employed the telescope that he had constructed; Smith used a reflecting telescope lent by Thomas Penn; Lukens used one constructed from lenses which Franklin had purchased for Harvard College but which had arrived too late to be forwarded to Cambridge in time for the transit. Sellers, who evidently had no telescope, perhaps assisted by "counting" the clock. The party also employed a quadrant belonging to the proprietors of East New Jersey, lent by the Surveyor General of the province, William Alexander, son of James Alexander, the old friend of Cadwallader Colden. The second site was in Philadelphia, where the Reverend Mr. John Ewing, a Presbyterian clergyman later to become Professor of Natural Philosophy at the College of Philadelphia, led a group who observed from the observatory set up at public expense in State House Square. Ewing used the new telescope provided by the Pennsylvania Assembly. He was associated with Joseph Shippen, who used a small reflecting telescope lent by the proprietors, Thomas Prior, who had his own telescope and supplied a timepiece lent him by the proprietors, and Dr. Hugh Williamson, who used a large refracting telescope borrowed from Miss Polly Norris. The third site was Cape Henlopen on Delaware Bay, where Owen Biddle, a member of the old American Society, assisted by Joel Bailey, a surveyor, made observations. Biddle employed an excellent telescope owned by the Library Company and newly fitted with a screw elevation at the company's expense; but Bailey had only a four-and-a-half-foot refracting telescope. Altogether, it was an impressive lot of apparatus which the committee of the American Philosophical Society had assembled for their purposes.

Outside the American Philosophical Society, a number of colonials also made observations of the transit. At Providence, Rhode Island, Joseph Brown, a wealthy merchant with an interest in science, in conjunction with Benjamin West, a bookseller and almanac-maker, and others in the province, purchased fine new equipment in England for the project. Abraham Redwood, founder of the Redwood Library at Newport, Rhode Island, teamed up with Ezra Stiles, the scholarly Librarian. Tristram Dalton, of Newburyport, Massachusetts, enlisted the aid of a Harvard tutor, the Reverend Mr. Samuel Williams. Three separate observations were made in New York City, two of them by staff members of King's College, the third by a former professor of mathematics and natural philosophy there. In New Jersey, William Alexander, the self-styled Earl of Stirling, employed his own telescope at his estate at Baskenridge, and William Poole, at Wilmington, Delaware, did likewise. John Leeds, Surveyor General of Maryland, guessed at his position at Talbot City and set his pocket watch by the sun—but wrote up his observations as if they were of value; amusingly enough, they were subsequently published in the *Philosophical Transactions*. At Rosewell, in Virginia, John Page did little better; he evidently understood what he was about, but his equipment was inferior, and his description of his work was so muddled that it was impossible for anyone not thoroughly familiar with astronomy to comprehend. Others, too, observed the transit in the colonies. Publicity of it had aroused the interest, or the curiosity, of hundreds of people. Crowds assembled to watch the observers at Philadelphia and at Norriton, and all over the colonies countless persons, supplied with anything they could turn up by way of magnifying equipment, or merely with smoked glasses, sought to view with their own eyes this extraordinary event. How many of them recorded the times of contacts is unknown. Hindle has reported that at least twenty-two supposedly independent contacts were ultimately published in the newspapers and in the journals of learned societies.[395]

At the Royal Society in London, where records of the observations were collected, studied, and compared by Nevil Maskelyne, the first of the American colonial accounts to be received was that of the Reverend William Smith. Dated at Philadelphia, July 19, 1769, it was transmitted by Thomas Penn to Maskelyne, who communicated it to the Society on November 23. Entitled

> Account of the Transit of Venus over the Sun's Disk, as observed at Norriton, in the County of Philadelphia, and Province of Pennsylvania, June 3, 1769. By William Smith, D.D. Provost of the College of Philadelphia; John Lukens, Esquire, Surveyor-General of Pennsylvania; David Rittenhouse, A.M. of Norriton; and John Sellers, Esquire, one of the Representatives in Assembly for Chester County; the Committee appointed for that Observation, by the American Philosophical Society, held at Philadelphia, for promoting useful Knowledge. Communicated to the said Society, in Behalf, and by Direction, of the Committee, by Doctor Smith; and to the Royal Society of London by Nevil Masklyne, B.D. Astronomer Royal

it was a long, detailed account, with a description of the instruments employed, accompanied by a diagram and tables of data recorded. It was published in full in the *Philosophical Transactions* soon afterward.[396]

John Winthrop's was the second American colonial report to be received by the Royal Society. Directed to Maskelyne, it was dated at Cambridge, September 5, 1769, and read to the Royal Society on December 7. Entitled "Observations of the Transit of Venus over the Sun, June 3, 1769. In a Letter to the Reverend Nevil Maskelyne, F.R.S. Astronomer Royal, from John Winthrop, Esquire, F.R.S. Hollisian Professor of Mathematics at Cambridge, in New England," it was more succinct and less detailed than that of William Smith, although it included an account of the instruments used and

---

[395] Hindle, *The Pursuit of Science*, p. 156. I have followed Hindle's account closely, finding little in the primary accounts to add to it.

[396] *Phil. Trans.*, LIX, Pt. II (1769), 289–326.

the data collected. It, too, was published in the *Philosophical Transactions.*[397]

"The Observations of the Transit of Venus over the Sun, June 3, 1769; made by Mr. Owen Biddle and Mr. Joel Bayley, at Lewestown, in Pennsylvania," was dated at Philadelphia, June 9, communicated by Benjamin Franklin to the Royal Society, where it was read on December 21. The Astronomer Royal, after consultation with Jeremiah Dixon, corrected the latitude and longitude for Lewestown as presented by Biddle, and the account, which also described the instruments employed and included the data collected, likewise was published in the *Philosophical Transactions.*[398]

John Ewing's account was neither communicated to the Royal Society nor published in the *Philosophical Transactions.* Hindle has revealed the reason. William Smith, after having sent a copy of his report to Thomas Penn (who transmitted it to Maskelyne), argued in the American Philosophical Society that all of the accounts of the observations made under the auspices of the Society should be reserved for publication first in its forthcoming *Transactions.* Accordingly, Ewing did not forward his account to London until it was too late to be included in the volume of the *Philosophical Transactions* which printed the principal observations from various parts of the world. A copy of Biddle's account fell into the hands of Dr. Cadwalader Evans, a partisan Quaker member of the old American Society and an inveterate enemy of Smith, and it was Evans who had forwarded Biddle's "Observations" to Franklin.[399] Thus, the effects of the personal intrigues of Philadelphians, left over from previous years before the Philosophical Society had united with the American Society, cheated Ewing out of the reputation which he might have won had his account been published by the Royal Society of London.

At least one more of the American colonial accounts of the transit of Venus reached the Royal Society, although it was not published in the *Philosophical Transactions.* This was *An Account of the observation of Venus upon the Sun the 3d day of June 1769 at providence in New England, with some account of the use of*

[397] *Ibid.*, pp. 351–358.
[398] *Ibid.*, pp. 414–421.
[399] Hindle, *The Pursuit of Science*, pp. 159–160.

those observations: by Benjamin West. Printed at Providence, Rhode Island, in 1769, a copy of this book was presented to the Royal Society and referred to the Astronomer Royal on May 3, 1770.[400]

Before Nevil Maskelyne could assemble and study the results of the many observations of the transit of Venus, John Winthrop raised a technical point which demonstrated his superiority over other American astronomers. In a letter to Franklin (September 6, 1769), Winthrop protested that an article in the *Philosophical Transactions* by the late Professor Nathaniel Bliss and Thomas Hornsby (1733–1810), both of Oxford, was in error when it argued that the phases of the transit of Venus would be accelerated in time by the aberration of light. On the contrary, said Winthrop, the transit would be retarded. The letter was read to the Royal Society (April 5, 1770) and referred to Maskelyne for a decision. It was published in the *Philosophical Transactions* for 1770, together with an article by Maskelyne in which the Astronomer Royal agreed with Winthrop, and a further letter from Richard Price (presented to the Society December 20, 1770) in which Price, too, agreed with Winthrop, calculating that the net effect would cause the apparent transit to begin five minutes and fifty-five seconds after the instant of its actual beginning. But Price also pointed out circumstances in which the effect would be reversed.[401] No other colonial astronomer, however, appears to have gone beyond the raw data gathered in his observations.

After Maskelyne had compared and sifted all the data from the wide variety of observations submitted, he had no success in determining the solar parallax with finality. In the most important observations the skies had been clear, but human errors, together with flaws in the equipment, had led to wide variations in the results obtained, although the range of difference was smaller than it had been in 1761. One of the most uncertain was William Smith's report of the observations made at Norriton. The three observers not only disagreed in their reports of the time of contact, but also David Rit-

---

[400] *Journal-Book,* XXVII, 342–343. I. Lorimer's account, dated at Mobile, June 24, 1769, was presented to the Society Feb. 13, 1772. It included observations made by Lorimer at Pensacola. *Ibid.,* XXVII, 577.

[401] *Phil. Trans.,* LX (1770), 358–362 (Winthrop), 536–540 (Maskelyne and Price); *Journal-Book,* XXVII, 335.

tenhouse had become so excited that he fainted momentarily before he could report his sight of the first contact, thereby rendering his observations of dubious value.[402] Measurement of the solar parallax is a very difficult matter. The sun is very large and very bright, and the smallness of the solar parallax is almost impossible to determine with great accuracy by direct observations of the transit of Venus. Astronomers continued in their efforts until late in the nineteenth century, with observations of the transits of Venus, Mercury, and Mars. Then scholars began to observe the transits of minor planets, which occur more frequently, and with the aid of better instruments have obtained values sufficiently alike to make possible the international acceptance of 8.80" for the solar parallax—the value employed in computing materials for various almanacs and ephemerides today.

On November 9, following the transit of Venus in 1769, there was a transit of Mercury across the sun. Both John Winthrop and William Smith made observations of it and sent their results to the Royal Society, which published them in the *Philosophical Transactions*.[403] Using the same apparatus at Norriton which had been employed in viewing the transit of Venus, Smith, in association with John Lukens, David Rittenhouse, and Owen Biddle, "agreeable to an appointment of the American Philosophical Society, held at Philadelphia for promoting useful knowledge," cooperated in the enterprise. Smith prepared the "Account" of it and sent it to Franklin (December 19, 1769), who presented it to the Royal Society on December 13, 1770.[404] Winthrop's report, also sent to Franklin (December 7, 1769), was presented to the Society on January 10, 1771.[405] Smith reported that his observations had been partially obstructed by clouds; Winthrop found that the sun had set before the transit had been completed. Both reports were referred to the Astronomer Royal for study.

Two further communications from John Winthrop reached the Royal Society before the American Revolution. The first, published

[402] Hindle, *David Rittenhouse*, pp. 54–59; Woolf, *The Transits of Venus*, p. 190.

[403] LX (1770), 504–507 (Smith); LXI, Pt. I (1771), 51–52 (Winthrop).

[404] *Journal-Book*, XXVII, 405–406.

[405] *Ibid.*, XXVII, 419–420.

in the *Philosophical Transactions* as "Remarks upon a Passage in Castillione's Life of Isaac Newton,"[406] was a complaint against an Italian Professor of Mathematics at Utrecht, J. F.–M. M. Salvemide Castillon (1709–91), whose *L'Arithmétique universelle de Newton* was published at Amsterdam in 1751. Winthrop held that Castillon had mistranslated Sir Henry Pemberton's *View of Sir Isaac Newton's Philosophy* (London, 1728), and in consequence erroneously presented Newton as if he had censured himself for handling geometrical subjects by algebraic calculations, for calling his algebra by the name of "universal Arithmetic," and for having commended Descartes for having done better in calling the subject "geometry." Winthrop sprang to Newton's defense, quoting Pemberton to show that Newton did not censure himself, but rather had censured Descartes for his *"injudicious* title of *Geometry."* Newton, said Winthrop, had attempted to follow the Euclidean *"geometrical* form of demonstration in preference to that of algebraic calculation; which is of modern invention." As a great admirer of Newton, Winthrop insisted that writers should be very exact in treating his works, even in the most trivial matters.[407] It was a curious echo of an old dispute between Cartesians and Newtonians, but it revealed Winthrop's deep admiration for the works of Sir Isaac Newton.

The last recorded communication of Winthrop to the Royal Society was an echo of another continuing dispute. Arising from interest in American Indian relics and artifacts, it referred to the puzzling inscription on Dighton Rock, near the shore of the Taunton River in eastern Massachusetts. Long since reproduced by the Reverend Mr. John Danforth in 1680 and brought to the attention of the Royal Society by Cotton Mather in 1714 and by Isaac Greenwood in 1730,[408] a number of other colonials had also speculated about the inscription. Ezra Stiles had made four drawings of it in 1767 and 1768; and Stephen Sewall, Hancock Professor of Hebrew and Oriental Languages at Harvard College, made a life-size drawing of it which was employed by Court de Gébelin (1725–84), a French antiquarian and linguist, who asserted in *Le Monde primitif,*

[406] LXIV, Pt. I (1774), 153–157.
[407] Franklin presented Winthrop's "Remarks" to the Society on Jan. 20, 1774. *Journal-Book,* XXVIII, 273–274.
[408] See pp. 452ff. above.

analysé et comparé avec le monde moderne (9 vols. in 4, Paris, 1773–82; the first three books were published at Paris in 1773) that the figures were Punic or Carthaginian in origin. Winthrop sent a reduced copy of Sewall's drawings with a letter to Thomas Hollis, F.R.S., dated at Cambridge, November 14, 1774, asking for help in deciphering the inscription. He advanced the opinion that it was Indian in origin, although very ancient—an opinion that appears closer to the truth than the notion that it stemmed from the ancient or Oriental world. The Royal Society, to whom Hollis presented Winthrop's communication on May 25, 1775, offered no opinion regarding the matter.[409]

The Crowning Point: The American Philosophical Society, 1768–69

The formation, in the winter of 1768–69, of the American Philosophical Society, held at Philadelphia, for Promoting Useful Knowledge was an appropriate climax to scientific developments during the Old Colonial Era. Clearly modeled after the Royal Society of London insofar as its functions were concerned, the American Philosophical Society was nevertheless an American institution, the culmination of an impulse toward formal organization that had been evident among American colonial scientists for nearly forty years. Cadwallader Colden had proposed something of the kind to Dr. William Douglass in 1728; Cromwell Mortimer, Secretary of the Royal Society, in the mid-1730's, had invited Paul Dudley to organize at Boston "a Company . . . subservient to the Royal Society"—ostensibly a colonial branch of the latter; John Bartram had broached the matter in a letter to Peter Collinson in 1739 and was advised by Collinson that "As to the Society that thee hints at, had you a set of learned well-qualified members to set out with, it might draw your neighbours to correspond with you. Your Library Company I take to be an essay towards such a Society. But to draw learned strangers to you, to teach sciences, requires salaries and good encouragement; and this will require public, as well as pro-

---

[409] *Journal-Book*, XXVIII, 599–600. See also Edmund Burke Delabarre, "Early Interest in Dighton Rock," Publications of the Colonial Society of Massachusetts, *Transactions, 1915–1916*, XVIII (1917), 235–299.

prietary assistance,—which can't be at present complied with, considering the infancy of your colony."[410] Collinson's words proved prophetic in view of the short life of the first American Philosophical Society, which, founded by Bartram, Thomas Bond, and Franklin at Philadelphia in 1743, broke down for lack of support within three or four years.[411]

In the meantime, however, American colonials were organizing other societies of one kind and another. As Collinson had said, the Library Company of Philadelphia, founded by Franklin in 1731, was "an essay" toward further organizations. During the previous year, a "Society for the Promotion of Knowledge & Virtue by a Free Conversation" had been set up at Newport, Rhode Island. Dr. William Douglass formed a medical society at Boston in 1736. The Charles Town Library Society was organized in 1748. The Royal Society of Arts, founded in London in 1754, led to a number of similar organizations in the colonies. Dr. Alexander Garden's efforts in its behalf in South Carolina proved abortive, but a society in imitation of the English model began at Williamsburg, Virginia, in 1759, at Boston in 1763, at New York in 1764, and the American Society held at Philadelphia for Promoting and Propagating Useful Knowledge—to which reference has been made above—came into being between 1750 and 1766, its organization being tightened and its cumbersome name adopted in the latter year. Some of these arose, in part at least, in response to colonial reactions to the Sugar Act, the Stamp Act, and other imperial acts that aroused the colonists to opposition against the mother country; but all of them bespoke as well the new opportunities which had arisen out of urbanization and the societal development of the colonies generally. Medical societies also multiplied. One was organized in New York in 1749, one in Charles Town in 1755, a new, enlarged one in Boston in 1765, and others in Philadelphia, New Jersey, and Litchfield, Connecticut, all in 1766. Furthermore, led by physicians who, for the most part, had had their medical training abroad, especially at Edinburgh, a medical school was organized in conjunction with the

---

[410] Collinson to Bartram, July 10, 1739, in Darlington, *Mems.*, pp. 131–132; Francis D. West, "John Bartram and the American Philosophical Society," *Pennsylvania History*, XXIII (1956), 463–466.

[411] Hindle, *The Pursuit of Science*, pp. 66–74.

College of Philadelphia in 1766 and in conjunction with King's College, New York, in 1767. In the meantime, too, the Philadelphia Hospital had been set up in 1752. A hospital was organized in New York in 1771, and a mental hospital was created in Virginia two years later. Obviously, shortly after the mid-century, the organization of science and of scientists, especially at grass-roots levels, was a widespread activity in the American colonies. The time was ripening for successful wider association on an intercolonial level. Ezra Stiles sensed this when, in 1765, he drew up a plan for an "American Academy of Sciences." Stiles's plan, which he revised repeatedly in 1766 and 1767, was distorted somewhat by his religious sectarianism and, as the tension rose in the imperial relations between the colonists and Great Britain, by his blatant patriotism. Moreover, it never advanced beyond the paper stage. But it conformed in general outline with the Royal Society of London and other great European scientific organizations in that it was intended to "collect all the curious Things in Science, especially in America, and maintain a Correspond[ence] over all the World."[412]

The merging of the American Society held at Philadelphia for Promoting and Propagating Useful Knowledge with the resuscitated American Philosophical Society in the winter of 1768–69 was a culmination of all these efforts. The new American Philosophical Society, held at Philadelphia, for Promoting Useful Knowledge soon came to embrace a membership not only of intercolonial proportions, but also sufficiently interlarded with men of political, economic, and financial prestige and power to insure its continued existence. The contemporaneous accident of the transit of Venus in 1769 cemented its membership—in spite of lingering personal animosities—by giving it something important to do in which the members took much pride, and at the same time enabling the Society successfully to approach the Pennsylvania Assembly for financial aid at a critical moment in the Society's existence. The Assembly increased its appropriations in support of the transit of Venus (and the publication of the results) until they totaled £450. In addition, the Assembly contributed £1,000 to match another £1,000 raised by public subscription to support a project of the

---

[412] *Ibid.*, pp. 120–121.

Society to encourage the development of silk culture, and both proprietary and private donations of money, scientific apparatus, books, natural curiosities, and the like further enhanced the purposes of the Society. Like the Royal Society, the American Philosophical Society levied upon its members admission fees and annual dues (of ten shillings each), but these proved difficult to collect and the Society was sustained during its early years by grants from the Assembly and by private contributions.[413]

Like the European scientific societies, the American Philosophical Society published reports of its scientific work for the scientific world. The first volume of its *Transactions* appeared in February, 1771, and thereafter, for the publication of American scientific materials, it generally took precedence over the Royal Society's *Philosophical Transactions*. Divided into four sections, including agriculture, medicine, and inventions, the most important section of the first volume of the Society's *Transactions* consisted of descriptions and calculations relating to the American observations of the transit of Venus in 1769. It included the principal American observations of the transit (with those made at Philadelphia by John Ewing), thereby supplementing those published in the *Philosophical Transactions*, the *Mémoires* of the Académie des Sciences, and others published in Europe. William Smith and David Rittenhouse attempted to calculate the solar parallax. Comparisons of the American observations with those of European observatories resulted in widely differing values ranging from 6.5″ to 9.2″. The mean value of a comparison of the Norriton and Philadelphia observations with ten European observations was found to be 8.4764″. The Norriton data compared with that of Greenwich resulted in 8.805″, a figure close to those calculated by Leonhard Euler at St. Petersburg (8.82″) and Joseph-Jérôme de Lalande (1732–1807), Director of the Paris observatory (8.53″ to 8.63″)—and a figure close to that accepted internationally today.[414]

The elaborate and painstaking detail of the accounts of observa-

---

[413] *Ibid.*, pp. 140–141. Professor Hindle's doctoral dissertation dealt with "The Rise of the American Philosophical Society, 1766–1787" (University of Pennsylvania, 1949).

[414] *Transactions of the American Philosophical Society*, I (1771), Appendix, pp. 54–70.

tions of the transit as published in the *Transactions* of the American Philosophical Society brought commendations from Europeans. Nevil Maskelyne stated that the American observations were "*excellent* and *compleat*, and do honour to the gentlemen who made them*,*" and Jean Bernoulli (1744–1807), astronomer at the Berlin Academy, commented that the fullness of the American accounts initiated "a practice well worthy of imitation by those European astronomers who are so sparing of detail and who speak only in general terms of their instruments and their observations."[415] Indeed, the appearance of the first volume of the *Transactions* of the American Philosophical Society excited widespread exclamations of surprise and acclaim in Europe. The Royal Society of London, to whom the American Philosophical Society presented the first volume of its *Transactions* on July 11, 1771, received the volume with pleased satisfaction and hastened to reciprocate by sending the American Philosophical Society copies of the *Philosophical Transactions* for 1769 and 1770.[416] The American *Transactions* served to announce to Europeans a new stage of maturity in American science. The volume of activity was impressive and the quality of the work compared favorably, in considerable part at least, with that of Europeans. The mere fact that the American Philosophical Society had come into being, with the capacity to publish a volume of *Transactions* (and the promise of more to come), was a further indication of the maturation of American scientific activity. All these factors were noted by Europeans, and they were matters of significance for American scientists and doubtless for American national independence when it was asserted only five years afterward.[417] During the War for Independence, the American Philosophical Society was formally chartered (1780) by the Pennsylvania Assembly, and it continues to be a foremost promoter and patron of the sciences in America to the present day.[418]

[415] Quoted in Hindle, *The Pursuit of Science*, p. 161.

[416] *C.M.*, VI, 106; *Journal-Book*, XXVII, 524.

[417] Publication of the second volume of the *Transactions of the American Philosophical Society* was delayed by the War for Independence until 1786.

[418] The American Academy of Arts and Sciences was organized at Boston in 1780, the second nationwide scientific society in the new nation. See Chapter 12.

TWELVE

Some Observations and Conclusions

THE HISTORIAN OF SCIENCE IN THE BRITISH colonies of America during the Old Colonial Era cannot ignore the enormous debt owed to the Royal Society of London. By its promotions, including those of some of its members, particularly Henry Oldenburg, James Petiver, Sir Hans Sloane, Peter Collinson, John Ellis, Dr. John Fothergill, and Dr. John Coakley Lettsom, the Royal Society constantly kept before the eyes of colonials the fundamental principles of the new experimental science, both as conceived by the Society in 1660 and as *developed* over more than a century of time. Moreover, the Society and its members supplied material aids in the form of books, scientific instruments, and occasionally financial support, usually in the form of subscriptions collected in England for the maintenance of specific scientific enterprises in the colonial field. It offered informed criticism and corrections to colonial men of science; it uniquely served as a clearing house for scientific data collected from nearly all parts of the world; it maintained the *Philosophical Transactions* as an outlet for the publication of the fruits of colonials' scientific discoveries; and frequently it bestowed recognition in the form of fellowship in the world's foremost scientific body of the time to those colonials who appeared to qualify for the honor. Such generous and open-handed support from the mother country must be set over against the more restrictive and occasionally selfish mercantilistic policies of the Old Regime, even admitting that these, too, sometimes infected the objectives of the Royal Society—or at least of some of its Fellows. On balance, however, one is inclined to envision the Royal Society as a kind of magnetic influence, unremittingly attracting the colonial mind from the old science to the new, from medievalism to the Age of Enlightenment. It continued to function after the American Revolution, of course, but thereafter international barriers dampened its enthusiasm and reduced its influence. For more than a century, however, the Royal Society promoted the new science in the American colonies with unflagging zeal.

Of course there were other influences simultaneously at work. These were principally societal developments within the colonies themselves. Until the second third of the eighteenth century, nearly every colonial who entered into scientific correspondence with the Secretaries of the Royal Society or others in the Old World complained, as John Winthrop, Jr., had complained in 1668, against the handicaps offered to all who sought to engage in scientific work in the colonies: the lack of leisure time, the unavailability of books and instruments, the absence of like-minded, stimulating associates, readily at hand, and general public apathy. Before the end of the Old Colonial Era these complaints had ceased, and in the larger urban communities the conditions which had provoked them had largely disappeared. For the American colonies, at least, the development of science was a *social* achievement, dependent upon a matrix which embraced far more than individual talents and the promotions of the Royal Society of London. This matrix was embraced within the general cultural maturation of colonial society, a subtle union of social forces originating in economic, demographic, and urban growth. Without it, it appears reasonable to assume that the promotions of the Royal Society would have continued to produce in the colonies little more than the lonely field collectors of scientific data which, as a whole, were the sole fruit of the Society's appeals until the early eighteenth century. With the matrix, however, there arose communities of colonial scientists, some of whom were able to advance beyond the mere collection of scientific data upon which Europeans constructed scientific ideas and to become a source of scientific ideas in themselves. The creation of the matrix, a product of collective forces in the growth of colonial society, transcended in importance the achievements of any individual colonial scientist, for it assured the future of science in America as no individual could provide.

The men who formed the scientific communities were a part—and an early product—of the cultural maturation of colonial society. It does not detract from their merits to point out that as they contributed to the maturation process, they also drew upon it for those factors which their predecessors lacked, factors which enabled them to reach new heights of scientific achievement during the last generation of the Old Colonial Era. But how great were these heights? Answers to this query have tended toward extremes and to polarize about either the contemptuous remark of the French Abbé Guillaume Raynal (1774) to the effect that "It is astonishing that America has not yet produced a good poet, an able mathematician, a man of genius in a single art, or a single science," or the equally absurd statement of Ezra Stiles, the patriotic President of Yale College, who claimed (1785) that Americans had contributed to science as much in "the last half century, as in all Europe."[1] Neither of these statements exemplifies balanced and informed judgment. Nor is it wholly true to state that American colonials served merely as field agents to gather *data*, empty of *ideas*, for European scholars, who creatively employed the data to construct organized science. There is an element of truth in the statement, of course. Until the early years of the eighteenth century, the statement will generally prove acceptable, although a few colonials, most notably John Banister, must be excepted. Beginning with the scientific contributions of Cotton Mather in the eighteenth century, however, the verdict loses validity rapidly, inasmuch as Mather and many of his colonial successors clearly contributed ideas as well as raw data. To be sure, some of the ideas proved faulty; but others were sound—and a few, unappreciated at the time, have won favorable responses from scientists who have read them in the light of later scientific developments.

A more discriminating judgment of the contributions of colonial scientists has been made by Professor I. Bernard Cohen.[2] Professor

[1] Both quotations are cited in Brooke Hindle, *The Pursuit of Science in Revolutionary America, 1735–1789* (Chapel Hill, 1956), p. 255.

[2] "The New World as a Source of Science for Europe," *Actes du IXᵉ Congrès International d'Histoire des Sciences (Collections de Travaux de l'Académie Internationale d'Histoire des Sciences*, No. 12 (UNESCO), Barcelona and Paris, 1960), I, 95–130. Cf. also Cohen, *Franklin and Newton (Memoirs of the American Philosophical Society*, Vol. XLIII, Philadelphia, 1956), pp. 590–591.

Cohen distinguishes between those colonials who merely contributed *data* and those who contributed *ideas*. With regard to ideas, he sets forth a severe test, admitting only those fundamental ideas of value to *pure* science. Employing these criteria, Professor Cohen concludes: "With the sole exception of Franklin's work in electricity, there is no contribution to pure science—whether concept, theory, law, or effect—made in any of the Americas prior to 1800 by a native or by a resident that is of such major importance that it is worthy of being recorded in every general history of scientific thought."[3] By the refined (and somewhat unrealistic?) standards of the twentieth-century concept of "pure" science (i.e., presumably, "theoretical science" as opposed to "applied science"), Professor Cohen is probably close to the mark. Indeed, somewhat similar criteria occasionally have been employed in this book to evaluate colonial men of science. But the question arises, is it historical—or is it not anachronistic—to evaluate eighteenth-century scientific achievements in terms of twentieth-century standards? Perhaps the question is given greater relevance when it is recalled that science in the eighteenth century, particularly that of the English-speaking world, was seldom "pure science" in the twentieth-century sense. English science, especially that set forth by the Royal Society of London—and, more broadly, that of the Enlightenment— was strongly Baconian in flavor, with a persistent strain of utilitarianism. Both in England and in the English colonies of America, this utilitarian emphasis, after the mid-eighteenth century, was fortified by the activities of the Royal Society of Arts. No doubt there were scientists in the eighteenth century—perhaps even in the colonies—who had at least momentary visions of "pure science" and of seeking knowledge only for its own sake; but for the vast majority the concept of science was mixed with utilitarianism, embracing both "pure" and "applied" science in varying proportions. And, admitting gladly that Franklin contributed to "pure" science, was he, therefore, only a "pure" scientist? Or, was not his simultaneous recognition of the possible use of pointed rods as a means of testing the identity of lightning and electricity *and* their probable utility as lightning rods an example of the mixed nature of science

[3] "The New World," p. 121.

EXPERIMENTS

AND

OBSERVATIONS

ON

ELECTRICITY,

MADE AT

Philadelphia in America,

BY

Mr. BENJAMIN FRANKLIN,

AND

Communicated in feveral Letters to Mr. P. COLLINSON, of *London*, F. R. S.

LONDON:

Printed and fold by E. CAVE, at *St. John's Gate.* 1751.
(Price 2s. 6d.)

31. The title page of the most significant single work by a colonial scientist (see p. 624 above).

Courtesy of the American Philosophical Society Library.

XLI. *Account of the Tranfit of* Venus *over the* Sun's *Difk, as obferved at* Norriton, *in the County of* Philadelphia, *and Province of* Pennfylvania, June 3, 1769. *By* William Smith, D. D. *Provoft of the College of* Philadelphia; John Lukens, *Efquire, Surveyor-General of* Pennfylvania; David Rittenhoufe, *A. M. of* Norriton; *and* John Sellers, *Efquire, one of the Reprefentatives in Affembly for* Chefter *County*; *the Committee appointed for that Obfervation, by the* American Philofophical Society, *held at* Philadelphia, *for promoting ufeful Knowledge. Communicated to the faid* Society, *in Behalf, and by Direction, of the Committee, by Doctor* Smith; *and to the* Royal Society *of* London *by* Nevil Mafkelyne, B. D. *Aftronomer Royal.*

GENTLEMEN,

Read Nov. 23, 1769. A MONG the various public-fpirited defigns, that have engaged the attention of this Society fince its firft inftitution, none does them more honour than their early refolution to

VOL. LIX.　　　　　P p　　　　　appoint

32. The opening page of the report of the Norriton observers for the new American Philosophical Society, held at Philadelphia, for Promoting Useful Knowledge, on the transit of Venus in 1769 (see p. 665 above).

Phil. Trans., LIX, Pt. II (1769), 289.

in the mind of Franklin? And does not this suggest, also, the answer Franklin expected when he posed the rhetorical question for young Polly Stevenson in London: "What signifies philosophy that does not apply to some use?"

If the criteria be altered to conform more closely to the mixed nature of eighteenth-century standards, the number of contributions to science made by colonials becomes more numerous. Disregarding all contributions of mere specimens and raw data, which continued to flow at a great rate throughout the Old Colonial Era, the following colonials set forth scientific ideas illustrative of rationalizations and hypotheses based upon observations of data but transcending the data themselves. If they were generally of a less fundamental nature than Franklin's ideas regarding electricity, nonetheless they were ideas which appear to have been of sufficient significance in the less rarified air of the eighteenth century to justify cataloguing the names of their authors with that of Franklin as contributors to the scientific knowledge of the day. Franklin's name heads the list, of course, as the foremost colonial scientist; the order of the names that follow is roughly chronological, with no attempt to rank them after Franklin:

BENJAMIN FRANKLIN, for his fundamental contributions to the knowledge of electricity.

COTTON MATHER, for his pioneer observations regarding plant hybridization, his contributions to the theory of disease, and his ground-breaking attention, unappreciated at the time, to psychosomatic medicine.

ZABDIEL BOYLSTON, for his early use of medical statistics to justify his inoculations for smallpox.

PAUL DUDLEY, for his enlargement of knowledge concerning the hybridization of plants and his identification of the source of ambergris.

ISAAC GREENWOOD, for his ingenious plan to construct ocean charts in unconscious anticipation of Maury's later work in the nineteenth century.

THOMAS GODFREY, for his quadrant to ascertain latitude at sea, a clear case of multiple invention in which the colonial nearly took precedence over the Englishman, Hadley, for whom the quadrant was named.

JAMES LOGAN, for his experimental demonstrations of the functions of the plant organs in the sexual reproduction of maize.

DR. WILLIAM DOUGLASS, for his clinical description of scarlet fever.

DR. JOHN LINING, for his experiments on metabolism, the relations between weather conditions and human sickness, and his description of yellow fever.

DR. LIONEL CHALMERS, for his careful studies to integrate weather conditions and the incidence of diseases in South Carolina.

DR. ALEXANDER GARDEN, for his corrections to Linnaeus in the classification of American flora and fauna.

EBENEZER KINNERSLEY, for his refinements and additions to Franklin's electrical discoveries, especially that lightning and electricity melt metals by hot rather than cold fusion.

JOHN CLAYTON, for his botanical studies in Virginia, which, though never published in their final form, placed him high in the ranks of systematic botanists.

JOHN WINTHROP, for his pioneer studies with regard to the mass and density of comets and his early perception of the undulatory character of earthquakes.[4]

Possibly the above list could reasonably be enlarged. No doubt the true measure of any contribution to science is the degree to which it can be found to have exerted a beneficial influence upon the progress of human knowledge. Science is more than a mere accumulation of data and ideas, yet specific data and ideas acquire scientific value only after an accumulation of related data and ideas clothe them with new meaning. In the process of synthesis, even errors can contribute in a negative sense to fortify the formulation of a scientific idea. The historian of science, especially the historian

[4] The list omits MARK CATESBY, DR. JOHN MITCHELL, and DR. PATRICK BROWNE as not permanent residents in the colonies. Had these men been colonials, their names would be added to the above list: Catesby, for his theories regarding the migration of birds; Mitchell, for his suggestions to establish generic relationships in taxonomy; and Browne, for his remarkable natural history of Jamaica.

of scientific theory, generally and of necessity ignores the minor steps taken in the final formulation of fundamental scientific ideas. Yet it was to this accumulation of minor steps, especially in terms of specimens, descriptions, and observations, that the bulk of the contributions of colonial scientists belonged. If the scientific horizon in the colonies contained few lofty structures fashioned by colonial architects, many of those erected by Europeans contained vital building materials supplied by colonial men of science. In this connection, it is useful to recall that many of the most significant European scientific works, from those of John Ray and Isaac Newton in the seventeenth century to those of Carl Linnaeus and Joseph Priestley in the eighteenth, rested, in part, upon specimens, descriptions, and observations supplied by colonials.

Too much has been expected of colonial scientists. It is unreasonable to denigrate their achievements simply because they failed to formulate as many significant basic scientific concepts as Europeans. With a total population less than half that of the United Kingdom, a fluid wealth far below that of the mother country, a new urban culture scarcely a generation removed from the deprivations of life in a virgin wilderness, and with no central clearing house for scientific data until the very end of the Old Colonial Era, is it reasonable to expect that colonials could compete on equal terms with the long-established cultural centers of Europe? Rather, is it not more reasonable to anticipate that the major scientific concepts would be generated at or near the centers of scientific collection, at such clearing houses for scientific data as London and Paris, where the broad synthesis could most readily be constructed? Indeed, it may well be considered remarkable that the colonial society, remote from these crossroads, was able to generate *one* fundamental contribution to pure science, together with a dozen or so lesser contributions, not to mention the thousands of specimens and observations fed into the maws of Europe for study and digestion.

By the close of the Old Colonial Era, American scientists had become a recognized part of the international scientific community. Throughout the colonial period they had followed the lead of Europeans, pursued the same range of subjects. Newtonian studies and natural history had dominated the scene and, as taxonomy came

681

to the fore with the Linnaean system, the colonists adjusted to it readily. Perhaps it was true, as patriots asserted during the War for Independence, that learning, including scientific learning, was more widely diffused among the people in America than it was in Europe.

Reservations must be noted, however. Much of the basic scientific research in America was inferior to that of Europe, and its total quality fell far short of that of the Old World. There were wide areas of scientific knowledge very little explored and almost wholly undeveloped in America. Amateurs played a prominent role in science on both sides of the Atlantic, but they more nearly dominated the scene in America, and there was little attempt on the part of professionals, except, perhaps, among medical men, to dissociate themselves from the amateurs and to professionalize scientific endeavors. American scientists remained heavily dependent upon European sources for precision instruments used in scientific work, and although some public support had been found for the observations of the transits of Venus, this support was so sporadic and uncertain that to all intents and purposes it was nonexistent. These factors, among others to be noted below, contributed to prevent the realization of the visions of the "Prophets of Glory" who, during and after the American Revolution, predicted that in an atmosphere of liberty and freedom, science and the arts would blossom unfettered in America as they had never blossomed in the Old World.

The Revolution itself, as Brooke Hindle has shown, swept away much of the scientific momentum which colonial scientists had created—a momentum which was not regained for many years. The number of colonial scientists of the last generation of the Old Colonial Era was reduced during the war. John Clayton, John Bartram, Cadwallader Colden, and John Winthrop died before its conclusion. Alexander Garden, the promising young Benjamin Thompson, and several lesser figures were British loyalists who left the American states; and Benjamin Franklin returned from France an aging man whose health was broken. Still, in the midst of the war, there was one gain to American science. Members of America's oldest scientific community at Boston launched a second scientific society which soon rivaled the American Philosophical Society at Philadelphia. Stung by French adulation of the Philadelphia society at Paris, John Adams returned to Boston in 1779 and rallied the scientific community there to give reality to plans long promulgated by the late Professor Winthrop. Assisted by James Bowdoin, Dr. Samuel Cooper (1725–83), a Harvard graduate and philosophical-minded pastor of the Brattle Street Church in Boston, Manasseh Cutler (1742–83), an equally philosophical-minded Yale man who, at the moment, was minister of Ipswich Hamlet, various members of the Harvard College staff, prominent Boston merchants, and other political leaders, Adams persuaded the Massachusetts General Court to charter (1780) the American Academy of Arts and Sciences. Created to lend support to arts and sciences as "the foundation and support of agriculture, manufacture and commerce; as they are necessary to the wealth, peace, independence, and happiness of the people; as they essentially promote the honor and dignity of the government which patronizes them; and as they are most effectually cultivated and diffused through a State by the forming and incorporating of men of genius and learning into public societies,"[5] the American Academy of Arts and Sciences quickly emerged as America's second national society for the promotion of science. Like its predecessor at Philadelphia, it published its proceedings, and the first volume of its *Memoirs*, to the end of the year 1783, was proudly presented to the Royal Society of London on June 15, 1786, by John Adams, now "Minister from the United States of America at the Court of London."[6]

Although the United States of America began with two well-organized national societies for the promotion of science, none of the accomplishments of its first generation of scientists equaled the importance of Benjamin Franklin's previous experiments in electricity. In fact, while a host of technological contrivances testified to the inventiveness of the early Americans, they also testified to the overwhelming utilitarian bent of the American mind. Commerce, manufacturing, and the settlement of the "New West"

---

[5] Dirk J. Struik, *Yankee Science in the Making* (Boston, 1948), p. 44.

[6] *Journal-Book*, XXXII, 427–428. This volume, as has been noted, included (pp. 396–397) Manasseh Cutler's "An Account of some of the vegetable Productions naturally growing in this Part of America, botanically arranged," a factor in reviving interest in botany in New England since the days of Paul Dudley and Dr. William Douglass.

engaged the principal attention of Americans. They were strongly committed to science, or what they believed to be science, and patriots confidently extolled political freedom as a release to scientific energies. Soon, they prophesied, American scientific attainments would surpass those of Europe and confound the world.[7] Post-revolutionary scientific accomplishments, however, failed to realize such proud hopes. The pursuit of happiness was believed to depend upon the pursuit of science, but science was too frequently confused with technology, and happiness was equated with utilitarian objectives embracing the augmentation of wealth and prestige. Indeed, the emphasis upon utilitarianism, that siren which had confused the Fellows of the Royal Society of London, was accompanied by a considerable amount of anti-intellectual sentiment in early America. Even such a profound observer as Samuel Miller (1769–1850) warned against giving too much attention to "useless sciences" in the schools. In 1803 he wrote that "the spirit of our people is *commercial*. It has been said, and perhaps with justice, that the *love of gain* peculiarly characterizes the inhabitants of the United States. The tendency of this spirit to discourage literature is obvious."[8] Its tendency to discourage basic scientific work was equally obvious, and only the conviction that basic scientific studies might ultimately turn up knowledge of utilitarian value prevented them from possible extinction. In such an intellectual atmosphere, and bereft of further close association with the Royal Society of London, Americans failed to keep abreast of European scientific developments.

However, the Royal Society, too, had fallen upon evil days. Torn by internal dissension during the 1780's, and forced to compete with new English scientific organizations, such as the Linnean Society (1788) and the Royal Institution (1799), it suffered a decline from which it did not emerge until after the Napoleonic Wars. In fact, English science as a whole, in spite of remarkable achievements in some areas, was partially eclipsed for a time by French and German accomplishments in chemistry, zoology, bot-

---

[7] Hindle, *The Pursuit of Science*, Chap. 12, devoted wholly to "The Prophets of Glory."

[8] Samuel Miller, *A Brief Retrospect of the Eighteenth Century* (2 vols., New York, 1803), II, 406–407.

any, theoretical astronomy, and even in electricity. American scientists contributed little to the new European attainments, and it was the second quarter of the nineteenth century before they began to assert themselves and to regain a position in the world of science comparable to that which their colonial forebears had established in the eighteenth century.

It is impossible, of course, to assert that Americans would have failed to reassert themselves in the world of science without the American Philosophical Society and the American Academy of Arts and Sciences. To be sure, these organizations rested upon societal foundations beyond science itself, and the demographic, economic, and social developments out of which they had been created were all present in post-revolutionary American society. However, with the exception of state and local scientific organizations, most of them ephemeral, the American Philosophical Society and the American Academy of Arts and Sciences were the only institutions of consequence dedicated to the preservation and promotion of science during the early years of the new nation. In spite of strong utilitarian forces unleashed by the expansive energies of the new republic, they kept alive the torch of the new science. Thus, if one looks to the future of American science after the War for Independence, the triumph of colonial science appears to have rested, in the long run, upon scientific organizations rather than upon the specific achievements of any particular colonial scientist.

In the broader realm of science outside the United States, it constituted only a minor triumph. But during the Old Colonial Era no "schools" of science had arisen; neither the colleges nor the individual scientists themselves had created focal points with a scientific tradition of continuing research that extended beyond the lives of individual men. The scientific organizations supplied this want by institutionalizing the manifold scientific interests and endeavors of colonials. They survived the Revolution and after independence had been won they were the principal—if not the sole—custodians and promoters of science in the new republic. Thus, although the War for Independence was primarily a military and political event—hardly a criterion in itself upon which to justify the conclusion of a study devoted to the history of science—the scientific organizations were both intellectual and cultural achievements marking the high point of scientific development during the Old Colonial Era. The "new science," which had been born during the distant years of early European exploration and settlement in the Americas and in part nurtured upon these events, had become institutionalized and made a permanent part of the learning and culture of the first independent nation of the New World.

Appendixes

I. "Directions for Sea-men, bound for far Voyages"

[The original copy of these "Directions," drawn up in the winter of 1662–63 by "Dr. Rook, late Professor of Geometry at Gresham College," lies in the Royal Society Archives (*Classified Papers, 1660–1740*, XIX, "Questions and Answers," No. 7). They were printed, in part, in the *Philosophical Transactions*, I, No. 8 (Jan. 8, 1665/66), 140–143, and, with some few modifications, in *ibid.*, II, No. 24 (April 8, 1667), 433–448. The latter copy was evidently printed separately for distribution to ship captains. The first printed version omitted Question 6 of the original, combined Questions 8 and 9, and omitted Questions 10–15 and the directions "To sound ye depth of ye Sea without a Line." But the first printed version prefaced the queries with a paragraph of some significance. What follows is a combined version, giving the Preface from the first printed version followed by the "Directions" from the manuscript copy. It seems likely that the copy given to John Winthrop, Jr., in the spring of 1663 before he returned to New England was this manuscript version.]

It being the Design of the *R. Society*, for the better attaining the End of their Institution, to study *Nature* rather than *Books*, and from the Observations, made of the *Phaenomena* and Effects she presents, to compose such a History of Her, as may hereafter serve to build a Solid and Useful Philosophy upon; they have from time to time given order to several of their Members to draw up both *Inquiries* of things Observable in forrain Countries, and *Directions* for the Particulars, they desire chiefly to be informed about. And considering with themselves how much they may increase the *Philosophical* stock by the advantage, which England injoyes of making Voyages into all parts of the World, they formerly appointed that Eminent Mathematician and Philosopher *Master Rook*, one of their Fellowes, and *Geometry* Professor of *Gresham Colledge* (now deceased to the great detriment of the Commonwealth of Learning) to think upon and set down some *Directions* for *Seamen* going into the *East* & *West-Indies*, the better to capacitate them for making such observations abroad, as may be pertinent and suitable for their purpose, of which the said Sea-men should be desired to keep an exact *Diary*, delivering at their return a fair copy thereof to the *Lord High Admiral of England*, his Royal Highness the *Duke* of *York*, and another to *Trinity-house* to be perused by the *R. Society*, which *Catalogue* of *Directions* having been drawn up accordingly by the said Mr. *Rook*, and by him presented to those, who appointed him to expedite such an one, it was thought not to be unreasonable at this time to make it publique, the more conveniently to furnish Navigators with Copies thereof. They are such, as follows:

[Thus far the printed version. What follows is the manuscript "Directions," identical with the printed copy except as noted above:]

1. To observe ye declination of ye Compass, or its variation from ye meridian of ye place, frequently marking withall ye latitude and longitude of ye place, where ever such observation is made, as exactly as may be, and setting down ye method by wch they made them.

2. To carry dipping needles with them, and observe ye inclination of ye needle in like manner.

3. To remark carefully ye Ebbings and Flowings of ye Sea in as many places as they can, together with all ye accidents ordinary and extraordinary of ye Tides, as their precise time of Ebbing and Flowing in rivers, at Promontories and on Coasts, which way their current runs, what perpendicular distance there is between the highest tide and lowest Ebbe, during ye Spring-tides and Neaptides; what day of ye moons age ye highest and lowest tides fall out. And all other considerable Accidents, they can observe in ye

Tides, chiefly neer ports and about islands, and in St. Helena's island, and ye 3 rivers there, at the Bermudas &c.[1]

4. To make plots and draughts of ye prospect of considerable Coasts, Promontories, islands, and Ports, marking ye bearings and distances as neer as they can.

5. To sound and mark ye depth of coasts and Ports, and such other places as they shall thinke fit.

6. To sound ye depth of ye ocean in severall places, as ye Bay of Biscay, &c.

7. To take notice of ye nature and ye soyle of ye bottom of ye Sea in all soundings, whether it be Clay, sand, rock, &c.

8. To keepe a register of all changes of wind and weather at all houres, by night and by day, showing ye point ye wind blows from, whether strong or weak; ye raines, Haile, Snow, &c. ye precise times of their beginnings and continuance, especially Hurricanes and Spouts: but above all

9. To take exact care, to observe ye Trade-winds, about what degrees of latitude and longitude they first begin, where and when they cease, or grow stronger or weaker, and how much, as neer and exact as may be.

10. To observe and record all extraordinary meteors, Lightenings, Thunders, ignes fatui, comets &c. marking still ye place and times of their appearing, continuing &c.

11. To carry with you good scales and Glass-vials of a pint or so with very narrow mouths, wch are to be filled with Sea-water in different degrees of latitude, as often as they please, and ye weight of ye viall full of water taken exactly at every time, and recorded, marking withall ye degree of Latitude and ye day of ye month.

12. To take up Sea-water in severall places, 2, 3, or 400 fathom deep, to compare ye weight and saltness thereof with ye water upon ye surfaces. Ye way to doe this is mentioned below.

13. To keep an exact dyary containing these things mentioned, and all other ordinary observations relating to the ships course, and all such extraordinary things as occurre; and at their return into England to deliver one copy of their diary, fairly written, and their

[1] "at the Bermudas &c." is added from the printed copy.

Course prickt out upon their Card, to his R. Highness ye duke of York, and another to Trinity House to be perused by ye Society at Gresham Colledge.

14. To observe ye end of ye Eclipse ye night after ye 8. of August 1663.

15. To observe ye totall darkness of ye Moons Eclipse that will be ye night after ye 27. July 1664. Ye houre it begins to be tolerably dark, and ye duration of ye totall darkness.

To sound ye depth of ye Sea without a Line.

Take a Globe of firre, or other light wood, a ring of brasse fastened to it, some 2 inches wide, let ye globe be secured altogether from soaking in water by varnish or otherwise: then take a piece of lead like the figure of 7. considerably heavier than will sink it; noting ye gravity as well of ye lead, as Globe, severally, both in aire and water, where you are to make ye Experiment: put ye short end of ye lead into ye ring, and so let globe and all sink gently into ye water, and as soon as ye point of ye lead touches ye ground, it will slip out of ye ring, and ye globe will immediately ascend, and note ye time of ye Globes stay under water, wch you may observe either by a watch that hath minutes and seconds, or by a good minute glass: for thereby by ye help of some tables and answers wch shall be given hereafter, ye depth of ye water may be known.

To take up water neer ye bottom of ye Sea

Take a vessel of mettal, like a pint-pot, or Cylinder, and having 2. valves: both opening upwards, one in ye bottom, ye other in ye top of ye vessel, so that, while it descends, both valves being open, leaves free passage for ye water to passe through it; then when it is neer ye bottom, as soon as it is pulled up ye valves will shut, and ye vessel bring ye water that is in it up.

II. *John Winthrop, Jr.'s Major Gift to the Royal Society, Sent October 4, 1669; Received February 10, 1669/70*

[The following is an excerpt from the *Journal-Book* of the Royal Society of London, Feb. 10, 1669/70.]

The Curator being absent, the Company, in stead of Experiments, was entertained with the reading of some Letters and the View of divers Curiosities of Nature, sent partly out of New England by Mr. Winthrop, partly out of Hungary and Transylvania by Dr. Edward Brown, partly out of Warwickshire by Mr. [Francis] Willoughby; all directed to Mr. Oldenburg for the Society.

The Letter of Mr. Winthrop was dated at Boston in New England Octob. 4th 1669, and the Curiosities accompanying it were these following:

1. Three dwarf oaks, with cupps of Acorns on them, the Acorns being fallen off; of which yet there were some found in the box, whereof two were given to Mr. Charles Howard, and one to Mr. Oldenburg for Mr. Evelyn, to plant them here; and two reserved for the Repository.

2. Two broad and one narrower Girdle of the Indian money.

3. One white and one blew string of Indian money, and one pair of Bracelets for the Aged, and one Childs bracelet.

4. A Small Pail made of the bark of a Birch.

5. Four Sorts of Sand, one found on the Sea-Shore neer Newhaven in New England; the other in St. Christophers Island, both metallin; the third very fine, of Saco in New England, taken out of a pond there; the fourth of Virginia.

6. Some Winter and Summer-Wheat, produced in those parts.

7. A Sort of Snakeweed.

8. Hartford-Earth, like Terra Sigillata, which the Cattle eat.

9. A Curious sort of Mosse growing on the Trees beyond Virginia towards Florida.

10. Black and Speckled Beans.

11. Red Beans and White in two papers.

12. A Lyme-stone lately found out in New England.

13. Three stones found in Clay deep in the ground at Hartford in New England.

14. Earth, which being put into common water, Swimms, as Wood or Cork, for a time; found about Patomack-River in Virginia.

15. The Horns of a kind of Beetle.

16. Flyes like moths, which engender the worms that Spoyle Apple-trees.

17. The Shape of those worms or Caterpillars which spoile the Apple-trees; into which forme they turne, after they have crept into the Earth from the tree, and in this forme they lye in the Earth, and onely in the Spring a Flye is bred of them like a Moth, and they come out of the Earth, and from them again the Caterpillars are produced. These two last in one Paper.

18. Exceedingly small Pismires.

19. Some Walnuts, whereof a couple were given to Mr. Charles Howard and another to Mr. Oldenburg for Mr. Evelyn, and one reserved in the Repository.

20. Three Indian purses or Bags; in one whereof are New England Chesnuts. Two of these purses are made of Porcupine's quills splitt.

21. A Small dish or Porenger of the bark of a Tree.

22. A Tray, made of the Root of a Tree.

23. Ten peices of Candle-wood, which being lighted burne with a good flame, and are used by many Planters in sted of Candles: they are Split out of the knots of Pitch-pine, And Tarr is made out of such knots.

24. A bag of Hasle-nuts grown in New England.

25. A bag of granat-stones, with a little paper of very small granats [garnets?].

26. Two unusual Shells.

27. A Black stone of Panirhall [or Paniwhall?].

28. Two flying-squirrils.

29. Some few Granat-stones by themselves, said to have been sent by one Thomas Edwards.

30. Two pretty big Shells.

31. Some ears of Indian Corne.

32. A Comb of the Indian Ladies.

33. Shreds of stuffe made by the English Planters, of Cotton and

Wooll: put up to shew the Colour, which is onely died with the bark of a kind of Walnut-tree, called by the Planters the Butter-nut-tree, the kernel of that sort of Walnut being very oyly, whence they call them Butter-nuts. They dye it onely with the decoction of that bark, without Allum or Coperas, as they say.

34. A Branch of the said Butter Walnut-tree.

35. Some minerals of New-England.

36. A mineral stuff, found in a Vein of the like kind of Sandy Stones, in digging a Trench at Hartford in New-England.

37. A Branch of a Tree, call'd the Cotton-tree, bearing a kind of Down, which yet is not fit to spin. The trees grow high and great; at the bottom of some of the leaves, next to the stalke of the leaf, is a knob, which is hollow, and a certain Flye, somewhat like a Pismire-fly, is bred in those knobs, in some years, more leaves have them knobs than in other years.

38. The Bottle-herb, alias Hedera trifolia Canadensis.

39. Lapides albi minerales.

40. Speckled gray and black Beans.

41. The Matrices, in which those shells are bred, of which the Indians make their white Wampan-peage, which is their money. They grow on the bottom of Sea-bayes; and the shells are like Periwinckles, but greater. Whilst they are very small and first growing, many are within one of the Concave receptacles of these matrices. If any of the litle cases be opened, there will be found the Primordia of those great shells of which the Indians make their Wampan; and many small shells will be found in one Case. They may be better discerned as to their full shape by a Microscope. These Cases are thin and separate from one another, yet one, that they are all fastened one close by another to a membrane like skin.

In the Second Box, marked B.

1. A Fish, full of prickles, called the Sea-hedge-hog.

2. A Flying Fish.

3. Pods of Silk-grasse.

4. Peices of the bark of a Tree, growing in Nova Scotia, and the more Easterly parts of New England; upon which barks they are little knobs, which opened yeild a Liquid matter like Turpentine, said to be of a very sanative and Balsamick nature.

5. Five Ears of Indian Corn of a special kind, said to ripen a month at least before other kinds.

In the third Box, Marked C.

1. Some Pine-apples, in number five.
2. Stones full of litle-holes, in every of which there was a litle creature, like a worme. They are taken from under the water.

In the fourth Box, Marked D.

This is a round Box, containing an extraordinary kind of Fish, somewhat resembling a Starr-fish, but yet differeing from it in divers particulars, and exceeding curiously wrought.

Besides these things there came along the head of a Dear which seems not an ordinary head: It was brought (saith the Presenter) far out of the Country by some Indians.

There came also with these Curiosities, Two Bibles in the Indian tongue; Three books of the Practice of Piety, translated into the Indian tongue. Two Astronomical Descriptions of the Comet of 1664. One book of Mr. Baxters Call of the Converted, turn'd into the Indian Language. One Indian Grammar. Of these Mr. Oldenburg desired one Copy of the Practice of Piety, and one of the Description of the Comet; which was granted him.

After the Company had viewed these Curiosities, it was ordered, that Mr. Oldenburg should returne Mr. Winthrop many thanks from the Society, and desire him to continue such Communications.

III. "Directions and Inquiries Concerning Virginia recommended to Edw. Diggs Esq. July 22.69."

[Several copies of "Inquiries Concerning Virginia" exist in manuscript. One, addressed to both Virginia and the Bermudas, is found both in the *Classified Papers, 1660–1740* of the Royal Society (XIX, No. 65) and among the *Sloane Manuscripts* in the British Museum (*Sloane 2903*, fols. 112–113). The following is one of the most complete and has the added feature of being addressed to a specific person at a specific time. On July 22, 1669, the Royal Society recorded (*Journal-Book*, IV, 78) that "The Society being made acquainted by Mr. Oldenburg that Mr. Edward Diggs intended to go shortly for Virginia and offred his service for philosophicall purposes, it was ordered that the Inquiries formerly drawn up for that Country should be recommended to him, and be thanked for his respect to the Society." Evidently, then, these inquiries had been previously drawn up, but if they had been directed to anyone in Virginia prior to Mr. Digges's departure there appears no record of it. Digges had made communications to the Society from Virginia, but considering the extent of the demands made upon the time, skill, energy, and financial outlay of one on the point of colonizing, it is not surprising if the Society received no detailed reply to its inquiries. The original of the following appears in the Royal Society Archives, *Classified Papers, 1660–1740*, XIX, No. 48.]

1. To endeavor the composure of a good History of the Virginian Plantation, concerning its Beginning, Increase, misfortunes and the present state thereof.

2. To give a perfect account of ye Planting and ordering of Tobacco; and its physical uses, made in Virginia of ye several preparations of that plant. And particularly to inquire of Mr. [Alexander] Moray, a minister living upon Ware river in Mockjack-bay, after ye improvement of yt new sort of sweet-sented Tobacco he mentioned in a letter of his written to Sir R[obert] Moray in Febr. 1665.[1]

3. To give a full account of the progress of the Silkwork; and to inquire of ye same Mr. Moray after ye success of yt extraordinary way he spoke of in ye same letter, of advancing Mulberry trees 2. or 3. years growth, by sowing them in seed, then cutting ym with a sith, and keeping them always under; or by planting ym as if they were Currans or Gooseberrys for to cut ym in great quantitys with dispatch. Further, to learn his way of serving many worms with few hands, and more clearly than before, as also his way of killing ym with expedition they now lying sometimes 3. or 4. days in the Sun before they dye.

---

[1] Moray's letter was dated Feb. 1, 1665/66. An extract of it is in *Letter-Book*, I, 241–242, and in *Guard-Book*, M-1, No. 36a. The extract was published in *Phil. Trans.*, I, No. 12 (May 7, 1665), 201–202, and in *Wm. & Mary Qty.*, 2nd Ser., II (1922), 157–161.

4. To try the raising of rice, Coffee-berries, Olives, Vines, and the like in that Country.

5. Concerning the Varieties of Earth, where as tis affirm'd, that there is a kind of Gummy consistence, white and cleer; another, white, and so light yt it swims upon water; another, red, called *Wapeigh*,[2] like *Terra sigillata*, to inquire after ym, and what other considerable kinds of it are there; and to send over a parcel of each.

6. What considerable Minerals, Stones, Bitumens, Tinctures, Druggs? To inquire after ye several sorts of iron-ore, to try wch of ym is kindest to make good and tough iron, and to encourage iron-mills for iron-work, for saving the wast of wood in England, fuell being much more plentifull and work much cheaper[?] there, than here.

7. What hot Baths, and of what Medicinal use?

8. What is the Original of those large Navigable Rivers, wch empty ymselves into ye bay of Chesapeak? And whether on ye other side of yt ridge of Mountains from wch they are supposed to proceed, there be not other Rivers yt flow into ye South-Sea; and if so, what search made after their current and use?

9. How the various sorts of Silkgrasse are prepared.

10. To give account of yt Vutnerary[?] root called *Wichacan:* of *Pocone*, a root of red Juyce a good tincture: of *Musquaspen*, an other root of red tincture; of ye plant *Maricock*, whose fruit is said to be fashioned like a Lemmon, exceeding pleasant to ye taste, and of a blossom most beautifull; of the *Chincomde-tree*, whose fruit is said to have a huske like a Chesnut, delicious and hearty meat, both raw and boiled.

11. Whether it be true that Virginia all over abounds wth oaks that are at least 50. or 60. foot high of clear timber, without boughs or branches, being also of yt largeness as to be very fit to make plank of any size, and very tough, and induring ye water exceeding well?

Whether the Cypres-trees there, as well as Pines, are tall enough and fit for Masts, and of as tough a nature as Yew, much lighter than firr, and so well lasting both in wet and dry, that it seems rather to polish than perish by long durance in the weather?

[2] Hariot had mentioned "Wapeigh," a molding clay. John Winthrop, Jr., had displayed a "swimming earth" before the Royal Society, from Virginia.

Whether there be abundance of Old Pines fit for making Rosin, Pitch, and Tarre?

Whether Virginia hath in many places, very proper soyle for planting hemp for Cordage and Sailcloths?

To send all sorts of Berries, Grasses, Grains and Herbs, growing in Virginia, and to wrapp up the Seeds very dry in paper, to send Seeds or Berrys, when they are ready to dropp off, wth as much husk and skins upon ym, as may be. to wrap up Roots in Mosse or light Earth, and to keep them from any dashing of Sea-water in the voyage; to gather ye smaller fruits, and dry ym in ye Air and in the Shade, to open fruits of a larger kind and ye Stones and Kernels being taken out, to dry ym; to set Plants or young Trees in half Tubbs of Earth, arched over with hoops, and cover'd with matts, to preserve ym from the dashing of Sea-water, giving them Air every day, ye weather being fair, and watering ym with fresh water every day.

12. What Animals the Contry is stored wth, both wild and tame? Whether Deer have there generally two or 3. Fawns at a brood? And whether any of the Catle transported thither from hence, become there more fruitfull, then they were here?

13. Whether at the bottom of ye Bay of Chespeak Nordward, the Natives be still of such gigantick Stature, as hath been reported? And whether there be another people, not far from these, Eastwardly, of a Dwarfish Stature?

14. To observe and informe us, How expert ye Indians are there in Hunting, Fishing, Swimming, and especially Diving, and to know the Uttmost, how long any Savages may be train'd to endure ye water Diving or Swimming; ye use of such men being very great for Merchants, Ship-Masters and others.

15. How their towns and Stocks of Catle are increased in Virginia? How their tillage and Pasture in the several parts of it have been proved or been improved? What care of gardens for the Kitchen or of orchards for wholesome fruit? What transplanted herbs prosper there?

16. What proceeding there is in making Pot-ashes and the way of making ym?

17. Whether there be really so great an alteration as to the health in

few years by opening ye woods of ye Country as is commonly reported?

18. Whether there be a foot-passage from Virginia to New England through Mary-land and yt not above an hundred miles; and if so, what plantations there are by ye way, and what distance from one another.

19. To observe the Variations of ye Needle from the Meredian exactly, in as many places as they can in their voyage.

20. To mark carefully ye Ebbings and Flowing of ye Sea; and therein: 1. Ye precise times of ye Flood and Ebb. 2. wch way currents run. 3. what perpendicular distance there is between the highest reach of ye Tide, and lowest of ye Ebb, both of all their Spring-tides and Neap-tides. 4. What day of ye Moons age and what times of ye year ye highest and lowest tides fall out. 5. The position of ye wind at every observation of the Tides.

21. To make a good Map of Virginia, and especially of ye coast thereof, wth the Longitude, Latitude; and to sound the depth near the Coast in ye shallow places, roads etc.

21 [sic]. To sound ye deeper Seas wthout a line.[3]

22. To keep a register of all changes of wind and Weather at all hours by night and day, shewing ye Point, ye wind blows from; as also the snows and ye Hurricanes, especially what season of ye year the latter happen most; and What are their prognosticks, concomitants, and consequences.

23. To observe and record all Extraordinary Meteors, lightnings, Thunders, lignes fatues, Comets.

24. To Carry wth ym good Scales and Glas-viols of a pint or so, wth very narrow mouths, wch are to be filled with Sea-water in different degrees of Latitude, and ye weight of ye water to be taken exactly at every time, and recorded, marking wthall ye degrees of Long. and Latitude of ye place; and yt as well of water near ye top, as at a greater depth.

---

[3] For the method, see Appendix I above.

IV. *"Inquiries Recommended to Colonel [Sir Thomas] Linch going to Jamaica, London, Decem. 16, 1670."*

[The Royal Society Archives contain several sets of inquiries addressed to the West Indies. One of the earliest of these, entitled "Inquiries for ye Antiles: out of ye French Naturall History of those iles¹ and Lygon's hist. of the Barbadoes"² (*Classified Papers, 1660–1740*, XIX, No. 64, undated) was published in the *Philosophical Transactions*, III, No. 33 (March 16, 1667/68), 634–639. Another, labeled "Inquiries Concerning Bermudas recommended to Mr. Hotham going thither. March 7, 1669/70," is very similar to that addressed to Edward Digges for Virginia (see Appendix III above), with some adaptations to the Bermuda scene. The following "Inquiries Recommended to Colonel Linch going to Jamaica, London, Decem. 16, 1670," though endorsed in Henry Oldenburg's hand, "A Memoriall recommended to ye favour and care of Colonell Linch going to Jamaica," appear, not in the Royal Society Archives, but among the *Sloane Manuscripts* in the British Museum, possibly used by Sloane as a guide to his own activities when he went to Jamaica a few years later. See *Sloane 3984*, fols. 194–195v. Sir Thomas Lynch served as Governor of Jamaica, and the Royal Society's "Inquiries" brought at least one response from him, noted in the *Journal-Book*, IV, 252 (May 27, 1672), as follows: "Sir Robt Moray brought in an Account Concerning Cacao-Trees, their Planting & Culture, the way of cureing them, the observables in their fruit &c transmitted to him by Sir Thomas Linch from Jamaica."]

1. Whether in Jamaica every night it blows off ye Island every way at once, so that no ship can any where come in by night; nor goe out but early in the morning, before the Sea-brise come in?

2. Whether it be true, that old Seamen can tell you each island towards Evening by ye shape of ye Cloud over it; vpon this account,

---

¹ Charles de Rochefort, *Histoire Naturelle et Morale des Isles Antilles de l'Amerique* . . . (2nd ed., Rotterdam, 1665).

² Richard Ligon, *A True & Exact History of the Island of Barbadoes* (London, 1657).

yt, as ye Sun declines, ye Clouds gather, and shape according to ye Mountains?

3. Whether in the harbour of Iamaica there grow many Rockes, shaped like Bucks- and Staggs-horns; as also divers Sea-plants whose roots are stony? And whether of those Stone-trees (if I may so call ym) some are insipid, others perfectly nitrous?

4. Whether the Observation, whereby ye Seasons of ye year betwixt ye Tropicks are divided by ye Rains and Faire weather, and six months are attributed to each Season, hold true at the Point in Jamaica? Or whether it be so, as some relate, that at ye sd Point there scarce fall 40 showers in a year, beginning in August and falling to October inclusively; but six miles from ye Point, towards Port morant, there be scarce an afternoon for 8. or 9. months, (beginning from April) in wch it rains not?

5. Whether at ye Point of Iamaica, whereuer you dig five or Six feet, ye water yt appears does ebb and flow? And whether it be true, that brackish water, though it be Vnwholesome for men, prove wholesome for Hoggs?

6. Whether it be true, yt at ye laymans the brackish water be wholesome for men, insomuch yt many recover there by feeding on Tortises, and drinking no other water?

7. Whether the Bloud of Tortises be colder than any water there; and yet their Heart beat as vigorously, as yt of any Animal?

8. Whether it be true, yt the Urine of those, yt haue eaten the fat of Tortises, wch is said to be green, looks of a yellowish green and oily after eating it?

9. Whether in some parts of Iamaica you ride through woods yt are full of very large Timber, and yet seem to haue nothing of Earth, only firm rock to grow in?

10. Whether in some ground in Iamaica, yt is full of Salt-peter, yr Tobacco flasheth as it smoaketh? And whether such nitrous Tobacco is subject to putrefaction? As also, whether the Potato's, yt grow in ye Salt-peter-grounds there, ripen two months sooner than elsewhere; but if they be not spent presently, they rot?

11. Whether it be true, yt ye flowers of ye fruit, called the Sowersop, when they open, doe giue an extraordinary crack?

12. Whether Ants have been observed to eat Brown Sugar White, and at last reduce it to an insipid powder?

13. Whether most brute creatures drink litle or nothing in Iamaica? And whether Horses in Guanaboa neuer drink? Nor Cows in some places of ye island for six months? Parrots neuer; nor Civet-cats, but once a month?

14. Whether it be true, that in ye midst of Iamaica there is a Plain, called *Magotti Savanna*, in wch whensoeuer it rains, ye rain as it setles vpon the seams of any garment turns in half an hour to maggots; yet is yt plain healthfull to dwell in?

15. Whether ye Sugar in Iamaica cures faster in ten days, than that at Barbados in six months?

16. Whether ye wood of ye Tree, call'd ye Bastard-Cedar, be so porous (though close to via) yt being turn'd into Cupps, wine and brandy will soak through at ye bottom in a short time?

17. Whether the Tree, call'd *White-wood*, in Iamaica, neuer breeds any worms? wch if so, it would be good, among other uses, to build ships of, if large enough.

18. Whether ye Berries of ye *Soape-tree*, wthout any proportion of Salt Lixiviate, or Sulphur, or Oyle, wash better than any Castile-soape; but rot ye Linnen in a short time?

19. Whether the several sorts of Tanning-barks, yt are in Iamaica, doe tann better than in England, and in six weeks the Leather there tanned is ready to work into shoes?

20. Whether ye Palma Christi yields such an exceeding great quantity of oyle, yt, if it were minded, it might be made a Staple-Commodity? And whether it be true, yt ye Indians and Negro's make the leaues of it, applied to the Head, ye only remedy for their Head-ache?

21. To observe, whether the Shining or Fire-flyes can contract and expand their Light as they fly? And especially, whether their Light continues some days after they are dead?

22. Whether there be any Hurricans about Iamaica? And whether those, yt haue been in such winds, haue found it exceeding cold? Whether in Hurricans, ye wind varieth all the points of ye Compasse; and ceaseth, when it comes East?

23. To observe the variation of ye Needle at Iamaica?

24. What is observable there in the Tydes? The precise times of ye beginnings of ye Floud and Ebb? What perpendicular distance there is between the highest reach of ye Tyde, and lowest of ye Ebb,

both of Spring-Tydes and Neap-Tydes, wth their Irregularities? What day of ye moons age, and what times of ye year ye highest and lowest Tydes fall out? Wch way ye Tyde setts etc?

24 [*sic*]. To send ouer some specimens of ye Roots, Seeds, and Fruits of Iamaica. In ye doing of wch, 'tis desired, yt ye smaller fruits, being gathered, may be exposed in the Air and Shade, till they are dry as raisins or figgs are usually made: That fruits of a larger kind may be open'd, and dryed, ye stones and kernels being first taken out: That Seeds and Berries may be sent, when they are ready to drop off, wth as much husk and skins about ym as may be: That Roots may be wrapt vp in mosse or light Earth, and kept from any dashing of Sea-water in ye voyage: That Plants or young Trees may be set in half tubbs of Earth, arched over with hoops, and cover'd wth matts to preserve ym also from ye dashing of sea-water; giuing ym Air every day, when the weather is fair, and watering ym wth fresh water once a day: To send all sorts of Pota-toes in Earth; and all sorts of Berries, grasses, grains and Herbs, wrapping vp ye seeds very dry in paper.

25. To observe, what considerable minerals, stones, Bitumens, Tinc-tures, and Druggs there are in Iamaica?

26. To try the raising of rice, olives, Coffee-berries, Currants, and the like, in Iamaica: As also to try, whether our late-ripe fruit here in England, as all sorts of winter-pears and the like, will not ripen in Iamaica much sooner etc?

The Answers, wch Col. Linch shall favour us wth, he is desired to direct them to me, Henry Oldenburg, at my House in the Pal-mal, inclosing them in the pacquet, he shall send to Mr. Slingsby, or Sr Robert Moray, or whom else he writes to at Court.

V. "*Queries proposed to and answered by Captaine Guilleaume, and Mr. Baily, Concerning the voyage and Country of ye bottom of East-Hudson-bay, one of ye chief places for the Beaver-trader. By H[enry] Old[enburg]. Read at ye Society Apr. 18, 16[72], and order'd to be enterd.*"

[Volume XIX of the *Classified Papers, 1660–1740* in the Royal Society Archives contains fifty-three sets of "Questions and Answers," dated between December 24, 1662, and March 14, 1670/71. Of these only a few relate to the British colonies and trading posts in North America, but two of them (Nos. 19 and 78) contain "Questions and Answers" regarding the Hudson's Bay Company's posts on Hudson's Bay. The two contain essentially the same information, although it is organized somewhat differently; No. 19 extends the questions and answers to twenty-two, containing the same (or very nearly the same) data that No. 78 encompasses within sixteen paragraphs. It would appear that the two were notes taken by the two Secretaries of the Society in a common interview with Captain Zachariah Gillam and Charles Bailly, both in the service of the Hudson's Bay Company. Captain Gillam commanded the *Rupert* and served the company well until, in 1683, his son, Captain Benjamin Gillam, sailed out of Boston in the *Batchelors Delight*, to engage in trade in the Hudson's Bay region as an interloper. This episode, which resulted in a considerable amount of official correspondence involving both King Charles II and the Governor of the Massachusetts Bay Colony,[1] caused the company to suspect that Captain Zachariah Gillam was in collusion with his son and to suspend him from their service. Charles Bailly was the first Governor of the Hudson's Bay Company forts in Hudson's Bay, having been released from the Tower of London (December 23, 1669) on condition that he would "betake himselfe to the Navigation of Hudsons Bay . . . which Sir John Robinson, Lieutenant of the Tower [and one of the adventurers in the Hudson's Bay Company] hath un-

[1] See *Hudson's Bay Company Archives*, A-6, I, 32v, 37–39, 40 (Hudson's Bay House, London).

dertaken that he shall doe. . . ."[2] Bailly died in 1680 and was handsomely buried in St. Paul's Church, Covent Garden.[3] The interview reported below probably took place in the early spring of 1672, shortly before it was read before the Royal Society on April 18. Though one of several "Questions and Answers" preserved by the Society, it is the only one relating to the North American scene in which both the "Answers" and the "Questions" have remained extant—and it is the only one based upon a personal interview as opposed to written (or printed) "Queries" more or less cast upon the waters in the hope that answers would be forthcoming.[4] The original manuscript is in *Classified Papers, 1660–1740*, XIX, No. 78, endorsed "Enquiries for Hudsons Bay, with Answers. Enterd." It was previously published in Thomas Birch, *The History of the Royal Society of London* . . . (4 vols., London, 1756–57), III, 43–46.]

1. What time of ye year they set out from hence, and when they arrived at the place intended? They set out june 5th, and landed Aug. 22th in the bottom of East-Hudson-bay, being 50°. 45′. North latitude, and distinct from the West-bay, ye place of C. James's wintering, wch was in 52°. 30′. latitude.

2. In what degree of Latitude they met wth the first ice, and at what time of ye year? in '59. in the beginning of August they met wth icy ilands moving.

3. How farr North they sailed? First to the Entry of the Straits yt let them into Hudsons bay, wch Entry is at 61½ deg. N. Lat. whence they run vp higher to ye lat. of 63 deg. (ye most Nordward place they went to;) and thence they run down again near 300. leagues due South to about the lat. of 51. deg; and longitude about 307 d.

4. In what months ye most Northern parts, wch they must make, are most convenient to passe? In August and September; and they

---

[2] *Acts of the Privy Council of England, Col. Ser.*, I, 540; see also *State Papers, Domestic*, CCII, No. 82 (Public Record Office, London).

[3] *Registers of St. Paul's Covent Garden*, Vol. IV, *Burials, 1653–1752*, in *Harleian Society Publications*, XXXVI (London, 1908), 86; *Hudson's Bay Company Archives*, A-6, I, 3.

[4] For an earlier voyage by Gillam, see Joseph Robson, *An Account of Six Years Residence in Hudson's Bay* . . . *1733 to 1736 and 1744 to 1747* (London, 1752), pp. 4–6.

hope, they shall be able hereafter to goe and come in one and ye same Summer, by ordering their Voyage so, as to be there about the midle of August, and by coming away in the beginning of September, the commodities of ye place being, vpon agreement wth the natives, ready to be shipped immediately vpon their arrival.

5. What depth of water they had, where their Ship anchored and they winter'd? About 9. or 10. foot: but in the straight's mouth 'tis so deep, yt they found no ground at 300 fathoms; And all along wthin those Straights deep enough; though many ilands everywhere.

6. What they observ'd as to ye Variation of the Needle? At 53 deg. they reckon'd no variation; at 54 d. they reckon'd about one degrees' variation west-ward, and thence the variation increased very considerably, so yt at ye entry of the Straights in 61½ deg. lat. at Cape Worsenam, ye needle varied 32 deg. west-ward, and at 63 deg. at Digs island it varied about 36 degrees: but running downe to the South for about 300 leagues, to ye lat. of 50°. 45'. it varied 26. degrees. Captain James reckon'd 29 d. variation where he winter'd.

7. What observable about the Tides? In those Straights there runs a constant Tyde East-South-East, and West-South-West; but in the Bay it runs North and South. Entring into the Straights, a South-East and by South Moon maketh a full Sea; but farther wthin the Straights, a South-moon doth it: And where they winter'd, a South-south-west moon maketh a full Sea. The Tydes commonly rise not above 8. foot perpendicular hight; though they are much gouern'd by the winds, wch are very variable there, and being high from the North-west raise the Tydes to ye hight of 12. foot in Ruperts riuer.

8. Whether the thick foggy Air did make their compasse move so dully, yt it would not traverse? This they apprehended would come to passe, but they prevented it by using Muscovy-glasse.

9. What kind a people the Natives are, where they winterd? They are tawny, living in Tents, wch they remove from place to place, according to the seasons of hunting, fowling, fishing. Their arms are bows and arrows. Their meat is venison, wild foule, as geese, partridges and rabbets, all wch are as white as snow, and in great abundance; ye Captain affirming, yt he had kill'd above 700. such white partridges. They haue also store of fish as Sturgeon, large pikes, Salmon-trouts, taken by ym wth nets. Great fish they had

seen none, but some Sea-horses and Seales, going into the Bay: no Cod, nor Whales. As to their drink, they use much the broth of their boyled venison, no crude water. Concerning their physick, they use chiefly sweating, not by taking any thing inwardly, but by making a kind of stove, in heating many stones red hot in their Tents, and then powring water vpon ym, whereby they are made to sweat excessively, in wch condition when they haue sat a while, they run out into ye snow, whereby they say their pores are presently closed again, as they were open'd by the heat; laughing at the Europeans, yt cause themselues to be rubbed and dryed wth clothes. They liue many of ym to a great age, to 120 years. From the Southwest of Carleton iland about 50. d. Latit. there came many Indians to ym, yt were 6. foot 9. inch tall, living among ye freshes, and much vpon fish, on the riuer Mousibi, yt is, the riuer of Elks, so called from the store of Elks, yt are to be found there.

The commodities they delight in, are course cloath, iron hatchets, hammers, ketles, pins, needles and such like; very ready to exchange them for Beavors.

10. What kind of soyle it is, and what it produceth? 'Tis most Clayground, plain, and very mossy; bearing no grain at all: only fetches, goos-berries, straw-berries, cran-berries. it abounds in Wood, especially in Birch, Willow, and Firr-trees, wch last kind of tree hath an excellent Turpentin (as they call it) on its buds, wch boiled in their bier they found very wholsom, and restoring them to strength and vigor, when they looked pale, and were sick and weak.

11. What Animals the country affords? Store of Dear, hares; Elks, and Beavor: all wch are very good meat. For other Beasts, there are White foxes, white bears, white cats; all yeelding excellent furr, wch is exceeding thick.

12. What observable chiefly about Beavors? Re/. They said, they had not been vp so far into ye freshes (for upon them they only liue) as to see themselues their manner of living and breeding; but they had been told by the Indians, yt they build their lodges two stories high; cutting pieces of wood from the neighbouring Trees, of yt length and bigness as is requisite for their purpose, and then meeting a competent number of ym together, whereof ye one half place ymselues on one side of ye piece to be shov'd away, and thrust their tailes under ye wood to ye other side, where the other half

standing ready doe fasten their teeth into the tailes of ye other, and so shove away the wood to ye place designed to build in, where they raise two stories, to ye end yt when the water swells, they may goe vp to ye upper story, in wch they also breed their young ones. Beavors liue not vpon fish, but rinds of Trees.

13. How the Captain of our Ship and his Company orderd ymselues as to their manner of living, whilst they stay'd there? Re/. When they came a shore, they built ymselues a house of wood, and dug a caue some 8. or 10. foot deep, into wch they put some barrels of good bier, wch at the time of their coming away being taken vp again, after it had remained there 8. or 9. months. prov'd very excellent liquor. Mean time they brew'd all the winter long of the provision of malt, they had taken wth them. And for their meat, they went a hunting, and wth their guns killd store of dear and fowle.

14. What temper they found that contry of, in the months of May and June? Re/. The spring began in May; in June they found it pretty hot in ye day time, and store of muskitos; but frost in the night.

15. How they had their health there? Re/. Reasonably well; only in returning they found some trouble of the Scurvy, and yt chiefly in their mouths.

16. What gouernmt and religion they haue amongst ymselues? Re/. They haue some chief person, yt is above the rest; yet working wth them. They found no quarreling amongst ym. They love keeping ones word; are very sensible of love and kindness; and they expresse their hearty forgivenes by a gesture of throwing ye arms behind their back; wch when they doe you may rely vpon them as perfectly reconciled. They acknowledge some supream power, wch they call Maneto, and they haue a Pawaw, by whom they addresse themselues to their God, and acquaint him wth their necessities, wch Pawaw returns them answers of help and relief, and yt commonly vpon conditions of giuing such and such commodities, amng wch Tobacco is one of the chief. [The copy in *Classified Papers*, *1660–1740*, XIX, No. 19, adds in the margin: "No worship but to ye sun, at whose appearance they rejoiced, and brought out when it shone all their wealth, and expos'd it to their reputed deity."]

VI. A Check-List of the Colonial Fellows of the Royal Society of London, 1661–1783[1]

Name	Location	Date of Election	Date of Formal Admission
Barham, Henry, c. 1650–1726	Jamaica	Nov. 14, 1717	Nov. 21, 1717
Boylston, Zabdiel, 1680–1766	Massachusetts	July 7, 1726	July 7, 1726
Brattle, William, 1662–1717	Massachusetts	March 11, 1713/14	Declined as "unqualified"
Burnet, Gov. William, 1688–1729	New York, New Jersey, Massachusetts	Feb. 13, 1705/06	Feb. 27, 1705/06
Butt, Dr. John Martin, d. 1769	Jamaica	Feb. 26, 1767	March 18, 1767
Byrd, William, II, 1674–1744	Virginia	April 29, 1696	April 29, 1696
Calvert, Gov. Benedict Leonard, 1700–1732	Maryland	March 25, 1730/31	no record[2]
Calvert, Gov. Charles, fifth Lord Baltimore, 1699–1751	Maryland	Dec. 9, 1731	Jan. 27, 1731/32
Campbell, Colin, d. 1752	Jamaica	Dec. 10, 1730	Nov. 7, 1734
Cuming (Cumming, Comyns), Sir Alexander, c. 1690–1775	South Carolina	June 30, 1720	July 7, 1720[3]
Douglas, John, d. 1743	Antigua	Nov. 30, 1720	Dec. 8, 1720
Douglas, Gov. Walter, fl. 1695–1716	Leeward Islands	Nov. 30, 1711	no record

[1] Adapted from *Col. F.R.S.*, with omission of non-English Fellows.

[2] It is apparent that the absence of record indicates that no formal admission was made, usually because the colonial Fellow never got to London for the ceremony.

[3] Ejected June 9, 1757, for nonpayment of fees.

Name	Location	Date of Election	Date of Formal Admission
Dudley, Paul, 1675–1751	Massachusetts	Nov. 2, 1721	*no record*
Ellis, Gov. Henry, 1721–1806	Georgia, Nova Scotia, etc.	Feb. 8, 1749/50	Feb. 22, 1749/50
Fauquier, Lt. Gov. Francis, c. 1704–68	Virginia	Feb. 15, 1753	*no record*
Franklin, Benjamin, 1706–90	Pennsylvania	April 29, 1756	Nov. 24, 1757
Fuller, Dr. Rose, d. 1777	Jamaica	April 20, 1732	May 4, 1732
Garden, Dr. Alexander, 1730–91	South Carolina	June 10, 1773	May 15, 1783
Glenie, Lt. John, 1750–1817	Canada	March 18, 1779	*no record*
Gray, John, fl. 1730	Cartagena	March 16, 1731/32	March 22, 1732/33
Greg (Gregg), John, fl. 1765	South Carolina, Dominica	July 9, 1772	Feb. 10, 1785
Houstoun, Dr. William, 1695–1733	Georgia	Jan. 18, 1732/33	*no record*
Hoy, Dr. Thomas, 1659–c. 1725	Jamaica	Dec. 1, 1707	*no record*
Hughes, Rev. Griffith, c. 1707–50	Barbados	June 9, 1748	*no record*
Hunter, Gov. Robert, d. 1734	New York	May 4, 1709	*no record*
Kuckhan (Kukhan), Tesser Samuel, d. 1776	Jamaica[4]	June 4, 1772	*no record*
Lashley, Dr. Thomas, d. 1807	Barbados	Nov. 24, 1768	Nov. 25, 1784
Lee, Dr. Arthur, 1740–92	Virginia	May 29, 1766	Nov. 10, 1768[5]

[4] Elected on the foreign list.

[5] Fellowship terminated in his own mind by the American Revolution; later formally terminated by action of the President and Council of the Royal Society, Jan. 17, 1788.

709

Name	Location	Date of Election	Date of Formal Admission
Lettsom, Dr. John Coakley, 1744–1815	Virgin Islands	Nov. 18, 1773	Nov. 18, 1773
Leverett, John, President, Harvard College, 1662–1724	Massachusetts	March 11, 1713/14	*no record*
Livius, Peter, 1727–95	New Hampshire, Canada	April 29, 1773	July 1, 1773
Lloyd, Philemon, fl. 1725——	Maryland	Nov. 9, 1727	*no record*
Macfarlane, Alexander, d. 1755	Jamaica	Feb. 19, 1746/47	*no record*
Mather, Dr. Cotton, 1663–1728	Massachusetts	July 27, 1713[6]	*no record*
Mathew, Gov. William, d. 1751?	Leeward Islands	March 10, 1719/20	March 10, 1719/20
Mitchell, Dr. John, 1690?–1768	Virginia	Dec. 15, 1748	Dec. 22, 1748
Morgan, Dr. John, 1735–89	Pennsylvania	March 7, 1765	*no record*
Morris, Gov. Robert Hunter, c. 1700–1764	New York, New Jersey, Pennsylvania	June 12, 1755	Feb. 15, 1759
Nicholson, Gov. Francis, 1655–1728	Virginia, South Carolina, etc.	Dec. 4, 1706	Dec. 4, 1706
Oglethorpe, James Edward, 1696–1785	Georgia	Nov. 9, 1749	Nov. 16, 1749[7]
Penn, William, 1644–1718	Pennsylvania	Nov. 9, 1681	*no record*
Pownall, Gov. Thomas, 1722–1805	New Jersey, Massachusetts, South Carolina	April 9, 1772	May 7, 1772[8]

[6] Election confirmed, April 11, 1723. See *Col. F.R.S.*, pp. 199–203.
[7] Ejected June 9, 1757, for nonpayment of fees.
[8] Withdrew Aug. 14, 1789.

Name	Location	Date of Election	Date of Formal Admission
Riz, David, fl. 1765	Jamaica	June 5, 1766	June 19, 1766[9]
Robie (Roby), Dr. Thomas, 1689–1729	Massachusetts	April 15, 1725	*no record*
Taylor, John (later Bart.), d. 1786	Jamaica	May 9, 1776	June 20, 1776
Tennant, Dr. John, fl. 1750–70	New York	June 13, 1765	June 20, 1765
Thompson, Benjamin (later Sir Benjamin, Count von Rumford), 1753–1814	New Hampshire	April 22, 1779	May 6, 1779
Wales, William, 1734?–98	Hudson's Bay	Nov. 6, 1776	*no record*
Winthrop, John, 1606–76	Connecticut	original Fellow, Jan. 1, 1661/62; confirmed under second charter, May 20, 1663	
Winthrop, John, 1681–1747	Connecticut	April 4, 1734	April 25, 1734
Winthrop, Prof. John, 1714–79	Massachusetts	Feb. 20, 1766	*no record*
Wright, Dr. William, 1735–1819	Jamaica, Barbados	March 12, 1778	Feb. 3, 1780
Yale, Elihu, 1649–1721	Massachusetts, East Indies, Connecticut	Nov. 30, 1717	*no record*

James Bowdoin (1726–90), of Massachusetts, was elected Fellow of the Royal Society on the foreign list, April 3, 1788, and David Rittenhouse (1732–96), of Pennsylvania, also was elected on the foreign list, April 16, 1795. These men, the first citizens of the United States of America to be so honored, were also the last North Americans to be elected to the Society in the eighteenth century.

[9] Ejected May 1, 1783, for nonpayment of fees.

711

Bibliographical Note

Inasmuch as all published materials consulted in the preparation of this book, both primary and secondary in nature, have been listed either in the text or in the footnotes with full bibliographical information, it appears both tedious and unnecessary to repeat them here. Nearly all of the rare sixteenth- and seventeenth-century works were consulted in the Library of the British Museum, London, where they are available under one roof. A few were consulted in the Library of the Royal Society of London, and some were found at the Bodleian Library, Oxford. It seems unlikely that any large proportion of them will be found in any one library in the United States, although the various libraries of Harvard University contain many of them. Practically all of them can be found in the United States, however, by searching in a variety of collections.

What follows relates solely to unpublished manuscript sources cited, arranged with reference to their geographical locations.

I. London

1. Archives of the Royal Society of London

Boyle Letters. Letters, largely to Sir Robert Boyle.

Boyle Papers. Miscellaneous papers of Sir Robert Boyle.

Certificates. Copies of the original signed certificates submitted to recommend persons to fellowship in the Society under the statutes as revised in December, 1730.

Classified Papers. Bound volumes of original papers received by the Society, including many reported to the Society but not published in the *Philosophical Transactions,* and some received but neither reported to the Society nor published.

Council Minutes. Minutes of the meetings of the Council of the Royal Society.

Guard-Books. Volumes of bound letters and other papers from correspondents of the Society, labeled alphabetically according to the surname of the correspondent (although the volumes occasionally contain other papers, sometimes in a rather confused manner).

Journal-Books. Minutes of the meetings of the Royal Society of London. The Society's Secretaries appear to have made notes at each meeting (a few of these rough notes are still extant). From these notes, often with the addition of papers or letters read at the meeting, the Secretaries prepared formal notes of the proceedings. A beautiful "fair copy" of these minutes was made in the nineteenth century. They make for easier reading, but they occasionally contain omissions and embellishments by the copyists. The originals prepared by the Secretaries are those cited.

Letter-Books. Primarily incoming letters from correspondents of the Society, although other papers are occasionally bound in. Arranged alphabetically under the surnames of the correspondents.

Letter-Book, Supplement. Miscellaneous leftovers from the above.

Miscellaneous Manuscripts. Odds and ends of various kinds.

Original Minutes. Minutes of meetings held by the Society before it became a chartered corporation.

Register-Book. Incomplete lists of members and other data.

2. British Museum (Division of Manuscripts)

Additional Manuscripts. Numbered volumes of bound letters, papers, and other materials relating to the Royal Society and its correspondents.

Egerton Manuscripts 2381. Reports of dissent in the Royal Society, 1759.

Sloane Manuscripts. Letters, papers, and miscellaneous items from the collections of Sir Hans Sloane.

Stowe Manuscripts 747. The Royal Society's troubles with artisans.

3. British Museum (Natural History)

Sloane Herbarium. Folio volumes of mounted plant specimens and seeds, occasionally with notations and other information, sometimes by the field collectors and sometimes by the persons who mounted the specimens. There is some confusion and error in labeling, especially with regard to the places from which the specimens derived. The herbarium has been well catalogued by J. E. Dandy, ed., *The Sloane Herbarium* (London, 1958).

4. Apothecaries Hall

Court-Book, I, Sept. 1, 1651–April 6, 1680. Relates to apprentices and visitations.

5. Fulham Palace Library

Fulham Palace Papers 14, S.R. 652. The papers of Bishop Henry Compton.

6. Guildhall Library

Hooke's Papers Ms. 1757. The papers of Robert Hooke.

7. Library of the Royal College of Physicians

Horsman Manuscripts, I. Sir Hans Sloane's troubles over the licensing of his *Prodromus*.

8. Hudson's Bay House

Hudson's Bay Company Archives, A-6. Contains manuscript materials relating to voyages to Hudson's Bay and cooperation between the Hudson's Bay Company and the Royal Society, especially with reference to the transits of Venus in the 1760's.

9. Public Record Office

C.O. 324/52, British State Papers. Relates to John Ellis in West Florida. The *Colonial Office Papers* are of some value.

10. Wellcome Medical Library

Wellcome Library Manuscripts 67457. Describes a curious amphibian from the American colonies. The library contains little relating to the colonies.

II. Oxford

1. Bodleian Library

Ashmolean Manuscripts. Contain letters between Edward Lhwyd, Martin Lister, and colonial field workers in natural history.

Bradley Manuscripts. Include astronomical letters and papers from the American colonies.

Lister Manuscripts. Contain Dr. Martin Lister's scientific correspondence to Virginia, Maryland, South Carolina, and the West Indies.

Additional materials from the same period (c. 1680–1720), principally pertaining to scientific relations between Oxford scientists and American colonies as well as relations between Oxonians and the Royal Society of London, are in the *English History Manuscripts*, the *Radcliffe Trust Manuscripts*, and the *Sherard Letters* (the latter supplemented materially by the Sherard materials in the Royal Society Archives in London).

2. Library of the Oxford Botanical Garden

The Sherardian Pinax, Ms. 32, c. 3. Contains materials relating to William Sherard's patronage of Mark Catesby.

III. Boston

1. Library of the Massachusetts Historical Society

Excepting the six-volume manuscript copy of Cotton Mather's *Biblia Americana* and the Gay transcriptions of Mather's *Curiosa Americana* (*Ms. C. 61.2.6*), I believe that I have cited no other manuscript materials from this library. I have looked at other manuscripts, especially materials relating to the Winthrop family, but found nothing for this book which the society has not published in its *Collections*, as cited.

IV. Cambridge, Massachusetts

1. Libraries of Harvard University

Other scholars had preceded me in winnowing manuscript materials relating to Professor John Winthrop, Isaac Greenwood, and others associated with Harvard College, and as I found nothing more to cite in this book, I have referred the reader to materials already published by my predecessors.

V. Worcester, Massachusetts

1. Library of the American Antiquarian Society

The manuscript copy of Cotton Mather's *Angel of Bethesda*. Although technically not unpublished manuscripts, this library's unique collection of rare colonial almanacs and newspapers was of great value.

VI. Philadelphia

1. Library of the Historical Society of Pennsylvania

I am deeply indebted to Professor George Frederick Frick, of the University of Delaware, who has shared with me manuscript materials hitherto unpublished gleaned from the *Bartram Papers* and the *Gratz Manuscripts*.

Obviously, the manuscript materials pertaining to science in the British colonies in America are widely scattered. The subject itself is relatively new, and its sources are not yet clearly delineated. Small amounts of the source materials hidden among the *Sloane Manuscripts* have long been recognized, but their total riches have been ignored hitherto (I have cited in this book eighty-two separate volumes of the *Sloane Manuscripts*, and I cannot guarantee that I have found all the pertinent materials relating to colonial science, although I searched many more volumes in vain). I have attempted to make a thorough search of all the major repositories cited above. But I suspect that there may be useful manuscript materials which I have not seen in some of the state archives, libraries, and historical society libraries as well as in local repositories. This may be equally true of county and other local collections in the West Indies and in Great Britain. I hope that this study will spur others to continue the search and to publish informative materials that I may have overlooked. With reference to many of the relatively obscure persons whose names appear in this book, it seems likely that local archives and records in the United States, in the West Indies, and in Great Britain may enrich the materials set forth herein. I have sought to open the field, not to close it.

To the discerning reader it will be obvious that there are many subjects referred to in this book which merit further study. Some of these, such as the Boston Philosophical Society organized by Increase Mather in the 1680's, may prove fruitless unless and until new sources are revealed in the future. Clearly, however, more satisfying biographical information is likely to be found by more thorough searches into the British and European backgrounds of such colonials as Dr. William Douglass, the activities in England of

716

Dr. Zabdiel Boylston, and the relationships between Isaac Greenwood and J. T. Desaguliers. Indeed, the activities of the latter, in relation to colonial science and scientists as a whole, might prove a rewarding subject for further study.

Several important colonial scientists merit further study in depth with regard to their scientific work—as apart from other biographical details. These include John Winthrop, Jr., the Governor of Connecticut; Cotton Mather; Thomas Clap; John Winthrop, the Hollis Professor of Mathematics and Natural Philosophy at Harvard; Dr. Zabdiel Boylston (whose medical and surgical career has been largely ignored, excepting his work as an inoculator for smallpox); Dr. John Mitchell; and Cadwallader Colden.

Doubtless, there are broader subjects as well. I make no claim to being exhaustive regarding these, but a few come to mind that may deserve attention: the intellectual interconnections between English scientists and English explorers in the New World prior to the foundation of the Royal Society of London; land survey in the American colonies—the Surveyors General and their deputies, their instruments, methods, and competence; colonial lecturers on science, their subjects, apparatus, numbers, and possible influence; American colonial cartography and science on the colonial frontier, with attention to the scientific explorations and observations of British military personnel stationed in Canada, Detroit, the Illinois country, and the Floridas. Both the relation between colonial science and the Enlightenment in America and the colonial response to English agricultural reform movements are subjects already under study by other scholars, although it is unlikely that they will exhaust the possibilities of such rich topics.

The condition of American science during the early federal period of American national history is inadequately explored as a whole. Between the end of the Old Colonial Era and about 1830, when patriots struggled to "Americanize" science and such important undertakings as the Lewis and Clark Expedition (to mention only one) marked the formation of scientific endeavor with a peculiarly American stamp, comparatively few significant studies have been made. I suspect that this is an era worthy of greater attention, especially if scholars will study it with an eye to European scientific accomplishments of the time.

Index

Abaco Island: Mark Catesby visits, 318
Abies balsamea, L. (balsam fir), 145
An Abstract from Dr. Berkeley's Treatise on Tar-Water, by Cadwallader Colden, 562
Académie Royale des Sciences: experiments to determine shape of earth, 391–393; Benjamin Franklin elected to, 638; *Mémoires*, 673; mentioned, 449, 461
Acadia, 458
"An Account of the Gymnotus Electricus or Electric Eel," by Alexander Garden, 615
An Account of the observation of Venus upon the Sun, by Benjamin West, 666–667
An Account of the Weather and Diseases of South-Carolina, by Dr. Lionel Chalmers, 598
Acosta, José d' (1539?–1600): *De Natura novi orbis*, 10, 35; *Historia Natural y Moral de las Indias*, 35; translations of, 35–36; views on cosmography, 36–37; anti-Copernican bias, 37; denies temperatures equal in equal latitudes, 37–38; on origin of Indians, 38; compared with Oviedo and Monardes, 39; views on novity of New World flora and fauna, 39; on metals, 39–40; on flora, 40–41; on fauna, 41; on origin of New World fauna, 41–42; on Indian history and institutions, 42; works known in England, 61, 65; on Northwest Passage, 248; mentioned, 21, 141
Acta Physico-Medica Academiae ... Leopoldina ... Ephemerides: publishes John Mitchell's work, 541

Acta Societates Regiae Scientiarum Upsaliensis, 565
Adams, John (1735–1826): and organization of American Academy of Arts and Sciences, 683
Admiralty, English Board of, 659
Africa, 6, 24, 37, 49, 82, 334, 417, 449, 528
Agawam (Ipswich, Mass.): John Winthrop, Jr., and, 118
Agricola (1490–1555), 11, 31
Agriculture: Indian, 69; colonial attention to, 72, 77–78, 133, 145, 181, 182, 192–193, 213–214, 216, 226, 229, 311, 317, 329, 335, 489–490
Alabama, 584
Albany, N.Y.: Capt. John Rutherford, 566–567; mentioned, 414, 462, 503
Albemarle: Duke of (1653–88), 236, 237, 241; Duchess of, 237, 238, 259
Albertus Magnus, 31
Albin, Eleazar (fl. 1713–59), scientific illustrator: *A Natural History of English Insects*, 520; *A Natural History of Birds*, 520
Alcalá de Henares, Spain, 29
Aleppo, Syria, 473
Alexander, James (1691–1756), of New York and New Jersey: and Cadwallader Colden, 493, 563, 568, 570, 574; mentioned, 497, 508, 647, 648, 663
Alexander, James, of Pennsylvania, 592
Alexander, William, 663
Alexander, William (self-styled Earl of Stirling), 664
Alfrey, George, 383
Allen, Mrs. Elsa G.: on American ornithology, 141, 193
Alligators: confused with crocodiles, 26

Index

745

Index

752

A Note on the Author

After more than thirty years at the University of Illinois—twenty-two of them as professor of history—Raymond P. Stearns became professor of history at Illinois State University, Normal, in September, 1970. He received his B.A. from Illinois College in 1927 and his Ph.D. from Harvard University in 1934. Since then he has received numerous honors, including the Frank S. Brewer Prize of the American Society of Church History (1940), the Medal of the University of Ghent (where he was Fulbright lecturer, 1959), and a National Science Foundation grant (1967–68). He has served on many academic committees and councils, among them the Council of the Institute for Early American History and Culture (Williamsburg, Virginia, 1947–53), the board of editors of the *William and Mary Quarterly* (1948–51) and of *The Historian* (1955–61), and the Council of the History of Science Society (1954–58). In 1966 he was an associate member of the Center for Advanced Study at the University of Illinois.

His publications include *James Petiver, Promoter of Natural Science, c. 1663–1718* (1953); *The Strenuous Puritan: Hugh Peter, 1598–1660* (1954); *Pageant of Europe* (1957, revised 1961); *Mark Catesby: The Colonial Audubon* (with G. F. Frick, 1961); and *A Journey to Paris in the Year 1698 by Martin Lister* (1967).

UNIVERSITY OF ILLINOIS PRESS